3rd International Conference on Modeling, Simulation, Optimization and Algorithm (ICMSOA 2021)

Journal of Physics: Conference Series Volume 2173

Sanya, China
12-14 November 2021

ISBN: 978-1-7138-5144-8
ISSN: 1742-6588

Printed from e-media with permission by:

Curran Associates, Inc.
57 Morehouse Lane
Red Hook, NY 12571

Some format issues inherent in the e-media version may also appear in this print version.

Institute of Physics
Dirac House, Temple Back
Bristol BS1 6BE UK

Phone: 44 1 17 929 7481
Fax: 44 1 17 920 0979

techtracking@iop.org

Additional copies of this publication are available from:

Curran Associates, Inc.
57 Morehouse Lane
Red Hook, NY 12571 USA
Phone: 845-758-0400
Fax: 845-758-2633
Email: curran@proceedings.com
Web: www.proceedings.com

TABLE OF CONTENTS

2021 The 3rd International Conference on Modeling, Simulation, Optimization and Algorithm[ICMSOA2021], 12-14 November, 2021

1. About ICMSOA2021

Organized by Yaseen Academy, **2021 The 3rd International Conference on Modeling, Simulation, Optimization and Algorithm (ICMSOA2021)** was held during 12-14 November, 2021 at Crowne Plaza Sanya City Center, Sanya, Hainan Province, China. Due to the different restrictions caused by Covid-19 according to particular provinces and cities, some of the participants participated in the physical conference at Sanya, but others had to join the conference online via Tencent Meeting. The Conference looks for significant contributions to all major fields of Computational Modeling, Simulation, Optimization and Algorithm in theoretical and practical aspects. The aim of the conference is to provide a platform to the researchers and practitioners from both academia as well as industry to meet and share cutting-edge development in the field.

Organizer: *Yaseen Academy* is located at Huangjiahu Lake University Town, Wuhan, Hubei, China, focusing on international academic conferences organizing and international academic journals corperation in areas of science and engineering, aiming to provide a high-level and open academic platform for global scholars, researchers and practioners. It has gathered an academic committee composed of more than 200 experts and scholars, and has successfully held many international academic conferences. Facing the vast number of experts and scholars, Yaseen Academic is a team with strong professional quality and professional ethics, it has creating a mature, deep and complete academic exchange platform, not only allowing experts and scholars to show and share their latest academic research results and realizing face-to-face academic exchanges, but also allowing the most cutting-edge scientific information, ideas, and viewpoints to be fully communicated and exchanged, which will benefit all mankind.

2. Call for papers

The conference topics include, but are not limited to the following areas:
Simulation modeling theory and technology, Computational modeling and simulation, System modeling and simulation, Device/VLSI modeling and simulation, Control theory and applications, Military Technology Simulation, Aerospace technology simulation, Information engineering simulation, Energy Engineering Simulation, Manufacturing Simulation, Intelligent engineering simulation, Building engineering simulation, Electromagnetic field simulation, Material engineering simulation, Visual simulation, Fluid mechanics engineering simulation, Manufacturing simulation technology, Simulation architecture, Simulation software platform and Intelligent Optimization Algorithm, Dynamic Programming, Ant Colony Optimization, Genetic Algorithm, Simulated Annealing Algorithm, Tabu Search Algorithm, Ant Colony System Algorithm, Hybrid Optimization Algorithm in other related areas.

3. Conference Committee Members

CONFERENCE GENERAL CHAIR
Prof. Yichuan Zhao, Georgia State University, USA

PROGRAM CHAIR
Prof. Yilun Shang, Northumbria University, United Kingdom

CONFERENCE COMMITTEE CHAIR MEMBERS
Prof. Zongwei Luo, Southern University of Science and Technology (SUSTech), China
Prof. Ying Ma, Xiamen University of Technology, China
Dr. Hui Sun, Academy of OPTO-Electronics, Chinese Academy of Science, BeiJing, China
Prof. Peng Lu, Central South University, China
Prof. Jinyang Xu, Shanghai Jiao Tong University, China

INTERNATIONAL TECHNICAL COMMITTEE MEMBERS
Prof. Bin Liu, Nanjing Medical University, China
Prof. Quanxin Zhu, Hunan Normal University, China
Prof. Zhixiao Wang, Xi'an University of Technology, China
Prof. Yingxue Zhao, University of International Business and Economics, China
Prof. Lulu Wang, Shenzhen Technology University, China
Dr. José Joaquim De Moura Ramos, Department of Computer Science, University of A Coruna, Spain
Dr. Pisutti Dararutana, Royal Thai Army Chemical Department, Thailand
Dr. Altan Onat, Eskisehir Technical University, Turkey
Dr. Zul-Atfi Ismail, School of Environmental Engineering, Universiti Malaysia Perlis, Malaysia
Dr. Ismail Yenilmez, Department of Statistics, Science Faculty, Eskisehir Technical University, Eskisehir, Turkey

INTERNATIONAL REVIEWERS
Prof. Kyaw Zaw Ye, Ministry of Defense, Myanmar
Prof. Gorbunov Vladimir, National Research University of Electronic Technology, Russia
Prof. Yu Xue, Nanjing University of Information Science and Technology, China
Prof. Goharik Petrosyan Razmiki, Armenian State Pedagogical University, Armenian
Prof. Ning Chen, Xian Polytechnic University, China
Prof. Nina Popva, St.Petersburg Polytechnic University, Russia
Prof. Anastasios Doulamis, National Technical University of Athens (NTUA), Greece
Dr. Jie Hu, University of Massachuttes, USA
Dr. LO, Man-fung, The Hong Kong Polytechnic University, China
Dr. Anand Nayyar, Duy Tan University, Da Nang, Vietnam
Prof. Peng Lu, Central South University, China
Dr. José Carlos Metrolho, Polytechnic Institute of Castelo Branco, Portugal
Dr. Izhar Mithal Jiskani, China University of Mining and Technology, China
Dr. Larisa Ivancu, Politehnica University of Timisoara, Romania
Dr. Ma Quanjin, Universiti Malaysia Pahang, Malaysia
Dr. Wondaferaw Yohannese Chubato, Southwest JiaoTong University, China

Dr Noraida Abdul Ghani, Universiti Sains Malaysia, Malaysia
Dr. Imran Memon, Zhejiang University ,China
Dr. Pratibhamoy Das, Department of Mathematics, Indian Institute of Technology, Patna, India
Prof. Subhabrata Banerjee, Department of Electronics and Communication Engineering, Institute of Engineering and Management, Kolkata, India
Assoc. Prof. Weili Zeng, College of Civil Aviation, Nanjing University of Aeronautics and Astronautics, Nanjing, China

4. Paper Submission

The total count of paper submission was 154. A preliminary review was conducted by the main editor and only the approved papers were sent for peer-reviewed. The peer review for selection of papers was conducted then, each paper was reviewed by 1-3 reviewers(average two) and finally 89 papers were accepted for paper presentation and publication in the conference. The acceptance rate is 58%. Please see the "Peer review declaration" file for more information.

5. Programs

In November 12, 2021, from 13:00-18:00, participants who participated in the physical conference registered and sign-in at the Conference Hall. In November 13-14, the conference was begun at 9:00 am at Qingshuiwan Meeting Room and also it is on air via Tencent Meeting. We are very honored to have Assoc. Prof. Jinyang Xu of Shanghai Jiao Tong University hosted the conference. There were 27 participants in total, inlcuding 5 on-site participants. 3 invited keynote speakers delivered their speech, each speech cost about 35 minutes and 5 minutes for questions&discussion. And the other two invited speakers delivered the oral speeches, each sharing cost about 25 minutes.

Group photo of on-site participants

Keynote Speeches

Assoc. Prof. Jinyang Xu is delivering the speech

Speech Title：Numerical Modeling and Experimental Study of Machining CFRP/Titanium Stacks

Assoc. Prof. Jinyang Xu
School of Mechanical Engineering, Shanghai Jiao Tong University, Shanghai, China

Biography: Dr. Jinyang Xu, MASME, MSCS, AMIMechE, is currently an Associate Professor and a Doctoral Supervisor of Mechanical Engineering at Shanghai Jiao Tong University, China. His main research interests concern the machining of fibrous composites utilizing both experimental and numerical methodologies as well as the micromachining of functional surfaces for dental dentures. His key skills and expertise lie in the multi-scale analysis of machining fibrous composites in relation to subsurface imperfections and the investigations addressing the mechanisms of heat transfer and issues of heat-induced damage for composites subjected to mechanical machining. He has published over 40 articles in highly-ranked JCR-referenced journals as a first/corresponding author, co-authored more than 50 peer-reviewed papers, and edited 8 special issues, 1 monograph & 2 book chapters, being cited over 1400 times with an h-index of 20 according to the Google Scholar Citation database. He is currently serving as the Editor-in-Chief of the Journal of Coating Science and Technology (JCST) and the International Journal of Product Sound Quality (IJPSQ, INDERSCIENCE PUBLISHERS, UK), the Associate Editor of SIMULATION - Transactions of the Society for Modeling and Simulation International (indexed by SCIE/EI), and the Academic Editors/Board Members of Green Materials (indexed by SCIE/EI), International Journal of Aerospace Engineering (indexed by SCIE/EI), LUBRICANTS, Journal of Composites Science (indexed by ESCI/SCOPUS), Journal of Engineering (indexed by ESCI/SCOPUS), and Current Materials Science (indexed by EI/SCOPUS). Moreover, Dr. Xu is also serving as a scientific reviewer for over 50 international scholarly journals indexed by SCI/SCIE/SCOPUS, such as IJMTM, JMPT, IJAMT, IJMS, COST, WEAR, etc. He has been invited as Keynote/Invited Speakers, Section Chairs, and TPC members for a number of international conferences (e.g., APMAS conference series, FAIM conference series, ICMEM conference series, EM2022, MD2021, AMSE2021, etc.) with themes in materials science and manufacturing processes. He is the principal investigator of some national and provincial projects, including the NSFC grants, Shanghai Pujiang Talents Program, Shanghai Academy of Spaceflight Technology projects, State Key Laboratory of Mechanical System and Vibration fund projects, as well as a number of industrial projects. For his notable and outstanding research in the field of Materials Processing & Manufacturing, he was honored with the 2020 IAAM Scientist Medal by the world-renowned scientific research organization in the fields of materials science - The International Association of Advanced Materials (IAAM) and received the Best Paper Awards at the CJUMP2017 and the ISGMA2015

Journal of Physics: Conference Series **2173** (2022) 011001 doi:10.1088/1742-6596/2173/1/011001

conferences.

Assoc. Prof. Harco Leslie Hendric Spits Warnars is delivering the speech online

Speech Title：System intelligent with Unstructured Data Text

Assoc. Prof. Harco Leslie Hendric Spits Warnars

Doctor of Computer Science (DCS), Bina Nusantara University, Jakarta Barat 11530, Indonesia

Biography: Harco Leslie Hendric Spits Warnars is Head of Concentration of Information Systems at Doctor of Computer Science (DCS) Bina Nusantara University (http://dcs.binus.ac.id) and supervisor some Ph.D. students in Computer Science. He did Bachelor's degree in Computer Science in the Information Systems field from STMIK Budi Luhur (http://fti.budiluhur.ac.id/id/), Jakarta Selatan, Indonesia with title S.Kom (Sarjana Komputer) between 1991-1995 with a bachelor thesis topic about information systems using Budi Luhur scholarship. Between 2004-2006 he continued his Master's degree in Computer Science with major field Information Technology in University Indonesia, with degree title M.T.I. (Magister Teknologi Informasi) (http://mti.cs.ui.ac.id/indonesia/?mod=content&type=home) with master thesis topic about Datawarehouse which was funded by Budi Luhur university. His Ph.D. Computer Science was done at The Manchester Metropolitan University, Manchester, United Kingdom (http://www2.mmu.ac.uk/science-engineering/), with a Ph.D. Thesis topic about Data Mining between 2008-2012. His Ph.D. Was funded by the Directorate General of Higher Education, Ministry of Education and Culture, Republic of Indonesia (DIKTI) scholarship.

He has been teaching computer science subjects since 1995. He has an Indonesian national lecturer degree Lektor Kepala (550) since 2007 which is recognized as Associate Professor. He had awarded some research grants such as Program of research incentive of national innovation system (SINAS) from the Ministry of Research, Technology and Higher Education of the Republic of Indonesia and Incentives article in the International Journal from the directorate of research and community service, Ministry of Research, Technology and Higher Education of the Republic of Indonesia.

He is a member of some professional membership such as IEEE (Institute of Electrical and Electronics Engineers) since 2011, member number: 92305834, (www.IEEE.org), SDIWC (The Society of Digital Information and Wireless Communication) member ID:3518 since March 2014, (www.SDIWC.net), IAENG (International Association of Engineers) member number: 140849 since April 2014, (www.IAENG.org), IACSIT (International Association of Computer Science and Information Technology) senior member since Jan 2014, (www.IACSIT.org), INSTICC (the Institute for Systems and Technologies of Information, Control and Communication) member number 5279 since June 2014, (http://www.insticc.org/myINSTICC/). He is active as a reviewer/program committee for some International journals or conferences and active as general chair, program chair, the general committee for some international conferences, active as the advisory and editorial board for 6 journals. His review activities can be seen at https://publons.com/researcher/1620795/harco-leslie-hendric-spits-warnars/metrics/. He had success running as general chair for 2 international conferences such as 2017 cyberneticsCOM (https://ieeexplore.ieee.org/document/8311679), 2018 Indonesian Association for Pattern Recognition International (INAPR) conference (https://ieeexplore.ieee.org/xpl/conhome/8612461/proceeding). Technical Program Chairs for 2019 IEEE International Conference on Teaching, Assessment, and Learning for Engineering (TALE) (https://ieeexplore.ieee.org/xpl/conhome/9220968/proceeding). His research publications can be reached at https://www.researchgate.net/profile/Harco_Leslie_Hendric_Spits_Warnars2, Or google scholar at https://scholar.google.co.id/citations?user=pplO3mEAAAAJ&hl=id or SCOPUS at https://www.scopus.com/authid/detail.uri?authorId=57219696428. He has published more than 261 papers which were cited by 1265 papers with h-index=20 and i10-index=51 as can be seen in his google scholar link. He has SCOPUS publication as can be seen his SCOPUS link account with 156 documents which were cited by 244 documents with h-index=12.

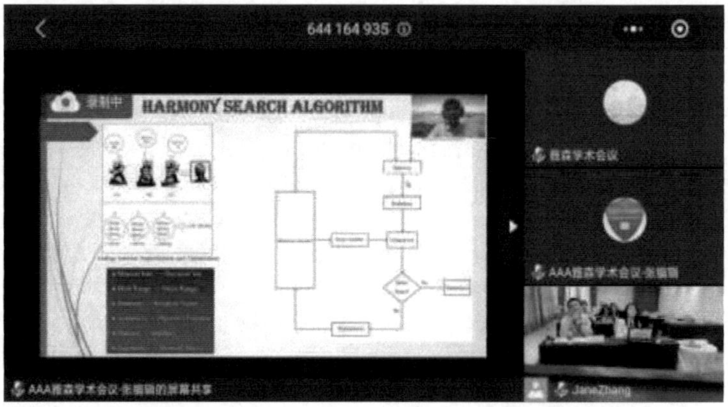

Dr. Subhabrata Banerjee is delivering the speech online

Speech Title：Nature based Meta-heuristic Algorithm and its impact on Modern Communication System

Dr. Subhabrata Banerjee

Department of Electronics and Communication Engineering, Institute of Engineering and Management, Kolkata, India

Biography: Dr. Subhabrata Banerjee is a Professor in the department of Electronics and Communication Engineering, Institute of Engineering and Management, Kolkata, India. He was awarded National Scholarship by WBED, West Bengal Government. He received his M.Tech. and PhD degrees in Electronics and Telecommunication Engineering from renowned Jadavpur University, Kolkata, India. He initially worked as Programmer Analysist in Cognizant Technology Solution and then is working as Assistant Professor in the Department of Electronics and Communication Engineering at Future Institute of Engineering and Management, Kolkata, India and as Associate Professor in the same institution thereafter. He is now working as Professor in Institute of Engineering and Management, Kolkata, India since 2021. He was authored numerous articles published in prestigious journals including Wireless Personal Communication (Springer), Personal and Ubiquitous Computing (Springer), International Journal of Communication Systems (Wiley) and Institution of Engineering and Technology (IET, UK). He has published more than 30 papers in International Conferences and International Journals of repute. He also published book chapter under IET,UK publication. His current research interests include Communication and Coding Theory, Machine Learning, Robotics, IOT, Control System, and Evolutionary

Algorithms, Soft Computation.

Oral Speeches

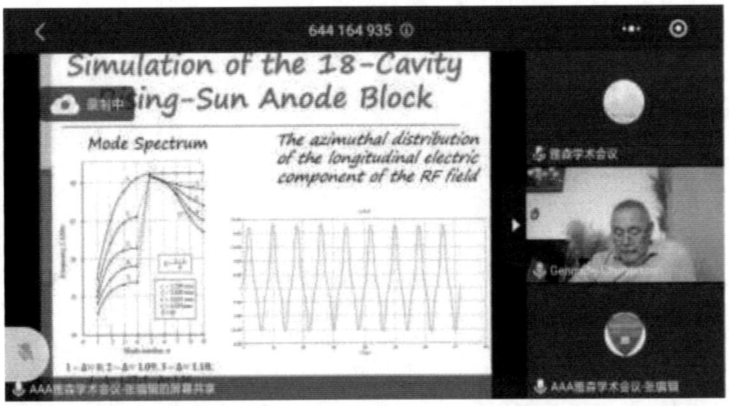

Dr. Gennadiy Churyumov is delivering the speech online

Speech Title：Simulation and Design of Rising-Sun Anode Blocks for Magnetrons with Two Outputs of Energy

Dr. Gennadiy Churyumov
O.Ya. Usikov Institute for Radiophysics and Electronics of National Academy of Science of Ukraine, Kharkiv, Ukraine

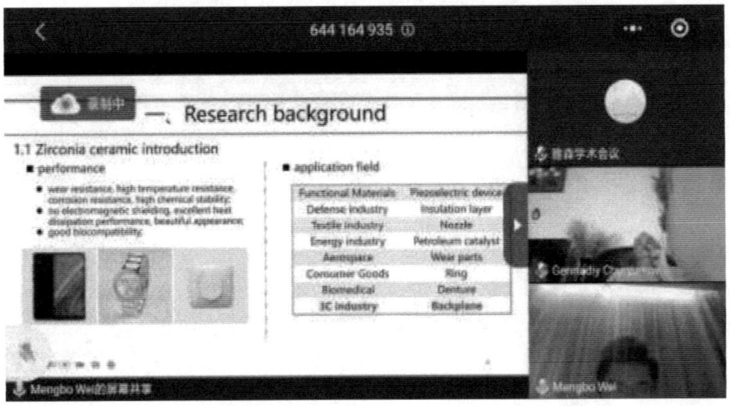

Mr. Wei Mengbo is delivering the speech online

Speech Title：System intelligent with Unstructured Data Text

Mr. Wei Mengbo
Tsinghua University, Xili University Town, Nanshan District, Shenzhen, Guangdong, China

Peer review declaration

All papers published in this volume of IOP Conference Series: Materials Science and Engineering have been peer reviewed through processes administered by the Editors. Reviews were conducted by expert referees to the professional and scientific standards expected of a proceedings journal published by IOP Publishing.

- **Type of peer review:** Double-blind review

First of all, the editor Iris will take a simple check to know whether the paper topics are included in the call-for-paper areas of this conference. Only the included submissions will then be numbered and sent to the conference committee members(reviewers) for peer-review, the author names, emails and affiliations were removed constantly in this porocess. Each reviewer should fill in the reviewing form and 9 criterias in total are to be scored (5-point scale, with 1=lowest and 5=highest), including "Originality: Is the work relevant and novel", "Motivation: Does the problem considered have a sound motivation", "Repetition: Have significant parts of the manuscript already been published", "Length: Is the content of the work of sufficient scientific interest to justify its length", "Conclusion: Does the paper contain a carefully conclusion, summarizing what has been learned and why it is useful", "Reference: Is the reference adequate and necessary", "Structure: Is the manuscript structure(including the title, abstract and the content) rational", "Language: Is the language accurate and readable", "Diagrams, figures, tables and captions: Are they essential and clear". Then the reviewer will give an overall evaluation or Detailed reviewing result(If any). In the end of the form, the final decision of Accept/Revision/Reject should be selected. No matter which selection the paper was marked, we send the whole reviewing form to the authors. When it is marked by Revision, we will have a double check of the revised papers before acceptation. The whole reviewing process can be tracked.

- **Conference submission management system:** Yaseen Academy

- **Number of submissions received:** 154

- **Number of submissions sent for review:** 128

- **Number of submissions accepted:** 89

- **Acceptance Rate (Number of Submissions Accepted / Number of Submissions Received X 100):** 58%

- **Average number of reviews per paper:** 2

- **Total number of reviewers involved:** 44

- **Any additional info on review process:** Selected papers were checked by Tuinition

- **Contact person for queries:** Iris Zhang (Yaseen Academy, Hongshan District, Wuhan, Hubei, China, iriszhang@yaseen.org.cn)

Research on Modeling and Simulation Technology of Equipment Maintenance Support System Based on Agent

Qing Xu*, **Hong-lin Liu, Xing-ming Hou**

Space Engineering University , Beijing 102206, China

*corresponding author's e-mail: xu950559@sina.com

Abstract. Aiming at the problem that the equipment maintenance support system is difficult to describe with highly stylized mathematical model, a framework model of equipment maintenance support agent is constructed. The paper introduces the Agent modeling method, analyzes the structure and workflow of the equipment maintenance support system, and abstracts the Agent framework structure according to the equipment maintenance guarantee and the modeling characteristics of the Agent. The internal modules of each Agent are designed and the agents are designed. The task environment between them has been studied, and the designed structural model can provide reference for other similar studies.
Keywords. Equipment maintenance support, Agent , Modeling and simulation.

1. Introduction

With the wide application of high-tech weapons and equipment, equipment maintenance and support work faces new opportunities and challenges. As the emergence of intelligent behavior, equipment maintenance support system has the research characteristics that traditional mathematical models are difficult to describe. Agent Based Modeling and Simulation(ABMS) is an emerging modeling thinking and methodology in the field of modeling and simulation. Since its introduction in the 1990s, it has rapidly spread in the study of complex phenomena in various fields such as economy, politics, transportation, disease transmission, and military, and has become an effective way to study complex systems. Practice has proved that the method based on Agent modeling and simulation has played a supportive role in studying the equipment maintenance support activities.

2. Agent related concepts of modeling and simulation

Base on Agent modeling and simulation technology originated from the theory of complex adaptive systems（Complex Adaptive System, CAS）.The CAS theory was put forward by Professor Holland in the 1990s on the basis of many years of research on complex systems. The basic idea is that the complexity of complex adaptive systems originates from the adaptability of Active Agents. It is these individuals and the environment and others. The interactions between individuals constantly change their own and at the same time change the environment. The most important feature of complex adaptive systems is adaptability. [1]This theory can be understood as abstracting the individual entities of the system into agents with active behaviors. The top-level phenomena and behaviors emerge from the bottom-up through the interaction between Agents and Agents and the environment. This is a way of generating modeling. Through the interaction and communication between the various underlying individuals, new structures, phenomena and more complex behaviors can be emerged on the whole level, such as the emergence of new levels, the emergence of differentiation and diversity, the formation of new aggregates, and greater The appearance of an individual, etc. Agent-based modeling and simulation is based on the analysis of the system. The system object is abstracted into an agent with autonomous behavior. The simulation system encapsulates the internal state and behavior of the agent. After the agent receives the external stimulus, the agent Self-acting behavior has complete control. [2]。 At present, agent has become the most important method and means to study complex adaptive systems, and has strong expressive ability for the emergence, nonlinearity and uncertainty of complex systems. Through the agent, not only can the simulation of the cooperation mode and the emerging effect of the equipment maintenance support force in combat, but also the environment similar to the battlefield can be simulated. At the same time, the dynamic cooperative effect of Agent reflects the advantages and disadvantages of equipment configuration in confrontation, which provides a better method and means for the research of optimal allocation of equipment resources.

3. characteristics analysis of the agent model

Agent is characterized by autonomy, responsiveness, initiative, sociality and intelligence. Agent-based modeling and simulation is applicable to systems with the following characteristics:

1) Individual behavior is difficult to describe with simple mathematical equations;

2) Individuals cannot perceive global information and can only act according to behavioral rules in a local information environment;

3) There is a dynamic relationship between individuals;

4) There are many random factors in the system.

Agent modeling and simulation are autonomous, cooperative and adaptive. They are suitable for the construction of distributed maintenance systems with autonomous, interactive and receiving information for the equipment maintenance support command and control system. The equipment maintenance support system can be regarded as a distributed system in which each information element realizes cooperation through information exchange to complete a specific task. For systems where such elements are physically or logically distributed and need to collaborate to solve problems, Agent technology provides a very natural modeling method, which maps the components of the actual security system to Agents. The resources, knowledge, goals, and capabilities of these components are encapsulated as the attributes of the Agent itself. Agents communicate, coordinate, and process information through events and activities. The relationship structure is mapped from the organizational structure in the actual system. The actual system runs the behavior of each agent and the interaction between the agents is reflected.

4. Research on Equipment Maintenance Support Command and Control System Based on Agent Modeling Method

The underlying behavior and the top-level phenomenon are organically linked through dynamic interaction. The individual behavior and interaction are the driving force of the system operation, reflecting the overall phenomenon and ability of the system in the dynamic interaction, and the bottom-level behavior is “bottom-up” dynamics. “Generate” the top-level phenomenon, through statistical analysis and other mechanisms to get the connection between the bottom layer and the top layer of the system. The traditional modeling and simulation based on "reduction theory" provides a way of decomposing the top-level phenomenon and simplifying the "top-down" decomposition of the underlying causes. The two have fundamental differences.

Stochastic factors make the model more expressive. Agent simulation systems usually contain many random factors. From the perspective of Agent's individual analysis of the underlying behavior, from the perspective of the system as a whole, the system is combined to facilitate better grasp of the system law, while the random factors make the disturbance of the underlying behavior produce a change in the top layer phenomenon, and strengthen the contrast between the bottom layer and the top layer. Enhance the expressive ability of the model.

The ABMS and artificial intelligence agents can be complemented by the factors that are difficult to consider in traditional modeling. The research results in the field of artificial intelligence can be applied to ABMS without barriers. Although the behavior of Agent is not much, it can show very high. Intelligence, including the intelligence, emotions, emotions, intentions, beliefs, etc. of people who cannot be considered by traditional modeling, such as the modeling and learning ability of commanders in combat command. With the continuous development of system theory, people's understanding of the system is no longer limited to the view of closed systems, but more and more attention to the study of open, uncertain and dynamic system environment. In the field of artificial intelligence, the external environment is classified into deterministic and random, fully observable and partially observable, static and dynamic, discrete and continuous, known and unknown, etc. Traditional modeling and simulation methods are difficult to express, and Agent-based modeling not only abstracts the Agent model, but also models the environment. Agents survive and interact in the environment model. By modeling the environment separately, you can fully examine the impact of the environment on the system.

Highly modular and extensible The Agent behaves as a completely closed whole in the environment. From the outside, it can only understand the external behavior of the Agent. Its internal mechanism is completely tightly encapsulated and is "externally invisible". This modular idea makes the Agent highly replaceable. That is, according to different research purposes and research levels, the same object can be abstracted into Agent models of different granularity, which can be easily replaced and modified. The scalability.

Applicable to large-scale simulation calculations It is concluded that the Agent modeling process generally follows the principle of "Keep It Simple, Stupid" (KISS), that is, the behavior of the Agent should be as simple as possible, so the calculation amount of the Agent simulation program is usually small and can be performed. Large-scale simulation calculations have important value in practical research.

5. Design of Agent-based equipment maintenance support system model and simulation steps

5.1. Design model structure

The components of the Agent Model System for Designing Equipment Maintenance Support Systems include:

1) Agent collection: The agent collection includes not only the command and control system of our military equipment maintenance support, the base level power agent, the force level agent, and the information platform agent, but also the enemy's confrontation force agent.

2) Agent relationship and its interactive topology network: Determine the interaction rules of the agent through the specific analysis of the maintenance support activities. The interactive topology network covers the interaction criteria such as management, deployment, and collaboration of the Agent, and the interconnection relationship between the agents.

3) Agent environment: According to the characteristics of equipment maintenance and support activities in the task, the agent environment of equipment maintenance support is called the task environment model, all agents exist in the task environment, and the agent's data, models, criteria, and plans pass the environment. Perceive to understand the state of the environment and interact dynamically in the environment.

Through this design idea, the structural framework of the Agent-based equipment maintenance support system is constructed as shown in Figure 1.

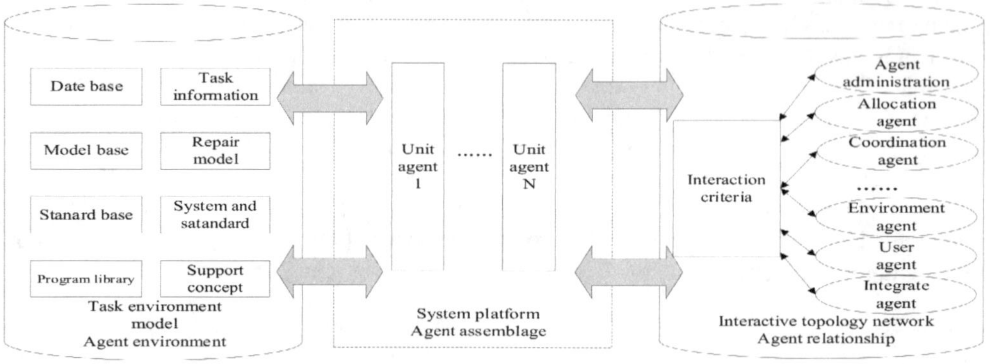

Figure 1. Structural frame of equipment maintenance support simulation system

5.2. *Task Environment and Maintenance Support Unit Agent Modeling*

Agent environment is a special entity in a multi-agent system. [3] It is also the place where the Agent exists and interacts. In order to conveniently express the characteristics of the task environment and interaction relationship, this paper proposes a battlefield environment model for the anti-missile system Agent modeling based on the three-dimensional Euclidean space model. It consists of three elements, which are expressed as:

$$Env = < Con, Prj, Opr >$$

Among them, Con is the environment state, Prj is the Agent relationship set, and Opr is the Agent operation set, which together constitute the Agent battlefield environment model, as shown in Figure 2.

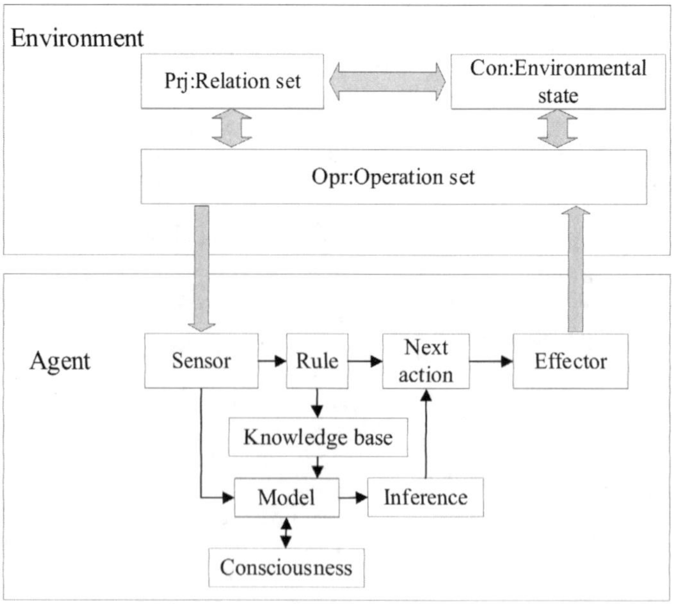

Figure 2. Interaction between Agent and task environment

According to the simulation target of the equipment support system maintenance support simulation system and the actual available computing power, the abstraction level of the simulation system is defined to the maintenance unit that can complete a certain degree of guarantee task. According to the ABMS method rules, the models are divided into three categories in the simulation system: entities, objects, and agents. Among them, Agent is divided into stimulus-response type, deliberate type and mixed type. This paper adopts Hybrid Agent, which can combine the characteristics of both reactive and deliberate Agents, both reactive and deliberate, with strong flexibility and rapid response.

5.3. Model Simulation

The agent-based maintenance support system simulation system architecture introduced in this paper is mainly composed of simulation engine, basic service component, environment model and Agent of various equipments. The Engine is the manager and driver of the whole simulation. The basic service component is some support libraries added for the convenience of simulation. The core of the simulation system is a multi-agent system consisting of various weapon equipment agents and environment model Environment. The basic organizational structure is described as: each class of agents is implemented as a class, the class object is an agent individual; the simulation engine class has a singleton object, a singleton object of the management environment model and all object agent individuals, providing a simulation environment for all objects And the active behavior of the Agent is invoked in each simulation step to ensure that each Agent object performs its active behavior at each step; the singleton object of the environment model has an interactive interface with all Agents, responding to the perceived behavior of the Agent and Effect behavior; the core of the Agent class is the member method/function step, which performs the agent's perceptual behavior, internal behavioral reasoning, state update, and effect behavior sequentially in each simulation step.

Through the equipment maintenance guarantee system performance simulation, it can carry out extensive analysis and calculation, obtain a large number of task processes and result data, and scientifically and effectively perform system performance index calculation and analysis. In the Agent-based simulation system, the behavior and interaction of the underlying individuals of the system is the driving force of the system operation.[4]In the dynamic interaction, the overall phenomenon and ability of the system are reflected. The underlying behavior dynamically "generates" the top-level phenomenon "bottom-up", and the relationship between the bottom layer and the top layer can be obtained by statistical analysis. An important advantage of Agent-based modeling and simulation is that if an expected behavior occurs in the simulation results, the researcher can quickly get a description of the top-level behavior and underlying dynamics of the system.[5]The Agent associates the underlying behavior of the system with the top-level behavior by means of "generation". Through this connection, it can be examined which underlying behaviors lead to the top-level phenomenon of a certain model, although this connection is sometimes very concealed and dynamically changed. This characteristic is reflected in the simulation calculation. It can be seen that by modifying the underlying rules or parameters, the changes of the simulation results can be viewed, and the influence of different factors on the overall effect of the system operation is judged from the statistical laws of a large number of results. The research method is based on the agent modeling and simulation to provide the "generation" of the underlying object and the top-level behavior, and provides a method guide for the agent simulation modeling of the maintenance support system.

6. Conclusion

Through the analysis of the simulation results of the maintenance support system, it is proved that the multi-agent simulation method is a feasible and effective method to study the equipment maintenance support effectiveness evaluation. In the next step, the wartime maintenance support activities can be simulated on the basis of the mission maintenance guarantee, considering the war damage situation in the confrontation, and providing feasible methods and suggestions for improving the overall maintenance support capability of the military.

References

[1] TAN Y.J,DENG H.Z. (2001) Research on Complex Adaptive System Theory and Its Application[J]. Systems Engineering.
[2] Huang H.B. (2009) Analysis of multi-agent system based on emerging perspective [D] .Changsha: National University of Defense Technology.
[3] CM Macal and MJ North. (2010) Tutorial on agent-based modelling and simulation [J]. Journal of Simulation, 4.

[4] Li X.(2013) Agent-based combat modeling [M]. Beijing:National Defence Industry Press.
[5] Zhang Z.X,Gao C.R, etc. (2010) Artificial Warfare: Multi-Agent Based Combat Simulation [M]. Artificial War: agent-Based Simulation of Combat. Beijing: Publishing House of Electronics Industry.

Kinematics and Simulation Analysis of Longitudinal-Torsional Ultrasonic Assisted Grinding of Zirconia Ceramics

Mengbo Wei[1], Pingfa Feng[1,2], Huiting Zha[1,*], Chao Xu[1], Yuan Ma[1] and Feng Feng[1]

[1]International Graduate School at Shenzhen, Tsinghua University, Shenzhen 518055, China
[2]Department of Mechanical Engineering, Tsinghua University, Beijing 100084, China

[*]corresponding author's e-mail: zhahuiting123@sz.tsinghua.edu.cn

Abstract. Zirconia ceramics is a typical hard and brittle material, which is very difficult to process. In recent years, with the rapid development of the 3C industry, zirconia ceramic materials have achieved a wide range of applications. Longitudinal-torsional ultrasonic assisted grinding (LTUAG) is efficient for processing hard and brittle materials. However, the mechanism of LTUAG of zirconia ceramic is not studied sufficiently. In this study, the spatial trajectory of the grinding grain in three different grinding methods is established and the critical cutting speed of LTUAG is calculated. Simulation model of LTUAG of zirconia ceramic is established. The results show that the longitudinal-torsional ultrasonic vibration increases the spatial trajectory length of the grinding grain, thereby reducing the thickness of the maximum removal layer and reducing the cutting force. When the cutting speed is less than the critical cutting speed, LTUAG can significantly reduce the Mises stress, the subsurface damage and the cutting force.
Keywords. Zirconia ceramics; Longitudinal-torsional ultrasonic assisted grinding; Finite element analysis; Spatial trajectory; Cutting force

1. Introduction

Zirconia ceramics have excellent properties such as high hardness, high strength, wear resistance, and less influence on signals, which is widely used in the fields of electronic products and communications [1]. The high hardness and brittleness of zirconia ceramics make its processing difficult. It has low processing efficiency. The processing cost is also relatively high, accounting for almost half of its production cost [2-4].The machining accuracy and surface quality of zirconia ceramics are difficult to guarantee [5]. This severely limits the application of hard and brittle materials.

In recent years, many scholars have done a lot of research on the grinding simulation of hard and brittle materials. In the grinding simulation of hard and brittle materials, JH-2 is the most frequently used constitutive model [6].When the hard and brittle material is in the brittle removal stage, the cutting force is closely related to the size of the indentation crack [7].In the material removal process, the cutting speed is the most critical factor affecting the integrity of the machined surface, and reducing the cutting speed is conducive to reducing the occurrence of cracks [8].In summary, the current grinding simulation is mainly aimed at stress and subsurface damage. There are relatively few grinding simulations for ultrasonic machining, and there is little discussion of process boundaries. In order to make better use of the advantages of ultrasonic machining, this research uses simulation to explore the technological boundaries of ultrasonic machining.

First, this study establishes the spatial kinematics equation of the grinding grain, and then establishes the spatial trajectory equation of the grinding grain by means of space curve integration. Through spatial trajectory analysis, the relationship between the maximum removal layer thickness and the spatial trajectory displacement of abrasive particles is established. On this basis, by analyzing the relationship between the cutting force and the maximum removal layer thickness, the critical speed of ultrasonic machining is found. Finally, the critical speed of ultrasonic machining is verified by grinding simulation, and the Mises stress of the tool and the subsurface damage of the workpiece are analyzed.

2. Kinematics analysis of the grinding grain

The motion trajectory equations of the grinding grain in the three processing methods of common grinding (CG), longitudinal ultrasonic assisted grinding (LUAG), and longitudinal torsional ultrasonic assisted grinding (LTUAG) can be expressed by formulas (1), (2),and (3),respectively.

Figure 1 is the schematic diagram of processing.

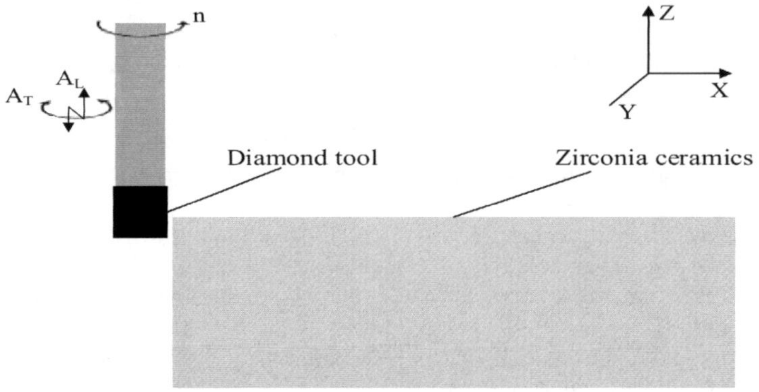

Figure 1. Grinding schematic.

$$\begin{cases} X_{CG} = R\cos(\omega t) + V_f t \\ Y_{CG} = R\sin(\omega t) \\ Z_{CG} = Z_0 \end{cases} \tag{1}$$

$$\begin{cases} X_{LUAG} = R\cos(\omega t) + V_f t \\ Y_{LUAG} = R\sin(\omega t) \\ Z_{LUAG} = Z_0 + A_L \sin(2\pi f t) \end{cases} \tag{2}$$

$$\begin{cases} X_{LTUAG} = R\cos(\omega t + \theta\cos(2\pi f t)) + V_f t \\ Y_{LTUAG} = R\sin(\omega t + \theta\cos(2\pi f t)) \\ Z_{LTUAG} = Z_0 + A_L \sin(2\pi f t) \end{cases} \tag{3}$$

where, R is the radius of the tool, n is the spindle speed, ω is the angular speed of the tool, V_f is the feed speed, A_L is the ultrasonic longitudinal amplitude, A_T is the ultrasonic torsion amplitude, f is the ultrasonic frequency, θ is the ultrasonic torsion amplitude angle.

Figure 2 shows the spatial trajectory diagram of the grinding grain in the three processing methods. It can be found that the ultrasonic vibration has obviously changed the motion trajectory of the grinding grain, making the motion of the grinding grain complex.

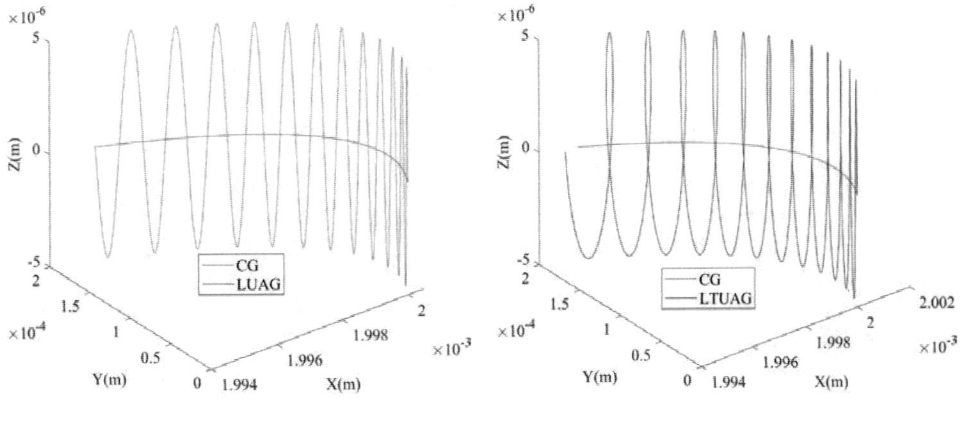

(a) LUAG (b) LTUAG

Figure 2. The spatial trajectory diagram of the cutting grain. (a) LUAG; (b) LTUAG.

3. The spatial trajectory length of the grinding grain and the critical cutting speed

The spatial trajectory length of the grinding grain directly affects the maximum removal layer thickness in the material removal process. The cutting force in the grinding process is directly related to the maximum removal layer thickness.

According to the kinematic analysis of the grinding grain, the spatial trajectory length of the grinding grain can be expressed by formula (4):

$$L = \int_{t_1}^{t_2} \sqrt{V_x^2 + V_y^2 + V_z^2} \tag{4}$$

The equation of the maximum removal layer thickness [9-12] in the CG and the UAG can be expressed by formulas (5) and (6) ,respectively:

$$a_{\mathrm{CG}_{\max}} = \left[\frac{V_f}{V_s} a_p\right] \left[\frac{N_d b L_{\mathrm{CG}}}{2}\right]^{-1} \tag{5}$$

$$a_{\mathrm{UAG}_{\max}} = \left[\frac{V_f}{V_s} a_p\right] \left[\frac{N_d b L_{\mathrm{UAG}}}{2}\right]^{-1} \tag{6}$$

where, V_s is the cutting speed, a_p is the cutting depth, N_d is the number of dynamic effective grinding grains in the grinding arc zone, b is the width of the grinding arc zone, L_{CG} is the spatial trajectory length of the grinding grain in the common grinding, L_{UAG} is the spatial trajectory length of the grinding grain in the ultrasonic compound grinding process.

The cutting force of grinding grain [11,13] can be expressed by the formula (7):

$$F_{nb} = \frac{1}{2}\xi \cdot (a_{\mathrm{CG}_{\max}/\mathrm{UAG}_{\max}})^2 \cdot \tan^2\frac{\alpha}{2} \cdot H_v \tag{7}$$

where, ξ is the geometrical factor of the indenter, α is the apex angel of the grinding grain, H_v is the hardness of the workpiece.

According to formula (7), it can be seen that the cutting force is proportional to the maximum removal layer thickness. The maximum removal layer thickness is inversely proportional to the spatial trajectory length of the grinding grain. Therefore, the spatial trajectory length of the grinding grain can effectively predict the cutting force during the machining process.

Figure 3(a) shows the relationship between the spatial trajectory length of the grinding grain and the cutting speed. The spatial trajectory length of the grinding grain increases with the increase of the cutting speed. The spatial trajectory length of the grinding grain is the longest in LTUAG. The spatial trajectory length of the grinding grain in CG is the smallest.

Figure 3(b) shows the decrease rate in the spatial trajectory length of the grinding grain between different processing methods. It can be seen that the decrease rate becomes smaller as the cutting speed increases. In the comparison between CG processing and LTUAG, the reduction rate dropped from the largest 67.2% to the smallest 4.7%. In the comparison between LUAG and LTUAG, the reduction rate dropped from the largest 16.9% to the smallest 0.2%. In the comparison between CG processing and LUAG, the reduction rate dropped from the largest 43.1% to the smallest 4.4%. Therefore, it can be seen that with the continuous increase of the cutting speed, the length of the grinding grain space trajectory of the three processing methods is continuously approaching. When the cutting speed reaches a certain value, the length of the grinding grain space trajectory of the three processing methods is close to the same. The length of the grinding grain space trajectory is closely related to the cutting force, so when the cutting speed reaches a certain value, the cutting force of the three processing methods is close to the same.

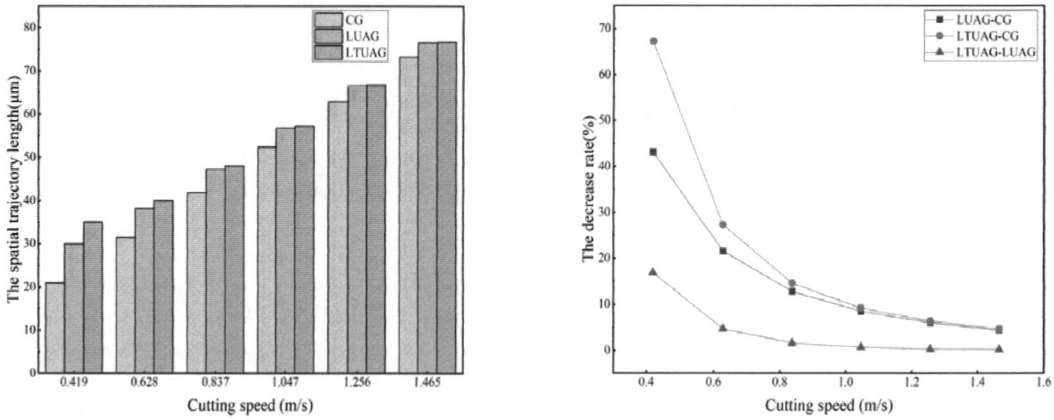

(a)The spatial trajectory length (b)The decrease rate

Figure 3. Changes in the space trajectory of grinding grain. (a) the spatial trajectory length;(b) the decrease rate.

4. Modeling of zirconia ceramic grinding simulation

The grinding simulation model includes two parts, the zirconia ceramic workpiece and the tool.The tool is set as a rigid body, so it does not need to be meshed, which reduces the amount of simulation calculations. The size of the grinding grain 74 μm, and its rake angle and back angle are both 30°.The size of the zirconia ceramic workpiece is 1000×50μm.This grinding simulation uses the JH-2 constitutive model, and the parameters [14] in the model are given in Table 1.According to the previous research, it is found that the cutting force is closely related to the cutting speed, so a single factor experimental method is adopted to verify. In the setting of simulation parameters, in order to ensure the consistency of the study, the same cutting speed as when studying the length of the abrasive particle space trajectory is adopted.

Table 1. Parameters of zirconia ceramic.

Parameters	Unit	Value	Parameters	Unit	Value
Density	Kg/m3	5850	T	GPa	0.2
Shear modulus	GPa	95.31	HEL	GPa	2.79
Poisson's ratio	/	0.22	PHEL	GPa	1.46
A	/	0.93	D1	/	0.005
B	/	0.31	D2	/	1.0
C	/	0	K1	GPa	130.95
M	/	0.6	K2	GPa	0
N	/	0.6	K3	GPa	0
EPSI	/	1.0	Beta	/	1.0

The simulation parameters are shown in Table 2. Figure 4 shows the orthogonal cutting geometry model of zirconia ceramics.

Table 2. Simulation parameters.

Cutting speed (m/s)	Cutting depth (μm)	Frequency (Hz)	AL (μm)	AT (μm)
0.419	15	20000	5	5
0.628	15	20000	5	5
0.837	15	20000	5	5
1.047	15	20000	5	5
1.256	15	20000	5	5
1.465	15	20000	5	5

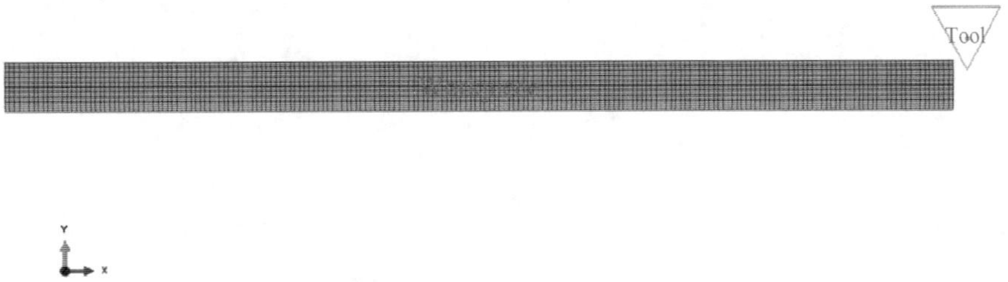

Figure 4. The orthogonal cutting geometry model of zirconia ceramics.

5. Results and discussion

Figure 5 shows the maximum Mises stress experienced by the grinding grain in an ultrasonic cycle under the three processing methods. Among them, in the processing of CG, LUAG, and LTUAG, the maximum Mises stress received by the grinding grain is 1.407×10^3MPa, 3.478×10^2MPa, and 3.31×10^2MPa, respectively. In the processing of CG, LUAG, and LTUAG, the average Mises stress of the grinding grain in an ultrasonic cycle is 7.947×10^2MPa, 3.064×10^2MPa, 2.655×10^2MPa, respectively. From the above analysis results, it can be seen that LTUAG can obviously reduce the Mises stress on the tool during the machining process, thereby reducing the wear of the tool.

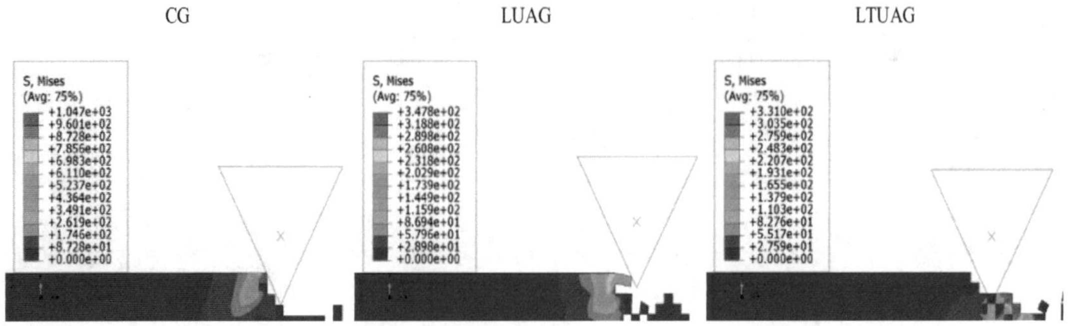

Figure 5. The maximum Mise stress of the grinding grain in an ultrasonic cycle.

Figure 6 shows the subsurface damage of zirconia ceramics under three processing methods. Among them, the maximum subsurface damage depth values of CG, LUAG, and LTUUAG are 26 μm, 25 μm, and 20 μm, respectively. The subsurface damage of zirconia ceramics directly determines its service life and safety, so LTUAG is beneficial to improve the subsurface damage when processing zirconia ceramics, and improve the strength and service life of the product.

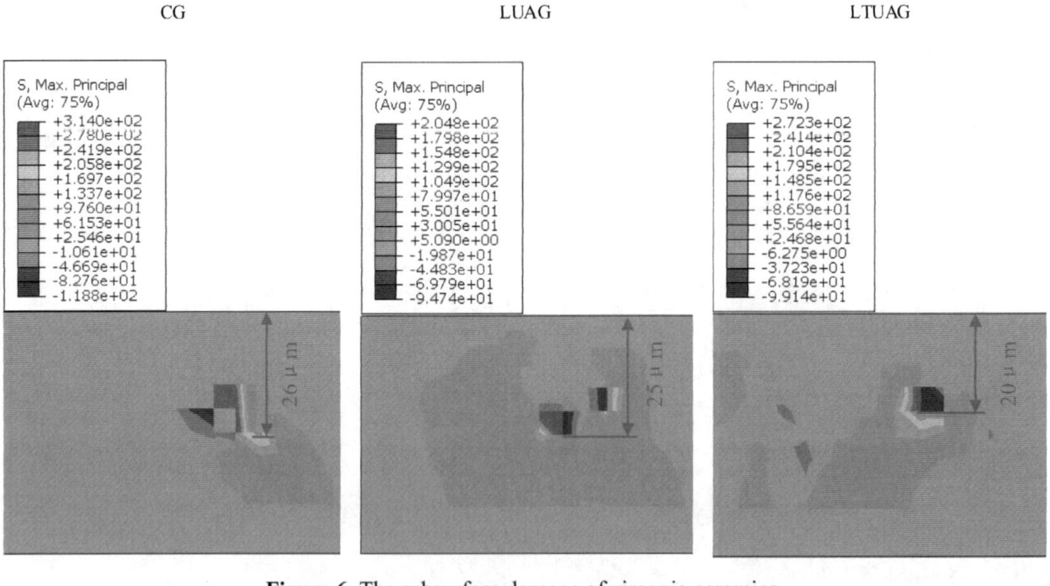

Figure 6. The subsurface damage of zirconia ceramics.

Figure 7 shows the cutting force in the three machining modes. It can be seen that the cutting force decreases with the increase of cutting speed and the cutting force of LTUAG is always smaller than that of CG and LUAG, but with the continuous increase of cutting speed, the cutting forces of the three processing methods are constantly approaching. Compared with CG, the cutting force of LUAG is reduced by 5.3%-41%.Compared with CG, the cutting force of LTUAG is reduced by 6.1%-73.7%. Compared with LUAG, the cutting force of LTUAG is further reduced by 0.8%-55.4%.Based on the above results, it is found that LTUAG has great advantages in reducing cutting force and extending tool life. However, as the cutting speed increases, the advantages of LTUAG are gradually decreasing, so it can be found that there is a critical speed in the application of LTUAG, which is consistent with the previous study on the length of the grinding grain space trajectory.

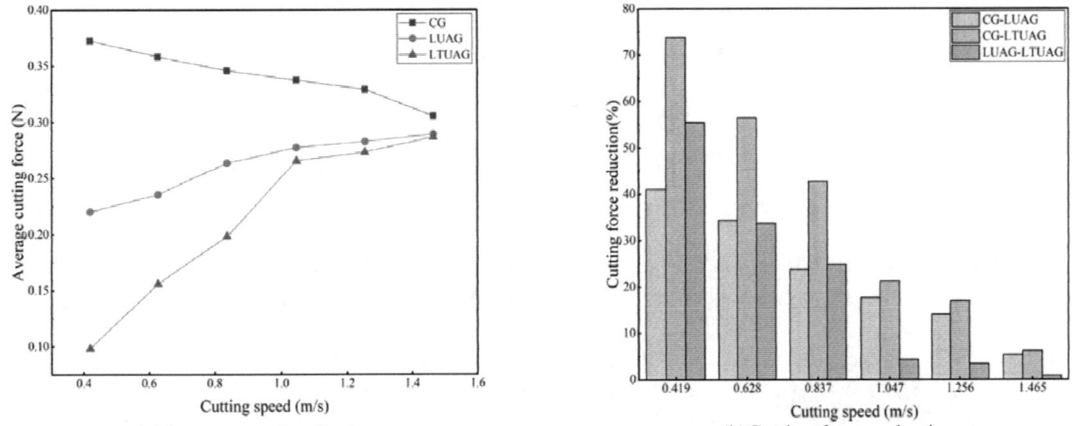

(a)Average cutting force

(b)Cutting force reduction

Figure 7. Changes in cutting force. (a) Average cutting force; (b) Cutting force reduction;

6. Conclusion

In order to make full use of the advantages of LTUAG processing zirconia ceramics, the kinematic equations of CG, LUAG and LTUAG are first established. Then, the spatial trajectory length equations of the abrasive particles under the three processing methods are established by means of curve integration. The critical speed of LTUAG is predicted by the mathematical relationship between the cutting force and the maximum removal layer thickness. By analyzing the simulation results of zirconia ceramic grinding, the following conclusions can be drawn:

(1) The length of the abrasive particle space trajectory of LTUAG is longer than that of CG and LUAG, but with the continuous increase of cutting speed, the length of the abrasive particle space trajectory of the three processing methods keeps getting closer. According to the formula of cutting force, it can be found that when the cutting speed increases to a certain value, the cutting force of the three processing methods tends to be the same. Therefore, there is a critical speed. When the cutting speed is less than the critical speed, LTUAG can significantly reduce the cutting force.

(2) Compared with the CG and LUAG, the average Mises stress on the tool in LTUAG is reduced by 199.32% and 15.4%, respectively. Therefore, LTUAG has obvious advantages in reducing tool wear and improving tool life.

(3) LTUAG can also significantly reduce the subsurface damage of zirconia ceramics during processing, thereby improving the processing quality of zirconia ceramics, and increasing the service life and safety.

(4) When using LTUAG to process zirconia ceramics, it is necessary to ensure that the cutting speed is less than the critical speed, which can reflect the processing advantages of LTUAG, thereby reducing the cutting force and increasing the tool life

Acknowledgments

This research was funded by Shenzhen Foundational Research Project (General program) (Grant No. JCYJ20190813173607172), Shenzhen Foundational Research Project (Discipline Layout) (Grant No.JCYJ20180508152128308) and Shenzhen Science and technology plan (Grant No.JSGG20180507181638381).

References

[1] BIAN R, FERRARIS E, HE N, et al. 2014. Process investigation on meso-scale hard milling of ZrO2 by diamond coated tools. Precision Engineering-Journal of the International Societies for Precision Engineering and Nanotechnology [J], 38: 82-91.

[2] GONG H, FANG F Z, HU X T 2010. Kinematic view of toot life in rotary ultrasonic side milling of hard and brittle materials. International Journal of Machine Tools & Manufacture [J], 50: 303-307.
[3] LIU J W, BAEK D K, KO T J 2014. Chipping minimization in drilling ceramic materials with rotary ultrasonic machining. International Journal of Advanced Manufacturing Technology [J], 72: 1527-1535.
[4] DING K, FU Y, SU H, et al. 2014. Experimental studies on drilling tool load and machining quality of C/SiC composites in rotary ultrasonic machining. Journal of Materials Processing Technology [J], 214: 2900-2907.
[5] KROLCZYK G M, NIESLONY P, LEGUTKO S 2015. Determination of tool life and research wear during duplex stainless steel turning. Archives of Civil and Mechanical Engineering [J], 15: 347-354.
[6] JOHNSON G R, HOLMQUIST T J 1994. An improved computational constitutive model for brittle materials. AIP Conference Proceedings [J]: 981-984.
[7] LAMBROPOULOS J C, XU S, FANG T, et al. 1996. Twyman effect mechanics in grinding and microgrinding. Applied Optics [J], 35: 5704-5713.
[8] ZHU D, YAN S, LI B 2014. Single-grit modeling and simulation of crack initiation and propagation in SiC grinding using maximum undeformed chip thickness. Computational Materials Science [J], 92: 13-21.
[9] CHEN Y, SU H, HE J, et al. 2021. The Effect of Torsional Vibration in Longitudinal-Torsional Coupled Ultrasonic Vibration-Assisted Grinding of Silicon Carbide Ceramics. Materials [J], 14.
[10]LI C, ZHANG F, MENG B, et al. 2017. Material removal mechanism and grinding force modelling of ultrasonic vibration assisted grinding for SiC ceramics. Ceramics International [J], 43: 2981-2993.
[11]XIAO X, ZHENG K, LIAO W, et al. 2016. Study on cutting force model in ultrasonic vibration assisted side grinding of zirconia ceramics. International Journal of Machine Tools & Manufacture [J], 104: 58-67.
[12]LIANG Y, CHEN Y, CHEN B, et al. 2019. Feasibility of Ultrasonic Vibration Assisted Grinding for Carbon Fiber Reinforced Polymer with Monolayer Brazed Grinding Tools. International Journal of Precision Engineering and Manufacturing [J], 20: 1083-1094.
[13]FENG J, The Theoretical and Experimental Studies on Ultrasonic Aided High Effificiency Lapping With Solid Abrasive of Engineering Ceramic (Ph.D. thesis), Shanghai Jiao Tong University, 2008.
[14]DENG B, YANG M, ZHOU L, et al. 2019. Smoothed particle hydrodynamics (SPH) simulation and experimental investigation on the diamond fly-cutting milling of zirconia ceramics [M] //E. OZTURK, T. MCLEAY, R. MSAOUBI, 17th Cirp Conference on Modelling of Machining Operations: 202-207.

Journal of Physics: Conference Series **2173** (2022) 012003 doi:10.1088/1742-6596/2173/1/012003

Intelligent Evaluation Method of Marine Engine Simulator Based on Combination Weighting

Shiyuan Liang, Hui Cao* and Henglong Shen

School of Marine Engineering, Dalian Maritime University, Dalian, Liaoning, China

*corresponding author's e-mail: lustersoft@gmail.com

Abstract. In order to solve the problems of objectivity and inaccuracy in the current evaluation of marine engine simulator, an evaluation model is established based on C-OWA (combination ordered weighted averaging)-the Coefficient of Variation method-minimum discriminant information principle combination weighting method and Vague set. Taking standby of marine generator project as an experiment example, the evaluation index system is established, and the evaluation model is applied to evaluate each evaluation team, and the final operation score of each team is obtained. The results show that (1) the index system is comprehensive and targeted for the operation evaluation of generator standby project; (2) the combination weighting method can improve the rationality of index weighting; (3) Vague set fuzzy decision-making can improve the accuracy of fuzzy evaluation of simulator operation; (4) the model shows rationality and reliability in evaluation results and can provide a reasonable and feasible method for intelligent evaluation of marine simulator.

1. Introduction

The automation systems of modern ships are developing in the direction of intelligence. Correspondingly, the overall quality of management and operators at all levels also needs to be further improved[1]. It has become a trend to train the crew by using a marine simulator. At present, it is widely used to carry out teaching and training for marine engineers, and has achieved good results[2]. The students who participate in the training can get the experience of long-term navigation work by training and practicing the marine engine simulator in a short time. The use of simulation training can also avoid the potential safety hazards caused by the crew's mistakes and significantly improve safety[3]. As a professional simulation operation platform, the modern marine engine simulator[4] can fully meet the requirements of marine engineering training and assessment.

Simulator evaluation, as an essential way to assess marine engineers, has been paid more and more attention by relevant researchers of marine engineering[1]. Now many experts and scholars have put forward many evaluation methods for marine engine simulator. DUAN [2] extracted evaluation rules based on expert experience and evaluation criteria, and used the weight optimized by genetic algorithm to make multiple fuzzy comprehensive evaluation on the data; NIE [3] put forward the establishment method of evaluation criteria, and combined with fuzzy comprehensive evaluation method and expert system theory, realized the automatic scoring of evaluation system; WANG [5] established the pingaji index system, and used the C-OWA operator to realize the evaluation index weighting; ZHU [6] used the coefficient of variation of evaluation indexes to calculate the objective weight instead of the ratio of the importance of adjacent evaluation indexes determined by the decision maker; LIU [7] analyzed the existing vague set method to solve the multi-objective decision-making problem under fuzzy conditions.

Based on the analysis of the above views, combined with the actual operation data of the examiners on the simulator, an intelligent evaluation method combining evaluation index weighting and Vague set[7] is proposed. The combination method of subjective and objective is used to weight each evaluation index, so as to consider the actual operation ability of the team and realize the intelligent evaluation of the marine engine simulator.

2. Methodology

2.1. C-OWA method

C-OWA (combination ordered weighted average) operator is an average ordered weighted operator that is optimized by the OWA method and calculated by binomial coefficient[5]. This method is widely used in the subjective weighting of evaluation index, which needs to consider expert experience. The process of empowerment is as follows:

(1) Suppose that there are n experts in related fields invited to evaluate the criticality of the evaluation indexes according to their personal experience and knowledge (For the convenience of calculation, the score interval is generally set to [0,10]), rearrange their scoring values $A=(a_1,a_2,\cdots,a_j,\cdots,a_n)$ in descending order, and renumber $A=(a_1,a_2,\cdots,a_j,\cdots,a_n)$ from 0, that is $B=(b_0,b_1,\cdots,b_j,\cdots,b_{n-1})$.

(2) Using the combination number C_{n-1}^{j} to calculate the weight η_{n+1} of B,

$$\eta_{n+1}=\frac{C_{n-1}^{j}}{\sum_{k=0}^{n-1}C_{n-1}^{k}}=\frac{C_{n-1}^{j}}{2^{n-1}},j=0,1,2,\cdots,n-1 \tag{1}$$

Where n is the number of experts participating in the evaluation, $C_{n-1}^{j}=\frac{(n-1)!}{j!(n-1-j)!}$, $\sum_{0}^{n-1}\eta_{n+1}=1$.

(3) The absolute weight of each evaluation index is ω_i,

$$\omega_i=\sum_{j=0}^{n-1}\eta_{j+1}b_j,i=1,2,\cdots,m \tag{2}$$

Where m is the number of evaluation indexes, $\eta_{n+1}\in[0,1]$.

(4) Calculate the subjective weight of each evaluation index λ_i.

$$\lambda_i=\frac{\omega_i}{\sum_{i=1}^{m}\omega_i},i=1,2,\cdots,m \tag{3}$$

Where $\sum_{i=1}^{m}\lambda_i=1$.

2.2. Coefficient of Variation method

Coefficient of Variation method[6] is mainly used to characterize the differences between data, which can eliminate the interference of the expert's preferences. Further processing of the data by normalization can eliminate the difference of unit and order of magnitude.

The basic steps are as follows:

(1) Assuming that there are m evaluation indexes and n samples, the initial evaluation matrix A is constructed as follows:

$$A=\begin{bmatrix} x_{11} & \cdots & x_{1m} \\ \vdots & x_{ij} & \vdots \\ x_{n1} & \cdots & x_{nm} \end{bmatrix} \tag{4}$$

Where x_{ij} represents the value of the jth evaluation index in the ith element to be evaluated.

(2) Calculate the standard deviation S_j to indicate the absolute variation degree:

$$S_j = \sqrt{\frac{\sum_{i=1}^{n}(x_{ij} - \overline{x_j})^2}{n}} \tag{5}$$

Where $\overline{x_j}$ is the mean value, $\overline{x_j} = \frac{1}{n}\sum_{i=1}^{n} x_{ij}, j = 1,2,\cdots,m$.

(3) Calculate the coefficient of variation v_j to indicate the degree of relative variation:

$$v_j = \frac{S_j}{\overline{x_j}} \tag{6}$$

(4) Normalized processing, and calculate the objective weight of each evaluation index w_j:

$$w_j = v_j \Big/ \sum_{j=1}^{m} v_j \tag{7}$$

2.3. Combination weighting model

When determining the weight of each evaluation index, C-OWA method overemphasizes the subjective experience and knowledge of experts, while Coefficient of Variation method relies too much on objective data and lacks the expert's experience. The combination of these two methods based on the Minimum Discriminant Information Principle[8] can effectively consider both subjective and objective factors, and calculate the combination weight δ_i of each evaluation index.

Assume that $\lambda_i = (\lambda_{i1}, \lambda_{i2}, \cdots, \lambda_{im})$ is the weight vector result of C-OWA method, the weight vector result of Coefficient of Variation method is $w_i = (w_{i1}, w_{i2}, \cdots, w_{im})$, then the combined weight vector can be measured as follows:

$$\delta_i = \delta_1 \lambda_i + \delta_2 w_i \tag{8}$$

Where δ_1, δ_2 represents the ratio of combination.

The target function is as follows:

$$\begin{cases} \min J(\delta) = \sum_{i=1}^{n}\left(\delta_i \ln\frac{\delta_i}{\lambda_i} + \delta_i \ln\frac{\delta_i}{w_i}\right) \\ s.t.\ \ \sum_{i=1}^{n}\delta_i = 1, \delta_i \geq 0 \quad i = 1,2,\cdots,n \end{cases} \tag{9}$$

By solving this model, calculate combination weight:

$$\delta_i = \frac{\sqrt{\lambda_i w_i}}{\sum_{i=1}^{n}\sqrt{\lambda_i w_i}} \tag{10}$$

2.4. Vague set

Compared with the traditional fuzzy set, Vague set has made some development and progress[9].

Assume that x_i is an element of the universe $U=\{x_1,x_2,\cdots,x_m\}$, $t_A(x_i)$ is the true membership degrees of x_i relative to Vague set A and $f_A(x_i)$ is the false membership degrees, membership degree interval $[t_A(x_i),1-f_A(x_i)]$ is constructed by $t_A(x_i)$ and $f_A(x_i)$, where $t_A(x_i),f_A(x_i)\in[0,1]$, and the hesitation degree of x_i is defined as $\pi_A=1-t_A(x_i)-f_A(x_i)$.

2.4.1. Vague set representation of evaluation index. Each evaluation index in the set $A=\{A_1,A_2,\cdots,A_n\}$ can be divided into two types: qualitative and quantitative. The former should be converted to Vague value according to Table 1.

Table 1. Evaluation criteria of Vague set

Comment sets	Vague set	Hesitation degree
Absolutely good	[1,1]	0
Very good	[0.9,1]	0.1
Relatively good	[0.75,0.9]	0.15
Good	[0.6,0.75]	0.2
General	[0.5,0.6]	0.1
Bad	[0.25,0.5]	0.2
Relatively bad	[0.1,0.25]	0.15
Very bad	[0,0.1]	0.1
Absolutely bad	[0,0]	0

Quantitative type is divided into two types according to the actual connotation: the smaller the better is called cost type, and the corresponding larger the better is called benefit type (Each quantitative indicator type in the experiment analysis below is cost type.)

Conversion equation of cost index:

$$t_{ij}=\frac{\max_i x_{ij}-x_{ij}}{\max_i x_{ij}-\min_i x_{ij}},\quad f_{ij}=\frac{x_{ij}-\min_i x_{ij}}{\max_i x_{ij}-\min_i x_{ij}} \tag{11}$$

Conversion equation of benefit index:

$$t_{ij}=\frac{x_{ij}-\min_i x_{ij}}{\max_i x_{ij}-\min_i x_{ij}},\quad f_{ij}=\frac{\max_i x_{ij}-x_{ij}}{\max_i x_{ij}-\min_i x_{ij}} \tag{12}$$

Where t_{ij} and f_{ij} respectively represents the true and false membership degree of scheme A_i under index X_j, x_{ij} is the initial value of scheme A_i under index X_j.

The Vague set of scheme A_i under index set $X=\{X_1,X_2,\cdots,X_m\}$ is defined as:

$$A_i=\{(X_1,[t_{i1},1-f_{i1}]),(X_2,[t_{i2},1-f_{i2}]),\cdots,(X_m,[t_{im},1-f_{im}])\} \tag{13}$$

2.4.2. Fitness of Vague set. Vague set can deal with the information fuzziness of each evaluation index and get its fitness, so as to achieve the purpose of quantifying and describing each evaluation index. Equation (14) has good robustness[10-11], and the calculation result is reasonable by using the true and false membership degree of each evaluation index. The fitness matrix $Q=(q_{ij})_{n\times m}$ of each evaluation index of marine engine simulator team operation is calculated by equation (14):

$$q_{ij}=\frac{t_{ij}-f_{ij}}{1+\pi_{ij}} \tag{14}$$

Where q_{ij} is the fitness of the scheme A_i under the index X_j.

Normalization Q, $Q'=(q_{ij}')_{n\times m}$:

$$q_{ij}'=\frac{q_{ij}-\min_i q_{ij}}{\max_i q_{ij}-\min_i q_{ij}} \tag{15}$$

3. Experiment analysis

3.1. Establishment of comprehensive evaluation index system

The evaluation tasks of the marine engine simulator are usually set up by professionals, and the marine engineers need to complete the learning and assessment of relevant tasks before obtaining the corresponding competency certificate. Generally, the simulator evaluation tasks need the cooperation of the team to complete.

The number of evaluation indexes in standby of marine generator set task is moderate and the evaluation results are intuitive. It is a typical simulator evaluation task so that is selected as the experimental task. The main contents of the task include: preparation of starting air system (r_1), preparation of lube oil system (r_2), preparation of fuel oil system (r_3), preparation of low-temperature cooling water system (r_4), generator starting and parallel operation (r_5), etc. The task of generator paralleling includes: phase difference, frequency difference, voltage to be paralleled, operation duration, Load transfer, shut off synchrometer and other evaluation indexes. The specific index information is shown in Table 2.

Table 2. Index system and standard of generator unit standby

Rule layer	Index layer	Type	Standard
Starting air system r_1	Air pressure r_{11} p/MPa	Quantitative	0.7~1.05
Lube oil system r_2	Inlet temperature r_{21} T/°C	Quantitative	54~76
	Inlet pressure r_{22} p/MPa	Quantitative	0.3~0.8
Fuel oil system r_3	Inlet pressure r_{31} p/MPa	Quantitative	0.6~1.0
	Inlet viscosity r_{32} γ/cst	Quantitative	3~14
Cooling water system r_4	Inlet pressure r_{41} p/MPa	Quantitative	0.6~0.96
	Inlet temperature r_{42} T/°C	Quantitative	76~91
Parallel operation r_5	Phase difference r_{51} Δφ/°	Quantitative	0~2
	Frequency difference r_{52} Δf/Hz	Quantitative	0.1~0.3
	Voltage to be paralleled r_{53} U/V	Quantitative	448~452
	Operation duration r_{54} t/min	Quantitative	1~5
	Load transfer r_{55}	Quantitative	0.01~10

	ΔP/KW		
	Shut off synchrometer r_{56}	Qualitative	Yes

3.2. Calculate combination weight of each index

1) Based on the C-OWA method, five experts with rich working experience are invited to score the evaluation indexes according to Table (2). The subjective weight of each index is λ = (0.2602 0.1301 0.1301 0.1301 0.1301 0.0688 0.0688 0.021 0.0273 0.0141 0.004 0.0077 0.0077) according to equation (1) ~ (3).

2) At present, there are four training teams A_1, A_2, A_3, A_4, each training team consists of five operators and they independently complete tasks r_1, r_2, r_3, r_4, r_5. The operation data of the four teams are shown in Table 3 as follows:

Table 3. Operation data

Index	A_1	A_2	A_3	A_4
r_{11}	0.71	0.96	0.85	0.77
r_{21}	63	59	66	69
r_{22}	0.54	0.63	0.43	0.69
r_{31}	0.88	0.79	0.73	0.83
r_{32}	5	7	6	7
r_{41}	0.73	0.76	0.81	0.65
r_{42}	83	85	79	81
r_{51}	1	0	1	0
r_{52}	0.15	0.17	0.21	0.18
r_{53}	449	450	450	451
r_{54}	2	2.3	2.5	3.1
r_{55}	4	3.8	5.6	4.3
r_{56}	1	1	1	0

According to the index information in Table 2 and the team operation data in Table 3, the qualitative data of A_1 , A_2 , A_3 , A_4 are quantified according to Table 1. The quantitative data is quantified by equation (9)~(10), as shown in Table 4.

Table 4. Team operation data Vague set quantification

Rule layer	Index layer	A_1	A_2	A_3	A_4
r_1	r_{11}	[0.971,0.971]	[0.257,0.257]	[0.571,0.571]	[0.8,0.8]
r_2	r_{21}	[0.591,0.591]	[0.773,0.773]	[0.455,0.455]	[0.318,0.318]
	r_{22}	[0.52,0.52]	[0.34,0.34]	[0.74,0.74]	[0.22,0.22]
r_3	r_{31}	[0.3,0.3]	[0.525,0.525]	[0.675,0.675]	[0.425,0.425]
	r_{32}	[0.818,0.818]	[0.636,0.636]	[0.727,0.727]	[0.636,0.636]
r_4	r_{41}	[0.639,0.639]	[0.556,0.556]	[0.417,0.417]	[0.861,0.861]
	r_{42}	[0.533,0.533]	[0.4,0.4]	[0.8,0.8]	[0.667,0.667]
r_5	r_{51}	[0.5,0.5]	[1,1]	[0.5,0.5]	[1,1]
	r_{52}	[0.75,0.75]	[0.65,0.65]	[0.45,0.45]	[0.6,0.6]
	r_{53}	[0.75,0.75]	[0.5,0.5]	[0.5,0.5]	[0.25,0.25]
	r_{54}	[0.75,0.75]	[0.675,0.675]	[0.625,0.625]	[0.475,0.475]
	r_{55}	[0.601,0.601]	[0.621,0.621]	[0.44,0.44]	[0.571,0.571]
	r_{56}	[1,1]	[1,1]	[1,1]	[0,0]

Combined with Table 4, the evaluation index matrix Q' is obtained by using equation (14) ~ (15).

$$Q'=\begin{bmatrix} 1 & 0.602 & 0.59 & 0 & 1 & 0.483 & 0.329 & 0 & 1 & 1 & 1 & 0.887 & 1 \\ 0 & 1 & 0.248 & 0.605 & 0 & 0.313 & 0 & 1 & 0.643 & 0.489 & 0.7 & 1 & 1 \\ 0.423 & 0.293 & 1 & 1 & 0.463 & 0 & 1 & 0 & 0 & 0.489 & 0.524 & 0 & 1 \\ 0.735 & 0 & 0 & 0.333 & 0 & 1 & 0.646 & 1 & 0.483 & 0 & 0 & 0.758 & 0 \end{bmatrix} \quad (16)$$

Based on the Coefficient of Variation method, the objective weight w= (0.0694 0.0787 0.0822 0.0761 0.1133 0.0811 0.0755 0.1005 0.0680 0.0719 0.0657 0.0595 0.0580) is calculated by equation (5) ~ (7).

According to the Minimum Discriminant Information Principle, the index combination weight δ = (0.1516 0.1142 0.1167 0.1123 0.1370 0.0843 0.0813 0.0518 0.0486 0.0359 0.0183 0.0242 0.0238) is calculated by equation (8) ~ (10).

3.3. Comprehensive analysis

To make the team's evaluation results more intuitive, we can calculate the score of each team by $Q' \times \delta^T$.

$$Q' \times \delta^T = \begin{bmatrix} 1 & 0 & 0.423 & 0.735 \\ 0.602 & 1 & 0.293 & 0 \\ 0.59 & 0.248 & 1 & 0 \\ 0 & 0.605 & 1 & 0.333 \\ 1 & 0 & 0.463 & 0 \\ 0.483 & 0.313 & 0 & 1 \\ 0.329 & 0 & 1 & 0.646 \\ 0 & 1 & 0 & 1 \\ 1 & 0.643 & 0 & 0.483 \\ 1 & 0.489 & 0.489 & 0 \\ 1 & 0.7 & 0.524 & 0 \\ 0.887 & 1 & 0 & 0.758 \\ 1 & 1 & 1 & 0 \end{bmatrix}^T \times \begin{bmatrix} 0.1516 \\ 0.1142 \\ 0.1167 \\ 0.1123 \\ 0.1370 \\ 0.0843 \\ 0.0813 \\ 0.0518 \\ 0.0486 \\ 0.0359 \\ 0.0183 \\ 0.0242 \\ 0.0238 \end{bmatrix} = \begin{bmatrix} 0.659 \\ 0.5134 \\ 0.4664 \\ 0.3846 \end{bmatrix} \quad (17)$$

Increase the team score by 100 times. Then the evaluation scores $B_i (i=1,2,3,4)$ of the four teams (percentage system) are as follows:

B_1=65.90,

B_2=51.34,

B_3=46.64,

B_4=38.46.

It can be seen from the results that each member of team 1 has a relatively good command of the operation of standby of marine generator set, while each member of team 2 has a relatively general command of the operation and needs more practice. Each member of team 3 and team 4 is not familiar with the operation of generator standby, and many operations are not up to standard, which leads to the low overall evaluation result of the team, Team members need to focus on learning this operation task.

4. Conclusion

1) When a single weighting method is used to determine the weight of evaluation indicators, there will be a problem that the consideration is not comprehensive and the results will be limited;

Combining the subjective and objective weighting methods, the weight value obtained is in line with the actual situation, and the evaluation results also verify this conclusion.

2) Vague set is used to deal with various evaluation indexes of standby of marine generator set task operation, which can effectively solve the fuzzy problem in the evaluation process. Verified by the marine generator standby task project, the combination of vague set and combined weighting is reasonable and feasible, and can intelligently evaluate the marine engine simulator.

Acknowledgments

This study is supported by the Liaoning general higher education undergraduate teaching reform research project (2019 / 3), the Intelligent ship comprehensive test and verification research project (No. 473 [2018] of Ministry of Industry and Information Technology).

References

[1] JIA BZ, LIN YJ, CAO H, ZHANG JD 2013 *Navigation of China* (in Chinese), 36(03) 28-33
[2] DUAN ZL, REN G, ZHANG JD, CAO H 2016 *J. Traffic and Transportation Engineering* (in Chinese), 16(06) 82-90
[3] NIE W, WU Y, HU DB, LI YD 2013 *J. Wuhan University of Technology (Transportation Science & Engineering)* (in Chinese), 37(04) 834-8
[4] CAO H, ZHANG JD 2012 *Maritime Education Research* (in Chinese), 29(01) 33-6
[5] WANG JB, YOU WB, LIU FM, YIN XZ 2018 *J. Safety and Environment* (in Chinese), 18(03) 900-6
[6] ZHU ZC 2017 *Statistics & Decision* (in Chinese), 12 78-81
[7] LIU HW 2004 *Systems Engineering-Theory & Practice* (in Chinese), 05 103-9
[8] YANG XY, LIU H, ZHANG B, XIAO YQ 2014 *Electric Power Automation Equipment* (in Chinese), 34(09) 118-22
[9] QIU RS, PAN JF, LI WS ZHAO J 2021 *Fire Control & Command Control* (in Chinese), 46(01) 62-6+72
[10]GAO JW, LIU HH, GU YD 2014 *Systems Engineering-Theory & Practice* (in Chinese), 34(12) 3175-81
[11]YE YT, LIANG LL, CAO Y, ZHOU ZH, GONG JG, ZHAO HL, JIANG YZ 2017 *Systems Engineering-Theory & Practice* (in Chinese), 37(07) 1926-36

Application of Uncertainty Analysis in Evaluation of Marine Engine Simulator

Henglong Shen, Hui Cao* and Shiyuan Liang

Marine Engineering College, Dalian Maritime University, Dalian, Liaoning, China

*Corresponding author's e-mail: lustersoft@gmail.com

Abstract. Aiming at the problem of fuzziness and randomness in the evaluation process of marine engine simulator, a comprehensive evaluation model based on subjective and objective combination weighting method and cloud model method is established on the basis of single subjective and objective combination weighting method. Firstly, the combined weighting method is used to determine the comprehensive weight; Then, the standard cloud model is generated by using the cloud model method, and the digital characteristics of the evaluation cloud are obtained through the sample data and cloud computing rules. The digital characteristics of the evaluation cloud are obtained by the backward cloud generator and comprehensive weight; Finally, the similarity between the generated evaluation cloud and the standard cloud model is calculated, so as to evaluate the operation of the student marine engine simulator. The example analysis shows that the evaluation method can scientifically and reasonably evaluate the crew's operation ability, and is feasible and effective.

1. Introduction

In the context of the implementation of the new international convention standards, higher-level requirements have been put forward for the training quality of ship engine room personnel. At the same time, the scientific and effective intelligent evaluation model of marine engine simulator (MES) has also become a hot topic for many people in the industry. Therefore, it is urgent to build a reasonable and effective method model in the intelligent evaluation of MES training operation. The introduction of intelligent models into the training and evaluation category of MES can not only weaken the randomness and ambiguity of evaluation, but also enrich the evaluation methods of engine room simulator, making them continue to move towards the goal of intelligent evaluation.

At present, models based on fuzzy comprehensive evaluation, inference engine, analytic hierarchy and neural network are commonly used in MES training evaluation [1]-[5]. These models are a single assignment analysis for the most important part of the evaluation index weight, which often makes the results too one-sided and unconvincing. Therefore, this article establishes an evaluation model based on the subjective and objective combination weighting- cloud model evaluation method for the training operation of the MES, and visualizes the qualitative and quantitative conversion of the evaluation results.

2. Marine engine simulator

MES is a simulation teaching facility and device based on software simulation and artificial intelligence technology to build a mathematical model of the operating conditions of ship engine room equipment and operating system, and finally transmit the operating data, system status and other information to the human-computer interaction media. Students can use this device to maximize the virtual action and process simulation of the main propulsion control system, cabin control console, cabin power station and cargo management. The operation scenes of various aspects of the MES are modeled based on the equipment environment of the real ship's engine room, etc., which keeps the consistency with the real ship's operation interface to the greatest extent. The MES interfaces are shown in Figure 1 and Figure 2.

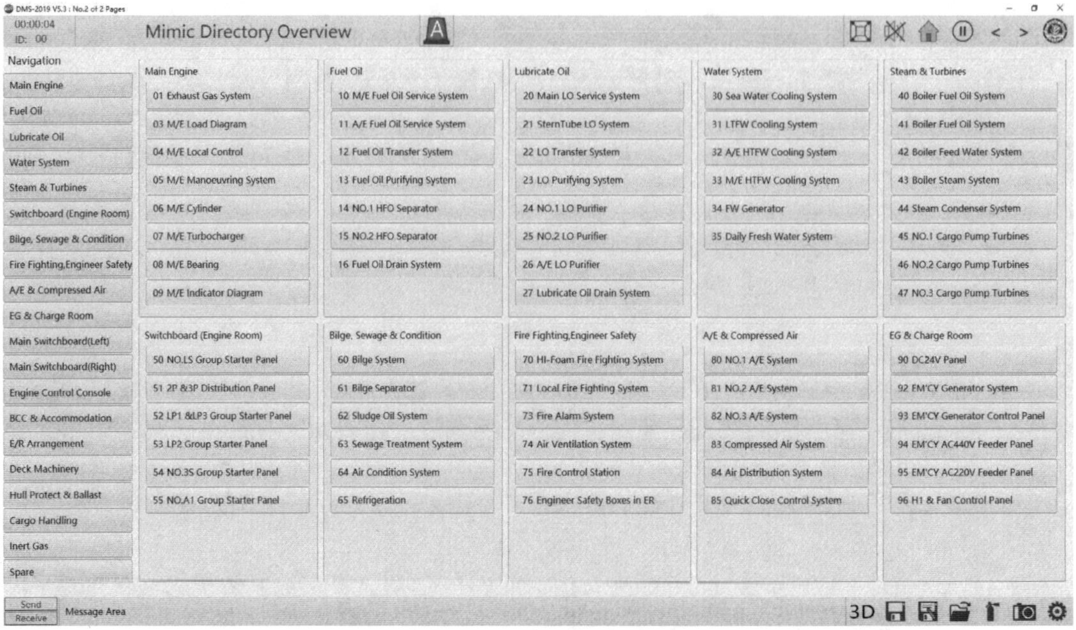

Figure 1. Main interface of marine engine simulator.

Figure 2. Synchro & Bus Tie CB Panel interface of marine engine simulator.

3. Combination weighting and cloud barycenter gravity evaluation

3.1. Combination weighting

3.1.1. Analysis of order relation. As an optimized subjective weight assignment method, order relation analysis method (G1 method) [6][7] has the advantages of less computation, without constructing judgment matrix and consistency check in determining the weight of each evaluation index, in addition to the characteristic that AHP can reflect the experience and knowledge of experts. The basic principle is as follows:

1). Determine the order relationship of evaluation indexes

According to the work experience of relevant experts, it is determined that each evaluation element $\{b_1, b_2...b_n\}$ has an order of $b_1 > b_2 > ... > b_n$ relative to the degree of influence of the evaluation object, and it is said that a " > " sequence is established among the evaluation elements grade.

2). Adjacent index importance judgment

If the expert comprehensively judges the ratio of the influence of the k-1 and k-th evaluation elements b_{k-1} and b_k on the evaluation object is r_k, that is to say:

$$r_k = \frac{w_{k-1}}{w_k}, k = n, n-1, n-2, ..., 3, 2 \tag{1}$$

The values of r_k mentioned above are shown in Table 1 w_{k-1}, w_k represents the weight of b_{k-1} and b_k.

Table 1. r_k assignment table.

	Instruction
1.0	Factor b_{k-1} is just as significant as factor b_k
1.2	Factor b_{k-1} is a bit more significant than factor b_k
1.4	Factor b_{k-1} is obviously more significant than factor b_k
1.6	Factor b_{k-1} is strongly more significant than factor b_k
1.8	Factor b_{k-1} is exceedingly significant than factor b_k
1.1,1.3,1.5,1.7,1.9	Between the above levels

3). Determine the weight of the n-th evaluation index w_n

$$w_n = (1 + \sum_{k=2}^{n} \prod_{i=k}^{n} r_k)^{-1} \tag{2}$$

4). Determine the index weight set $\boldsymbol{w_s}$

$$w_{k-1} = r_k w_k, k = n, n-1, ..., 3, 2 \tag{3}$$

Therefore, the weight set of each index calculated by G1 method is $\boldsymbol{w_s}=(w_1, w_2, ..., w_n)$.

3.1.2. CRITIC method. CRITIC (Criteria Importation Through Interrogation Correlation) method [8] is an objective weight assignment method. Compared with objective weighting methods such as standard deviation method and entropy weight method, Critic method can complete the assignment of elements more comprehensively and accurately. The calculation principle is as follows:

1). Construct evaluation matrix $\boldsymbol{B}$

If there are m objectives and n elements, the original data evaluation matrix $\boldsymbol{B}$ is constructed

$$\boldsymbol{B} = \left(x_{ij}\right)_{m\times n} = \begin{bmatrix} x_{11} & x_{12} & \cdots & x_{1n} \\ x_{21} & x_{22} & \cdots & x_{2n} \\ \cdots & \cdots & \cdots & \cdots \\ x_{m1} & x_{m2} & \cdots & x_{mn} \end{bmatrix} \tag{4}$$

Where, x_{ij} represents the *j*-th element value of the *i*-th objective.

2). Data preprocessing

Formulas (5) and (6) are used to normalize the transformation of the above matrix $\boldsymbol{B}$, and the new matrix $\boldsymbol{B'}=(x'_{ij})_{m\times n}$ is obtained.

For the smaller the better type indicators:

$$x'_{ij}=\frac{\max(x_{ij})-x_{ij}}{\max(x_{ij})-\min(x_{ij})} \tag{5}$$

For the bigger the better type indicators:

$$x'_{ij}=\frac{x_{ij}-\min(x_{ij})}{\max(x_{ij})-\min(x_{ij})} \tag{6}$$

3) Each index parameter calculation

The standard deviation δ_j, correlation coefficient λ_{ij} and conflict coefficient c_j of each element was calculated by using the matrix $\boldsymbol{B'}$ obtained in the 2) step.

$$\delta_j=\sqrt{\frac{1}{m}\sum_{i=1}^{m}(x'_{ij}-\overline{x'_j})^2} \tag{7}$$

$$\lambda_{ij}=\frac{\sum_{k=1}^{m}(x'_{ki}-\overline{x'_i})(x'_{kj}-\overline{x'_j})}{\sqrt{\sum_{k=1}^{m}(x'_{ki}-\overline{x'_i})^2}\sqrt{\sum_{k=1}^{m}(x'_{kj}-\overline{x'_j})^2}} \tag{8}$$

$$c_j=\sum_{i=1}^{n}(1-\lambda_{ij}) \tag{9}$$

Where, $\overline{x'_j}$ is the average value of element *j*, $\overline{x'_j}=\frac{1}{m}\sum_{i=1}^{m}x'_{ij}$.

4) Determine the index weight set $\boldsymbol{w}_o$

Figuring out the weight of the elements *j*

$$w_j=\frac{\delta_j\times c_j}{\sum_{j=1}^{n}\delta_j\times c_j},\ j=1,\dots,n \tag{10}$$

Therefore, the weight set of each index calculated by CRITIC method is $\boldsymbol{w}_o=(w_1, w_2, \dots, w_n)$.

3.1.3. Combination of empowerment. The basic principle of combinatorial weighting in game theory is applied to fully combine the internal distribution information of subjective consciousness and objective data to increase the scientific rationality of weights to a certain extent [9]. The steps to construct the combinatorial optimization weighting method are as follows:

1). Solve the linear combination of the weight set $\boldsymbol{W}$

Constitute the set of weight vectors $\boldsymbol{W}=\{w_s, w_o\}$, then the linear combination of the two weight vectors is

$$\boldsymbol{W}=\alpha_1 w_s^T+\alpha_2 w_o^T \tag{11}$$

2). Construct the objective function

The optimal linear combination of the above *two* weights are carried out, and the objective is to minimize the deviation to obtain the comprehensive weight, namely, the objective function is

$$\min\left\|\boldsymbol{W}^T - w_i^T\right\|_2, \quad i=1,2 \tag{12}$$

3). Solve the optimal first derivative condition

Based on matrix differential properties, the optimal conditions for the first derivative in Equation (12) above can be obtained, and the linear formula can be expressed as

$$\begin{pmatrix} w_s w_s^T & w_s w_o^T \\ w_o w_s^T & w_o w_o^T \end{pmatrix}\begin{bmatrix} \alpha_1 \\ \alpha_2 \end{bmatrix}=\begin{bmatrix} w_s w_s^T \\ w_o w_o^T \end{bmatrix} \tag{13}$$

4). Normalization of coefficients

The weight coefficient obtained by the above formula (13) is normalized

$$\alpha_1^* = \frac{|\alpha_1|}{|\alpha_1+\alpha_2|}, \alpha_2^* = \frac{|\alpha_2|}{|\alpha_1+\alpha_2|} \tag{14}$$

5). Determine the comprehensive weight set $\boldsymbol{W}$

$$\boldsymbol{W} = \alpha_1^* w_s^T + \alpha_2^* w_o^T \tag{15}$$

3.2. Cloud barycenter model evaluation method

3.2.1. Cloud model theory. The cloud model [10][11] uses three digital quantities, expected *Ex*, entropy *En* and hyper entropy *He*, to describe the qualitative language value, and realizes the conversion between natural language expression and quantitative data through calculation principles.

If $\boldsymbol{U}$ is a domain representing quantity, $\boldsymbol{C}$ represents a natural language expression in $\boldsymbol{U}$, and if the number $x \in \boldsymbol{U}$, and x is a random realization of $\boldsymbol{C}$, and $\mu(x) \in [0,1]$, x is called a cloud droplet. Where $\mu(x)$ is the membership of x to $\boldsymbol{C}$.

$$\mu\text{:}\ \boldsymbol{U} \to [0,1], \forall x \in \boldsymbol{U}, x \to \mu(x) \tag{16}$$

Expectation *Ex* represents the central quantity of uncertain factors transformed into quantitative evaluation; Entropy *En* describes the fuzziness of natural language, that is, the size of allowable value space expressed by qualitative language in the domain interval. The hyper entropy *He* describes the stability level of the cloud droplet, as shown in Figure 3.

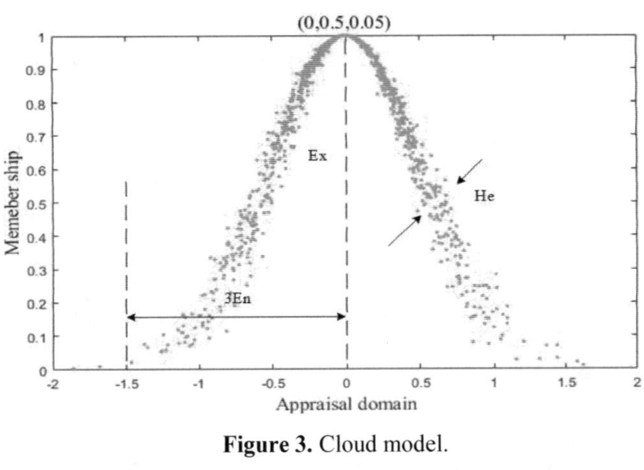

Figure 3. Cloud model.

Cloud Generator [12] is a logical tool used for cloud generation, mainly consisting of two types: Forward Cloud Generator (FCG) and Backward Cloud Generator (BCG), which can complete the transformation between natural language description and fixed value data. The forward cloud generator can parse characteristics such as the value range of quantitative data from the qualitative expression information. The calculation principle is as follows:

1). Enter *Ex*, *En*, *He* and cloud droplet number *N*;

2). Generating normal random numbers En^*, and $En^* \sim NORM(En, He^2)$;

3). Generating normal random numbers x_i, where $x_i \sim NORM(En, En^{*2})$;

4). Generate the cloud droplets e(x_i, $\mu(x_i)$), and $\mu\left(x_i\right) = \exp\left(-\left(x_i - Ex\right)^2\left(2\left(En^*\right)\right)^{-2}\right)$

5). Repeat the above 2), 3) and 4) until $i = N$ ($i = 1, 2, ..., N$) to generate N cloud droplets.

The implementation of reverse cloud generator algorithm is the reverse process of forward cloud generator, which is mainly used in the production of integrated cloud. The function in this paper is to transform the training evaluation value of the Marine engine simulator into the cloud digital feature of the training evaluation, and realize the transformation from quantitative value to qualitative concept. The algorithm is as follows:

1) Calculate the average

$$Ex = \bar{x} = \frac{1}{n}\sum_{i=1}^{n} x_i \tag{17}$$

2) Calculate the variance

$$S^2 = \frac{1}{\text{n-1}}\sum_{i=1}^{n}(x_i - \bar{x})^2 \tag{18}$$

3) Cloud model digital feature

$$\begin{cases} E_x = \bar{x} \\ E_n = \sqrt{\frac{\pi}{2}}\frac{1}{n}\sum_{i=1}^{n}\left|x_i - \bar{x}\right| \\ H_e = \sqrt{S^2 - E_n^{\,2}} \end{cases} \tag{19}$$

3.2.2. Setting up a judgment set. The evaluation set $\boldsymbol{V}$ described by natural language is established, and the evaluation values are divided into the interval $[x_{\min}, x_{\max}]$. All the evaluation values have a one-to-one corresponding interval, and all the evaluation values are represented by the normal cloud model.

$$\boldsymbol{V} = \{v_1, v_2, \ldots, v_m\} \tag{20}$$

Among them, the v_i ($i = 1, ..., m$) represents the various possible evaluation results.

3.2.3. Calculate the cloud digital characteristics of evaluation index. Several experts in the industry were invited to score the results of the students' operation indicators in accordance with the bilateral constraint principle, and give the highest and lowest scores in the score range [0,1].Then, the cloud digital features with the highest and lowest scores are calculated according to the reverse cloud algorithm ($Ex^{\min}$, $En^{\min}$, $He^{\min}$)、($Ex^{\max}$, $En^{\max}$, $He^{\max}$), according to the comprehensive cloud generator, the digital characteristics of the comprehensive cloud model of each operation result are calculated (Ex_z, En_z, He_z). The calculation principle is as follows:

$$\begin{cases} Ex_z = \dfrac{Ex^{\min} En^{\min} + Ex^{\max} En^{\max}}{En^{\min} + En^{\min}} \\ En_z = En^{\min} + En^{\max} \\ He_z = \dfrac{He^{\min} En^{\min} + He^{\max} En^{\max}}{En^{\min} + En^{\max}} \end{cases} \tag{21}$$

3.2.4. Generate comprehensive index evaluation cloud map of training operation. In order to accurately evaluate the students' operation training, according to the cloud digital characteristics of each operation result and the comprehensive weight $\boldsymbol{W}$ weighted operation, the cloud digital characteristics of the comprehensive evaluation of the operation of turbine simulation training was obtained, and then the cloud map was obtained by the forward cloud generator. The calculation formula is as follows:

$$\begin{cases} Ex = \dfrac{\sum_{i=1}^{n} Ex_z W}{\sum_{i=1}^{n} W} \\ En = \dfrac{\sum_{i=1}^{n} En_z W^2}{\sum_{i=1}^{n} W^2} \\ He = \dfrac{\sum_{i=1}^{n} He_z W^2}{\sum_{i=1}^{n} W^2} \end{cases} \tag{22}$$

3.2.5. Comprehensive similarity calculation. By comparing the digital characteristics of students' comprehensive cloud with the comprehensive evaluation set $\{v_1, v_1, \ldots, v_m\}$, the comprehensive similarity with each evaluation cloud is calculated, so as to determine the evaluation results of students' MES operation training.

Assuming that the two clouds are $C_1(Ex_1, En_1, He_1)$ and $C_2(Ex_2, En_2, He_2)$, respectively, the shape similarity and distance similarity are calculated. Then the comprehensive similarity is calculated. The steps are as follows:

Step 1: Shape similarity calculation $\boldsymbol{S}_{s12}$

The shape similarity of cloud model is only related to entropy *En* and super-entropy *He*, and the change of expected *Ex* does not affect the shape similarity.

$$S_{S12}(C_1, C_2) = \frac{\min\left((En_1 + 3He_1)/En_1, (En_2 + 3He_2)/En_2\right)}{\max\left((En_1 + 3He_1)/En_1, (En_2 + 3He_2)/En_2\right)} \tag{23}$$

Step 2: Position similarity calculation $\boldsymbol{S}_{p12}$

Position similarity refers to the similarity difference caused by the lateral displacement of the cloud model on the cloud map.

$$S_{P12}(C_1, C_2) = p_{\text{over}} / p_{\text{total}} \tag{24}$$

Where

$$p_{\text{over}} = \begin{cases} \min\left(Ex_1 + 3En_1, Ex_2 + 3En_2\right) - \max\left(Ex_1 - 3En_1, Ex_2 - 3En_2\right) \\ 0 \quad if \quad p_{\text{over}} < 0 \end{cases} \tag{25}$$

$$p_{\text{total}} = \max\left(Ex_1 + 3En_1, Ex_2 + 3En_2\right) - \min\left(Ex_1 - 3En_1, Ex_2 - 3En_2\right) \tag{26}$$

Step 3: Shape - position comprehensive similarity calculation

Based on shape similarity and the effect of distance between clouds, the comprehensive shape-position similarity of cloud model is modified.

$$S_{12}\left(C_1, C_2\right) = S_{\text{S12}}\left(C_1, C_2\right) \cdot S_{\text{P12}}\left(C_1, C_2\right), \quad S_{12} \in [0,1] \tag{27}$$

3.3. Project Evaluation Process

In summary, the block diagram of the project evaluation program of the MES is shown in Figure 4.

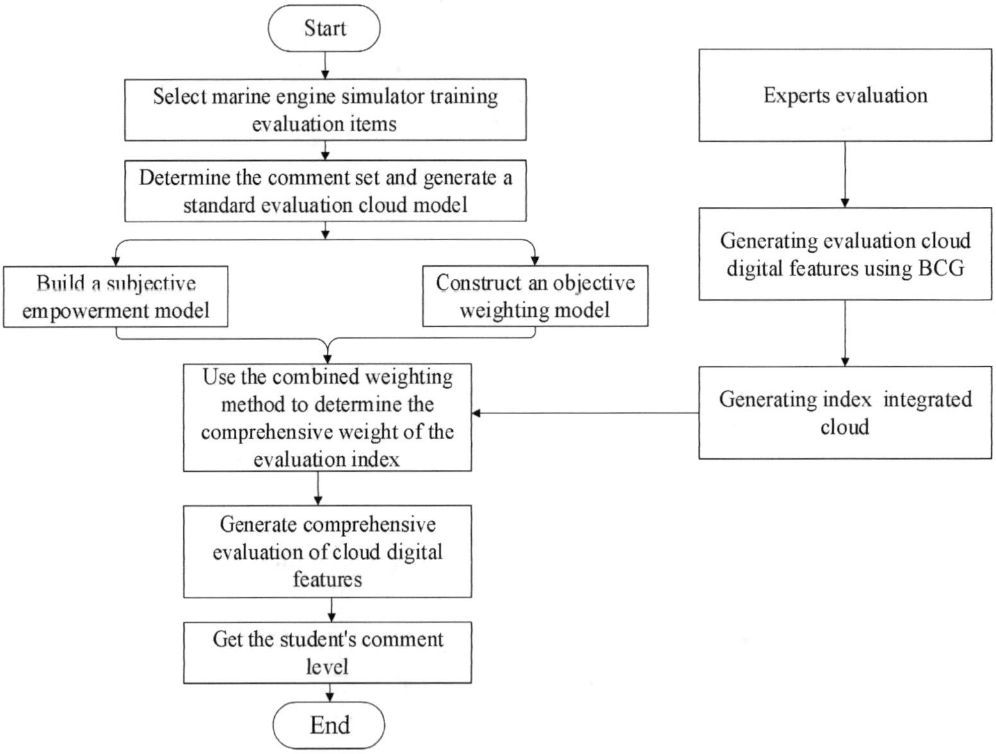

Figure 4. Visual representation of the proposed methodology.

4. Examples of application

4.1. Evaluation item selection

In this paper, the assessment project of "Manual Parallel and Load shift operation of Main Generator" is selected as an example for analysis. The project requires students to manually start the No. 2 generator and integrate it into the power grid to conduct power distribution operation. After completion, all the load of the No. 1 generator is transferred to the No. 2 generator and the No. 1 generator is shut down. In this project, 9 factors such as voltage difference, frequency difference and synchronizing meter state between two generators were selected as evaluation indexes. With this project as the background, the performance data of four representative students from a certain period of evaluation training examination of MES were selected for evaluation, i.e., the operational data of "manual paralleling and disassembling of power station" of the four students are shown in Table 2.

Table 2. Assessment data of students' project of "manual parallel and load shift operation of main generator".

Evaluation element	Student 1	Student 2	Student 3	Student 4
I01 Evaluation operation on duration /Min	5.97	4.77	6.59	5.97
I02 Frequency deviation /Hz	0.20	0.15	0.14	0.20
I03 Voltage deviation /V	1.11	0.07	1.31	1.11
I04 Phase deviation /Rad	2.34	2.45	1.43	2.34
I05 Frequency value /Hz	60.18	60.20	60.07	60.18
I06 Voltage value /V	449.06	450.01	450.73	449.06
I07 Close synchronous-meter	1.00	1.00	1.00	1.00
I08 Shift load /Kw	1.58	0.86	0.89	1.58
I09 No alarm	1.00	1.00	0.00	1.00

4.2. Build a review book

In this paper, 7-level comments are used to form the comment set $V=\{v_1, v_2, v_3, v_4, v_5, v_6, v_7\}$, and the comments are divided into continuous domain intervals [0,1] by the golden section method, as shown in Table 3. In MATLAB, the standard level comprehensive cloud model of turbine simulator is output according to FCG algorithm, as shown in Figure 5 below.

Table 3. Cloud assessment level.

Level cloud	Cloud digital feature
v_1	(0.000,0.090,0.015)
v_2	(0.167,0.056,0.028)
v_3	(0.373,0.034,0.017)
v_4	(0.500,0.021,0.011)
v_5	(0.627,0.034,0.017)
v_6	(0.833,0.053,0.028)
v_7	(1.000,0.090,0.055)

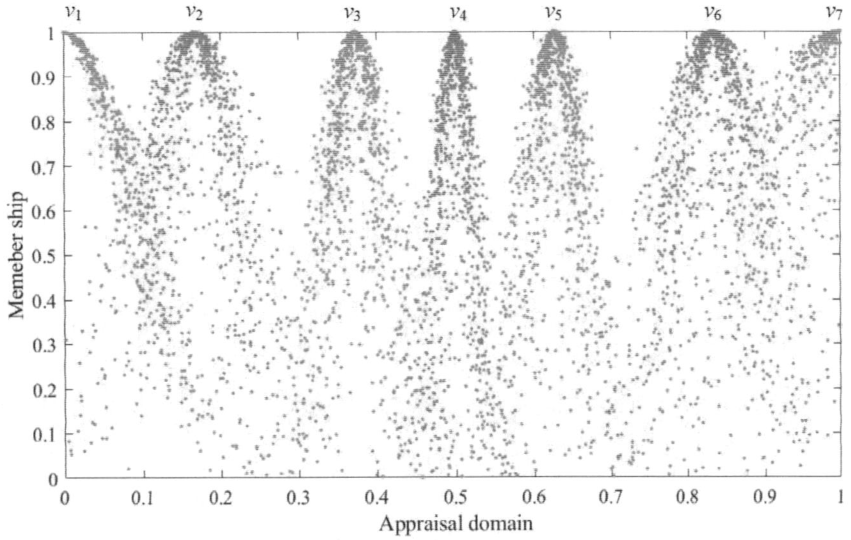

Figure 5. Visual distribution of the appraisal grade clouds.

4.3. Evaluation index weight calculation.

G1 theory gets used to figure out the subjective weight. Authorities in the industry were invited to make an independent sequential comparison of the above 9 evaluation elements according to their knowledge and experience, and the relative importance ratio r_k among the evaluation elements was given as follows:

Element importance ranking: I02>I03>I01>I04>I05>I08>I06>I07>I09.

Importance of indicators: r_2=I02/I03=1 ; r_3=I03/I01=1.4 ; r_4=I01/I04=1.4 ; r_5=I04/I05=1 ; r_6=I05/I08=1; r_7=I08/I06=1.4; r_8=I06/I07=1; r_9=I07/I09=1. Then, according to formulas (2) and (3), the subjective weight set $\boldsymbol{w}_s$ can be calculated. $\boldsymbol{w}_s$=(0.1338,0.1873,0.1873,0.0956,0.0956, 0.0683, 0.0683,0.0956,0.0683).

CRITIC method calculates objective weights. Firstly, the index elements are classified according to the original operation data matrix $\boldsymbol{B}$ of students 1~4, and the standardized matrix $\boldsymbol{B'}$ is obtained through normalization processing of matrix $\boldsymbol{B}$ by using formulas (5) and (6).

According to the data in matrix $\boldsymbol{B'}$, the δ_j, λ_{ij} and c_j values of each index parameter are obtained by using formulas (7) ~ (9). Then, according to the difference factors of evaluation indexes, the objective weight set $\boldsymbol{w}_o$ of the project of " manual parallel and load shift operation of main generator " is solved by formula (10). $\boldsymbol{w}_o$= (0.0913,0.1125,0.0913,0.1080,0.0856,0.1542,0.1382,0.1207,0.0982).

Finally, based on the equilibrium strategy of game theory and combining the advantages and disadvantages of subjective and objective assignment methods. According to formulas (11) ~ (14), the subjective and objective weight coefficients are α^*_1=0.8242 and α^*_2=0.1758 respectively. Furthermore, the combination weight set $\boldsymbol{W}$ is obtained according to formula (15). $\boldsymbol{W}$=(0.1263,0.1742,0.1704,0.0978, 0.0938,0.0834,0.0806,0.1000,0.0736).

4.4. Expert evaluation.

Taking student 1's operation result as an example, the above model is used to evaluate his operation. Provide MES evaluation data, scoring rules and standard requirements of Maritime Safety Administration to the above 5 experts, and invite them to score the results of operation indicators within [0,1]. In the evaluation process, the actual situation of the case and the professional experience and knowledge of the experts were integrated, so the scoring results were relatively scientific and reliable. The evaluation results of each index are shown in Table 4.

Table 4. Evaluation form of operation results of Student 1.

Evaluation element	Value type	Expert 1	Expert 2	Expert 3	Expert 4	Expert 5
I01	Max	0.92	0.95	0.90	0.94	0.93
	Min	0.86	0.88	0.87	0.89	0.90
I02	Max	0.97	0.93	0.89	0.95	0.92
	Min	0.90	0.88	0.87	0.89	0.88
I03	Max	0.98	0.95	0.94	0.92	0.96
	Min	0.90	0.87	0.89	0.88	0.91
I04	Max	0.97	0.95	0.99	0.92	0.94
	Min	0.92	0.88	0.91	0.89	0.90
I05	Max	0.88	0.94	0.91	0.96	0.90
	Min	0.85	0.87	0.90	0.91	0.88
I06	Max	0.94	0.90	0.96	0.93	0.97
	Min	0.89	0.86	0.91	0.88	0.87
I07	Max	0.42	0.37	0.30	0.35	0.27
	Min	0.29	0.31	0.25	0.27	0.20
I08	Max	0.87	0.94	0.91	0.85	0.92
	Min	0.81	0.85	0.84	0.82	0.87
I09	Max	0.90	0.98	0.95	0.94	0.95
	Min	0.87	0.91	0.85	0.89	0.92

The cloud digital characteristics of each index of student 1 calculated by BCG algorithm are shown in Table 5. According to formula (19), the final evaluation comprehensive cloud result E = (0.861,

0.062, 0.008) of students' training on two turbine simulator was calculated. FCG algorithm is used to generate the cloud map of the evaluation index, as shown in Figure 6.

Table 5. Student 1 evaluation index training results cloud digital characteristics table.

Evaluation element	The lowest cloud digital feature	The highest cloud digital feature	Integrate cloud digital features
I01	(0.880,0.150,0.005)	(0.928,0.018,0.007)	(0.885,0.168,0.005)
I02	(0.884,0.011,0.003)	(0.932,0.028,0.012)	(0.919,0.039,0.010)
I03	(0.890,0.015,0.005)	(0.950,0.020,0.010)	(0.924,0.035,0.006)
I04	(0.900,0.015,0.005)	(0.954,0.026,0.007)	(0.934,0.041,0.006)
I05	(0.882,0.023,0.006)	(0.918,0.032.0.003)	(0.903,0.055,0.004)
I06	(0.882;0.018;0.007)	(0.940,0.025,0.011)	(0.916,0.043,0.009)
I07	(0.264,0.039,0.016)	(0.342,0.057,0.014)	(0.310,0.096,0.015)
I08	(0.838,0.023,0.006)	(0.898,0.038,0.009)	(0.875,0.061,0.008)
I09	(0.888,0.028,0.006)	(0.944,0.024,0.016)	(0.914,0.052,0.011)

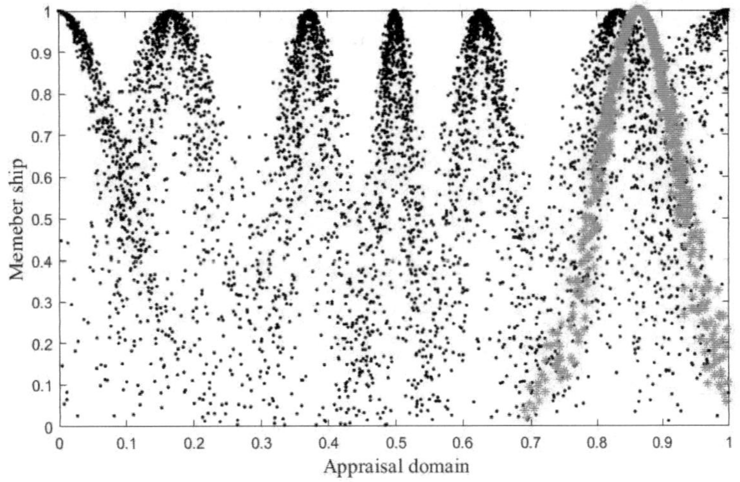

Figure 6. Cloud picture of MES training evaluation.

4.5. Determine student comment level

Comprehensive cloud in order to further analysis and evaluation results will be final evaluation result and benchmark set $\{v_1, v_2, v_3, v_4, v_5, v_6, v_7\}$ seven evaluation cloud similarity comparison, realize the evaluation results from the quantitative to qualitative analysis of the objective, according to the formula (20) - (24) calculations can get the result as shown in table 6.

Table 6. Comparison table of comprehensive similarity calculation.

Evaluation level	Shape similarity	Location similarity	Comprehensive similarity
v1	0.9247	0	0
v2	0.5548	0	0
v3	0.5548	0	0
v4	0.5394	0	0
v5	0.5548	0.0971	0.0539
v6	0.5366	0.8351	0.4481
v7	0.4896	0.5405	0.2646

It can be seen from Figure 6 and table 6 that cloud droplets are dense and the evaluation result is stable; The qualitative result is " v_6" according to the principle of maximum similarity. Based on the cloud digital characteristics of the training results of student 1, student 1 is proficient in the project training of "manual parallel train splitting of power station". The next training focus should be on the operation details to ensure the orderly completion of the whole training project, so as to meet the training requirements of the general program of the Maritime Safety Administration for crew members.

5. Conclusions

The research on the training evaluation of MES is carried out, and a training evaluation model based on subjective and objective combined weighting- cloud model evaluation method is established in this paper. In the MES training, the combined weighting method is used to quantify the comprehensive weight of the evaluation elements, which dilutes the unilateral influence of human latent consciousness or data characteristics, maximizes strengths and avoids weaknesses, and enhances the persuasiveness of comprehensive weights; it is more appropriate to introduce the cloud model evaluation method it solves the numerical processing problem of natural language expression, and at the same time weakens the uncertain and random factors in the evaluation process, so that the students' operation evaluation level is more realistic. Finally, the actual operation data of the students are used to analyze and calculate examples, the applicability of the combined weighting cloud model evaluation method in the training evaluation of MES students is verified, and the results are reasonable and intuitive.

Acknowledgments

This work was supported by the Liaoning general higher education undergraduate teaching reform research project (2019/3), the Intelligent ship comprehensive test and verify cation research project (No.473[2018] of Ministry of Industry and Information Technology).

References

[1] Nie W, Wu Y, Hu DB and Li YD 2013 Ship Science and Technology 35(12):78-82. (in Chinese)
[2] Cao H, Ma YX and BZ 2015 Journal of Dalian Maritime University 41(1):104-108. (in Chinese)
[3] Duan ZL, Ren G and Zhang JD 2016 Journal of Traffic and Transportation Engineering 16(6):82-90. (in Chinese)
[4] Gong HY and Cao H 2020 Ship & Ocean Engineering 49(01):111–114+119. (in Chinese)
[5] Duan ZL, Cao H and Ren G 2017 Journal of Marine Science and Technology 25(5): 71-580.
[6] Bao XY, Chai NJ and Wang QC 2018 Journal of the China Railway Society 40(10):15-22. (in Chinese)
[7] Yang XQ, Zhang ZH, Jiang HR and Yang CG 2020 China Safety Science Journal 30(03):1-7. (in Chinese)
[8] Lin GQ, Mo WT, Ye XJ, Han C, Lin ZZ and Wang ZK 2018 High Voltage Engineering 44(10):3383-3389. (in Chinese)
[9] Wang JQ and Zhang M 2018 China Safety Science Journal 28(11):142-148. (in Chinese)
[10]Li DY, Liu CX and Du W 2004 Journal of Software 15(11):1583-1594. (in Chinese)
[11]Li DY 1995 Journal of Computer Research and Development 32(6):15-20. (in Chinese)
[12]Liu GH, Song CX and Liu H 2007 Application Research of Computers (01):46-48. (in Chinese)

Modeling and Simulation of Layered Water Tank Based on MATLAB/SIMULINK

Xin Tan, Zhenjing Zhu*, Guoxin Sun, Li Diao and Jingyu Cao

College of Mechanical Engineering, Inner Mongolia University of Science and Technology, Baotou, Inner Mongolia, China

*Corresponding author's e-mail: qnananana@163.com

Abstract. In order to improve the accuracy of the stratified hot water storage tank model and simplify the modeling process of the stratified water tank, a modelling method of the stratified water tank based on MATLAB/SIMULINK is proposed. By comparing with TRNSYS platform simulation, the correctness and accuracy of this method are verified. At the same time, this method is used to establish a six-node stratified hot water storage tank. Through the simulation of 8 different working conditions, the influence of different inlet and outlet water flow rates and temperature of the inlet and outlet water on the stratification effect of the hot water storage tank is analyzed.

Keywords. Hot Water Storage Tank; Stratified Water Tank; MATLAB/SIMULINK

1. Introduction

Clean heating is one of the important measures to achieve my country's low-carbon environmental protection, energy saving and emission reduction development goals, and to meet the needs of sustainable development strategies[1]. With the steady progress of clean energy heating in the northern region, solar heating and combined heating have become the main alternatives for "heating by coal"[2].

Regardless of whether the hot water storage tank is used in the heating system or in the daily production process, as a key equipment for storing heat, its design and performance directly affect the overall performance of the system[3~5]. The hot water storage tank is an indispensable part of the composite heating system. The heat collector stores the heated hot water in it for users to use, which to a certain extent compensates for the discontinuity of the solar heating system caused by weather problems[6~7]. Due to the influence of temperature on water density, the phenomenon of temperature stratification occurs in the hot water storage tank. Making full use of this phenomenon not only improves the heating effect of the solar heating system, but also improves the heat collection efficiency[8].

At present, Wang Dengjia[8] uses the CFD platform to simulate the temperature stratification in the hot water storage tank, and obtains the stratification of the water tank under different working conditions. Yang Xianliang[9] used the Gambit platform to build a hot water storage tank model, and used Fluent to simulate the heating mode. Tay[10], Arslan[11], Baek[12] and others all use CFD software to build a hot water storage tank model to facilitate their research.

On the basis of the above research, this paper gives a simpler way of modeling a layered water tank based on MATLAB/SIMULINK for reference.

2. Mathematical model of stratified water tank

This paper chooses the current mainstream modeling method multi-node water tank as the model. Its characteristic is that the water tank is divided into several layers from top to bottom. Each layer is represented by a node. The uppermost node has the highest temperature and the lowermost node has the lowest temperature. The temperature of the nodes is different. The point at which the outlet water of the solar collector and the return water from the heating end enters the water tank is determined by the control function. The more nodes, the better the layering effect. When the number of nodes is 3~4, a better layered system can be expressed[17]. Therefore, this paper selects a hot water storage tank with 3 nodes for modeling. The node (layer) water tank model is shown in Figure 1[18].

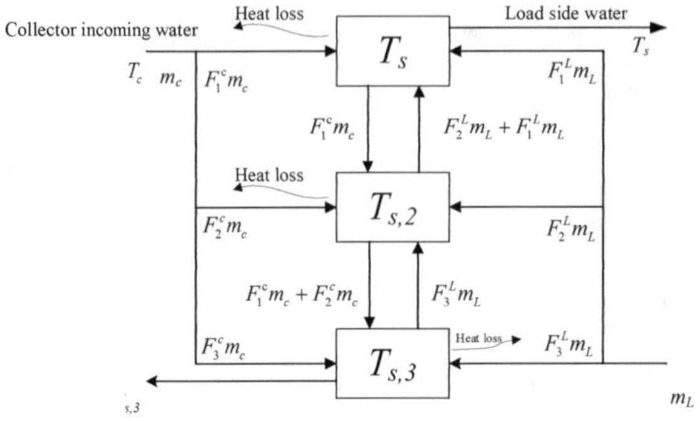

Figure 1. Three-node stratified water tank

2.1. Design control function

When the temperature of the incoming water from the collector is greater than or equal to the temperature of the first layer of the water tank, the incoming water from the collector enters the first layer of the water tank; when the incoming water temperature of the collector is between the temperature of the first layer and the second layer of the water tank, The incoming water from the collector enters the second layer of the water tank; when the incoming water temperature of the collector is between the temperature of the second and third layers of the water tank, the incoming water from the collector enters the third layer of the water tank;

The mathematical method to describe the above-mentioned heat collector incoming water into position-controlled heating is as follows[8]:

1)Define a function F_i^c to describe at which layer the water from the collector enters

$$F_i^c = \begin{cases} 1, & if \ \ i=1 \ \ and \ \ T_c \geq T_s \\ 1, & if \ \ T_{s,i-1} \geq T_c > T_{s,i} \\ 0, & else \end{cases} \tag{1}$$

2)In the same way, define a function F_i^r to describe at which floor the return water from the heating terminal enters

$$F_i^r = \begin{cases} 1, & if \ \ i=3 \ \ and \ \ T_{r,3} \leq T_{r,3} \\ 1, & if \ \ T_{r,i-1} \geq T_r > T_{r,i} \\ 0, & else \end{cases} \tag{2}$$

2.2. Calculate the heat exchange between each layer

When the heating system is working, the water in the water tank enters the load side from the uppermost layer and enters the solar collector from the lowermost layer. The flow of water causes the water exchange between the layers, and the water exchange causes the heat exchange between the layers. The greater the flow, the greater the degree of exchange. The current temperature of each layer is determined by the law of conservation of energy.

The energy balance equation expression of node[8]:

$$M_i \frac{dT_{s,i}}{dt} = \left(\frac{UA}{C_\mathrm{p}}\right)_i \left(T_a - T_{s,i}\right) + F_i^c m_c (T_{c,0} - T_{s,i}) + F_i^L m_l (T_{L,r} - T_{s,i}) + \begin{cases} m_{m,i}(T_{s,i-1} - T_{s,i}) \ \ if \ \ m_{m,i} > 0 \\ m_{m,i+1}(T_{s,i} - T_{s,i+1}) \ \ if \ \ m_{m,i+1} < 0 \end{cases} \tag{3}$$

Where: M_i — is the mass of the node in the hot water storage tank; m_c — is the average mass flow of water in the outlet pipe of the collector; m_l — is the average mass flow of water in the heating supply and return pipes; UA — is the heat loss coefficient of the water tank; C_p — is the water Specific heat; the net flow from the $i-1$ node to the i node can be expressed by the blended flow $m_{m,i}$, which only considers the vertical exchange, and does not consider the flow directly entering the node from the collector or heating load.

$$m_{m,1} = 0$$
$$m_{m,i} = m_c \sum_{j=1}^{i-1} F_j^c - m_L \sum_{j=i}^{N} F_j^L \tag{4}$$
$$m_{m,N+1} = 0$$

In the expression of the energy balance equation of the i node, the first term on the right represents the heat loss of the node to the surrounding environment; the second term represents the heat provided by the collector; the third term represents the heat supplied by the heating system; the fourth term is the above formula It represents the net heat exchange between the previous node and this node due to the vertical mixing, and the following formula represents the net heat exchange between the next node and this node.

3. SIMULINK model of stratified water tank

With the continuous deepening of research on solar heating systems and composite heating, they have basically met the needs of daily use. The main problems and challenges at present are the optimal design of the control system.

When designing and optimizing complex control systems, the solar heating modules and components included in TRNSYS and other software cannot meet actual needs, so it is necessary to use SIMULINK to develop heating modules and components.

3.1. Temperature model of each node

The first, second, and third layers of the water tank are independently modeled, and the heat change of each layer is represented by Q_i.

$$Q_i = Q_1 + Q_2 + Q_3 + Q_4 \tag{5}$$

In the formula: Q_1 is the heat of incoming water on the collector side; Q_2 is the heat of return water on the load side; Q_3 is the heat of exchange water; Q_4 is the heat lost.

The temperature of each layer of the water tank is determined by water exchange, and the quality of each layer of water is constant, that is, without considering the heat loss, the heat gain and loss are realized through water exchange, so the gains and losses of each layer of the water tank. Analyzing the water situation, the temperature of each node can be calculated. Define a function to determine whether each layer is in a state of water gain or loss (only for the uppermost or lowermost layer). In the figure, the solid line indicates that it must happen, and the dashed line indicates that it may happen.

$$m_i = m_{l/c} - m_c \cdot F_i^c - m_l \cdot F_i^r \tag{6}$$

1) Node one

The gains and losses of node one are shown in Figure 2.

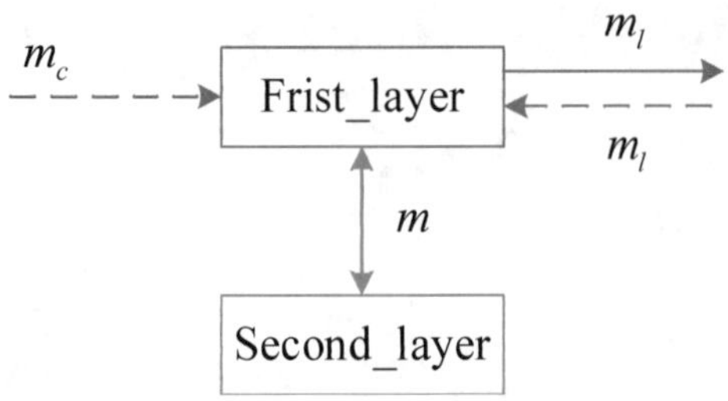

Figure 2. Gain and loss of water on the first floor

The heat of node one indicates:

$$Q_1 = m_c \cdot F_1^C (T_c - T_s) \cdot C_\rho \tag{7}$$

$$Q_2 = m_l \cdot F_1^r (T_r - T_s) \cdot C_\rho \tag{8}$$

$$Q_3 = \begin{cases} m_1 \cdot (T_{s,2} - T_s) \cdot C_\rho & m > 0 \\ 0 & m < 0 \end{cases} \tag{9}$$

$$Q_4 = UA \cdot M_1 \cdot (T_a - T_s) \tag{10}$$

$$Q_{frist} = Q_1 + Q_2 + Q_3 + Q_4 \tag{11}$$

Among them, if Q_3 in $m > 0$ indicates that the water from the second layer enters the first layer, heat exchange occurs; if $m < 0$ indicates that the water from the first layer enters the second layer, the resulting heat loss is counted as the heat gain of the second layer , calculated in the second layer.

2) Node three

The situation of node three is similar to that of node one. When $m < 0$ indicates that the water in the third layer enters the second layer, the heat exchange generated thereby is also calculated in the second layer. The situation of node three's gains and losses is shown in Figure 3.

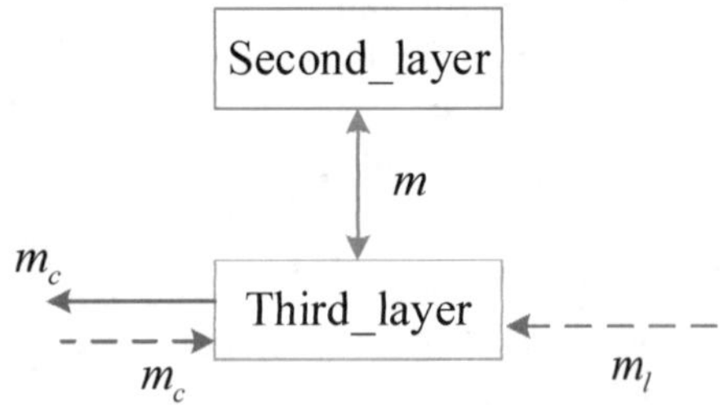

Figure 3. Gains and losses of the three layers of the water tank

The heat of node three indicates:

$$Q_1 = m_c \cdot F_3^C (T_c - T_{s,3}) \cdot C_\rho \tag{12}$$

$$Q_2 = m_l \cdot F_3^r (T_r - T_{s,3}) \cdot C_\rho \tag{13}$$

$$Q_3 = \begin{cases} m_3 \cdot (T_{s,2} - T_{s,3}) \cdot C_\rho & m > 0 \\ 0 & m < 0 \end{cases} \tag{14}$$

$$Q_4 = UA \cdot M_3 \cdot (T_a - T_s) \tag{15}$$

$$Q_{third} = Q_1 + Q_2 + Q_3 + Q_4 \tag{16}$$

3) Node second

The gains and losses of node two are shown in Figure 4.

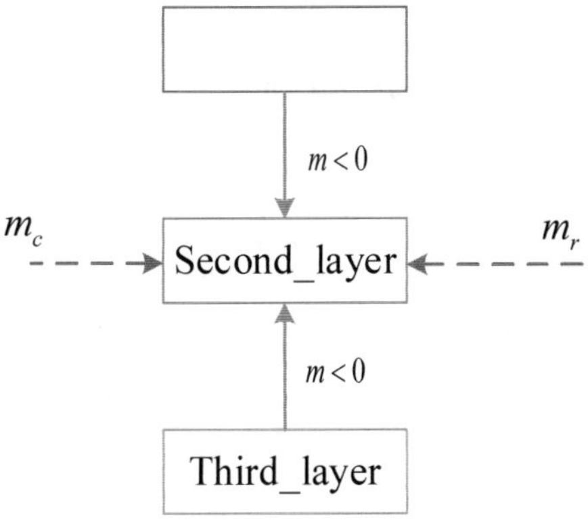

Figure 4. Water gains and losses on the second floor of the water tank

The heat expression of node second:

$$Q_1 = m_c \cdot F_2^C (T_c - T_{s,2}) \cdot C_p \tag{17}$$

$$Q_2 = m_l \cdot F_2^r (T_r - T_{s,2}) \cdot C_p \tag{18}$$

$$Q_3 = m_a \cdot (T_s - T_{s,2}) \cdot C_p + m_b \cdot (T_{s,3} - T_{s,2}) \cdot C_p \tag{19}$$

$$Q_4 = UA \cdot M_2 \cdot (T_a - T_s) \tag{20}$$

Among them, function m_a, m_b is used to judge whether m in node one and node three is less than zero.

3.2. SIMULINK Model

According to the method introduced in the previous section, the mathematical formula is converted into a code language to build a three-layer hot water storage tank as shown in Figure 5.

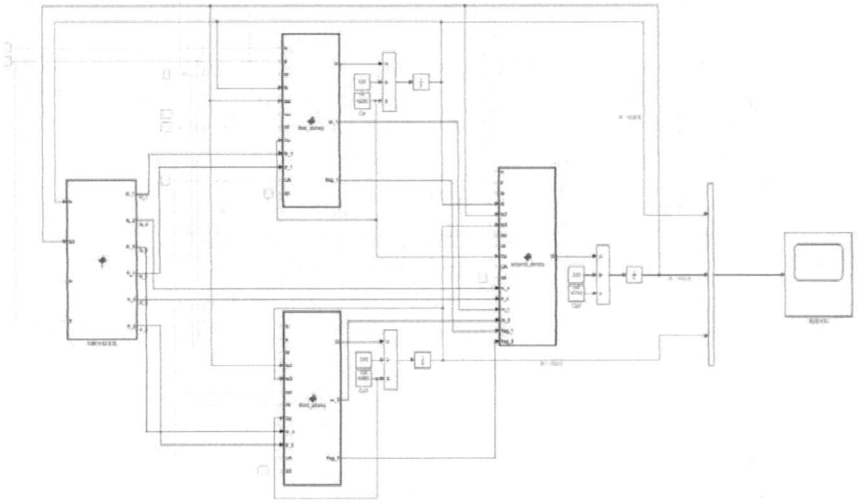

Figure 5. SIMULINK model of three-node layered hot water storage tank

4. Simulation analysis

4.1. Accuracy analysis
In order to verify the accuracy of the model established by this method, compare it with the currently recognized simulation software TRNSYS. Under the same working conditions, the layering effect of the three-node water tank is shown in Figure 6.

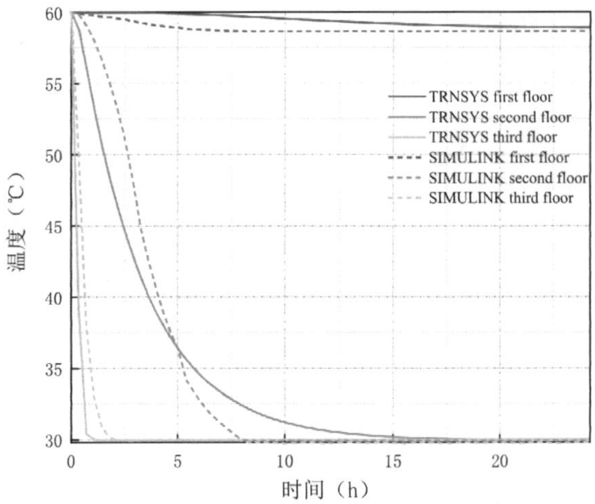

Figure 6. Comparison of the effect of stratification

It can be seen from Fig. 6 that compared with TRNSYS[19~21] which is widely used and accurate in the field of HVAC, the instantaneous maximum temperature difference is 2℃, which verifies the correctness and accuracy of the modeling method.

4.2. Flow analysis
The size of the inlet and outlet flow of the water tank will have a greater impact on the overall heating system. The larger the inlet and outlet flow of the water tank, the more obvious the mixing effect between the layers of the water tank. If the layering effect of the water tank is better, the heat utilization rate will be increased, and the efficiency of the heating system will also increase accordingly.

In this paper, with the same inlet and outlet water temperature and different inlet and outlet water flow rates, a 6-node layered hot water storage tank model is established according to the modeling method shown above. The volume of the water tank is 2m^3 and the ambient temperature is 20°C. Simulate 4 kinds of working conditions, and the simulated working conditions are shown in Table 1. The layering effect of the hot water storage tank is shown in Figure 7~ 10.

Table 1. Flow condition table of hot water storage tank

Working condition	Inlet water temperature on collector side℃	Load side return water temperature℃	Inlet water flow on the collector sidekg/s	Backwater flow on load sidekg/s
One	65	30	0.3	0.4
Two	65	30	0.1	0.4
Three	65	30	0.4	0.3
Four	65	30	0.4	0.1

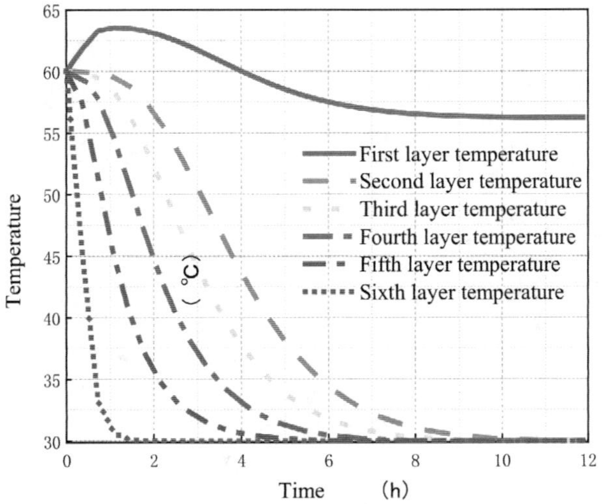

Figure 7. Working condition one

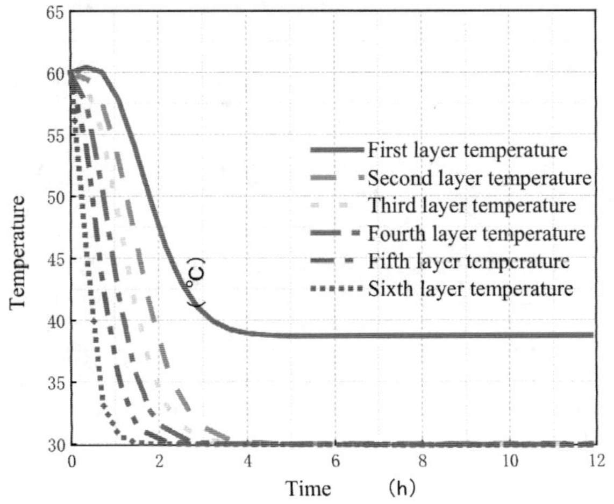

Figure 8. Working condition Two

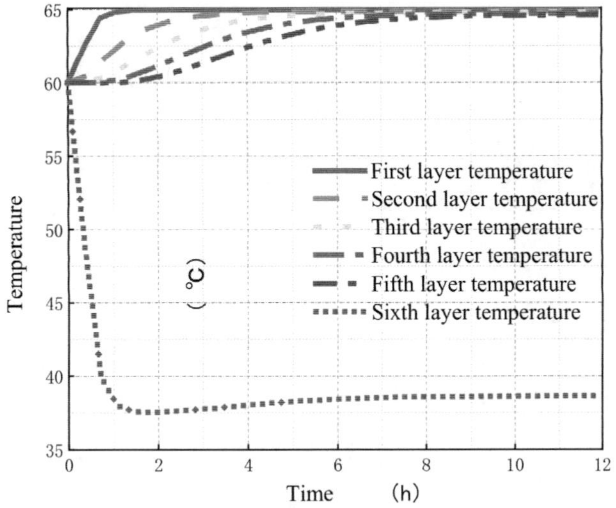

Figure 9. Working condition Three

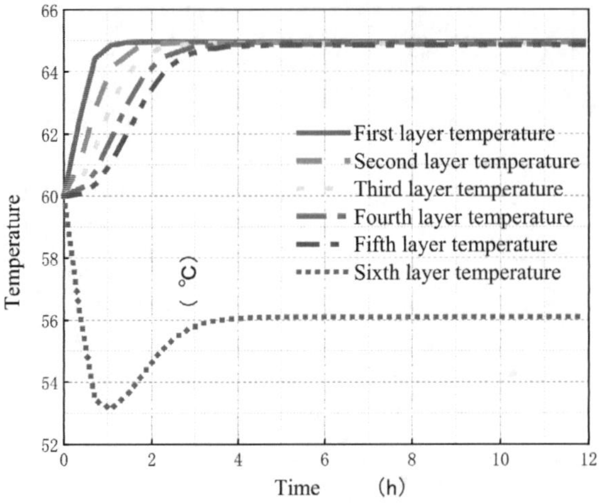

Figure 10. Working condition Four

It can be seen from the above four working conditions that when the inlet water flow rate on the collector side is greater than the return water flow rate on the load side, it is more conducive to the stratification of the hot water storage tank. When the backwater flow rate on the load side is 0.4kg/s, when the inlet water flow rate on the collector side is 0.3kg/s, it is more conducive to the stratification of the water tank.

4.3. Temperature analysis

In addition to the influence of flow on the stratification of the hot water storage tank, the inlet water temperature on the collector side and the return water temperature on the load side also have a greater influence on the stratification effect of the water tank. Choose the inlet and outlet flow with obvious stratification effect, and simulate four different working conditions with different inlet water temperature and return water temperature, as shown in Table 2. The layering effect of the hot water storage tank is shown in Figure 11~14.

Table 2. Heat storage tank temperature condition table

Working condition	Inlet water temperature on collector side℃	Load side return water temperature℃	Inlet water flow on the collector sidekg/s	Backwater flow on load sidekg/s
One	65	30	0.3	0.4
Two	65	20	0.3	0.4
Three	80	30	0.3	0.4
Four	80	20	0.3	0.4

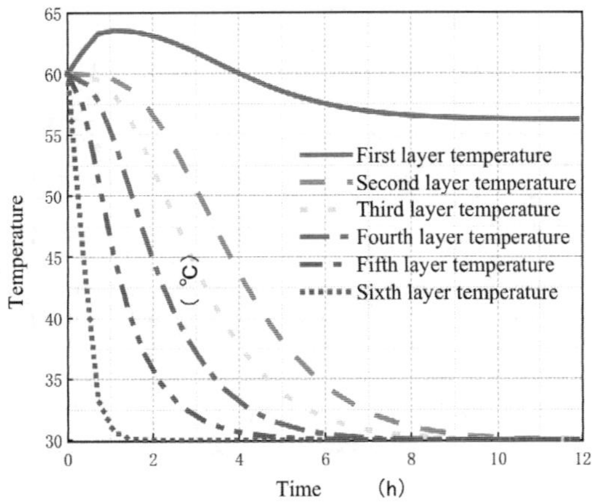

Figure 11. Working condition Five

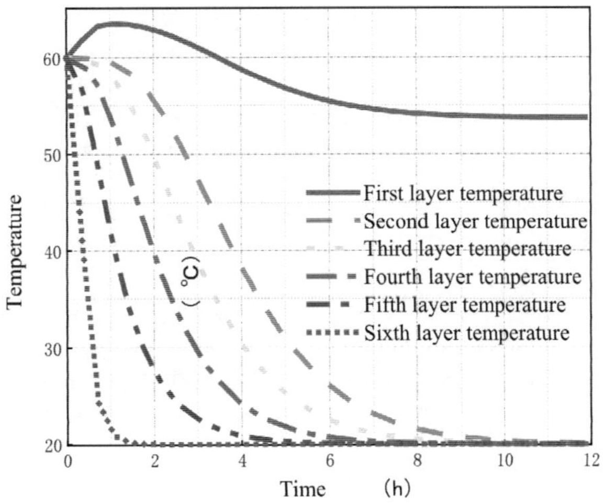

Figure 12. Working condition Six

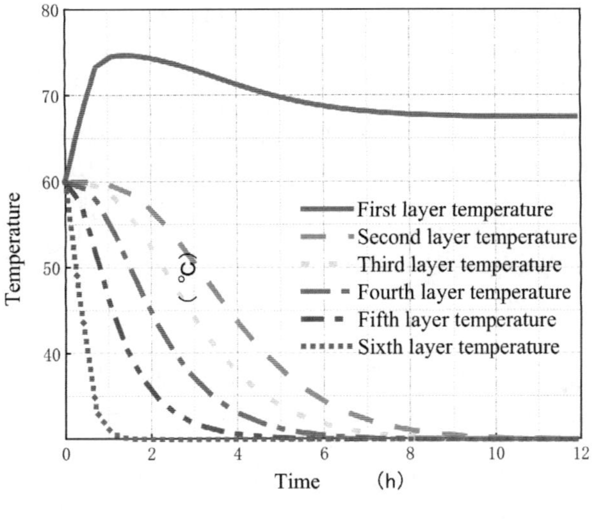

Figure 13. Working condition Seven

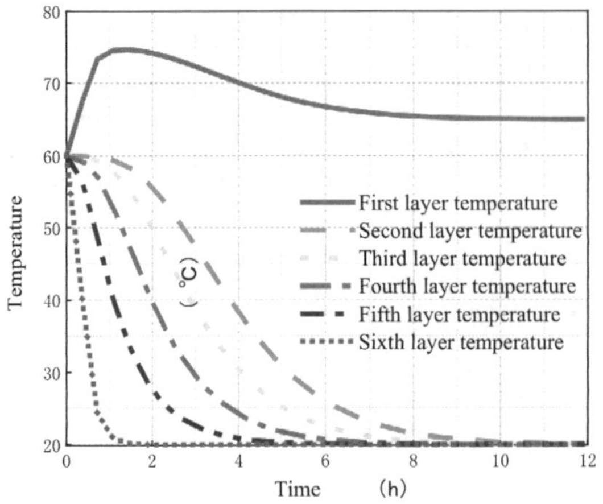

Figure 14. Working condition Eight

It can be seen from the above four that when the inlet water flow rate on the collector side and the return water flow rate on the load side are constant, the greater the temperature difference between the inlet water temperature on the collector side and the return water temperature on the load side, the more conducive to the stratification of the hot water storage tank. That is, the inlet water temperature of the water tank should be increased as much as possible, and the return water temperature should be reduced.

5. Conclusion

(1) A relatively simple MATLAB/SIMULINK layered hot water storage tank modeling method is given, and compared with TRNSYS, the maximum instantaneous temperature difference of each layer is 2℃, which proves the correctness and accuracy of the method .

(2) When the inlet water flow on the collector side is greater than the return water flow on the load side, it is more conducive to the stratification of the hot water storage tank. When the backwater flow rate on the load side is 0.4kg/s, when the inlet water flow rate on the collector side is 0.3kg/s, it is more conducive to the stratification of the water tank.

(3) When the inlet water flow rate on the collector side and the return water flow rate on the load side are constant, the greater the temperature difference between the inlet water temperature on the collector side and the return water temperature on the load side, the more conducive to the stratification of the hot water storage tank, that is, it should be increased as much as possible The inlet water temperature of the water tank reduces the return water temperature.

Acknowledgements

This research was supported by the Natural Science Foundation of China, Project No. 61765012, the Natural Science Foundation of Inner Mongolia, Project No. 2021LHBS05005, Science and Technology Research Project of Inner Mongolia Autonomous Region Higher Education, Project No. 2021SHZR0620, Inner Mongolia Autonomous Region 2017 science and technology innovation guidance award funding projects, Project No. 2017CXYD-2. "The funders had no role in the design of the study; in the collection, analyses, or interpretation of data; in the writing of the manuscript, or in the decision to publish the results".

References

[1] Yuan T H, Zhu N, Shi Y F, Chang C, Yang K, Ding Y 2018 Sample data selection method for improving the prediction accuracy of the heating energy consumption *Energy & Buildings* p158
[2] Li Q R 2018 Study on the Application of Solar-air source Combined Heating System in PlateauCold Regions (Chongqing: Chongqing University)
[3] Jiang H M, Wang Y, Ma Z L, Yao Y 2006 Design and Experimental Analysis on a Hot-Water

Storage Tank *Buiding Energy and Envionment* **4**(02) 74-8
[4] Yu H T, Gao J M, Chen Y 2015 Heat release performance of heat storage water tank with phase-change material in solar drying system *Transactions of the Chinese Society of Agricultural Engineering (Transactions of the CSAE)* **24** 207-11
[5] Elgamal M, Kriaa K, Farouk M 2021 Drainage of a Water Tank with Pipe Outlet Loaded by a Passive Rotor *Water* **13**(13) 1872
[6] Kaygusuz K, Ayhan T 1999 Experimental and theoretical investigation of combined solar heat pump system for residential heating *Energy Conversion and Management* **40**(13) 1377-96
[7] Carboni C, Montanari R 2007 Solar thermal systems: Advantages in domestic integration *Renewable Energy* **33**(16) 1364-73
[8] Wang D J, Liu Y F 2010 Temperaure stratification studying of heat storage Tank *Buiding Energy and Envionment* **29**(01) 16-9
[9] Yang X L, Feng H, Shi G H 2021 Study on temperature stratification of heat storage tank with new equalizer *Acat Energies Solaris Sinica* **42**(2) 19-24
[10] Tay N H S, Bruno F, Belusko M 2012 Experimental validation of a CFD and an ε-NTU model for a large tube in-tank PCM system *International Journal of Heat and Mass Transfer* **55**(21-22) 5931-40
[11] Arslan M, Igic A A 2015 Thermal performance of a vertical solar hot water storage tank with a mantle heat exchanger depending on the discharging operation parameters *Solar Energy* **116** 184-204
[12] Baek S M, Nam J H, Hong H K et al. 2011 Effect of brine flow rate on the performance of a spiral-jacketed thermal storage tank used for SDHW systems: A computational fluid dynamics study *Applied Thermal Engineering* **31** 2716-25
[13] Solkane P M 2008 *Refrigeration and air-conditioning technology* Canada: Delmar Cengage Learning p56-59
[14] Gao Y J et al. 2020 Mathematical modeling and simulation of solar finned tube evaporator. *Acat Energies Solaris Sinica* **41**(7) 57-63
[15] Sudhakar K, et al. 2019 Modelling of a solar desiccant cooling system using a TRNSYS-MATLAB co-simulator: A review *Journal of Building Engineering* **24** 100749
[16] [16]Zhang W B, Du Z M, Jin X Q, Fan B, Yang C 2009 *2009 Annual Conf. Proc. of Shanghai Society of Refrigeration* (Shanghai: Shanghai Society of Refrigeration)
[17] Duffie J, Beckman W, Worek W 1994 Solar Engineering of Thermal Processes, 2nd ed. *Journal of Solar Energy Engineering* **116**(1)
[18] Yu G Q, Tang J H, Zou Z J 2007 Research on the Effect of Temperature Stratification in Water Tank in Solar Domestic Hot Water System *Building Science* **4**(04) 70-3
[19] Ran S Y, Li X T, Xu W, Wang B L 2020 A solar-air hybrid source heat pump for space heating and domestic hot water *Solar Energy* **199** 347-59
[20] Xu W, Liu C P, Li A G, Li J, Qiao B 2020 Feasibility and performance study on hybrid air source heat pump system for ultra-low energy building in severe cold region of China *Renewable Energy* **146** 2124-33
[21] Huang B J, Lee C P. 2004 Long-term performance solar assisted heat pump water heater *Renewable Energy* **29**(4) 633-9

Research on an Anechoic Wind Tunnel in the Design Phase by Numerical Simulation

Hanqin Li[1], Bin Fang[2] and Yongming Zhang[1,*]

[1]Department of Operational Research and Programming, Naval University of Engineering, Wuhan, Hubei, China
[2]Department of Naval Architecture and Marine Engineering, Naval University of Engineering, Wuhan, Hubei, China

[*]Corresponding author's e-mail: 1243964100@qq.com

Abstract. As an important facility for acoustic measurement and research in air medium, anechoic wind tunnel requires high acoustic performance. In this paper, the flow field and acoustic characteristics of anechoic wind tunnel were studied by modeling and simulation. The velocity and pressure distribution of the whole flow field in the wind tunnel are obtained by simulation calculation, and the noise source distribution in the anechoic wind tunnel is studied by the broadband noise source method, which provides a reference for the design, construction and optimization of the anechoic wind tunnel.
Keywords. Anechoic wind tunnel; Broadband noise source method; RANS

1. Introduction

With the development of transportation industry, people pay more and more attention to the noise problem out of the process when aircraft, automobile and other means of transportation are working. Anechoic wind tunnel plays an important role in the experimental study of acoustics, especially aeroacoustics. Anechoic wind tunnel can carry out a variety of acoustic tests in the aerodynamic environment, and can be used in the test and research of vehicle flow noise and aircraft flow noise. For example, the Audi wind tunnel in Germany is one of the most modern wind tunnels in the world, which is mainly used to test the aerodynamic and acoustic performance of cars.[1]The DNW wind tunnel built jointly by Germany and the Netherlands is a reflux wind tunnel with a total length of 20m, which can measure different test models of different size. [2]NASA Ames wind tunnel is the largest anechoic wind tunnel in the world for now, which is mainly used for the measurement of aerodynamic noise in aerospace industry.[3]

An anechoic wind tunnel needs good flow field and acoustic characteristics, especially the latter one. Flow field of good quality can reduce the turbulence noise caused by the unstable flow, thus reducing the background noise. Researchers from all over the world have been researching actively on the aerodynamic and acoustic characteristics of different anechoic wind tunnels to improve the quality of them. Therefore, before the design and construction of an anechoic wind tunnel, the flow field and acoustic performance of an anechoic wind tunnel is an important research direction. The numerical simulation method can be used to estimate the flow field and acoustic characteristics of the wind tunnel, as well as to make contribution to the design, modification and optimization. Cong took the method of modeling and numerical simulation to study how the fan section influence on the performance of the aero-acoustic wind tunnel.[4]Ismail carried out application of the computational fluid dynamic method with several turbulence models to obtain the ideal position of the test object in the test section as well as the performance of the wind tunnel.[5]S. Redonnet assessed acoustic installation effects through numerical simulation, using Computational Aeroacoustics, which shows potential in the advanced design of optimized aeroacoustic wind tunnels.[6]

The Reynolds Average Navier Stokes equation and broadband noise source method were adopted in this paper for numerical simulation of a certain type of anechoic wind tunnel in the period of preliminary design and demonstration. And the flow field characteristics and acoustic characteristics of the anechoic wind tunnel were studied .

2. Fundamental principles of numerical calculation

2.1. Reynolds Average Navier Stokes equation

Due to the existence of turbulence fluctuation, it will cost a lot of computing resources and time to simulate the turbulence directly in practical engineering. The Reynolds Average Navier Stokes equation (RANS) is used to calculate the turbulence fluctuation after time averaging by statistical averaging method. The term of small-scale motion after time averaging is eliminated and combined into the large-scale motion equation, so as to replace the original instantaneous control equation, which will reduce the computational cost.

For a certain physical quantity f , the average value is defined as $\bar{f}$,and the pulsating value is defined as f' , then it can be obtained that:

$$f = \bar{f} + f' \tag{1}$$

$$\bar{f}(x, y, z, t) = \frac{1}{T} \int_{t-T/2}^{t+T/2} f(x, y, z, \tau) d\tau \tag{2}$$

The Reynolds Average Navier Stokes equation can be obtained by averaging the flow variables in the transient continuity equation and momentum equation.

$$\frac{\partial \rho}{\partial t} + \frac{\partial}{\partial x_i}(\rho u_i) = 0 \tag{3}$$

$$\frac{\partial}{\partial t}(\rho u_i) + \frac{\partial}{\partial x_j}(\rho u_i u_j) = -\frac{\partial p}{\partial x_i} + \frac{\partial}{\partial x_j}\left[\mu\left(\frac{u_i}{u_j} + \frac{u_j}{u_i} - \frac{2}{3}\delta_{ij}\frac{\partial u_1}{\partial x_1}\right)\right] + \frac{\partial}{\partial x_j}\left(-\rho\overline{u_i' u_j'}\right) \tag{4}$$

The Reynolds Average Navier Stokes equation is consistent with the Navier Stokes equation in form. It is worth noting that the variables in the equation represent the mean quantity. And an additional term in the rightmost of the equation is the Reynolds stress term, which represents the turbulence effect.

The equations above are not closure. It is necessary to introduce the eddy viscosity coefficient into the equations. Express the turbulent stress as a function of the eddy viscosity coefficient and then solve the equations.

2.2. Broadband noise source method

In frequency domain, broadband noise has continuous spectrum, in which sound energy is continuously distributed in a certain frequency range. The location and intensity of the main noise sources related to the aerodynamic characteristics can be calculated by the broadband noise source model. Due to RANS simulation method taken by broadband noise source model, it is adopted to reduce the amount of calculation and shorten the calculation time.

By this method the accurate noise spectrum value of the monitoring point can't be obtained, but it can predict the location and strength of the noise source. Boundary layer noise model is a common method to study the noise caused by turbulent boundary layer on solid surface. Curle equation is as follows:

$$p'(\vec{x},t) = \frac{1}{4\pi c_0} \int_s \frac{(x_i - y_i) n_i}{r^2} \frac{\partial p}{\partial t}(\vec{y},\tau) dS(\vec{y}) \tag{5}$$

x is vector at the measuring point, y is vector at the sound source, c_0 is acoustic speed in the air, τ is time delay, S is the surface。

The far-field sound density is as follows:

$$\overline{p'^2} \approx \frac{1}{16\pi c_0^2} \int_s \frac{\cos^2\theta}{r^2} \overline{\left[\frac{\partial p}{\partial t}(\vec{y},\tau)\right]^2} A_c(\vec{y}) dS(\vec{y}) \tag{6}$$

The total noise power is as follows:

$$SAP = \frac{1}{12\rho_0 \pi c_0^3} \int A_c(y) \overline{\left[\frac{\partial p}{\partial t}\right]^2} dS(y) \tag{7}$$

3. Modeling and simulation

The Star CCM software was adopted to carry out modeling and simulation on the anechoic wind tunnel. The model consists of contraction section, diffuser section and test section. The size of the test section is designed to be 4m×2.4m×1.6m.

The wall thickness of the contraction section is neglected. While the wall thickness of the diffuser section can not be ignored due to the filling of sound-absorbing materials in it. The model of diffuser section contains inner wall and outer wall. The geometric model is established as shown in the figure 1 below.

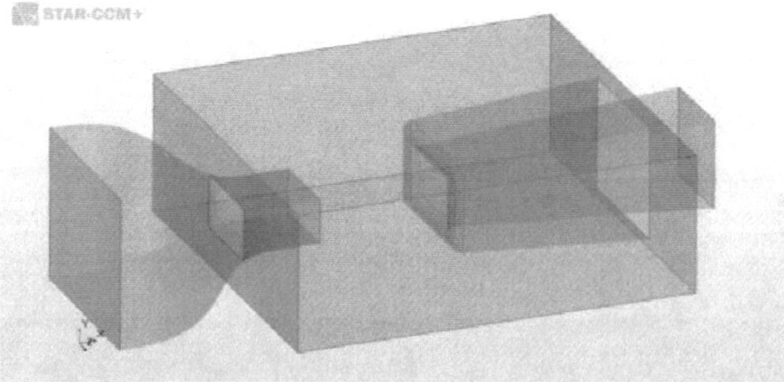

Figure 1. Geometry model of anechoic wind tunnel

The mesh of anechoic wind tunnel is divided. When the wind tunnel is running, the turbulent boundary layer is prone to be generated on the wall of contraction section and diffuser section. The mesh of these parts needs to be defined smaller and denser for calculation, so as to obtain the information of pressure fluctuation in turbulence and capture the movement of small vortices.

The velocity at the entrance is determined according to the theorem of momentum.

$$v_0 S_0 = v_T S_T \tag{8}$$

The velocities at the inlet and outlet of the contraction section are represented as v_0、v_T .The area at the inlet and outlet of the contraction section are represented as S_0、S_T .

4. Result of calculation and analysis

4.1. Flow field characteristics

The simulation calculation of the anechoic wind tunnel is carried out under two conditions of $V_T = 24$

m/s and V_T = 60 m/s. Here, the velocity nephogram and pressure nephogram of V_T = 24m/s and V_T = 60m/s are taken for observation and comparison.As is shown in figure 2 and figure 3.

In the condition of V_T = 24 m/s, from the velocity nephogram in the longitudinal section plane, it can be obtained that:

- The velocity gradient at the junction of contraction section and test section is large.
- After the airflow enter the test section, there is a little variaty in velocity gradient at the beginning of the test section, Then the airflow in the test section become uniform and stable at the velocity of 25m / s. From the velocity nephogram, it can be seen that the velocity of airflow in the test section changes little. And the flow field of the test section is relatively stable and uniform.
- It is found that there is a small circle of velocity gradient near the edge of the test section. Within the circle, the velocity of airflow in the test section remains uniform and stable, and there is no obvious velocity gradient changing.

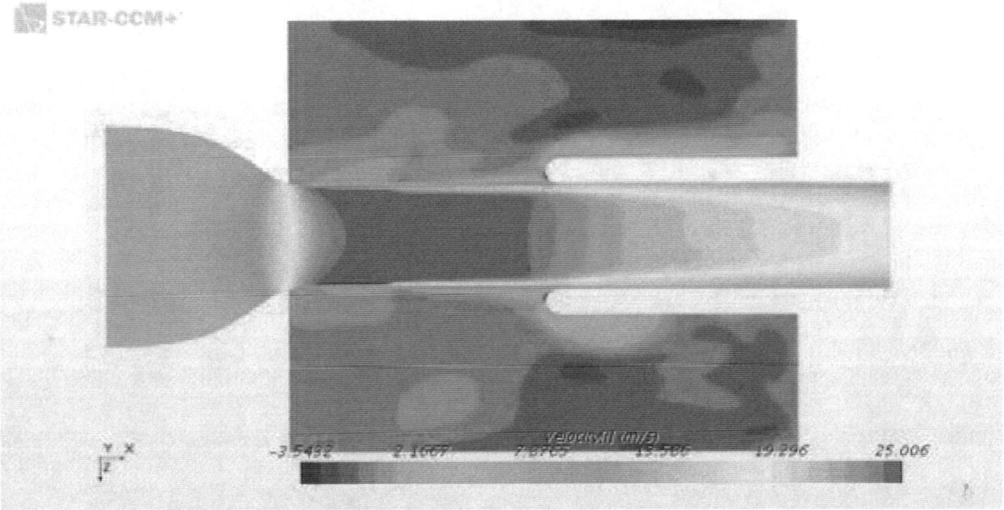

Figure 2. Longitudinal section of velocity nephogram when V_T = 24 m/s

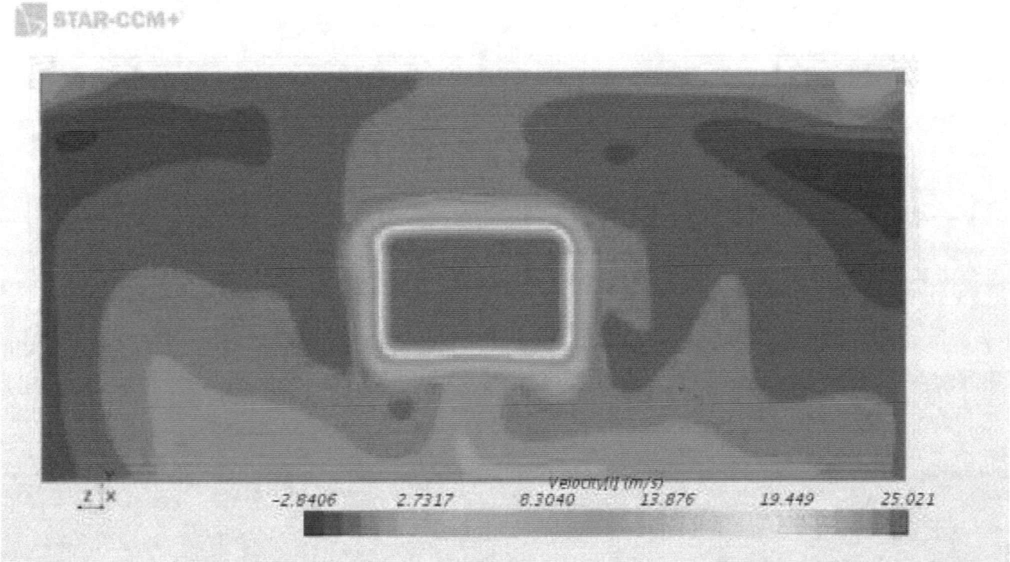

Figure 3. Middle section of the velocity nephogram when V_T = 24 m/s

When V_T = 60m / s, choose the longitudinal section and middle section of the velocity nephogram for observation. From figure 4 and figure 5, it can be observed that the velocity field of contraction section and test section is basically uniform. While in the diffuser section, from the longitudinal section plane we can see that the velocity of airflow in the end of the diffuser section becomes a little non uniform. The cause may be that the increase of velocity caused the increase of airflow instability.

Therefore, in general, the velocity field of the test section in the wind tunnel is relatively uniform, but with the increase of velocity, the stability of velocity variation in the diffuser section decreases.

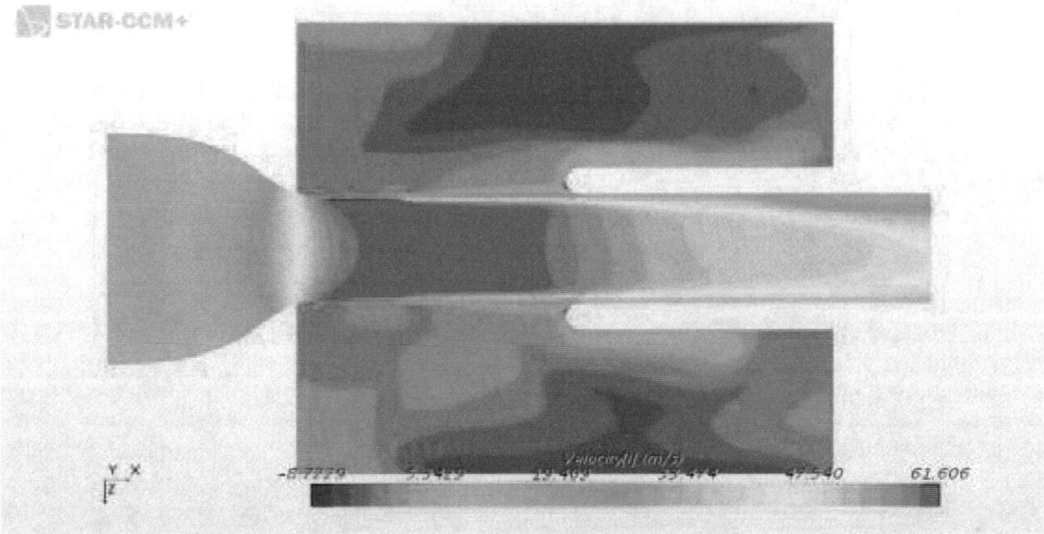

Figure 4. Longitudinal section of velocity nephogram when V_T = 60m / s

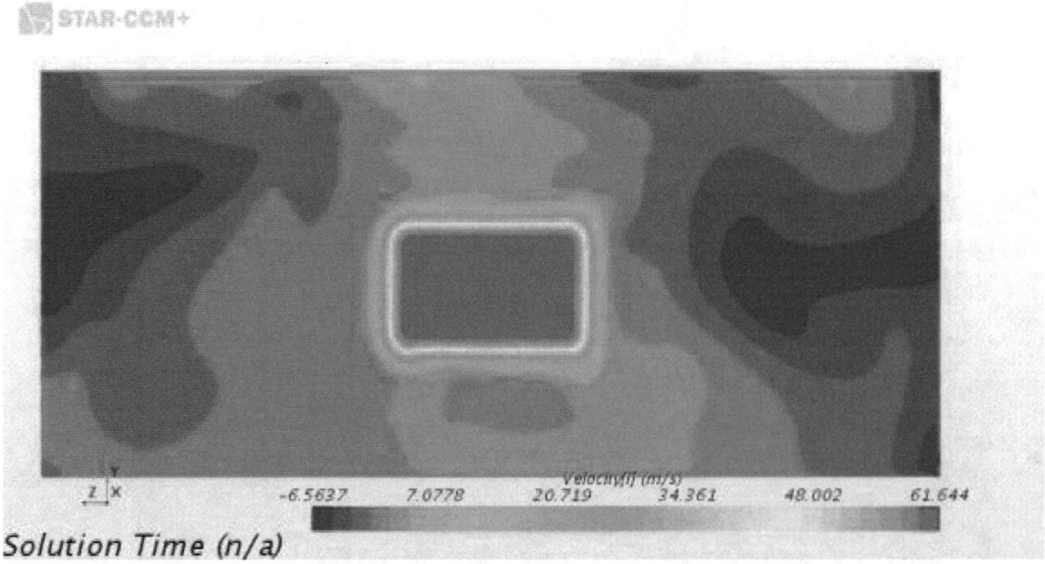

Figure 5. Middle section of the velocity nephogram when V_T = 60m / s

Turbulence intensity is an important index for the flow field quality assessment of an wind tunnel. Turbulence intensity of an wind tunnel should be less than 0.5%.When it comes to the value less than 0.2%, flow field quality of an wind tunnel can be regarded as excellent.

Turbulence intensity in the test section of the wind tunnel was calculated at the velocity of 60m/s.There are 9 virtual measuring points, which are distributed with the same distance of 0.4m in the central axis of the test section. According to the sequence of direction from the contraction to the diffuser, the points are numbered from 1 to 9.The result of calculation for the turbulence intensity is showed in the table below.

Table 1. Turbulence intensity of the measuring points when VT = 60m/s

Number of the points	1	2	3	4	5	6	7	8	9
Turbulence intensity(%)	0.24	0.30	0.21	0.19	0.15	0.16	0.18	0.22	0.28

From the table above, it can be seen that the turbulence intensity is low, which basically meet the requirement for the flow field quality of a wind tunnel. Turbulence intensity at some measuring points are less than 0.2%,which has reached to good flow field quality.

4.2. Acoustic characteristics

By observing the Curle nephogram in figure 6, it can be found that the dipoles are distributed on the walls of different sections, including contraction section, diffuser section and anechoic chamber. Among them, the dipole distribution in the region of contraction is the largest. And it is the same with dipole intensity. The contribution of the radiation source is the largest.

It is found from figure 7 that the quadrupole represented by Proudman nephogram is distributed near the outer edge of the test section, forming a hollow cuboid shape. It is also observed that quadrupoles are distributed in the diffuser section.

Figure 6. Curle nephogram of sound sources

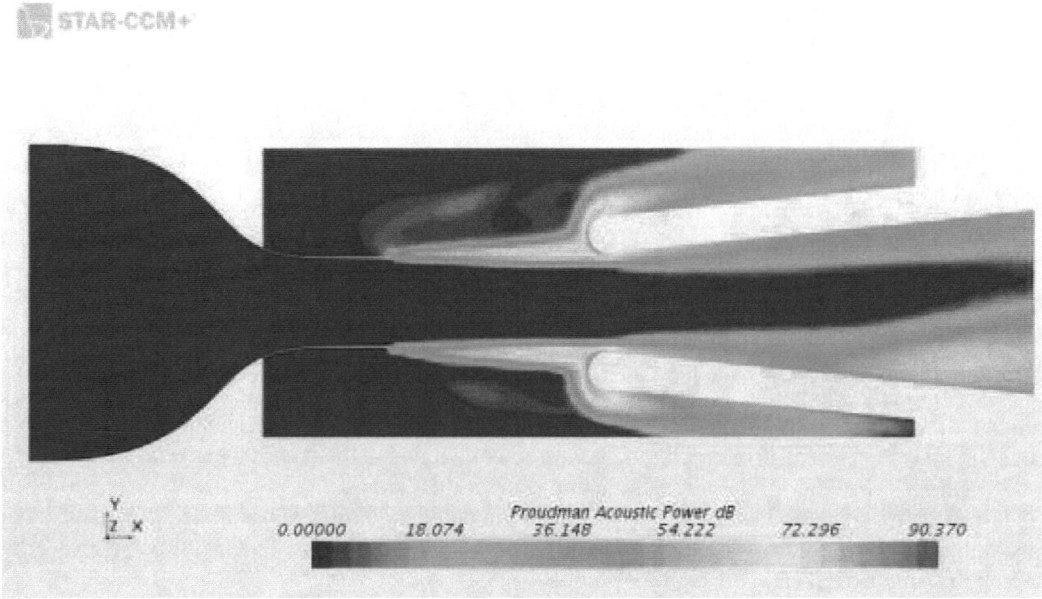

Figure 7. Proudman nephogram of sound sources

5. Conclusions

In this paper, flow field and acoustic characteristics of an anechoic wind tunnel in the period of preliminary design was carried out by modeling and numerical simulation. Through calculation and analysis, the following conclusions can be obtained:

- Under different working conditions, the flow field in the test section is uniform, the flow field characteristics is good, and the turbulence intensity is low. The preliminary design can meet the construction standard of wind tunnel.
- Under different working conditions, the velocity gradient in the diffuser changes slowly, which can meet the requirement for a wind tunnel.
- There are some dipole noise sources on the wall of the exit of the contraction section, which may increase the background noise of the anechoic wind tunnel.

In view of the noise source problem mentioned above, it is suggested that sound absorption materials can be laid on the contraction wall in the following construction process of the wind tunnel to strengthen the noise reduction effect.

On the other hand, the linear design of the contraction section can be optimized in the future work. And further study should be carried out on the acoustic characteristics of the anechoic wind tunnel.

References

[1] Wiedemann J,Wickern G,Ewald B,et al.Audi Aero-Acoustic Wind Tunnel[C].International Congress & Exposition.1993.
[2] Brouwer H H.Anechoic wind tunnels[R].Amsterdam:National Aerospace Laboratory NLR,1997.
[3] Stephen J M,Christopher A S.Reduction of background noise in the NASA Ames 40 by 80 foot wind tunnel[C].1st Joint CEAS/AIAA Aeroacoustics Conference,1995:1-4.
[4] Cong Chenghua, Liao Daxiong, Qu Xiaoli. Numerical study on characteristics of aero-acoustics pilot wind tunnel fan section[J]. Journal of Propulsion Technology,2012, 30(4):450-455
[5] Ismail, Johanis John, Erlanda A. Pane. Computational fluid dynamics simulation of the turbulence models in the tested section on wind tunnel [J]. Ain Shams Engineering Journal,2020,11(4),1201-1209.
[6] S. Redonnet. Investigation of the acoustic installation effects of an open-jet anechoic wind tunnel using computational aeroacoustics [J].Applied Acoustics,2020,169, 107469.

Resource Allocation for Cloud-Edge-End Simulation System

Guoqiang Yan[1,*], Shaojie Mao[1] and Yuping Li[1]

[1]Science and Technology on Information System Engineering Laboratory, Nanjing, Jiangsu, China

[*]corresponding author's e-mail: misu.yan@qq.com

Abstract. Aiming at the problem of simulating service requests and optimal scheduling during the operation of the cloud computing/edge computing environment, the real-time optimization scheduling technology of computing resources is studied, and elastic resource optimization scheduling is realized through data feature mining analysis, service load prediction, and collaborative resource management. Ensure that the simulation service quality meets the mission requirements and provide support.

1. Introduction

The high-quality resources based on the cloud computing center combined with the resources of the edge computing environment (edge cloud) can provide support for computing-intensive and strong real-time weapon equipment system/system simulation. However, due to differences in resources and capabilities between cloud computing center computing nodes and edge cloud computing nodes, as well as the limitation, heterogeneity and instability of edge cloud resources, the resource management and scheduling mechanism in traditional cloud computing centers is difficult to apply Multi-granularity simulation services respond to demands in real time under the "cloud computing/edge computing" system architecture[1-3].

Therefore, in response to the demand for unified management, control and scheduling of computing resources that support simulation operations under the "cloud computing/edge computing" architecture, the theoretical methods based on cloud-edge collaborative computing, big data analysis technology, resource topology analysis technology, and machine learning are proposed based on Data-driven intelligent computing resource integration technology, through the analysis of cloud and edge service requirements and computing resource distribution characteristics, establishes the dependence of simulation services on resources and data, breaks through the intelligent analysis technology of system resource situation based on perception, and forms system network communication Model, establish a cloud-edge collaborative resource management and allocation mechanism based on service demand perception, and an active resource optimization framework based on multi-dimensional resource balance, to achieve environment-aware adaptive task scheduling and resource integration; target cloud computing/edge computing environment operation processes In the simulation service request and optimization scheduling problem, the real-time optimization scheduling technology of computing resources is studied, and the simulation service load prediction model based on load big data analysis is constructed through data feature mining analysis and service load prediction. Sharing mode, breakthrough network-aware cloud-edge collaborative resource dynamic configuration method, and real-time creation algorithm of service replicas under unstable edge cloud environment, realize load-aware flexible resource optimization scheduling, and provide support to ensure that the simulation service quality meets the task requirements[10-11] .

2. Formatting the title, authors and affiliations

With the rise of edge computing models in recent years, and the need for stronger data analysis in information systems, although traditional cloud data centers can provide powerful data analysis capabilities, data transmission will consume a lot of time, resulting in longer Service delays and other issues. For real-time simulation systems, the QoS (Quality of Service) caused by long service delays is unacceptable. Therefore, cooperation with edge computing is an option to reduce service delays. The author of literature [4] proposes a service recommendation method based on collaborative filtering to be applied in the mobile edge computing environment, and to predict QoS based on user mobility. This method first calculates the similarity of users or edge servers and selects the most similar k neighbors to predict service QoS, and then performs service recommendation. Literature [5] studied the task allocation problem of reducing service delay in cloud-edge system, and used W-DAG (weighted directed acyclic graph) to model data-intensive services or business logic, and analyzed the tasks that constitute integrated services. Data and resource requirements, and resource allocation between cloud data centers and edge nodes, a data-driven task allocation algorithm is proposed to reduce business latency. Literature [6] studied how to select services based on user needs to construct a multimedia file transmission path from the source node to the destination node. In previous studies, the transmission is always based on a fixed functional path.

3. Formatting the text

3.1. Conception

The cloud side simulation system architecture is a new type of simulation system architecture, which is a further development of the cloud simulation architecture. It is based on cloud-side joint application requirements, based on cloud computing and edge computing concepts, and integrates various technologies such as network technology, virtualization technology, artificial intelligence technology, and realizes the safe sharing and reuse of various resources in the system on demand , To achieve on-demand collaborative and interoperability between the cloud and the edge, to achieve dynamic optimization and scheduling of the system, and to support the construction, operation and evaluation of complex simulation systems. The cloud edge simulation system architecture is shown in Figure-1.

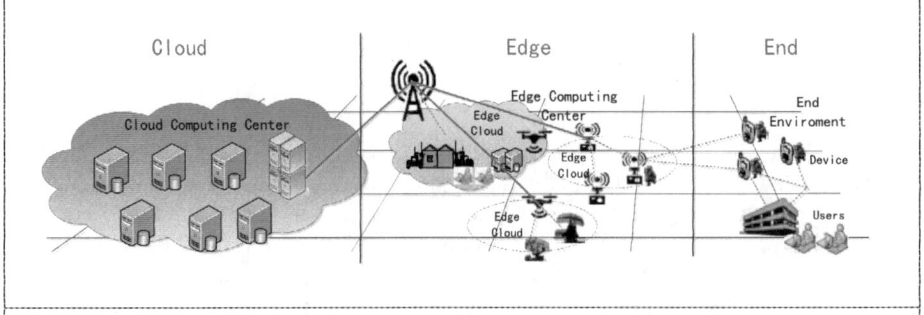

Figure 1. Cloud-edge-end simulation system architecture.

Cloud refers to the cloud center, which has massive on-demand computing/storage capabilities, and encapsulates various computing, model, data and other resources through virtualized service technology to provide services to the edge.

Edge refers to the edge information service center, the local edge computing environment built by the units that develop and use various equipment systems for their own tasks, and have domain-related simulation capabilities. The "edge" is closer to the user, and can process a large amount of data locally, thereby reducing the amount of data transmitted on the network, and also giving users a faster real-time response.

The end is the user of various equipment system simulations, or various existing simulation subsystems, which are dynamically connected as a special link of simulation to form a complete simulation system. As an integral part of the simulation system, the "end" is used for system input, such as system intervention commands. The second is the output of the system, such as the situation display of the simulation process.

3.2. Application Architecture

There are three main types of users of cloud-side simulation system architecture. The first category is the design and development personnel of the simulation system, who is responsible for designing the simulation system based on the architecture. The second category is cloud computing and edge computing operation and maintenance personnel, responsible for operating and maintaining various computing and simulation resources based on this architecture, and deploying and operating simulation systems. The third category is the users of the simulation system, who use the simulation system in accordance with the application mode of the architecture design. The new simulation system architecture is divided into infrastructure service IaaS, platform service PaaS, and software service SaaS according to the service mode. The roles of the three types of users are shown in Figure-2[9].

The computing resource scheduling of the cloud-side simulation system is aimed at the dynamic and heterogeneous nature of resource requirements for the operation of the simulation system, and the dynamic allocation of resources is used to meet the resource requirements of the simulation service and ensure the quality of the simulation service. The computing resource scheduling of the cloud-side simulation system can be divided into three aspects: intelligent perception of resource situation, collaborative resource management and utilization, and real-time optimized scheduling of computing resources.

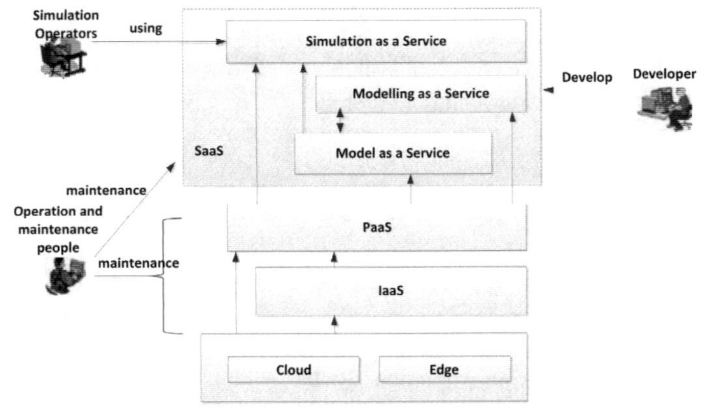

Figure 2. Application architecture.

4. Automatic Resource allocation for Simulation Systems

Taking into account the dynamic loading and migration of simulation services between the cloud and the edge, the computing resource scheduler can perform resource scheduling and allocation in accordance with designated calculations, automatic calculations, and restricted conditions. The following is to test the three scheduling methods to observe the ability of computing resource scheduling function.

4.1. Intelligent perception of resource situation

Intelligent perception of resource situation is based on simulation load analysis to grasp the utilization status of platform resources and locate potential resource bottlenecks. Based on resource data detection, a simulation service load prediction model based on LSTM is proposed to provide a basis for real-time resource scheduling. The packet loss model during communication uses the Gilbert model, and the delay model uses the token bucket filter model. Based on these system network communication models, the system network communication can be modelled and analysed, the current system network status can be grasped, and the calculation basis for calculating the data transmission cost and delay can be provided.

Monitoring and real-time detection and perception of resource information in the system realizes real-time feedback of various alarm information and resources in the cluster, and provides a forecast model of future load based on current operating data, which can effectively avoid resource bottlenecks..

4.2. Collaborative resource management

Collaborative resource management and utilization is to combine the resource requirements of simulation services to realize environment-aware collaborative resource management through cloud-edge collaboration, multi-dimensional resource balance and other means. The resource requirements and data dependence model of simulation services are studied, and the task scheduling mechanism algorithm of service quality perception in the cloud-side fusion environment is proposed, which can adapt to batch scenes and online scenes. The resource requirements of simulation services can be analysed. The labelling data processing method based on hybrid heterogeneous memory is also studied. Based on a hybrid storage system containing DRAM, NVM and disk, this method can effectively increase the probability of data in the memory and save the running time of the task. Improve calculation efficiency. Finally, a task scheduling mechanism algorithm for the perception of service quality in a cloud-side convergence environment is proposed to balance the network bandwidth utilization load between edge clouds. Whether in batch or online scenarios, this algorithm is in terms of completion time and average response time. Both are better than traditional methods to achieve the purpose of improving service quality and reliability.

4.3. Real-time scheduling of computing resources

Real-time optimal scheduling of computing resources is based on the dynamic analysis of the simulation service runtime, and real-time optimal scheduling of computing resources is realized by means of elastic configuration and resource reuse. Aiming at the runtime load changes of services, a simulation service composition algorithm based on data location awareness is proposed, which can reduce the average delay of service chain requests while ensuring the balance of host load, and finally achieve the effect of reducing the response delay of key simulation services.

5. Experiments

5.1. Experiments Environment

Constructed cloud computing, edge computing and user terminal environments in the laboratory environment, and simulated the connection between the actual edge to the cloud data center through bandwidth controllable network equipment, and adopted the local area network of computer equipment components connected to the edge environment as the end environment, The entire experimental verification environment is shown in the figure.

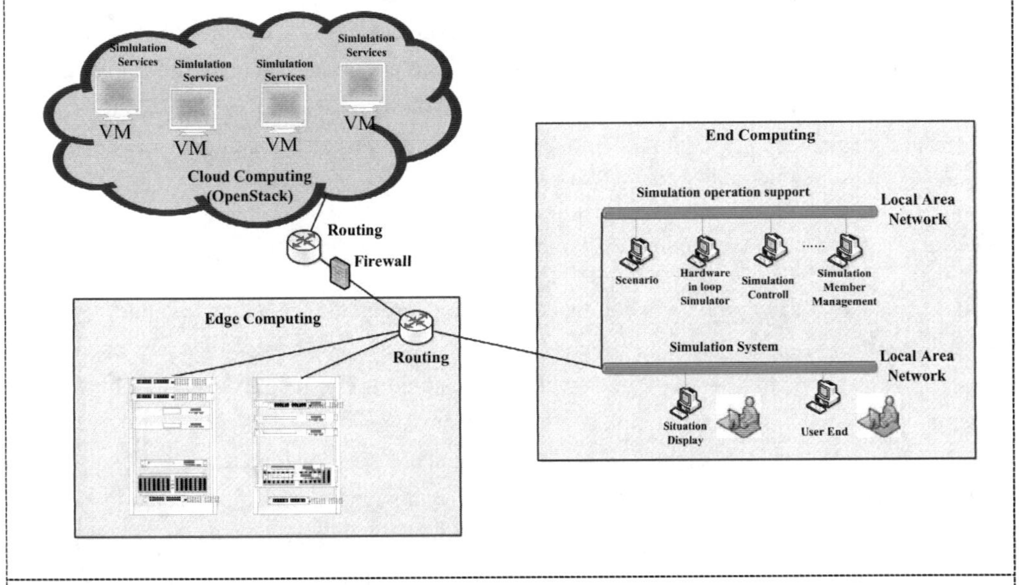

Figure 3. Experiments Environment.

5.2. Automatic Resource Allocation for Simulation Service

Taking into account the dynamic loading and migration of simulation services between the cloud and the edge, the computing resource scheduler can perform resource scheduling and allocation in accordance with designated calculations, automatic calculations, and restricted conditions. The following is to test the three scheduling methods to observe the ability of computing resource scheduling function.

5.2.1. *Designated computing resource allocation* Deploy the service by specifying the computing node to verify that the simulation service can be correctly allocated to the specified resources for execution. By specifying the LT-server (the brigade service) to the edge node edge06 and executing the deployment command, it is observed that the LT-server (the brigade service) is deployed to the specified edge node edge06. Modify the configuration file to deploy LT-server (Brigade Service) to node edge04, execute the deployment command again, and observe that LT-server (Brigade Service) is deployed to the designated edge node edge04.

5.2.2. *Automatic calculation of resource allocation* For automatic computing resource allocation, set the service installation location and the number of copies, and use the cloud platform to automatically allocate computing resources for service deployment. It is verified that under certain conditions, the service may be distributed on each computing node according to the principle of equal distribution. Open the simulation deployment configuration file, the configuration file is designated as automatic allocation, and the configuration of LT-server (Brigade Service) is 6. To execute the deployment command, the service may be evenly distributed to 6 edge nodes, or one or several edge nodes may be randomly assigned.

5.2.3. *Computing resource allocation under restricted conditions* By setting the memory size limit of the service deployment container, the verification service will automatically select the appropriate cloud node or edge node for deployment under limited conditions. There are 6 edge nodes during execution, among which the three edge nodes edge1, edge2, and edge3 have a memory size of 16G, and the three edge nodes edge4, edge5, and edge6 have a memory size of 4G. Open the simulation deployment configuration file, configure LT-server (Brigade Service) to have a memory limit of 3G, and configure LT-server (Brigade Service) to have a memory requirement of 2.5G. Execute the deployment command and observe the deployment of LT-server (Brigade Service). The three edge nodes of edge1, edge2, and edge3 may each deploy 3 services; the three edge nodes of edge4, edge5, and edge6 each deploy 1 service;

5.3. Real-time scheduling of computing resources

5.3.1. *Average service response time* For the computing resource scheduling test under restricted conditions, by configuring and deploying typical common service deployment files, the response time of typical common simulation services under restricted conditions is verified. Open the terminal of the cloud center work node, run the simulation deployment command, execute the test script, and perform 12 request tests on the common simulation service, and calculate the average request response time to be 0.025 seconds. Open the edge work node terminal, execute the test script, and perform 12 request tests in the same way, and the average request response time is calculated to be 0.026 seconds.

5.3.2. *Edge cloud throughput* The scheduling data is generated in advance by publicizing the task running data in the data set, and the data set is applied to the scheduler constructed based on the optimization algorithm, and the result is compared with the scheduler based on the greedy strategy. First, use the greedy scheduling strategy to deploy 100 simulation services, and the completion degree of statistical execution is 85%; then the simulation service is removed, the optimized scheduling strategy is replaced, the simulation service is deployed again, and the simulation service completion degree is calculated to be 98.5%.

6. Conlusions

The characteristics of cloud edge simulation system architecture are analysed, and the resource scheduling problem in the dynamic deployment of simulation system based on this architecture is analysed, and the analysis and solution framework is given from the aspects of resource situation awareness, collaborative resource management, and real-time optimization scheduling. Designed and implemented a resource scheduling system for a cloud-side joint environment facing resource-constrained environments. Through specific experiments, it verified the dynamic on-demand automatic allocation of simulation resource services and the ability to handle more fragmented computing resources. When the total amount of resources is the same, it can significantly increase the carrying capacity of the workload and have a higher resource utilization efficiency.

References

[1] MERKEL D. Docker: lightweight linux containers for consistent development and deployment[J]. Linux journal,2014(239):76-90 .

[2] Kozhirbayev, Zhanibek, and Richard O. Sinnott. A performance comparison of container-based technologies for the cloud[J]. Future Generation Computer Systems 2017,68:175-182.

[3] TAO, YE, et al. Dynamic resource allocation algorithm for container-based service computing[C].// 2017 IEEE 13th International Symposium on Autonomous Decentralized System (ISADS). IEEE, 2017.

[4] Xin Li, Zhen Lian, Xiaolin Qin et al. Delay-aware Resource Allocation for Data Analysisin Cloud-Edge System[Z].Nanjing:IEEE,2018.

[5] Zhuzhong Qian,Minyi Guo,Sheng Zhang etal. Service-Oriented Multimedia Delivery in Pervasive Space[C]. Budapest: WCNC 2009:IEEE Wireless Communications & Networking

Conference,2009
[6] Xin Li, Jie Wu, Sanglu Lu. QoS-Aware Service Selection in Geographically Distributed Clouds[Z].Nanjing:IEEE,2013.
[7] AL-HAIDARI F, SQALLI M, SALAH K. Impact of CPU utilization thresholds and scaling size on autoscaling cloud resources[C]//Proceedings of 2013 IEEE International Conference on Cloud Computing Technology and Science. Bristol:IEEE,2013:256-261.
[8] Monsalve, Jose, Aaron Landwehr, and Michela Taufer. Dynamic cpu resource allocation in containerized cloud environments[C].// 2015 IEEE International Conference on Cluster Computing. IEEE, 2015.
[9] Cayirci E., " Modelling and Simulation as a Cloud Service: A Survey," In Proceedings of the 2013 Winter Simulation Conference, edited by R. Pasupathy, S.-H. Kim, A. Tolk, R. Hill, and M. E. Kuhl, Washington DC, December 2013.
[10]CHENG Wendi; MAO Shaojie; DING Ran; "Generation and Deployment Technology for Cloud Simulation Test Environment, Command Information System and Technology, 2019Vol10(3):37-40
[11]CHU Wei, ZHOU Fang, DING Ran, MAO Shaojie; Parallel Simulation System Construction Method for Battlefield Behavior Prediction, Command Information System and Technology, 2019Vol10(3):26-33

Optical Simulations on the Surface Plasmon Polaritons Loss & Power Dissipation in Organic Light Emitting Diodes

Guorui Chen[1], Jian Duan[1], Shuai Liu[1], Guiwen Yue[1] and Min Qian[1,2,*]

[1]Department of Electronic Science and Technology, Soochow University, Suzhou, Jiangsu 215006, China
[2]City College of Suzhou, Suzhou, Jiangsu 215104, China

[*]Corresponding author's e-mail: qianmin@suda.edu.cn

Abstract. An optical energy loss mechanism including the surface plasmon polaritons (SPPs) loss in organic light-emitting diodes (OLEDs) is introduced based on CPS theory. The theoretical calculations of power dissipation spectrum (PDS), power dissipation map (PDM) and the out-coupling efficiency (OCE) of OLEDs are proposed. MATLAB tools are applied to simulate the optical model and shows that a considerable proportion of SPP loss leads to the low EQE of the OLEDs especially when the distance from the light-emitting boundary to cathode is short, it can even account for up to 40% of the total energy. Wave guide (WG) loss and SPPs loss go aganist *vs.* distance from the light-emitting boundary to cathode, and there exists a optimized distance of OCE. Study of SPPs losses and power dissipation mechanism provides the theoretical basis to enhance OLED light-emitting efficiency. The OCE calculation results are ≤20%.

1. Introduction

OLEDs (Organic light-emitting diodes) based panel display are the high-end product compared to LCD, owing to their fast response, broad viewing-angle, low driving voltage and so on[1]. Now, they have been applied in digital electronic devices, especially in mobile phones. Higher luminous efficiency is the important indicator in OLEDs technology. Previous studies have revealed that phosphorescence OLEDs[2] have 100% IQE (the internal quantum efficiency) and the EQE (external quantum efficiency) is mainly decided by the OCE (out-coupling efficiency) which is generally less than 20%, including the WG (wave guide) modes losses on ITO anodes, TIR (total internal reflection) on the glass substrate, SPPs loss on the interface of metal electrode, and so on[3]. Therefore, improving the OCE is significant for attaining high-efficiency OLEDs. Certainly, first of all, we must make clear of the energy dissipation mechanism. In the past many years, many empirical light out-coupling methods have been proposed and some theoretical studies on the optical energy loss mechanism have been presented[3]. However, few clear simulation algorithms are reported. Therefore, the study on the clear physical mechanism is essential to further improve the OCE and EQE. In this paper, we investigate and simulate the energy loss physics in detail based on the relevant model with MATLAB tools.

2. Theoretical model

OLED is a flat light source of near radiation within two electrodes, anode and cathode. Normally, one electrode is made of metal thin film or both, based on its light outgoing mode, bottom-emitting, top-emitting, or transparent. According to classic electromagnetic field theory, the exciton in OLEDs can be treated as a bound harmonically oscillating dipole including two types: vertical and parallel, see it in Figure 1 (left). When the dipole is very close to the metal electrodes, the radiation of the exciton will be affected by the metal. Due to the loss property of metal, the exciton will lose its energy, even be quenched.

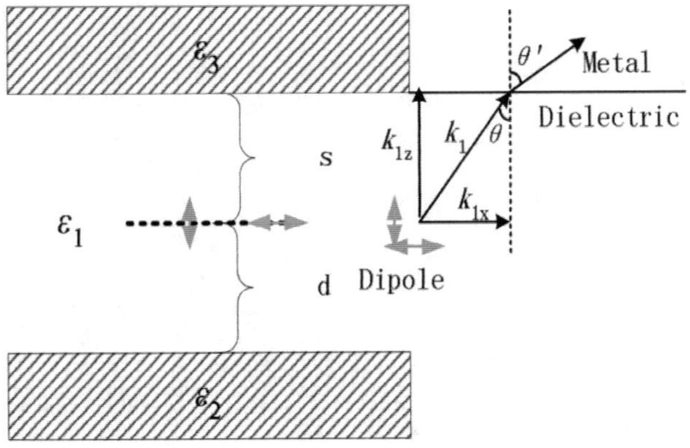

$$l_1 = -i\sqrt{1-u^2} = -i\sqrt{1-\frac{k_{1x}^2}{k_1^2}}$$
$$= -i\frac{k_{1z}}{k_1}, (near\_field)$$
$$= -i\cos\theta, (far\_field)$$

$$l_2 = -i\sqrt{\frac{\varepsilon_2}{\varepsilon_1}-u^2} = -i\sqrt{\frac{k_2^2}{k_1^2}-\frac{k_{1x}^2}{k_1^2}} = -i\sqrt{\frac{k_2^2}{k_1^2}-\frac{k_{2x}^2}{k_1^2}}$$
$$= -i\frac{k_{2z}}{k_1}, \quad (near\_field)$$
$$= -i\sin\theta ctg\theta', (far\_field)$$

Figure 1. Left: The wave-vector and components on the interface of a 3-layers OLED (two kinds of dipoles, the distances to electrodes are *s* and *d*, dielectric constants are ε_1, ε_2, ε_3); Right: parameters in far-field/near-field.

2.1. Property of near-field radiation in OLED

As is known, the near-field radiation of a dipole is much different from its far field radiation, because the reflected electric field will impact the radiation dipole in near-field situation and it can be neglected in far-field situation. The total reflected electric field is the superposition of waves of all directions reflected by the metal surface. It is very tedious to compute the value. Luckily, predecessor researchers had done a lot of work and gotten the result based on classic electrodynamics.

The incidence angle of far field radiation is confined in the range of 0° to 90°. The approximate spherical wave of near-field radiation is a Fourier series sum of plane waves. Figure 1 shows the parameters in far field incidence. Considering the completeness in mathematics, the in-plane wave-vector values of the incident light k_{1x} (*x* direction) vary from 0 to ∞. Equivalently, the incidence angle is more than 90°. It just means that the component in the *z* direction, $k_{1z}(=\sqrt{k_1^2-k_{1x}^2})$ is an imaginary number although the intrinsic wave-vector of incident light k_1 $(=2\pi/\lambda_1)$ is a constant. From the viewpoint of optics, this concept refers to the evanescent wave. When interacting, energy transfer/loss will occur on the lossy metal surface. As shown in figure 1 (right), the normalized in-plane wave-vector (u=k_1/k_{1x}), corresponding to $\sin\theta$ of the incident angle in the case of far field incidence, ranges from 0 to ∞ in the case of near-field incidence.

2.2. Surface Plasmon Polaritons (SPPs) loss in OLED

According to classic electrodynamics theory, when the incident angle is greater than the TIR critical angle, the evanescent wave will appear and interacts with the metal surface. If the wave-vector or momentum condition is matched, SPPs will occur. SPPs are trapped electromagnetic surface mode on the interface of metal and dielectric.[4] They are a combined collective oscillation of electric field and the metal's surface charges, which can only be activated by transverse magnetic (TM) wave. The dispersion relation is illustrated in inset equation of figure 2(a) [5].

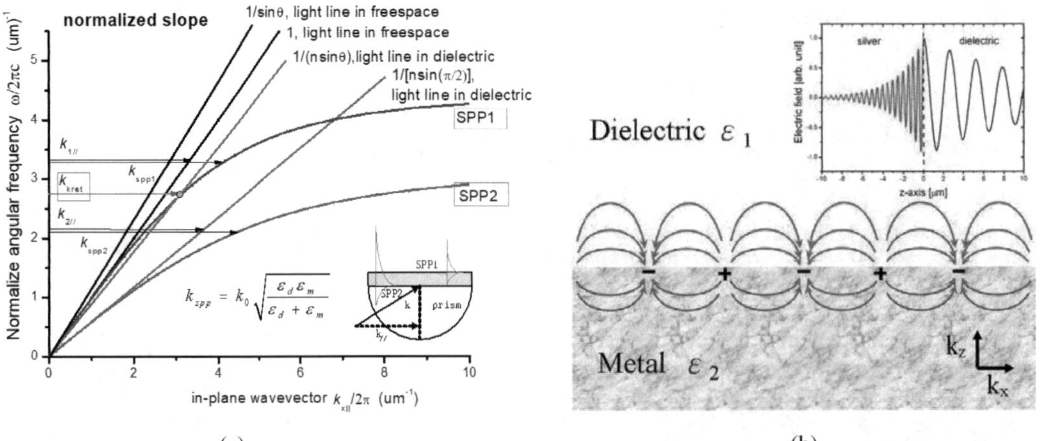

(a) (b)

Figure 2. (a) The dispersion relation of SPPs, Kretschmann method principle, Kretschmann method diagram ; (b) Diagram of SPPs' electric field on the interface

Where, ε_m: the dielectric function of the metal, ε_d : that of the adjacent dielectric. Research demonstrated that, due to the loss property of metal, the amplitude of SPPs decays negative exponentially along with normal and parallel direction on both sides from the interface. They can only propagate along the direction of the interface within a limited distance, causing non-radiative loss, see figure 2(b). In figure 2(a), commonly, $k_{Spp}>k_0$, so the far-field light can't activate the SPPs directly. But we still can activate it by some special means such as Kretschmann method, Otto method, *etc.*[4], which is based on photon tunneling effect, see figure 2(a) inset diagram.

Another means to activate SPP is the near-field method. The near-field excited SPPs exist in the OLEDs widely, an important factor leading to the low quantum efficiency of OLEDs. In OLEDs, due to the near-field radiation field can be expressed as a sum of plane waves and each plane wave has its in-plane wave-vector component k_x, it can be coupled into SPPs on a near metal surface directly when the momentum condition is matched, as discussed in section *2.1.*

2.3. PDS & PDM of OLED

Many works have been done about the power dissipation in OLED. A related key creative work was fulfilled by R. R. Chance, A. Prock and R. Silbey (CPS theory[6]) and proposed the following equation of power dissipation spectrum (**PDS**), $I(u)$. It represents the power dissipation distribution *vs* normalized in-plane wave-vector of near-field u and has two forms for two dipole orientations [6], see equation (1). Here, β: the phase shift; r^s/r^p (functions of u): the reflectance of s-polarized and p-polarized component. Based on Fresnel's equation, they are expressed in equation (2).

$$I_{\perp}(u)=\mathrm{Im}\{\frac{3}{2}\frac{u^3}{l_1}\frac{(1-r^p_{1,2}\,e^{-\beta_{1,2}})(1-r^p_{1,3}\,e^{-\beta_{1,3}})}{[1-r^p_{1,2}r^p_{1,3}\,e^{-(\beta_{1,2}+\beta_{1,3})}]}\}$$

$$I_{\parallel}(u)=\mathrm{Im}\{\frac{3}{4}\frac{u}{l_1}\{\frac{(1+r^s_{1,2}\,e^{-\beta_{1,2}})(1+r^s_{1,3}\,e^{-\beta_{1,3}})}{[1-r^s_{1,2}r^s_{1,3}\,e^{-(\beta_{1,2}+\beta_{1,3})}]}+(1-u^2)\frac{(1+r^p_{1,2}\,e^{-\beta_{1,2}})(1+r^p_{1,3}\,e^{-\beta_{1,3}})}{[1-r^p_{1,2}r^p_{1,3}\,e^{-(\beta_{1,2}+\beta_{1,3})}]}\}\} \quad (1)$$

$$\beta_{1,2}=2l_1k_1d=\frac{4\pi}{\lambda_1}l_1d,\ \ \beta_{1,3}=2l_1k_1s=\frac{4\pi}{\lambda_1}l_1s,$$

$$r^s(u)=\frac{l_1-l_2}{l_1+l_2} \qquad r^p(u)=\frac{\varepsilon_1 l_2-\varepsilon_2 l_1}{\varepsilon_1 l_2+\varepsilon_2 l_1} \quad (2)$$

In small molecular OLEDs, the orientations of the dipoles are random. Therefore, 2/3 of the excitons are in parallel orientation and 1/3 is vertical,

$$I(u)=\frac{2}{3}I_{\parallel}(u)+\frac{1}{3}I_{\perp}(u) \tag{3}$$

PDS aims at a specific wavelength and power dissipation map (PDM) focuses on all wavelengths in the light spectrum. We use color depth to represent the degree of the energy loss. In fact, the profile of PDM just comprises the dispersion relation with more information.

The out-coupling light refers to the part of the light that emerges out of the device, whose incidence angle is less than the critical angle. The out-coupling efficiency is obtained by integrating the power of a certain range compared to the total integrated power. As a result, the OCE can be represented as

$$\xi(d,s,\lambda)=\frac{\int_0^{n_0/n_1} I(u)du}{\int_0^{\infty} I(u)du} \qquad \xi(d,s)=\int_{\lambda}\xi(d,s,\lambda)S_{PL}(\lambda)d\lambda \tag{4}$$

$S_{PL}(\lambda)$ is the spectrum. When u is in the following ranges, the energy couples into corresponding mode: [0 -- k_0] Out-coupling mode, [k_0 -- n_1k_0] TE, TM wave guide mode, substrate mode, [n_1k_0 --∞] SPPs mode, lossy surface wave (LSW) mode. LSW is related to the electron-hole pair activation near the Fermi surface; k_0 is the wave-vector in air. By normalizing the in-plane wave-vector to the total wave-vector ($u=k_{1x}/k_1$), corresponding ranges are obtained as [0 -- n_0/n_1], [n_0/n_1 -- 1], and [1 -- ∞].

For a practical multi-layer OLED device, there exists many reflection interfaces as shown in Figure 3(a). To obtain the PDS, we can use the recursion method by converting the multi-layer reflection and transmission interfaces into two layers reflective surface structure based on the amplitude-splitting

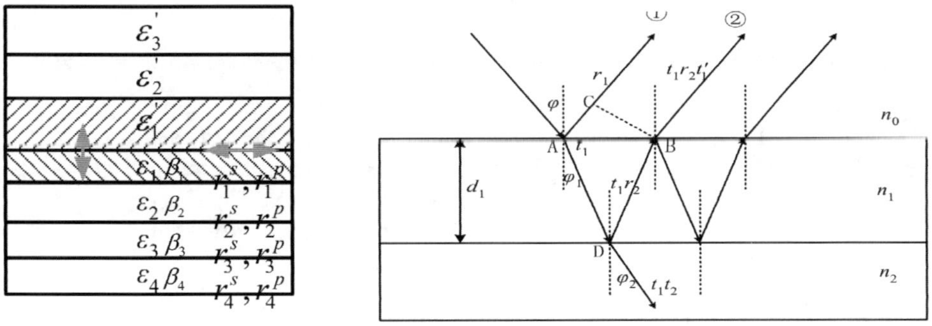

(a) (b)

Figure 3. (a) Reflection interfaces in a practical multi-layer OLED device; (b) multi-beam interference theory of multi-beam optical interference.

The recursion is done from the bottom and top on both sides of the dipole. The recursiveλs are expressed as follows

$$r_{i-1}^{p,s}=\frac{r_{i-1}^{p,s}+r_i^{p,s}e^{-\beta_{i-1}}}{1+r_{i-1}^{p,s}r_i^{p,s}e^{-\beta_{i-1}}} \qquad \beta_{i-1}=2l_{i-1}k_1d_{i-1}=\frac{4\pi}{\lambda_1}l_{i-1}d_{i-1} \tag{5}$$

where β_{i-1} is the phase shift, l_{i-1} is defined as aforementioned. k_1 is contained in the formula only.

The refractive index of the metal is obtained by fitting the experimental data from the handbook of solid optical constant of Edward D. Palik's[7]. The relative dielectric function of metal of Lorentz-Drude model[8] can be described as follows, the meaning of every parameter can be found in the reference.

$$\varepsilon_r(\omega)=1-\frac{f_0\omega_p^2}{\omega^2+i\gamma_0\omega}+\omega_p^2\sum_{j=1}^{K}\frac{f_j}{\omega_j^2-\omega^2-i\gamma_j\omega} \tag{6}$$

3. Optical simulations

Next, according to Equations (3-4), MATLAB programming language was used to calculate the optical loss of the OLED, including OCE *vs.* distance from the light-emitting position to cathode. The simulation was performed for a normal bottom-emitting OLED. As illustrated in figure 4(a), all the refractive indexes are also listed. In figure 4(b), we provide a PDS of a BEOLED via numerical simulations. The leaky mode can couple out from emitter and contributes the main part of the quantum efficiency. However the WG and SPP modes are lossy modes. As shown in figure 5(a), from blue to red, it demonstrates the degree of deepened energy losses (in logarithm).

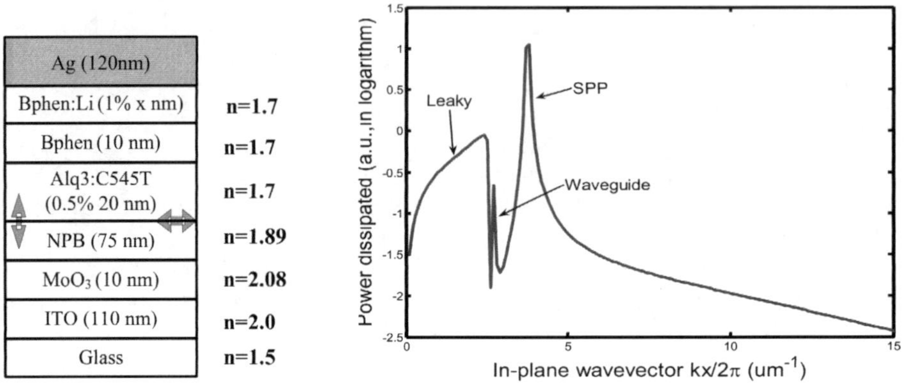

(a) (b)

Figure 4. (a) BEOLED structure; (b) PDS of BEOLED with Ag reflecting cathode @ λ=528nm.

The power dissipation *vs.* distance is shown in figure 5(b). SPPs loss ratio is relatively small when the distance is greater than 100 nm but it accounts for more than 40% when the distance is shorter than 40 nm. OCE has an optimal thickness of ~120 nm. Waveguide mode and the substrate mode increased with increasing of the distance, which agrees with optical waveguide theory. The success of CPS dipole theory lies in that various physical mechanisms have been included in a unified calculation equation. The OCE calculation results are ≤20%.

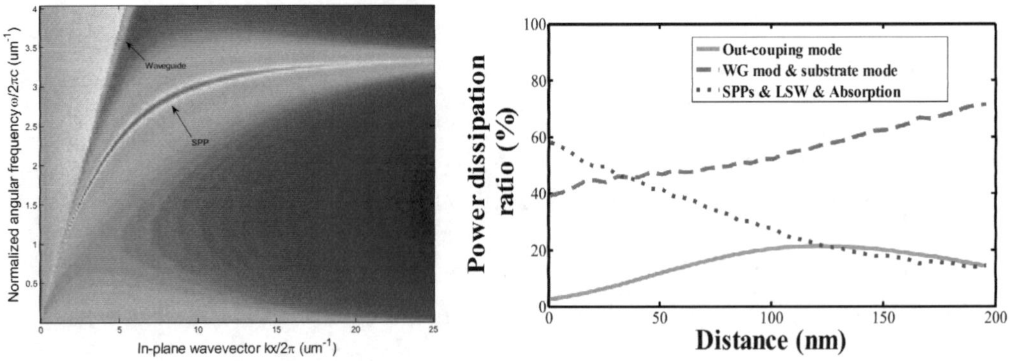

(a) (b)

Figure 5. (a) PDM of BEOLED with Ag reflecting cathode; (b) Energy loss of bottom-emitting OLED *vs.* the distance between the emission interface and the Ag cathode.

4. Conclusion

Following CPS theory, we calculate PDS, PDM and OCE in OLED devices. The near-field property of the dipole in OLED activates the SPPs. Simulation results shows that a considerable proportion of SPP loss leads to the low EQE of the OLEDs especially when the distance from the light-emitting boundary to cathode is too short. The relationss between energe loss of WG and SPPs and distance take on a opsite trend and the OCE has a optimum about 20%. Study on the power dissipation mechanism of OLED provides the theoretical basis to enhance the luminous efficiency.

Acknowledgements
This work is support by the Natural Science Foundation of the Jiangsu Higher Education Institutions of China (19KJD510006)

References

[1] Templier F 2014 *OLED Microdisplays: Technology and Applications (John Wiley & Sons, Inc.)* 35-51
[2] Adachi C, Baldo M A, Thompson M E, Forrest S R 2001 Nearly 100% internal phosphorescence efficiency in an organic light-emitting device *Journal of Applied Physics* **90**(10) 5048-51
[3] Hong K, Lee J L 2011 Review Paper:recent Developments in light extraction Technologies of Organic Light Emitting Diodes. *Electronic Materials Letters* **7**(2) 77-91
[4] Raether H 1988 Surface plasmons on smooth surfaces (Springer)
[5] Barnes W L, Dereux A, Ebbesen T W 2003 Surface plasmon subwavelength optics. *Nature.* **424**(6950) 824-30
[6] Chance R, Prock A, Silbey R 1978 Molecular fluorescence and energy transfer near interfaces. *Advances in Chemical Physics* 1-65
[7] Prucha E J, Palik E D 1998 Handbook of optical constants of solids **1-2** + 3 1998-01-01
[8] Eldlio M, Che F, Cada M 2014 Drude-Lorentz Model of Semiconductor Optical Plasmons. IAENG Transactions on Engineering Technologies (Springer) 41-9

Simulation and Design of Rising-Sun Anode Blocks for Magnetrons with Two Outputs of Energy

Gennadiy Churyumov[1,*], Shuang Qiu[2], Nan-nan Wang[3], Wei Li[3], Volodymyr Gerasimov[4] and Tetyana Frolova[4]

[1]O.Ya. Usikov Institute for Radiophysics and Electronics of National Academy of Science of Ukraine, Kharkiv, Ukraine
[2]Tsinghua University, Beijing, China
[3]Harbin Institute of Technology, Harbin, China
[4]Kharkiv National University of Radio Electronics, Kharkiv, Ukraine

[*]corresponding author's e-mail: g.churyumov@ukr.net

Abstract. This paper presents the simulation results of the oscillation modes excited in the interaction space of an anode block of a magnetron and the azimuthal distributions of a longitudinal electric component of the RF field for either of the excited modes. Using CST code, there were considered available anode blocks with different configurations of the resonators: a rising sun anode block and a rising sun hole-and-slot anode block. The dispersion characteristics and distributions of the azimuthal components of the RF field for different geometrics of their cavities as applied to the *X*- and *Ku*-bands are given. The condition of acceptable separation both in frequency and in phase velocity between the operating $N/2$ -mode (or π-mode) and neighboring $(N/2\ -1)$-mode has been determined. It suggested that the mechanism of forming the total RF field connects with possible interference of the RF fields excited severally in the resonant system consisting of the small and large cavities. The features design of the magnetrons with two energy outputs is presented, and their application in practice is determined. The results of the simulation are in good agreement with the experiment.

1. Introduction

One of the main challenges of modern vacuum microwave electronics is creating effective, reliable, and relatively simple fabrication of vacuum microwave sources [1]. Among available microwave sources, it is necessary to stand out the magnetrons whose parameters are the closest to such requirements. The attainment of the high values of the energy parameters (efficiency, output power) of the magnetrons had been become a result of the multiyear studies, starting with a moment of creating a magnetron by Hall in 1921 [2]. The existing designs of the magnetrons cover a range of frequencies from 330 MHz to 300 GHz and electronic efficiency from ~ 80 % up to several percent, respectively. The output power of the magnetrons at the continued regime changes from units of W up to several decades of kW, and peak power at the pulse regimes equals from ~ 10 W up to ~ 20 MW. In addition to the investigations associated with improving power characteristics of the magnetrons (such as output power and efficiency), improvements in anode and cathode technologies as well as the technology of fabrication of the magnetic focusing systems, there were carried out investigations concerning the betterment of frequency properties of the magnetrons, including enhancement frequency stability, upgrading of spectrum quality, lowering the levels of spurious oscillations, reduction amplitude, and phase noises, and so on [3-10]. As a result, the magnetrons have been used in the different ground-based and boarded electronic and radar systems, industrial and household systems of microwave heating, physical experiments for heating a plasma, etc. Also, there are different studies directed to the application of the magnetrons in linear accelerators, phased arrays, and space systems for transforming the energy of electromagnetic waves into an energy of the continue electric current (so-called system of wireless transmission of power to the Earth) [11-13]. The miniaturization of the magnetron designs allows to decrease their mass up to ~ 100 g provided a value of output power is ~ 1 kW, and efficiency is ~ 50 %. Those values of parameters of the magnetron generators correspond to the best prototypes of the microwave generators based on the transistors.

It should recognize that modern development of the magnetrons connected with improving their output characteristics and directed on understanding the complex non-linear processes of electron-wave interaction depends on the progress of their theory, especially the 3-D non-linear one.

Even though a magnetron has a relatively simple design, the 3-D simulation of electron-wave interaction is a complicated challenge that requires significant computer time and resources of the hardware used. In this connection, searching the fresh approaches directed towards speeding up the simulation process provided conservation of a typical self-consistent setting up a problem is a topical and complicated challenge.

The result of such work will be creating novel designs of the magnetrons possessing improved performances. One such design is a magnetron with two outputs of energy: by active output matched with a load and by reactive one needed for controlling an operation of a magnetron (for example, for frequency tuning (or adjustment) [14]). The creation of such designs requests additional investigations to get novel knowledge about wave processes inside a magnetron.

This paper presents the 3-D simulation results of cold electrodynamic characteristics of the rising-sun systems used in the X and Ku-bands magnetrons with two energy outputs.

It is worthy to note that a magnetron's design process is a challenge that consists of several steps. Among these steps, it is necessary to cause two essential moments connected with a simulation of a cold resonant system, including an output of energy and a simulation of electron-wave interaction and calculation of the main output parameters (the output power, efficiency, etc.). As a resonant system of the cm-band short-wave part of the magnetrons, let's consider the rising sun systems that give good separation of the resonance frequencies between the operating $N/2$ - mode (or π-mode) and the neighboring $(N/2 \pm 1)$ – modes, where N – the number of cavities.

In Fig. 1, several examples of the anode blocks having cavities with different shapes are presented. The analysis carried out was shown that a definition of the frequency's separation of the oscillation modes in the rising sun anode blocks essentially depends on the choice of the shape of the cavities.

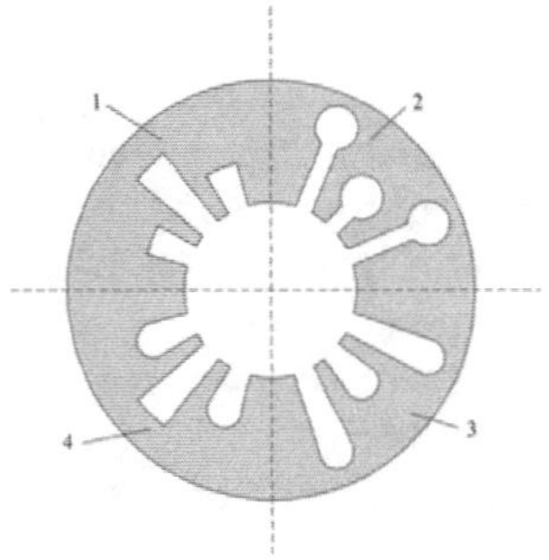

1 – vane type; 2 – slot-and-hole type; 3 – drop-like type; 4 – a big resonator – vane type, a small resonator – drop-like type

Figure 1. The possible views of anode blocks.

Fig. 2 schematically presents the two samples (types) of the most-used anode blocks of the magnetrons which are usually applied into the magnetrons operating in the short-wave part of the cm-band. The main geometric dimensions of the given resonant systems are shown below. The CST code is used for the simulation of their dispersion characteristics.

2. 3-D Simulation of the cold resonant systems

2.1. An 18-cavity rising-sun system

Fig. 3 presents the cold resonant frequencies of modes of oscillations for various sizes (radiuses) of the big cavities [14]. As can be seen, a case $\boldsymbol{r_s = r_b}$, when $\Delta = \mathbf{0}$ corresponds to a vane type of the resonant structure (a curve 1 in Fig. 5). Increasing the radiuses of the big cavities lead to the rise of separation

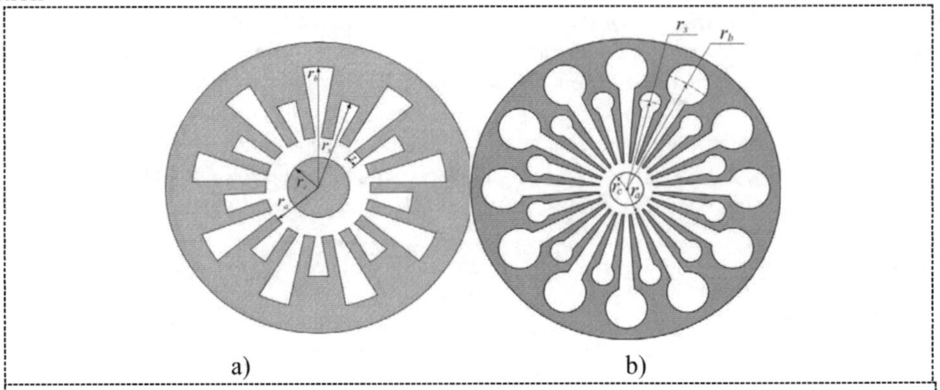

Figure 2. Cross-section schematically views of the resonant systems of the rising sun anode blocks having various configurations of the cavities: a) – "rising sun" type, and b) – "hole-and-slot" type.

between an operating $N/2$ – mode ($\pi -$ mode) and its neighboring mode of oscillation $n = N/2 - 1$. The case, when $\Delta = 0.325$ mm (the curve 5) for a frequency of the π –mode corresponds to a frequency f = 23,8 GHz, it correlates well with the experiment [15]. Thus, changing a relationship between radiuses the big and small resonators we may vary the cold resonant frequency of the π –mode in wide limits controlling the value of separation between the π –mode and neighboring low-voltage and high-voltage modes of oscillations.

Fig. 4 shows a distribution of the azimuthal electric component of an RF field in the 18-cavities rising sun system (Fig. 2, a). As it is seen, the maximum values of the RF field under larger and smaller resonators are different. This outgrowth agrees with the data obtained in [15].

2.2. A 24-cavity rising-sun hole-and-slot system

Fig. 5 shows a spectrum mode for different values of the ratios between the radiuses of large and small resonators. The mode spectrum for different values the diameters of the large resonator is shown in Fig. 6. As it is seen, increasing the radiuses of the large resonators leads to varying not only the $\boldsymbol{N/2}$ - mode frequency and increasing the separation between neighboring modes (specifically, between $\boldsymbol{N/2}$ – mode and neighboring $(\boldsymbol{N/2 - 1})$ –mode, but also changes the behaviour of the dispersion curve of the large cavities.

The distributions of the azimuthal electrical components of the RF field at $r = r_a$ for an operating mode of oscillation $n = N/2 = 12$ (π –mode) are presented in Fig. 7. It is seen that for the given resonant system, the RF field amplitudes under small and large resonators have the same values.

2.3. A 20-cavity rising-sun hole-and-slot system

Fig. 8 shows a spectrum mode for the 20-cavity rising-sun slot-and-hole resonator system. The main geometric dimensions of the resonant system and their values are also represented in Fig. 8. As indicated in the picture, the resonant frequency a cold resonant frequency of the $\boldsymbol{N/2}$ – mode is 9.65 GHz. As with the case of the 24-cavity rising-sun system, the given resonant system also has good separation between the operating mode and neighboring $(\boldsymbol{N/2 - 1})$ - mode both a frequency and phase velocity.

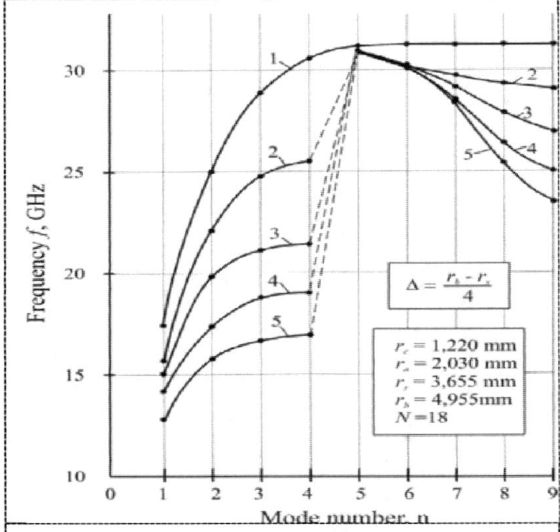

1 – Δ= 0; 2 – Δ= 1.09; 3 – Δ= 1.18;

4 – Δ= 1.27; 5 – Δ= 1.36

Figure 3. Mode spectrum for the cold 18-cavity rising-sun structure [15].

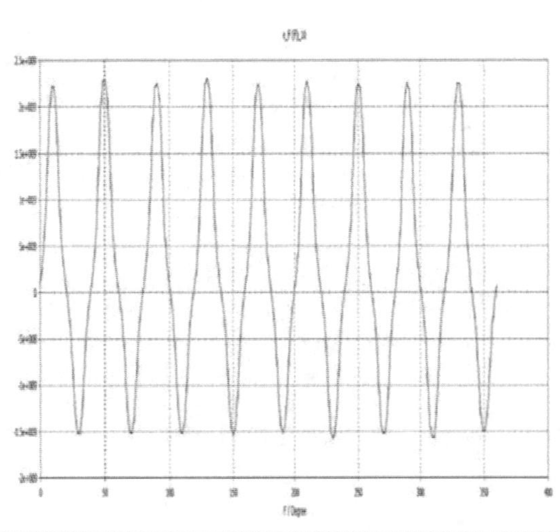

Figure 4. Azimuthal distribution of the $N/2$ - mode of the cold 18-cavity rising-sun system.

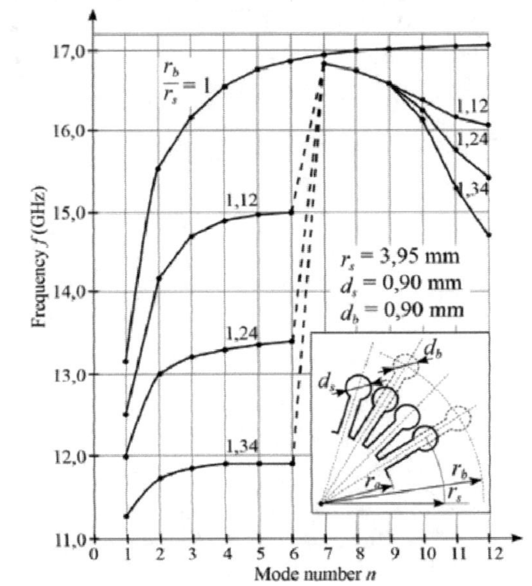

Figure 5. Mode spectrum for the cold 24-cavity rising-sun structure [14].

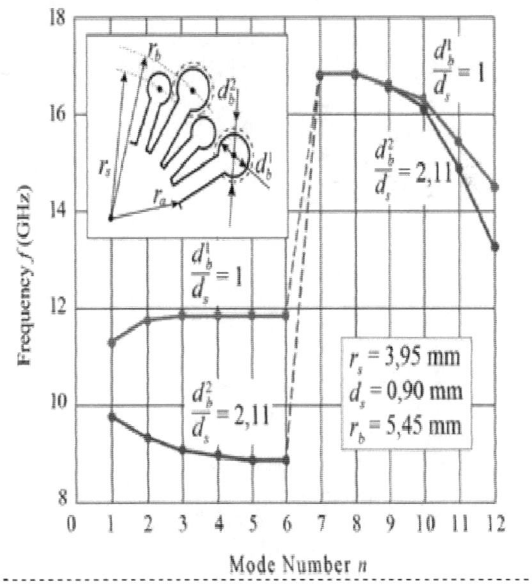

Figure 6. Schematic view and the mode spectrum for the cold 24-cavity hole-and-slot rising-sun system [14].

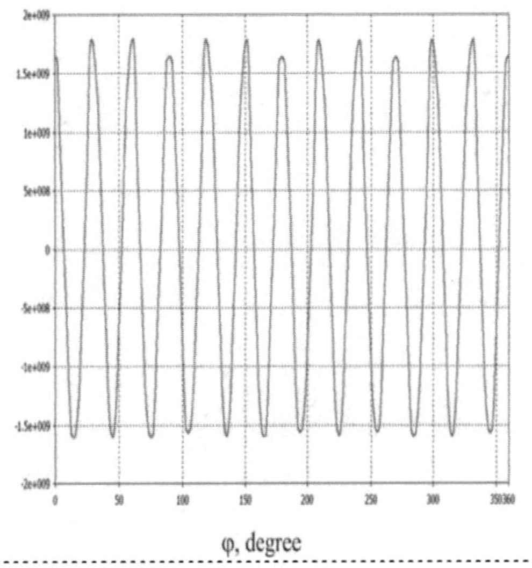

Figure 7. Distribution of the azimuthal electrical component of RF field for the operating mode of oscillation n = N/2 ($\pi -$ mode).

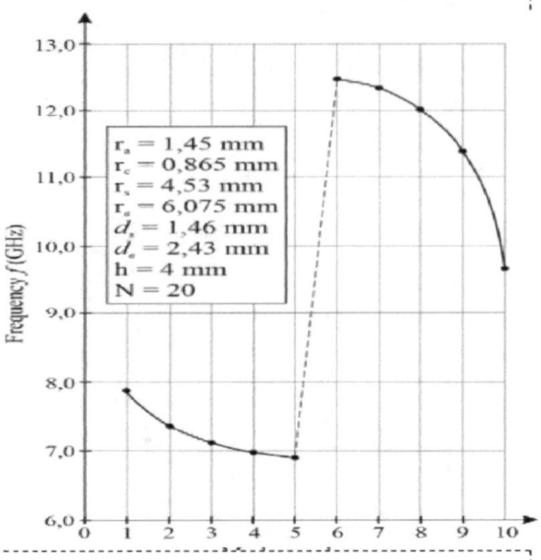

Figure 8. Mode spectrum for the cold 20-cavity rising sun system.

2.4. *Designs of the X and Ku-Bands Magnetrons with Two Energy Outputs*

The obtained data of a simulation of the 20- and 24-cavity rising sun systems can be used to create the designs of the magnetrons with two energy outputs in *X* and *Ku*-bands (Fig. 9). The main advantages of the given magnetrons are the additional possibilities associated with the frequency tuning, including frequency agility, as well as improving the stability of a frequency [14, 16].

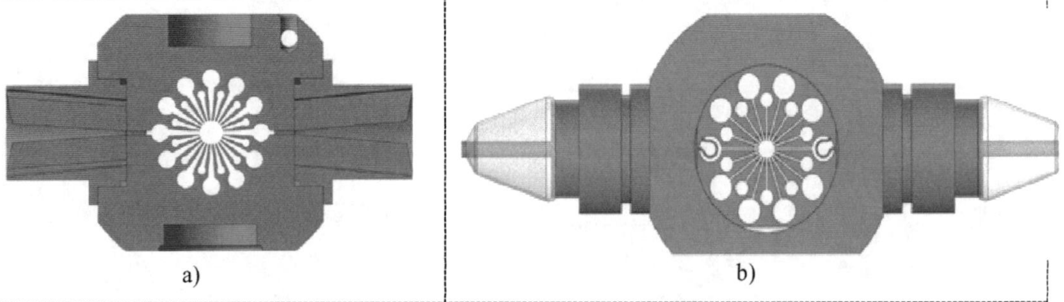

a) b)

Figure 9. Designs of the magnetrons with two outputs of energy.

Thus, the paper presents several designs of the resonant systems having a different configuration and shape of the cavities for applying in the short-wave part of the cm-band. A simulation of the electrodynamic characteristics of the given resonant systems has been carried out by using the CST code. It showed that a "hole-and-slot" rising sun system possesses the broader possibilities associated with by choosing a resonant frequency of the operating $N/2$ - mode of oscillation ($\pi -$ mode) as well as a separation between the operating $N/2$ - mode and the neighboring $(N/2-1)$ - mode both in frequency and in phase velocity. An explanation mechanism of forming a total RF field in the interaction space of the rising sun resonant systems is given. The obtained results of the 3-D simulation are in good agreement with experimental data. Based on obtained simulation data, it developed the designs of the anode blocks for the low-voltage *X* - and *Ku* -bands magnetrons with two energy outputs.

Acknowledgments
The authors wish to thank Prof. J. Qiu from the Harbin Institute of Technology for taking an active interest and support of the given work.

References
[1] Special Issue on New Vistas for Vacuum Electronics / Proceedings of the IEEE, May 1999. - Vol. 87, No 5. – 935 C.
[2] Hull A.W. The effect of a uniform magnetic field on the motion of electrons between co-axial cylinders // Phys.Rev. – 1921. – V. 18, № 1. – P. 13.
[3] K. Yamamoto, H. Kuronumo, T. Koinumo, and N. Tashiro, A Study of Magnetron Noise. *IEEE Trans. On Electron Devices*. Vol. ED-34, No 11, May 1987, pp. 1223 -1226.
[4] Chen X. Computer Simulation of Magnetrons // Vacuum Electronics and Displays / X. Chen, M. Esterson, P. A. Lindsay // Springer-Verlag. - 1992. — P. 99-104.
[5] Lindsay P. A. Simulation of some transient effects in a magnetron using the MAGIC code / P.A. Lindsay, J. Watkins, X. Chen // MAGIC Users' Group Newslett. - 1995. - N. 4. – P. 1-4.
[6] Wang, Z. Characterization of resonance modes of a magnetron cavity using MAGIC-an FD-TD code. Effective Microwave CAD Tools / Z. Wang, X. Chen, M. Esterson, P.A. Lindsay // IEE Colloquium on. – 1999. - No1999/064. – P. 2/1 -2/4.
[7] G. Churyumov, T. Frolova, A. Gritsunov. The State-o-the-Art of Computer Modeling and Design of the Vacuum Microwave Devices. 13th International Conference "Modern Problems of Radio Engineering, Telecommunications and Computer Science" (TCSET'2016), Lviv-Slavske (Ukraine), 23-26 February, 2016. – PP. 1114-1116.
[8] G. I. Churyumov, A. I. Ekezli. Investigating the frequency tuning mode in an impulse magnetron with two energy outputs. Electronnaya Technika, Ser. 1, Technika SVCh. - 2014. - No. 2 (521). – PP. 39 - 45. (in Russian).
[9] G. I. Churyumov, T. I. Frolova, and K. M. Basrawi, "The Two-Stage Magnetron For Radar Applications," 2006 International Radar Symposium, Krakow, 2006, pp. 1-3, doi: 10.1109/IRS.2006.4338148.
[10]Churyumov G.I., Frolova T.I., Gritsunov A.V., Nikitenko O.M., Zin'kovsky V.N.. The magnetrons - EMI sources: computer modeling and experimental investigations // Proceedings International Symposium on Electromagnetic Compatibility (EMC Europe 2004). - Vol. 1. - Eindhoven. - 2004, September 6-10. - P. 327-331.
[11]Microwave Power Engineering. Edited by E.C. Okress. V. 1, 2. Academic press, New-York & London. 1968.
[12]Edgar R. H. Consumer, commercial, and industrial microwave ovens and heating systems / R. H. Edgar, J. M. Osepchuk. – Eds. New York : Marcel Dekker, 2001. – PP. 215–278.
[13]Brown W. C. Beamed microwave power transmission and its application to space / W. C. Brown, E. E. Eaves // IEEE Trans. on MTT. – 1992. – Vol. 40, No. 6. – PP. 1239-1250.
[14]Gennadiy I. Churyumov, Nan-nan Wang, Volodymyr P. Gerasimov, Wei Li, Low-Voltage K_u-Range Magnetron With Two Outputs of Energy: Design Features and Main Advantages. IEEE Transaction on Electron Devices. Vol. 67, # 12, December 2020. pp. 5743-5749
[15]Microwave Magnetrons. Edited by George B. Collins. Department of Physics University of Rochester. – New York-Toronto-London. 1948. – 806 p.
[16]Patent of Ukraine № 98574, Magnetron generator with frequency tuning from pulse to pulse. G.I. Churyumov, A.I. Ekezly, a 201104603, 14.03.2011 (in Ukraine).

ICMSOA 2021 IOP Publishing
Journal of Physics: Conference Series **2173** (2022) 012010 doi:10.1088/1742-6596/2173/1/012010

Agile Modeling and Simulation Technology of Heterogeneous Systems Based on Meta-model Modeling Method

Yunfei Liu[1] and Tao Ma[1,*]

[1] China Academy of Engineering Physics, Institute of Computer Application, Mianyang, Sichuan, China

[*]corresponding author's e-mail: whumatao@163.com

Abstract. Modeling and simulation of heterogeneous systems are faced with the problems of long modeling and simulation cycle, complex and changeable modeling fields and complex domain knowledge. In order to support the requirements analysis and initial design of the system, build the system model as the only authoritative truth source throughout the whole life cycle, and quantitatively characterize the requirements, functions, structure and performance of the system, this paper proposes an agile modeling and simulation technology for heterogeneous systems based on the meta-model modeling method, and on this basis, constructs an agile modeling and simulation system framework for heterogeneous systems.

1. Introduction

Many interactive elements can form complex systems, such as the Internet, economic systems, public departments and administrative organizations, cities, living bodies, brains, power grids, transportation, ecological systems, etc. Complex systems exist in all areas of our lives[1].

Various complex systems can often be abstracted as a group of units, which are connected through a network and interact with each other according to certain dynamic laws. However, because there are many elements and subjects in a complex system, and there are complex functions such as dependence, competition and association among the system components and between the system and the environment, special properties such as nonlinearity, emergence, spontaneous order, adaptability and feedback loop are produced, which makes it difficult to describe the interaction mode and influence, and the increase of the number of parameters may be faster than the increase of the system size. We can neither simply understand the whole system as the sum of individuals based on reductionism nor be effective from the overall behavior. Therefore, it is extremely difficult for us to model complex systems. At the same time, with the further development of research and design, the interaction between design rules and scientific mechanisms in different fields increases the complexity of system modeling and simulation.

Model-based systems engineering (MBSE) is an effective modeling method for complex systems at present[2], which can effectively support the requirements, design, analysis, verification and validation activities of complex systems in the whole life cycle through formal modeling methods. At present, MBSE has become a hot technology concerned by academia and industry, and has the following characteristics: 1) Implement model-based design and research and development based on system engineering methods; 2) Adopt formal modeling method to support system requirements, design, analysis, verification and confirmation in the whole life cycle; 3) Requirements tracing, system architecture driving, behavior and performance expression, simulation and testing are realized through relevant technologies. Since MBSE engineering is a good solution for complex systems, it is an effective research technology path for complex systems.

Figure 1. Aircraft error analysis case

However, due to the long learning cycle of MBSE method, the learning cycle for domain design experts is long[3], the modeling cost is huge, and the cycle is often old. When the system design changes, it often cannot keep up with the system changes. Can't adapt to the current demand of customization and intelligent manufacturing. At the same time, this method is difficult to simulate and cannot be used for agile modelling[4].

In view of the above difficulties, we have done the following work: 1) Standardize the expression of design variables (inherent attributes, parameters, structures, etc.) and design constraints (functional constraints, coupling constraints, process constraints, etc.), and establish a system meta-model. Standardize the expression of design variables (inherent attributes, parameters, structures, etc.) and design constraints (functional constraints, coupling constraints, process constraints, etc.), and establish a system meta-model. 2) Based on the system model, the quantitative tracing of requirements, indicators and parameters in the design process is realized, and various model views of the system are automatically generated. 3) According to the above design method, we designed an agile modeling and simulation system framework for heterogeneous systems based on meta-model modeling method.

2. Background and Motivation

2.1. Meta-modeling method[5]

Modeling is to establish a meta-model to describe a modeling language, and to provide modeling tools to support the modeling language. Meta-modeling is to build a collection of "concepts" (events, conditions, etc.) in a specific field. The model is an abstraction of the phenomenon in the real world, and the meta-model is another abstraction, paying attention to the attributes of the model itself, so a meta-model can be regarded as an abstraction of the model. The model should be subject to the definition of its meta-model, just as a computer program is subject to the syntax definition of this programming language.

M3 MOF	Defines a language for specifying a metamodel Example: MOF
M2 UML	Defines a language for specifying models Example: UML
M1 User Models	Defines a language that describe semantic domains Example: model of a problem domain
M0 Instance Models	Contains run-time instances of the model elements defined in a model

Figure 2. OMG meta-model hierarchy

2.2. Meta-object facility

The meta-object facility (MOF)[6] provides an open and platform-independent metadata management framework and related metadata service set to support the development and interoperability of model-driven and metadata-driven systems. Systems using MOF include modeling and development tools, data warehouse systems, metadata repositories and so on. MOF has made great contributions to the core principles of OMG model-driven architecture. On the basis of modeling established by UML[7], MOF introduces the concepts of formal meta-model and platform-independent model (PIM) of metadata (including several standard OMG meta-models including UML and MOF itself).

3. Solution

In this chapter, from the expression of design variables and constraints to the design of system architecture, the agile modeling system of heterogeneous systems is described in detail.

3.1. Formal expression of design variables and constraints

Meta-modeling can be formalized by tree structure[3]. In order to ensure that all constraints can be

Journal of Physics: Conference Series **2173** (2022) 012010 doi:10.1088/1742-6596/2173/1/012010

expressed by tree structure, we deform the constraints as follows.Functional constraints can be expressed directly by nodes, process constraints can be expressed sequentially by transforming nodes, and coupling constraints can be transformed into functional constraints by adding coupling constraint nodes. Figure 3 is a representation of structural load constraints, including functional constraints, coupling constraints and process constraints.

Figure 3. Constraint expression case. red: process constraint, green: functional constraint, beige: coupling constraint.

3.2. Meta model construction

Constructing meta-model is the main part of modeling expression in agile modeling system. Heterogeneous systems need real-time meta-model expansion and expression activities.

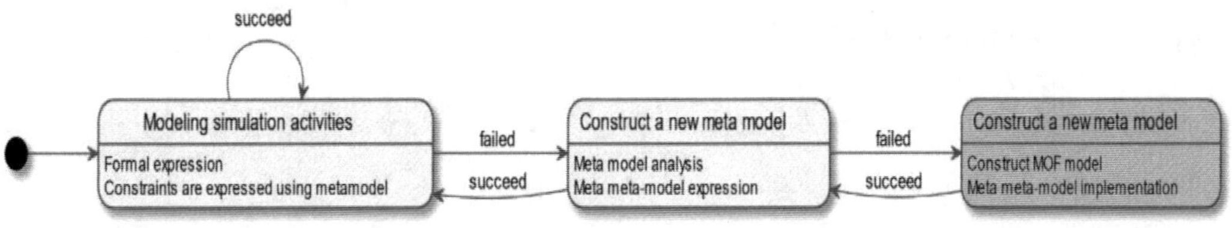

Figure 4. Schematic diagram of meta-model construction process.

3.3. System design

For agile modeling of heterogeneous systems, when the system structure changes dynamically, the modeling system needs to dynamically characterize the expression elements and meta-models. The specific steps are as follows. Element reconstruction: the elements in the business model of domain characteristics are extracted, matched with the standard system modeling metamodel, and instantiated; Relationship traceability: extract the relationships in the domain feature business model, match the relationship header and tail object meta-models, and iteratively obtain the global traceability relationship; Relationship generation: match the corresponding relationship meta-model and instantiate it; Pre-processing: In the stage, all nodes are stored in the stack, edges are simplified to undirected edges, and artificially set fixed edges are eliminated. The system architecture is shown in Figure 5.

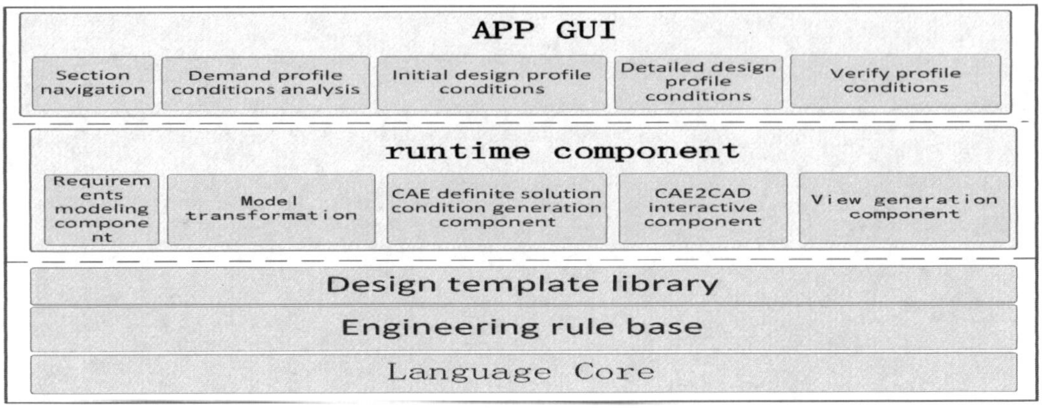

Figure 5. System architecture

4. Conclusion

In this paper, an agile modeling and simulation technology for heterogeneous systems based on meta-model modeling method is proposed, and on this basis, an agile modeling and simulation system framework for heterogeneous systems is constructed, which supports the requirements analysis and initial design of the system, and the system model is constructed as the only authoritative truth source throughout the whole life cycle to quantitatively characterize the requirements, functions, structure and performance of the system.

References

[1] Meyers, Bart, et al. A Framework for Temporal Verification Support in Domain-Specific Modelling. (2020): 1-1.
[2] Wolny, Sabine, et al. Thirteen years of SysML: a systematic mapping study. 19 (2020): 111-169.
[3] Peng, ZHAO, et al. Research on Application of Ontology Modeling using Object-oriented Technology. (2008): 64-68.
[4] Forootan, Amin, et al. Methods to determine limit of detection and limit of quantification in quantitative real-time PCR (qPCR). 12 (2017): 1-6.
[5] Xu, P. , Deng, Z. , Cui, C. , Zhang, T. , Choi, K. , Suhang, G. , Wang, J. , & Wang, S. . (2019). Concise Fuzzy System Modeling Integrating Soft Subspace Clustering and Sparse Learning. 27(11), 1-1.
[6] Yang, Yilong, et al. "Real-time System Modeling and Verification through Labeled Transition System Analyser (LTSA)." abs/1803.05362 (2018).
[7] Holliman, N. , Antony, M. , Charlton, J. , Dowsland, S. , James, P. , & Turner, M. . (2020). Petascale Cloud Supercomputing for Terapixel Visualization of a Digital Twin.. abs/1902.04820..

Mathematical Model of the Integrated Filtering Reactor Windings Transformer

Shijun Li[1,*], Jiayi Yang[2], Fumei Wu[3], Longfu Luo[4], Wendi Liao[4] and Jinwen Xiang[4]

[1]College of Electrical & Information Engineering, Hunan Institute of Engineering; Xiangtan, Hunan, China
[2]School of Electrical Engineering ShaoYang University; ShaoYang, Hunan, China
[3]Swan college of Central South University of Forestry and Technology, Changsha, Hunan, China
[4]College of Electrical and Information Engineering, Hunan University, Changsha, Hunan ,China

[*]corresponding author's e-mail: lsj181627@qq.com

Abstract. The integrated filtering reactor windings transformer is designed by a special winding arrangement design scheme. The air-core reactors of the harmonic filtering devices are integrated as the transformer windings in the transformer.so as to reduce the number of the equipments and covering area, The harmonic model of this kind of winding have been shown in this paper, and the harmonic current of each winding are deduced on the basis of the self-inductance and mutual-inductance voltage equation of multi-winding transformers. At last, this paper took a 300 KVA integrated filtering reactor winding rectifier transformer prototype as the experimental object to verify the correctness of the mathematical model.

1. Introduction

Integrated filtering reactor winding transformer [1,2], to integrate traditional filtering device of air-core reactor in transformer winding in special winding arrangement scheme, demonstrates the following electromagnetic properties: 1) the mutual inductance between integrated filtering reactor windings in the transformer and other power windings in the transformer is 0, or close to 0; 2) mutual inductance of multiple Integrated filtering reactor windings is 0; 3)the inductance of integrated filtering reactor winding remains constant within a certain range in large current range. Therefore, the voltage and current ratio curve of the winding keeps linear.

2. The integrated approach of filtering reactor winding of transformer

Fig 1 is the structure diagram for the integrated filtering reactor winding[3,4]. Winding W_A is a common-used power-supply winding, and winding W_B was series connected by W_{B1} and W_{B2}, with same number of turns. In addition, in the space vision winding W_{B1} and winding W_{B2} are of level mirror symmetrical relationship to the winding W_A, and their winding directions are in contrast. Current and induced electromotive force through the two windings is showed in Fig. 1.

The mutual inductance between winding W_A and winding W_{B1} or W_{B2}:

$$\begin{cases} M_{AB1} = -M_{AB2} \\ M_{B1A} = -M_{B2A} \end{cases} \tag{1}$$

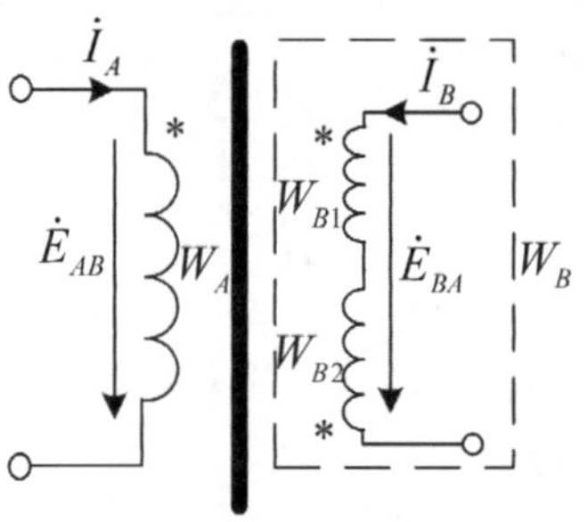

Figure 1. Integrated filtering reactor windings

From the relationship between the induction electromotive force and the current, we have:

$$\begin{cases} \dot{E}_{B1} = -j\omega M_{B1A}\dot{I}_A \\ \dot{E}_{B2} = -j\omega M_{B2A}\dot{I}_A \end{cases} \tag{2}$$

So,

$$\dot{E}_B = \dot{E}_{B1} + \dot{E}_{B2} = -j\omega(M_{B1A} + M_{B2A})\dot{I}_A = 0 \tag{3}$$

Similarly, we can know the induction electromotive force of winding WA:

$$\dot{E}_A = -j\omega(M_{AB1} + M_{AB2})\dot{I}_A = 0 \tag{4}$$

From the derivation of the integrated filtering reactor winding decoupling theory above, it is known that the electromagnetic coupling does exist between integrated filter inductor winding and the power-supply winding, but the mutual induced electromotive force is close to 0 because of special winding arrangement, Therefore, integrated filter inductance winding realized power decoupling in the same field direction and winding in one iron core, so as to live up to the requirements for the integrated filtering reactor windings. With the special winding design and the configuration method of winding[5-8].

Mentioned above, an integrated filtering reactor winding transformer was developed, the structure of which can be seen in fig.2. The transformer utilizes 3-phase structure; each phase integrates multiple integrated filtering windings, clinging to the iron core of the transformer. The peripheries are respectively arranged low-voltage winding and the high-voltage winding.

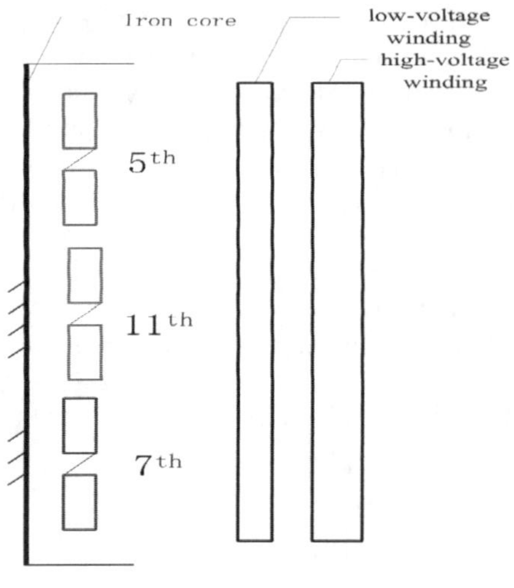

Figure 2. Integrated filter reactor windings transformer winding arrangement structure

3. Calculation with mathematical model

3.1. The harmonic model

Fig. 3 demonstrates the harmonic circuit model of integrated filtering reactor transformer. Winding W_1 is short circuited. Winding W_3 is in series connection with filter capacitor C, consisting of the n times filter branch. The n times harmonic current source connects the winding W_2 and the n times filter branch respectively. Assuming the harmonic impedance of the 3 windings are Z_{n1}, Z_{nl} and Z_{n1}, the capacitor harmonic impedance is ZnC[9-11].

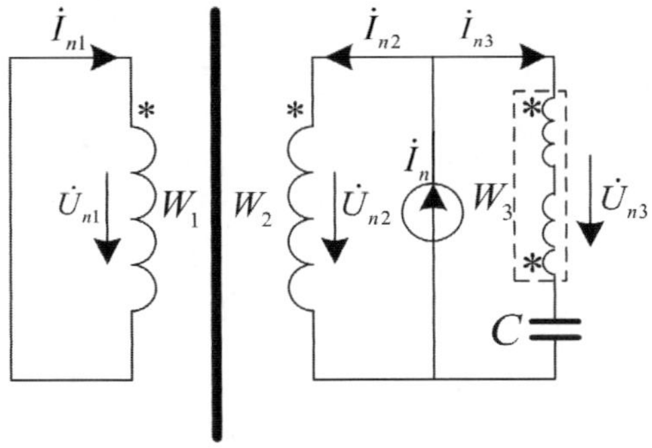

Figure 3. Integrated filter reactor windings transformer harmonic circuit model

$$\begin{cases} Z_{n1} = j\omega_n L_1 + R_1 \\ Z_{n2} = j\omega_n L_2 + R_2 \\ Z_{n3} = j\omega_n L_3 + R_3 \\ Z_{nC} = \dfrac{1}{j\omega_n C_f} + R_C \end{cases} \tag{5}$$

ω_n is the n times harmonic's angular velocity.

Based on the transformer harmonic circuit model, the voltage equation of the winding's self-inductance and mutual inductance matrix are:

$$\begin{cases} \dot{U}_{n1} = Z_{n1}\dot{I}_{n1} + j\omega_n M_{12}\dot{I}_{n2} + j\omega_n M_{13}\dot{I}_{n3} \\ \dot{U}_{n2} = j\omega_n M_{21}\dot{I}_{n1} + Z_{n2}\dot{I}_{n2} + j\omega_n M_{23}\dot{I}_{n3} \\ \dot{U}_{n3} = j\omega_n M_{31}\dot{I}_{n1} + j\omega_n M_{32}\dot{I}_{n2} + Z_{n3}\dot{I}_{n3} \end{cases} \tag{6}$$

Port conditions for transformer harmonic circuits

$$\begin{cases} \dot{U}_{n1} = 0 \\ \dot{I}_n = \dot{I}_2 + \dot{I}_3 \\ \dot{U}_{n3} = \dot{U}_{n2} - \dot{I}_{n3} Z_C \\ M_{13} = M_{31} = 0 \\ M_{23} = M_{32} = 0 \end{cases} \tag{7}$$

Put formulas (7) into formulas (6), we have:

$$\begin{cases} \dot{I}_{n1} = \dfrac{-\dot{I}_n j\omega_n M_{12}(Z_{n3}+Z_{nC})}{Z_{n1}Z_{n2}+\omega_n^2 M_{12}^2+Z_{n1}(Z_{n3}+Z_{nC})} \\ \dot{I}_{n2} = \dfrac{\dot{I}_n X_{n1}(X_{n3}+X_{nC})}{Z_{n1}Z_{n2}+\omega_n^2 M_{12}^2+Z_{n1}(Z_{n3}+Z_{nC})} \\ \dot{I}_{n3} = \dfrac{\dot{I}_n (Z_{n1}Z_{n2}+\omega_n^2 M_{12}^2)}{Z_{n1}Z_{n2}+\omega_n^2 M_{12}^2+Z_{n1}(Z_{n3}+Z_{nC})} \end{cases} \quad (8)$$

Due to the all-tuned design of the n times filter branch and the small resistance of it can be neglected, then:

$$Z_{n3}+Z_{nC}=(j\omega_n L_3+\frac{1}{j\omega_n C_f})+(R_3+R_C)\approx 0 \quad (9)$$

Put formulas (9) into formulas (8):

$$\begin{cases} \dot{I}_{n1}\approx 0 \\ \dot{I}_{n2}\approx 0 \\ \dot{I}_{n3}\approx \dot{I}_n \end{cases} \quad (10)$$

4. Experimental and Simulation verification

Takes a 300 KVA integrated filtering reactor winding rectifier transformer prototype and its filtering system as the experimental object. Fig.4 shows the systematical experimental scheme.

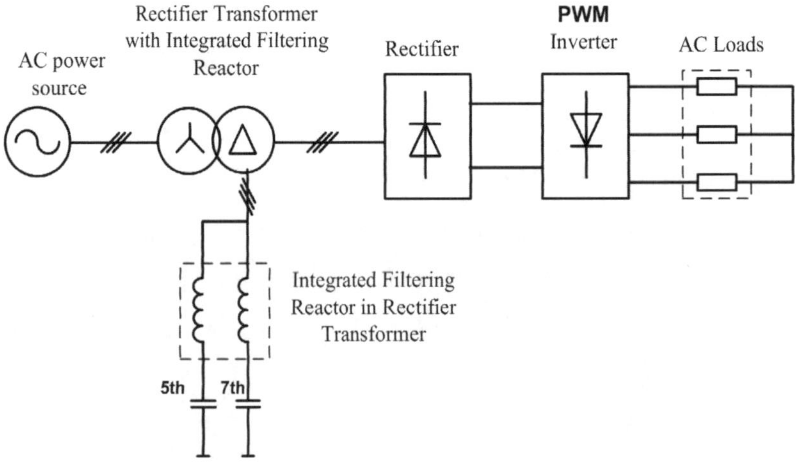

Figure 4. The integrated filtering reactor rectifier transformer filtering, power supply experimental

In MATLAB/simulink circuit simulation software, the circuit simulation model was established on the basis of mathematical model of integrated filtering reactor winding transformer. Fig.5 is the simulation waveform for the current mathematical model of the transformer high-voltage side when the transformer is disconnecting the filtering branch; Fig.5 is the simulation waveform for the current mathematical model of the transformer high-voltage side when the transformer is accessing to the filtering branch.

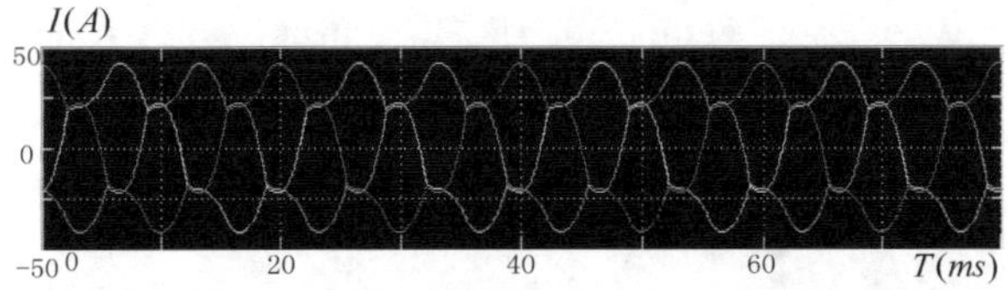

Figure 5. High-voltage side three phase current waveforms with the tuning branch disconnected

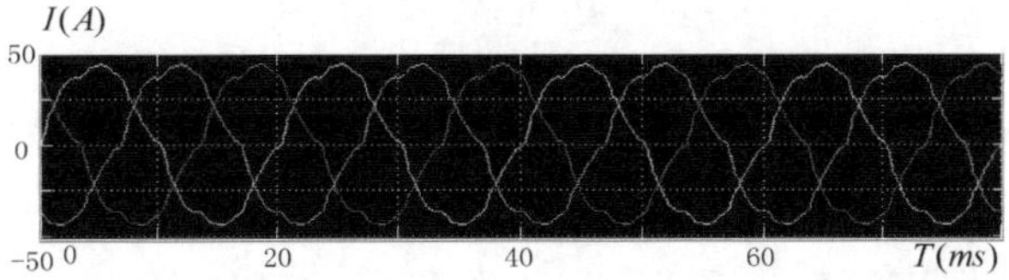

Figure 6. High voltage side three phase current waveforms with the tuning branch connected

From fig.5 and fig6, the current simulation waveform of the transformer high-voltage side calculated on the basis of the mathematical model is suggests the smooth operation and sound filtering effect of the filtering branch consisting of filtering reactor winding.

5. Conclusion

For this transformer with special arrangement mode, this paper takes advantage of the self-inductance and mutual inductance data of the winding, and derivate and calculated its harmonic model. At last, this paper testifies the mathematic models' correctness.The result shows: the mathematic models of the integrated filtering reactor winding transformer can reflect the mathematic structure of the transformer well, as well as calculate the power-supply performance of the transformer accurately.

Acknowledgments

This paper is supported Project supported by Scientific Research Fund of Hunan Provincial Education Department(19A108), and provincial natural science foundation of Hunan(2019JJ40059) AND collaborative education project of industry university cooperation of the Ministry of Education(201902143006) (2019003020)(201902145014), and Hunan teaching reform project (HNJG-2021-1350), and Doctoral Talent Research Fund(20RC002), and Xiangtan enterprise science and technology special expert project(CG-TPZJ20211019), and Xiangtan science and technology innovation project(CG-YB20211019).

References

[1] Hunan University 2009 Integrated filtering reactors transformer, China, CN200910043186.8 [P].2009.12-9
[2] Luo Long-fu, Li Yong, Xu Jia-zhu, et al 2007 The New-Type Converter Transformer's Mathematical Model Based on Phase-Coordinate *Transactions of China Electrotechnical Society* **22**(1) 34-40
[3] Corasaniti V F, Barbieri M B, Arnera P L, et al 2009 Hybrid power filter to enhance power quality in a medium-voltage distribution network *IEEE Transactions on Industrial Electronics* **56**(8) 2885-93
[4] Tziouvaras D A, McLaren P, Alexander G, et al 2000 Mathematical models for current, voltage, and coupling capacitor voltage transformers Power Delivery *IEEE Transactions* **15**(1) 62-72
[5] Deng M S, Wang X Z, Huang J 2013 New Transformer's Longitudinal and Asymmetric Faults Calculation Based on Additive Process *Proceedings of the Chinese Society of Universities for Electric Power System and its Automation* **1** 011
[6] Wang J, Witulski A F, Vollin J L, et al 1999 Derivation, calculation and measurement of parameters for a multi-winding transformer electrical model[C]//Applied Power Electronics Conference and Exposition, 1999. APEC'99. Fourteenth Annual. IEEE **1** 220-6
[7] L. Luo, Y. Li, J. Xu, J. Li, B. Hu, and F. Liu 2008 "A new converter transformer and a

corresponding inductive filtering method for HVDC transmission system," *IEEE Trans. Power Del.* 23(3) 1426–31 Jul. 2008

[8] Zhanxin Zhu, Dexin Xie, Gang Wang, Yanli Zhang, and Xiuke Yan 2012 "Computation of 3-D Magnetic Leakage Field and Stray Losses in Large Power Transformer," IEEE Transactions on Magnetics **48**(2) 739-742 February 2012

[9] Luo L, Li H, Li Y, et al 2008 Study on the electromagnetic transient state in the new converter transformer based on coupled field-circuit method[C]Electrical Machines and Systems, 2008.International Conference on. IEEE, Wuhan, China 4291-5

[10]Kumbhar G B, Kulkarni S V 2007 Analysis of Short-Circuit Performance of Split-Winding Transformer Using Coupled Field-Circuit Approach Power Delivery, IEEE Transactions on **22** 936-43

[11]Peterson M, Singh B N 2009 Multipulse controlled ac-dc converters for harmonic mitigation and reactive power management. IET on Power Electronics **2**(4) 443–55

Design and Development of Virtual Simulation Teaching Resources of "Safe Electricity" Based on Unity3D

Qing-Sheng Zheng* and Ke Li

Mechanical and Electrical Engineering Department, Nanchang Institute of Technology, Nanchang, Jiangxi ,China

*corresponding author's e-mail: 46561461@qq.com

Abstract. While electricity brings convenience to people's production and life, it also has certain hazards. It is imperative to popularize common sense of safe electricity use. For non-electrical majors, offering electrical and electronic technology courses is one of the ways. Electrical and electronic technology is a technical basic course with high practical requirements. It is mainly taught through a combination of theory and experiment. However, in actual teaching, there are still some teaching contents that are not experimental and experience teaching difficulties, which give teachers and students brings teaching problems. This article takes the teaching content of "Safe Electricity" as an example. Explain the design, development and application of virtual simulation teaching resources. Experimental research results show that virtual simulation teaching resources play a positive role in improving the shortcomings of traditional teaching, promoting students' interest in learning, and improving learners' learning effects.

1. Introduction

The electrical and electronic technology course is an important technical basic course for non-electrical majors, with strong theoretical and practical nature. The teaching purpose of the course is to enable students to acquire the basic theoretical knowledge and skills of electrical and electronic technology, and to lay a solid foundation for the subsequent professional courses. There are many concepts and formulas in the course. For courses with many concepts and formulas, students will feel that the content is more abstract and boring, so it is difficult for students to master the knowledge. Therefore, a combination of theoretical teaching and experimental teaching is usually used to help students better understand and consolidate knowledge points. However, for some experimental or experiential teaching content that is restricted by experimental conditions or environment, such as the content of electric shock in safe use of electricity, most teachers use the "full classroom" teaching method, which is led by teachers and passively accepted by students. In the process of learning, students lack experimental learning situations and experience platforms, leading to low learning interest and learning efficiency. Their understanding of knowledge stays at the theoretical level, and there are obvious deficiencies in deep understanding of knowledge and skills development. [1]

The application of virtual simulation technology provides brand new teaching resources and teaching methods for electrical and electronic technology teaching. With the help of virtual simulation development software, the teaching content that cannot be experimented or experienced in traditional teaching is designed as a virtual simulation teaching resource. In the teaching process, students use the virtual platform to learn or practice close to the real environment, which is convenient to achieve a learning experience of embodied cognition. The application of virtual simulation teaching resources has opened up a broad space for the development of electrical and electronic technology teaching in colleges and universities. It has very important practical significance for solving the teaching difficulties in traditional teaching, promoting the reform of electrical and electronic technology teaching, and improving students' learning interest and learning effect. [2]

2. Analysis of learner characteristics

Electrical and electronic technology courses usually start in the second semester of the freshman year. For freshmen in this period, they have basically adapted to the study and life of the university. The individual's physical and mental development tends to become stable and mature, and the intellectual development reaches its heyday. The method is mainly based on dialectical and logical thinking. Individual self-awareness has formed obvious characteristics. The content has been greatly enriched and differentiated. Self-concept is more multi-dimensional, multi-layered psychological structure, and self-evaluation is more independent, unique, and trending. Because of stability, the inner experience is deeper and more sensitive. [3] Therefore, in the teaching process of various subjects, teachers should grasp the characteristics of students during the period, take advantage of the new learning opportunities and environment of the university, and adopt more effective teaching mode and methods to re-stimulate students' interest in learning and establish self-confidence in learning, So that they have a strong desire for knowledge and motivation to become talents.

3. Analysis of learning content

The teaching content of "safe electricity use" belongs to the second semester of electrical engineering and electronic technology teaching content for freshman students. It is one of the main content of electrical technology. It mainly introduces the general knowledge of power supply and distribution and the basics of safe electricity use. Have an overall understanding of power transmission, distribution and safe use of electricity. Through the study of this content, learners will understand the composition of the general power system, understand the common sense of daily safe electricity use, standardize daily electricity use behavior, cultivate students' logical thinking ability, improve students' ability to use knowledge to solve some problems. Lay the necessary foundation for students' future study and work life . [4] Electricity is harmful, and the experiment of electric shock cannot be experienced by students. These contents are a big problem in the teaching of electrical and electronic technology courses. When students' understanding and analysis skills are not strong, learners will find it difficult. If virtual simulation teaching resources are used for auxiliary teaching in teaching, not only teaching problems can be solved, but also the learning enthusiasm and initiative of learners can be mobilized. In the end, the teaching effect will be doubled with half the effort.

4. Design of virtual simulation teaching resources for "safe electricity "

The author chooses Unity3D, which is popular and powerful in the market, as the virtual simulation teaching resource development platform. The teaching content of "Safe Electricity" is designed and developed into a virtual simulation teaching resource. The learners learn the general knowledge of electric power system, electric shock simulation experiment and safe electric behavior code in the created virtual scene as the third person.

4.1. Design Ideas

Use virtual simulation technology to study subject knowledge, and pay attention to reasonable choices in content. Not all content is suitable for virtual simulation methods, mainly for knowledge points that are highly abstract and inconvenient to experiment or experience in real teaching conditions; The resource design needs to conform to the learner's learning characteristics, and the operation should not be too complicated to reduce the cognitive load; the content should match the learner's cognitive ability, and the difficulty of the simulation experiment should be in line with the learner's current skill level. The teaching resource design concept of this case is shown in 'Figure 1'.

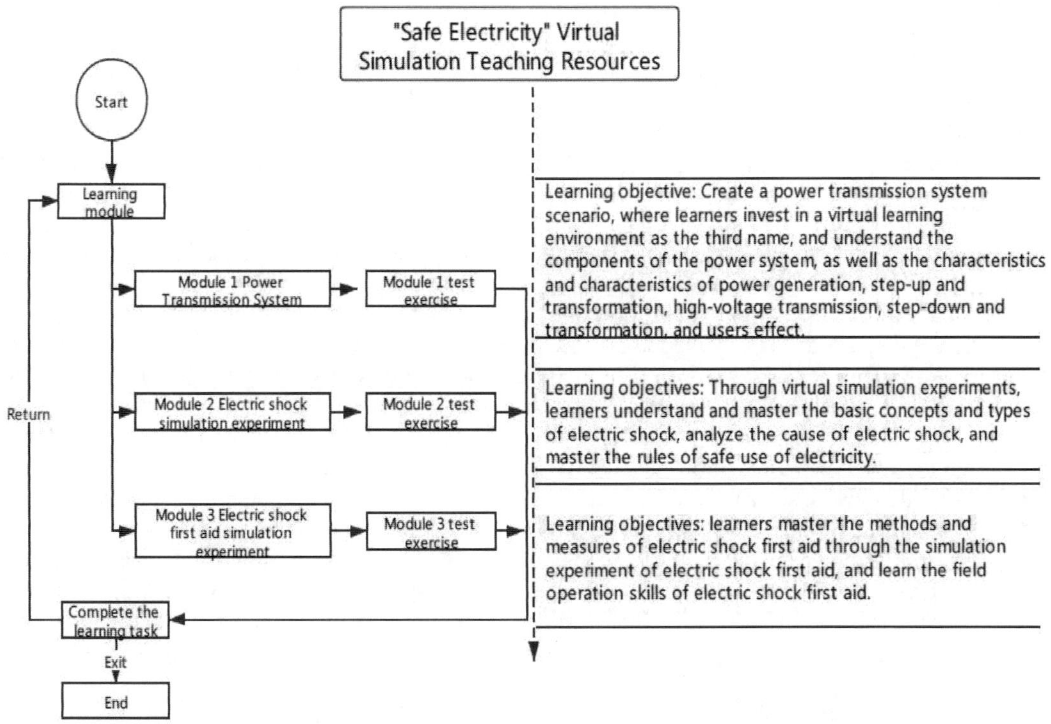

Figure 1. Design concept drawing.

4.2. Teaching resource module and function design

This case uses the teaching content of the electrical and electronic technology course "Power Supply and Safe Use of Electricity" as the design content, and follows the requirements of the teaching objectives. This teaching resource is divided into three modules, namely, the power transmission system, the electric shock simulation experiment, and the electric shock first aid. Simulation. After the learner enters the virtual simulation software, he chooses the corresponding module to study. Each module is equipped with corresponding test questions, and the learner can pass the test link to test the learning effect. Through learning, students can understand the basic concepts and composition of the electric power system, understand and master the basic concepts of electric shock, the types of electric shock, and the basic knowledge of electric shock prevention, and master the field operation skills of electric shock first aid. This case resource can be used as an auxiliary teaching resource in the classroom or as an after-school learning resource.

5. Development of virtual simulation teaching resources for "safe electricity "

5.1. Development tools

Unity 3D, also known as Unity, is derived from a multi-platform comprehensive game development tool developed by Unity Technologies in Copenhagen, Denmark that can easily create interactive content such as 3D video games, architectural visualization, and real-time 3D animation. It has the advantages of easy-to-implement physical collision effects, convenient development of a virtual environment, and free access to personal editions. Therefore, the author chose Unity3D as the development tool for this case. At the same time, in order to increase the learning interest of learners, in the process of resource development, use Unity3D Fungus dialogue plug-in to design the dialogue link between learners and the system reasonably, which will attract learners' attention to a certain extent and improve learners'learning interest.[5]

5.2. Development process and technical realization

5.2.1. "Power System" Virtual Learning Module. (1) Scene creation and modeling. Create a power system scene in Unity, add building models for each link of the power system to the scene, and place power stations, step-up substations, high-voltage transmission lines, step-down substations (substations), and low-voltage distribution in the scene according to the power system transmission process. Objects such as electrical cabinets, factory premises, residential areas, etc. (2) Role and function design. The third-person character object is imported into the scene, and the learner controls the character to learn in the virtual power transmission system scene by controlling the keyboard direction keys. Use Unity's physical collision effects to set trigger detection functions for each link model of the power system, and realize that the mouse pointer moves into the object and the object explanation is displayed. The Fungus plug-in is used to set up a dialogue system containing teaching content for roles and power systems. When the role collides with power stations, step-up substations, high-voltage transmission lines, step-down substations (substations), low-voltage power distribution cabinets, factory buildings, When objects such as residential areas, enter the corresponding dialogue link. After the content learning of this module is over, set up test questions, and test the learner's learning situation by answering the questions. Through the dialogue system, question answering system and virtual power system scenarios, learners can simulate and experience the various links of the real power system, understand the composition of the power system and the characteristics and functions of each component, and finally achieve the learning goal of the power system. (3) Main function scripts and renderings. Mouse ray detection, display text: realize the function of displaying explanatory text when the mouse pointer moves into each component object of the power system.

```
public class tip : MonoBehaviour
{   public Text text;
    public string info;
    private void OnMouseEnter()
    {text.text = info;text.transform.parent.gameObject.SetActive(true);}
    private void OnMouseExit()
    {text.text = "";text.transform.parent.gameObject.SetActive(false);}}
```

The effect of the virtual learning environment of "power system" is shown in ‘ Figure 2’, ‘ Figure 3’.

Figure 2. Panoramic effect picture .

Figure3. Effect drawing of power station.

5.2.2. "Electric shock simulation experiment" module.

(1) Scene creation. This module is mainly to set up electric shock simulation experiment scenes, including single-phase electric shock, two-phase electric shock, contact voltage electric shock and step voltage electric shock. The learner selects each type of electric shock by clicking the corresponding label button with the mouse to conduct an electric shock simulation experiment. The objects in the scene are mainly high and low voltage poles (towers) and character models required for various types of electric shocks.

(2) Role and function design. The learner chooses the corresponding electric shock form, then enters the simulation experiment scene of this electric shock form, controls the character model through the mouse operation to complete the electric shock experiment. When the electric shock conditions are met, electric shock special effects will appear in the scene. Learners can observe various electric shock conditions and electric shock phenomena that are difficult to see or experience in reality in the simulation experiment, understand and master the basic concepts, types and reasons of electric shock from them, and master the basic behaviors of safe electric use rule. Through the learning experience of the electric shock simulation experiment module, the learning effect of students will be significantly better than the traditional teaching effect.

(3) Program control flow and effect diagram. The program control function and process of the "Electric Shock Simulation Experiment" module is shown in ' Figure 4'.

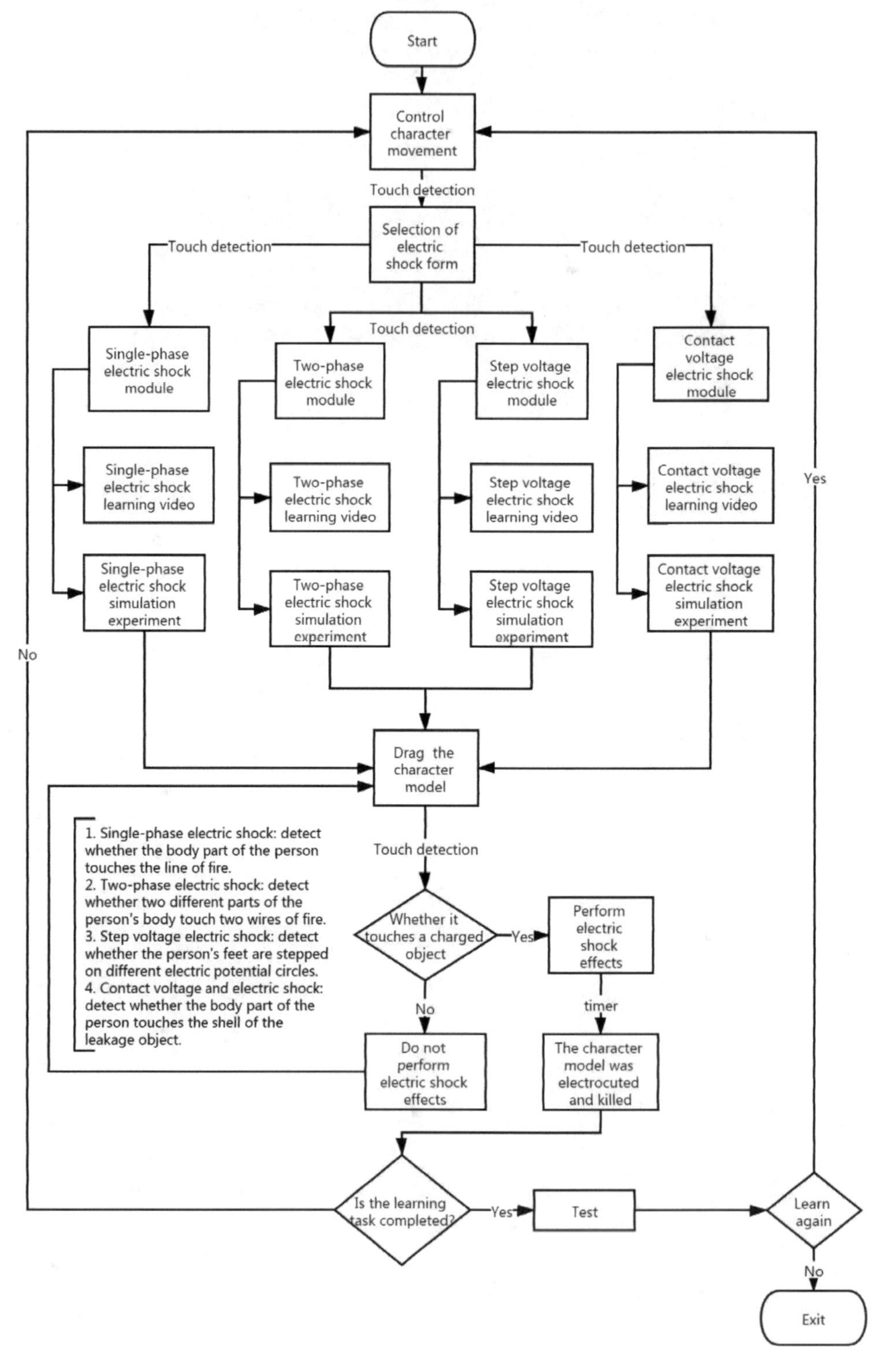

Figure 4. Program control flow chart.

The single-phase electric shock and step voltage electric shock effects in the "Electric Shock Simulation Experiment" module are shown in ' Figure 5', ' Figure 6'.

Figure 5. Single-phase electric shock.

Figure 6. Step voltage electric shock.

5.2.3. "Electric Shock First Aid" Simulation Module.

(1) Scene creation**.** The scene creation of this module is mainly to add video learning components, electric shock first aid animation, electric shock first aid simulation operations, and character models to the scene.

(2) Role and function design. Set the operation buttons in this scene, which are "Watch video learning", "Electric shock simulation", and "Start rescue". The learner first learns the related knowledge of electric shock first aid by watching the electric shock first aid video, and masters the methods and steps of electric shock first aid. Subsequently, the learner enters the electric shock simulation state, the scene presents the character model electric shock animation special effect, simulating the electric shock scene. Finally, the learner is required to complete the "Electric shock first aid" task. when the learner arrange the rescue measures in a correct and orderly manner within a limited time, the rescue is successful, and the animation effect of the ambulance rescuing the electrocuted person will be played, otherwise the rescue will fail. Through the simulation experiment of electric shock first aid in this module, learners master the correct methods and measures of electric shock first aid, and improve the teaching goal of on-site operation skills of electric shock first aid.

Program control flow and effect diagram. The program control function and process of the "Electric Shock First Aid" simulation module is shown in ' Figure 7'.

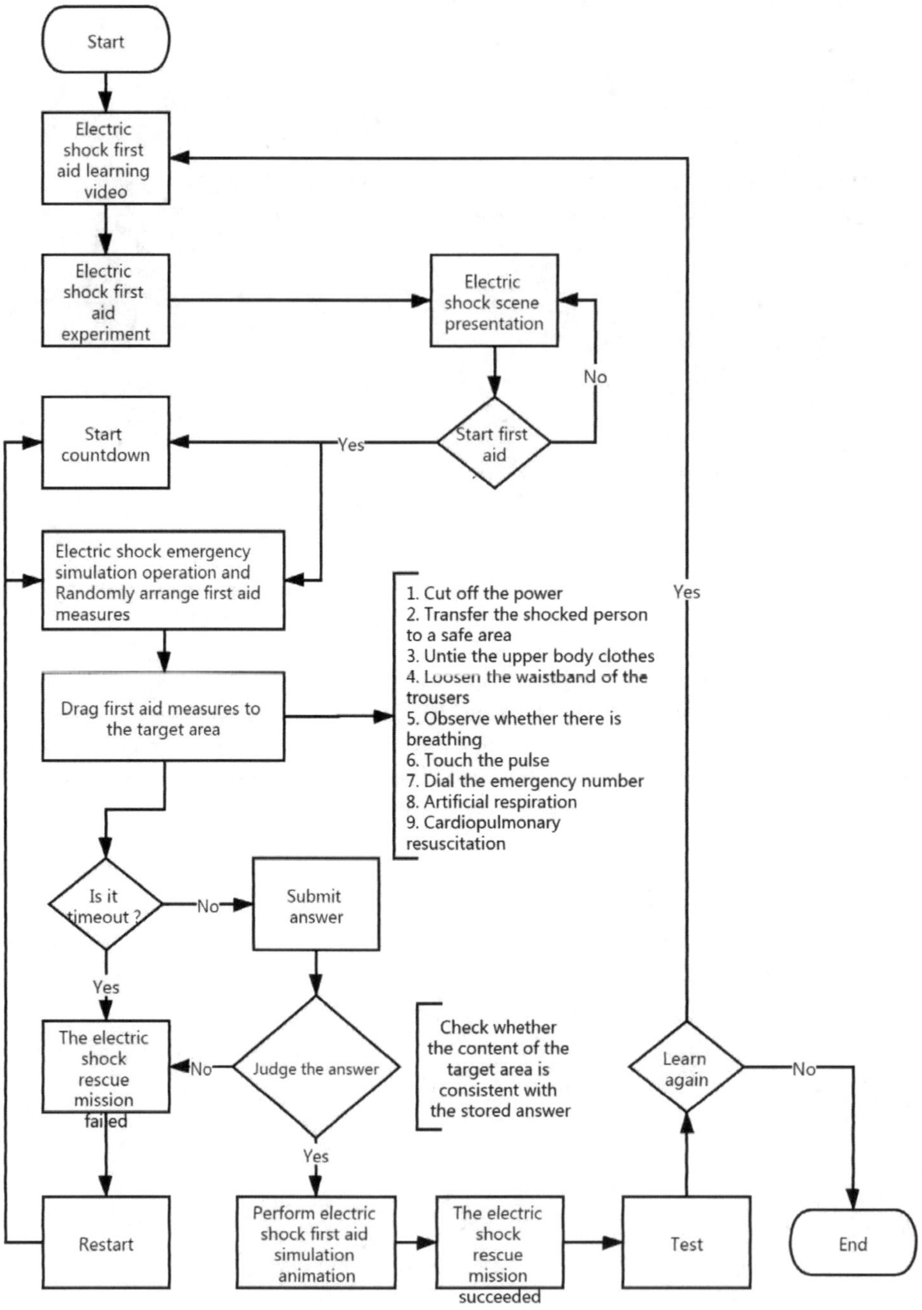

Figure 7. Program control flow chart.

The effect of the "Electric Shock First Aid" simulation scene is shown in ‘ Figure 8’ ,‘ Figure 9’.

Figure 8. Video learning.

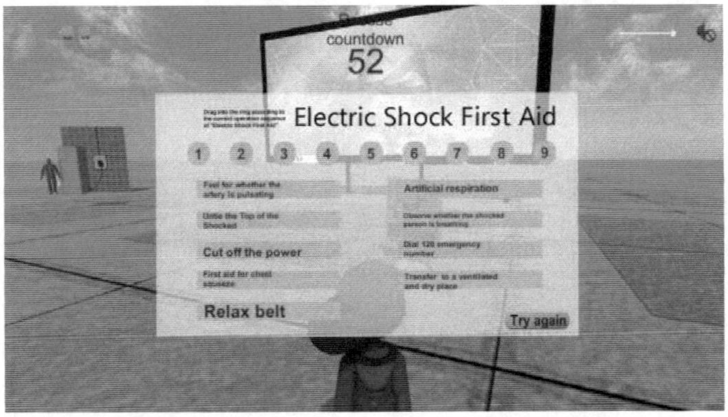

Figure 9. Simulation effect.

5.3. Test and release of virtual simulation software

After the above-mentioned design and development, the test of this case work shows that the interactive functions and control logic of the software meet the design requirements, so the executable program is packaged to generate the work and the work is released. The work developed by the author is used in the computer room, so select the "PC, Mac&Linux Standalone" type in the "Platform" publishing platform list, then select "Player Settings...", and fill in the publisher name and work name, and finally click The "Build" button generates an executable file, and the user only needs to double-click to open the .exe file to run it.

5.4. Application and analysis of resources

5.4.1. Implementation Plan. In order to study the application of virtual simulation teaching resources in the classroom teaching of electrical and electronic technology, the virtual simulation software of this project "safe electricity use" is used as a teaching resource and used in teaching. Divide students into two groups: experimental class and control class according to the natural class (as shown in 'Table 1.'). And use the experimental method to explore the influence of using virtual simulation teaching resources to assist teaching in the teaching of electrical and electronic courses on the learning effect. In the teaching process, the experimental class uses "safe electricity" virtual simulation software for auxiliary teaching, while the control class uses traditional teaching resources and methods for teaching. Before the experiment, a pre-test was performed on the experimental group and the control group. After the experiment, the experimental group and the control group were post-tested. The test questions were the same as the pre-test. Finally all students were asked to fill in the satisfaction survey questionnaire.

Table 1. Introduction of the experimental group and control group

Group	Number of people	Experiment description
experimental group	Freshman in a college Specialty Mechanical and Electrical Class 1 48	Teaching is organized in the computer room, and the teacher spends 10 minutes introducing the use and functions of the software. Under the guidance of the teacher, use virtual simulation software to learn the knowledge of safe electricity use in 2 classes.
control group	Freshman in a college Specialty Mechanical and Electrical Class 2 46	According to the traditional teaching mode, teaching is organized without using virtual simulation software.

5.4.2. Data collection. After the experiment, collect the research sample data. The experimental group is a freshman majoring in mechanical and electrical engineering, a total of 48 students, the effective sample size is 48; the control group is a parallel class of 46 students, the effective sample number For 46 people. The test scores include the pre-test and post-test scores of the experimental group and the control group, as shown in 'Table 2.'. The questionnaire results are the results of the questionnaire survey on the learning satisfaction of the experimental group and the control group after the experiment, as shown in 'Table 3.'.

Table 2. Before and after experiment result statistics between the experimental group and control group

score	experimental group（n=48）	control group（n=46）	t	P
Pre-test	20.0000±6.28414	19.6042±6.70117	0.258	0.795
Post-test	85.3840±8.04756	76.3542±8.96444	2.323	0.033
t	-62.544	-93.412		
P	0.000	0.000		

Table 3. Learning satisfaction scores after experiment between the experimental group and control group

factor	After the experiment		t	P
	control group	experimental group		
Learning satisfaction	3.1888±0.68310	3.9814±0.62336	-5.875	0.000
teaching methods	3.1596±0.71556	4.2021±0.71216	-7.080	0.000
learning result	3.4255±0.89064	3.8617±0.76401	-2.548	0.012
learning environment	3.1702±0.78211	4.0426±0.69023	-5.058	0.000
Learning participation	3.0000±0.80757	3.8191±0.76204	-5.875	0.000

5.4.3. Data analysis and conclusion. (1) Analysis of learning effect. The learning effect of the subjects is shown in 'Table 2.'. The pre-test and post-test results of the experimental group and the control group are in accordance with the law of normal distribution. The results in 'Table 2.' show that the average scores of the experimental group and the control group are not significantly different, which are 20 points and 19.6 points, respectively. P=0.795>0.05, indicating that the pre-test scores are not significant at the 0.05 level of significance. difference. Under the premise that the selected experimental group and the control group are homogeneous, different teaching methods are adopted, and the learning performance of the learners has been significantly improved. The average learning performance of the experimental group is 85.384 points, which is compared with the control group

(76.3542 points) There is a significant difference (P=0.0000<0.05), indicating that the experimental group and the control group have significant differences in their academic performance. It shows that the use of virtual simulation resources to assist teaching can significantly improve the academic performance of electrical and electronic technology. (2) Analysis of the factors of learning satisfaction. By comparing the results of the satisfaction survey questionnaire between the experimental group and the control group after the experiment, it is found that the satisfaction degree of the experimental group after the experiment is 86%, while the satisfaction degree of the control group is 77% after the study. The above is divided into two categories: positive learning satisfaction (mean>=3.00) and negative learning satisfaction (mean<3.00). The results in 'Table 3.' show that there are significant differences between the experimental group and the control group in terms of learning satisfaction, teaching methods, learning effects, learning environment, and learning participation. In terms of being positive. Each factor is significantly different at a significant level of 0.05 (all P<0.05), indicating that after the experimental group uses virtual simulation teaching resources, the learners' learning satisfaction is significantly improved in all dimensions of learning satisfaction compared with the control group. Moreover, the differences in the dimensions of teaching methods, learning environment, and learning participation are more significant than those of learning effectiveness.

The above analysis shows that the application of virtual simulation teaching resources as auxiliary teaching resources for electrical and electronic technology courses not only optimizes the learning effect of learners, but also provides a guarantee for students' main status; while solving teaching problems under realistic experimental conditions, It improves learners' learning interest and learning participation, and provides a platform for learners to develop their learning potential. Through personal experience in virtualized experimental scenes, students form practical perceptions, realize the context transfer of knowledge, complete the learning and consolidation of knowledge, and finally achieve the goal of completing the hidden teaching goals.

6. Conclusion

In accordance with the general design and development principles of teaching resources, the author elaborated on the design, development and application of virtual simulation software for "safe electricity use" in electrical and electronic technology courses. Through the application of cases, the teaching application effect of virtual simulation teaching resources is studied. The experimental results show that the virtual simulation software of "safe electricity use" as a teaching resource to assist teaching has a significant positive effect on learners' learning effect and learning satisfaction. It is hoped that the design and development of this case can provide a reference for the research and application of virtual simulation teaching resources, and play a role in improving and making up for the lack of traditional teaching resources and teaching modes.

References

[1] Fu G D, Jiang W, Ren P F and Liu H 2015 Discussion on the Teaching Method of Electrical and Electronic Technology Course *Decision and Information Magazine* **2**

[2] Dong Y 2021 Application of Virtual Simulation Technology in Electrical and Electronic Technology Teaching in Secondary Vocational Schools [J]. *Electronic Technology and Software Engineering*, **16** 39-40

[3] Lin C D 2018 *Developmental Psychology* [M] (Beijing: People's Education Press) 401-415

[4] Cheng X F 2014 On the Visual Teaching of "Safe Use of Electricity". *Prose 100 Ten-day Journal*, **000 (010)** 170-170

[5] Zhang X L 2016 Research on game design and application of junior middle school physics education based on unity3d [D]. East China Normal University

Construction and Exploration of Virtual Simulation Experimental Teaching Platform for Network Security and Computer Technology

Jin Zhang

Teachers' Work Department, Nanchang Institute of Technology, Nanchang, Jiangxi 330088, China

*corresponding author's e-mail:63999230@qq.com

Abstract. Today, with the arrival of informatization, the application of education informatization in the field of education and teaching and the management process is being gradually deepened. The application mode has effectively promoted the profound reform of education mode, teaching methods and learning mode. The main content of the construction of higher education informatization and experimental teaching demonstration center is to establish a virtual simulation experimental teaching center, which has had a positive and important impact on the improvement of higher education quality and the deepening of experimental teaching reform. Virtual simulation technology uses a variety of emerging science and technologies to make otherwise challenging and dangerous experiments simple. According to the main characteristics of virtual simulation teaching, we discusses the application of virtual simulation teaching in computer network teaching and proposes corresponding suggestions for relevant problems.
Keywords. Computer technology; Cyber security; Experimental teaching

1. Computer Virtual Simulation Experimental Teaching

Computer virtual simulation experimental teaching based on the principle of "unified planning, fast sharing, unified integration", closely combined with the specific situation of the virtual simulation laboratory and the future development, according to the laboratory information interaction portal, virtual simulation, practice teaching, experimental resources sharing access to promote the information, intelligent construction of the laboratory. As an important course of this computer network engineering major, the computer network course should not only familiarize students with the theoretical knowledge about the network, but also let students master the ability of practical operation. Due to the influence of network equipment, training conditions and other factors, the school just lets students learn according to books, but did not really do hands-on practice, which makes students lack of hands-on practice opportunities. Over time, more and more students find the course boring and lose some of their interest in the course. We should focus on cultivating students' interest in learning, so as to cultivate talents that meet the needs of the society.

Computer virtual simulation system to establish laboratory scientific, intelligent, management, with effective management monitoring, complete experimental teaching resources integrated management and open service as the purpose, give full play to the efficiency of experimental resource utilization, serving teachers and students as the goal, further meet the needs of users and managers, thought all aspects of three-dimensional intelligent management. Its construction objectives mainly include virtual simulation resource management, the main idea of virtual simulation resource management of each major, unified display, unified login. Teaching experimental management is mainly to monitor the comprehensive information management in the process of experimental teaching arrangement. Can implement the unified management platform and implement intelligent, scientific and rational management of the arrangement of simulation teaching.

2. The development course and construction of computer virtual simulation experimental teaching center

From 1994 to 2013, it was the integration and development stage of the computer virtual simulation teaching center. From 2013 to 2016, it was the completion and development stage of the computer virtual simulation system demonstration center. Currently, the computer virtual simulation system center covers a total area of 3,000 square meters and has 16 laboratories. Computer virtual simulation experimental teaching for the whole school, 30 virtual simulation experimental projects have been built, realizing the sharing of experimental teaching resources and achieving good teaching results. The main part of the computer virtual simulation experiment teaching construction is to build the computer virtual simulation experiment teaching platform. Computer virtual simulation experimental teaching platform is composed of platform and resources combination. The platform has a virtual experimental center website, and has built an open virtual simulation experimental teaching management and sharing platform, followed by the resources of computer virtual experimental courses on the reform platform for management. Through the research of key technologies, a virtual cognitive experimental teaching platform with comprehensive functions, advanced technology and in line with the characteristics of computer discipline has been successfully built. It can support to some extent the open virtual experimental teaching and teaching materials sharing, and also has certain benefits for every user using this platform.

3. Necessity of the construction of computing computer virtual simulation test teaching center

The basic theoretical knowledge of computer undergraduates in universities in China is relatively solid, but there is still a large gap in practical ability compared with computer undergraduates in world-class universities. How to close the gap between the two has become a difficult problem for computer students in universities in China. In view of the differences of students 'personalities, it is the unremitting efforts direction of college computer teachers in recent years to cultivate students' innovation consciousness, improve their practical ability, and explore and improve the results of computer experimental teaching. The necessity of computer virtual simulation experimental teaching center can be divided into the following: first, it can provide effective ways to improve students' interest and motivation in computer practice courses; second, it is the best opportunity to build computer simulation model system, which helps students to consolidate professional foundation and promote practical innovation ability, it can broaden the only way of experimental teaching, can effectively promote the function of higher education from "talent trainer" to "talent designer". Therefore, experimental teaching has now fully entered the information age, and the establishment of computer virtual simulation experimental teaching center is the inevitable trend of the development and reform of experimental teaching.

4. Characteristics of the virtual simulation experiments

As a product of the deep combination of discipline and information technology, virtual simulation experiment is a computer-aided teaching software using virtual reality to simulate things technology. Its main characteristics include the following points:

4.1. Strong interactivity

The traditional teaching mode generally adopts the mode of teachers teaching students 'learning. Under this mode, students' learning is often too passive and students can learn the knowledge that is less and not comprehensive, which has certain obstacles to the all-round development of students. Under the virtual simulation teaching system, the teaching mode of two-way interaction between teaching and learning pays more attention to the teaching content, which is conducive to students and teachers to independently express their learning opinions and views, so as to further improve students 'interest in the curriculum and improve students' initiative in learning.

4.2. Incentive

In the usual course teaching, teachers can show simulation examples to students, guide students to participate in practice, can stimulate students 'learning desire, help students to fully understand and master the learning content, stimulate students' innovative thinking and creative motivation. Compared to the traditional teaching, the virtual simulation experimental teaching has more advantages at the moment.

4.3. The whole sex was higher

Some more dangerous projects if the student teacher is injured by operation negligence, is not worth the loss. Virtual simulation experiments cover the equipment and knowledge used in various aspects, such as toxic and harmful substances, highly destructive experiments can be completed in the virtual laboratory, while ensuring the safety of the experimenter while also obtaining the desired experimentation results.

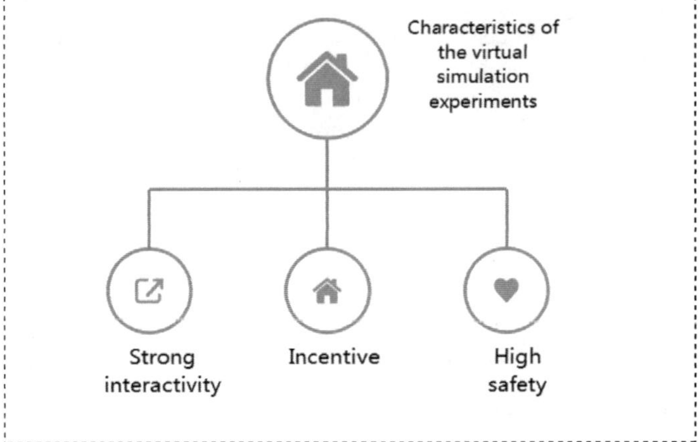

Figure 1. Characteristics of the virtual simulation experiments

5. Suggestions for the construction of virtual simulation experimental teaching center

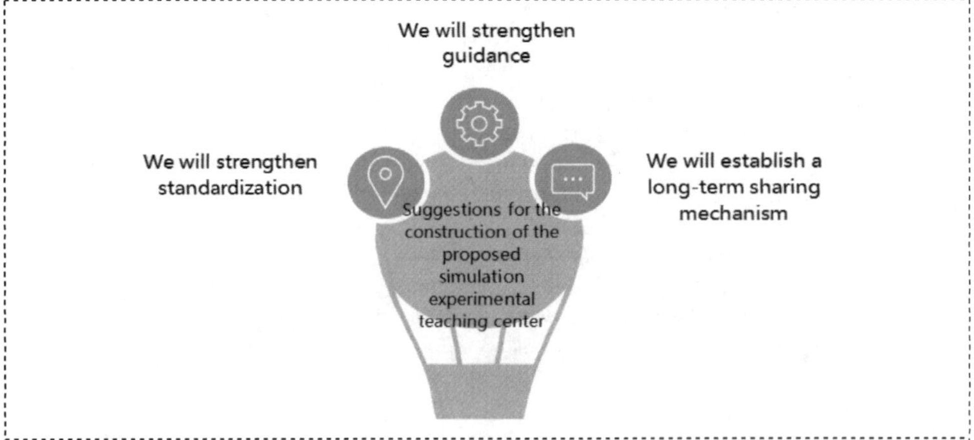

Figure 2. Suggestions for the Construction of the Virtual Simulation Experimental Teaching Center

5.1. We will strengthen standardization

In the construction of virtual simulation experiment teaching center, should take research, transformation, secondary development, for different experiments targeted virtual simulation teaching technology, at the same time, establish the corresponding functional institutions, perfect the experimental teaching information standard research, testing, inspection and certification, so as to better standardize and strengthen the virtual simulation experiment technology facilities resources, information standardization construction.

5.2. We will strengthen guidance
Formulate policies and regulations on experimental teaching information work, build the ability system of virtual simulation experimental teaching center under the guidance of education information development planning, establish the construction and management provisions of virtual simulation experimental teaching center, and strengthen the orientation of virtual simulation experimental teaching center.

5.3. We will establish a long-term sharing mechanism
Statistical data show that the network laboratories with more than nearly 600 class hours in colleges and universities only account for 27% every year. Due to the imperfect management system and operation mechanism, high-quality teaching resources are limited, which leads to great waste, and the phenomenon of resource imbalance is still very serious. Therefore, virtual simulation experiment teaching should establish and improve the long-term mechanism of long-term resource sharing, so as to form and promote the sharing of high-quality experimental teaching resources as the core.

6. Conclusion
Computer network course is a practice and theory course, using virtual simulation experiment in computer network course, can effectively solve the problems encountered in computer network teaching, to improve students' learning enthusiasm and practical operation ability, computer technology, virtual simulation technology, multimedia technology, make the experiment is not limited by time and space, can effectively solve problems and reduce the lack of sufficient equipment, further improve the results of computer network teaching. With the growth of time, China's virtual experimental teaching technology will also show different improvement and progress. I believe that in the near future, China will gradually narrow the gap with other countries in virtual simulation experimental technology.

Acknowledgement
Science and Technology Research Project of Jiangxi Provincial Department of Education,GJJ202115, Research on the Performance Model of National Musical Instruments based on IT audio technology

References
[1] Li L, Wu C 2021 The Design and Application of the Network Teaching System Based on Virtual Simulation *Modern electronics technology* pp 1-4 (2021-09-23) http://kns.cnki.net/kcms/detail/61.1224.TN.20210910.1454.002.html
[2] Xiao L, Li L, Liu S H, Wang M, Shan L X 2021 Design of an embedded system virtual experimental simulation platform based on Proteus and Arduino *Scientific And Technological Innovation And Application* **11**(24) 57-9
[3] Deng Y, Xiao H, Wang M, Wu S, Li C, Wei X, Xu W 2021 The Application of Virtual Simulation Technology in Modern Instrument Analysis *Guangming TCM* **36**(15) 2630-3
[4] Yu Dong 2021 Application of Virtual Simulation Technology in Secondary Vocational Electrical and Electronic Technology Teaching *Electronic Technology and Software Engineering* **16** 39-40
[5] Lu H, Li M M, Zhang Y, Gu F 2021 Application of Virtual Simulation Technology in Teaching of Computer Network in the Background of "Internet +" *China New Communications* **23**(13) 177-8
[6] Xu X, Zhao J, Liu C, Hui W 2020 Research on Virtual Simulation Experimental Development and Its Technology Application *Technology Entrepreneurship Monthly* **33**(09)160-2
[7] Zhang X 2020 Research on Practical Teaching System Reform Combined with Virtual Simulation *Electronic Devices and Information Technology* **4**(09) 167-8
[8] Song Y, He P, Pan J, Xu F, Zhao Y 2020 The Application of Virtual Simulation Technology in Classroom Teaching *Journal of Shijiazhuang Vocational and Technical College* **32**(04) 78-80
[9] Huang Y, Liu Z, Zhang L 2020 Construction of a Virtual Simulation Laboratory Based on Cloud

Computing *Times Agricultural Machinery* **47**(05) 88-9
[10] Lin J 2020 Research on Virtual Simulation of Computer Technology *Computer Products and Circulation* **06** 256
[11] Wang Y, Zhou W 2020 The Application of Virtual Simulation Technology in the Experimental Teaching of Computer Network *Computer Products and Circulation* **07** 190
[12] Liu X, Zhang S, Cui F, Zhou A 2019 The Application of Virtual Simulation Experiments in Teaching **26** 115-7

Simulation of Wide Band BAW Filter for N78 Application

Lu Lv, Yao Shuai*, Wenbo Luo, Chuangui Wu and Wanli Zhang

State Key Laboratory of Electronic Thin Films and Integrated Devices, University of Electronic Science and Technology of China, Chengdu, Sichuan, China

* corresponding author's e-mail: yshuai@uestc.edu.cn

Abstract. Based on the requirements of large bandwidth filters for 5G communications, finite element models for material analysis and filter design are established. Characteristics of lithium niobate under different cut angle are simulated, and 36° Y cut lithium niobate is used in resonator with SiO_2/Mo acoustic reflector. The influence of top electrode is evaluated and the area of the electrode is designed. A method of co-simulation using finite element model and modified Butterworth van Dyke model is introduced. A wide band filter with relative bandwidth of 10.1% is designed while the insertion loss is less than 2 dB, which can satisfy the requirement of N78 band.

1. Introduction

The 5th generation wireless communication systems (5G) put forward an urgent demand for broadband filter at higher frequency. Acoustic filter, which is the key component in RF front-end system due to its compact size, also faces challenges with and wide bandwidth at high frequency. Bulk acoustic wave (BAW) filter [1] and surface acoustic wave (SAW) filter [2] are the two categories of acoustic filters. The SAW filter is difficult to use at the frequency higher than 3 GHz because of performance degradation at high frequencies. On the contrary, BAW filter seems a better choice for 5G communication. However, the widely used AlN-BAW filter suffers from the disadvantage of low acoustic coupling. Since filter bandwidth is related to acoustic coupling ability of the piezoelectric materials, the relative bandwidth of AlN based BAW filter is typically less than 4.1% [3,4]. A way to increase bandwidth is to use Sc doped AlN, but the bandwidth is still limited [5]. Another way is to replace the AlN with other piezoelectric materials with high acoustic coupling ability. Recently, the anisotropic single crystal lithium niobate (LN) appears to be a promising candidate. The electromechanical coupling coefficient (K_t^2) of longitudinal wave in LN varies greatly by changing the crystal orientation [6], As a result, it is possible to obtain wide band filters if an appropriate orientation is chosen, along which the longitudinal wave with high K_t^2 is excited.

In this work, the finite element method (FEM) model is built up to investigate the bulk acoustic characteristics of the LN. It is demonstrated that the 36° Y-cut LN shows a large K_t^2 exceeding 26%. A 2D model of solidly mounted resonator (SMR) BAW filter is built, and a filter satisfying N78 band is simulated with co-simulation of modified Butterworth van Dyke (MBVD) model [7].

2. Resonator design

For a resonator, the effective electromechanical coupling coefficient (K_{eff}^2) is the main factor limiting the bandwidth of the BAW filter. The relationship between K_{eff}^2 and bandwidth can be estimated by equation 1 [7,8].

$$K_{eff}^2 > \frac{\pi^2}{4} \cdot \frac{BW}{f_0} \tag{1}$$

Where BW is the bandwidth of filter and f_0 is the center frequency. The N78 band has a center frequency of 3450 MHz with 8.7% bandwidth, which means the K_{eff}^2 of the piezoelectric layer should be larger than 21.75% according to equation 1. For this purpose, a 2D FEM model of LN film is built to acquire the electromechanical coupling coefficient and sound velocity under different cut angle, as well as the impedance characteristic. The cut plane to build the 2D-model is shown in figure 1.

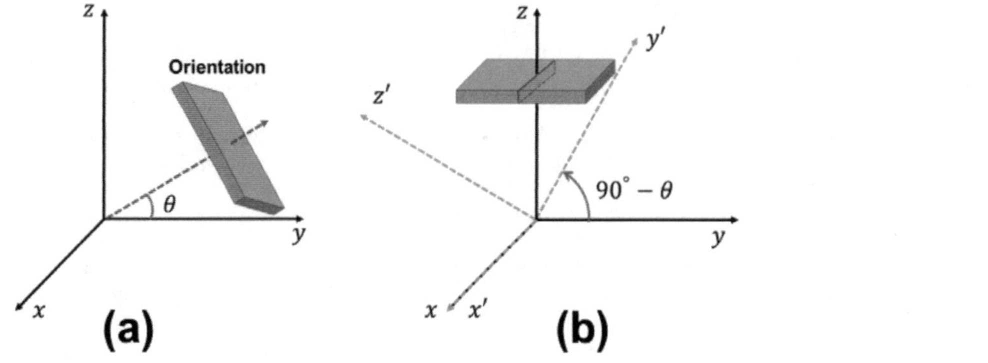

Figure 1. (a) Purpose orientation of θ angle Y cut LN, (b) rotation of material coordinate relative to geometry coordinate.

The purpose orientation is shown in figure 1(a), the z-axis is the thickness direction (figure 1(b)), the material coordinate is set as x'-y'-z' to keep the LN θ angle Y cut orientation. To get different cut angle, Euler angle (α, β, γ) is set to rotate material matrix of LN with z-x-z order. For example, a 30° Y cut LN wafer, the material coordinate system should be rotated 60° counterclockwise alone the x-axis if c-axis orientated LN is used as reference. According to this rule, The Euler angle is $(0,90° - \theta, 0)$ for θ angle Y cut LN wafer, and the material matrix of LN is transformed by equations below. The rotation matrix is denoted by R.

$$R = \begin{bmatrix} 1 & 0 & 0 \\ 0 & \cos\beta & \sin\beta \\ 0 & -\sin\beta & \cos\beta \end{bmatrix} \tag{2}$$

Where $\beta = 90° - \theta$,
Then the stress tensor transformation matrix is described as

$$[M] = \begin{bmatrix} 1 & 0 & 0 & 0 & 0 & 0 \\ 0 & \cos^2\beta & \sin^2\beta & -2\sin\beta\cos\beta & 0 & 0 \\ 0 & \sin^2\beta & \cos^2\beta & 2sin\beta cos\beta & 0 & 0 \\ 0 & \sin2\beta & -\sin2\beta & \cos^2\beta - \sin^2\beta & 0 & 0 \\ 0 & 0 & 0 & 0 & \cos\beta & \sin\beta \\ 0 & 0 & 0 & 0 & -\sin\beta & \cos\beta \end{bmatrix} \tag{3}$$

The relative permittivity matrix [ε], elastic matrix [c] and piezoelectric coupling matrix [e] can be transformed with equations below, respectively.

$$\varepsilon' = [R]\varepsilon[R]^{-1} \tag{4}$$

$$c' = [M]c[M]^t \tag{5}$$

$$e' = [R]e[M]^t \tag{6}$$

Angle θ is set from 0° to 180° during the simulation. As plotted in figure 2, in the region of 18°~ 44°, the impedance ratio (Zr) and K_t^2 are both high, which means these cut angles are suitable for preparing wide band filter. The electromechanical coupling coefficient is larger than 22% in the range of 18°~ 44°, and is larger than 26% between 25°~36°, which is high enough for the design of N78 band filter. Considering the design margin, finally the 36° Y cut LN is chosen as piezoelectric material.

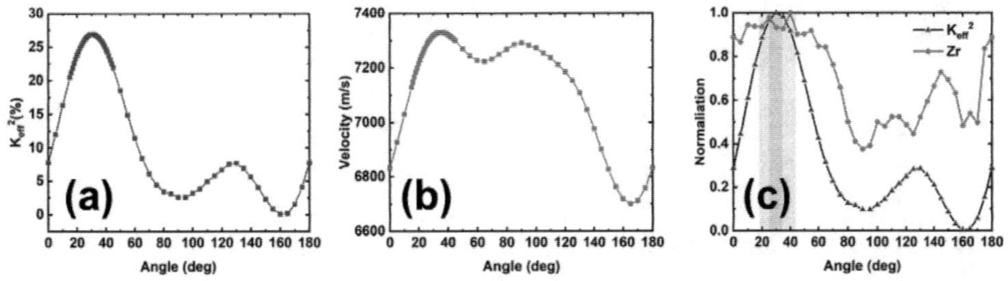

Figure 2. (a) Effective electromechanical coupling coefficient of LN film along different orientation; (b) longitudinal wave velocity along different orientation; (c) normalization K_{eff}^2 and Zr along different orientation.

For a SMR resonator, the reflector should be taken into consideration to get high Q-factor. Here, SiO_2 and Mo are used as low and high acoustic impedance materials, respectively, due to their large impedance ratio. The reflector is designed using the 1/4 λ principle at the center frequency of 3.45 GHz. FEM method is adopted to evaluate the reflection efficiency. As figure 3 shows, the transmissivity between 3.3 GHz to 3.6 GHz is less than -35 dB in a seven layers reflector consisting of SiO2/Mo, and the minimum point is located at 3.45 GHz. The reflection window is wide enough for covering the 300 MHz band of N78.

Then a model of SMR is built as illustrated in figure 4 (a). Al serves as the top and bottom electrodes. Seven layers of SiO_2 and Mo is alternately stacked between the bottom Al electrode and the Si substrate. In order to reduce the reflection from boundary, perfect match layer (PML) is set beside the bottom and lateral boundary to simulate half infinite zone. A step is set up at the edge of the top electrode to inhibit lateral energy leakage [9]. The bottom boundary and lateral boundary are set as fix boundary. A voltage of 1 V under different driving frequency is applied to the top electrode and the bottom electrode is set as ground.

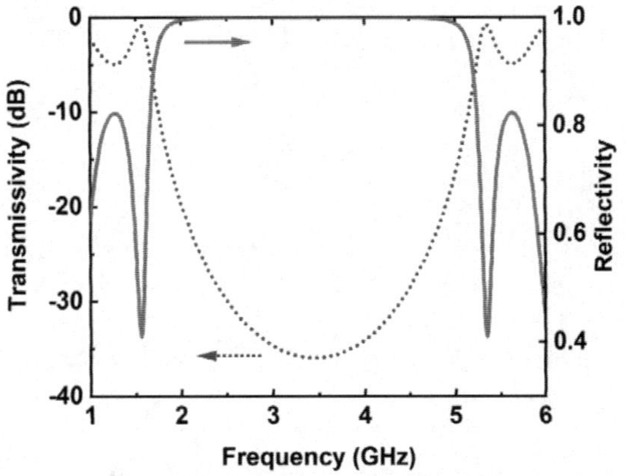

Figure 3. Transmissivity and reflectivity of seven layers SiO_2 and Mo combination.

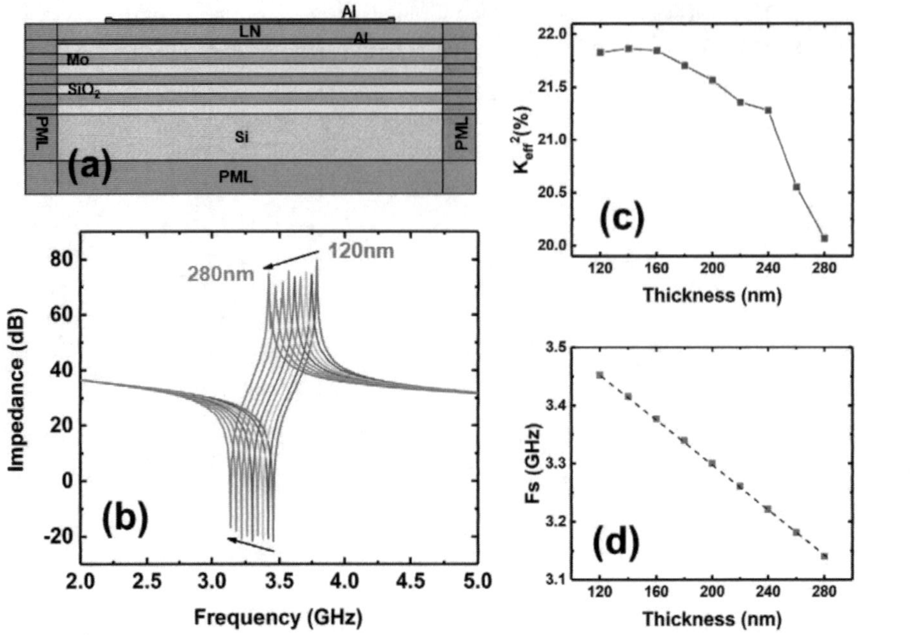

Figure 4. (a) 2D model of SMR; (b) impedance of resonators under different thickness of top Al electrode; (c) K_{eff}^2 under different top electrode thickness; (d) resonant frequency under different thickness of top Al electrode.

The initial thickness of the LN layer can be calculated from the velocity of longitudinal wave by d=v/(2f). the velocity of 36° Y cut is 7330 m/s, thus the calculated thickness of LN is 1.062 μm. However, the mass loading effect of the electrode reduces the resonant frequency, thus the thickness of LN should be thinner than initial thickness. Here, the bottom electrode is 110 nm, the LN is 800 nm, and the top electrode is 120 nm. Detail thickness is listed in table 1.

Table 1. Thickness parameters of SMR

Layer	Thickness	Unit
Al	120	nm
LN	800	nm
Al	110	nm
SiO_2	430	nm
Mo	450	nm
Si	500	μm

As figure 4(b) shows, there is only a pure longitude mode in the resonator when top Al electrode is 120 nm. The resonant frequency is 3452 MHz, and anti-resonant frequency is 3787 MHz. The corresponding K_{eff}^2 is 21.8%. The top Al electrode is used as mass load to adjust resonant frequency. To further design the filter, resonators with upper electrode thickness between 120 and 280 are simulated. The resonant frequency decreases almost linearly by increasing the electrode thickness as shown in figure 4(d). The resonant frequency decreases by 1.95 MHz when the electrode thickness increases by 1 nm, which is convenient for filter design. But the Zr decreases with thickness increase, and parasitic mode appears between 240 nm to 280 nm (figure 4(c)). Meanwhile, the K_{eff}^2 also changes with electrode thickness, a relatively large K_{eff}^2 is obtained when the electrode thickness is near 140 nm.

3. Filter design

The filter is designed as a ladder type topology. The model of the filter is cascade by resonator model.

Since the resonators are separated, the potential of the electrodes that need to be connected is set to be the same in order to realize the electrical connection between different resonators. The area of resonator and thickness of electrodes are variables that need optimization. There are three goals for the designed filter. (a) bandwidth covers the 3.3 GHz ~ 3.6 GHz. (b) Attenuation big than 40dB outside the passband. (c) Insertion loss less than 2 dB.

Since the FEM method need considerable computing resources, modified Butterworth van Dyke (MBVD) model is used to optimize these parameters. The MBVD parameters of simulated resonators are extracted by fitting method. These parameters are basic design parameters for filter optimization. Before optimizing, relationship between resonator variables and MBVD parameters need be built up. From the expression for MBVD equivalent circuit [10], brief equations $C_0 \propto A$, $C_m \propto A$ and $L_m \propto (a \cdot d_{top} + b)^2/A$ can be derived for a resonator with specified thickness. Where a and b are the coefficient between f_s and top electrode thickness d_{top}, and A is the area of resonators. Here, the influence of each parameter is simplified, and other parameters are assumed to be unaffected when adjusted in a small norm dimension.

The d_{top} and A can be set as variables at MBVD model, and the most common ladder type topological structure is adopted. The final filter is 11 order with 6 series resonators and 5 parallel resonators as shown in figure 5(a). Combined with simulated annealing algorithm, the optimized results are listed in table 2.

Table 2. Parameters of eleventh order filter

Resonator number	Area(um^2)	Top electrode thickness(nm)
Series 1	2503.8	140
Series 2	1468.1	140
Series 3	1551	140
Series 4	1472.1	140
Series 5	1418.8	140
Series 6	2452.9	140
Parallel 1	2630.1	270
Parallel 2	2123.8	270
Parallel 3	2148.6	270
Parallel 4	2176.2	270
Parallel 5	2627.3	270

To further verify the design, these parameters are reused in FEM filter model. As figure 5(b) shows, a simulated S21 of the N78 bandpass filter is obtained. The max insertion loss in the pass band is 1.92dB, while the center insertion loss is 1.24 dB. In the region of 3280 MHz to 3632 MHz, the insertion loss is less than 2 dB. The attenuation from 2000 MHz to 3184 MHz is more than 40 dB. From 3690 MHz to 5000 MHz, the attenuation is more than 40 dB. The overall stopband attenuation is about 46 dB.

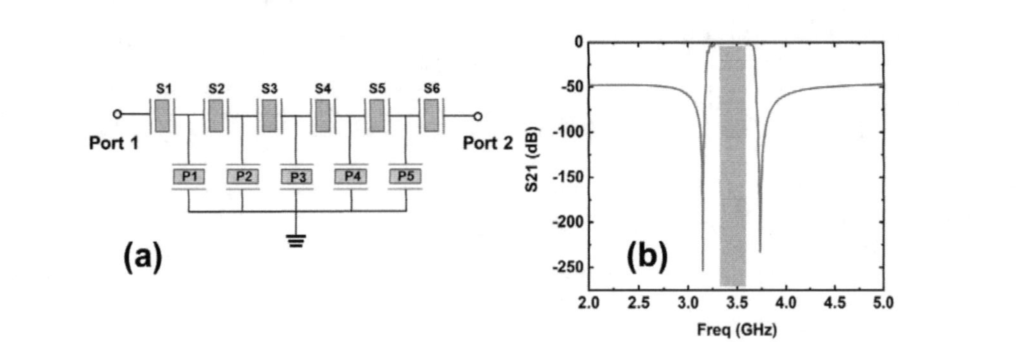

Figure 5. (a) Topological structure of 11th order filter; (b) S21 curve of 11th order filter simulated by FEM.

4. Conclusion

In this work, FEM simulation model of materials and devices is established. Based on the simulation model, it is found that 25°~36° Y cut LN can have large K_{eff}^2 and Zr simultaneously, which is suitable for wideband filter design. 36° Y cut LN is chosen to design filter for N78 band. A SMR type filter with relative bandwidth 10.1% is designed with FEM method and MBVD model co-simulated, which can meet the requirements of N78 band. This work is of great significance to solve the large bandwidth requirements of 5G communication in the future.

Acknowledgements

This work is supported by the China National Key Research and Development Plan Project(2017YFB0406402), the National Natural Science Foundation of China(No. 51772044).

Reference

[1] Volatier A, Fattinger G, Dumont F, Stoyanov P and Aigner R 2013 Technology enhancements for high performance BAW duplexer *2013 IEEE International Ultrasonics Symposium (IUS)* 2013 IEEE International Ultrasonics Symposium (IUS) pp 761–4

[2] Sun J-H and Yu Y-H 2016 A SAW filter using SiO2/LiNbO3 layered-structure phononic crystals *2016 IEEE International Frequency Control Symposium (IFCS)* 2016 IEEE International Frequency Control Symposium (IFCS) pp 1–3

[3] Shealy J B, Vetury R, Gibb S R, Hodge M D, Patel P, McLain M A, Feldman A Yu, Boomgarden M D, Lewis M P, Hosse B and Holden R 2017 Low loss, 3.7GHz wideband BAW filters, using high power single crystal AlN-on-SiC resonators *2017 IEEE MTT-S International Microwave Symposium, IMS 2017, June 4, 2017 - June 9, 2017* IEEE MTT-S International Microwave Symposium Digest vol 0 (Honolulu, HI, United states: Institute of Electrical and Electronics Engineers Inc.) pp 1476–9

[4] Vetury R, Kim D, Fallon K, Kwon J, Winters M, Gibb S R, Patel P, Hodge M D, Grundman E, McLain M A, Shen Y, Houlden R and Shealy J B 2018 Low loss, wideband 5.2GHz BAW RF filters using single crystal AlN resonators on silicon substrates *2018 International Conference on Compound Semiconductor Manufacturing Technology, CS MANTECH 2018, May 7, 2018 - May 10, 2018* CS MANTECH 2018 - 2018 International Conference on Compound Semiconductor Manufacturing Technology (Austin, TX, United states: CS Mantech) p et al.; Plasma-Therm, LLC; Revasum; Virginia Diodes, Inc.; WIN Semiconductors Corp.; Wolfspeed, A Cree Company

[5] Wang J, Zheng Y and Ansari A 2021 Ferroelectric Aluminum Scandium Nitride Thin Film Bulk Acoustic Resonators with Polarization-Dependent Operating States *Phys. Status Solidi - Rapid Res. Lett.* **15** 2100034

[6] Bai X, Shuai Y, Luo W, Wu C and Zhang W 2020 The Simulation of Resonant Mode and Effective Electromechanical Coupling Coefficient of Lithium Niobate Crystal with Different Orientations *J. Phys. Conf. Ser.* **1637** 012064

[7] Gevorgian S S, Tagantsev A K and Vorobiev A K 2013 *Tuneable film bulk acoustic wave resonators* (London: Springer) p 68

[8] Jia L, Gao Y and Han C 2017 Design of a bulk acoustic wave filter for wi-fi band *AOPC 2017: Optical Sensing and Imaging Technology and Applications* Optical Sensing and Imaging Technology and Applications ed Y Jiang, H Gong, W Chen and J Li (Beijing, China: SPIE) p 64

[9] Fattinger G G, Marksteiner S, Kaitila J and Aigner R 2005 Optimization of acoustic dispersion for high performance thin film BAW resonators *IEEE Ultrasonics Symposium* IEEE Ultrasonics Symposium **2** 1175–8

[10] Reddy P R and Mohan B C 2012 Design and Analysis of Film Bulk Acoustic Resonator (FBAR) Filter for RF Applications *International Journal of Engineering Business Management* January-2012

Modeling and Analysis of Spike-Train Based on Poisson-GML

Yuanpeng Li[1], Wenge Yang[2] ,Yile Fan[1], Yu Zhang[1,*]

[1]Space Engineering University, Beijing, China
[2]Department of Electronics and Optics, Space Engineering University, Beijing, China

[*]corresponding author's e-mail: fanyile028@qq.com

Abstract. Mathematical modeling is of great significance to the study of brain function. In this paper, an analysis method based on Poisson's general linear model is adopted to model the brain impulse sequence. First, the ideas of Hodegkin-Hulex model, STA and Poisson's general linear model are introduced. Secondly, the implementation of the general linear model is introduced. Finally, Poisson GLM Fit Spike-train is simulated and compared with Linear Gaussian-GLM. The AIC tradeoff estimation model fits well. The superiority of Poisson GLM Fit Spike-Train is demonstrated

1. Introduction

With the development of brain neuroscience, scholars believe that neuron activity pulse signals are of great significance in analysis. It is helpful to analyze the interaction between brain molecules. In the past research process, scholars have adopted various and rich modeling methods. One of the analysis methods is based on the mechanical thought of various formulas, and the other is based on the thinking method of statistics.[1]

In an organism, each neuron is connected to many other neurons through synapses, and the whole neuron forms a large network. Neurons in the case of stimulation will produce reflection, stimulus can be physical in nature, such as to stimulate the light of retinal neurons, or used to stimulate the auditory cortex neural activity, also can be abstract or cognitive, such as in working memory task, it will be in front of the hippocampus and frontal cortex to stimulate neural activity. After the stimulation, the membrane potential of nerve cells will change. It is believed that neurons transmit signals mainly through the information content of spiketrain.

After obtaining the original signals from neurons, researchers can use this method to analyze the issuing characteristics of impulsive nerve signals from different brain regions when the organism completes a specific task, and build corresponding mathematical statistical models to qualitatively or quantitatively analyze the functional maps of different brain regions.

It is a common method for researchers to analyze and study brain function qualitatively and quantitatively by using mathematical statistical model to model neuron impulse signals of organisms. Characterizing the characteristics of neuronal impulse signals through the mathematical characteristics of the model and verifying the reasonable assumptions made by these characteristics can help researchers deepen their cognition of the functional connections between different types of neurons in the brain and reveal the functional connections among neurons [2,3,4].Exploring the hidden information inside the neuronal pulse signals [5,6,7,8] can also build the corresponding neural prosthesis according to the pulse signal issuing characteristics of specific brain regions [9], which can be used to make up for the functional loss or abnormality of specific brain regions caused by brain damage and other reasons. With the continuous development of microelectrode array technology and the continuous improvement in the spatial and temporal accuracy of the original data of neuron impulse signals collected by related instruments, it has become very common to use data to drive the construction of statistical models, so as to assist researchers to complete the exploration of functional connections between neurons in different brain regions.

In this paper, the probabilistic model of Poisson process is established and the statistical characteristics are analyzed by combining the knowledge of random process point process.The difference between the predicted pulse emission probability sequence and the real pulse emission sequence was compared by mathematical statistics, and the modeling quality of the model was further improved by using the corresponding mathematical statistics.

2. Model analysis

2.1. The simulation model

Hodgkin-Huxley model [10] belongs to one of the models with high imitation degree.A class of neuron simulation models represented by this model attempt to model the neuron signal from the Angle of voltage and current inside and outside the neuron cell, so as to simulate the dynamic characteristics of the level inside and outside the cell before and after the neuron cell sends the pulse signal.

In terms of implementation, Hodgkin-Huxley model uses a set of nonlinear partial differential equations to simulate a series of electrophysiological characteristics of neuron cells before and after sending pulse signals. Therefore, the model itself has the ability to simulate the sending and spreading process of impulse signals. However, in the theoretical research of the encoding and decoding field of neural impulse signals, the flexibility of Hodgkin-Huxley model in practical application is slightly insufficient due to the large amount of calculation and the complexity of the model, which also attracts many scholars to make some modifications on the basis of Hodgkin-Huxley model for further study.

2.2. Linear model

This article framework based on Poisson model adopted by the general linear model (Poisson Generalized Linear Model, Poisson GLM), [11,12,13,14].Thought is cut along the time axis pulse signal is divided into several isometric time slot, the timeline of each time slot contains the pulse signal number to poisson distribution model for modeling, the different time slots of the poisson model are independent of each other, so that the whole pulse sequence can be according to the non homogeneous poisson point process (Inhomogeneous Poisson Point Process) to model it. Compared with the traditional STA and STC models, the Poisson-GLM model has no special assumptions on the distribution of input signals received by neurons, and the corresponding parameters of Poisson-GLM model in each time slot on the non-uniform Poisson point process can be directly related to the independent variables of the model through GLM.GLM is a general term for a class of methods. According to the typical link function selected in GLM, the corresponding concrete model will have different properties.

GLM itself is a special case of linear-nonlinear Poisson framework (LNP framework for short). Its idea is to expand the input variable into a vector and carry out convolution operation with the parameter vector of GLM model itself. The convolution result obtained is then combined with the typical link function of GLM in each time slot of the non-uniform Poisson point process. The coefficients corresponding to the Poisson model are connected, and this coefficient is the expected probability value of pulse signal in this time slot, so as to complete the operation of the nonlinear part of the LNP framework.

Compared with other models, the advantage of Poisson-GLM model is that it can directly introduce any input independent variable into the model itself, and use the nonlinear mapping inside the GLM model to complete the nonlinear simulation of the nervous system. According to the difference of typical link function, GLM model can simulate a variety of nonlinear characteristics of impulse neural signals. In addition, the model itself has very good interpretability to the input variables, each independent variable has a certain physical interpretation of the meaning, the background of this feature in the biomedical neurons modeling task is particularly important, because the ultimate goal of neuron model is to explain the pulse signal of the connections between neurons, so the model good explanatory Gui help to research Scholars use models to generalize the relationship between input and output. Aimed at the upper limit of Poisson-GLM's ability to express neuron characteristics, Alison et al. [15] used Poisson-GLM model to simulate several impulse signaling characteristics of neurons in real situations, which demonstrated the representational ability of Poisson-GLM model to a certain extent.

2.3. STA

The main idea of STA is to solve the optimal projection direction by taking the mean value of the input signal variables corresponding to the pulse signal.

Specifically, STA sets a time window on the input variable sequence for the neuron impulse signal in any time slot on the time axis, and intercepts the input independent variables related to the impulse signal corresponding to the current time slot through this time window. Then, STA extracts the input variables corresponding to the time slot of pulse signal on the time axis, sums the extracted input variables and takes the mean value, and the final result is the projection direction obtained by STA.

3. Poisson's general linear model

The general Linear model (GLM) is a kind of flexible neural impulse signal modeling model. The processing of the input signal can be summarized into three steps, namely Linear mapping, Nonlinear transformation and constructing the impulse sending model, which is linear-nonlinear -PoissA special case of the ON cascading model (LNP cascading model).

Due to the different transformation functions selected for nonlinear transformation, GLM can be subdivided into several different categories, and these different categories of GLM correspond to different types of modeling frameworks for pulse emission models in each time slot of the pulse signal point process. Taking Poisson-GLM as an example, the pulse emission model in a single time slot of Δ the pulse signal point process is modeled by Poisson model. The Poisson models contained in different time slots are independent of each other, so the whole point process sequence can be regarded as a non-uniform Poisson point process.

Specifically, Poisson-GLM measures the probability $P(n(t))$ of the number of impulses emitted in the r time slot of the neuron impulse signal:

$$p(n(t)) = e^{-\lambda(t|x(t))\Delta} * (\lambda(t|x(t))\Delta)^{n(t)}/n(t)! \quad (1)$$

$\lambda(t|x(t))$ is Poisson GLM model of neuron pulse sequence in t moment corresponding conditional probability density function, $x(t)$ includes all independent variables involved in the model calculation, Δ time slot is the length of the window, in a single time slot $\lambda(t|x(t))$ can be regarded as a constant constants. $\lambda(t|x(t))$ formally defined as follows:

$$\lambda(t|x(t)) = \lim_{\Delta\to 0} P(N(t+\Delta) - N(t) = 1|x(t))/\Delta \quad (2)$$

N(t) represents the sum of the number of impulses sent by the neuron from the first time slot to the t-th time slot.

Under the assumptions of poisson, $\lambda(t|x(t))$ value is the model of neurons t a time slot in the first issue of the expected value of pulse number.In order to combine the parameters in the Poisson model; $\lambda(t|x(t))$ and the neurons of the independent variable x(t), at this moment, in different typical response function can be achieved by both. If the exponential function is selected to connect the two, the relationship between them is as follows:

$$\lambda(t)\Delta = e^{x(t)^T\theta} \quad (3)$$

Combined with the above formulas, the logarithm likelihood function of Poisson-GLM at a given time t can be obtained

$$logP(N(t)) =- \lambda(t)\Delta + n(t)log(\lambda(t)\Delta) - logn(t)! \quad (4)$$

$$logP(N(t)) =- e^{x(t)^T\theta} + n(t)(x(t)^T\theta) + const \quad (5)$$

Const represents a constant independent of the independent variable of the objective function, and the joint likelihood function is calculated for the whole D formed by the Poisson model on all time slots of the neuronal impulse signal point process, with the form as follows:

$$logP(D) = \sum_t (- e^{x(t)^T\theta} + n(t) * (x(t)^T\, \theta)) + const \quad (6)$$

In order to use the gradient descent method to optimize the solution, its derivative with respect to θ should be calculated as:

$$\frac{\partial P(D)}{\partial \theta} = \sum_t (n(t) -- e^{x(t)^T\theta} x(t)) \quad (7)$$

After obtaining the derivative information of the model, we can use all kinds of gradient descent methods in optimization calculation to solve the optimization parameters of the objective function of the model.

4.The experimental simulation

Data source: Uzzell & Chichilnisky 2004 spike train produced by retinal cells stimulated by white noise.

First, the raw data is loaded, and the following diagram is obtained by drawing:

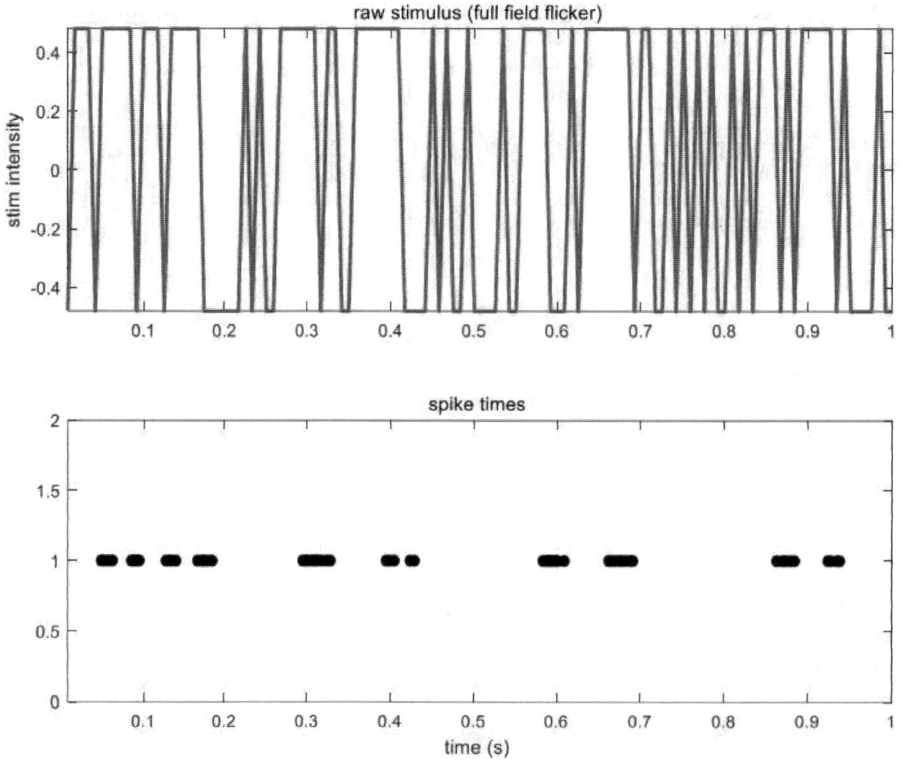

Figure 1. Uzzell & Chichilnisky 2004 spike train produced by retinal cells stimulated by white noise.

The data set contains 49954 spikes, and the original data is separated into the same size to obtain the following pulse sequence data:

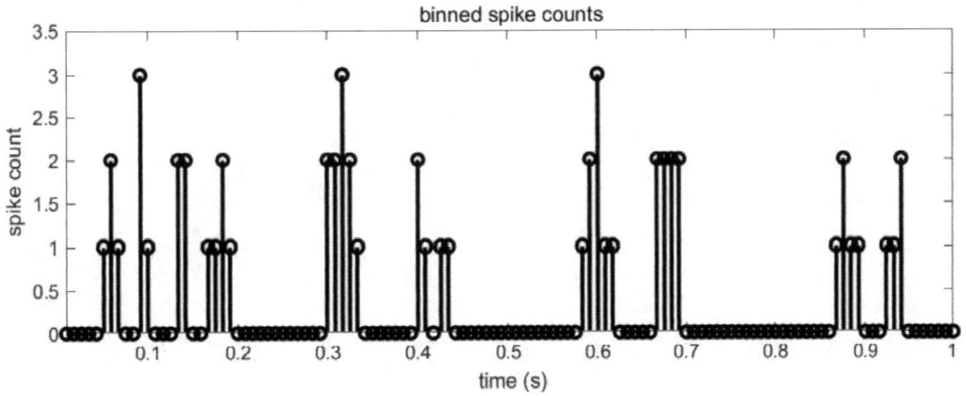

Figure 2. binned spike counts

To calculate STA

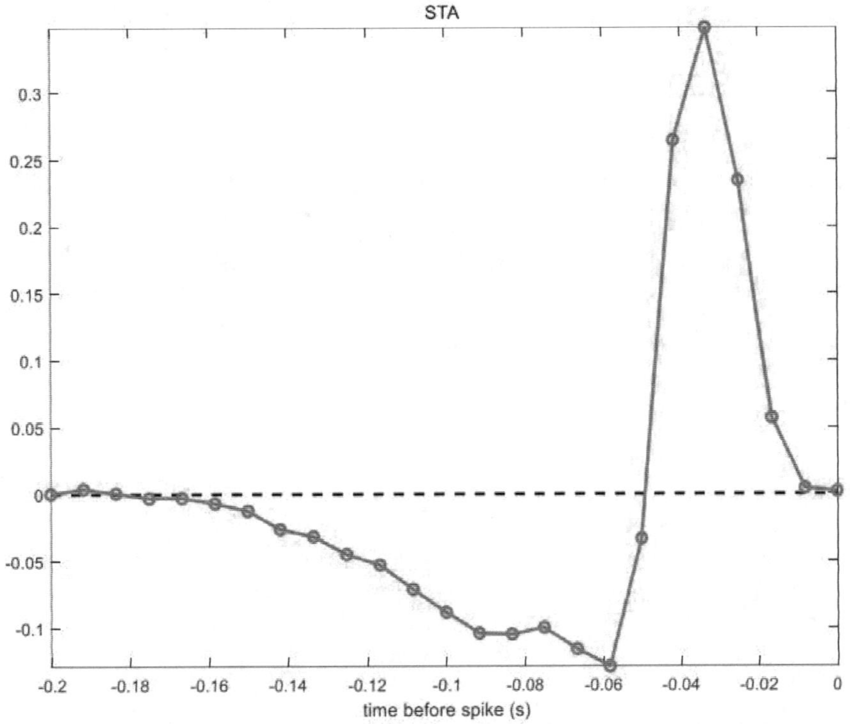

Figure 3. STA

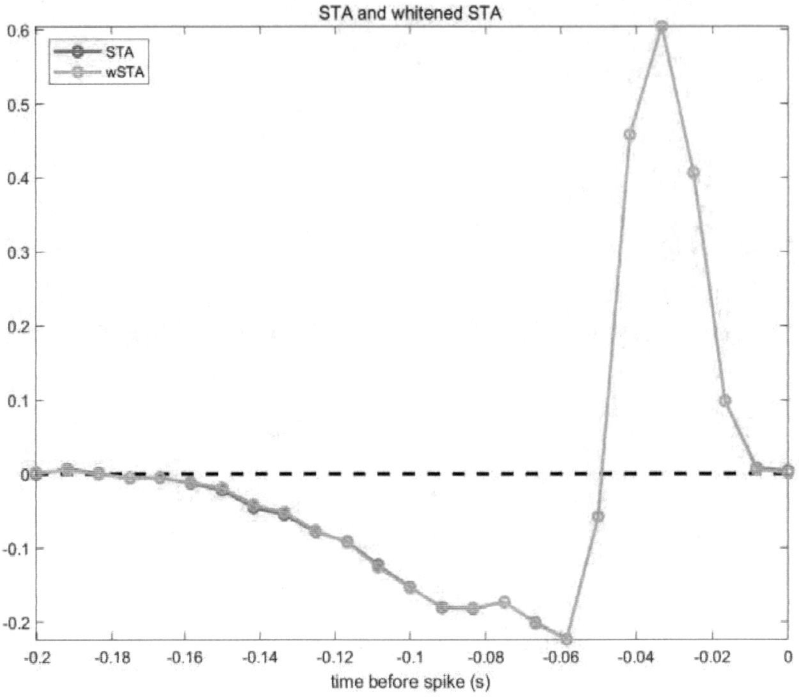

Figure 4. STA and whitened STA

Comparing with Poisson GLM fit spikes-train and linear-Gaussian-GLM:

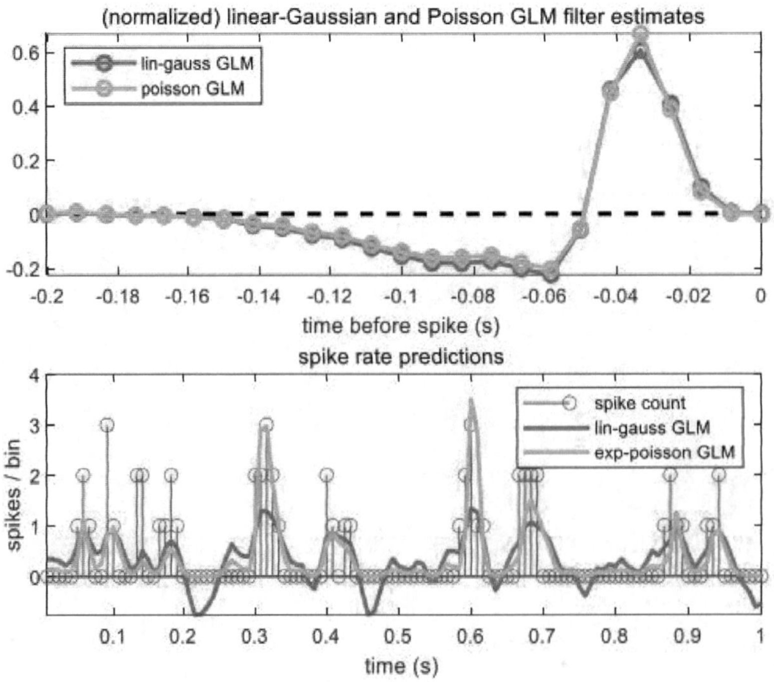

Figure 5.(a) (normalized) linear-Gaussian and Poisson GLM filter estimates

Figure 5.(b) spike rate predictions

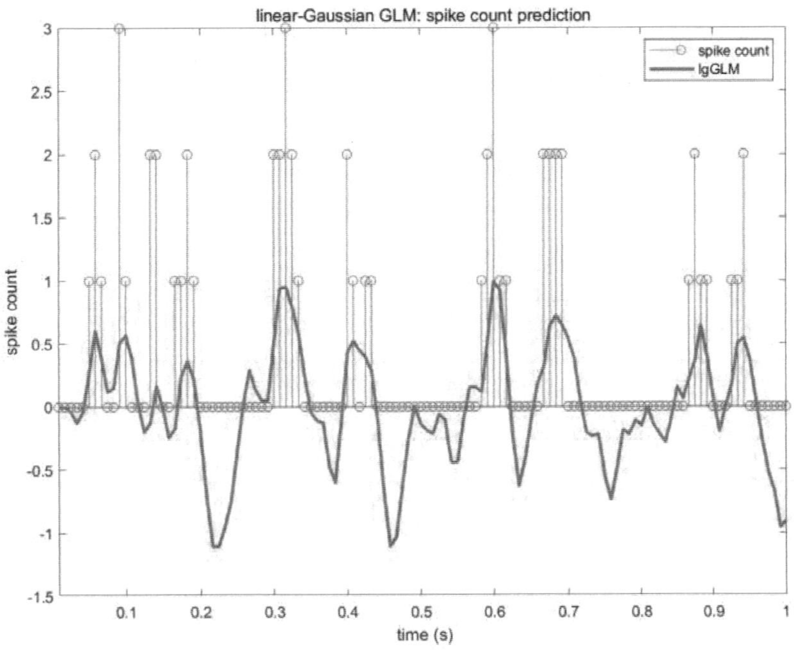

Figure 6. linear-Gaussian GLM: spike count prediction

5. Conclusion

This paper uses Uzzell & Chichilnisky 2004 retinal cells in white noise An analysis method based on Poisson's general linear model is established, and Poisson GLM Fit Spike-Train is used to compare with Linear Gaussian-GLM. AIC tradeoff estimation model fits optimally. We demonstrate the superiority of Poisson GLM Fit Spike-train and its high fitting degree to Spike Train.

Reference

[1] Computational Neuroscience: Mathematical and Statistical Perspectives
[2] Leutgeb S, Leutgeb JK,Treves A,et al Distinct ensemble codes in hippocampal areas CA3 and CA1[J] Science 2004,305(5688):1295-1298
[3] Pillow JW,Shlens J,Paninski L,,et al. Spatio-temporal correlations and visual signalling in a complete neuronal population[J].Nature,2008,4 54(707):995-999
[4] So K,Koralek A C, G anguly K, et al. Assessing functional connectivity of neural ensembles using directed information[J]. Journal of neural engine ering,2012,9(2):026004.
[5] Glaser C, Joublin F. Firingrate homeostasis for dynamic neural field formation[J].IEEE Transactions on Autonomous Mental Development,2011,3(4):285-299
[6] LiC, LiY. A spike-based model of neuronal intrinsic plasticity[J],IEEE Trans actions on Autonomous Mental Development, 2013, 5(1): 62-73.
[7] Miao J,Qing L,Zou B,et al. Top-down gaze movement control in target s earch using population cell coding of visual context[J]. IEEE Transaction s on Autonomous Mental Development, 2010,2(3): 196-215.
[8] VellisteM Perel S, Spalding M C, et al Cortical control of a prosthetic arm for selffeedind [J]. Nature,2008,453(7198): 1098-1101.
[9] Song D, Chan R H M, Marmarelis V Z, et al .Non-linear dynamic model ing of spike train transformations for hippocampal-cortical prosthes es[J]. IEEE Transactions on Biomedical Engineering, 2007, 5 4(6): 1053-1066.
[10] Hodgkin A L, Huxley A F. Aquantitative description of membrane current and its application to conduction and excitation in nerve[J]. The Journal of physiology, 1952, 1 17(4)500.
[11] Lawhem V, Wu W, Hat sopoulos Net al. Population de coding of motor cortical activity using a

generalized linear model with hidden states[J]. Journal of neuro science methods, 2010,1 89(2): 267-2 80.
[12] Panins ki L. Maximum likelihood estimation of cascade point-process neural encoding models [J]. Network: Computation in Neural Systems, 2004,1 5(4): 243- 262.
[13] ParkI M, Meister MLR, Huk AC, et al. Encoding and decoding in parietal cortex during sensorimotor decision-making[J]. Nature neuroscience, 20 14, 17(10)1395- 1403.6
[14] Truccolo W,Eden U T, Fellows MR et al. A point process framework for elating neural spiking activity to spiking history, neural ensemble, and ex trinsic covariate efifects[J]. Journal of neurophysiology, 2005, 93(2): 1 0 74-108 9.
[15] Weber A I, Pillow J W. Capturing the dynamic al repertoire of s ingle neurons with generalized linear models[J]. arXiv preprint arXiv: 1602.073 89, 2016

Discrete Population Competition Model of Poisonous Weed Invasion and Spatial Distribution

Hua Liu[1,*], Yexuan Li[1], Yumei Wei[2], Ming MA[1] and Gang Ma[1]

[1]School of Mathematics and Computer Science, Northwest Minzu University, LanZhou, GanSu, China

[2]Department of Experimental Teaching, Northwest Minzu University, LanZhou, GanSu, China

[*]corresponding author's e-mail: 7783360@qq.com

Abstract. Poisonous weeds invade ecosystems and compete with edible forage grassland, changing the population spatial distribution. In this paper, a discrete population competition model for the interaction between poisonous weeds and edible forage grass is established, and a diffusion coefficient to describe the invasion and diffusion effect of poisonous weeds is introduced. By using cellular automata methods, the influence of invasive diffusion of poisonous weeds on the spatial distribution of poisonous weed and edible forage populations is simulated. These findings indicated that the invasion of poisonous weeds affected the ecological distribution of grassland, and that the spatial distribution of poisonous weeds also changed from initial uniform distribution to aggregated distribution.

1. Introduction

In recent years, the spread of poisonous weeds has caused degeneration of many types of grasslands [1]. Studying the control of the breeding and spreading of poisonous weeds through mathematical modelling is of increasing interest to theoretical ecologists and mathematical biologists due to the rich dynamics and wide applications of this platform. When two species grow independently, the population is hampered by environmental resources, so the species follow a logistical growth pattern. However, in changeable ecological environments, many species are often found in the same habitats. Therefore, when poisonous weeds and edible grasses inhabit the same pasture, the resources consumed by the weeds can also affect the growth of edible grasses. In this context, the Lotka-Volterra model of edible forage and poisonous weed growth interaction was applied, resulting in a continuous model [2]. Mathematical biologists also studied other continuous-time systems and stochastic models and obtained relevant results [3,4].

The reproduction cycles of poisonous weeds and edible grasses do not overlap and belong to discrete species. Therefore, it is necessary establish a discrete competitive model of poisonous weeds and edible grasses. The simplest population growth model is the Malthus model, with the form $\frac{dx(t)}{dt}=rx(t)$, and its initial condition is $x(0)=x_0$. The solution to this equation is $x(t)=x_0\exp(r)$. This equation is transformed into a discrete iterative equation as $x(t+1)=x(t)\exp(r)$. Considering population growth consistent with logistical growth, the discrete population growth model is $x(t+1)=x(t)\exp(r-x(t)/K)$. Therefore, a discrete competitive model of poisonous weeds and edible grasses is established as follows [5]:

$$\begin{cases} x(t+1) = x(t)\exp\left(r_1 - \frac{r_1}{K_1}\left(x(t)+\alpha y(t)\right)\right), \\ y(t+1) = y(t)\exp\left(r_2 - \frac{r_2}{K_2}\left(y(t)+\beta x(t)\right)\right), \end{cases} \tag{1}$$

Where $x(t)$ is the edible grass population size in generation t, $y(t)$ is the poisonous weed population in generation t, r_1 is the growth rate of edible grass, r_2 is the growth rate of poisonous weeds, K_1 and K_2 are the carrying capacities of edible forage and poisonous weeds, respectively, in the environment, α is the competitive coefficient of poisonous weeds to edible grass, and β is the competitive coefficient of edible grass to poisonous weeds.

Poisonous weeds are an exotic species that compete with edible forage in distinct patches, and migrate from dense patches to low-density patches during growth. Moreover, considering that poisonous weeds are invasive species, toxic weeds are thought to have proliferative capacity during the invasion process, while edible grasses, as local species, do not have the ability to diffuse in space. In light of these considerations, a spatial diffusion model for the invasion of single patches of weeds was established as follows:

$$\begin{cases} x(t+1)_{(i,j)} = x(t)_{(i,j)} \exp\left[r_1 - \frac{r_1}{K_1}\left(x(t)_{(i,j)} + \alpha y(t)_{(i,j)}\right)\right], \\ y(t+1)_{(i,j)} = y(t)_{(i,j)} \exp\left[r_2 - \frac{r_2}{K_2}\left(y(t)_{(i,j)} + \beta x(t)_{(i,j)}\right)\right] + \mu(\sum y_{(i',j')} - 4y_{(i,j)}), \end{cases} \tag{2}$$

where $i' \in \{i+1, i-1\}$, $j' \in \{j+1, j-1\}$, and $\mu(\sum y_{(i',j')} - 4y_{(i,j)})$ is the diffusion effect of the invasive population.

The second formula in (2) is a space diffusion model for the (i,j) th patch of poisonous weed invasion. The model is designed using Neumann cellular automata, wherein each cell (patch) has four neighbors, as shown in Fig. 1, in which the dark color represents the cell and the light color represents its neighbors. $y_{(i,j)}$ is the number of organisms in the (i,j) patch in a two-dimensional mesh. The numbers of neighbors of the patch are $y_{(i-1,j)}$ (upper neighbor), $y_{(i+1,j)}$ (lower neighbor), $y_{(i,j\text{-}1)}$ (left neighbor), and $y_{(i,j+1)}$ (right neighbor). Plant species invade and spread from patches with a high number of biomass to those with a low biomass. The difference in the number of biological quantities between the (i,j) patch and the adjacent patches is $y_{(i-1,j)} - y_{(i,j)}$ (taking the left-hand patch as an example). In the model, $\sum y_{(i',j')} - 4y_{(i,j)}$ reflects the sum of the invasion and diffusion effects of the four neighbors of the (i,j) patch, and μ is the diffusion coefficient [6].

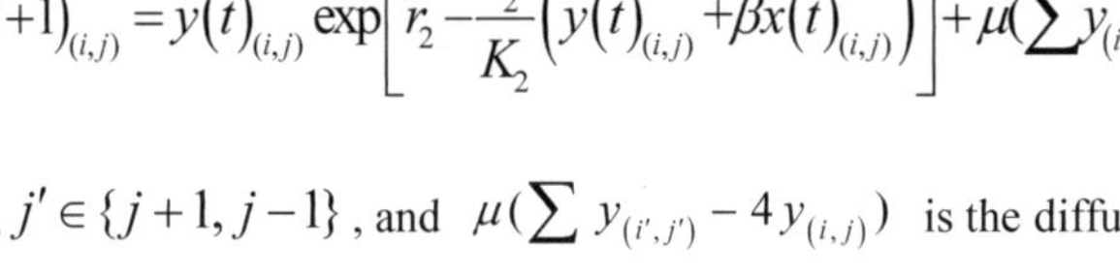

Figure 1. Neumann cellular automata diagram

2. Numerical simulation

In the two-dimensional 100×100 grid plane, the population dynamics of poisonous weeds and edible herbages are simulated, and the time scale t is the unit time step.

At the initial time, the numbers of edible grasses and poisonous weeds were randomly generated, and the number of poisonous weeds in single patches was assumed to be no greater than the number of edible grasses. In two-dimensional grids, the species is migrated between the grids, assuming that the species migrate only in the Neumann neighborhoods, which only considers the migration of four patches adjacent to each patch. Migration occurs from high density to low density, with a migration rate of 0.1 times the density difference.

Assuming that $r_1 = 0.1$, $r_2 = 0.12$, $\beta = 0.5$, and $\alpha = 0.7$, and that the survival ability of poisonous weeds is slightly stronger than that of edible grasses, the maximum environmental capacity of K_1 is 30, and that of K_2 is 40. Finally, Neumann-type cellular automata were used to simulate the programming using MatLab software (MathWorks, USA). The spatial distribution of poisonous weeds and edible grasses is represented as follows in Figs. 2–5, where white denotes survival and black represents extinction.

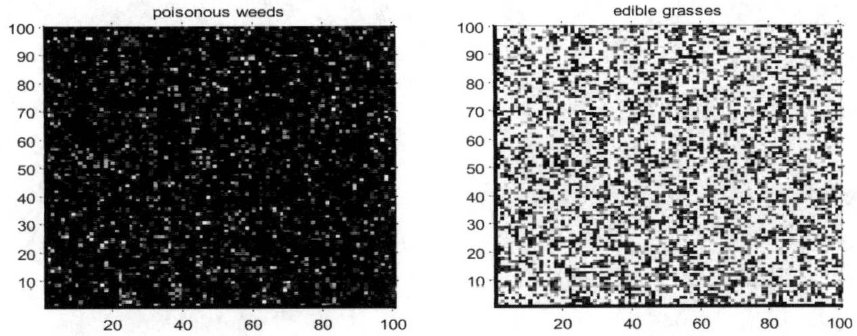

Figure 2. Spatial distribution of 10th-generation poisonous weeds and edible pasture

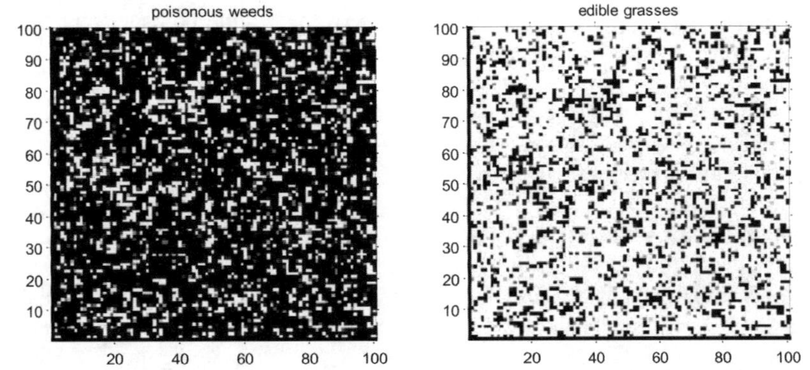

Figure 3. Spatial distribution of 40th-generation poisonous weeds and edible pasture

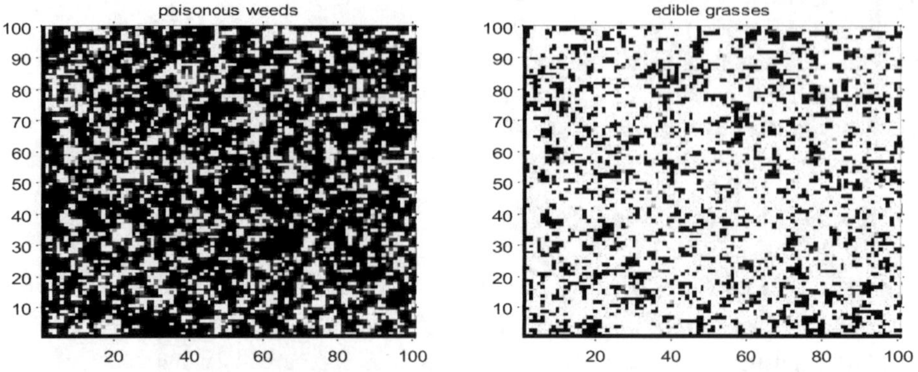

Figure 4. Spatial distribution of 60th-generation poisonous weeds and edible pasture

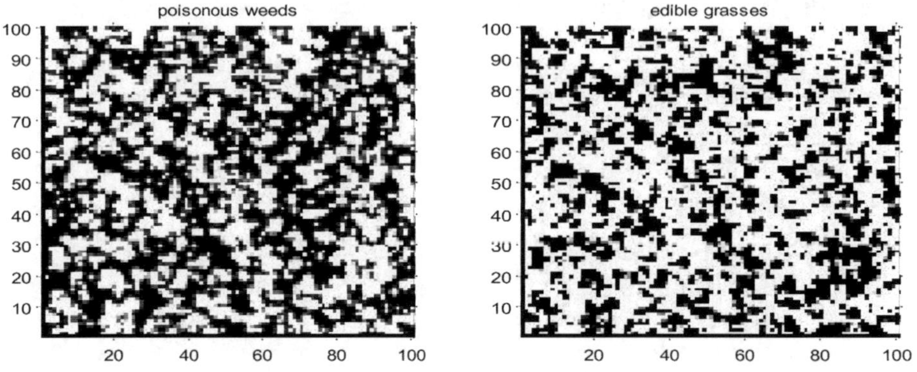

Figure 5. Spatial distribution of 80th-generation poisonous weeds and edible pasture

By observing the spatial distribution of poisonous weeds and edible herbage, one can find, by the color of poisonous weed images from the 10th to the 80th generation, that white has been increasing and black has been decreasing, showing the growth trend of poisonous weeds intuitively. The edible herbage image color of the 10th generation shows less white and is mostly black up to the 60th generation, while white gradually increased and black gradually decreased. From the 60th to 80th generations, black is quite clustering.

3. Test of the spatial distribution hypothesis

Spatial distribution structure is a basic characteristic of a population. The spatial distribution structure of poisonous weed invasion was studied, which has practical application value in understanding poisonous weeds and controlling their breeding and spreading. The distribution types of population spatial structures can be divided into three types: random, aggregation, and uniform distribution [7].

The hypothesis of spatial distribution structure was tested using data of the 80th generation of poisonous weeds as an example. Using the computer simulation data and 80th generation of data testing, the David-Moore method, Iwao $\overset{*}{m}-m$ regression method, average congestion index Lloyd M, and other methods were used. The results demonstrated that poisonous weeds invade edible herbage, and with strong competitive ability, the spatial distribution pattern gradually evolves into the final aggregation distribution from the initial uniform distribution. The data were statistically analyzed and tested as follows.

3.1. Index of Dispersion Test

We define an index of dispersion C to be

$$C = \frac{S^2}{\overline{x}} = \frac{\sum (x_i - \overline{x})^2}{\overline{x} \cdot (n-1)} = 17.0984 \cdot \tag{3}$$

For the theoretical Poisson distribution, the variance equals the mean, so the expected value of C is always 1.0 in a Poisson world. Therefore, if the C is close to 1, the distribution type is a random. Using the data of 80th generation of poisonous weeds, we calculate $C = 17.0984$. The confidence interval for C obtained with a probability of significance of 90% is

$$1 \pm 2\sqrt{\frac{2n}{(n-1)^2}} = 1 \pm 2\sqrt{\frac{20000}{99980001}} = 1 \pm 0.028287 , \tag{4}$$

and

$$17.0984 > 1.028287 \ , \ 17.0984 > 0.9717129 . \tag{5}$$

C falls outside of the range, so the simulation data belong to aggregation-type distribution.

3.2. Using the David-Moore method

Using the David-Moore method to test the data, it is first assumed that the analog data is randomly distributed, and it follows that

$$\chi^2_{0.975} < \chi^2 < \chi^2_{0.025} \cdot \tag{6}$$

The above analysis shows that $I = \frac{S^2}{\overline{x}} = C = 17.0984$, and

$$\chi^2 = I(n-1) = 17.0984 \times 9999 = 170966.9 \tag{7}$$

is obtained. Letting the freedom $i = 10000 - 1 = 9999$, by using the Excel (function CHIINV), the below is obtained

$$\chi^2_{0.975} = 9723.732237, \quad \chi^2_{0.025} = 10278.05632 . \tag{8}$$

It could be deduced that $\chi^2 > \chi^2_{0.025}$ and $\chi^2 > \chi^2_{0.975}$, but obviously the assumption is not accepted, so the simulation data have clustered distribution.

3.3. Using average congestion index Lloyd M

Using average congestion index Lloyd M to test the data, the average congestion index parameter is expressed as $\overset{*}{m} = \frac{\sum_{j=1}^{n} x_j (x_j - 1)}{\sum_{j}^{n} x_j}$, and the number of neighbors is used to determine the corresponding data. With $m = \overline{x} = 7.760174$, the following is obtained:

$$\frac{\overset{*}{m}}{m} = 3.0743 > 1 . \tag{9}$$

This demonstrates that the simulation data exhibit an aggregation-type distribution.

3.4. Using the Iwao $\overset{*}{m} - m$ regression method

Using the Iwao $\overset{*}{m} - m$ regression method to test the data, and considering that this study is focused on individual distributions, the basic component of distribution is a single individual. Thus, using

$\overset{*}{m} = \alpha + \beta m$,

$$\beta \approx 3.0743 > 1. \tag{10}$$

Here, the simulation data exhibit an aggregation-type distribution.

According to the simulation data, poisonous weeds were uniformly distributed before the 14th generation, from the 15th generation to the aggregation distribution. After the addition of the diffusion term, the poisonous weeds were uniformly distributed before the 20th generation and exhibited aggregated distribution from the 21st generation onwards. In conclusion, the invasion and spatial diffusion of alien plant species will affect the spatial distribution of plant species.

4. Conclusions

Poisonous weeds comprise an invasive species that invades edible pasture grasses and competes with these grasses for grassland resources. Poisonous weeds will spread from a region of high density to one of low density, and, using the theory of cellular automata, which is a dynamic evolution process, each cell interaction and mutual diffusion establishes a spatial diffusion model for the invasion of a single patch of weeds.

The simulation demonstrate that in the initial stage and in conditions of the survival resources sufficient, the competition between poisonous weeds and edible herbage was not obvious, with both species remaining in their own growth states. When survival conditions were limited, the survival ability of the poisonous weeds exhibited an obvious advantage, and gradually eliminated the edible herbage. The invasion of poisonous weeds affects the ecological distribution of grassland, and the spatial distribution of poisonous changes from initial uniform distribution to aggregated distribution.

The results presented in this paper theoretically prove that poisonous weeds with strong survivability will eliminate edible herbage in a certain amount of time. Thus, in order to protect grassland resources, measures to prevent further breeding and spread of poisonous weeds should be taken as early as possible.

Acknowledgements

This work was supported by the Gansu Science and Technology Fund (20JR5RA512), Research Fund for Humanities and Social Sciences of the Ministry of Education (20XJAZH006), the Leading Talents Project of State Ethnic Affairs Commission of China, Central Universities Fundamental Research Funds for the Graduate Students of Northwest Minzu University (Yxm2021067) and Innovation Team of Intelligent Computing and Dynamical System Analysis and Application of Northwest Minzu University.

References

[1] Cui J and Takeuchi Y 2006 A predator–prey system with a stage structure for the prey. Mathematical & Computer Modelling pp 1126-1132
[2] Jiang X Song Q and Hao M 2010 Dynamics behaviors of a delayed stage-structured predator–prey model with impulsive effect. Applied Mathematics & Computation pp 4221-4229
[3] Lei S Hua L Wei Y M Ming M and Ye J.H 2018 The permanence and periodic solution of a competitive system with infinite delay, feedback control, and Allee effect. Advances in Difference Equations 400 pp 1-8
[4] Shang Y 2013 The Limit Behavior of a Stochastic Logistic Model with Individual Time-Dependent Rates. Journal of Mathematics, pp 1-7
[5] Chen Y and Zhou Z 2003 Stable periodic solution of a discrete periodic Lotka–Volterra competition system. Journal of Mathematical Analysis & Applications pp 358-366
[6] Matis J H Zheng Q and Kiffe T R 1995 Describing the spread of biological populations using stochastic compartmental models with births. Mathematical Biosciences pp 215-247
[7] Yuan G Zhang Z and Cao Z Spatial distribution and spatial association of Stellera chamaejasme population in the different altitude in in degraded alpine grassland. Acta Ecologica Sinica pp 605-612.

Journal of Physics: Conference Series **2173** (2022) 012017 doi:10.1088/1742-6596/2173/1/012017

The (Vertex) Gutman Index of One Vertex Union of Two Cycles

Kelun Wang[1] and Haili Guo[2,*]

[1]School of Science, Dalian Maritime University, Dalian, Liaoning, P.R. China
[2]Department of Basic Sciences, Beijing City University, Beijing, P.R. China

[*]corresponding author's e-mail: guohaili@bcu.edu.cn

Abstract. The Gutman index is a variant of the Wiener index, which related to the degree and the distance. In this paper, we study the vertex Gutman indices of a special kind of graphs, one vertex union of two cycles. And the maximum and minimum of vertex Gutman indices are studied. The results show that the vertex Gutman indices has the maximum at the center and has minimum at some vertices nearest the center.

1. Preliminaries

The following concepts are cited in the literature [1-6].

Let G = (V, E) be a connected simple graph with the vertex set V and the edge set E, and the degree of the vertex u is denoted by d(u), the distance between vertices u and v of G is denoted by d(u,v). The Gutman index of a graph G is defined as $Gut(G)=\sum_{\{u,v\}\in V(G)}(d(u)\cdot d(v))d(u,v)$. $\forall$u$\in$G, the degree distance of a vertex u to G is defined as $Gut(u,G)=\sum_{v\in V(G)}d(v)d(u,v)$; Call $d(u)\cdot Gut(u,G)$ the vertex Gutman index of u, denoted by $Gut(u)$. Namely, The Gutman index of G is the sum of the products of the distance of pairs of unordered vertices and their vertex degrees, and

$$Gut(G)=\sum_{\{u,v\}\subseteq V(G)}\left(d(u)\cdot d(v)\right)d(u,v)=\frac{1}{2}\sum_{u\in V(G)}d(u)Gut(u,G)=\frac{1}{2}\sum_{u\in V(G)}Gut(u). \quad (1)$$

Definition 1. Let C_m ($m \geq 3$) be an m-cycle and C_n ($n \geq 3$) be an n-cycle: $V(C_m)=\{u_0,u_1,u_2,\text{L},u_{m-1}\}$, $V(C_n)=\{v_0,v_1,v_2,\text{L},v_{n-1}\}$. If $|V(C_m)\cap V(S_n)|=1$, then the union graph of C_m and C_n is called one vertex union of two cycles (denoted as $C_{m,n}$, see figure 1). The common intersection vertex c is the center of $C_{m,n}$.

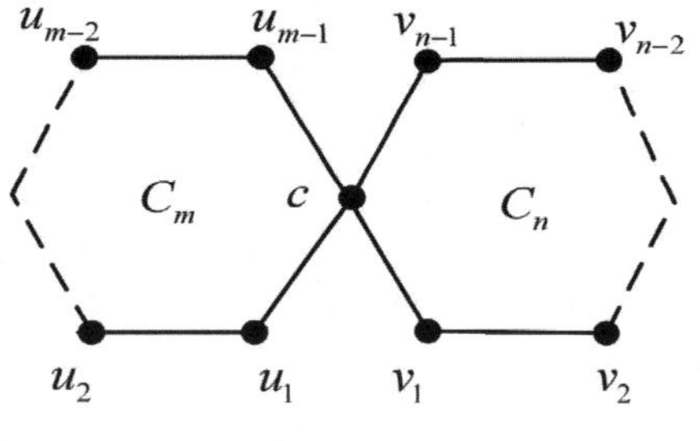

Figure 1. One vertex union of two cycles $C_{m,n}$

This paper focuses on the vertex Gutman indices of a special kind of graphs, one vertex union of two cycles. The maximum and minimum of vertex Gutman indices are studied.

2. Main results

2.1. The vertex-Gutman indices

Theorem 1. Let c be the center of $C_{m,n}$. Then the vertex-Gutman index of c is as follows:

$$Gut(c)=\begin{cases} 2(m^2+n^2-2), & m \text{ and } n \text{ are odd;} \\ 2(m^2+n^2), & m \text{ and } n \text{ are even;} \\ 2(m^2+n^2-1), & m+n \text{ is odd.} \end{cases} \quad (2)$$

Proof. From definition of the degree distance of a vertex to a graph, we get

$$\begin{aligned} Gut(c,C_{m,n}) &= \sum_{v\in V(C_{m,n})} d(v)d(c,v) = \sum_{i=1}^{m-1} d(u_i)d(c,u_i) + \sum_{j=1}^{n-1} d(v_j)d(c,v_j) \\ &= 2\sum_{i=1}^{m-1} d(c,u_i) + 2\sum_{j=1}^{n-1} d(c,v_j). \end{aligned} \quad (3)$$

(1) If m and n are odd, then

$$Gut(c,C_{m,n}) = 2\sum_{i=1}^{m-1} d(c,u_i) + 2\sum_{j=1}^{n-1} d(c,v_j). = 4\sum_{i=1}^{\frac{m-1}{2}} i + 4\sum_{j=1}^{\frac{n-1}{2}} j - \frac{1}{2}(m^2+n^2-2). \quad (4)$$

(2) If m and n are even, then

$$Gut(c,C_{m,n}) = 2\sum_{i=1}^{m-1} d(c,u_i) + 2\sum_{j=1}^{n-1} d(c,v_j). = 2\left(2\sum_{i=1}^{\frac{m}{2}-1} i + \frac{m}{2} + 2\sum_{j=1}^{\frac{n}{2}-1} j + \frac{n}{2}\right) = \frac{1}{2}(m^2+n^2). \quad (5)$$

(3) If m is odd and n is even, then

$$Gut(c,C_{m,n}) = 2\sum_{i=1}^{m-1} d(c,u_i) + 2\sum_{j=1}^{n-1} d(c,v_j). = 2\left(2\sum_{i=1}^{\frac{m-1}{2}} i + 2\sum_{j=1}^{\frac{n}{2}-1} j + \frac{n}{2}\right) = \frac{1}{2}(m^2+n^2-1). \quad (6)$$

(4) If m is even and n is odd, then

$$Gut(c,C_{m,n}) = 2\sum_{i=1}^{m-1} d(c,u_i) + 2\sum_{j=1}^{n-1} d(c,v_j). = 2\left(2\sum_{i=1}^{\frac{m}{2}-1} i + \frac{m}{2} + 2\sum_{j=1}^{\frac{n-1}{2}} j\right) = \frac{1}{2}(m^2+n^2-1). \quad (7)$$

So, the equation (2) holds.

Theorem 2. For the vertex u_k $(1 \le k \le m-1)$ in the cycle C_m (of $C_{m,n}$), the vertex-Gutman index of u_k is as follows:

$$Gut(u_k)=\begin{cases} m^2+n^2+4kn-2, & m \text{ and } n \text{ are odd, } 1\le k\le \frac{m-1}{2};\\ m^2+n^2+4(m-k)n-2, & m \text{ and } n \text{ are odd, } \frac{m+1}{2}\le k\le m-1;\\ m^2+n^2+4kn, & m \text{ and } n \text{ are even, } 1\le k\le \frac{m}{2};\\ m^2+n^2+4(m-k)n, & m \text{ and } n \text{ are even, } \frac{m}{2}+1\le k\le m-1;\\ m^2+n^2+4kn-1, & m+n \text{ is odd, and } 1\le k\le \lfloor \frac{m}{2}\rfloor;\\ m^2+n^2+4(m-k)n-1, & m+n \text{ is odd, and } \lfloor \frac{m}{2}\rfloor+1\le k\le m-1. \end{cases} \quad (8)$$

Proof. Denote c as u_0 or v_0, then the vertex-Gutman index of u_k $(1\le k\le m-1)$ is that

$$\begin{aligned} Gut(u_k,C_{m,n}) &= \sum_{i=1}^{m-1} d(u_i)d(u_k,u_i)+d(c)d(u_k,c)+\sum_{j=1}^{n-1} d(v_j)d(u_k,v_j)\\ &=2\sum_{i=1}^{m-1} d(u_k,u_i)+2d(u_k,u_0)+2d(u_k,v_0)+2\sum_{j=1}^{n-1} d(u_k,v_j)\\ &=2\sum_{i=0}^{m-1} d(u_0,u_i)+2k+2\sum_{j=1}^{n-1}(d(v_0,v_j)+k). \end{aligned} \quad (9)$$

(1) If m and n are odd, and $1\le k\le \frac{m-1}{2}$, then

$$Gut(u_k,C_{m,n})=4\sum_{i=0}^{\frac{m-1}{2}} i+2k+4\sum_{j=1}^{\frac{n-1}{2}}(j+k)=\frac{1}{2}(m^2+n^2)+2kn-1. \quad (10)$$

And, it is clear that, for $\frac{m+1}{2}\le k\le m-1$, $Gut(u_k,C_{m,n})=\frac{1}{2}(m^2+n^2)+2(m-k)n-1$.

(2) If m and n are even, and $1\le k\le \frac{m}{2}$, then

$$Gut(u_k,C_{m,n})=4\sum_{i=0}^{\frac{m}{2}-1} i+2\frac{m}{2}+2k+4\sum_{j=1}^{\frac{n}{2}-1}(j+k)+2\left(\frac{n}{2}+k\right)=\frac{1}{2}(m^2+n^2)+2kn. \quad (11)$$

And, it is clear that, for $\frac{m}{2}+1\le k\le m-1$, $Gut(u_k,C_{m,n})=\frac{1}{2}(m^2+n^2)+2(m-k)n$.

(3) If m is odd, n is even, and $1\le k\le \frac{m-1}{2}$, then

$$Gut(u_k,C_{m,n})=4\sum_{i=0}^{\frac{m-1}{2}} i+2k+4\sum_{j=1}^{\frac{n}{2}-1}(j+k)+2\left(\frac{n}{2}+k\right)=\frac{1}{2}(m^2+n^2)+2kn-\frac{1}{2}. \quad (12)$$

And, it is clear that, for $\frac{m+1}{2}\le k\le m-1$, $Gut(u_k,C_{m,n})=\frac{1}{2}(m^2+n^2)+2(m-k)n-\frac{1}{2}$.

(4) If m is even and n is odd, similarly, we can get

$$Gut(u_k,C_{m,n})==\begin{cases} \frac{1}{2}(m^2+n^2)+2kn-\frac{1}{2}, & m+n \text{ is odd, and } 1\le k\le \frac{m}{2};\\ \frac{1}{2}(m^2+n^2)+2(m-k)n-\frac{1}{2}, & m+n \text{ is odd, and } \frac{m}{2}+1\le k\le m-1. \end{cases} \quad (13)$$

So, the equation (3) holds.

From Theorem 2, we can get the following results similarly.

Corollary 1. For the vertex v_k $(1\le k\le n-1)$ in the cycle C_n (of $C_{m,n}$), the vertex-Gutman index of v_k is as follows:

$$Gut(v_k)=\begin{cases} m^2+n^2+4km-2, & n \text{ and } m \text{ are odd, } 1\le k\le \frac{n-1}{2};\\ m^2+n^2+4(n-k)m-2, & n \text{ and } m \text{ are odd, } \frac{n+1}{2}\le k\le n-1;\\ m^2+n^2+4km, & n \text{ and } m \text{ are even, } 1\le k\le \frac{n}{2};\\ m^2+n^2+4(n-k)m, & n \text{ and } m \text{ are even, } \frac{n}{2}+1\le k\le n-1;\\ m^2+n^2+4km-1, & m{+}n \text{ is odd, and } 1\le k\le \left\lfloor \frac{n}{2}\right\rfloor;\\ m^2+n^2+4(n-k)m-1, & m{+}n \text{ is odd, and } \left\lfloor \frac{n}{2}\right\rfloor+1\le k\le n-1. \end{cases} \tag{14}$$

2.2. The extremal of vertex Gutman indices

From Theorem 1, Theorem 2 and Corollary 1, it's easy to get the following results:

Corollary 2.

$$\max\{Gut(u)|u\in\{u_1,u_2,\cdots,u_{m-1}\}\}=\begin{cases} Gut(u_{\frac{m-1}{2}})=Gut(u_{\frac{m+1}{2}}), & \text{if } m \text{ is odd},\\ Gut(u_{\frac{m}{2}}), & \text{if } m \text{ is even}.\end{cases}$$
$$\max\{Gut(v)|u\in\{v_1,v_2,\cdots,v_{n-1}\}\}=\begin{cases} Gut(v_{\frac{n-1}{2}})=Gut(v_{\frac{n+1}{2}}), & \text{if } n \text{ is odd},\\ Gut(v_{\frac{n}{2}}), & \text{if } n \text{ is even}.\end{cases} \tag{15}$$

Corollary 3.

$$\min\{Gut(u)|u\in\{u_1,u_2,\cdots,u_{m-1}\}\}-Gut(u_1)=Gut(u_{m-1}),$$
$$\min\{Gut(v)|v\in\{v_1,v_2,\cdots,v_{n-1}\}\}=Gut(v_1)=Gut(v_{n-1}). \tag{16}$$

Theorem 4.

$$\max_{u\in V(C_{m,n})}\{Gut(u)\}=\begin{cases} Gut(c)=Gut(u_{\frac{m}{2}})=Gut(v_{\frac{n}{2}}), & \text{if } m \text{ and } n \text{ are even}, m=n,\\ Gut(c), & \text{others}.\end{cases} \tag{17}$$

Proof. (1) If m and n are odd, then

$$Gut(c)-Gut(u_k)\ge Gut(c)-Gut(u_{\frac{m-1}{2}})=(m-n)^2+2(n-1)>0,$$
$$Gut(c)-Gut(v_k)\ge Gut(c)-Gut(v_{\frac{n-1}{2}})=(n-m)^2+2(m-1)>0. \tag{18}$$

So, for odd m and n, $\max_{u\in V(C_{m,n})}\{Gut(u)\}=Gut(c)$.

(2) If m and n are even, then

$$Gut(c)-Gut(u_k)=\begin{cases} 0, & k=\frac{m}{2}, m=n,\\ (m-n)^2>0, & k\ne\frac{m}{2}, m\ne n;\end{cases}$$
$$Gut(c)-Gut(v_k)=\begin{cases} 0, & k=\frac{n}{2}, m=n,\\ (m-n)^2>0, & k\ne\frac{n}{2}, m\ne n.\end{cases} \tag{19}$$

So, for even m and n,

$$\max_{u\in V(C_{m,n})}\{Gut(u)\}=\begin{cases} Gut(c)=Gut(u_{\frac{m}{2}})=Gut(v_{\frac{n}{2}}), & m=n,\\ Gut(c), & m\ne n.\end{cases} \tag{20}$$

Similarly, we can get that, if m+n is odd, $\max_{u\in V(C_{m,n})}\{Gut(u)\}=Gut(c)$. So, the equation (4) holds.

Theorem 4 shows that the vertex Gutman indices always get the maximum the center c, or at the vertices farthest from the center c, in addition, if two cycles have the same length.

Theorem 5.

$$\min_{u\in V(C_{m,n})}\{Gut(u)\}=\begin{cases} Gut(v_1)=Gut(v_{n-1}), & m>n; \\ Gut(u_1)=Gut(u_{n-1})=Gut(v_1)=Gut(v_{n-1}), & m=n; \\ Gut(u_1)=Gut(u_{n-1}), & m<n. \end{cases} \tag{21}$$

Proof. From Theorem 2, Corollary 1, we get that $Gut(u_1)=Gut(u_{m-1}),\ Gut(v_1)=Gut(v_{n-1})$. And since $Gut(u_1)-Gut(v_1)=4(m-n)$. So, from Corollary 3, we get the equation (5).

Theorem 5 shows that the vertex Gutman indices have the minimum at some vertices nearest the center.

2.3. Gutman index of one vertex union of two cycles

Theorem 6. The Gutman index of one vertex union of two cycles is as follows:

$$Gut(C_{m,n})=\begin{cases} \frac{1}{2}\left(m^3+n^3+2m^2n+2mn^2-3m-3n\right), & \text{If } m \text{ and } n \text{ are odd,} \\ \frac{1}{2}\left(m^3+n^3+2m^2n+2mn^2\right), & \text{If } m \text{ and } n \text{ are even,} \\ \frac{1}{2}\left(m^3+n^3+2m^2n+2mn^2+m^2+n^2+2mn-m-2n+1\right), & \text{If } m \text{ is odd and } n \text{ is even,} \\ \frac{1}{2}\left(m^3+n^3+2m^2n+2mn^2+m^2+n^2+2mn-2m-n+1\right), & \text{If } m \text{ is even and } n \text{ is odd.} \end{cases} \tag{22}$$

Proof. Since

$$Gut(C_{m,n})=\frac{1}{2}\sum_{u\in V(G)}Gut(u)=\frac{1}{2}\sum_{u\in V(C_m)}Gut(u)+\frac{1}{2}Gut(c)+\frac{1}{2}\sum_{v\in V(C_n)}Gut(v). \tag{23}$$

(1) If m and n are odd, then,

$$\begin{aligned} Gut(C_{m,n})&=\frac{1}{2}\cdot 2\sum_{i=1}^{\frac{m-1}{2}}(m^2+n^2+4in-2)+\frac{1}{2}\cdot 2(m^2+n^2-2)+\frac{1}{2}\cdot 2\sum_{i=1}^{\frac{n-1}{2}}(m^2+n^2+4im-2) \\ &=\frac{1}{2}\left(m^3+n^3+2m^2n+2mn^2-3m-3n\right). \end{aligned} \tag{24}$$

(2) If m and n are even, then,

$$\begin{aligned} Gut(C_{m,n})&=\frac{1}{2}\cdot 2\sum_{i=1}^{\frac{m}{2}-1}(m^2+n^2+4in)+\frac{1}{2}\cdot(m^2+n^2+2mn)+\frac{1}{2}\cdot 2(m^2+n^2) \\ &\quad+\frac{1}{2}\cdot 2\sum_{i=1}^{\frac{n}{2}-1}(m^2+n^2+4im)+\frac{1}{2}\cdot(m^2+n^2+2mn) \\ &=\frac{1}{2}\left(m^3+n^3+2m^2n+2mn^2\right). \end{aligned} \tag{25}$$

(3) If m is odd and n is even, then,

$$
\begin{aligned}
Gut(C_{m,n}) &= \frac{1}{2}\cdot 2\sum_{i=1}^{\frac{m-1}{2}}(m^2+n^2+4in-1)+\frac{1}{2}\cdot 2(m^2+n^2-1) \\
&+\frac{1}{2}\cdot 2\sum_{i=1}^{\frac{n}{2}-1}(m^2+n^2+4im-1)+(m^2+n^2+4\cdot\frac{n}{2}\cdot m) \\
&=(m^2+n^2-1)\frac{m-1}{2}+4n\cdot\frac{1}{2}\left(1+\frac{m-1}{2}\right)\frac{m-1}{2}+(m^2+n^2-1) \\
&+(m^2+n^2-1)\left(\frac{n}{2}-1\right)+4m\cdot\frac{1}{2}\left(1+\frac{n}{2}-1\right)\left(\frac{n}{2}-1\right)+(m^2+n^2+2mn) \\
&=\frac{1}{2}\left(m^3+n^3+2m^2n+2mn^2+m^2+n^2+2mn-m-2n+1\right).
\end{aligned}
\tag{26}
$$

(4) If m is even and n is odd, similarly, we get

$$
Gut(C_{m,n})=\frac{1}{2}\left(m^3+n^3+2m^2n+2mn^2+m^2+n^2+2mn-2m-n+1\right). \tag{27}
$$

So, the equation (6) holds.

3. Conclusions

The one vertex union of two cycles is a typical bicyclic graph. This paper focused on the (vertex) Gutman indices of the graph, and the maximum and minimum of vertex Gutman indices were studied. We got that the vertex Gutman indices had the maximum at the center and minimum at some vertices nearest the center.

Acknowledgement

This research was financially supported by the Fundamental Research Funds for the Central Universities (3132019400).

References

[1] Wang D, Feng M and Wang K 2017 2nd International Congference on Conputational Modeling, Simulation and Applied Mathematics The extremal values and rankings of the Gutman index of k-caterpillar 204.
[2] Gutman I 1994 Chem. Inf. Comput. Sci Selected properties of the Schultz molecular topological index 34 1087.
[3] Gutman I, Araujo O, Morales D A. 2000 Chem. Inf. Comput. Sci. Bounds for the Randić connectivity index 40 572.
[4] Gutman I, Furtula B, Das K C 2016 Applied Mathematics and Computation On some degree-and-distance-based graph invariants of trees 289 1.
[5] Ilić A, Stevanović D, Feng L, Yu G, Dankelmann P 2011 Discrete Appl. Math. Degree distance of unicyclic and bicyclic graphs 159 779.
[6] Klavzar S, Gutman I 1997 Discrete Appl. Math. Wiener number of vertex-weighted graphs and a chemical application 80 73 .

Heart Sound Classification Method Based on Complete Empirical Mode Decomposition with Adaptive Noise Permutation Entropy

Quanyu Wu*, Meijun Liu, Sheng Ding, Lingjiao Pan and Xiaojie Liu

Institute of Biological Information and Medical Engineering, School of Electrical Information and Engineering, Jiangsu Institute of Technology, Changzhou, Jiangsu Province, China

*Corresponding author's e-mail: WU Quanyu@jsut.edu.cn

Abstract. This paper proposes a method of complete empirical modal decomposition with adaptive noise (CEEMDAN) arrangement entropy as a characteristic vector of heart sound signal and support vector machine (SVM) as a classifier to classify heart sounds. Firstly, PCG is decomposed into a few intrinsic mode functions (IMF) from high frequency to low frequency with CEEMDAN. Secondly, this method uses the correlation coefficient, energy factor and signal-to-noise ratio of the IMF and the original signal to optimize IMFs for Hilbert transform getting instantaneous frequency. Making use of the instantaneous frequency calculates arrangement entropy of each IMF. These arrangement entropy values form a feature vector. Finally, the extracted features were classified by help of support vector machine (SVM) with the mark of normal and abnormal heart sounds. The 100 heart sound samples from the 2016 PhysioNet /CinC Challenge were classified between normal and abnormal, and experimental results show that this method can effectively improve the recognition accuracy. The classification results of support vector machine reach 87%, and are better than Fisher discrimination methods.

1. Introduction

Heart sound signal is a sound signal formed by fibrillation caused by blood colliding with the ventricle and aortic wall during systole and diastole [1-2], which can reflect a large amount of cardiovascular physiological information and pathological information, and is useful for clinical evaluation of cardiac function status [3-5]. Due to the non-linearity and non-stationarity of heart sound signals [6-7], the main methods in the analysis of heart sound signals are time-frequency analysis and energy analysis. The two methods are usually combined with each other to analyze physiological signals [8-10]. Liu Liping [11] and others used C-W distribution time-frequency analysis combined with the time-frequency and energy analysis method of AR model power spectrum to extract the feature vector of the heart sound signal. Yin Ming [12] in-depth study based on HHT, heart sound signals after noise reduction using EMD decomposition, the HT conversion and normalization Shannon theorem to extract energy from a plurality of angles heart sounds characteristic values. Huang Linzhou [13] proposed that EMD and approximate entropy constitute the eigenvector of the heart sound signal. Li Hongquan [14] proposed a method combining EMD and Mel-frequency cepstrum to identify diastolic heart murmurs. Marton [15] extracted the frequency characteristics of the systolic and diastolic phases and the resampled wavelet envelope features, and used a support vector machine to classify records with several training set configurations, and achieved an accuracy of 80.28%. Iga [16] proposed a machine learning algorithm based on neural networks. The segmentation of PCG signals uses an algorithm based on hidden Markov models, focusing on statistical features representing signal features, and the best overall score obtained is 0.79. Therefore, empirical mode decomposition and time-frequency and energy analysis are of great significance to the analysis and processing of complex physiological signals [17,18].

Wu [19]and others proposed lumped empirical mode decomposition (EEMD) with the empirical mode decomposition (EMD) method as the core, adding Gaussian white noise to the original signal many times, and the Gaussian white noise signal is a kind of easy analysis. The ideal noise signal uses the uniform distribution of its spectrum to make the reconstructed signal have continuity on different time scales, and solves the problem of EMD modal aliasing. Due to the low decomposition efficiency of EEMD, Yeh [20] proposed complementary lumped empirical mode decomposition (CEEMD) based on EEMD, adding a white noise to the original signal, and then subtracting a white noise. The residual noise in the reconstructed signal is eliminated, and the number of lumped averages is reduced, which improves the decomposition efficiency, but there is false IMF [21]. The adaptive noise complete empirical mode decomposition (CEEMDAN) [22] improves the shortcomings of the above methods, adding adaptive Gaussian white noise to each IMF, making the reconstruction error close to 0, and the decomposition efficiency is extremely high and complete.

This paper proposes a method of extracting the instantaneous frequency permutation entropy of the CEEMDAN modal component of heart sound signal as a feature vector, classifying and identifying normal and abnormal heart sounds through a support vector machine. Finally, to reflect the advantages of this method, it is also compared with the Fisher discriminant method. The experimental results show that this method can be effectively applied to heart sound classification.

2. Method

2.1. Data Source and Preprocessing

The heart sound data in this article comes from a large database provided by the "Classification of Heart Sound Records-2016 PhysioNet/CinC Challenge". Each record lasts from 5 seconds to 120 seconds, with a sampling frequency of 2000 Hz, and is provided in *.wav format. They are divided into two types: normal and abnormal heart sound records. Normal records are from healthy subjects, and abnormal records are from patients with diagnosed typical heart valve defects and coronary artery disease (CAD).

This paper selects 300 heart sounds for experiment. The heartbeat period of a normal adult is 0.8s, while the heart sound data sampling point of the heart sound database is 0.0005s, a heartbeat period is 1600 sampling points, for convenience analysis, so we took 7.5 heartbeat cycles for research. The typical records of normal and abnormal heart sounds are shown in Figure 1. Since the heart sound signal collection environment is not controlled, many recordings will contain various noises, such as conversation, stethoscope movement, breathing, and intestinal sounds. Therefore, this article uses a Butterworth low-pass filter to filter out the noise above 300Hz caused by improper operation in the original signal; a notch filter is used to filter out 50Hz power frequency interference, and the heart sound signal is preprocessed.

 2173 (2022) 012018 doi:10.1088/1742-6596/2173/1/012018

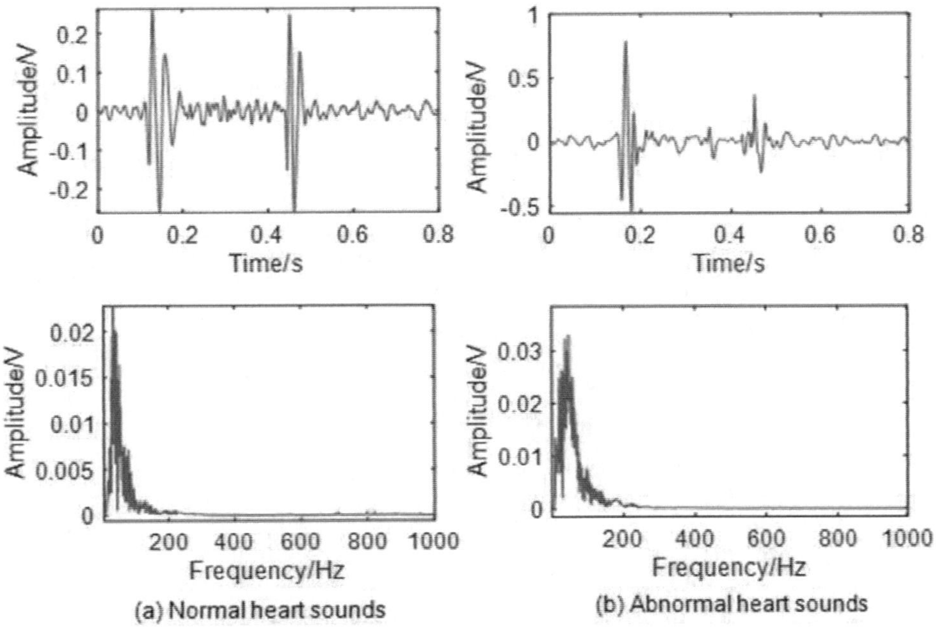

Figure 1. Typical records of normal and abnormal heart sounds.

2.2. Heart sound signal feature extraction

The flowchart of heart sound classification based on CEEMDAN permutation entropy and SVM is shown in Figure 2. CEEMDAN can obtain a series of frequency-layered IMFs, and only a part of the modal components can reflect the physiological and pathological information of the heart. Therefore, the correlation coefficient, signal-to-noise ratio and energy factor of each modal component and the original heart sound signal are calculated, and the energy is selected. Selecting larger IMFs, and with reference to the correlation between the component signal and the original signal, the effectiveness of a certain order of modal components is analyzed. The energy factor in this method is to select DB6 wavelet and perform 5-layer wavelet decomposition of the original heart sound signal, and construct the energy value of the original signal through high-frequency coefficients and low-frequency coefficients as an indicator of the frequency domain distribution of the original signal energy. Perform Hilbert transform on the effective modal components to obtain the frequency distribution of each modal component in the original time domain, that is, the instantaneous frequency, and calculate the instantaneous frequency permutation entropy of each IMF to form a one-dimensional eigenvector. Input to the SVM classifier for classification training to achieve heart sound classification diagnosis.

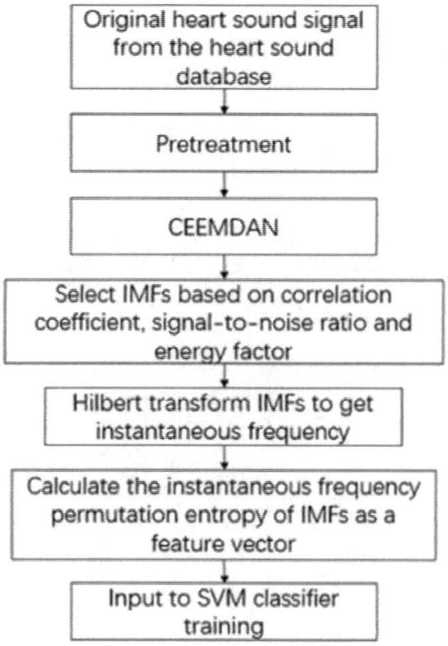

Figure 2. Heart sound classification flowchart.

3. Results

The pre-processed normal heart sounds are decomposed by CEEMDAN. The noise standard deviation of the modal components is 0.2, and the total number of averaging is 50 times. Figures 3 and 4 shows that the normal heart sound signal of a heartbeat cycle is decomposed into 12 IMFs, and the frequency spectrum of each modal component decreases from large to small, which clearly indicates that the main information of the heart sound signal is concentrated in 0~300Hz. Meanwhile, the frequency distribution of abnormal heart sounds is more prone to high-frequency distribution than the frequency distribution of normal heart sounds, and the component frequencies of abnormal heart sounds are mainly 200Hz~300Hz.

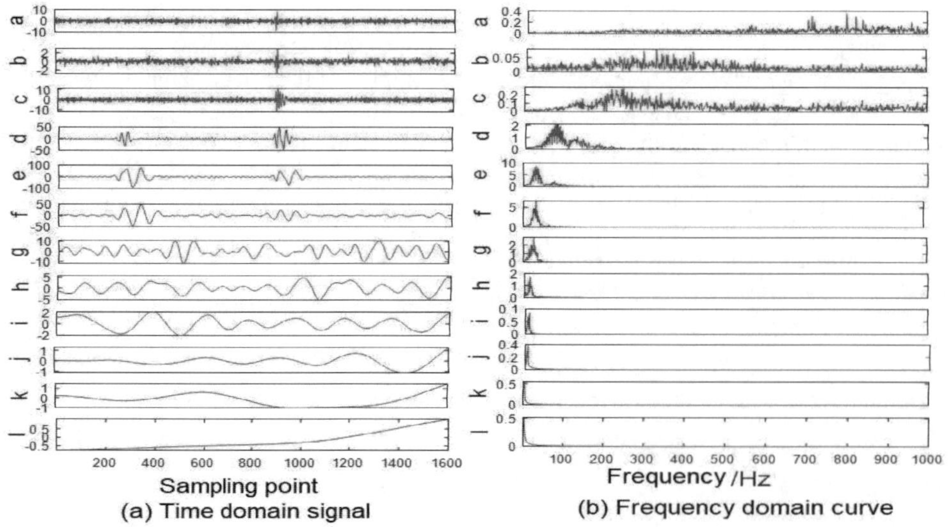

Figure 3. CEEMDAN decomposition result of normal heart sounds signal. a—IMF1;b—IMF2;c—IMF3;d —IMF4;e —IMF5;f —IMF6;g —IMF7;h —IMF8;i—IMF9;j —IMF10;k —IMF11;l —IMF12.

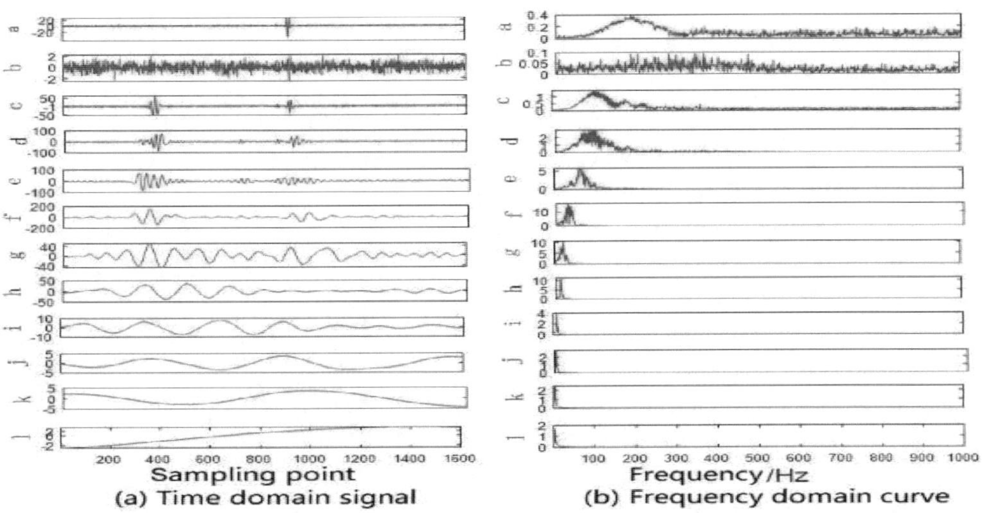

Figure 4. CEEMDAN decomposition result of abnormal heart sounds signal. a—IMF1;b—IMF2;c—IMF3;d —IMF4;e —IMF5;f —IMF6;g —IMF7;h —IMF8;i—IMF9;j —IMF10;k —IMF11;l —IMF12.

Here, we only show the correlation coefficient and energy factor value of each modal component of the typical normal heart sound signal decomposition and the original input heart sound signal. Table 1 shows the distribution of ratios, which shows a trend of small at both ends and large in the middle. Considering that the energy of the final residual signal of CEEMDAN decomposition is small, the degree of correlation with the original heart sound signal is low, and the signal-to-noise ratio is small, so this article selects the first nine-order modal components which are subjected to Hilbert transform proceed to the next step. Their instantaneous frequency is used to calculate the permutation entropy to form a one-dimensional feature vector.

Table 1. Correlation coefficient, signal-to-noise ratio, and energy factor of each IMF.

IMFs	Correlation coefficient	SNR / dB	Energy factor
IMF1	0.0469	-27.0017	0.0123
IMF2	0.0330	-26.0017	0.0010
IMF3	0.0885	-25.2831	0.0104
IMF4	0.2534	-18.5387	0.0209
IMF5	0.5631	-11.7590	0.1673
IMF6	0.8862	-3.1142	1.6327
IMF7	0.5928	-10.8264	0.0690
IMF8	0.1998	-17.9982	0.0009
IMF9	0.0473	-26.2056	0
IMF10	0.0038	-34.1481	0
IMF11	0.0005	-35.6126	0
IMF12	0.0004	-36.3620	0

The comparison of CEEMDAN permutation entropy between typical normal heart sounds and abnormal heart sounds is shown in Figure 5. The permutation entropy of abnormal heart sounds is higher than that of normal heart sounds, indicating that the higher the permutation entropy, the more complex and more information the content contains.

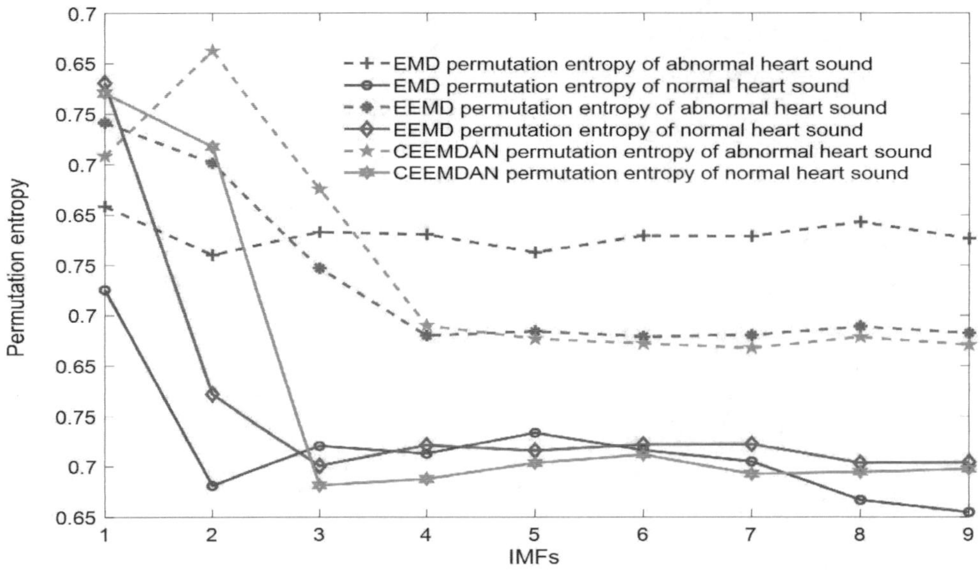

Figure 5. Comparison of permutation entropy of EMD、EEMD and CEEMDAN of normal and abnormal heart sounds.

3.1. CEEMDAN permutation entropy combined with SVM

The 200 original heart sounds signals were selected as training samples, including 100 normal signals and 100 abnormal signals. SVM selects bi-classification trainer, label the normal signal to "1" and the abnormal signal to "-1". The Gaussian radial basis function (RBF) is selected as the kernel function, and a cross-validation method is used to determine the penalty factor C and the kernel function parameter Y, to improve the prediction accuracy. The test sample included 100 original data, half normal, half abnormal. The SVM classification result accuracy reached 87%.

3.2. Comparison of EMD and EEMD permutation entropy combined with SVM

To verify the advantages of this method, the 300 samples were decomposed using EMD and EEMD respectively with the same steps. Figure 5 shows the comparison of EMD and EEMD permutation entropy of normal heart sound and abnormal heart sound signals. The discriminability of the first order and second-order IMFs of the heart sound signal CEEMDAN is significantly better than that of EEMD and EMD, while the difference in the entropy value of the three higher-order IMFs is not significant. In general, the entropy value of the high-order modal component of CEEMDAN is lower than that of EMD and EEMD, while the entropy value of the low-order modal component of CEEMDAN is higher than that of the low-order modal component of EMD and EEMD. Therefore, CEEMDAN shows excellent performance. The decomposition ability is consistent with the theory. Then SVM classification results are 66% and 71% respectively, as shown in Table 2. Therefore, the classification result based on CEEMDAN permutation entropy as the feature vector is higher than that of EMD and EEMD.

3.3. Comparison of EMD、EEMD and CEEMDAN permutation entropy combined with Fisher discriminant method

To reflect the advantages of SVM classifier, the 300 samples establish a classification method of Fisher discriminant. The discrimination function of CEEMDAN-Fisher is:

$$Y = 0.0229H(IMF1) - 0.0178H(IMF2) + 0.0310H(IMF3) - 0.0337H(IMF4) + 0.0076H(IMF5) + 0.0172H(IMF6) + 0.0038H(IMF7) - 0.0151H(IMF8) - 0.0012H(IMF9) - 0.1071$$

The result was 66.28%. Similarly, the classification results of EMD-Fisher and EEMD-Fisher are 65.61% and 65.95%, as shown in Table 2. The analysis shows that Fisher discriminants are insensitive to the permutation entropy classification of the empirical modal decomposition, it cannot fit the discriminant function well. The classification results of support vector machine are better than Fisher discrimination methods.

Table 2. Accuracy comparison of classification results.

Decomposition method	Fisher accuracy /%	SVM accuracy /%
CEEMDAN	66.28	87
EEMD	65.95	71
EMD	65.61	66

4. Discussion

The characteristic extraction of cardiac voice signal is an important part of the classification and recognition of cardiac voice signal. Since the cardiac voice signal is one of the important weak physiological signals of the human body, the information contained has complexity, nonlinear and non-stationary, and there is no specific characteristic extraction standard. Combining the time-frequency analysis method and the energy analysis method, this method avoids the problem of modal mixing and false components after EMD decomposition in [13]. At the same time, after adding adaptive white noise, the reconstruction error of the signal is almost 0, which improves the correlation coefficient and decomposition effect of the component signal and the original signal. This study proposes that the component signal as Hilbert transform obtains the instantaneous frequency arrangement entropy, it calculates the probability of a certain sequence of instantaneous frequency in each component. It extracts the information more accurately than in the original signal, and more accurately reflects the essential characteristics of the cardiac signal. However, there are certain pseudo frequency components in the first two order mode components of CEEMDAN decomposition, compared with the higher arrangement entropy of other order mode components.

The classification of cardiac sounds must combine the extracted feature vectors with the classifier to efficiently classify the identification. Because Fisher discriminant analysis constructs a discriminant function with the idea of variance analysis, which makes the standardization distance between the two groups of arrangement entropy after the projection, which is a high dimensional to low dimensional map, and cannot effectively distinguish the two categories of normal and abnormal cardiac tones. In this study, we select the SVM classifier to project two sets of eigenvectors into a higher dimensional space, dividing the optimal hyperplane for the feature space, very suitable for the small sample, nonlinear classification. This method, although using time-frequency and energy techniques, still needs to manually extract eigenvalues and set threshold, so this study is still room for improvement and future work will be devoted to the study of convolutional neural network classifiers. Moreover, the training and test sets of this study are recorded from different individuals, ensuring the rationality of the sample distribution.

5. Conclusion

In this paper, we propose a cardiac tone classification method combining adaptive noise complete empirical mode decomposition arrangement entropy with SVM, it proves the effectiveness by performing experiments on 300 sample data. The analysis results show that the CEEMDAN arrangement entropy can effectively show the intrinsic information arrangement characteristics and energy distribution characteristics of the cardiac sound signal, which can be used as the characteristic vector of cardiac sound classification, and can distinguish between normal and abnormal cardiac sound, and the classification results are significantly better than the EMD and EEMD algorithms.

Acknowledgement

This work was supported in part by the Jiangsu key R & D plan (No.BE2020648, No. BE2019317), National Natural Science Foundation of China (No.62001196) and Teaching Reform Project of Jiangsu University of Technology (No.11610211943).

References

[1] Ma Jing, Cai Wenjie and Yang Li 2017 Cardiac Signal Analysis Chinese Journal of Medical Physics chapter 34 pp 72-75
[2] Singh-Miller NE and Singh-Miller N 2017 Using Spectral Acoustic Features to Identify Abnormal Heart Sounds 2016 Computing in Cardiology Conference IEEE pp 557-560
[3] Xu Dongdong, Zhou Jing, Ying Dong and Long Qinghua 2019 Optimization of MFCC feature vector extraction method for heart sound signals Signal Processing pp 410-418
[4] Xing Suxia and Chen Tianhua. 2011 Cardiac Signal Acquisition and Analysis Based on DSP Journal of Biomedical Engineering chapter 28 pp 273-276
[5] Zabihi M, Rad A B, Kiranyaz S, Gabbouj M.and Katsaggelos A K 2017 Heart sound anomaly and quality detection using ensemble of neural networks without segmentation. 2016 Computing in Cardiology Conference IEEE pp 613-619
[6] Nogueira D M, Ferreira C A and Jorge A M 2017 Classifying Heart Sounds Using Images of MFCC and Temporal Features EPIA Conference on Artificial Intelligence Springer Cham.
[7] Ortiz JJG and Phoo CP Wiens J 2017 Heart Sound Classification Based on Temporal Alignment Techniques 2016 Computing in Cardiology Conference IEEE pp 589-592
[8] Zhang Xiaolan, Fang Yu, Liu Dongbo and Wang Haibin 2020 Research on the Algorithm for Extracting and Recognizing the Time-frequency Clutter Feature of Heart Sounds in Obese Heart Disease Journal of Electronic Measurement and Instrument chapter 34 pp 20-26
[9] Guo Chunli and Yue Xiaobing 2016 Automatic recognition of cardiac signal Based on Support Vector Machine Computer and Modernization pp 36-39
[10]Quan Xuefeng 2015 Heart sound signal recognition based on S transform and local singular entropy Electronic Technology and Software Engineering chapter 15 pp 260-264
[11]Liu Liping and Yuan Gang. 2016 Characteristic analysis and classification and identification of cardiac tone signal Information and Communication chapter 10 pp 26-29.
[12]Yin Ming, Wang Yan and Liu Xin 2013 Heart sound denoising preprocessing and HHT-based feature analysis research Electroacoustic Technology chapter 37 pp 58-63
[13]Huang Linzhou, Guo Xingming and Ding Xiaorong 2012 Research on Heart Sound Signal Recognition Based on EMD Approximate Entropy and Support Vector Machine Vibration and Shock chapter 31 pp 21-25.
[14]Li Hongquan, Guo Xingming and Zheng Yineng 2017 Classification and recognition of diastolic heart murmurs based on EMD and MFCC Journal of Vibration and Shock chapter 36 pp 8-13
[15]Goda MA and Hajas P 2017 Morphological Determination of Pathological PCG Signals by Time and Frequency Domain Analysis 2016 Computing in Cardiology Conference IEEE pp 1133-1136
[16]Grzegorczyk I, Katarzyna S, Gieratowski J, Rymko J and Perka 2017 PCG Classification Using a Neural Network Approach 2016 Computing in Cardiology Conference IEEE pp 1129-1132.
[17]Zheng Yineng, Guo Xingming, Qin Jian and Xiao Shouzhong 2015 Computer-assisted diagnosis for chronic heart failure by the analysis of their cardiac reserve and heart sound characteristics Comput Methods Programs Biomed chapter 122 pp 372-383
[18]Vernekar S, Nair S, Vijayasenan D and Ranjan,R 2017 A Novel Approach for Classification of Normal/Abnormal Phonocardiogram Recordings using Temporal Signal Analysis and Machine Learning 2016 Computing in Cardiology Conference IEEE pp 1141-1144
[19]Wu Zhaohua and Huang Ne 2009 Ensemble empirical mode decomposition:a noise assisted data analysis method Advances in Adaptive Data Analysis chapter 1 pp 1-41
[20]Yeh JR, Shieh JS and Huang NE 2011 Complementary Ensemble Empirical Mode Decomposition: A Novel Noise Enhanced Data Analysis Method Advances in Adaptive Data Analysis chapter 2 pp 135-156
[21]Dong Lichao, Guo Xingming and Zheng Yineng 2019 Research on wavelet packet denoising algorithm of heart sound signal based on CEEMD Vibration and Shock chapter 38 pp 192-198
[22]TORRES M E, Colominas M A, Schlotthauer G and Flandrin P 2011 A complete ensemble empirical mode decomposition with adaptive noise IEEE International Conference on Acoustics pp 4144-4147

Analysis of Regularized Poisson GLM Spike-Train Modeling

Yile Fan[1,*], Yuanpeng Li[1], Naiyang Xue[1] and Dan Ding[2]

[1]Graduate School, Aerospace Engineering University, Beijing China
[2]Department of Electronics and Optical Engineering, Aerospace Engineering University, Beijing China

[*]Corresponding author's e-mail: fanyile028@qq.com

Abstract. This paper introduces a method for modeling and analyzing neural impulse sequences. In this paper, we define the response value of a scale-independent neuron and construct the correlation graph of the neuron under the response value. The minimum cut algorithm is applied continuously to obtain the maximum group of neurons. According to the characteristics of the firing of neurons, a Poisson-process based model is proposed to mathematically model the neural coding, and the gradient descent method is used to optimize it. Through the modeling analysis method, information such as maximum neuron group and Inter-spike-Interval (ISI) can be effectively analyzed according to neuron impulse sequence.

1. Introduction

The mechanism of brain work, learning and memory has long been the focus of research and exploration.Since the end of last century, people's research on the brain has not only been limited to the molecular level, cellular level and behavioral level. Researchers pay more attention to the establishment of the overall model of the brain, revealing how physical and chemical processes are connected with behavioral patterns, and how the information processing processes hidden behind behavioral patterns work.

How is sensory information integrated into information inside the brain in order to understand the external world and how does the formation of learning and memory relate to the firing of neurons?We are only just beginning to understand these complex problems.As a highly complex, huge and dynamic system, it is difficult and one-sided to study the brain science itself, which promotes the emergence of a large number of interdisciplinary subjects, especially the integration of mathematics, informatics and computer science, which provides new viewpoints and means for the brain research.In recent years, the computational methods of machine learning, pattern discovery, probability model and graph model have been increasingly introduced into the analysis of neural data.The development of computer science, especially artificial intelligence, also needs the help of brain research to open up ideas.The brain's high versatility, high concurrency, high redundancy, and extremely fast response time are incomparable to the current computer.Researchers are trying to find good computer algorithms by studying the brain.Among them, the development of artificial neural network is inseparable from the inspiration of brain structure, and the adjustment strategy of weight in artificial neural network comes from the inspiration of prominent plasticity.In recent years, the modeling of visual processing by neural mechanisms has also begun to develop, in which the non-classical sensory model simulates the structure of human retina and optic nerve, and abstracts the representation of image information.

The study of information encoding can be traced back to the 1920s, when the theory of neural action potential was first proposed.In the decades that followed, various scholars proposed various hypotheses: the 1949 hypothesis, the single neuron coding hypothesis in 1972, and the cluster coding theory of neurons in 1996. Recent years have favored the hypothesis that clusters of neurons encode information because the firing of individual nerves is highly random even under phasic stimulus conditions.To a question: How often is the nerve firing or the exact time at which the nerve firing is encoding information?The academic world is still debating.Many experiments, however, suggest that the former may be closer to the truth.In a specific neural cluster, individual neurons have their own firing base frequency, and work together through temporary correlation to transmit and process information.A single neuron has a high degree of uncertainty and can obtain reliable information.

2. Processing mechanism through cluster coding

With the development of multi-electrode synchronous recording technology, various experiments have been designed to study the relationship between neural emission sequence and external stimulus. Among them, neural decoding is a study of the relationship between the temporal and spatial characteristics of neurons observed and recorded synchronously and the external events experienced by animals in the course of animal behavior experiments. In terms of theory of probability and statistics, the bayesian rule as one of the basic theory of statistics, also in the early introduction to the study of neural coding, in the end of last century the probability that the encoding and decoding model is put forward, and detailed analysis of the performance of various neurons distributed model, including model, the variation of model and model.Bayes' rule has always been considered as the optimal rule for brain inference and has been applied successfully in the field of neural computing. On the other hand, neuron firing sequence can be seen as discrete point sequence, namely point process.Therefore, the multi-channel records can be regarded as multi-dimensional point processes, and these point processes are dynamic and random.These point processes can be described by probabilistic models that describe the probability density generated at a given moment, et al. studied the point process model of a single neuron.Another widely studied problem is how to tell if a neuron is functionally involved in a particular task. By studying changes in neuron correlation in a group of thought neurons can be deduced to play a more precise role in certain types of tasks. In recent years, many techniques have been used to study the pair correlation of neurons during the task and then analyze the clustering characteristics of neurons. However, direct correlation-based methods are difficult to examine the correlation of neurons at multiple scales. The motivation for the study of neuronal groups in this paper is: since the intrinsic cognitive processes that induce the firing of neurons are different in different experiments, if a neuron is closely involved in the task, the biological signals that induce the firing of neurons will be consistent with the changing trend of the cognitive process. The cognitive processes of each trial are not observable, so it is difficult to tell from a single neuron whether the differences between trials are caused by internal cognitive processes or by other unrelated factors. However, if large numbers of neurons act in correlated ways in different experiments, the synergistic fluctuations are most likely due to internal cognitive processes. In this paper, the response value of a scale-independent neuron is defined, and the correlation graph of the neuron under the response value is constructed. The minimum cut algorithm is applied continuously to obtain the maximum neural group

3. A mathematical model of three neural codes

We assume that each action potential depends on an internal continuous signal that varies over time and reflects the instantaneous discharge frequency. And we also assume that each action potential is independent of each other, which is the independent action potential hypothesis. If the independent action potential hypothesis holds, it can be completely described by a nonhomogeneous Poisson process, which is also the mathematical model that this study relies on, and this model is not valid in the following cases: when the transmission frequency is very high, it is close. Because the effect of absolute and relative refractory periods increases significantly in the case of high firing frequency, the neuron will not generate the next action potential for a very short period of time after one action potential is generated. In addition, when neurons fire at a high frequency, there is often a phenomenon associated with the firing of several nerve impulses at a high frequency in succession. However, the vast majority of neurons observed in the brain satisfy or approximately satisfy the independent action potential hypothesis

3.1. Neural response function and instantaneous emission frequency

In order to accurately describe the whole Spiketrain, we introduce the neural response function ρ(t), which represents the generation time of each action potential:

$$\rho(t) = \sum_{i=1}^{k} \delta(t - t_i) \quad (1)$$

Where k is the number of action potential in Spiketrain, and T_i represents the generation moment of the ith action potential.

Unit response signal

$$\delta(t) = \begin{cases} 1 & t = 0 \\ 0 & t \neq 0 \end{cases} \quad (2)$$

We can calculate that the integral of delta of t is equal to one.The neural response function can be thought of as a random process.And the neural response function is equivalent to the list of all the action potentials in Spiketrain. To represent the number of spikes at moments T1 to T2, we can integrate the neural response function,

$$n = \int_{t_1}^{t_2} \rho(t) \quad (3)$$

So the instantaneous rate at which a neuron fires can define the expected value of the neural response function, which is the mean of the same stimulus for an infinite number of times,

$$\lambda(t) = E(\rho(t)) \quad (4)$$

In practical calculation, the instantaneous transmission frequency is often obtained by averaging finite number of tests, and its estimation is written.

$$r_M(t) = 1/M \sum_{j=1}^{M} \rho_j (t) \quad (5)$$

Where M is the number of tests and ρ_j (t) is the neural response function of each test. Poisson process is an important class of stochastic process, which has a wide range of applications in various fields. Intuitively, the problem described by it is as follows: a particle (customer) arrives at (service station) at a certain average speed, and the dynamic probability distribution N(t) of the arrival times N with time t is a Poisson process. The increment N(t) -n (T_0) is written as $N(T_0, t)$, $0 \leq T_0 \leq t$ which represents the number of particles occurring in the time interval (T_0, t)."There are k particles in (T_0, t)", that is, $N(T_0, t) = k$ is an event, and its probability is denoted as:

$$P_k(t_0, t) = PN(t_0, t) = k \quad (6)$$

It is assumed that the following conditions are satisfied :

(1) Independence in the non-overlapping interval;

(2) For sufficiently small Δt, $p_1(t, t + \Delta t) = P\ N(t_0, t) = 1 = \lambda\Delta t + o(\Delta t)$Where the constant λ is called the intensity of process N(t);

(3) For sufficiently small Δt, the probability of two particles appearing in $(T_0, t)$is negligible compared with the probability of one particle;

(4) The counting process (N(t),t≥0) satisfying the above conditions is called the Poisson process with the intensity of λ, and the corresponding particle flow or the random moment when the particle appears is called the Poisson flow with the intensity of λ. When λ changes dynamically with time, it is called a nonhomogeneous Poisson process.

When we randomly place an action potential into a time period of length of, and then randomly select a sub-time period $(T_1, T_2), \Delta T = T_2 - T_1$, the probability of the action potential during this time period is "p" T/T.If you put in k action potentials, the probability of n of them falling on Δ time t follows a binomial distribution,

$$P\Delta T = k!/(k - n)!\,n!\,p^n q^{k-n} \quad (7)$$

Where P=Δ T /T and Q = 1-P.When we increase k and n, and keep λ=k/T constant, we can think of it as the emission frequency, the number of action potentials per second. When k approaches infinity, the probability of n action potentials with internal parameters obeys the Poisson distribution, that is,

$$P\{\Delta \text{ T parameters n action potential}\} = e^{-\lambda\Delta t}(\lambda\Delta t)^n/n! \tag{8}$$

This expression is the Poisson distribution probability distribution function. For homogeneous Poisson process, without considering the temporal correlation of action potential before and after action potential, the mean value of action potential generated within time is the expected value of Poisson distribution with parameter $\lambda\Delta t$, i.e.,

$$E(n) = \int_{t1}^{t2} \lambda dt = \lambda\Delta t \tag{9}$$

For the Poisson distribution with an intensity of $\lambda\Delta t$, the variance is the same as the expected value, and the variance is also λ square t.

When we know that a random process is a Poisson process, the probability of an action potential occurring within $(T_0, T_0 + \tau)$ is,

$$P\{(t\ 0,t\ 0{+}\tau) \text{ is an action potential generated}\} = 1 - e^{-\lambda\tau} \tag{10}$$

This expression is the cumulative probability function of the generation of action potential in $(T_0, T_0 + \tau)$.When τ is 0, p is 0, and when τ is large, p approaches to 1.However, we would like to obtain the probability distribution of the time to generate the next action potential, or the distribution of waiting time. In computational neurobiology, this value is called the Interspike-interval (ISI).Obviously the distribution is the derivative of the cumulative probability function,

$$p(\tau) = d/dt(1 - e^{-\lambda\tau}) = \lambda e^{-\lambda\tau} \tag{11}$$

It can be seen that the ISI of the Poisson process follows exponential distribution, and the probability density is higher with the decrease of t. Statistical histogram of ISI can be made by counting the number of times of ISI in different length time Windows in Spike-Train. For the mean length of exponential distribution ISI is,

$$E[t] = \int_0^{+\infty} \tau p(\tau) d\tau = 1/\tau \tag{12}$$

For an exponential distribution, the variance is going to be,

$$\sigma_\tau^2 = \int_0^{+\infty} \tau^2 p(\tau) d\tau - E[\tau] = 1/\lambda^2 \tag{13}$$

The joint likelihood function is calculated for the whole M constituted by Poisson's model on all time slots of the neuronal impulse signal point process, and its form is as follows:

$$\log P(M) = \sum_t \left(-e^{x(t)^T\theta} + n(t) * \left(x(t)^T\theta\right)\right) + \text{const} \tag{14}$$

In order to optimize it using the gradient descent method, its derivative with respect to θ needs to be calculated as

$$\frac{\partial P(M)}{\partial \theta} = \sum_t \left(n(t)x(t) - -e^{x(t)^T\theta}x(t)\right) \tag{15}$$

After obtaining the derivative information of the model, we can use all kinds of gradient descent methods in optimization calculation to solve the optimization parameters of the objective function of the model.

4. Model and Simulation

Load data and reinforce it

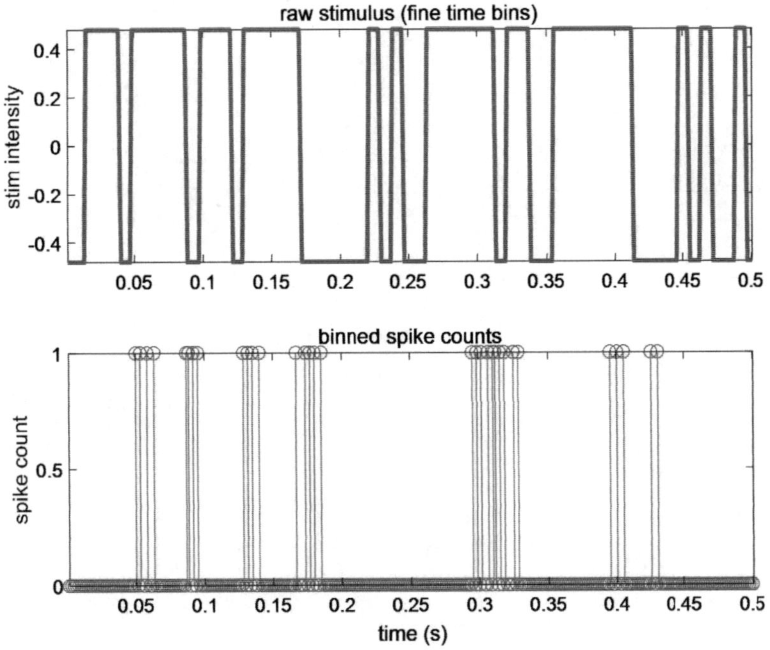

Figure 1. Spike-Train Data (Raw)

Gradient descent algorithm was used to fit Poisson-GLM
FMINUNC is used to calculate the maximum likelihood estimate
Calculate the STA

$$STA = \frac{1}{n_{sp}} \sum_{t=1}^{T} y(t) * x(t) \tag{16}$$

Wherein, N_{sp} represents the number of impulse signals sent by neurons in T time slots, and the value range of y(T) is [0,1].

Iteration	f(x)	Norm of step	First-order optimality	CG-iterations
0	-5655.08		1.77e+03	
1	-5655.08	4.42674	1.77e+03	4
2	-7323.19	1.10668	484	0
3	-7747.9	1.10668	159	8
4	-7792.28	0.45665	41.2	6
5	-7797.19	0.553691	7.08	15
6	-7797.44	0.165432	0.302	21
7	-7797.44	0.0110023	0.0196	20

Figure 2. Model Parameters

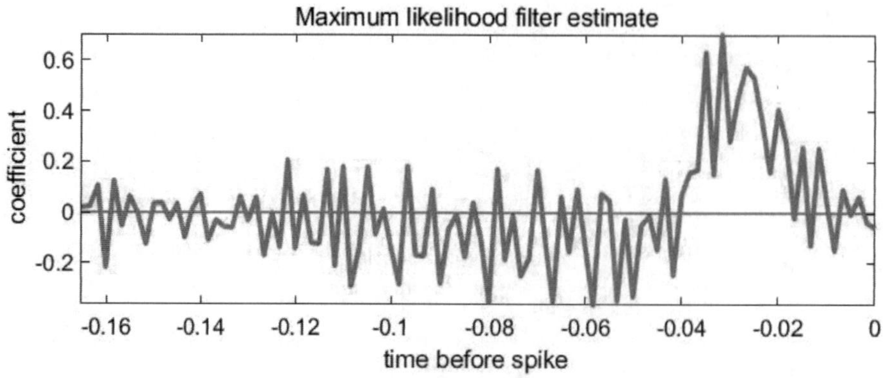

Figure 3. Coefficient of Maximum Likelihood Filter Estimate

The above results of lack of regularization are relatively poor, and the regularization operation is carried out below.

Ridge regression optimization:

$$\text{penalty}(\lambda) = \lambda * \left(\text{sum}_i \text{w}(\text{i}).^2\right), \tag{17}$$

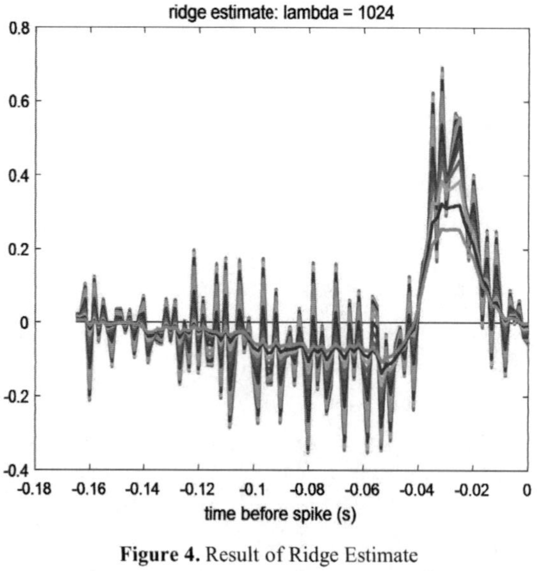

Figure 4. Result of Ridge Estimate

It can be seen that with the increase of λ, the data curve gradually flattens out:

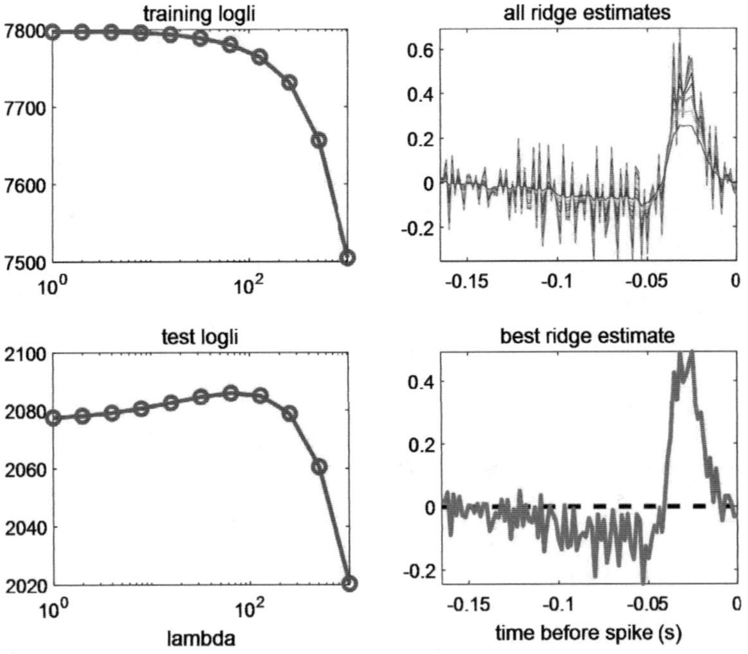

Figure 5. Data Curve of Ridge Estimate

5. Summarizes

Traditional Poisson-GLM modeling is used to solve the problem of neuron impulse train modeling. It can be further optimized by ridge regression regularization or L2 average and other methods. Through this large-scale practice, the application of stochastic process in scientific research is more clear, and the understanding of Poisson's general linear model is deepened. In future scientific research, the idea and tool method of stochastic process will be used to complete the task.

References

[1] Kass R E, Amari S I, Arai K et al. 2018 Computational Neuroscience: Mathematical and Statistical Perspectives *Annu Rev Stat Appln* **5** 183-214

[2] Leutgeb S, Leutgeb JK, Treves A et al. 2004 Distinct ensemble codes in hippocampal areas CA3 and CA1 *Science* **305**(5688) 1295–98

[3] Pillow J W, Shlens J, Paninski L et al. 2008 Spatio-temporal correlations and visual signalling in a complete neuronal population *Nature* **454**(7207) 995–9

[4] So K, Koralek A C, Ganguly K et al. 2012 Assessing functional connectivity of neural ensembles using directed information *J. Neural Eng.* **9** 026004

[5] Glaser C, Joublin F 2011 Firingrate homeostasis for dynamic neural field formation *IEEE Transactions on Autonomous Mental Development* **3**(4) 285–99

[6] Li C, Li Y 2013 A Spike-Based Model of Neuronal Intrinsic Plasticity *IEEE Trans actions on Autonomous Mental Development*, **5**(1) 62–73

[7] Miao J, Qing L, Zou B et al. 2010 Top–Down Gaze Movement Control in Target Search Using Population Cell Coding of Visual Context *IEEE Transaction s on Autonomous Mental Development* **2**(3) 196–215

[8] Velliste M, Perel S, Spalding M C et al. 2008 Cortical control of a prosthetic arm for self-feeding *Nature* **453** 1098–101

[9] Song D, Chan R H M, Marmarelis V Z et al. 2007 Nonlinear Dynamic Modeling of Spike Train Transformations for Hippocampal-Cortical *Prostheses IEEE Transactions on Biomedical Engineering* **54**(6) 1053–66
[10] Hodgkin A L and Huxley A F 1952 Aquantitative description of membrane current and its application to conduction and excitation in nerve *J Physiol* **117**(4) 500–44
[11] Lawhem V, Wu W, Nicholas, Liam P 2010 Population Decoding of Motor Cortical Activity using a Generalized Linear Model with Hidden States *Journal of Neuro Science Methods* **189**(2) 267–80
[12] Panins L. 2004 Maximum likelihood estimation of cascade point-process neural encoding models *Network: Computation in Neural Systems* **15**(4) 243–62
[13] ParkI M, Meister M L R, Huk A C et al. 2014 Encoding and decoding in parietal cortex during sensorimotor decision-making *Nat Neurosci* **17** 1395–403
[14] Truccolo W, Eden U T, Fellows MR et al. 2005 A Point Process Framework for Relating Neural Spiking Activity to Spiking History, Neural Ensemble, and Extrinsic Covariate Effects *Journal of Neurophysiology* **93**(2) 1074–89
[15] Weber A I, Pillow J W 2017 Capturing the Dynamical Repertoire of Single Neurons with Generalized Linear Models *Neural Comput* **29**(12) 3260–89

Equivalences in Variables Refinement for Differential Semi-algebraic Hybrid Systems

Bai Liu[1,*], Yitong Wang[1] and Liming Nie[2]

[1]Hubei University of Technology, Wuhan, Hubei, 430068, China
[2]Zhejiang Sci-Tech University, Hangzhou, Zhejiang, 310018, China

*Corresponding author's e-mail: liubai@hbut.edu.cn

Abstract. Differential semi-algebra hybrid system is a new formalized hybrid system model; it is widely used in engineering technology. In this paper we proposed differential semi-algebraic event structure as the system model, and then build the variables refinement theory of differential semi-algebraic hybrid system. Based on the various equivalence relationship, we researched the keeping equivalence after variables refinement.
Keywords. Variable refinement; Differential semi-algebraic hybrid systems; Bisimulation equivalence; Trace equivalence

1. Introduction

The hierarchical design is a method to depict the complicated system into simple abstract criterion. The key operation of hierarchical design method is refinement, which is called bottom-up software design [1].

The conventional refinement has been investigated in lecture [2, 3]. But the conventional refinement method is discrete, and cannot depict or solve the data flow. Consequently, researchers begin to study the refinement in hybrid systems [4-7]. Generating the invariant by polynomials Gröebner base is proposed firstly in hybrid system in 2004[8]. Then the variables refinement and equivalences were proposed [9-12]. But in differential semi-algebra system, the refinement has not been concerned, so we study the refinement in differential semi-algebra systems that contain differential and semi-algebra.

In this paper we use differential semi-algebra event structure as the standard pattern that is according to the event structures[12], and refine the variables in differential semi-algebra system. After variables refinement, research the equivalence preservation problem. In the paper we proposed trace equivalence and bisimulation equivalence which is discussed in literature[9, 10, 13, 20]. Then we find which equivalences are kept after the variables refinement.

Symbolic computation is an formal method which can be used to standardize the system before processing the system[14]. At the international symposium Algorithms and computation conference(05), B. Buchberger had reported that: "Symbolic computation with the numerical computation are based on the logical method." The Wu-Ritt characteristic set [15] which has been formed a strict and complete framework in theoretical foundation [16] is used to calculate for differential algebraic symbols. Cylindrical algebraic decomposition[17]is used to calculate for semi-algebraic symbols computation. So we can use differential semi-algebraic symbols symbol computation and numerical computation to research the variables refinement and equivalence preservation in differential semi-algebraic hybrid system.

2. Differential semi-algebraic event structures

Let $K[X_1, X_2, ..., X_n]$ be the differential multivariate polynomials ring with the variables $\{X_1, X_2, ..., X_n\}$. An ideal is any set $I \subset \{X_1, X_2, ..., X_n\}$ which contains the null polynomial and is closed under addition and multiplication by any element of $K[X_1, X_2, ..., X_n]$. Let $F \subseteq K[X_1, X_2, ..., X_n]$ be a set of polynomials. The ideal generated by F is an finite set $(F) = \{k_i = P_iQ_i \in K[X_1, X_2, ..., X_n], Q_i \in F, k \geq 1\}$. A set of polynomials F is called a basis of an ideal I if $I = (F)$. By the Hilbert basis theorem, we know that all ideals have a finite basis.

Definition2.1. A differential semi-algebra event structure is a tuple $\varepsilon = \langle V, E, Init, \varphi, \prec, \#, |||\rangle$, where

V is variables set of $\{x_1, x_2, ..., x_n\}$, n is the number ;

$E = \{e_1, e_2, ..., e_n\}$ is a countable set, where e_i is a differential algebraic equation or inequality equation about variables $\{x_1, x_2, ..., x_n\}$, differential semi-algebraic event $e \in \varphi_e$.

Init is an initial-value, initial events are usually given the initial values before running;

$\varphi = \{\varphi_e \mid e \in E\}$ is the differential algebra equations set or inequality equation set φ_e;

$\leq \subseteq E \times E$ is the causal relation, it is an irreflexive relation and satisfy transitivity;

$\# \subseteq E \times E$ is the conflict relation, it is a symmetric relation;

$||| \subseteq E \times E$ is the independently relation which means the event are independently.

In the differential semi-algebra event structure, cause relation, independent relationship, and conflict relationship are three kinds of relationship. $(d, e) \in |||$ means the events are independent. The execution sequence of the events is arbitrary or parallel. $(d, e) \in \leq$ express causal relationship between the two events, that means the cause and effect relation between d and e. $(d, e) \in \#$ express the conflict relationships between the events, it indicate that the conflicting eventsn d and e cannot occur at the same time. The differential semi-algebra system can express by figure. The conflict relationship is represented by dotted line, the events are represented by circle, the cause relationship is represented by arrow.

Example2.1 A rail transit control system which is described by a differential semi-algebra event structure, here is abbreviated as the train control center (TCC)[7]. The TCC can send command according to the ground subsystem which transmits the information from the monitor (MA) to the train transmission system to ensure the safety of the train. If there is an obstacle ahead the moving train, the TCC will send the distance the safety driving distances s from the monitor (MA). According to the situation we can establish a differential semi-algebra event structure model as Figure1:

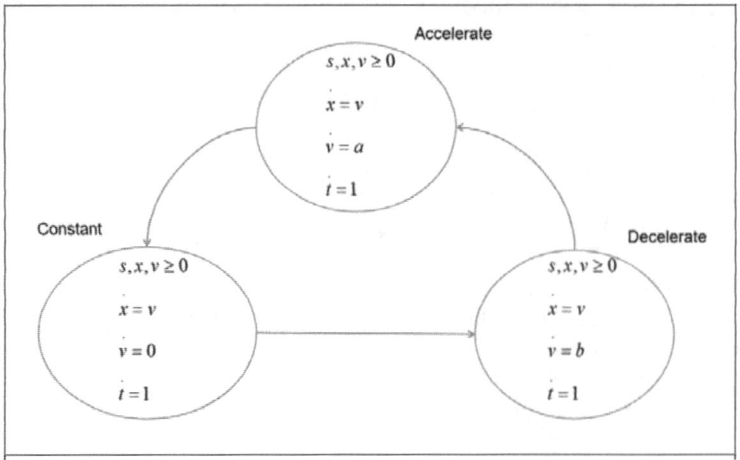

Figure 1. A rail transit control system

Differential semi-algebra systems is a complex physical systems, which includes discrete and continuous state. Continuous state is denoted by differential equations. In example2.1, the continuous movement (the train at distance x with the velocity v)is expressed by $\dot{x} = v$ ($\dot{x}$ is the time-derivative of x), we also can use the continuous variation of velocity v to get the time-derivative($\dot{v} = a$, a is the acceleration). Then we can model other behaviors by discrete states, for example, the instantaneous change of control variables like the acceleration (the changing of a by $a := -b$, $b > 0$ is braking force) or change of status information in discrete controllers. The continuous state will affect the decisions of discrete controllers (when v is too high that the train is on braking mode). Otherwise, the control variables of the continuous state will effect by the resulting of control choices (for example the changing control variable a can get from x twice derivative $x'' = a$). If exist continuous state with analog or discrete control superposition, then will cause more complex system behavior that can neither be verified by purely continuous reasoning nor by considering discrete change in isolation.

Hybrid systems[18], hybrid automat[19, 20] are the most widely research in control theory. These models contain discrete and continuous system. Figure1 is a simple train control system. Each node corresponds to a continuous system, and it is denoted by its differential algebra equations and the invariant region. For example, in brake node of Figure1, the differential algebra equations $\dot{x} = v$, $\dot{v} = a$, the invariant region $v \geq 0$ (in braking modes the train impossible move back). Edges is denotes the discrete switching behavior to the continuous modes. The edges can be expressed with conditions that need to hold and with discrete state transformations that take instantaneous effect. For example, the Figure1 can take an edge to leave node constant when the train is distance in less than s, then set the braking acceleration is $a := -b$, and enter the braking mode. So we can get the differential semi-algebra event structure model of rail transit control system in Figure2.

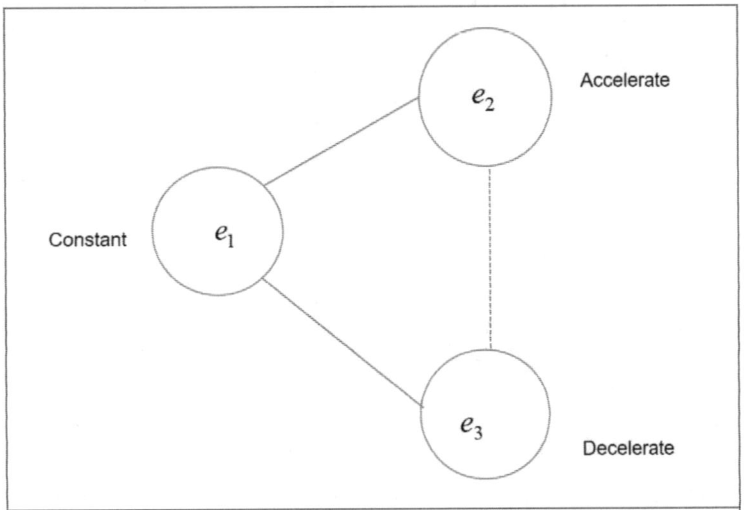

Figure 2. Differential semi-algebra event of rail transit control system

Events set $E = \{e_1, e_2, e_3\}$, where e_1 is the initial event.

Assume that the initial velocity is v_0 assuming that $v_0 = c$.

$$\phi_{e_1} = \begin{cases} x \geq 0, v \geq 0 \\ \dot{x} = v \\ \dot{v} = 0 \\ \dot{t} = 1 \end{cases} \quad \phi_{e_2} = \begin{cases} x \geq 0, v \geq 0 \\ \dot{x} = v \\ \dot{v} = a \\ \dot{t} = 1 \end{cases} \quad \phi_{e_3} = \begin{cases} x \geq 0, v \geq 0 \\ \dot{x} = v \\ \dot{v} = -b \\ \dot{t} = 1 \end{cases}$$

Cause set: $\{e_1 \leq e_2, e_1 \leq e_3\}$

Conflict set: $\{e_2 \# e_3\}$

Definition2.2 An evaluation events sequence $C = \{e_1, e_2, ..., e_n\}$, satisfying:

(1) $\{e_1, e_2, ..., e_n\}$ is free-conflict, i.e . $\forall e_i, e_j : \neg\{e_2 \# e_3\}$

(2) $\forall i, e : e_i \Rightarrow \exists j \prec i : e_j \leq e_i \wedge (e = e_j \vee e \# e_j)$

C is called a configuration .

Notice that in configuration C no events will be conflict with each other or cause each other. We use $C(\varepsilon)$ to denote the set of all configurations of E .

If a configuration C of ε is called maximal, iff $C \subseteq C' \in C(\varepsilon)$, implies that $C = C'$.

We called configuration C of ε is successfully terminated, iff $\forall d \in E : d \notin C \Rightarrow \exists e \in C$ with $d \# e$. We use $DAES$ to denoted the set of all differential semi-algebra event structures, $\varepsilon, \varepsilon_i \in DAES$ and $\varepsilon = \langle V, E, Init, \varphi, \prec, \#, ||| \rangle$, $\varepsilon_i = \langle V_i, E_i, Init_i, \varphi_i, \prec_i, \#_i, |||_i \rangle$. For two $DAES$ ε_1 and ε_2, we always assume that $E_1 \cap E_2 = \phi$.

3. Trace equivalence and bisimulation equivalence

3.1. The definition of equivalence

Based on the definition of differential semi-algebra event structure, it is easily to define the trace equivalence and bisimulation equivalence between two different differential semi-algebra event structures.

Definition3.1 There is a differential semi-algebra event structure $\varepsilon = \langle V, E, Init, \varphi, \prec, \#, ||| \rangle$, an events sequence $\{e_1, e_2, ..., e_n\}$ is a trace of ε iff $\{e_1, e_2, ..., e_n\} \in C(\varepsilon)$.

We use $trs(\varepsilon)$ to express all trace set of ε . If the event d and e are trace equivalence, then denote as $d \approx e$.

Definition3.2 There are two differential semi-algebra event structures ε_1 and ε_2 , $\varepsilon_1 = \langle V_1, E_1, Init_1, \varphi_1, \prec_1, \#_1, |||_1 \rangle$ and $\varepsilon_2 = \langle V_2, E_2, Init_2, \varphi_2, \prec_2, \#_2, |||_2 \rangle$, $\alpha \in trs(\varepsilon_1)$, $\beta \in trs(\varepsilon_2)$, if $\forall 1 \leq i \leq n$, $d_i \in E_1$, $e_i \in E_2 : d_i \approx e_i$, then $\alpha = d_1, d_2, ..., d_n$ and $\beta = e_1, e_2, ..., e_n$ are equivalence(denote as $\alpha \approx_{trs} \beta$)

Definition 3.3 There are two differential semi-algebra event structures ε_1 and ε_2 , $\varepsilon_1 = \langle V_1, E_1, Init_1, \varphi_1, \prec_1, \#_1, |||_1 \rangle$ and $\varepsilon_2 = \langle V_2, E_2, Init_2, \varphi_2, \prec_2, \#_2, |||_2 \rangle$, ε_1 and ε_2 are trace equivalence iff

(1) $\forall \alpha \in trs(\varepsilon_1), \exists \beta \in trs(\varepsilon_2)$, then $\alpha \approx_{trs} \beta$

(2) $\forall \beta \in trs(\varepsilon_2), \exists \alpha \in trs(\varepsilon_1)$, then $\alpha \approx_{trs} \beta$

In concurrent systems trace equivalence is widely used in non-deterministic equivalence verification.

Definition3.4 There are two differential semi-algebra event structures ε_1 and ε_2 , $\varepsilon_1 = \langle V_1, E_1, Init_1, \varphi_1, \prec_1, \#_1, |||_1 \rangle$ and $\varepsilon_2 = \langle V_2, E_2, Init_2, \varphi_2, \prec_2, \#_2, |||_2 \rangle$, A relation R is a bisimulation equivalent between ε_1 and ε_2 if and only if $(X_0, Y_0) \in R$ and if $(X, Y) \in R$, then

$\forall X \xrightarrow{d}_{\varepsilon_1} X', d \in E_1 \Rightarrow \exists e \in E_2, Y \xrightarrow{e} \varepsilon_2 Y' \wedge d \approx e \wedge (X', Y') \in R$;

$\forall Y \xrightarrow{e}_{\varepsilon_2} Y', e \in E_2 \Rightarrow \exists d \in E_1, X \xrightarrow{d} \varepsilon_1 X' \wedge d \approx e \wedge (X', Y') \in R$.

Differential semi-algebra event structure ε_1 and ε_2 are bisimulation equivalence (denote as $\varepsilon_1 \approx_{ib} \varepsilon_2$), if and only if differential semi-algebra event structures ε_1 and ε_2 are bisimulation equivalence.

3.2. The calculation method for differential semi-algebra event structure

Definition3.5: There are two differential semi-algebra event structures ε_1 and ε_2, $\varepsilon_1 = \langle V_1, E_1, Init_1, \varphi_1, \prec_1, \#_1, |||_1 \rangle$ and $\varepsilon_2 = \langle V_2, E_2, Init_2, \varphi_2, \prec_2, \#_2, |||_2 \rangle$, $e_1 \in E_1, e_2 \in E_2$ are the differential semi-algebra events, assume $\varphi_{e_1} = \{DF_1 = 0, DF_2 = 0, ..., DF_s = 0\}$ then the different semi-algebra events e_1 and e_2 are equivalence, if they satisfy the following:

(1) $V_1 = V_2$;

(2) $DAV(DF_1, DF_2, ..., DF_s) = DAV(DG_1, DG_2, ..., DG_t)$

where $DAV(.)$ means the differential semi-algebra variety of differential semi-algebra sets.

Corresponding to the two differential semi-algebraic events, the differential semi-algebraic varieties are equal, and then the zero set of the two differential algebraic events are equal. Then we can use Wu differential characteristic set method and columnar algebraic decomposition. Ritt-Wu ordering principle takes an important role in theory of Wu differential characteristic set; it shows that the zero of differential polynomials set and the zero relation of differential characteristic set.

Give the null empty differential polynomial set $DPS \subseteq R\{u_1, u_2, ..., u_m\}$ and the characteristic set DCS, according to Ritt-Wu ordering principle, then get

$$Zero(DPS) = Zero(DCS / J) \cup_i Zero(DPS \cup \{I_i, S_i\})$$

Here I_i, S_i refers to the initial type of DCS and separant respectively, J refers to IS-power product of DCS.

Then keeping adding the initial type and separant I_i, S_i, and keeping calculating differential characteristic set DCS_i of $DPS \cup \{I_i, S_i\}$, thus we can get the zero set decomposition of differential algebra collection DPS:

$$Zero(DPS) = \sum_i Zero(IRR_i / J_i)$$

Here IRR_i is an irreducible characteristic set, J_i denotes IS-power product of IRR_i. Because the irreducible characteristic set of differential algebra collection is existences and uniqueness, then we can use Wu differential characteristic set method to judge the equivalence of two differential algebra event structures.

Theorem3.1 There are two differential semi-algebra event structures ε_1 and ε_2, $\varepsilon_1 = \langle V_1, E_1, Init_1, \varphi_1, \prec_1, \#_1, |||_1 \rangle$ and $\varepsilon_2 = \langle V_2, E_2, Init_2, \varphi_2, \prec_2, \#_2, |||_2 \rangle$, $e_1 \in E_1, e_2 \in E_2$ are the differential semi-algebra events, and assume $\varphi_{e_1} = \{DF_1, DF_2, ..., DF_s\}$, $\varphi_{e_2} = \{DG_1, DG_2, ..., DG_t\}$, $Irrcharset(DF_1, DF_2, ..., DF_s)$ and $Irrcharset(DG_1, DG_2, ..., DG_t)$ are denote the irreducible characteristic set of differential algebra events e_1 and e_2 respectively. Then the differential semi-algebra events e_1 and e_2 are equivalence in the same regions if they satisfy the following:

$V_1 = V_2$;

$Irrcharset(DF_1, DF_2, ..., DF_s) = Irrcharset(DG_1, DG_2, ..., DG_t)$;

φ_{e_1} and φ_{e_2} have the same real solutions.

Then according to the DISCOVERER algorithm[21], the real solutions or interval solutions can be calculated.

Theorem3.2 There are two differential semi-algebra event structures ε_1 and ε_2, $\varepsilon_1 = \langle V_1, E_1, Init_1, \varphi_1, \prec_1, \#_1, |||_1 \rangle$ and $\varepsilon_2 = \langle V_2, E_2, Init_2, \varphi_2, \prec_2, \#_2, |||_2 \rangle$, $e_1 \in C_1, e_2 \in C_2$ are the differential semi-algebra events, C_1 and C_2 are configurations of ε_1 and ε_2, and assume $\varphi_{e_1} = \{DF_1, DF_2, ..., DF_s\}$, $\varphi_{e_2} = \{DG_1, DG_2, ..., DG_t\}$, if satisfy the fellowing:

$V_1 = V_2$;

φ_{e_1} and φ_{e_2} have the same real solutions of the two differential semi-algebra event structures ε_1 and ε_2, $e_1 \in C_1$ and $e_2 \in C_2$ are equivalence $e_1 \approx e_2$.

Trace equivalence and bisimulation equivalence are based on the differential semi-algebra events equivalence, then we can use the differential semi-algebra events equivalence in next section.

4. Variables refinement

4.1. The definition of variables refinement

For differential semi-algebra hybrid system, we use differential semi-algebra equation to express data flow exchange and data flow continuous evolution, so in hierarchical theory we attempts to refine variables like in literature [11].

After the variable refinement, the differential semi-algebra event structure is still a differential semi-algebra event structure. The definition of variables refinement is given below.

Definition 4. 1Variable refinement

Assume $DAES$ is a differential semi-algebra event structure domain, $\varepsilon \in DAES$, $ref : v \to DAES - \{\Phi\}$ is a variable refinement function for differential semi-algebra event structure ε, $ref(v)$ is the refinement for variable v, iff $\exists v \in V : ref(v)$ and it is non empty and finite. Thus, $ref(\varepsilon)$ is also the differential semi-algebra event structure, the definition is following:

ref is a refinement function. $V(e)$ is variables set of event e with differential semi-algebra event structure. The variable refinement in differential semi-algebra event structure can be defined as:

$ref(\varepsilon) = \langle V_{ref}, E_{ref}, \varphi_{ref}, \prec_{ref}, \#_{ref} \rangle$, where

$$V_{ref(\varepsilon)} = V_\varepsilon \cup V_{ref(v)} \tag{4.1}$$

$$Init_{ref(\varepsilon)} = Init_\varepsilon \cup Init_{ref(v)} \tag{4.2}$$

$$E_{ref} = E_\varepsilon \cup \{(e,e') \mid (e' \in E_{ref(v)}) \wedge v \in V(e)\} \tag{4.3}$$

$$\begin{cases} \varphi_e \quad if \quad e \in E_\varepsilon \\ \varphi(e,e') = \varphi_{e'} \quad if \quad v \in V(e), e' \in E_{ref(v)} \end{cases} \tag{4.4}$$

$$\begin{cases} d \le_{ref(\varepsilon)} e \quad if \ (d \le_\varepsilon e) \wedge (v \notin V(e)) \\ d \le_{ref(\varepsilon)} (e,e'') \quad if \ (d \le_\varepsilon e) \wedge (v \in V(e)) \wedge (e' \in E_{ref(v)})_b) \\ (e,e'') \le_{ref(\varepsilon)} e \quad if \ (v \in V(e)) \wedge (e'' \in E_{ref(v)})_{end}) \\ (e,e') \le_{ref(\varepsilon)} (e,e'') \quad if \ (v \in V(e)) \wedge (e' \le_{ref(v)} e'') \end{cases} \tag{4.5}$$

$$\begin{cases} d\#_{ref(\varepsilon)}e & if\ (d\#_{\varepsilon}e) \\ d\#_{ref(\varepsilon)}(e,e') & if\ (d\#_{\varepsilon}e)\wedge(v\in V(e))\wedge(e'\in E_{ref(v)})_b) \\ (e,e'')\#_{ref(\varepsilon)}(e,e'') & if\ (v\in V(e))\wedge(e''\#_{ref(v)}e'') \\ (d,d')\#_{ref(\varepsilon)}(e,e'') & if\ (d\#_{\varepsilon}e)\wedge(v\in V(d))\wedge(v\in V(e))\wedge(d',e'\in(E_{ref(v)}) \\ d\#_{ref(\varepsilon)}(e,e'') & if\ (d\leq_{\varepsilon}e)\wedge(v\in V(e))\wedge(e'\in(E_{ref(v)})\setminus E_{ref(v)})_b) \\ (d,d')\#_{ref(\varepsilon)}e & if\ (d\leq_{\varepsilon}e)\wedge(v\in V(d))\wedge(d'\in(E_{ref(v)}) \\ (d,d')\#_{ref(\varepsilon)}(e,e') & if\ (d\leq_{\varepsilon}e)\wedge(v\in V(d))\wedge(v\in V(e))\wedge(d',e'\in(E_{ref(v)}) \end{cases} \quad (4.6)$$

$$\begin{cases} d\,|||_{ref(\varepsilon)}e & if\ (d\,|||_{\varepsilon}e) \\ d\,|||_{ref(\varepsilon)}(e,e'') & if\ (d\,|||_{\varepsilon}e)\wedge(v\in V(e))\wedge(e'\in E_{ref(v)})) \\ (e,e')\,|||_{ref(\varepsilon)}(e,e') & if\ v\in V(e))\wedge(e'|||_{ref(v)}e'') \\ (d,d')\,|||_{ref(\varepsilon)}(e,e') & if\ (d\,|||_{\varepsilon}e)\wedge(v\in V(d))\wedge(v\in V(e))\wedge(d',e'\in E_{ref(v)}) \end{cases} \quad (4.7)$$

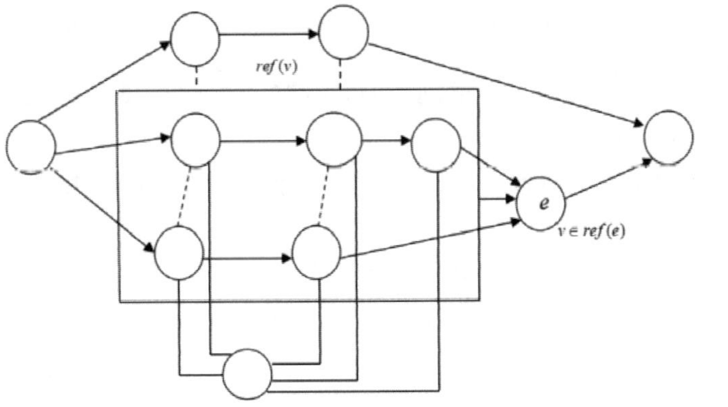

Figure 3. A refinement model of differential semi-algebra event structure

The Figure3 is described the variable refinement of differential semi-algebra event structure, and $ref(v)$ can get from refining variables v in differential semi-algebra event structure, $v\in V(e)$. The causal relationship $d\leq_{\varepsilon}e$ in original differential semi-algebra event structure is refined by the variable refinement function, and then the causal relationship exchange between d and $ref(v)$, and d and e are not any more causal relationship. The causal relation $e\leq_{\varepsilon}e''$ in original differential semi-algebra event structure is still maintains the causal relationship $e\leq_{ref(\varepsilon)}e''$ after the variable refinement. The conflict relations $d\#_{\varepsilon}e$ in original differential semi-algebra event structure will maintain the conflict between d and the refinement $ref(v)$ after the variables refinement. The independent relations $d\,|||_{\varepsilon}e$ in original differential semi-algebra event structure will maintain the independent relationship after variable refinement, means d will independ with $ref(v)$.

4.2. The calculation of variables refinement algorithm

Algorithm 4.2.1: The calculation algorithm of variables refinement $ref(v)$.

Input: differential semi-algebra event structure ε;

Output: differential semi-algebra event structure $ref(\varepsilon)$.

(1) Firstly, merging the variable set of ε and the variable set of $ref(v)$ as one variable set of $ref(\varepsilon)$.

(2) Let the event set of ε as the initial event set of $ref(\varepsilon)$, then loop computation the event $e \in E_\varepsilon$, and incorporate $\{\langle e,e'\rangle \mid x \in V(e), e' \in E_{ref(v)}\}$ into the event set of $ref(\varepsilon)$.

(3) Combine the initial event set of ε with the initial event set of subevent structure $ref(v)$ as the initial event set of $ref(\varepsilon)$.

(4) Copy the event $e \in E_\varepsilon$ to the differential semi-algebra set φ of $ref(\varepsilon)$. For the event $\langle e,e'\rangle$ which is in set $\{\langle e,e'\rangle \mid x \in V(e), e' \in E_{ref(v)}\}$, then change the differential semi-algebra equations $\varphi_{e'}$ as $\varphi_{\langle e,e'\rangle}$, and add the set $\varphi_{\langle e,e'\rangle}$ to the set of $ref(\varepsilon)$.

(5) Copy $Inv(e)$ of ε into differential semi-algebra event e of $ref(\varepsilon)$; for $\{\langle e,e'\rangle \mid x \in V(e), e' \in E_{ref(v)}\}$, change $Inv(e')$ as $Inv(\langle e,e'\rangle)$, and add $Inv(\langle e,e'\rangle)$ into the event $\langle e,e'\rangle$ of $ref(\varepsilon)$.

(6) Take out event e from set E_ε of ε, if e included the variables v of differential semi-algebra equations, then turn (6.1); otherwise turn (6.2), complete turn (7).

(6.1) Take out the event $d \in E_\varepsilon$ which is different from e orderly, and turn (6.1.1), complete turn (6).

(6.1.1) If $d \leq_\varepsilon e$, then d and the initial events of $ref(v)$ are formed new causal relationship, with the other events in $ref(v)$ are formed conflict relationship, and $d \#_{ref(\varepsilon)} e$, then turn (6.1.2).

(6.1.2) If $e \leq_\varepsilon d$, then d and all the events of $ref(v)$ are formed conflict relation, and $e \leq_{ref(\varepsilon)} d$, then turn (6.1.3).

(6.1.3) If $d \#_\varepsilon e$, then d and all the events of $ref(v)$ are formed conflict relation, and $d \#_{ref(\varepsilon)} e$, then turn (6.1.4).

(6.1.4) If $d \,|||_\varepsilon\, e$, then d and all the events of $ref(v)$ are formed concurrence relationship, and $d \,|||_{ref(\varepsilon)}\, e$, then turn (6.1).

(6.2) Take out $d \in E_\varepsilon$ which is different from e orderly, and $v \in V(d)$, turn（6.2.1）, complete turn (6).

(6.2.1) If $d \leq_\varepsilon e$， $d \leq_{ref(\varepsilon)} e$； ； then turn（6.2.2）.

(6.2.2) If $e \leq_\varepsilon d$， $e \leq_{ref(\varepsilon)} d$； then turn（6.2.3）.

(6.2.3) If $d \#_\varepsilon e$， $d \#_{ref(\varepsilon)} e$； then turn（6.2.4）.

(6.2.4) If $d \,|||_\varepsilon\, e$， $d \,|||_{ref(\varepsilon)}\, e$； then turn（6.2）.

(7）End.

5. Equivalences in variables refinement

The definition of interweave trace equivalence and interweave bisimulation equivalence have given in section3. Now we will investigate which equivalences are kept under the variables refinement.

Definition 5.1 Assume $\varepsilon \in DAES$, an events sequence $\{e_1, e_2, ..., e_n\} \in C(\varepsilon)$ satisfy $\{e_1, e_2, ..., e_n\} \in C(\varepsilon) : e_i \xrightarrow{e_i} e_{i+1}, i = 1, 2, ..., n$.

Here we use $trs(\varepsilon)$ to denotes all the traces of differential semi-algebra event structure ε.

Definition5.2 Suppose differential semi-algebra event structure ε_1, $\varepsilon_2 \in DAES$, $\alpha \in trs(\varepsilon_1)$, $\beta \in trs(\varepsilon_2)$ .We will get $\alpha = d_1, d_2, ..., d_n$ and $\beta = e_1, e_2, ..., e_n$ are equivalence(denotes $\alpha \approx_{trs} \beta$), if $\forall 1 \leq i \leq n, d_i \in E_{\varepsilon_1}, e_i \in E_{\varepsilon_2}, d_i \approx e_i$.

Definition5.3 Interweave traces equivalence, suppose differential semi-algebra event structure ε_1, $\varepsilon_2 \in DAES$, $\varepsilon_1, \varepsilon_2$ are called traces equivalence(denotes $\varepsilon_1 \approx_{trs} \varepsilon_2$)if and only if $trs(\varepsilon_1) \approx trs(\varepsilon_2)$.

Definition5.4 Suppose differential semi-algebra event structure ε_1, $\varepsilon_2 \in DAES$, a relation $R \subseteq X_{\varepsilon_1} \times Y_{\varepsilon_2}$ is called an interweave bisimulation equivalence between ε_1 and ε_2 if and only if $(X_0, Y_0) \in R$ and if $(X, Y) \in R$, then

$$\forall X \xrightarrow{d}_{\varepsilon_1} X', d \in E_1 \Rightarrow \exists e \in E_2, Y \xrightarrow{e} \varepsilon_2 Y' \wedge d \approx e \wedge (X', Y') \in R;$$

$$\forall Y \xrightarrow{e}_{\varepsilon_2} Y', e \in E_2 \Rightarrow \exists d \in E_1, X \xrightarrow{d} \varepsilon_1 X' \wedge d \approx e \wedge (X', Y') \in R.$$

Differential semi-algebra event structure ε_1, ε_2 are called interweave bisimulation equivalence (denotes $\varepsilon_1 \approx_{ib} \varepsilon_2$), if and only if ε_1 and ε_2 have an interweave bisimulation equivalence.

From the literature [13,18], we know that the traditional interweave trace equivalence and bisimulation equivalence are not kept.

Definition5.5 Suppose differential semi-algebra event structure $\varepsilon_1, \varepsilon_2 \in DAES$, $ct(\alpha)$ is the casual trace α, suppose the casual trace set $ct(\varepsilon_1) = \{ct(\alpha) \mid \alpha \in trs(\varepsilon_1)\}$ and $ct(\varepsilon_2) = \{ct(\beta) \mid \beta \in trs(\varepsilon_2)\}$, called ε_1 and ε_2 casual trace equivalence(denote $\varepsilon_1 \approx_{ct} \varepsilon_2$), if and only if

$\forall \mu \in ct(\varepsilon_1), \exists v \in ct(\varepsilon_2)$, make $u \approx v$;

$\forall v \in ct(\varepsilon_2), \exists \mu \in ct(\varepsilon_1)$, make $u \approx v$;

Theorem5.1 If differential semi-algebra event structure $\varepsilon_1, \varepsilon_2 \in DAES$, ref is a refinement function, and $\varepsilon_1 \approx_{ct} \varepsilon_2$, then $ref(\varepsilon_1) \approx_{ct} ref(\varepsilon_2)$.

Proof:

$\forall \tilde{u} = ct(ref(\varepsilon_1))$, $\forall \tilde{v} - ct(ref(\varepsilon_2))$, make $\tilde{u} \approx \tilde{v}$

For $\tilde{u} = ct(\tilde{\alpha})$, then $\tilde{\alpha} \in trs(ref(\varepsilon_1))$. By the definition of refinement, $\exists \alpha \in trs(\varepsilon_1)$, makes $\tilde{\alpha} = ref(\alpha)$. For $\varepsilon_1 \approx_{ct} \varepsilon_2$, then $\tilde{u} = ct(\tilde{\alpha})$, $\exists v \in ct(\varepsilon_2)$, makes "$u \approx v$". Assume $u = ct(\alpha)$, then $\exists \beta \in trs(\varepsilon_2)$, makes $ct(\alpha)$ and $ct(\beta)$ are isomorphism. Based on the definition of isomorphism, we can know that the variables in α and β are corresponding. Then get the trace after refinement the variables of β, satisfy $ct(\tilde{\alpha})$ and $ct(\tilde{\beta})$ are isomorphism. Then $v = ct(\tilde{\beta})$, so $\exists \tilde{v} = ct(ref(\varepsilon_2))$, makes $\tilde{u} \approx \tilde{v}$.

Proof $\forall \tilde{v} = ct(ref(\varepsilon_1)), \forall \tilde{u} = ct(ref(\varepsilon_2))$, makes $\tilde{u} \approx \tilde{v}$.

According to the symmetry, we can get that.

The interweave traces equivalence and the interweave bisimulation equivalence are not keeping after variables refinement. But if we restrict the properties of differential semi-algebra event structure, then we can get some new results. If there is not independent relationship in differential semi-algebra event structure, then the interweave traces equivalence and the interweave bisimulation equivalence are preserved after variables refinement.

Theorem5.2 Assume two differential semi-algebra event structures $\varepsilon_1, \varepsilon_2 \in DAES$, if $|||_{\varepsilon_1} = |||_{\varepsilon_2} = \phi$, and

ref is a refinement function, then

(1) $\varepsilon_1 \approx_{trs} \varepsilon_2 \Rightarrow ref(\varepsilon_1) \approx_{trs} ref(\varepsilon_2)$

(2) $\varepsilon_1 \approx_{ib} \varepsilon_2 \Rightarrow ref(\varepsilon_1) \approx_{ib} ref(\varepsilon_2)$

Proof:

(1) $\alpha \in trs(\varepsilon_1)$, for $|||_{\varepsilon_1} = \phi$, then we can get $ct(\alpha) = \{\alpha\}$, so $ct(\varepsilon_1) = \{ct(\alpha) \mid \alpha \in trs(\varepsilon_1)\} = \{\{\alpha\} \mid \alpha \in trs(\varepsilon_1)\}$. Similarly $ct(\varepsilon_2) = \{ct(\beta) \mid \beta \in trs(\varepsilon_2)\} = \{\{\beta\} \mid \beta \in trs(\varepsilon_2)\}$. Because $\varepsilon_1 \approx_{trs} \varepsilon_2$, if and only if

(a) $\alpha \in trs(\varepsilon_1)$, $\exists \beta \in trs(\varepsilon_2)$ makes $\alpha \approx \beta$, then get $\{\alpha\} \approx \{\beta\}$.

(b) $\beta \in trs(\varepsilon_2)$, $\exists \alpha \in trs(\varepsilon_1)$ makes $\alpha \approx \beta$, then get $\{\alpha\} \approx \{\beta\}$.

So $ct(\varepsilon_1) \approx ct(\varepsilon_2)$, get $\varepsilon_1 \approx_{trs} \varepsilon_2 \Rightarrow \varepsilon_1 \approx_{ct} \varepsilon_2$. Meanwhile, based on theorem5.2, get $\varepsilon_1 \approx_{ct} \varepsilon_2 \Rightarrow ref(\varepsilon_1) \approx_{ct} ref(\varepsilon_2)$, from the proposition5.1, we can get $ref(\varepsilon_1) \approx_{ct} ref(\varepsilon_2) \Rightarrow ref(\varepsilon_1) \approx_{trs} ref(\varepsilon_2)$, so $\varepsilon_1 \approx_{trs} \varepsilon_2 \Rightarrow ref(\varepsilon_1) \approx_{trs} ref(\varepsilon_2)$.

(2) Similarly (1), we can get the result.

6. Conclusion

Differential semi-algebra hybrid system is a new formalized hybrid system model; it is widely used in engineering technology. At present, the perfect theory of differential semi-algebra hybrid system is not established. In this paper we based on the concurrent theory, using differential semi-algebraic symbols and symbols mixed with numerical calculation method to integration and application research, and using the differential semi-algebraic event structure as the system model to build the variables refinement theory of differential semi-algebraic hybrid system. Then we study the problem of keeping equivalence after variables refinement. Concretely include:(1) the definition of variables refinement of differential semi-algebra event structure; (2)the definition of kinds of equivalence which are based on the dynamic behavior characteristics of the differential semi-algebraic event structure model, for example the trace equivalence and bisimulation equivalence; (3)equivalent determination: studied the determination of various equivalence of differential semi-algebraic event structure, and given the equivalent calculation method; (4)lastly research which equivalence are kept after the refinement.

Acknowledgments

This work was supported by the National Natural Science Foundation of China(62002105), Ph. D. Programs Foundation(BSQD2019024).

References

[1] Wirth N 1971 Program Development by Stepwise Refinement *Communications of the ACM* **14**(4) 221-7

[2] Gorrieri R, Rensink A 2001 Action Refinement, To appear as Chapter XVI of the *Handbook of Process Algebra* (Elsevier Science) 1047-1147

[3] Aceto L 1992 Action refinement in Process Algebra(Cambridge: Cambridge Univ. Press) 265-271

[4] Doyen L, Frehse G, Pappas G J, and Platzer A 2018 Verification of Hybrid Systems. In Edmund M. Clarke and Thomas A. Henzinger and Helmut Veith, editors, *Handbook of Model Checking* (Springer, Cham)1047-1110

[5] Khadim U, Cuijpers P J L 2012 Repairing time-determinism in the process algebra for hybrid systems ACPhssrt,*Theoretical Computer Science*(Elsevier Science Publishers Ltd)**446** 20-47

[6] Asarin E, Pace G, Schneider G 2008 Algorithmic analysis of polygonal hybrid systems, Part II: Phase portrait and tools *Theoretical Computer Science* **390**(1) 1-26

[7] Gan T, Chen M, Li Y, Xia B and Zhan N 2017 Reachability Analysis for Solvable Dynamical Systems *IEEE Transactions on Automatic Control* **63**(7) 2003-2018

[8] Sankaranarayanan S, Sipma H B, Manna Z 2005 Non-linear loop invariant generation using Gröbner bases *ACM SIGPLAN Notices* **39**(1) 318-329

[9] Yang H, He A, Zhang Z, Yang S, Liu Y 2013 Approximate Completed Trace Equivalence of Linear Algebra Transition Systems. In: Yin Z., Pan L., Fang X. (eds) Proceedings of The Eighth International Conference on Bio-Inspired Computing: Theories and Applications (BIC-TA), 2013. Advances in Intelligent Systems and Computing, (Berlin, Heidelberg: Springer) 233-237

[10] Deng H, Wu J Z, Zhou N 2011 Approximate Equivalence and Optimization for High-Level Datapath *Journal of Information & Computational Science* **8**(16) 4131-4142

[11] Pola G, Pepe P, Benedetto M.D D 2015 Symbolic models for time - varying time - delay systems via alternating approximate bisimulation *International Journal of Robust & Nonlinear Control* **25**(14) 2328-47

[12] Liu B, Wu J, Xie Z 2014 Approximate Bisimulation Equivalence and Variable Refinement

Applied Mathematics& Information Sciences **8**(4) 1959-1966
[13] Winskel G, Event structures 2001 in: W. Brauer, W. Reisig, G. Rozenberg(Eds.), Petri-Nets: Applications and Relationships toother Models of Concurrency, AdvancesinPetriNets Part II; *Proc. Advanced Course, Bad Honnef, September, LectureN otesin Computer Science* (Berlin: Springer) **255** 325-392
[14] Buchberger B 2005 A View on the Future of Symbolic Computation. ISSAC, (Proceedings of the 2005 international symposium on Symbolic and algebraic computation:ACM Press)**2005**:1-1
[15] Wu W J 2003 Mathematical mechanization (Beijing: Science press)
[16] Dulę B I 2009 Computational Algebra Support for the Chen-Fliess-Sussmann Differential Equation Robot Motion and Control 396 133-142
[17] Wilson D 2014 Advances in cylindrical algebraic decomposition *University of bath*:the degree of Doctor of Philosophy
[18] Wurdemann H A, Stilli A, Althoefer K 2015 Lecture Notes in Computer Science: An Antagonistic Actuation Technique for Simultaneous Stiffness and Position Control. In: Liu H., Kubota N., Zhu X., Dillmann R. (eds) Intelligent Robotics and Applications *Lecture Notes in Computer Science* (Cham: Springer) **9246** 164-174
[19] Benvenuti L, Bresolin D, Collins P, Ferrari A, Geretti L, Villa T 2012 Ariadne: Dominance Checking of Nonlinear Hybrid Automata Using Reachability Analysis. In: Finkel A., Leroux J., Potapov I. (eds) Reachability Problems. RP *Lecture Notes in Computer Science*(Berlin, Heidelberg: Springer) **7550** 79-91
[20] He A, Wu J, Yang S, et al 2014 Approximate Equivalence of the Hybrid Automata with Taylor Theory *Journal of Applied Mathematics* **1** 1-5
[21] Chen Y, Xia B, Yang L, Zhan N 2007 Generating Polynomial Invariants with DISCOVERER and QEPCAD. In: Jones C.B., Liu Z., Woodcock J. (eds) Formal Methods and Hybrid Real-Time Systems. *Lecture Notes in Computer Science* **4700** 67-82

Analysis of COVID-19 Based on SEQR Dynamic Model in Xiaogan City, Hubei Province

Jiayu Shen[1,2], Jiehao Zhang[1,2] and Yongxian Wen[1,2,*]

[1]College of Computer and Information, Fujian Agriculture and Forestry University, Fuzhou, China
[2]Institute of Applied Statistics, Fujian Agriculture and Forestry University, Fuzhou, China

[*]Corresponding author's email: Wen9681@sina.com

Abstract. In this study, we established a *SEQR* model for the spread of *COVID-19*. The impact of city lockdown measures and other factors on the spread of the epidemic was discussed in this paper through dynamic analysis and numerical simulation of the model. The research results showed that the input ratio of the exposed population should be controlled to at least 0.001 if the government wanted to effectively control the epidemic. Closing the city was the most effective way. The exposed population of *COVID-19* in Xiaogan City was hoping to be cleared on February 29 under strict control. The closed city could be released around March 28. But at the same time, we should pay attention to prevent the virus from entering it again. The peak number of existing patients would reduce by about 1,400 if city lockdown measures were taken in Xiaogan City four days in advance.

1. Introduction

The sudden outbreak of *COVID-19* posed a threat to people's health at the end of 2019. The virus is highly contagious and concealed, so it can spread rapidly in the population and has a certain incubation period.[1] It is difficult for patients to be detected in time because the people in the incubation period do not show symptoms, which leads to the spread of the virus on a large scale around the world, causing unpredictable losses. A total of 178,088,231 people were diagnosed worldwide and a total of 3,852,743 people died until June 18, 2021.[2] It is of great significance for human health to study the effectiveness of epidemic control measures.

In terms of epidemic control, obvious results have been achieved in China. A series of anti-epidemic measures were adopted by the Chinese government to minimize the number of infections as much as possible. One of the measures that had caused widespread concern was the city lockdown measures taken at the beginning of the epidemic. The concept of lockdown measures was put forward by Li Lanjuan on January 19, 2020. That content was to prevent entry and exit and control the epidemic in Wuhan.[3] Subsequently, many cities in Hubei were closed. As we all know, the closure of the city has a positive effect on the control of the epidemic. It can reduce population movement and control the input of exposed people, thus reducing the risk of mass infection. The quantification of the impact of the lockdown on epidemic control will help us take more immediate measures in future epidemic outbreaks. In the early stage of the domestic epidemic, the severity of the epidemic in Xiaogan City was second only to Wuhan City, and the city lockdown measures in Xiaogan City were taken at 0:00 on January 26.[4] Therefore, we selected fact data from Xiaogan City for simulation.

Relevant scientific researchers have successively carried out research, hoping to provide scientific evidence for the stability of the epidemic, help the government make decisions, effectively control the epidemic and protect public health. Aiming at the outbreak of *COVID-19* in Wuhan, Jonathan M Read et al[5] made the early estimation of epidemiological parameters and epidemic prediction. Tang et al[6] discussed the transmission risk estimation of *COVID-19* and its significance for public health interventions. The traditional epidemiological related documents were referenced in this paper.[7,8] Define the *COVID-19* infected quarantined as Q category. Use class E to indicate exposed people. The susceptible population is defined as the S class. It is defined removed people as R category that refers to those who healed during the incubation period, those who were cured after being infected, and those who received epidemic information and took corresponding protective measures to avoid infection.

2. Model

2.1. Model assumption

S, E, Q, R indicate respectively $S(t), E(t), Q(t), R(t)$ in the *COVID-19* transmission mechanism, which respectively represent the number of the four types people at time t. $N(t)$ represents the value of the total population at time t. And $N(t) = S(t) + E(t) + Q(t) + R(t)$. The study will establish a *COVID-19* transmission model based on the following assumptions.

- (1) The exposed patients carrying the virus are the main reason for the large-scale spread of the epidemic compared with the confirmed patients being admitted to the hospital for isolation and treatment because the symptoms of exposed patients are unknown and the virus has a strong spread. Therefore, we mainly consider the virus transmission effect of exposed patients in our model.
- (2) Since the exposed patients do not show obvious symptoms, we should also consider the incubation period patients among the urban importers. We assume that a is the input ratio of exposed patients, $0 \leq a \leq 1$. Then $1 - a$ is the input ratio of susceptible people, $0 \leq 1\text{-}a \leq 1$.
- (3) Assuming that the patients are cured and produce corresponding antibodies in the body so immune to the same virus.
- (4) The natural mortality rate is the same for the four categories (S, E, Q, R). Symbol d represents daily natural mortality. Infected persons have daily disease mortality p.
- (5) The constant input factor is out of consideration of the mobility of life and the expansion of the scope of development activities due to convenient transportation. Since the epidemic in Xiaogan City is mainly caused by imported cases, the number of people going to Xiaogan City is set as the input A of the model.($A > 0$, And the input includes birth)
- (6) The disclosure of disease information and the government's vigorous publicity to mobilize people not to gather, wash hands frequently, wear masks, and follow-up measures such as movement restrictions, which have reduced the number of susceptible people who may be infected. Let q be the proportion of people who are susceptible be removed from the infected population due to government propaganda.
- (7) Only consider the person-to-person spread of *COVID-19* in this model.

2.2. Model flow chart

Based on the above analysis of the epidemic background and model assumptions, the flow chart of the research and establishment of the model (the basic transmission mechanism of *COVID-19*) is shown in Figure 1 .

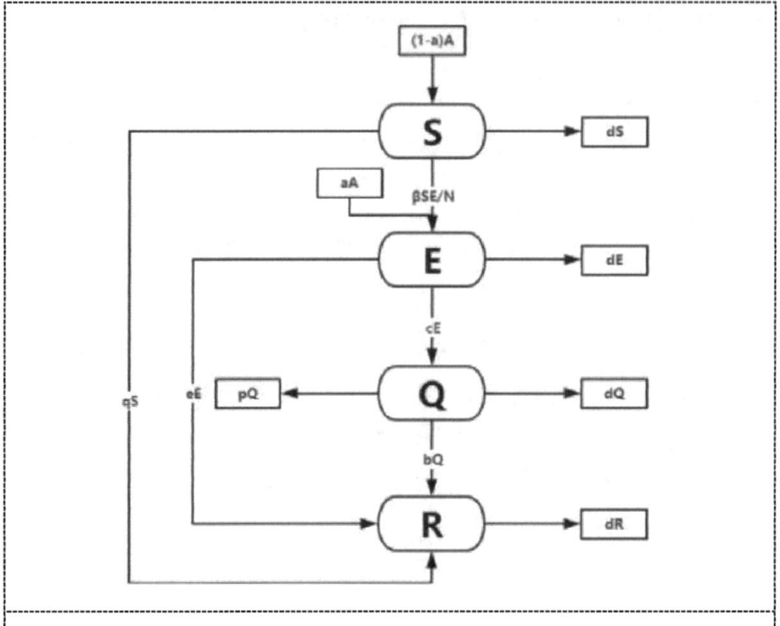

Figure 1. The transmission mechanism of COVID-19.

2.3. Model establishment

Build the model as follows.

$$\begin{cases} \dfrac{dS}{dt} = (1-a)A - \beta \dfrac{S}{N} E - qS - dS \\ \dfrac{dE}{dt} = aA + \beta \dfrac{S}{N} E - (d+e+c)E \\ \dfrac{dQ}{dt} = cE - (b+p+d)Q \\ \dfrac{dR}{dt} = qS + bQ + eE - dR \end{cases} \tag{1}$$

Where a is the input ratio of class E. A is the constant input rate of S category. q is the daily removed ratio of susceptible people. The coefficient of disease transmission rate is β. e is the daily natural healing rate of category E. b is the recovery rate of the confirmed patients. p is the daily disease mortality of Q category. c is the daily disease incidence of exposed patients. d is the daily natural death rate. It can be seen that $a, A, q, \beta, e, b, p, c, d$ are all non-negative constants.

The kinetic analysis shows that when $a \neq 0$, the system has the only disease equilibrium point $P^*(S^*, E^*, Q^*, R^*)$ and satisfies $S^* = N^* - E^* - Q^* - R^*$, $Q^* = \dfrac{A - dN^*}{\rho}$, $E^* = \dfrac{\delta}{c} Q^*$, $R^* = \dfrac{q(N^* - E^* - Q^*) + bQ^* + eE^*}{d+q}$. Where N^* is the only solution of the equation $\dfrac{aAp}{A-dN} - \beta \cdot \dfrac{\delta(d+e) + c(b+d)}{c(b+q)} \cdot \dfrac{\delta}{c} \cdot \dfrac{1}{p} \cdot (\dfrac{A}{N} - d) + \dfrac{\delta}{c}(\dfrac{\beta \cdot d}{d+q} - w) = 0$. And $w = d+e+c$, $\delta = b+p+d$. At $a = 0$, the model has a free equilibrium point $P_0 = (\dfrac{A}{d}, 0, 0, 0)$. When $\dfrac{\beta \cdot d}{(\mathrm{d}+q)(d+e+c)} > 1$ the model has a similar disease equilibrium point $P_1(S^*, E^*, Q^*, R^*)$.

Similar to the literature[8], the analysis results show that the disease equilibrium point p_1 is always locally asymptotically stable. When $a=0$, if $\frac{\beta \cdot d}{(d+q)(d+e+c)}>1$, the free equilibrium point p_0 is an unstable point. When $\frac{\beta \cdot d}{(d+q)(d+e+c)}<1$, and the free equilibrium point P_0 is locally asymptotically stable. The local asymptotic stability of the disease equilibrium point P^* means that disease may appear, so disease control is critical. So when $a=0$, decreasing β or increasing q or e is helpful to control the epidemic.

3. Simulation

The ODE module in MATLAB was used to simulate the construction of the above-mentioned *COVID-19* transmission model. First, the parameter values were estimated as shown in Table 1.

Table 1. Parameter values

Symbol	Unit	Value	Source
a	-	0.03	Assumption
A	People/day	23780	Estimation
b	Treatment healing rate/day	0.11	Estimation
$1/c$	Incubation period/day	7	[4]
β	-	0.0000000011886	[5]
d	Constant dead rate/day	0.0000136986	Population mortality in 2018
e	Natural healing rate/day	0.034821	[5]
p	Disease mortality/day	0.00000068331	[5]
q	Removal rate/day	0.0357	Assumption
$S(0)$	people	4920000	Permanent population in 2018

Theoretical analysis results showed that strict detection or input isolation made theoretically $a=0$ if the input of exposed people could be eliminated, then the system had a free equilibrium point, as shown in Figure 2. The value of parameter a reflected the strength and effect of government input control. The simulation showed that strict input monitoring could effectively control the epidemic. It could be seen in Figure 2, the value of a needed to be controlled at least below 0.001 to stabilize the epidemic at a low level. Otherwise, the number of infected people would increase sharply. It could be found that the number of existing patients with *COVID-19* had a stable value of about 1700 when $a=0.01$, which was about 700 more than when $a=0.005$, and about 1500 more than when $a=0.001$. And it was expected that the number of patients would be cleared around 70 days with strictly eliminating the input of exposed people.

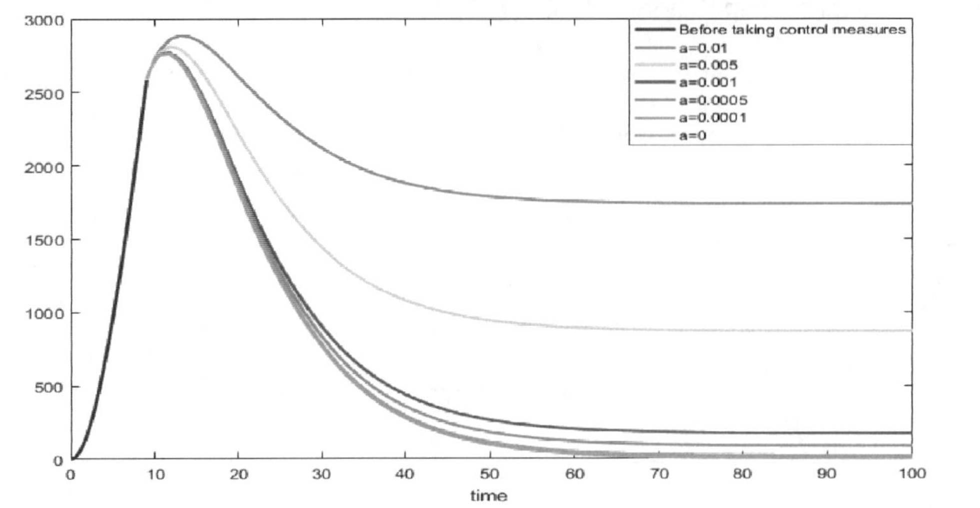

Figure 2. Adjust the value of a and observe the curve of the number of patients with time under different control measures.

The curve that simulated the number of exposed persons and the number of confirmed persons in Xiaogan City over time is shown in Figure 3. It could be seen that the lockdown measures were taken in Xiaogan City at 24:00 on January 24, and the daily peak of the number of residents was about 2,750, which was basically in line with the actual number of confirmed diagnoses. The simulation results showed that exposed people were cleared on February 29 and the number of existing quarantine patients was basically cleared on March 25. Taking into account that urban development required the operation of materials and personnel in various industries, at the same time, the confirmed patients received good isolation and treatment in the hospital. Therefore, it was recommended to consider lifting the lockdown around March 28 after the number of people in the incubation period was cleared for 28 days. Continue to strictly monitor the imported personnel in the city, such as concentrated medical observation for the possible incubation period of 14 days.

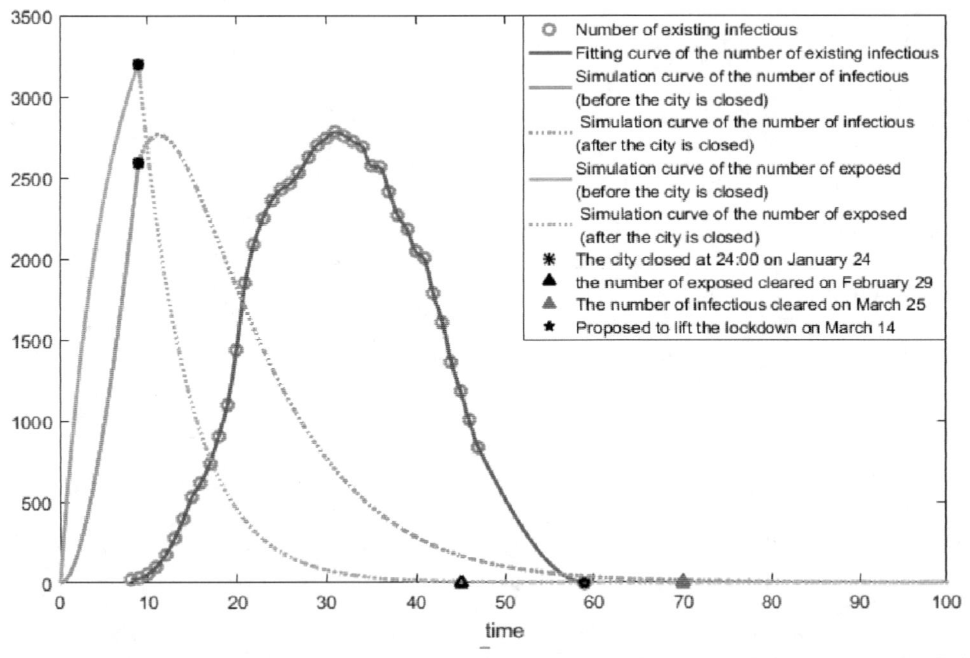

Figure 3. Simulate the changes of the exposed class and the number of existing patients over time.

The above analysis results showed that the value of a needed to be controlled at least below 0.001 to stabilize the epidemic situation at a low level. A notable feature of the *COVID-19* virus infection was the presence of asymptomatic infected persons, so the surveillance was not completely effective. We could only take measures to close the city to ensure that the input of exposed people was strictly eliminated to make $a = 0$. That is to say, lockdown measures were necessary effective measures to control the vicious spread of the epidemic. But making sure the time of the beginning of taking measures was also important in epidemic control. In Figure 5, we adjusted the parameter time in the system to compare the impact of the lockdown measures one day, two days, three days, and four days in advance on the epidemic. It is concluded that closing the city one day in advance could reduce the number of peak daily patients in Xiaogan City by about 350. Closing the city four days in advance could reduce the number of peak daily patients in Xiaogan City by about 1,400. Closing the city one day in advance in the face of emerging infectious diseases without medical protection was the guarantee of life.

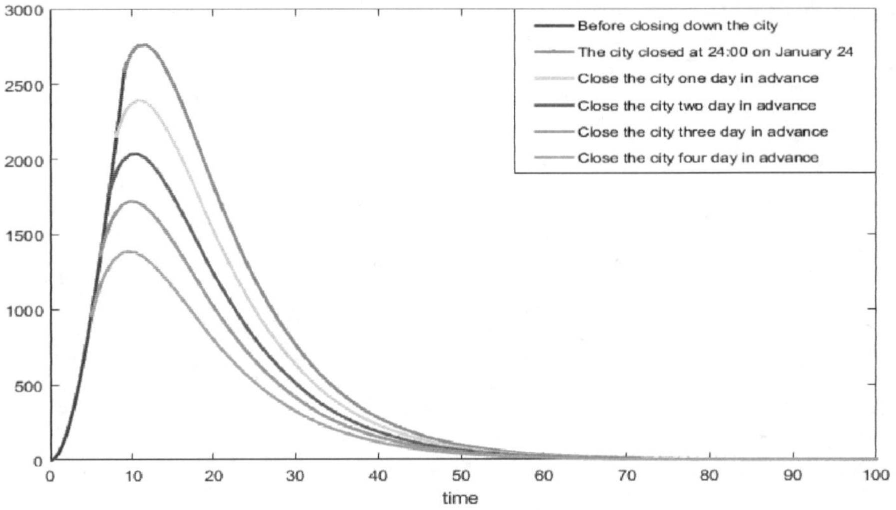

Figure 4. Adjust the lockdown time to simulate changes in the number of patients over time.

Of course, the closure of the city is an extreme method that will inevitably harm social and economic development. Therefore, theoretically studying the effects of other measures is also urgent. The publication of disease information by relevant government departments and the guidance of preventive measures enable people to understand the spread of the disease and increase awareness of prevention. Conscious action and good protective measures can effectively reduce the infection. The simulation in Figure 5 showed that the city was also closed at 24:00 on January 24, and the a value was reduced to 1/2 and 1/4 of the current parameter, indicating that the public reduced their activities due to the knowledge of the epidemic-related information, thereby reducing the proportion of imported virus carriers in the region. Adjusting the value of q before and after the closure of the city to $q = 0.0357, 0.0714$ respectively indicates that the self-awareness of part of the public and measures such as community closure and suspension of work and school after the closure of the city before the government takes control measures to make q increase. The simulation changed the value of 1/4 of the original value before closing the city and $q = 0.0714$. It could be found that the number of existing patients on the peak day might be reduced by about 2,000 people compared to before the parameter adjustment.

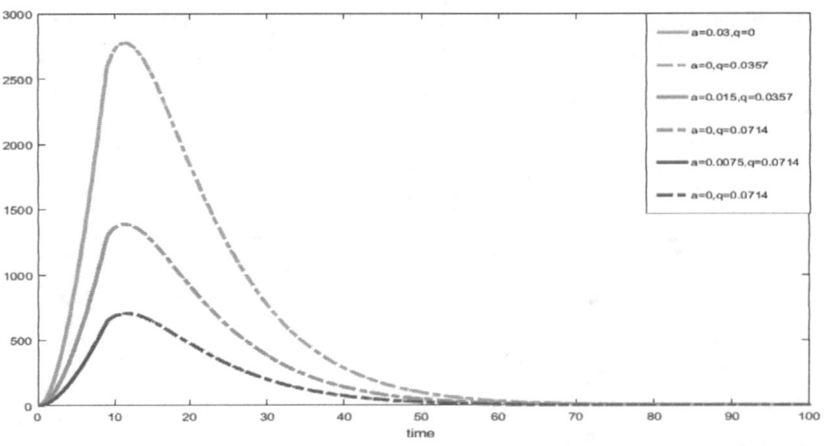

Figure 5. The public had different levels of understanding of public health knowledge when the epidemic broke out.

4. Conclusion

This paper established a *SEQR* infectious disease dynamics model that reflected the closure measures, the effect of information dissemination, and the restrictions on the movement of people, to study the spread and control of *COVID-19*. The analysis results drew the following conclusions.

In the epidemic control work of sudden infectious diseases, the key is to control the import of viruses carried by exposed people, especially viruses that are highly transmissible during the incubation period. Take this *COVID-19* virus as an example. Compared with a value was 0.01, the stable value of the number of patients would decrease by about 1500 if the control a value was below 0.001. And if necessary, the city lockdown policy was adopted to prohibit personnel input so that the theoretical $a = 0$, the epidemic situation could be stabilized around 70 days and patients could be cleared. The simulation showed that the number of existing patients on the peak day was about 2,750 in the past two months of the epidemic development in Xiaogan City. The incubation period might be cleared on February 29 under the influence of the city lockdown policy. The number of existing quarantine patients would be cleared to zero around March 25. It could be considered to lift the lockdown around March 28 after the number of people in the incubation period was cleared for 28 days. However, it was also necessary to continue monitoring and centralized medical observation of imported personnel in the city to prevent re-entry of viruses.

By summarizing the experience in the prevention and control of the epidemic, we found that the epidemic could be stabilized at a relatively low level if government departments took early control measures. The simulation found that the number of existing patients on the peak day may be reduced to about 1,350 if the lockdown policy was adopted in Xiaogan 4 days in advance, which is about 1,400 less than the number of existing patients in the actual situation. The early gatherings and activities of some people who did not understand the prevention of infectious diseases were important reasons for the spread of the epidemic. It could be found in the simulation that the number of existing patients on the peak day might be reduced by about 2000 people compared to the number of existing patients before the parameter adjustment by changing the value of a to 1/4 of the original value and setting $q = 0.0714$. In summary, we found that spreading the knowledge of infectious disease prevention to the public, adhering to the attitude of early detection and isolation, and strictly restricting the import of the virus carried by latent persons can effectively prevent the spread of the epidemic in the early stage of an outbreak of infectious disease. The city closure policy should be adopted immediately when necessary, supplemented by measures such as community closure, suspension of work and school.

Acknowledgments

This work was supported by National Natural Science Foundation of China (Grant/Award Number: 32071892) and Science and Technology Innovation Special Fund of Fujian Agricultural and Forestry University (Project No: KFA20109A, KFA19127A).

References

[1] National Health Commission Disease Control and Prevention Bureau, Chinese Center for Disease Control and Prevention 2020 *Guidelines for Public Protection of Pneumonia from New Coronavirus Infection(Xinxing Gaunzhuang Bingdu Ganrna de Feiyan* Gongzhong Fanghu Zhinan) (Beijing: People's Medical Publishing House) p

[2] *Novel coronavirus pneumonia epidemic trackin in real time* (2021-06-18) https:// sa.sogou.com/new-weball/page/sgs/epidemic?type_page=VR

[3] Sun M. 2020 Academician Li Lanjuan *Disclosed Details of Wuhan's "Closed City" for the First Time(Li Lanjuan Yuanshi Shouci Pilu Wuhan "Fengcheng" Xijie).* Party life (Dang De Shenghuo)**04** 30-2

[4] *Xiaogan City COVID-19 Prevention and Control Headquarters (Order No.7)* (Xiaoganshi Xinxing Guanzhuang Bingdu Ganran De Feiyan Fangkong Zhihuibu (7 Hao Ling)) 2020-01-25 http://www.xiaogan.gov.cn/gsgg/741990.jhtml

[5] Jonathan M R, Jessica R B, Derek A C, Antonia H, Chris P J 2020 *Novel coronavirus 2019-nCoV: early estimation of epidemiological parameters and epidemic predictions* Phil. Trans. R. Soc. B **376**

20200265
[6] Tang B, Wang X, Li Q, Nicola L B, Tang S Y, XiaoY N, Wu J H 2020 *Estimation of the Transmission Risk of 2019-nCov and Its Implication for Public Health Interventions* J Clin Med **9**(2) 462
[7] Ma Z E, Zhou Y C, Wang W D, et al 2004 *Mathematical Modeling and Research on Dynamics of Infectious Diseases* (Beijing: Science Press) p
[8] Chen Y X 2012 *Global Stability of an Avian Influenza Model* Mathematics in Practice and Theory **42**(03) 119-25

Soliton Solutions of Kaup-Kupershimdt Equation Using HPTM

Tiejun Chen

Yiyang Medical College, Yiyang, Hunan, China

Corresponding author's e-mail: 362392572@qq.com

Abstract. In this paper, we employ the idea of homotopy perturbation transform method(HPTM), which is a combination of Laplace transformation and homotopy perturbation method(HPM) for solving nonlinear ordinary and partial differential equations. The equations are Laplace transformed and nonlinear terms are represented by He's polynomials. The solutions are obtained in the series form which converges at fast rate with easily computable terms. Comparison of standard perturbation method shows that HPTM is appropriate even for system without any small/large parameters and therefore it can be applied more extensively than traditional perturbation techniques. HPTM is used to obtain soliton solution of Kaup-kupershimdt(KK) equation, which is modified form of fifth order KdV equation. Numerical results are presented in tabular form to show the accuracy, efficiency and reliability of this method.

Keywords. Nonlinear differential equation, Homotopy Perturbation Transform Method, Fourier series, Laplace transformed

1. Introduction

Nonlinear partial differential equations(NLPDEs) are used to describe the mathematical models of complex physical phenomena around US in the world. In recent years,study of nonlinear processes gains much interest due to their vast application. Therefore the number of concepts used by mathematicians to prove the accuracy and efficiency of their processes and determine the application of mathematical modelling. These concepts include Soliton, compactness,cusps,breathers,kinks and pecans have entered into many branches of natural sciences such as mathematics,biology,chemistry,communication and especially in different branches of physics like condensed matter physics,field theory. Plasma physicst fluid dynamics,hydrodynamics,nonlinear optics, solid state physics and acoustics. Nonlinear partial differential equations are studied in three different ways.which depends on the type of their solution. Mathematician and researchers face three major problems when they studied nonlinear partial differential equations (NLPDEs):

1. To find a solvable NLPDEs and study their properties.
2. To find the exact solution of NLPDEs.
3. To find special(particular)solution of nonlinear differential equations.

In recent years,many mathematicians and researchers studied NLPDEs using various methods to determine the exact and approximate solutions of NLPDEs such as Hirota method [1], Inverse scattering method [2], extended tanh method [3], The Tanh method [4], Lie transformations [5], Projective Riccati equation method [6],generalized projective Riccati method [7], Exp-function method [8], Jacobi elliptic function expansion method [9], G/G expansion method [10],Sine-Cosine method [11], Adomian decomposition method(ADM) [12], variational iteration method(VIM) [13], homotopy analysis method(HAM) [14], differential transform method(DTM) [15], reduced differential transform method(RDTM) [16], F-expansion method [17], parameterized perturbation method [18], Elzaki projected differential transform method(EPDTM) [18-20], and homotopy perturbation method(HPM)[21].

In this paper,we consider the well-known generalized fifth order equation of the form

$$\frac{\partial\phi}{\partial t}+\alpha\phi\frac{\partial^3\phi}{\partial x^3}+\beta\frac{\partial\phi}{\partial x}\frac{\partial^2\phi}{\partial x^2}+\gamma\phi^2\frac{\partial\phi}{\partial x}+\frac{\partial^5\phi}{\partial x^5}=0, \tag{1}$$

Where α, β and γ are arbitrary constants. We can assign different values to α, β and γ a,which directly affect the properties of Eq.(1). For example,when $\alpha=10$, $\beta=20$, and $\gamma=30$, than Eq.(1),becomes Lax equation and it becomes Sawada-Kotera(SK)equation if $\alpha=3$, $\beta=6$, and $\gamma=2$. . These two equations have N-soliton solution and an infinite set of conservation laws. The Kaup-kupershimdt (KK)equation with $\alpha=5$, $\beta=\frac{25}{2}$ and $\gamma=5$ is also known to be integrable and have bilinear representation,but its N-soliton solution is apparently not known.

In this paper,we use homotopy perturbation transform method(HPTM)to solve KK equation with different initial conditions and real constants to determine the accuracy and efficiency of HPTM.HPTM well suited to physical model problems because it does not require any Linearization,assumptions and restrictions which may amend the problem. HPTM essentially demonstrate how we can use the Laplace transform to calculate approximate solution of nonlinear differential equations using decomposition methods. HPTM present the solution in fast converging series,which may be appearing in closed form.In section II the basic idea of homotopy perturbation is discussed,while in section III, implementation of the proposed method of nonlinear problem is explained.Solution of the problems is given in section IV. Convergence analysis is drawn in section V.

2. Homotopy Perturbation transform method

To illustrate the idea of this modified method, we consider a nonlinear non-homogeneous partia differential equation with initial conditions of the form

$$D\phi(x,t)+R\phi(x,t)+N\phi(x,t)=f(x,t) \tag{2}$$

$$\phi(x,0)=g(x), \tag{3}$$

Where $g(x)\in C(R_+), D$ is the first order linear differential operator $D=\partial/\partial t$,R is the linear differential operator and N is nonlinear differential operator and f(x,t) is source term. Taking the Laplace transform (denoted by L) on both sides of Eq.(2)

$$L[D\phi(x,t)]+L[R\phi(x,t)]+L[N\phi(x,t)]=L[f(x,t)] \tag{4}$$

Using the differentiation property of Laplace Transform, we have

$$L[\phi(x,t)]=\frac{g(x)}{s}+\frac{1}{s}L[f(x,t)]-\frac{1}{s}L[R\phi(x,t)]-\frac{1}{s}L[N\phi(x,t)] \tag{5}$$

Operating Laplace inverse transform of both sides of Eq.(5), gives

$$\phi(x,t)=G(x,t)-L^{-1}\left[\frac{1}{s}L[R\phi(x,t)+N\phi(x,t)]\right] \tag{6}$$

Where G(x,t)represents the term arising from the source term and prescribed initial conditions. The homotopy perturbation method admits the solution in the form

$$\phi\ (x,t)=\sum_{n=0}^{\infty}p^{n}\phi_{n}(x,t) \tag{7}$$

and nonlinear terms can decompose as

$$\mathrm{N}\phi\ (x,t)=\sum_{n=0}^{\infty}p^{n}H_{n}(\phi) \tag{8}$$

using He'S polynomia

$$H_{\mathrm{n}}(\phi_{0},\phi_{1}.....\phi_{n})=\frac{1}{ni}\frac{\partial^{n}}{\partial p^{n}}\left[N\left(\sum_{i=0}^{\infty}p^{i}\phi_{i}\right)\right] \tag{9}$$

Substitute Eq.(7)and Eq.(8)in Eq.(6). According to HPTM,we construct a homotopy of the form

$$\sum_{\mathrm{n}=0}^{\infty}p^{n}\phi_{n}(x,\mathrm{t})=G(x,t)-p\left(L^{-1}\left[\frac{1}{s}L\left[R\sum_{n=0}^{\infty}p^{n}\phi_{n}(x,t)+\sum_{n=0}^{\infty}p^{n}H_{n}(\phi)\right]\right]\right) \tag{10}$$

Which is the coupling of Laplace transform and homotopy method using He'S polynomials. Equating the coefficients of like powers of P, we have

$$\mathrm{p}^{0}\ :\ \phi_{0}(x,t)=G(x,t) \tag{11}$$

$$\mathrm{p}^{1}\ :\ \phi_{1}(x,t)=L^{-1}\left[\frac{1}{s}L\left[R\phi_{0}(x,t)+H_{0}(\phi)\right]\right], \tag{12}$$

$$\mathrm{p2}\ :\ \phi_{2}(x,t)=L^{-1}\left[\frac{1}{s}L\left[R\phi_{1}(x,t)+H_{1}(\phi)\right]\right], \tag{13}$$

$$\mathrm{p}^{3}\ :\ \phi_{3}(x,t)=L^{-1}\left[\frac{1}{s}L\left[R\phi_{2}(x,t)+H_{2}(\phi)\right]\right], \tag{14}$$

Clearly,the general recursive relation is given by

$$p^{\mathrm{n+1}}:\phi_{n+1}(x,t)=L^{-1}\left[\frac{1}{s}L\left[R\phi_{n}(x,t)+H_{n}(\phi)\right]\right], \tag{15}$$

3. Implementation of the method

Consider generalized fifth-order nonlinear evolution equation known as Kaup-Kupershmidt equation of the form

$$\frac{\partial\phi}{\partial\mathrm{t}}+\alpha\phi\frac{\partial^{3}\phi}{\partial x^{3}}+\beta\frac{\partial\phi}{\partial x}\frac{\partial^{2}\phi}{\partial x^{2}}+\gamma\phi^{2}\frac{\partial\phi}{\partial x}+\frac{\partial^{5}\phi}{\partial x^{5}}=0 \tag{16}$$

$$\phi\ (x,0)=g(x), \tag{17}$$

First, we apply Laplace transform of both sides of Eq.(16)

$$\mathrm{L}\left[\frac{\partial\phi}{\partial x}\right]=-\alpha L\left[\phi\frac{\partial^{3}\phi}{\partial x^{3}}\right]-\beta L\left[\frac{\partial\phi}{\partial x}\frac{\partial^{2}\phi}{\partial x^{2}}\right]-\gamma L\left[\phi^{2}\frac{\partial\phi}{\partial x}\right]-L\frac{\partial^{5}\phi}{\partial x^{5}}$$

$$\mathrm{s}L[\phi(x,t)]-\phi(x,0)=-\alpha L\left[\phi\frac{\partial\phi^{3}}{\partial x^{3}}\right]-\beta L\left[\frac{\partial\phi}{\partial x}\frac{\partial\phi^{2}}{\partial x^{2}}\right]-\gamma L\left[\phi^{2}\frac{\partial\phi}{\partial x}\right]-L\left[\frac{\partial^{5}\phi}{\partial x^{5}}\right],$$

$$L\phi(x,t)=\frac{1}{s}g(x)-\frac{1}{s}L\left[\alpha L\phi\frac{\partial^3\phi}{\partial x^3}+\beta\frac{\partial\phi}{\partial x}\frac{\partial^2\phi}{\partial x^2}+\gamma\phi^2\frac{\partial\phi}{\partial x}+\frac{\partial^5\phi}{\partial x^5}\right], \tag{18}$$

Apply inverse Laplace transform of both sides of Eq.(18)and using Eq.(17), we get

$$\phi(x,t)=g(x)-L^{-1}\left[\frac{1}{s}L\left[\alpha L\phi\frac{\partial^3\phi}{\partial x^3}+\beta\frac{\partial\phi}{\partial x}\frac{\partial^2\phi}{\partial x^2}+\gamma\phi^2\frac{\partial\phi}{\partial x}+\frac{\partial^5\phi}{\partial x^5}\right]\right], \tag{19}$$

Now according to the homotopy perturbation method, we construct a homotopy of the form

$$\sum_{n=0}^{\infty}p^n\phi_n(x,t)=g(x)-p\left(L^{-1}\left[\frac{1}{s}L\left[\sum_{n=0}^{\infty}p^nH_{1n}(\phi)+\sum_{n=0}^{\infty}p^nH_{2n}(\phi)+\sum_{n=0}^{\infty}p^nH_{3n}(\phi)+\sum_{n=0}^{\infty}p^n\phi_n(x,t)\right]\right]\right) \tag{20}$$

Where $H_{1n}(\phi),H_{2n}(\phi) and H_{3n}(\phi)$ are He'S polynomials representing the nonlinear terms in the solution. Comparing the coefficients of like powers of P on both sides of Eq.(20), we get

$$p^0:\phi_0(x,t)=g(x) \tag{21}$$

$$p^1:\phi_1(x,t)=-L^{-1}\left[\frac{1}{s}L\left[H_{10}(\phi)+H_{20}(\phi)+H_{30}(\phi)+\phi_0(x,t)\right]\right], \tag{22}$$

$$p^2:\phi_2(x,t)=-L^{-1}\left[\frac{1}{s}L\left[H_{11}(\phi)+H_{21}(\phi)+H_{31}(\phi)+\phi_1(x,t)\right]\right], \tag{23}$$

$$p^3:\phi_3(x,t)=-L^{-1}\left[\frac{1}{s}L\left[H_{12}(\phi)+H_{22}(\phi)+H_{32}(\phi)+\phi_2(x,t)\right]\right], \tag{24}$$

$$p^4:\phi_4(x,t)=-L^{-1}\left[\frac{1}{s}L\left[H_{13}(\phi)+H_{23}(\phi)+H_{33}(\phi)+\phi_3(x,t)\right]\right], \tag{25}$$

$$p^{n+1}:\phi_{n+1}(x,t)=-L^{-1}\left[\frac{1}{s}L\left[H_{1n}(\phi)+H_{2n}(\phi)+H_{3n}(\phi)+\phi_n(x,t)\right]\right], \tag{26}$$

Where some of the He'S polynomial used are as follows

$$H_{10}(\phi)=\phi_0\frac{\partial^3\phi_0}{\partial x^3}, \tag{27}$$

$$H_{11}(\phi)=\phi_0\frac{\partial^3\phi_1}{\partial x^3}+\phi_1\frac{\partial^3\phi_0}{\partial x^3}, \tag{28}$$

$$H_{12}(\phi)=\phi_0\frac{\partial^3\phi_2}{\partial x^3}+\phi_1\frac{\partial^3\phi_1}{\partial x^3}+\phi_2\frac{\partial^3\phi_0}{\partial x^3}, \tag{29}$$

$$H_{13}(\phi)=\phi_0\frac{\partial^3\phi_3}{\partial x^3}+\phi_1\frac{\partial^3\phi_2}{\partial x^3}+\phi_2\frac{\partial^3\phi_1}{\partial x^3}+\phi_3\frac{\partial^3\phi_0}{\partial x^3}, \tag{30}$$

$$H_{20}(\phi)=\frac{\partial\phi_0}{\partial x}\frac{\partial^2\phi_0}{\partial x^2}, \tag{31}$$

$$H_{21}(\phi)=\frac{\partial\phi_0}{\partial x}\frac{\partial^2\phi_1}{\partial x^2}+\frac{\partial\phi_1}{\partial x}\frac{\partial^2\phi_0}{\partial x^2}, \tag{32}$$

$$H_{22}(\phi)=\frac{\partial\phi_0}{\partial x}\frac{\partial^2\phi_2}{\partial x^2}+\frac{\partial\phi_1}{\partial x}\frac{\partial^2\phi_1}{\partial x^2}+\frac{\partial\phi_2}{\partial x}\frac{\partial^2\phi_0}{\partial x^2}, \tag{33}$$

$$H_{23}(\phi)=\frac{\partial\phi_0}{\partial x}\frac{\partial^2\phi_2}{\partial x^2}+\frac{\partial\phi_1}{\partial x}\frac{\partial^2\phi_1}{\partial x^2}+\frac{\partial\phi_2}{\partial x}\frac{\partial^2\phi_1}{\partial x2}+\frac{\partial\phi_3}{\partial x}\frac{\partial^2\phi_0}{\partial x^2}, \tag{34}$$

$$H_{30}(\phi)=\phi_0^2\frac{\partial\ \phi_0}{\partial x}, \tag{35}$$

$$H_{31}(\phi)=2\phi_0\phi_1\frac{\partial\ \phi_0}{\partial x}+\phi_0^2\frac{\partial\ \phi_1}{\partial x}, \tag{36}$$

$$H_{32}(\phi)=2\phi_0\phi_2\frac{\partial\ \phi_0}{\partial x}+\phi_1^2\frac{\partial\ \phi_0}{\partial x}+2\phi_0\phi_1\frac{\partial\ \phi_2}{\partial x}+\phi_0^2\frac{\partial\ \phi_2}{\partial x}, \tag{37}$$

$$H_{33}(\phi)=(2\phi_0\phi_3+2\phi_1\phi_2)\frac{\partial\ \phi_0}{\partial x}+\phi_1^2\frac{\partial\ \phi_1}{\partial x}+2\phi_0\phi_2\frac{\partial\ \phi_1}{\partial x}+2\phi_0\phi_1\frac{\partial\ \phi_2}{\partial x}+\phi_0^2\frac{\partial\ \phi_3}{\partial x}, \tag{38}$$

and so on.

4. Illustration of method by numerical examples

4.1. Example

Consider a fifth order Kaup-Kupershmidt(KK)equation with

$$\alpha=5,\beta=\frac{25}{2},\gamma=5andg(x)=-2k^2+\frac{24k^2}{1+e^{kx}}-\frac{24k^2}{(1+e^{kx})^2},$$

So Eq.(16) becomes

$$\frac{\partial\phi}{\partial t}=5\phi\frac{\partial^3\phi}{\partial x^3}+\frac{25}{2}\frac{\partial\phi}{\partial x}\frac{\partial^2\phi}{\partial x^2}+5\phi^2\frac{\partial\phi}{\partial x}+\frac{\partial^5\phi}{\partial x^5}, \tag{39}$$

$$\phi(x,0)=-2k^2+\frac{24k^2}{1+e^{kx}}-\frac{24k^2}{(1+e^{kx})^2}, \tag{40}$$

Applying Laplace transform and homotopy perturbation method on Eq.(39) and using Eq.(40), we get

$$\sum_{n=0}^{\infty} p^n \phi_n(x,t) = -2k^2 + \frac{24k^2}{1+e^{kx}} - \frac{24k^2}{(1+e^{kx})^2}$$
$$+L^{-1}\left[\frac{1}{s}L\left(5\sum_{n=0}^{\infty} p^n H_{1n}(\phi) + \frac{25}{2}\sum_{n=0}^{\infty} p^n H_{2n}(\phi) + 5\sum_{n=0}^{\infty} p^n H_{3n}(\phi) + \sum_{n=0}^{\infty} p^n \phi_n(x,t)\right)\right] \quad (41)$$

Now, equating the like powers of P and using the values of He'S polynomial from Eq.(27) to Eq.(38) to obtain values $\phi_i(x,t)(i=0,1,2......)$

$$p^0 : \phi_0(x, t) = -2k^2 + \frac{24k^2}{1+e^{kx}} - \frac{24k^2}{(1+e^{kx})^2}, \quad (42)$$

$$p^1 : \phi_1(x, t) = \left(\frac{-480\mathrm{k}^8 e^{kx}(e^{kx}-1)}{(1+e^{kx})^3}\right)t, \quad (43)$$

$$p^2 : \phi_2(x, t) = \left(\frac{-1368\mathrm{k}^{14} e^{kx}(4e^{kx}-e^{2kx}-1)}{(1+e^{kx})^4}\right)t^2, \quad (44)$$

$$p^3 : \phi_3(x, t) = \left(\frac{-7248\mathrm{k}^{20} e^{kx}(1-15e^{kx}+15e^{2kx}-e^{3kx})}{(1+e^{kx})^5}\right)t^3, \quad (45)$$

$$p^4 : \phi_4(x, t) = \left(\frac{-11520\mathrm{k}^{26} e^{kx}(1-30e^{kx}+72e^{2kx}-30e^{3kx}+e^{kx})}{(1+e^{kx})^5}\right)t^4, \quad (46)$$

and so on. $\mu(x,t)$ is the approximate fourth order solution of the Eq. (6. 39) obtained by HPTM

$$\mu(x, t) = -2k^2 + \frac{24k^2}{1+e^{kx}} - \frac{24k^2}{(1+e^{kx})^2} - \left(\frac{-480\mathrm{k}^8 e^{kx}(e^{kx}-1)}{(1+e^{kx})^3}\right)t - \left(\frac{-1368\mathrm{k}^{14} e^{kx}(4e^{kx}-e^{2kx}-1)}{(1+e^{kx})^4}\right)t^2$$
$$-\left(\frac{-7248\mathrm{k}^{20} e^{kx}(1-15e^{kx}+15e^{2kx}-e^{3kx})}{(1+e^{kx})^5}\right)t^3 - \left(\frac{-11520\mathrm{k}^{26} e^{kx}(1-30e^{kx}+72e^{2kx}-30e^{3kx}+e^{kx})}{(1+e^{kx})^5}\right)t^4, (47)$$

Which is the solution closed form solution of Eq.(16).

4.2. Example

Consider Eq.(16),when, $\alpha = 15, \beta = 15, \gamma = 45 with g(x) = \frac{4a}{3} - \frac{4a}{p}\sec h^2(\sqrt{ax})$

$$\frac{\partial \phi}{\partial t} = 15\phi\frac{\partial^3\phi}{\partial x^3} + 15\frac{\partial\phi}{\partial x}\frac{\partial^2\phi}{\partial x^2} + 45\phi^2\frac{\partial\ \phi}{\partial x} + \frac{\partial^5\phi}{\partial x^5}, \quad (48)$$

$$\phi(x,0) = \frac{4a}{3} - \frac{4a}{p}\sec h^2(\sqrt{ax}), \quad (49)$$

Apply Laplace transform of both sides of Eq.(48)and using initial conditions given in Eq.(49), according to homotopy perturbation transform method. Construct ahomotopy of the form

$$\sum_{n=0}^{\infty} p^n \phi_n(x,t) = \frac{4a}{3} - \frac{4a}{p}\sec h^2(\sqrt{ax}) + L^{-1}\left[\frac{1}{s}L\left(15\sum_{n=0}^{\infty} p^n H_{4n}(\phi) + 15\sum_{n=0}^{\infty} p^n H_{5n}(\phi) + 45\sum_{n=0}^{\infty} p^n H_{6n}(\phi) + \sum_{n=0}^{\infty} p^n \phi_n(x,t)\right)\right] \quad (50)$$

Comparing the coefficients of like powers of P and using the values of He'S polynomials,we get the values of $\phi_i (i = 0,1,2......)$

$$\phi_0(x,t) = \frac{4a}{3} - \frac{4a}{p}\sec h^2(\sqrt{ax}), \quad (51)$$

$$\phi_1(x,t) = \frac{64a^{5/2}}{p}((3a-5p)\sec h^2(\sqrt{ax})\tanh^2(\sqrt{ax}))t, \quad (52)$$

$$\phi_2(x,t) = \frac{256a^3}{p}\begin{bmatrix} a(3a-5p)^2 \sec h^2(\sqrt{ax})\tanh(\sqrt{ax}) + a\sec h^2(\sqrt{ax}) \\ (-\frac{a}{2}(3a-5p)^2) + a(3a-5p)^2\tanh^2(\sqrt{ax}) \end{bmatrix} t^2, \quad (53)$$

$$\phi_3(x,t) = \frac{1024a^3}{p}\begin{bmatrix} (\frac{1}{8\sqrt{a}}(3a-5p)\ \sec h^2(\sqrt{ax})\tanh(\sqrt{ax}))(\frac{1}{4}(3a-5p)^2 + \frac{1}{2}(3a-5p)^2 \\ \tanh^2(\sqrt{ax})) + 2\sec h^2(\sqrt{ax})(-\frac{a^{9/2}}{2})(3a-5p)^3\tanh(\sqrt{ax}) \\ -\frac{1}{64a^{3/2}}(3a-5p)\tanh(\sqrt{ax})(-\frac{1}{16}(3a-5p)^2) + \frac{1}{8}(3a-5p)^2\tanh^2(\sqrt{ax}) \end{bmatrix} t^3, \quad (54)$$

In order to verify numerically whether the proposed method lead to higher accuracy, we can evaluate the numerical solution using fourth and eighth term of the approximation. Difference of absolute errors between exact and numerical solution is shown in Table 1 and 2. Numerical results obtained by proposed method show that a good approximation has been achieved with the exact solution of the KK equation. Absolute errors can be minimized by adding more approximation terms. Numerical results proved the accuracy and efficiency of the proposed method. Furthermore, EPDTM does not require any discretization of variables.

5. Convergence analysis

In this section convergence analysis of the Kaup-Kupershimdt equation is studied. Let US consider Hilbert space H which may be defined by

$H = L^2((\alpha,\beta)\times[0,T])$ The set of applications

$$\phi:(\alpha,\beta)\times[0,T] \to R\,with \int_{(\alpha,\beta)\times[0,T]} \phi^2(x,s)ds\,d\tau < +\infty \quad (55)$$

According to the assumption of Eq.(39)KK equation can be written as

$$L_1(\phi) = \frac{\partial}{\partial t} = 5\phi\frac{\partial^3\phi}{\partial x^3} + \frac{25}{2}\frac{\partial\phi}{\partial x}\frac{\partial^2\phi}{\partial x^2} + 5\phi^2\frac{\partial\phi}{\partial x} + \frac{\partial^5\phi}{\partial x^5}, \quad (56)$$

6. Hypothesis for convergence

HPTM is convergent if the following hypotheses are satisfied

- 1) $\left(L_1(\phi)-L_1(\psi),\phi-\psi\right)\geq k\|\phi-\psi\|^2\,\forall k>0,\phi,\psi\in H$ (57)
- 2) For M>O,there exist a constant C(M)>o,such that for $\phi,\psi\in H$ with $\|\phi\|\leq M,\|\psi\|\leq M$,we have for every $\eta\in H$

$$\left(L_1(\phi)-L_1(\psi),\eta\right)\leq C(M)\|\phi-\psi\|\eta \tag{58}$$

Theorem: (Sufficient condition of convergence for KK equation)The HPTM applied to KK equation without initial and boundary conditions converges towards particular solution.

$$L_1(\phi)=5\phi\frac{\partial^3\phi}{\partial x^3}+\frac{25}{2}\frac{\partial\phi}{\partial x}\frac{\partial^2\phi}{\partial x^2}+5\phi^2\frac{\partial\phi}{\partial x}+\frac{\partial^5\phi}{\partial x^5}, \tag{59}$$

Proof: In order to proof convergence of the above theorem,we will verify the Eq.(57)and Eq.(58). For $\mathrm{L}_1(\phi)$

$$\mathrm{L}_1(\phi)-\mathrm{L}_1(\psi)=5\phi\frac{\partial^3\phi}{\partial x^3}+\frac{25}{2}\frac{\partial\phi}{\partial x}\frac{\partial^2\phi}{\partial x^2}+5\phi^2\frac{\partial\phi}{\partial x}+\frac{\partial^5\phi}{\partial x^5}-5\psi\frac{\partial^3\psi}{\partial x^3}-\frac{25}{2}\frac{\partial\psi}{\partial x}\frac{\partial^2\psi}{\partial x^2}-5\psi^2\frac{\partial\psi}{\partial x}-\frac{\partial^5\psi}{\partial x^5}$$

$$=5\left[\phi\frac{\partial^3\phi}{\partial x^3}-\psi\frac{\partial^3\psi}{\partial x^3}\right]+\frac{25}{2}\left[\frac{\partial\phi}{\partial x}\frac{\partial^2\phi}{\partial x^2}-\frac{\partial\psi}{\partial x}\frac{\partial^2\psi}{\partial x^2}\right]+5\left[\phi^2\frac{\partial\phi}{\partial x}-\psi^2\frac{\partial\psi}{\partial x}\right]+\left[\frac{\partial^5\phi}{\partial x^5}-\frac{\partial^5\psi}{\partial x^5}\right]$$

Now

$$\begin{aligned}(\mathrm{L}_1(\phi)-\mathrm{L}_1(\psi),\ \phi\text{-}\psi)&=\frac{25}{4}(\frac{\partial}{\partial x}(\phi_x^2-\psi_x^2),\phi-\psi)+5(\frac{\partial}{\partial x}(\phi^3-\psi^3),\phi-\psi)\\&+\frac{5}{4}(\frac{\partial^3}{\partial x^3}(\phi^2-\psi^2),\phi-\psi)+(\frac{\partial^5}{\partial x^5}(\phi-\psi),\phi-\psi)\end{aligned} \tag{60}$$

But there exist a real constant χ_i for i=l,2,...9, such that

$$(\frac{\partial^5}{\partial x^5}(\phi-\psi),\phi-\psi)\geq\chi_1\|\phi-\psi\|^2 \tag{61}$$

and according to Schwartz inequality

$$(\frac{\partial^3}{\partial x^3}(\phi^2-\psi^2),\phi-\psi)\geq 3\chi_2M^2\|\phi-\psi\|^2 \tag{62}$$

similarly, we have

$$(\frac{\partial}{\partial x}(\phi^3-\psi^3),\phi-\psi) \tag{63}$$

and

$$(\frac{\partial}{\partial x}(\phi_x^2-\psi_x^2),\phi-\psi)\le \chi_4\|\phi_x^2-\psi_x^2\|^2\|\phi-\psi\|\le \chi_4\|\phi_x-\psi_x\|\ \|\phi_x+\psi_x\|\|\phi-\psi\|\le \varepsilon\|\phi-\psi\|^2 \quad (64)$$

Where $\exists\varepsilon > 0/_{\varepsilon} = \chi_4\chi_5\chi_6$. Substituting Equ.(61)to Eq. (63) in Eq.(60)

$$L_1(\phi)-L_1(\psi)\ge(\frac{25}{4}\varepsilon+10\chi_3 M+\frac{15}{4}\chi_2 M^2+\chi_1)\|\phi-\psi\|^2 = k\|\phi-\psi\|^2, \quad (65)$$

where $k=(\frac{25}{4}\varepsilon+10\chi_3 M+\frac{15}{4}\chi_2 M^2+\chi_1)$

Which shows that hypothesis in(57)is satisfied. Now in order to verify a hypothesis in(58)

$$\begin{aligned}L_1(\phi)-L_1(\psi) &= \frac{25}{4}p(\frac{\partial}{\partial x}(\phi_x^2-\psi_x^2),\eta)+5(\frac{\partial}{\partial x}(\phi^3-\psi^3),\eta)+\frac{5}{4}(\frac{\partial^3}{\partial x^3}(\phi^2-\psi^2),\eta)+(\frac{\partial^5}{\partial x^5}(\phi-\psi),\eta)\\ &= \frac{25}{2}pM\|\phi-\psi\|\|\eta\|+5M^2\|\phi-\psi\|\|\eta\|+\frac{5}{2}M\|\phi-\psi\|\|\eta\|+\|\phi-\psi\|\|\eta\|\\ &= (5M^2+\frac{25}{2}pM+\frac{5}{2}M+1)\|\phi-\psi\|\|\eta\|\\ &= C(M)\|\phi-\psi\|\|\eta\|\end{aligned} \quad (66)$$

where

$$C(M)=(5M^2+\frac{25}{2}pM+\frac{5}{2}M+1) \quad (67)$$

In this Chapter, we present a new application of HPTM by handling highly nonlinear model,namely Kaup-Kupershimdt equation with different initial conditions. HPTM does not require any perturbation or discretization, linearization and unjustified assumptions,and requires less computational work as compared with techniques and methods. HPTM gives solution in convergent power series form with easily computable coefficients. Numerical results obtained show HPTM is a powerful mathematical tool to handle nonlinear PDEs with higher nonlinearities. Numerical results presented in tabular form shows the accuracy of the proposed method.

7. Tables

Table 1. The absolute error $|\phi(x,t)-\mu(x,t)|$ for different values of t and x at k=0.25 for example IV.1

t/x	0.1	0.2	0.3	0.4	0.5
0.1	4.1233×10^{-20}	9.9008×10^{-20}	2.6730×10^{-19}	1.5481×10^{-19}	2.7797×10^{-18}
0.2	1.5576×10^{-18}	5.7760×10^{-18}	4.8915×10^{-17}	6.0794×10^{-17}	7.8304×10^{-17}
0.3	1.0358×10^{-16}	3.4548×10^{-16}	3.9752×10^{-16}	4.2065×10^{-16}	5.7977×10^{-16}
0.4	4.6099×10^{-15}	8.7253×10^{-15}	1.5354×10^{-15}	1.5273×10^{-15}	2.2783×10^{-15}
0.5	1.8313×10^{-15}	3.3261×10^{-15}	4.4021×10^{-15}	5.9421×10^{-15}	6.9063×10^{-15}

Table 2. comparison of exact solution $\phi(x,t)=\frac{4a}{3}-\frac{4a}{p}\sec h^2\left[\sqrt{a}x+8(3a^2-5ap)t\right]$ with approximate solution, when c=0.01 and p=5/2 for example IV.2

t/x	0.1	0.2	0.3	0.4	0.5
0.1	3.0634×10^{-15}	5.0131×10^{-15}	1.4232×10^{-14}	1.2781×10^{-14}	2.7900×10^{-13}
0.2	7.7857×10^{-14}	1.1227×10^{-14}	4.8915×10^{-17}	4.2734×10^{-13}	5.2594×10^{-12}
0.3	8.9842×10^{-13}	1.5338×10^{-13}	3.9752×10^{-16}	3.5259×10^{-12}	3.6696×10^{-11}
0.4	4.1964×10^{-12}	4.7246×10^{-12}	1.5354×10^{-15}	2.0495×10^{-11}	1.1530×10^{-10}
0.5	1.5036×10^{-11}	1.4122×10^{-11}	4.4021×10^{-15}	3.7838×10^{-10}	4.8341×10^{-10}

Acknowledgements
This paper was supported by the Scientific research project of hunan province department of education (Fourier Series Multiscale Method for Compututional analysis in Science Engineering, No: 19C1837).

References

[1] Tanoglu G 2006 Hirota Method for Solving Reaction-Diffusion Equations with Generalized Nonlinearity *International Journal of Nonlinear Science* **1**(1) 30-6
[2] Ablowitz M J, Clarkson P A 1991 *Solitons, Nonlinear Evolution Equations and Inverse Scattering* (New York: Cambridge University press)
[3] Zayed E M E, Abdelaziz M A M 2011 Exact solutions for the nonlinear Schrödinger equation with variable coefficients using the generalized extended tanh-function, the sine–cosine and the exp-function methods. *Applied Mathematics and Computation* **218**(5) 2259-68
[4] Khater A H, Melfliet W, Callebaut D K, Kamel E S 2002 The tanh method, a simple transformation and exact analytical solutions for nonlinear reaction–diffusion equations *Chaos, Solitons & Fractals* **14** 513-22
[5] Lakshmanan M, Kaliappan P 1983 Lie transformations, nonlinear evolution equations, and Painlevé forms *J. Math. Phys.* **24** 795
[6] Li B, Chen Y 2003 Nonlinear Partial Differential Equations Solved by Projective Riccati Equations Ansatz *Zeitschrift für Naturforschung A* **58a** 511-9
[7] Chen Y, Li B 2004 General projective Riccati equation method and exact solutions for generalized KdV-type and KdV–Burgers-type equations with nonlinear terms of any order *Chaos, Solitons & Fractals* **19**(4) 977-84
[8] Bekir A 2010 Application of the Exp-function method for nonlinear differential-difference equations *Applied Mathematics and Computation* **215**(11) 4049-53
[9] Liu S K, Fu Z T, Liu S D, Zhao Q 2001 Jacobi elliptic function expansion method and periodic wave solutions of nonlinear wave equations *Physics Letters A* **289**(1-2) 69-74
[10]Zhang S, Tong J L, Wang W 2008 A generalized -expansion method for the mKdV equation with variable coefficients *Physics Letters A* **372**(13) 2254-7
[11]Wazwaz A M 2004 A sine-cosine method for handlingnonlinear wave equations *Mathematical and Computer Modelling* **40**(5-6) 499-508
[12]Adomian G 1994 *Solving Frontier Problems of Physics: The Decomposition Method Kluwer* (Dordrecht: Springer)
[13]He J H 1999 Variational iteration method – a kind of non-linear analytical technique: some examples *International Journal of Non-Linear Mechanics* **34**(4) 699-708
[14]Liao S J 1992 *The proposed homotopy analysis technique for the solution of nonlinear problems* (Shanghai: Shanghai Jiao Tong university)
[15]Y Keskin, G Oturanc 2010 Reduced differential transform method for fractional partial differential equations *Nonlinear Science Letters A* **1**(2) 61-72
[16]Cai G L, Wang Q C 2007 A modified F-expansion method for solving nonlinear PDEs *J Comput Inf Sci Eng* **2**(1) 3-16

[17]Ghasemi S E, Zolfagharian A, Hatami M, Ganji D D 2016 Analytical thermal study on nonlinear fundamental heat transfer cases using a novel computational technique *Applied Thermal engineering* **98** 88-97
[18]Tarig M, Elzaki S M 2011 Application of New Transform "Elzaki Transform" to Partial Differential Equations Global Journal of Pure and Applied Mathematics **7**(1) 65-70
[19]Tarig M 2014 Projected Differential Transform Method and Elzaki Transform for Solving System of Nonlinear Partial Differential Equations World Applied Science Journal 32(9) 1974-9
[20]Elzaki T M 2012 A Solution for Nonlinear Systems of Differential Equations Using a Mixture of Elzaki Transform And Differential Transform Method International Mathematical Forum **7**(13) 625-30
[21]He J H 1999 Homotopy Perturbation Technique Computer Methods in Applied Mechanics and Engineering **178**(3-4) 257-62

Modeling Thermal Environment Responses to Terrain and Urbanization of Metropolis

Sulin Tao[1,3] and Yuhong Li[2,4,*]

[1]School of Applied Meteorology, Nanjing University of Information Science & Technology, Nanjing 210044, China
[2]Institute of Atmospheric Environment, China Meteorological Administration, Shenyang 110166, China
[3]Technology Development Center, Nanjing Daqiao Machine Co., Ltd, Nanjing 211101, China
[4]Liaoning Ecological Meteorology and Satellite Remote Sensing Center, Shenyang 110166, China

[*]corresponding author's e-mail: liyh0609@163.com

Abstract. Thermal environment responses to terrain and urbanization of Beijing, a metropolis of China, were simulated by the Weather Research and Forecasting model (WRF) with the single-layer Urban Canopy model (UCM). Four numerical experiments were designed: (1) the control experiment (Control exp), (2) no-terrain sensitive experiment (No-Terrain exp), (3) no-urban surface sensitive experiment (No-Urban exp), and (4) real land surface sensitive experiment (LU exp). Results indicate that better near-surface air temperature simulations were obtained after using MODIS land surface type product data MCD12Q1. The foehn effect due to northwestern mountains in Beijing were simulated to intensify the high temperature intensity in the plain area. Urban heat island effect existed in this area day and night, and the heat island intensity in the nighttime was obviously greater than that in the daytime. As a result of urbanization, the nighttime temperature had risen by 1.5~3.5 °C and the expansion range of nighttime high temperature areas was basically consistent with the urban expansion range.

1. Introduction

Urban thermal environment has become worse for life of citizens especially in metropolis. Frequent occurrence of sultry nights and heat waves threatens urban residents' health and daily activities, poses serious adverse impacts on urban power and water supply and safety in production. High night-time temperature accompanied with high humidity will further aggravate extreme heat events in metropolitan areas that posing a particular threat for the elderly. This deadly combination results in a significant public health problem responsible for more deaths annually than other natural hazards combined [1]. It is important to investigate the causes for frequent occurrence of urban high temperature in metropolis.

Global climate change and rapid urban growth might lead to the worsening urban thermal environment. Solomon et al. [2] point out that heat waves have become more frequent as the average temperature of the earth is continuously raised with the influence of global warming. The increasing variability among today's warming climate will also result in increased risk of frequent and severe weather [3]. In this context, urban heat island (UHI) effect which urban on near-surface air temperature appear higher than that of surrounding rural areas exacerbates the temperature increase in urban areas and thus the urban thermal environment will further deteriorate [4]. The development of urbanization, as an important aspect of human influence on natural environment and further the thermal environment, contributes to UHI due to the change of urban underlying surface (i.e., land use and land cover) through the replacement of vegetations by artificial facilities of impervious cover [5,6]. It is also pointed out that complex terrain and land use change may affect energy and momentum exchange between land surface and atmosphere the portion known as urban canopy layer and further affect near-surface air temperature [7,8]. The primary changes affecting the exchanges occur in this layer including the drag force resulting from the exertion of urban structures on wind field, the reduction of sky view-factor, the multiple reflection of short wave radiation, the interception of long wave radiation and the release of anthropogenic heat etc. [9].

To describe and simulate the complicated wind field distribution and heating processes in metropolis is difficult. The methods of statistical analysis and remote sensing analysis are still unsatisfied for the investigation of the effects of complex terrain and change of underlying surface on meteorological environment [10,11]. However, a lot of numerical simulation models has accounted for these processes in detail such as the WRF model coupled with the Noah land surface and urban canopy model. Lin et al. [12] successfully use this model to evaluate the UHI effect over northern Taiwan province of China and obtain great remarkable improvements in the simulation of this effect. Analogously, the perfect performances of this meso-scale modeling system are found by Kusaka et al. [13] with accurate capture of nocturnal UHI, diurnal variations and frequency distribution of surface air temperature.

The interest of this work is to simulate the effects of northwestern mountains and urbanization development on summer high temperature weather in Beijing by designing sensitive experiments using WRF model with the single-layer UCM.

2. Methodology

2.1. Study region

Beijing is a typical metropolis in China with a population of over 20 million residents, located at the junction of North China Plain and Taihang and Yanshan Mountains. Its northwest part is encompassed by mountains while the southeast part is plain. The meteorological environment is obviously affected by terrain effect. Great changes have taken place in urban underlying surface with the acceleration of urbanization process of Beijing in recent twenty years and the frequency of occurrence of summer high temperature and daily maximum temperature tends to increase [14]. One of them occurred on July 4 to 6, 2010 with continuous high temperature in Beijing, mainly caused by subsidence warming under the control of continental warm high pressure and foehn effect.

2.2. Numerical model and data

The WRF model coupled with the single-layer UCM was applied in this study for better description of physical processes involved in urban boundary layers. The advanced Noah land surface/hydrology model coupled in WRF model provides surface layer temperature in the lower boundary, surface sensible and latent heat fluxes [15] and the single-layer UCM estimates the heat fluxes from the roof, wall, and road as well as their surface temperatures. The effects of wind shear from urban structures are also considered [16,17].

The National Centers for Environmental Prediction (NCEP) reanalysis data (data set 083.2) with a resolution of $1°\times1°$ and 6-hour intervals were used as initial field. To analyze heat island intensity expressed by the difference between urban air temperature and suburban air temperature, observations were selected through preliminary quality control from seven urban stations, including Tiananmen, Beijing University Student Gymnasium, Siyuanqiao, Shibalidian, Lizeqiao, Zizhuyuan, Ancient Observatory, and six suburban stations, including Fangshan, Tongzhou, Shunyi, Huairou, Changping, Daxing, where the elevation above sea level is similar to that of urban area. The heat island intensity was defined by subtracting mean temperature of the six suburban stations from mean temperature of the seven urban stations.

The urban area of Beijing has been continuously expanded with the acceleration of urbanization process and physical characteristics of urban underlying surface have changed significantly. In this study, the land use type data in 2010 were provided by MODIS satellite remote sensing product MCD12Q1 based on International Geosphere Biosphere Programme (IGBP) for the comparison to the simulation results with the land use type data of USGS in 1993. Compared with the original USGS land use in 1993 (Figure 1a), the urban underlying surface of Beijing has changed obviously and the urban area has been expanded in all directions and nearly doubled in the MODIS land use in 2010 (Figure 1b).

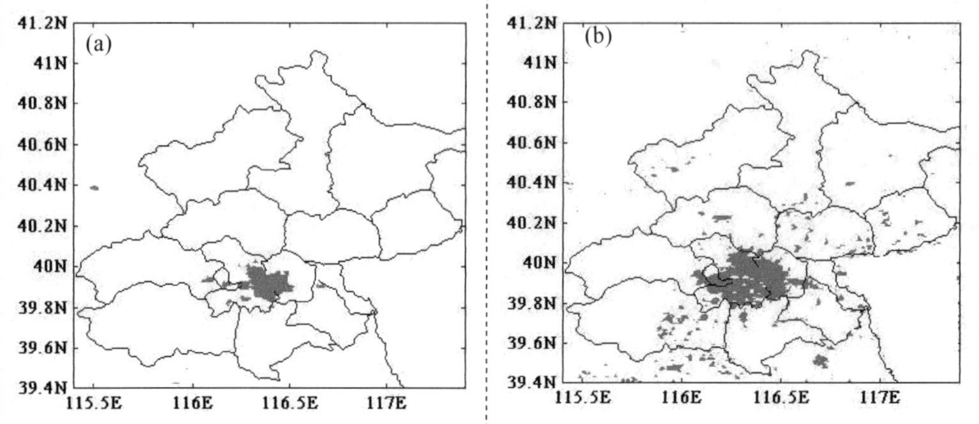

Figure 1. The urban area variations of Beijing from the year of (a) 1993 and (b) 2010.

2.3. Modeling scheme

Four numerical simulation experiments were conducted to analyze the influence of northwestern mountains in Beijing and urbanization on the thermal environment (Table 1). The WRF model was configured with triple-nested grids for the simulated region and the center point of our study region is 39.9°N, 116.4°E. Nested horizontal grid intervals were 18km, 6km and 2km. The number of grid points for the third layer nested domain was 121×121, covering the area with latitude ranging from 38.8°N to 41.2°N and longitude from 114.8°E to 117.9°E. The vertical structure consists of 37 terrain following coordinate layers with the model top air pressure at 50hPa. Resolutions of static terrain data for the triple-nested domains are 10 minutes, 2 minutes and 30 seconds, respectively. WSM 3-class simple ice scheme, RRTM long wave radiation scheme and Dudhia shortwave radiation scheme were adopted and Noah LSM parameterization scheme was selected for land surface process. Integration starts from 20:00, July 4 to 8:00, July 6.

Table 1. Details of four numerical simulation experiments.

Name	Details
Control exp	the control experiment, in which the original terrain and land surface type in the model were used
No-Terrain exp	no-terrain sensitive experiment, in which the mountain topography for triple-nested grids were removed and it was assumed that terrain of the simulated region was homogeneous
No-Urban exp	no-urban surface sensitive experiment, in which the urban surface was replaced by the farmland surface in suburban area
LU exp	real land surface sensitive experiment, in which MODIS land surface type product data MCD12Q1 in 2010 was used as real land surface

3. Results and discussion

3.1. The control experiment

The daily air temperature variation tendency from 8:00, July 5 to 8:00, July 6 in urban area was well simulated in Control exp (Figure 2). During the day, the maximum air temperature in urban area was 39.6 °C which appeared at 16:00. The observed maximum air temperature was 40.2 °C, 0.6 °C higher than the simulated value. The reason might be the utilization of air conditioners. During the high temperature period, the use of air conditioners increased in urban area and heat emissions from the air conditioners may heat the air. However, the effect of anthropogenic heat on near-surface air temperature hasn't been considered in simulation process.

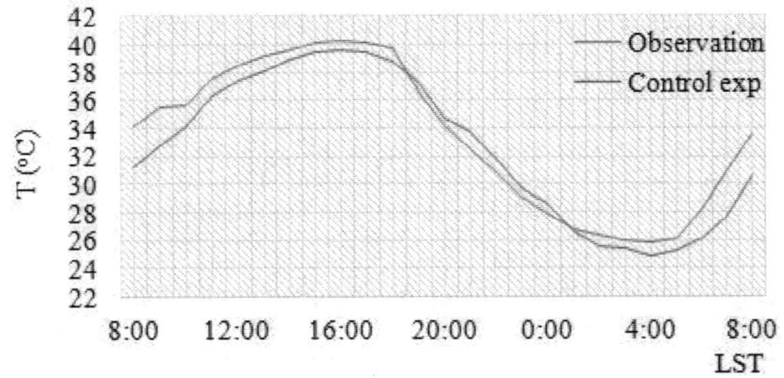

Figure 2. Diurnal variations of observed and simulated hourly urban temperatures in urban area in the control experiment.

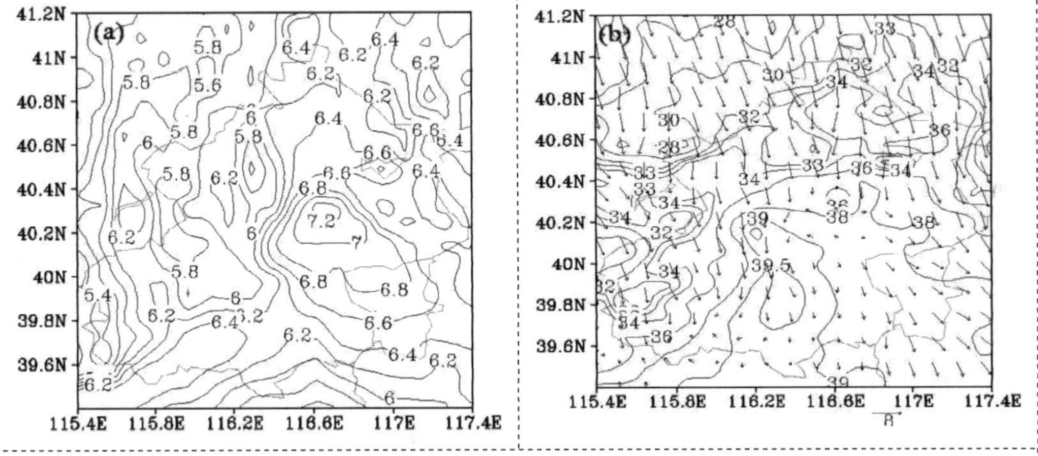

Figure 3. Simulated meteorological fields at 16:00 LST on July 5: (a) specific humidity (g/kg), and (b) temperature (°C) at 2m as well as wind (m/s) at 10m.

The specific humidity, temperature fields at 2m and wind field at 10m at 16:00 on July 5 were also simulated in Control exp. The air humidity in mountainous areas in the northwest of Beijing was lower than that in the eastern plain. The northwest wind was prevailing in Beijing area. Warm air with relatively low humidity descended and lead to an increase in air temperature after passing through the northwestern mountains (Figure 3a). Thus, foehn effect was produced to make air temperature of southeast plain on the leeward side rise and to enhance high temperature intensity. Air temperature at 2m in urban area and suburban area was mostly above 39 °C generally identical or slightly lower than the actual air temperature at 16:00 on July 5 (Figure 3b).

3.2. The thermal environment response to terrain

The surrounding terrain of Beijing is complex. The altitude difference between northwestern mountainous areas and urban area varies between 200 and 1800m, and the mountainous areas are near to the urban area of Beijing. The northwestern mountains were removed to investigate the influence of northwestern mountains on summer high temperature intensity in Beijing.

On the average, the temperature difference between the Control exp and No-Terrain exp was 1~2 °C, and the temperature difference between 10:00 and 17:00 was very small (Figure 4). From 17:00, the temperature difference tended to be much more obvious, and the maximum difference reached 2 °C at 4:00, July 6. After replacing the mountains with homogenous terrain, the foehn effect disappeared and the temperature deviation in urban area was small during the day due to the synoptic weather influence. At night, the increase of wind speed in urban area enhances turbulence intensity and weakens heat island intensity. Thus, urban air temperature was obviously reduced.

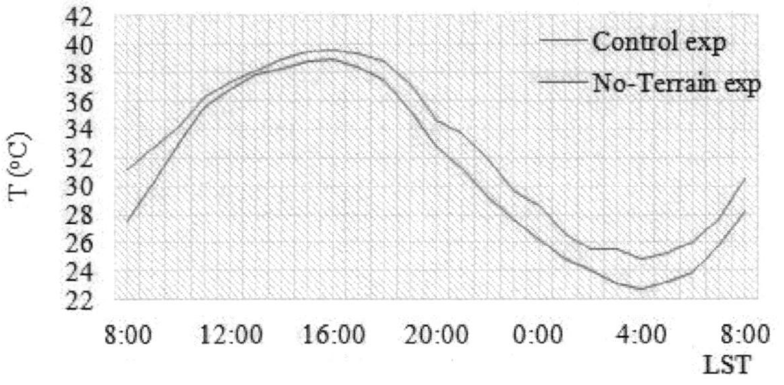

Figure 4. Time variations of simulated hourly temperature in the Control exp and No-Terrain exp.

3.3. The thermal environment response to urbanization

It was clear that urbanization could increase air temperature at 2m by 1.5~3.5 °C from the spatial distribution of the differences in simulated air temperature between LU exp and No-Urban exp (Figure 5a). With the development of urbanization, in the areas where suburban farmland has been covered by urban surface, the nighttime air temperature raised by 1.5~3.5 °C and the expansion range of nighttime high temperature areas was basically consistent with urban expansion range (Figure 5b).

The temperature differences between Control exp, No-Urban exp and LU exp was small during the daytime; and after 17:00, the temperature in the No-Urban exp began to drop significantly and the temperature drop reached the maximum value at 5:00, July 6 (Figure 6). The maximum temperature difference between the observed and No-Urban exp at this time was about 2.5 °C, which indicated a stronger urban heat island effect in nighttime than that in daytime during the high temperature period. When the urban area was replaced by farmland, high temperature intensity in daytime was affected by synoptic weather condition and the temperature difference was small; meanwhile, the temperature dropped obviously in nighttime due to the disappearance of heat island effect.

Using MODIS land surface type product data as real land surface could improve the simulate accuracy of the urban air temperature at 2m. The urban air temperature simulated by LU exp was slightly higher than that simulated in the Control exp. The maximum air temperature was 39.8 °C at 16:00, closer to the actual temperature (41.2 °C at 16:00) than in the Control exp.

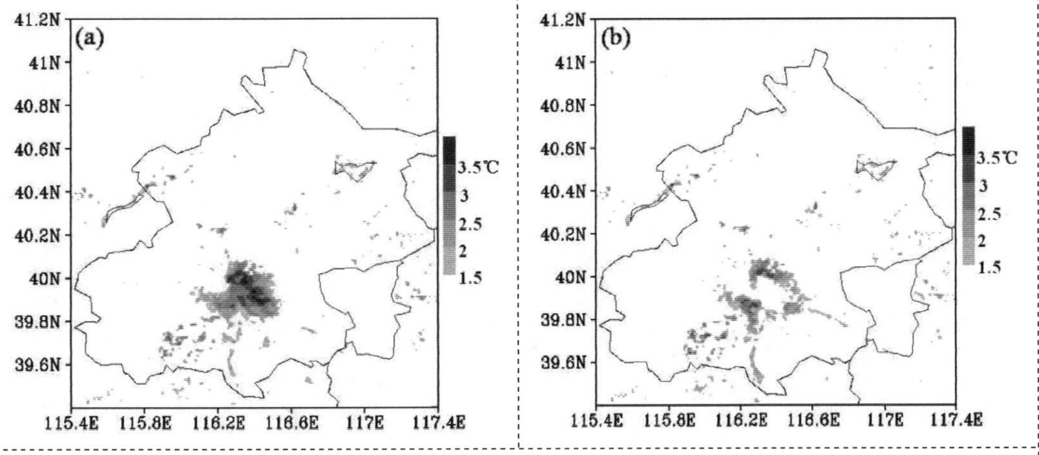

Figure 5. Differences of near surface air temperature (a) in LU exp and No-Urban exp and that (b) in LU exp and Control exp at 20:00 LST 5 July.

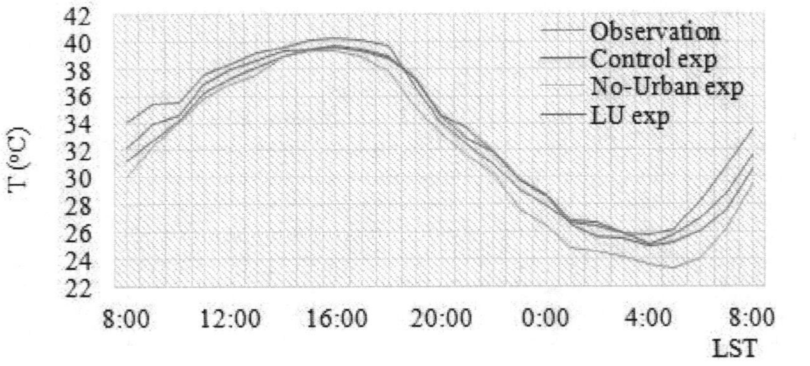

Figure 6. Simulated hourly temperature variations in the numerical experiments with different urban land surface.

3.4. The effect of terrain and urbanization on the nighttime heat island intensity

The effects of terrain and land surface type on the nighttime heat island intensity were further investigated by the simulations of the temperature field at 2m and wind field at 10m at 20:00, July 5. A high temperature center with temperature above 34 °C was simulated in urban area in Control exp, but the range was relatively small compared with actual situation (Figure 7a).

Northerly wind was prevailing in the west and north and southerly wind was prevailing in the south. Two types of air currents converged and raised in urban area and form urban heat island circulation. However in No-Terrain exp, northwest air current before the ridge of high pressure was not blocked by mountains; the wind speed was fast; foehn effect did not exist and urban air temperature dropped obviously (Figure 7b). Yet still there existed a high temperature center with temperature of about 32 °C.

In the No-Urban exp, a uniform temperature zone with temperature of about 32 °C existed over the plain area and urban heat island effect disappeared (Figure 7c). In the LU exp, a high temperature center of about 34 °C was simulated in urban area (Figure 7d). Compared with the Control exp (Figure 3b), the range of heat island was expanded and closer to the actual situation. Thus, foehn effect and urbanization development intensify nighttime urban high temperature.

 2173 (2022) 012023 doi:10.1088/1742-6596/2173/1/012023

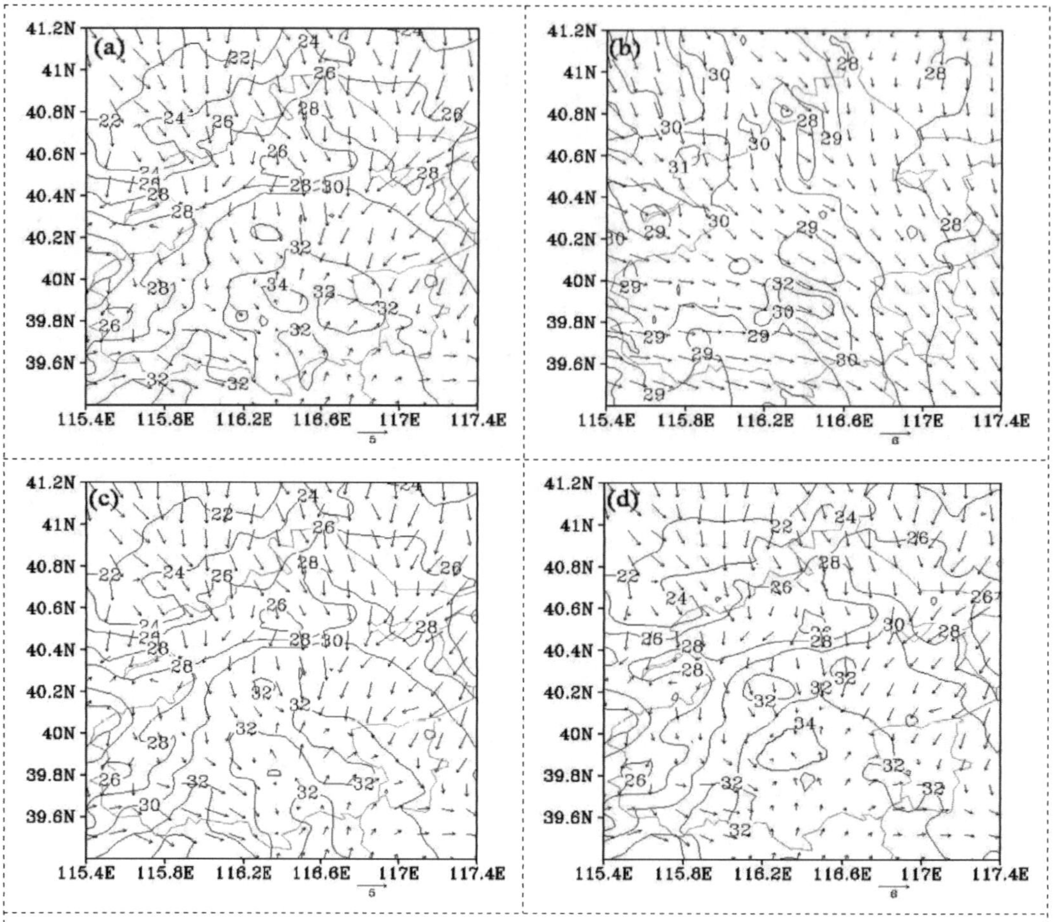

Figure 7. Simulated temperature at 2m (°C) and wind at 10m (m/s) in (a) the Control exp, (b) No-Terrain exp, (c) No-Urban exp, and (d) LU exp at 20:00 LST, July 5.

4. Conclusions

Detailed thermal environment responses to the terrain and urbanization of Beijing were simulated via several numerical schemes. Better near-surface air temperature simulations were obtained after using MODIS land surface type product data MCD12Q1. The foehn effect due to northwestern mountains in Beijing can intensify the high temperature intensity in the plain area. When the influence of mountains was eliminated, the foehn effect disappeared, the wind speed increased, the nighttime heat island intensity was weakened and near-surface air temperature dropped obviously. Also after eliminating urban underlying surface, the high temperature intensity was weakened, the nighttime urban heat island effect disappeared and the air temperature dropped. However, as a result of urbanization, the nighttime air temperature raised by 1.5~3.5 °C. The expansion range of nighttime high temperature areas was basically consistent with urban expansion range.

The main cause for the high-temperature weather process in Beijing area was the foehn effect caused by subsidence warming and northwest over mountains airflow. In this high temperature process, urban heat island effect existed in this area day and night, and the heat island intensity in the nighttime was obviously greater than that in the daytime. The heat island effect, in addition, intensified nighttime urban high temperature.

References

[1] Luber G and McGeehin M 2008 Climate Change and Extreme Heat Events *American Journal of*

Preventive Medicine **35** 429–35
[2] Solomon S, Qin D, Manning M, Chen Z and Miller H 2007 Climate change 2007: the physical science basis *Working Group I Contribution to the Fourth Assessment Report of the IPCC* (Intergovernmental Panel on Climate Change)
[3] Tao S, Shen S, Li Y, Wang Q, Gao P and Mugume I 2016 Projected crop production under regional climate change using scenario data and modeling: Sensitivity to chosen sowing date and cultivar *Sustainability* **8** 1–23
[4] Wua C, Lung S and Jan J 2013 Development of a 3-D urbanization index using digital terrain models for surface urban heat island effects *ISPRS Journal of Photogrammetry and Remote Sensing* **81** 1–11
[5] Grossman-Clarke S, Zehnder JA, Loridan T and Grimmond C 2010 Contribution of land use changes to near-surface air temperatures during recent summer extreme heat events in the Phoenix metropolitan area *Journal of Applied Meteorology and Climatology* **49** 1649–64
[6] Chow W, and Svoma B 2011 Analyses of nocturnal temperature cooling-rate response to historical local-scale urban land-use/land cover change *Journal of Applied Meteorology and Climatology* **50** 1872–83
[7] Im E, Coppola E, Giorgi F and Bi X 2010 Validation of a high-resolution regional climate model for the Alpine region and effects of a subgrid-scale topography and land use representation *Journal of Climate* **23** 1854–73
[8] Tao S, Li Y and Mugume I 2019 Model terrain correction using variational adjoint method with Tikhonov-total variation regularization *Journal of Physics Conference Series* **1176** 022034
[9] Ca V, Asaeda T and Ashie Y 1999 Development of a numerical model for the evaluation of the urban thermal environment *Journal of Wind Engineering and Industrial Aerodynamics* **81** 181–96
[10] Yuan S, Gu X, Miao Q, Kang W, Yu F and Qiu X 2009 Distributed simulation of monthly mean daily maximum temperature over complex terrains on Guizhou Plateau *Acta Geographica Sinica* **64** 888–95
[11] Zheng Q, Liu H and Chen Y 2009 Observational and analytical researches on the impacts of urban development on meteorological environment *Scientia Meteorologica Sinica* **29** 214–9
[12] Lin C, Chen F, Huang J, Chen W, Liou Y, Chen W and Liu S 2008 Urban heat island effect and its impact on boundary layer development and land–sea circulation over northern Taiwan *Atmospheric Environment* **42** 5635–49
[13] Kusaka H, Chen F, Tewari M, Dudhia J, Gil D, Duda M, Wang W and Miya Y 2012 Numerical simulation of urban heat island effect by the WRF model with 4-km grid increment: An inter-comparison study between the urban canopy model and slab model *Journal of the Meteorological Society of Japan* **90** 33–45
[14] Wang Y, Ge G and Tao Z 2003 Climatic characteristics of Estival Muggy Weather in Beijing related to 2008 Olympics *Meteorology* **29** 23–7
[15] Ek M, Mitchell K, Lin Y and Rogers E 2003 Implementation of Noah land surface model advances in the National Centers for Environmental Prediction operational mesoscale Eta model *Journal of Geophysical Research: Atmospheres* **108** 1984–2012
[16] Kusaka H and Kimura F 2004 Coupling a single-layer urban canopy model with a simple atmospheric model: Impact on urban heat island simulation for an idealized case *Journal of the Meteorological Society of Japan* **82** 67–80
[17] Kusaka H and Kimura F 2004 Thermal effects of urban canyon structure on the nocturnal heat island: Numerical experiment using a mesoscale model coupled with an urban canopy model *Journal of Applied Meteorology* **43** 1899–910

Deterministic and Stochastic Modeling Approaches for Saturation Nonlinearity

Pai Peng and ChunQing Huang*

Department of Automation, Xiamen University, Xiamen, Fujian, China

*corresponding author's e-mail: chqhuang@xmu.edu.cn

Abstract. Being different from the most traditional methods of nonlinearity modelling, non-parametric modelling approaches viz. deterministic and stochastic approaches for saturation nonlinearity are proposed in the sense of system impulse response. Based on deterministic approach, the closed-loop system involved saturation nonlinearity can be modelled as the response of the closed-loop system that is subject to impulse stimulation. Alternatively, the closed-loop system involved saturation nonlinearity can be modelled in stochastic manner, in which the impulse response coefficients are estimated by the FCOR algorithm. Moreover, it shows some linear relationship of the impulse response coefficients between different saturation ratios in both the deterministic and stochastic models. This is illustrated by three different numerical examples.

1. Introduction

Nonlinear phenomena and nonlinear systems are extremely ubiquitous in various aspects of society, such as biological systems, engineering practice, chemical processes, et al [1, 2]. In the real world, almost all control systems are nonlinear to a certain extent [3, 4]. Nonlinear models are somuch needed to improve control performance. Nonlinearities in systems can be roughly divided into two categories: the polynomial nonlinearity and the hard nonlinearity [2, 5]. Compared with the polynomial nonlinearity, the hard nonlinearity cannot be written as the analytic expression of the input and is more widely encountered in engineering practice [6].

There exist so many varying forms of nonlinear models presented in diverse literatures, including Volterra series, NARMAX (Nonlinear AutoRegressive Moving Average with eXogenous inputs) models, artificial neural network models, block-oriented nonlinear models, etc.

Volterra series is a powerful mathematical tool for analyzing nonlinear systems [7, 8]. In essence, It directly extends the description of the impulse response function of a linear system to a nonlinear system [9]. Volterra series offers a highly structured analysis platform for modeling the nonlinear systems, but the key issue utilizing the Volterra series is the identification of its kernel functions, which has demonstrated to be a great challenge [10].

NARMAX models describe nonlinear systems through expanding the nonlinear function of lagged input, output and prediction errors [11, 12]. It is a kind of nonlinear time series model. The NARMAX model realizes a uniform representation for a wide variety of nonlinear systems and includes several known nonlinear input-output models as special cases [13]. However, in practical applications, it is commonly difficult to obtain an accurate NARMAX model because of the system unknown nonlinear structure [14].

Artificial neural networks, as a non-parametric modeling method, have been applied vastly since it can approximate any nonlinear function with arbitrary accuracy in theory [9, 15]. Neural networks have good approximation capabilities and adaptive features and can also achieve parsimonious modeling [16, 17]. The main disadvantage is that for most neural networks, the parameters are too hard to explain in a physical sense, which makes it very difficult to understand the results of the nonlinear system modeling process [18].

Block-oriented models consist of two basic building blocks as a static nonlinear block and a linear dynamic block. They can combine in some different ways and form several different nonlinear systems as shown in Figure 1 [19]. Block-oriented nonlinear models are quite popular due to the advantages of being simple to understand and easy to use, and they provide such a highly structured form for different nonlinear systems [20]. The crucial problem is that plenty of effort and time may be wasted via choosing an improper candidate model structure at the beginning of the system modeling process [21].

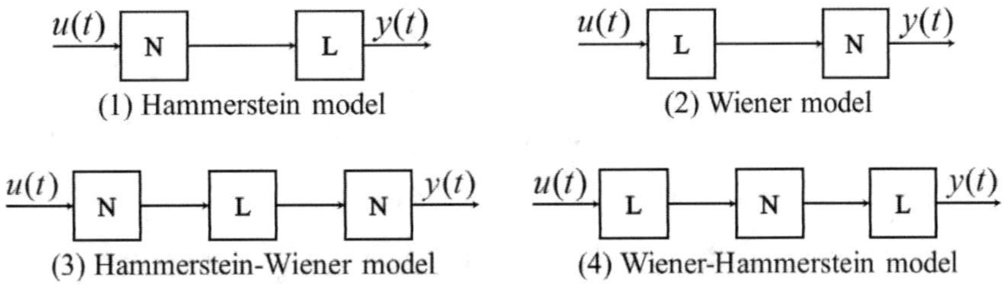

Figure 1. The different structures of block-oriented models

The above modeling methods always encounter many difficulties in industrial applications, therefore, a kind of simple and intuitive nonlinear model structure is required. The main contribution of this paper is to provide alternative approaches for modelling the closed-loop system with saturation nonlinearity by non-parametric models that are characterized by the impulse response of saturated nonlinear system in deterministic and stochastic sense respectively.

The rest of this paper is organized as follows. In Section 2, the nonlinear control system is depicted. Section 3 presented the recursive form of impulse response calculation. The FCOR algorithm is given in Section 4. And section 5 provides the results of three simulation examples. Finally, conclusions are presented in Section 6.

2. System description

Consider a SISO feedback nonlinear system shown in Figure 2 as

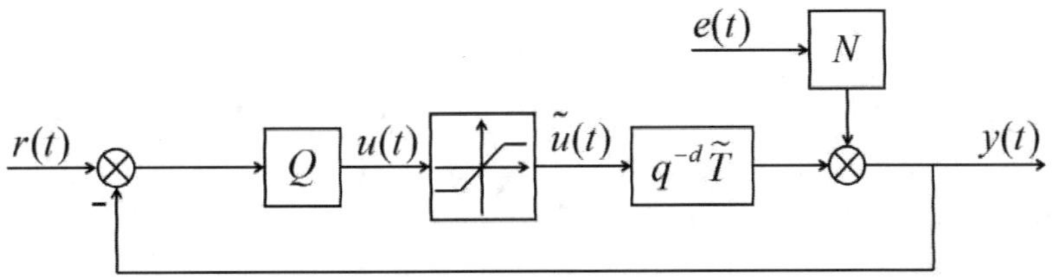

Figure 2. Schematic nonlinear closed-loop system

where d denotes pure time delay of the system, $\{r(t)\}$, $\{u(t)\}$ and $\{y(t)\}$ denote the external input sequence, controller output sequence and closed-loop output sequence, respectively. $\{e(t)\}$ stands for the unmeasured disturbance, which is commonly modeled as a white Gaussian noise sequence with zero mean and unit variance. $Q(q^{-1})$, $\tilde{T}(q^{-1})$ and $N(q^{-1})$ represent the controller transfer function, delay-free plant transfer function, and disturbance transfer function, respectively, with q^{-1} denoting the unit backward shift operator, i.e. $q^{-1}y(t)=y(t-1)$.

The saturation nonlinearity is one kind of the hard nonlinearity which is common in engineering. The characteristic of saturation nonlinearity presented in Figure 3 is a piecewise function and can be expressed as

$$\tilde{u}(t) = Sat\{u(t)\} = \begin{cases} \alpha & u(t) > \alpha \\ u(t) & |u(t)| \le \alpha \\ -\alpha & u(t) < -\alpha \end{cases} \tag{1}$$

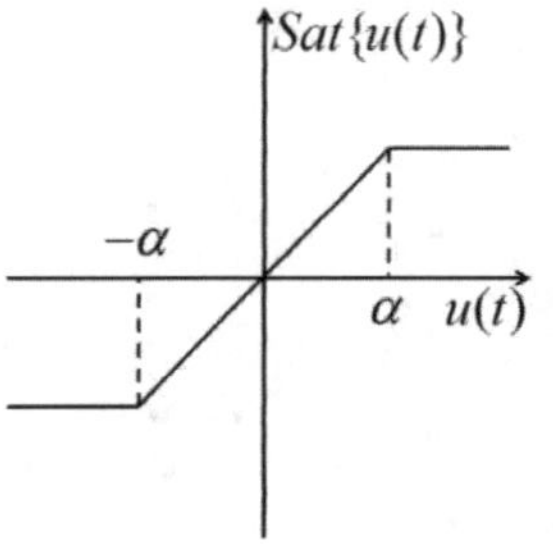

Figure 3. The saturation nonlinearity

Define the saturation ratio η (i.e. the percentage of the control signal $\{u(t)\}$ into saturation) as

$$\eta(\%) = \frac{number\{|\tilde{u}(t)| > \alpha\}}{number\{\tilde{u}(t)\}} \tag{2}$$

3. Deterministic modeling approach

In order to obtain the accurate closed-loop impulse response coefficients of the nonlinear system, let $\{e(t)\}$ be a unit impulse $\{\sigma(t)\}$, where

$$\sigma(t) = \begin{cases} 1 & t = 0 \\ 0 & t \ne 0 \end{cases} \tag{3}$$

Suppose that $y(t) = 0$ for $t \le 0$ and $\{r(t)\} = 0$, therefore, the closed-loop output $\{y(t)\}$ can be written as follows

$$y(t) = q^{-d}\tilde{T}(q^{-1})\tilde{u}(t) + N(q^{-1})e(t) \tag{4}$$

Then, we expand the controller transfer function $Q(q^{-1})$, delay-free plant transfer function $\tilde{T}(q^{-1})$ and disturbance transfer function $N(q^{-1})$ to finite order line model, namely FIR (Finite Impulse Response) model, respectively as

$$\begin{aligned} Q(q^{-1}) &= c_0 + c_1 q^{-1} + c_2 q^{-2} + \cdots c_m q^{-m} \\ \tilde{T}(q^{-1}) &= a_0 + a_1 q^{-1} + a_2 q^{-2} + \cdots a_m q^{-m} \\ N(q^{-1}) &= b_0 + b_1 q^{-1} + b_2 q^{-2} + \cdots b_m q^{-m} \end{aligned} \tag{5}$$

According to the definitions given above, the closed-loop output $\{y(t)\}$ can eventually be rewritten in recursive form as follows

$$y(t) = b_{t-1} + a_0\tilde{u}(t-d) + a_1\tilde{u}(t-d-1) + \cdots + a_{m-1}\tilde{u}(t-d-m+1) + a_m\tilde{u}(t-d-m) \tag{6}$$

where

$$\begin{gathered}\tilde{u}(t)=Sat\{u(t)\}\\ u(t)=-[c_0y(t)+c_1y(t-1)+c_2y(t-2)+\cdots+c_my(t-m)]\end{gathered}\tag{7}$$

Substituting (7) into (6) yields

$$\begin{aligned}y(1)&=b_0\\ &\vdots\\ y(d)&=b_{d-1}\\ y(d+1)&=b_d+a_0\cdot Sat\{-c_0y(1)\}\\ y(d+2)&=b_{d+1}+a_0Sat\{-c_0y(2)-c_1y(1)\}+a_1Sat\{-c_0y(1)\}\\ &\vdots\\ y(k)&=b_{k-1}+a_0\cdot Sat\{-c_0y(k-d)-c_1y(k-d-1)-\cdots-c_my(t-d-m)\}\\ &\quad+a_1\cdot Sat\{-c_0y(k-d-1)-c_1y(k-d-2)-\cdots-c_my(k-d-m-1)\}\\ &\quad+\cdots+a_m\cdot Sat\{-c_0y(k-d-m)-c_1y(k-d-m-1)-\cdots-c_my(k-d-m-m)\}\end{aligned}\tag{8}$$

The above is the deterministic modeling method. Based on the known conditions, it can be inferred that there may exist some linear relationship in impulse response coefficients between different saturation ratios.

4. Stochastic modeling approach

FCOR (Filtering and CORrelation analysis) algorithm, a relatively simple and stochastic modeling method to estimate the closed-loop impulse response coefficients from stochastic noise $\{e(t)\}$ to output $\{y(t)\}$ for the stable SISO closed-loop system, which is proposed by Huang in [22]. The output data of a stable closed-loop process can be written in the form of MA (moving-average) model with infinite order as

$$y(t)=(f_0+f_1q^{-1}+f_2q^{-2}+\cdots+f_{d-1}q^{-(d-1)}+f_dq^{-d}+\cdots)e(t)\tag{9}$$

Since $\{e(t)\}$ is a zero-mean white noise sequence, $e(t)$, $e(t-1)$, $e(t-2)$, ..., $e(t-n+1)$ are mutually independent of each other. Multiplying both sides of the above formula by $e(t)$, $e(t-1)$, $e(t-2)$, ..., $e(t-n+1)$ respectively and calculating their expectations yields a set of following equations as

$$\begin{aligned}r_{ye}(0)&=E[y(t)e(t)]=f_0\sigma_e^2\\ r_{ye}(1)&=E[y(t)e(t-1)]=f_1\sigma_e^2\\ r_{ye}(2)&=E[y(t)e(t-2)]=f_2\sigma_e^2\\ &\vdots\\ r_{ye}(d-1)&=E[y(t)e(t-d+1)]=f_{d-1}\sigma_e^2\\ r_{ye}(d)&=E[y(t)e(t-d)]=f_d\sigma_e^2\\ &\vdots\end{aligned}\tag{10}$$

Therefore, the impulse response coefficients of the closed-loop system transfer function can be computed through the above FCOR algorithm. Although the white noise sequence $\{e(t)\}$ is unknown, it can be represented by the estimated residual sequence $\{\hat{e}(t)\}$, which is obtained by pre-whitening the process output data $\{y(t)\}$ through fitting appropriate time series model.

5. Illustrative examples

In order to explore the inferred linear relation, S is defined as

$$S=\frac{S_2}{S_1} \tag{11}$$

where

$$S_1=\frac{f_i^{(2)}-f_i^{(0)}}{f_i^{(1)}-f_i^{(0)}},\ S_2=\frac{f_i^{(3)}-f_i^{(0)}}{f_i^{(2)}-f_i^{(0)}},\ (i=0,1,2\cdots) \tag{12}$$

In simulation, the external input sequence $\{r(t)\}$ is taken as zero, i.e. $\{r(t)\}=0$, and the stochastic signal $\{e(t)\}$ is a white noise sequence with mean of zero and variance of $\sigma_e^2=1$.

5.1. *Example 1*

Consider a closed-loop control system studied in [23], where

$$q^{-d}\tilde{T}(q^{-1})=\frac{0.1}{1-0.8q^{-1}}q^{-3},\ Q(q^{-1})=\frac{6.286-8.814q^{-1}+3.163q^{-2}}{1-q^{-1}},\ N(q^{-1})=\frac{0.1}{1-0.3q^{-1}-0.1q^{-2}} \tag{13}$$

The impulse response coefficients shown in Figure 4 is computed through equation (8). As can be seen from Figure 5, the ratio S between different impulse response coefficients is a linear constant.

Applying the FCOR algorithm to estimate the closed-loop impulse response coefficients of this nonlinear system, the results are presented in Figure 6. It shows that the trend of each estimated impulse response is roughly the same as the trend of each precise impulse response. There exists some kind of connection between each impulse response. As the saturation ratio gets higher and higher, the impulse response of nonlinear system gradually approaches the impulse response of the disturbance transfer function. Due to the randomness of the estimated impulse response coefficients from FCOR algorithm itself, the impulse response coefficients cannot be estimated completely accurately so S calculated by them cannot be a strictly linear constant just like the value computed by deterministic modelling approach. It is easy to note that S shown in Figure 7 fluctuates around between 0.7 and 0.8.

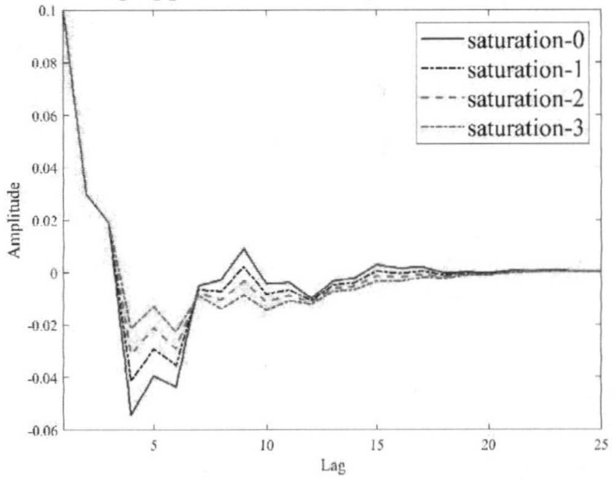

Figure 4. Impulse responses calculated by deterministic modeling approach for Example 1

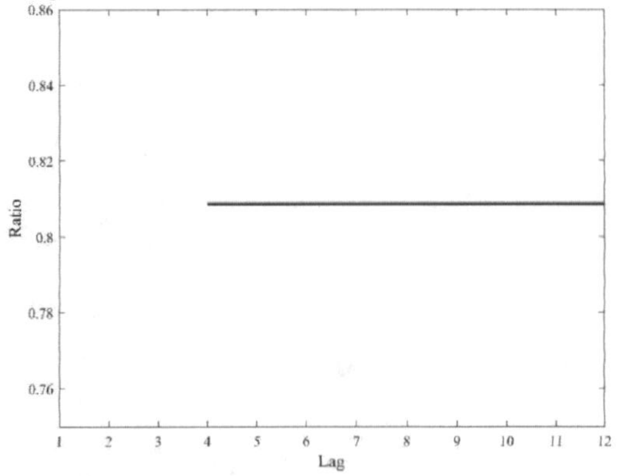

Figure 5. Relation of the impulse responses for Example 1

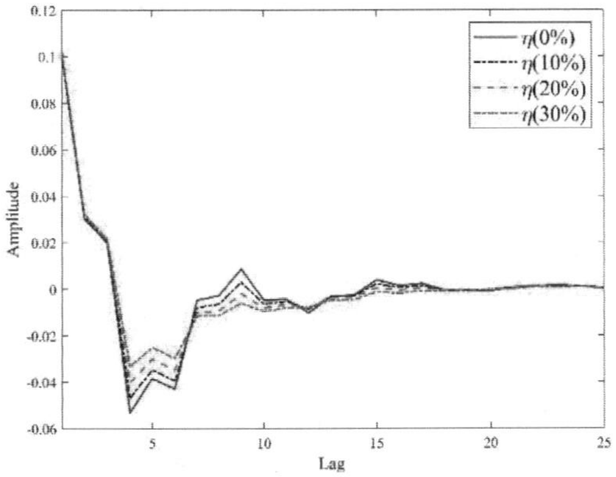

Figure 6. Impulse responses estimated by stochastic modeling approach for Example 1

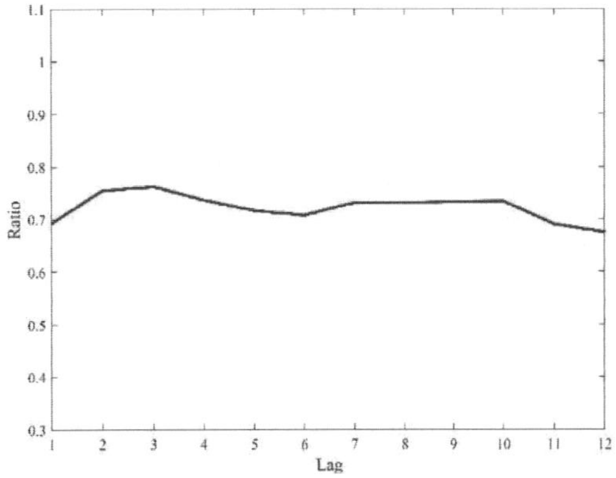

Figure 7. Relation of the estimated impulse responses for Example 1

5.2. Example 2

Consider a closed-loop numerical example given by [24] as

$$q^{-d}\tilde{T}(q^{-1})=\frac{0.2}{1-0.7q^{-1}}q^{-4},\ Q(q^{-1})=\frac{2.8047-4.4056q^{-1}+1.7485q^{-2}}{1-q^{-1}},\ N(q^{-1})=\frac{1}{1-0.1q^{-1}-0.2q^{-2}} \quad (14)$$

The impulse response coefficients presented in Figure 8 is computed through equation (8). It is easy to note that the ratio S shown in Figure 9 is a linear constant. The FCOR algorithm is applied to estimate the closed-loop impulse response coefficients of this nonlinear system. The characteristics of each impulse response presented in Figure 10 are roughly the same as those of each impulse response presented in Figure 8. As shown in Figure 11, the value of S fluctuates around 0.75 in this example.

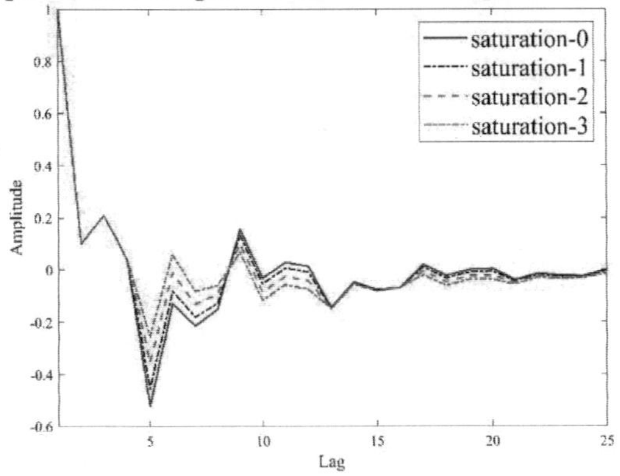

Figure 8. Impulse responses calculated by deterministic modeling approach for Example 2

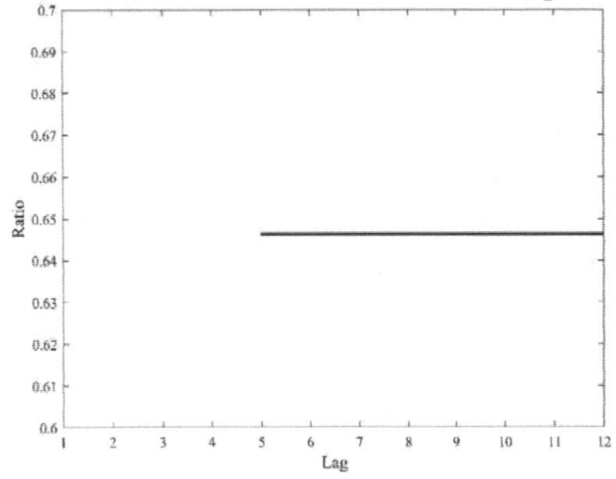

Figure 9. Relation of the impulse responses for Example 2

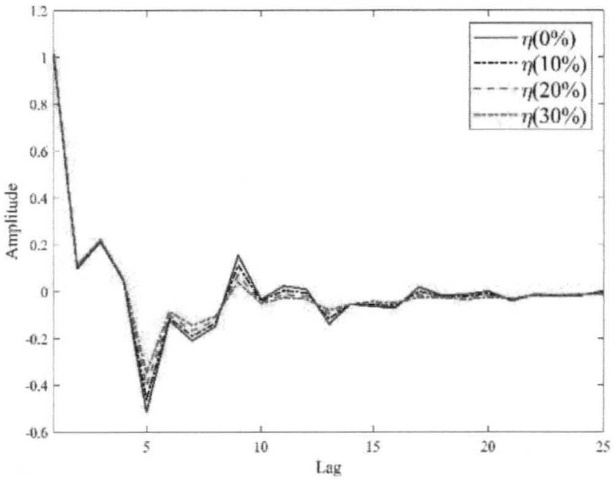

Figure 10. Impulse responses estimated by stochastic modeling approach for Example 2

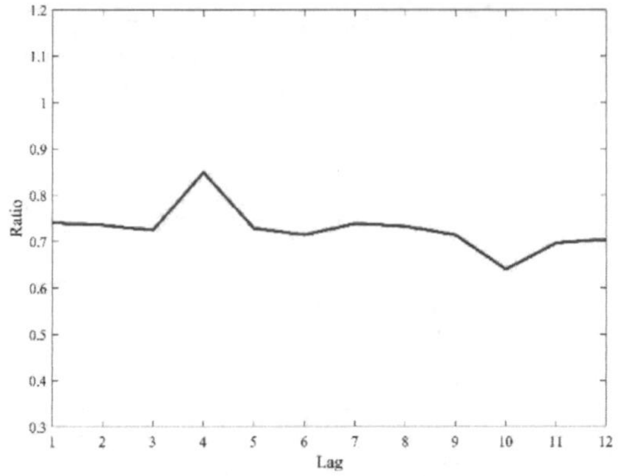

Figure 11. Relation of the estimated impulse responses for Example 2

5.3. *Example 3*

Consider the following closed-loop control system in [23], where

$$q^{-d}\tilde{T}(q^{-1})=\frac{0.1}{1-0.8q^{-1}}q^{-3},\quad Q(q^{-1})=\frac{6.5331-9.2362q^{-1}+3.3574q^{-2}}{1-q^{-1}},\quad N(q^{-1})=\frac{1}{1-0.5q^{-1}} \tag{15}$$

The impulse response coefficients displayed in Figure 12 is computed through equation (8). It is seen from Figure 13 that the ratio S between different impulse response coefficients is a linear constant. Similar to the phenomenon of the previous examples, the characteristics of each impulse response presented in Figure 14 is roughly the same as those of each impulse response presented in Figure 12. It can be seen from Figure 15 that S fluctuates around 0.7 in this example.

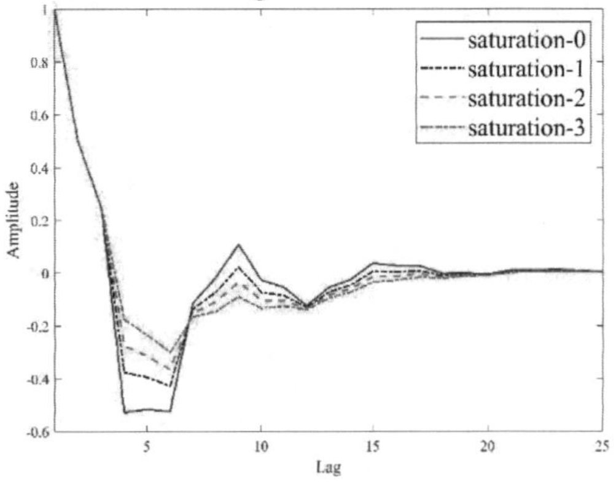

Figure 12. Impulse responses calculated by deterministic modeling approach for Example 3

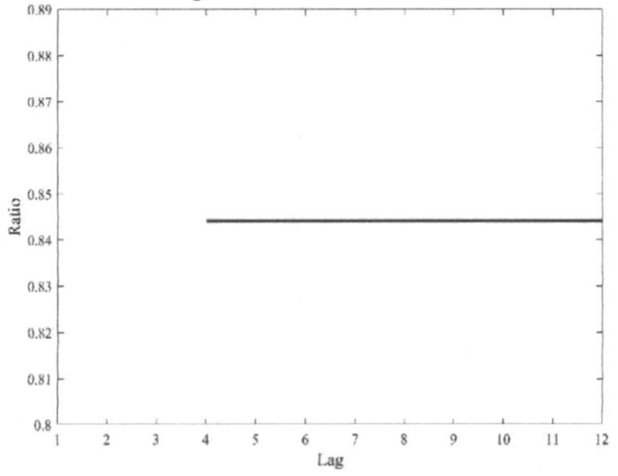

Figure 13. Relation of the impulse responses for Example 3

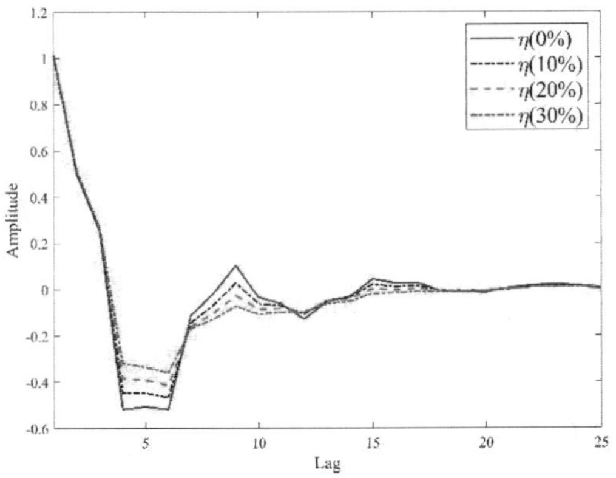

Figure 14. Impulse responses estimated by stochastic modeling approach for Example 3

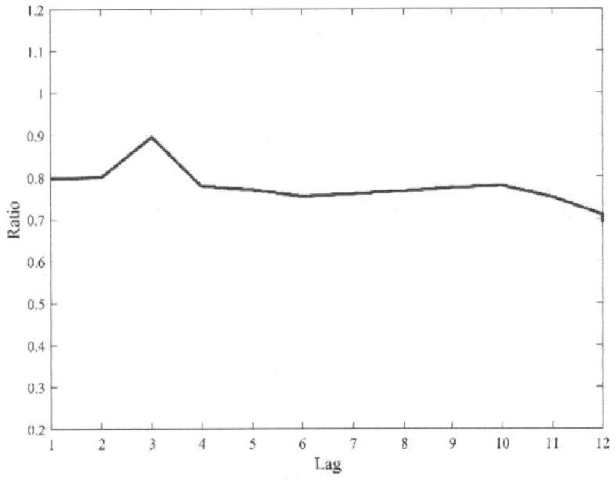

Figure 15. Relation of the estimated impulse responses for Example 3

6. Conclusions

The deterministic and stochastic approaches for non-parametric modelling are proposed for the closed-loop system involved saturation nonlinearity in this paper. Both non-parametric models are characterized by the impulse response of closed-loop nonlinear system. For the former, the impulse response is derived by the deterministic approaches under different saturation ratios; while for the latter, the impulse response coefficients are estimated by FCOR algorithm as the external disturbance of closed-loop is modelled as stochastic signal. In the presence of different saturation ratios, it shows the linear relationship among the resulting impulse response coefficients. The results of three numerical examples can be summarized as that the ratio relationship between the impulse response coefficients estimated by using the FCOR algorithm at different saturation ratios is not completely linear, which is a bit different from the ratio relationship of impulse response via deterministic modeling approach.

7. References

[1] B. Shen, F. Ding, A. Alsaedi, and T. Hayat 2016 *Circuits Syst. Signal Process.* Gradient-Based Recursive Identification Methods for Input Nonlinear Equation Error Closed-Loop Systems **36** 2166-83

[2] Y. Pu, Y. Q. Yang, and J. Chen 2020 *International Journal of Control Automation and Systems* Maximum Likelihood Iterative Algorithm for Hammerstein Systems with Hard Nonlinearities **18** 2879-89

[3] K. Deng, and F. Ding 2014 *Nonlinear Dyn.* Newton iterative identification method for an input nonlinear finite impulse response system with moving average noise using the key variables separation technique **76** 1195-202

[4] C. M. Cheng, Z. K. Peng, W. M. Zhang, and G. Meng 2017 *Mech. Syst. Sig. Process.* Volterra-series-based nonlinear system modeling and its engineering applications: A state-of-the-art review **87** 340-64

[5] J. Chen, X. Lu, and R. Ding 2012 *J. Vib. Control* Gradient-based iterative algorithm for Wiener systems with saturation and dead-zone nonlinearities **20** 634-40

[6] X. Wang, F. Ding, A. Alsaedi, and T. Hayat 2021 *Int. J. Robust Nonlinear Control* Auxiliary model - based iterative parameter estimation for a nonlinear output - error system with saturation and dead - zone nonlinearity **31** 4262-86

[7] L. Carassale, and A. Kareem 2010 *J. Eng. Mech.* Modeling Nonlinear Systems by Volterra Series **136** 801-18

[8] J. Xavier, S. K. Patnaik, and R. C. Panda 2021 *ChemBioEng Rev.* Process Modeling, Identification Methods, and Control Schemes for Nonlinear Physical Systems – A Comprehensive Review **8** 392-412

[9] Y. R. Wang, and C. M. Cheng 2020 *Int. J. Syst. Sci.* Volterra series identification and its applications in structural identification of nonlinear block-oriented systems **51** 1959-68
[10] C. M. Cheng, Z. K. Peng, W. M. Zhang, and G. Meng 2013 *Nonlinear Dyn.* Wavelet basis expansion-based Volterra kernel function identification through multilevel excitations **76** 985-99
[11] I. J. Leontaritis, and S. A. Billings 2007 *Int. J. Control* Input-output parametric models for non-linear systems Part I: deterministic non-linear systems **41** 303-28
[12] I. J. Leontaritis, and S. A. Billings 2007 *Int. J. Control* Input-output parametric models for non-linear systems Part II: stochastic non-linear systems **41** 329-44
[13] T. Baldacchino, S. R. Anderson, and V. Kadirkamanathan 2013 *Automatica* Computational system identification for Bayesian NARMAX modelling **49** 2641-51
[14] S. Chen, and S. A. Billings 2008 *Int. J. Control* Representations of non-linear systems: the NARMAX model **49** 1013-32
[15] S. L. Xie, Y. H. Zhang, C. H. Chen, and X. N. Zhang 2013 *Mech. Syst. Sig. Process.* Identification of nonlinear hysteretic systems by artificial neural network **34** 76-87
[16] Z. Ahmad, R. A. M. Noor, and J. Zhang 2009 *Asia-Pac. J. Chem. Eng.* Multiple neural networks modeling techniques in process control: a review **4** 403-19
[17] K. J. Hunt, D. Sbarbaro, R. Żbikowski, and P. J. Gawthrop 1992 *Automatica* Neural networks for control systems—A survey **28** 1083-112
[18] T. A. Tutunji 2016 *Appl. Soft Comput.* Parametric system identification using neural networks **47** 251-61
[19] M. Schoukens, and K. Tiels 2017 *Automatica* Identification of block-oriented nonlinear systems starting from linear approximations: A survey **85** 272-92
[20] M. Schoukens, A. Marconato, R. Pintelon, G. Vandersteen, and Y. Rolain 2015 *Automatica* Parametric identification of parallel Wiener–Hammerstein systems **51** 111-22
[21] J. Schoukens, R. Pintelon, Y. Rolain, M. Schoukens, K. Tiels, L. Vanbeylen, A. Van Mulders, and G. Vandersteen 2015 *Automatica* Structure discrimination in block-oriented models using linear approximations: A theoretic framework **53** 225-34
[22] B. Huang 1997 Thesis Multivariate statistical methods for control loop performance assessment Type University of Alberta (Canada)
[23] R. W. Fu, L. Xie, Z. H. Song, and Y. Cheng 2012 *J. Process Control* PID control performance assessment using iterative convex programming **22** 1793-9
[24] A. Y. Sendjaja, and V. Kariwala 2009 *J. Process Control* Achievable PID performance using sums of squares programming **19** 1061-5

Three-Layers Model of Pedestrian in 2D Complex Space

Fa Zhang

Zhuhai School, Beijing Institute of Technology, Zhuhai, Guangdong,China,519088

E-mail:richter2000@163.com

Abstract. Walking is the basic way of human movement. How to model pedestrian in complex space is a challenge. Various models such as social force and cellular automaton have been proposed. These models can reproduce some macroscopic characteristics of pedestrian, but are not very effective in complex space. This paper proposes a pedestrian model framework, including mental-planning-moving three layers. The mental layer depicts the mental state of pedestrians, selects the movement mode, destination, and speed. The planning layer performs route planning and determines the sequence of temporary target points. The moving layer could embed social force model to realize physical movement. The framework can introduce incomplete information and uncertainty, which makes the pedestrian's decision-making probabilistic, and the trajectory has a certain degree of randomness. A pedestrian simulation software has been developed, and multiple simulation experiments have been performed in a complex two-dimensional space. The results show that the model can describe the pedestrian in a complex space realistically.

1. Introduction

Walking is a commonly used mode of human movement. pedestrian models are of great value in the fields of traffic planning, emergency evacuation, animation design, and video games[1] [2]. A variety of pedestrian models have been proposed[3]. The most influential ones are social force model[4], potential field model [5], cellular automata (CA)[6], game theory[7], and multi-agent model[8] . The social force model regards pedestrians as self-driven particles that move under the action of attractive and repulsive forces[9]. Although there are some flaw[10], it can reproduce some important phenomena, such as “arched in front of the exit” and “fast is slow”. The potential field model introduces a variety of fields, and regards pedestrians as particles moving in the field. The cellular automata use simple rules to describe the movement of individuals. The game theory model uses game theory to describe the individual's decision-making.

These models adopt particle or rational assumptions. Theses simplified assumption cannot describe the characteristics of human individual. Pedestrians are complex individuals. The individual is affected by many factors, such as mental state, physiological condition, the distribution of buildings (obstacles) in the space, the location and speed of other pedestrians around. In this paper we regard individual as agent with mental states and capable of real time decision-making. We expect this model can describe pedestrian in complex spaces.

2. Framework of pedestrian model

In a complex space, individuals perform cycles of observation, orientation, decision-making, and action (OODA) while walking. Treating individuals as particles cannot portray pedestrians realistically. To fully describe walking involves three levels, which are the mental layer, the planning layer, and the moving layer. As shown in figure 1.

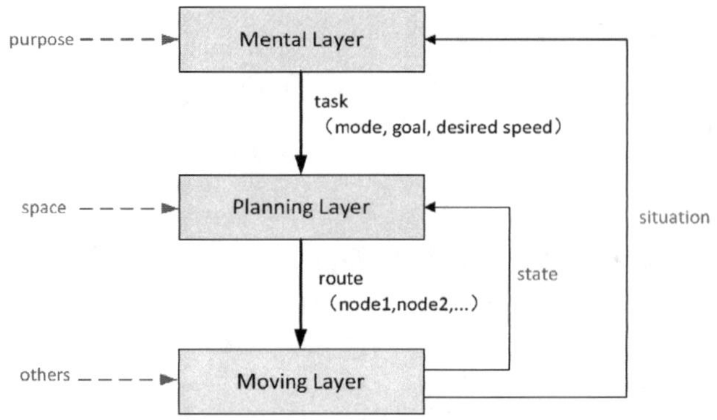

Figure 1. The three-layer architecture of pedestrian model

The mental layer selects the walking mode, determine the destination, and set the main parameters of walking. Human minds are very complex and need to be simplified from the perspective of building models. There are many modes of walking. Common modes include: approaching the target, following, searching, waiting, etc. Taking approaching the target as an example, the individual needs to determine the location of the desired target and determine the speed. Decision-making is affected by a variety of mental factors. Among them, the more important is the mental state. The mental state has an important influence on the choice of mode and parameter.

The planning layer generates the route according to the walking mode selected by the mental layer. In the approaching target mode, the optimal path needs to be found. We designed a heuristic path-finding algorithm.

The moving layer realizes the movement of individuals in the space according to the route made by the planning layer. A variety of micro-pedestrian models can be used, such as social force, CA, etc. Furthermore, during the walking the individual continuously collects information, feeds it back to the planning layer, and re-plans the route if needed. If there is major change in the situation, it will be directly fed back to the mental level and re-select the walking mode.

3. Three-layers model design

3.1. Mental layer

The mental layer determines the walking mode and destination based on the individual's purpose and the situation. In normal situations, individuals choose destinations based on purpose and spatial information. In the emergency, the safer exit is selected. The utility of exit i is

$$V_i = (1 - d_i)(1 - Risk_i) \tag{1}$$

According to the logit model, the probability of choosing exit i is:

$$P_i = \frac{exp(\beta V_i)}{\sum_{i=1}^{n} e\,xp(\beta V_i)} \tag{2}$$

After the target position is selected, the desired speed is determined. In emergency state, the desired speed is the maximum speed.

3.2. Planning layer

Scholars adopt different methods to solve the path planning problem. Hoogendoorn et al. established a dynamic programming model to find the optimal path [11]. The NOMAD chooses the direction that minimizes the cost [12]. Nasir et al. used a set of fuzzy rules to reason the direction and distance of travel [13]. Asano calculate the shortest path and obtains the desired walking direction[14]. Gao designed the LRP model to select the route [15].

There are many buildings (obstacles) in a complex space, and individuals do not fully know the spatial information, so they cannot make a theoretically optimal plan. We use heuristic methods for sequential route planning. Assuming that the pedestrian knows the location of the destination, when a building obscures the target, the pedestrian will choose to detour. Starting from the current position, find the closest building that occludes the target, determine the temporary target, and then move to the temporary target. After reaching the temporary target, the next temporary target is generated. The process is repeated until the target is finally reached.

Suppose that at time t, the pedestrian position is O, the target point is G (Goal), and there is a building W between OG. Assuming that W is a convex polygon (if it is a concave polygon, it can be convexized into a convex polygon, which does not have a substantial effect on the detour), W has n vertices, denoted as A_i, $i = 1,2, ..., n$.

The algorithm for determining the temporary target point is as follows:

- 1) For all vertices A_i, $i = 1,2, ..., n$ of W, calculate the rotation angle θ_i from $\overrightarrow{OG}$ to $\overrightarrow{OA_i}$, and the rotation angle of the vertex on the clockwise side of $\overrightarrow{OG}$ is negative, The one on the counterclockwise side is non-negative.
- 2) According to the sign of θ_i all vertices of W are divided into two sets: the counterclockwise side vertex set $CCW = \{A_k|\theta_k \geq 0\}$, the clockwise side vertex set $CW = \{A_j|\theta_j < 0\}$.
- 3) For the vertex sets CCW and CW, find the vertices with the largest $|\theta_i|$ and mark them as A^+ and A^-.
- 4)For A^+ and A^-, calculate the approximate distance $d^+ = |\overrightarrow{OA^+}| + |\overrightarrow{A^+G}|$, $d^- = |\overrightarrow{OA^-}| + |\overrightarrow{A^-G}|$, if the angle θ of A^+ (or A^-) is an obtuse angle, then the vertex is ignored (assuming that it will not go in the opposite direction).
- 5)Select A^+ or A^- as detour vertex in probability, where $p(A^+) = \frac{\exp(d^+)}{\exp(d^+)+\exp(d^-)}$, $p(A^-) = \frac{\exp(d^-)}{\exp(d^+)+\exp(d^-)}$, and mark it as A^*.
- 6)Use A^* to make the vertical line of the line segment OG, and extend the distance ε from A^* to the outside of W along the vertical line (ε is the distance margin of the detour), and the point R is obtained, which is the detour point.

An example of the algorithm for determining the temporary target point according to the near building is shown in figure 2:

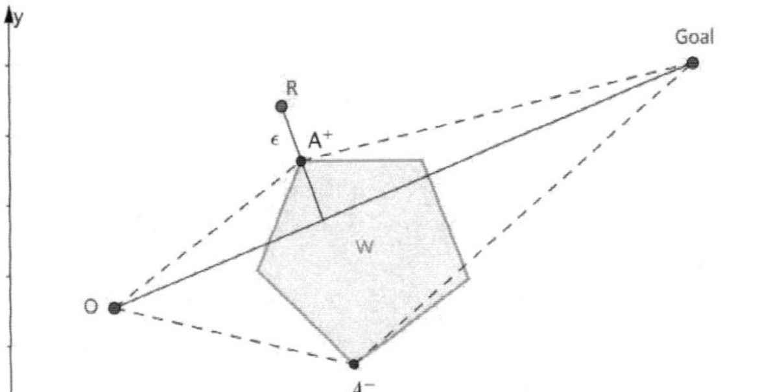

Figure 2. An example of finding temporary target

We have experimented with this algorithm in a variety of space configures, which can realize path planning in a complex space. For example, the path generated in a space scene is shown in figure 3:

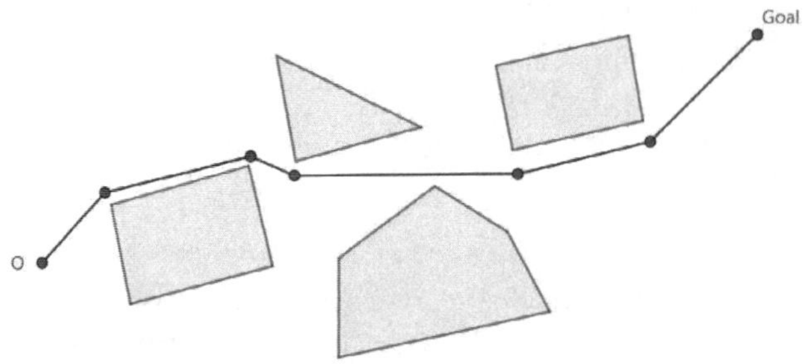

Figure 3. The route generated by the heuristic algorithm

3.3. Moving layer
The moving layer realizes pedestrian movement according to the temporary target generated by the planning layer. There are many algorithms that can be used, and the social force model is used here. The basic equation of the social force model is [7] [12]:

$$\frac{d\boldsymbol{v}_\alpha}{dt} = \boldsymbol{f}_\alpha(t) + \boldsymbol{\xi}_\alpha(t) \tag{3}$$

The acceleration of pedestrian α is mainly determined by social force $\boldsymbol{f}_\alpha(t)$, $\boldsymbol{f}_\alpha(t)$ is the generalized force, and $\boldsymbol{\xi}_\alpha(t)$ represents random fluctuations. Social force is the sum of three components:

$$\boldsymbol{f}_\alpha(t) = \frac{1}{\tau_\alpha}\left(v_\alpha^0 \boldsymbol{e}_\alpha^0 - \boldsymbol{v}_\alpha\right) + \sum_{\beta(\neq\alpha)} \boldsymbol{f}_{\alpha\beta}(t) + \sum_i \boldsymbol{f}_{\alpha i}(t) \tag{4}$$

The first term at the right end of the equation is the self-driving force, the current speed of the pedestrian is $\boldsymbol{v}_\alpha$, and it is hoped to travel at the desired velocity v_α^0 and the direction $\boldsymbol{e}_\alpha^0$ to generate the self-driving force. The pedestrian's expected speed unit vector $\boldsymbol{e}_\alpha^0$ points to the target, The expected velocity v_α^0 is given by pedestrian statistics, and the relaxation time τ_α.

The second term is the interaction force between pedestrians and other pedestrians, which is the repulsive force. The third item is the interaction force between pedestrians and obstacles, which is the repulsive force. The role of repulsion is to keep a certain distance from other pedestrians (or obstacles) to avoid getting too close or colliding.

The principle of the repulsive force between pedestrian α and pedestrian β and the repulsive force between pedestrians and obstacles is the same. The magnitude of the repulsive force is related to the distance between the two, and increases as the distance approaches. Generally, it adopts the form of negative exponential function, and the direction of force is from β to α, like:

$$\boldsymbol{f}_{\alpha\beta} = A_\alpha e^{-d_{\alpha\beta}/B_\alpha} \frac{\boldsymbol{d}_{\alpha\beta}}{\|\boldsymbol{d}_{\alpha\beta}\|} \tag{5}$$

where $\boldsymbol{d}_{\alpha\beta} = \boldsymbol{r}_\alpha - \boldsymbol{r}_\beta$, $\boldsymbol{r}_\alpha$, $\boldsymbol{r}_\beta$ are the position vector.

Using the social force model, we only need to set the temporary target generated by path planning layer as the current target of pedestrian α. After setting temporary target, pedestrians move according to the social force model. Perceive the gap between the current position and the temporary target point during moving. When the gap is small enough, it is considered that the temporary target point has been reached. At this time, the progressive path planning algorithm is reused at the planning layer to determine the next temporary target point, and then the social force model is used again to realize pedestrian movement after determination. When arriving at the destination or a special situation occurs, feedback to the mental layer to re-select the walking mode.

4. Simulation
We developed pedestrian simulation software using Java. Use simulation software to calculate the trajectory of pedestrians. The rectangular space is set to be 30m long and 20m wide, surrounded by

walls, and there is an exit on the right side with a width of 5m. There are multiple buildings distributed in the space. At the initial moment, the pedestrian appears on the left, and its initial position is randomly set, and the pedestrian's target point is set to a random position at the exit. Draw the position of each pedestrian at each moment to form a trajectory.

Given the different distribution of buildings in the two-dimensional space, the pedestrian trajectory is generated by simulation to verify the model. We have conducted a large number of simulation experiments, and now show a few scenes for demonstration in figure 4.

In figure 4(a), a rectangular obstruction is placed upright, and the target point cannot be seen through. After passing the obstacle, the pedestrian walks in a straight line to the target point.

In figure 4(b), two rectangular obstacles are placed alternately in the middle. After the pedestrian in the middle position bypasses the first obstacle, but still does not meet the visibility conditions, continue to bypass the second obstacle.

In figure 4(c) and (d), multiple obstacles are randomly arranged. The obstacles are messy and there are no obvious rules, but pedestrians can find a reasonable path.

It can be seen that pedestrian logic can be applied to complex space and produce natural trajectories.

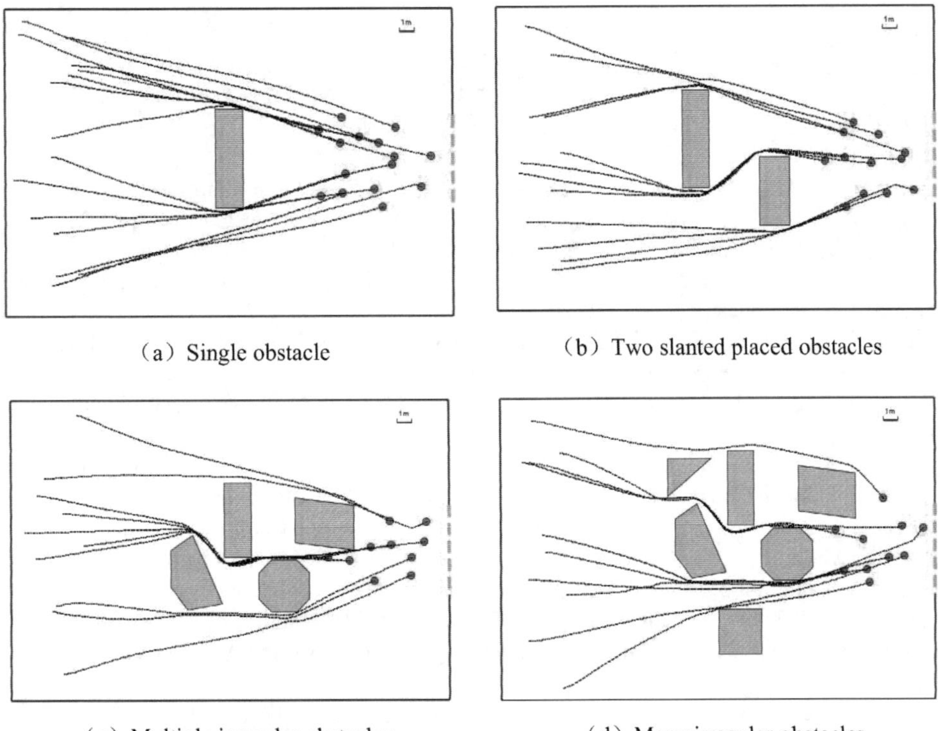

（a）Single obstacle （b）Two slanted placed obstacles

（c）Multiple irregular obstacles （d）Many irregular obstacles

Figure 4. The trajectory of pedestrians in scenarios with obstacles

5. Conclusions

Walking is very complicated, involving a large number of psychological, physical, spatial, and social factors. It is difficult to establish a pedestrian model suitable for complex spaces. We regard pedestrians as individuals with the ability to perceive, make decisions, and move. We build a three-layers model of mind, planning, and moving. The mental layer describes the mental state of the pedestrian, selects the movement mode, destination, and speed, and the planning layer performs route planning. The moving layer realizes individual moving. Incomplete information and uncertainty can be introduced into the model. In complex 2D space, a heuristic and progressive path planning method is proposed, and the algorithm is designed, which is combined with the social force model to realize movement in the complex space. We developed a pedestrian simulation model with Java to simulate pedestrians in a complex 2D space. Experiments show that the characteristics of pedestrians can be reproduced realistically.

Acknowledgments
This research was supported by National Natural Science Foundation of China (No. 71571190). The authors appreciate the supports deeply.

References

[1] S. Yang, T. Li, X. Gong, B. Peng, and J. Hu 2020 A review on crowd simulation and modeling Graphical Models, vol. 111, p. 101081.
[2] G. Kouskoulis, I. Spyropoulou, and C. Antoniou 2018 Pedestrian simulation: Theoretical models vs. data driven techniques International Journal of Transportation Science and Technology, vol. 7, no. 4, pp. 241-253.
[3] H. Vermuyten, J. Beliën, L. De Boeck, G. Reniers, and T. Wauters 2016 A review of optimisation models for pedestrian evacuation and design problems Safety Science, vol. 87, pp. 167-178.
[4] D. Helbing and A. Johansson 2010 Pedestrian, Crowd, and Evacuation Dynamics Encyclopedia of Complexity & Systems Science, vol. 16, pp. 697-716.
[5] R.-Y. Guo, H.-J. Huang, and S. C. Wong 2013 A potential field approach to the modeling of route choice in pedestrian evacuation, Journal of Statistical Mechanics: Theory and Experiment, vol. 2013, p. P02010.
[6] V. J. Blue, M. J. Embrechts, and J. L. 1997 Adler Cellular automata modeling of pedestrian movements, IEEE International Conference on Systems, Man, and Cybernetics. Computational Cybernetics and Simulation, vol. 3, pp. 2320-2323.
[7] D. Shi, W. Zhang, and B. Wang 2017 Modeling pedestrian evacuation by means of game theory, Journal of Statistical Mechanics, vol. 2017, no. 4, p. 043407
[8] N. Ronald, L. Sterling, and M. Kirley 2018 An agent-based approach to modelling pedestrian behaviour, SIMULATION, vol. 8, pp. 1473-8031.
[9] D. Helbing and P. Molnár 1995 Social force model for pedestrian dynamics. Physical Review E, vol. 51, pp. 4282-4286.
[10]T. Kretz 2015 On oscillations in the Social Force Model , Physica A, vol. 438, pp. 272-285.
[11]S. P. Hoogendoorn and P. H. L. Bovy 2004 Pedestrian route-choice and activity scheduling theory and models , Transportation Research Part B, vol. 38, pp. 169-190.
[12]M. Campanella, S. Hoogendoorn, and W. Daamen 2014 The Nomad Model: Theory, Developments and Applications, Transportation Research Procedia, vol. 2, pp. 462-467.
[13]M. Nasir, C. P. Lim, S. Nahavandi, and D. Creighton 2014 A genetic fuzzy system to model pedestrian walking path in a built environment , Simulation Modelling Practice and Theory, vol. 45, pp. 18-34.
[14]M. Asano, T. Iryo, and M. Kuwahara 2010 Microscopic pedestrian simulation model combined with a tactical model for route choice behaviour, Transportation Research Part C, vol. 18, pp. 842-855.
[15]Y. Gao, F. Chen, and Z. Wang 2019 Hybrid Dynamic Route Planning Model for Pedestrian Microscopic Simulation at Subway Station, Journal of Advanced Transportation, vol. 2019, p. 5207814.

A Detection Model of the Moving Direction in Vertical for UAV to Search Ground Infrared Targets

Yang Yang[1,*], Zhijun Li[2], Yunqiang Wu[1], Ke Li[1] and Jie Lu[2]

[1]School of Electronics and Information, Nanchang Institute of Technology, Nanchang 330044, China
[2]660 Design Institute, Hongdu Aviation Industry Group, Nanchang 330001, China

[*]corresponding author's e-mail: dzx@nut.edu.cn

Abstract. Regarding the relative detection of the moving direction in vertical between the unmanned aerial vehicle (UAV) and the ground target, we have been inspired by the functional advantages of flying insects in nature, such as lightweight, low computational complexity, low power, and the natural characteristics of high adaptability and high reliability in the detection and tracking or escape process of targets. Based on this, with the help of LGMD (lobula giant movement detector) neuron modeling, this paper proposes a detection model of the moving direction in vertical for UAV ground infrared targets based on LGMD neurons modeling (referred to as LGMD-UAVGIT model). In this paper, experiments and tests are mainly carried out on infrared imaging videos taken by UAVs, thus verifying the effectiveness of the new model proposed in the paper for detecting the two typical vertical movement directions of UAVs that are relatively far away from and close to the ground targets.
Keywords. UAV ground target; Target detection; Vertical moving; Modeling insect's visual processes

1. Introduction

Quickly and robustly extract the relative vertical motion (also called motion-in-depth) information of the unmanned aerial vehicle (UAV) on the ground from dynamic changes and complex scenes is the key to the UAV vision system to track targets and avoid obstacles and collisions. At the same time, this has always been a research hotspot in the field of computer vision technology for the two most basic motion modes of target (object) translation and depth motion in the three-dimensional space of the visual scene. In nature, the insects apply compound eye's vision to detect a depth moving target in highly variable and visual cluttered environments during flight, which are excellent paradigms to learn motion perception strategies. Compared with the visual system of primates, compound eyes have limited resolution and visual computing resources, but they have the natural characteristics of high adaptability and high reliability in the detection, tracking or escape process of targets in complex dynamic scenes. At present, simulating the optic nerve process of compound eyes has become an important model paradigm for the research and design of artificial motion perception visual systems[1-4]. In addition, this also provides a new approach to nonlinear signal processing for the relative vertical motion (motion-in-depth) and direction detection between the UAV and the ground target.

In recent decades, the research progress of the lobula giant movement detector (LGMD) system in locust optic lobe neuropile layers that can respond to emergent visual stimuli in a complex environment has resulted in LGMD-type neurons being significant for detecting target movement and collisions. The neuro-computer theory of sexual response is becoming clearer and clearer. Therefore, many scholars have proposed some artificial neural network models to clarify the process of LGMD neuron activity

[5,6]. However, these LGMD neuron-based collision detection models only have a significant response to moving objects that are gradually approaching, but do not respond to moving objects that are far away, which means that they cannot detect the distance and approach of the target movement simultaneously. This paper studies the two typical direction detection problems of the UAV's vertical movement of the ground target away from and approaching. Based on the bionic strategy combined with LGMD's improved collision detection model, this paper uses the zoom variable calculation of the target (object) movement to propose a detection model of the moving direction in vertical for UAV ground infrared targets based on LGMD neurons modeling (referred to as LGMD-UAVGIT model).

2. LGMD-UAVGIT modeling

The basic idea of LGMD-UAVGIT modeling is: firstly, with the help of the inter-frame difference method, the expansion or contraction change of the moving target between adjacent frames of the video is calculated to distinguish between static and moving targets. Then, combined with the improved collision detection model based on LGMD to complete the detection and estimation of the vertical movement direction of the target [5-7]. LGMD-UAVGIT modeling is divided into four continuous links, which mainly include: primary visual filtering, target motion scaling variable calculation, LGMD-based collision detection, and fusion calculation. The modeling framework is shown in Figure 1.

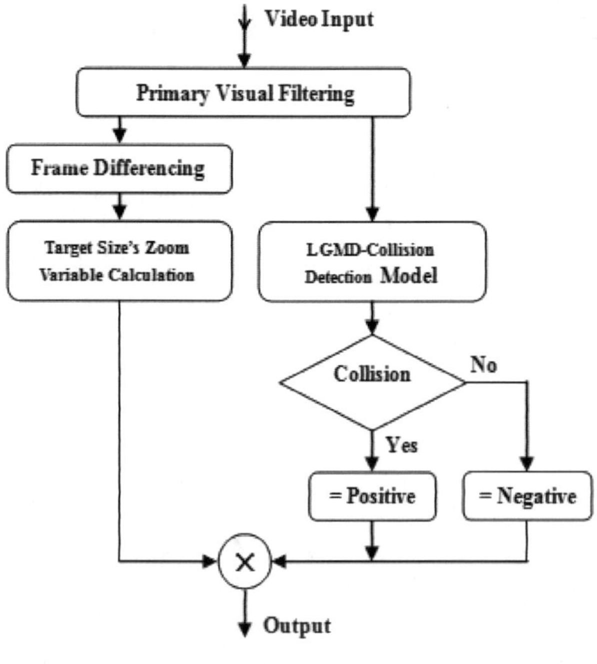

Figure 1. LGMD-UAVGIT model framework

(1) By simulating the mutual inhibition effect of centre-surrounding antagonism, which is the center of thin-lamellar monopolar cells in insect optic lobe neuropile layers, the bionic primary visual filtering is realized.

(2) The calculation of the zoom variable of the target motion is based on the following considerations, that is, when moving vertically, the target image size (the area surrounded by the edge of the image) will definitely change. Specifically, the target image size increases when it is close, and the image size remains unchanged or decreases when it is still or far away. This article defines the vertical movement of the target image size change as the expansion (enlargement) / contraction (reduction) variable, that is, the zoom variable. In this paper, the absolute value of the summed value after the difference between frames is used as the zoom variable of the target image size, and then it is judged whether the target has made a vertical motion away and close.

(3) LGMD's improved collision detection is based on the new biological discovery of lamina

monopolar cells (LMCs) that the motion visual information is separated into parallel ON and OFF channels [1,2,5,8,9]. In the detection, the half-wave rectifying method is used to decompose the ON and OFF channel signals. Then, it simulates the convergence and integration of the motion edge change information from the ON and OFF channels in the lobula complex layer, thereby forming the global membrane potential of the collision sensing neurons in the lobula complex.

(4) The collision detection model is combined with the added zoom variable calculation module, which is mainly used to distinguish between stationary and distant UAV ground targets. Multiplying the calculation result of the scaling variable of the relative vertical movement between the UAV and the ground target and the collision detection result can obtain the relative vertical movement direction between the UAV and the ground target. That is, the normalized output value of the model is greater than 0 means the target is close, less than 0 means the target is far away, and equal to 0 means no target is moving.

3. Simulation experiment and analysis

In the experiment, a four-rotor drone with an infrared camera was used, which can shoot infrared targets on the ground while flying vertically. Among them, the ground background is grass, the target is small rocks (about 6.5cm3), the infrared video was shot at a frame rate of 25 frames/s, and the video length is 120 frames. During the experiment, two videos of the quad-rotor UAV flying away from and approaching the target vertically were taken as test samples. The experimental scene is shown in Figure 2. The day was a sunny autumn day, with winds of level 3 to 4, and the target (small stone) was placed on the grass in the shadow of the sun next to the tall buildings. The representative frames intercepted in the infrared imaging video are shown in Figure 3 and Figure 4 respectively, and the target in the figure is marked with a red circle. It should be noted that due to the effect of wind, the UAV will drift when flying vertically.

Figure 2. Experimental scenario

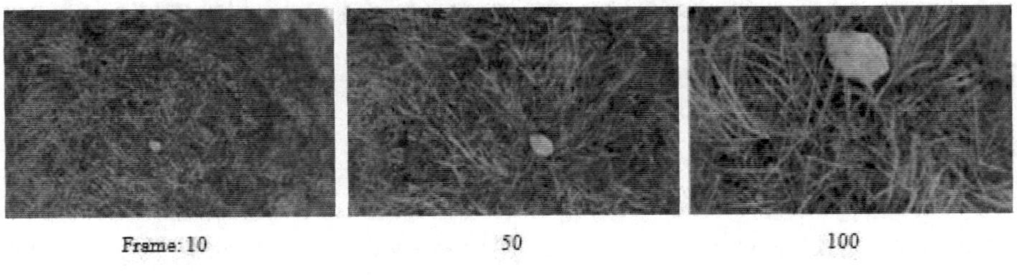

Figure 3. The target is approaching (the drone descends)

Figure 4. The target is far away (the UAV rises)

Insect compound eyes have low resolution, so the infrared imaging video is first converted and Gaussian smoothing is performed to simulate the visual effect of insect retina. Then input it to the LGMD-UAVGIT model. In the test of our new model, from the 10th frame, the target relative to the UAV movement is detected, and after the 110th frame, the drone stops ascending and descending. The direction detection and estimation output results of the new model are listed in Table 1. As can be seen from the table, the polarity of the new model represents the vertical motion (motion-in-depth) direction. Specifically, the normalized output value of the new model when the UAV is approaching is positive, and the normalized output value when the UAV is flying away is negative. This is consistent with the actual situation in the detection and estimation results of the two vertical movement directions of UAV flying close and away. In addition, interference caused by shaking or changes in flight speed during the flight of the UAV may cause the calculated target scaling variable to change. This will cause the output value of the new model to fluctuate between the 10th frame and the 110th frame, resulting in slightly lower than the ideal value after averaging.

Table 1. Normalized output value of LGMD-UAVGIT model

Infrared imaging video sample moving direction		Video frame number/normalized output average					
		1～9	10～40	41～70	71～110	111～120	Ideal value
Target	Close	0.05	0.81	0.90	0.83	0.07	1
	Away	-0.04	-0.84	-0.88	-0.86	-0.03	-1

4. Conclusion

Aiming at the two typical direction detection problems of the vertical movement of the UAV to the ground infrared target, the detection model of the moving direction in vertical for UAV ground infrared targets based on LGMD neurons modeling (LGMD-UAVGIT Model). This paper simulates and analyzes the neural calculation process of projecting light-sensitive signals from the retina of the compound eye to the downstream thin plate through the nonlinear signal processing of the insect optic lobe system. Then, experiments and tests are carried out on the sample video of infrared imaging taken by the drone to verify the effectiveness of our new model for the detection of the vertical movement direction of the drone on the ground target.

References

[1] A. Borst, J. Haag, A. S Mauss. How fly neurons compute the direction of visual motion[J]. Journal of Comparative Physiology A, 2020, 206: 109–124.

[2] Nathalie Neriec, Claude Desplan. From the eye to the brain: Development of the drosophila visual system[J]. Current Topics in Developmental Biology, 2016, 116: 247–271.

[3] Fu Q, Hu C, Peng J, et al. A robust collision perception visual neural network with specific selectivity to darker objects[J]. IEEE Transactions on Cybernetics, 2019, 1–15.

[4] Xu, M.X., Wu, X.B., Zhang, Z. and Lu, Y.Y. (2021) Compound-Eye Imaging Imitation-Based Whole-Field Flow Measurement. Computers and Electrical Engineering, 92, Article ID: 107141.https://doi.org/10.1016/j.compeleceng.2021.107141

[5] Blanchard M, Rind F C, Verschure P. Collision avoidance using a model of the locust LGMD

neuron[J]. Robotics & Autonomous Systems, 2000, 30(1-2):17–38.
[6] MA Xingzao, ZHAO Jiannan, ZHU Qiyuan, FU Qinbing, HU Cheng, LEI Fang, YUE Shigang. Study on obstacle avoidance method of UAV based on LGMD[J]. Computer Engineering and Applications, 2019， 55(15): 250–256. (in Chinese)
[7] Yang Xu. Research on information feature extraction method based on fly vision mechanism [D]. HoHai University, 2021.06. (in Chinese)
[8] Behnia, R., Clark, D.A., Carter, A.G., Clandinin, T.R. and Desplan, C. (2014) Processing properties of ON and OFF pathways for Drosophila motion detection. Nature, 512, 427–430.
[9] Strother J A, Wu S T, Wong A M, et al. The Emergence of Directional Selectivity in the Visual Motion Pathway of Drosophila[J]. Neuron, 2017, 94(1):168–182.

Feasibility analysis and Research on Intelligent trusted modeling of network architecture software

Kui Wang

School of Computer and Information Engineering, Nanchang Institute of Technology, Nanchang, Jiangxi 330044, China

*corresponding author's e-mail: 110557560@qq.com

Abstract. Mesh software is a new software form for network environment, and its construction relies on effective collaboration among software entities of network nodes in open, dynamic and changeable environments. According to the open, dynamic and changeable characteristics of network configuration software, this paper designs a software structure model of trusted network monitoring based on the model, with the network configuration software running at any time according to the related factors of network node state dynamic evaluation of internet ware credibility, and according to the monitoring results to improve the network configuration software and monitoring parameters provide support for dynamic, trusted Internet ware, and gradually improve the credibility of network configuration software.
Keywords. internet ware; trusted degree; software monitoring; model for tectonic software

1. Introduction

The concept of network configuration software was first proposed by Yang Fuqing of the Peking University and Hong Mei and several other experts [1], he Jifeng and other experts and scholars and Chen Yujian of Tsinghua University, the Chinese Academy of Sciences Jinzhi, Lv Jian of Nanjing University, East China Normal University, the in-depth study of the structure and content of software related network, related to software development methods, formal, model theory and platform framework [2]. Because of the importance of people gradually realize that the network configuration software on the software technology development and role can't be eignored, the model of internet are based on the sustained attention and become a research hotspot [3]. At present, network configuration software technology research mainly in the following aspects: based on Internet are model architecture, Agent based Internet are model and method, component oriented network software development method, Internet are middleware platform model, Internet are credibility theory and its measurement and evaluation system and network configuration software platform technology the standard system [4].

Software is a virtual reflection of objective things, is the solidification, concise and embodiment of knowledge [5]. The essence of software, driven by a basic model of software continue to pursue more skills, more in line with the thinking mode of human beings, more construction and evolution of the software structure [6]. With the rapid development of Internet, individual participation and independence, interactive information openness, group cooperation credibility, malicious deception, individual credit evaluation relative the local search of knowledge, has gradually formed a kind of social communication is different from the traditional users of social ideology, a new way of thinking requires a knowledge of new forms [7], Internet ware is a new form of software produced under this kind of environment. Specifically, network configuration software is a software entity group distributed in the Internet environment of each node, with the main character [8], and a set of software for supporting these entities in various interactive connection sub collaboration [9]. The presence of the Internet are to existing software technology research put forward a new topic, the existing software theory, method and corresponding technology of middleware system is difficult to effectively adapt to the basic network configuration software or can't dapt to this new form. The research and practice achievements of Chinese scholars in the national key basic research and development plan under the support of the Internet ware area in recent years, pointed out that the key work is to strengthen further the existing achievements and depth the breadth of emphasis in depth, focus on breakthrough of internet are model, intelligent trusted Internet ware middleware technology and autonomous management of network software development [10]. The degree of automation of law. The research on software trustworthiness is a hot issue at home and abroad in recent years. It mainly includes two aspects: identity trust and behavior trust. The former belongs to the static trust, relevant regulations promulgated China - the State Encryption administration regulations, the international series of trusted computing trusted computing organization to develop a trusted computing platform password technology both at home and abroad; the latter belongs to the dynamic trust.

2. Concept of trusted network and its properties

2.1. Concept of Trusted Network

Though the concept of the trusted network has been proposed for some time and the trusted computing to enhance the credibility of computer terminals is a research hotspot in recent years, the international research about trusted network has just begun for several years and the basic concepts and scientific issues are not in-depth. At present, the representative expositions about the credibility are mainly as follows. ISO/IEC 15408 indicates that the conduct of a trusted component, operation or process is predictable under any operation condition and has very good resistance to the damage caused by the application software, virus, as well as some physical interference. Microsoft's Bill Gates figured that trusted computing is a reliable safe computing ready to access, including the degree of human's trust in computers like the free and secure use of electric power systems or telephone communication systems. Algridas, Laprie and others would express the credibility as that the services provided by the system can be demonstrated to be trusted and the system can avoid unacceptable frequent or serious failure of service.

2.2. Properties of Trusted Network

To a trusted network, the network's behavior and its results can be expected, so the conduct state can be monitored, the results of the act can be evaluated, and the abnormal behavior can be controlled Trusted Computing Group. Specifically speaking, the credibility of the network should include a set of attributes: the security and survivability of services which should be protected from the user's point of view, as well as the network controllability which should be provided from the design point of view. Different from scattered and isolated connotation of the concept about the security, survivability and controllability in the traditional sense, the trusted network will integrate these three fundamental attributes in the goal of trusting in network, and form an organic whole around the maintenance of the trust between network components and behavior control.

At the beginning of the Internet design network security was not considered adequately. The early Internet protocol few took into account the safety of network architecture so the vulnerability of the Internet has led to the result that it is not possible to completely trust in Internet. The traditional network architecture focused on improving the efficiency of data transmission and formed the Internet model with simple core and complex edge. It is easy to expand for new application in this architecture, but difficult to detect problems in the application layer. The current design of network security has rarely touched the core of the architecture, most using the mode for information access control on the edge of the network with plugging loopholes, building high walls and preventing outside attack. Thus network security systems become increasingly complex and cumbersome and seriously reduce network performance. Considering the current complex status of heterogeneous network architecture, the design of trusted network architecture should provide the core security architecture in compliance with safety standards.

Controllability is an important attribute in the trusted network design and its main objective are following. Add authentication, authorization and other control mechanisms in the key part of network so that the network is more trusted. Maintain certain status information and take the necessary control in the network so that the network has some degree of connection-oriented features. Carry out network monitoring at different levels, provide the mechanism to collect and transmit network components confidence information and check network operation status and provide fast algorithm of abnormal behavior controlling and early warning of attacks.

Aiming at integration and complication of network attacks and acts of destruction, the corresponding mechanisms to fight against it should also be integrated. However, the current network security defense system is isolated and scattered, which requires the internal associated supervisory control system should be established when design the controllability of trusted network and effectively integrate all means of control in a trusted platform, such as network information collection and testing, attacks analysis, early warnings and so on. According to the concept of trusted network and its three basic attributes, behavior control, behavior monitoring, behavior authentication, behavior management and behavior combating can be provided in network and systems aiming at the behavior of business and technology and its results.

3. Trusted network architecture

Fault tolerance of internet ware credibility depends mainly on the network nodes, so this paper set two kinds of fault tolerance allocation method: (1) were randomly assigned to each fault tolerance function according with mean beta Poisson distribution; (2) priority allocation, call the number and weight of product fault tolerance function and it has the and the proportional relationship, as shown in (1):

$$\beta_i = E(\sum_{i=1}^{8} \frac{S_i}{S_{max}} \times \frac{k_i}{k_{max}}) \tag{1}$$

Among them, $S_i = \sum_{j=1}^{n} a_{ij}$ denotes the weight of node i, and k_i denotes the number of i callable relation of node, S_{max} denotes the weight of network and k_{max} denotes the maximum number of callable relation in network. In the process of internet ware system operation, if the function i fails, then the fault may be through inter function calls or dependencies with a certain probability to spread function j, and fault-tolerant ability decides whether the fault can lead to the failure of j function, and then affect the other function calls and dependent function j.

When the fault tolerance function j failure intensity function call i β for λ_j, considering the construction method of network configuration software this paper stipulates: when $\beta_i \geq \lambda_j$, j fault occurred whether the close degree infection function i depends on the two functions, as well as the interaction strength of fault and fault tolerant capability than the j at the time; the close degree, influence function i will be the fault depends only on the relationship between function and function call i j. In order to distinguish the two cases, a discrimination coefficient δ is set up, such as formula (2):

$$\delta_{ij} = \begin{cases} \dfrac{\lambda_j}{\beta_i}, \beta_i \ge \lambda_j \\ 1, \beta_i < \lambda_j \end{cases} \tag{2}$$

In summary, the fault infection probability P of node i in mesh software is shown in formula (3):

$$P_i = \sum_{j=1}^{m} (W_{ij} \times \delta_{ij}) \tag{3}$$

Where $W_{ij} = w_{ij} / w_{\max}$ represents the probability of the call relationship between function i and function j in mesh software.

Trusted Network Architecture (TNA) is a trusted network security technology architecture, which overall improves network security protection by effectively integrating and managing network available security products and security sub-systems, as well as combining the trusted network access control mechanisms and the internal information protection, encryption and transmission mechanisms. TNA includes Trusted Network Security Management System (TSMS), Trusted Network Security Access Control System (TSCS) and Trusted Network Protection Mechanisms (TNPM). The TNA is as figure1. The largest feature of this structure is the realization of the legitimate integration and management of existing resources, changing the past safety management system aiming at a kind of security incident, and implementing all-around, systemic, centralized security management.

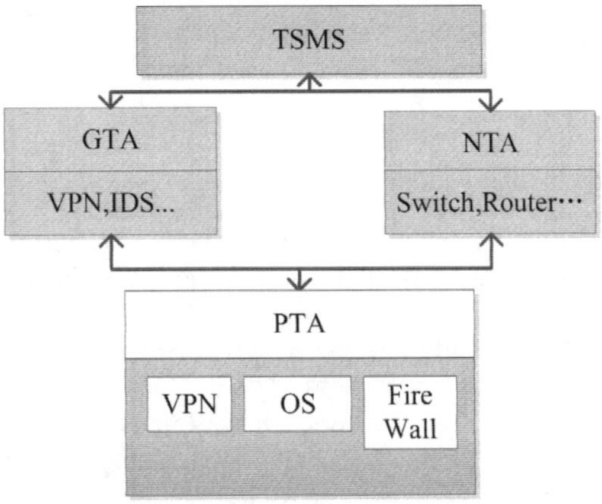

Figure 1. Trusted network architecture

TSMP is an open platform for network security management produced by Beijing Topsec Science and Technology Co., Ltd., which is based on their ideas of TNA. TSPM consist of five main components: TopAnalyzer, TopDesk, asset management system, authentication system and report forms management system. TopAnalyzer and TopDesk are the core components of TSMP. As basic components, the other three systems and database systems, as well as WEBbased portal management system constitute a support platform of TSMP. The TSMP is as figure 2.

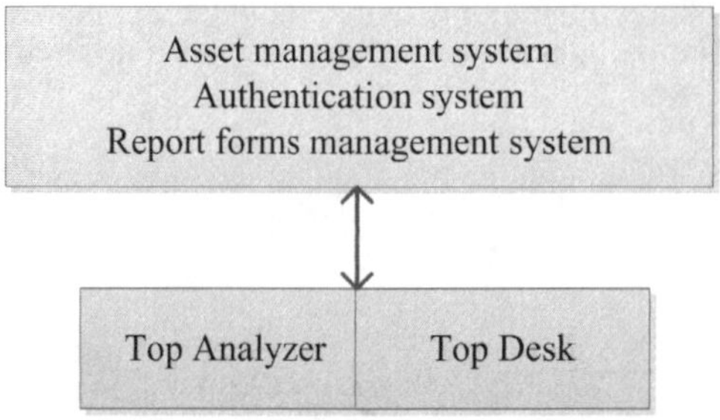

Figure 2. Trusted network management platform model

4. Dynamic adaptability and accuracy evaluation

In order to verify the performance of the model, in the Netlogo simulation platform to achieve the Trust system distributed trust model in Peer and typical MSCTrust multi service hierarchical trust model based on the time frame of this paper (Multiple service classification trust model) were compared. In the test set, simulation scenarios show that the system parameters of the NIF=40%, the busy degree is moderate; NDF=30%, complete with node dynamic strategy changes its strategy of frequency is 30%; SDF=10%, shows that the dynamic join the system and leave the number of nodes accounted for 10% of the total. Such scenario design can basically conform to the actual situation of a real network system. In order to test the accuracy of MSCTrust and dynamic adaptability, we designed two experiments scene, community is set to nodes in the system of goodwill accounting for 80% of the total, accounting for 20% of the total number of malicious entity nodes, which accounted for 5% of the malicious recommendation entity (the assumption here is that the malicious recommendation entity and malicious interactive entities in identity without overlapping). In the first experiment scene, malicious interactive node set 15% for class S DMI, these nodes take cooperation strategy in the interactive stage; in the "buy" mid take cooperative action, take collaboration on the "sell" behavior; later changed to take cooperation in "sell" behavior, take cooperation in "buy" behavior. The weight of interactive services between nodes is the same. In a second experiment in the scene, a malicious entity node set 15% for class V DMI, these nodes take cooperation strategy in the interactive stage, started in the middle of interaction in the service weight lower taken in cooperation, take cooperation strategy in the service and service of high weight, late again take cooperation strategy. The initial trust value of the node is set to 0.4. The dynamic adaptability and accuracy of the trust model are investigated by calculating the average trust values $trust_s^t$ and $trust_v^t$ of the nodes of the above two types of strategic change behavior. Figure 3 and figure 4 show the average trust value changes of malicious interaction nodes of the two dynamic behavior change strategies respectively.

It can be seen from Figure 3, the initial stage, the class S DMI cooperative nodes selection strategy, so the MSCTrust model and Peer Trust model of trust value approximates 1, the convergence rate of MSCTrust compared to the Peer model Trust model faster. With the increase of time, nodes began to use collaboration on the "buy" behavior, not collaboration on the "sell" behavior, the nodes in MSCTrust model, the Trust-CL value was not affected, as long as the good behavior in the service request node, its reputation request service value will remain at a higher value. At this time due to node selection not cooperation strategy in the provision of services, the Trust-SE value decreases rapidly and tends to zero, indicating that the node is a node of a malicious node in the provision of services, change the strategy again in the interaction stage, began to take the collaboration on the "sell" behavior, not cooperation in the "buy" behavior at this time, the node's Trust-CL value began to decline rapidly, the Trust-SE value began to rise, but it can be seen that the increase of Trust-SE value and Trust-SE value decreased when the speed not fast, which shows It takes longer for nodes to build trust values than to consume trust values. Peer and Trust model in the node began to use the "collaboration buy" behavior, not "sell" behavior in cooperative strategy, the average value of the node trust began to decline, but because the model does not distinguish between nodes as the service requester and service providers, so the trust values of nodes eventually converge to about 0.5. The average trust value of nodes does not really reflect the malicious collaboration strategy adopted by nodes.

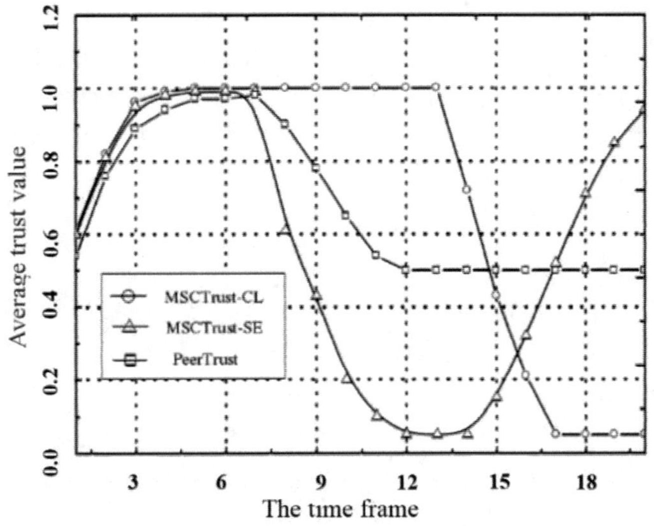

Figure 3. The change of average trust value

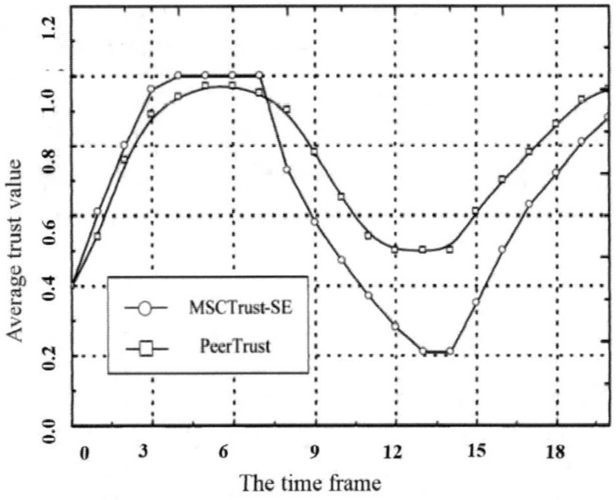

Figure 4. The change of average trust value

We can see from figure 4 that the initial stage, the class S DMI cooperative nodes selection strategy, so the MSCTrust model and Peer Trust model of trust value approximates 1, the convergence rate of MSCTrust compared to the Peer model Trust model faster. With the increase of time, nodes began to use the interaction in service weight low take cooperation, take cooperation strategy in the service and service of high weight, because both as a service requester node or service provider, its strategy is consistent, so the node Trust-SE values consistent with the changes of Trust-CL value, given only changes here the former value. Nodes in the interactive medium term strategy of cooperation led to two trust value began to decline, but because the node selection does not cooperate in higher weight services, so in the MSCTrust model the node trust value declined rapidly, and ultimately converge to 0.2. While the Peer Trust model is dynamic, it can't distinguish the node's service type, so the trust value is close to 0.5. Late again node change strategy, choose to cooperate strategy, two kinds of trust value are increased, but the MSCTrust model of the trust value of the rapid growth of Peer, the Trust model of the trust value of slower growth, indicating that the MSCTrust model is more sensitive to the change of the node, to do it faster. However, the trust value of the MSCTrust model increases faster or the trust value of the model does not drop faster, indicating that the establishment of trust value needs longer time than the consumption of trust value.

5. Conclusion

With the rapid expansion of Internet in the type of business, the number of users and the complexity, current scattered, isolated, single defense, external attached network security system has been unable to cope with attacks and acts of sabotage, which are of various, random and hidden features. Trusted Network borrows ideas from system credibility and syncretizes the traditional isolation research content into the object of network credibility, to provide security services. In this paper, an open and dynamic network environment Internet ware credibility problem in software systems is applied to analyze and research on the measurement of trust model and trust model of internet ware incentive mechanism and to ensure the credibility and reliability of the software. A time frame based multi service hierarchical trust model in mesh software system is proposed. The overall architecture evolution model of internet ware software online based on the measurement model of internet ware system trust has solved the existing trust measurement model granularity being rough due to the final results of the assessment is not accurate enough problems.

Acknowledgement

Project fund: Science and Technology Research Project of Jiangxi Provincial Department of Education, Project No. 202107, Research on MEC Smart Mall Information Management Scheme under the Background of Big Data.

References

[1] Alliez P, Yvinec M, Boissonnat J D 2015 CGALmesh: A Generic Framework for Delaunay Mesh Generation *ACM Transactions on Mathematical Software* **41**(4) 23

[2] Chang J S, Jeon Y H, Sim S, et al 2015 Information Security Modeling for the Operation of a Novel Highly Trusted Network in a Virtualization Environment *International Journal of Distributed Sensor Networks* (2015-9-27) **4** 1-12

[3] Abdel-Halim I T, Fahmy H M A, Bahaa-Eldin A M 2015 Agent-based trusted on-demand routing protocol for mobile ad-hoc networks *Wireless Networks* **21**(2) 467-83

[4] Koster J, Leckie G, Miller A, et al 2015 Multilevel modeling analysis of dyadic network data with an application to Ye'kwana food sharing *American Journal of Physical Anthropology* **157**(3) 507

[5] Gan C 2016 Modeling and analysis of the effect of network eigenvalue on viral spread *Nonlinear Dynamics* **84**(3) 1727-33

[6] Xie H, Lui J C S 2015 Modeling eBay-like reputation systems: Analysis, characterization and insurance mechanism design *Performance Evaluation* **91**(C) 132-49

[7] Yan S R, Zheng X L, Chen D R, et al 2011 User-centric trust and reputation model for personal and trusted service selection *International Journal of Intelligent Systems* **26**(8) 687–717

[8] Khabbaz M J, Alazemi H M K, Assi C M 2013 Modeling and Delay Analysis of a Retransmission-Based Bundle Delivery Scheme for Intermittent Roadside Communication Networks *IEEE Transactions on Intelligent Transportation Systems* **14**(2) 700-8

[9] Zhou W, Jia W, Haghighi M, et al 2015 A Sword with Two Edges: Propagation Studies on Both Positive and Negative Information in Online Social Networks *IEEE Transactions on Computers* **64**(3) 640-53

[10] Tu M, Yen I L 2014 Distributed replica placement algorithms for correlated data *Journal of Supercomputing* **68**(1) 245-73

The Application of Computer Technology in Mathematical Modeling

Jing Hu

Nanchang Institute of Technology, Nanchang, Jiangxi, China

E-mail: 69462301@qq.com

Abstract. Based on the analysis of mathematical modeling, this paper further expounds the application of computer technology in mathematical modeling. The specific role of computer technology mainly includes establishing mathematical modeling by using computer programming and using computer software. At present, the development of mathematical modeling cannot be separated from the support of computer technology. Through the specific analysis of computer technology model formula, the value of computer technology to mathematical modeling is clarified.
Keywords. Computer technology; Mathematics; Modeling

1. Introduction

Mathematical modeling has been deeply applied in all aspects of society. The mathematical modeling process is very complicated. In the current development of mathematical modeling, the application of information technology and computer technology has become the technical guarantee for the rapid development of mathematical modeling. As an important ideological subject, mathematics has a strong creativity for students' thinking. Mathematical modeling refers to the use of relevant mathematical knowledge to solve practical problems, in complex mathematical problems middle school students need to continuous analysis and demonstration, before and after data operations and drill, draw the conclusion in a simplified abstract processing, the practical problems into the first related mathematical problems, and then the systematic mathematical model construction. With the continuous development of computer technology, the application of computer technology in mathematics is also more active and extensive. The computer technology using the system programming function can quickly solve some mathematical computing formula, calculus, etc. Secondly, computer technology also contains a powerful drawing function, and in the process of learning mathematics, specific mathematical graphics are needed to further understand mathematical knowledge. Therefore, with the support of computer technology, mathematics can break through the previous modeling problems and further penetrate into the mathematical operation.

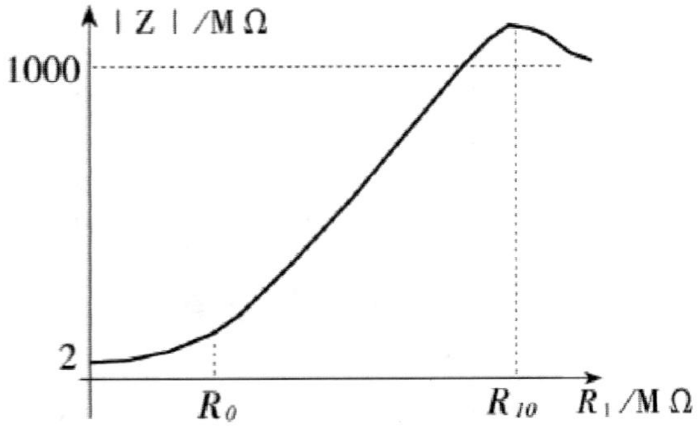

Figure 1. Coordinate equations for computer technology

2. Model building of computer software in mathematical modeling

The way of mathematical software used by mathematical modeling is relatively similar, because most of the mathematical software can draw the known mathematical formulas, and complete the calculation drill process. Mathematical software can not only greatly complete the conversion of various mathematical symbols, but also make accurate values of mathematical results, but still according to the actual mathematics is data. Data analysis method is the most common method of operation in business school modeling. It deals with some huge mathematical computing formulas, and the amount of mathematical knowledge designed is also very huge, which is also one of the difficult problems in mathematical modeling. Nowadays, many mathematical software provides powerful data processing functions, and introducing these powerful functions into the teaching curriculum of mathematical modeling can not only improve students 'thinking creative ability but also further inspire students' analytical processing ability. Therefore, the clustering analysis and time series analysis functions of mathematical software are based on computer technology, clarifying the role of computer technology in mathematical modeling.

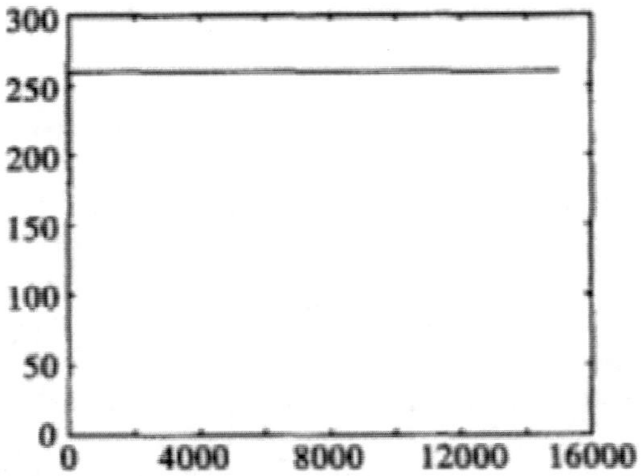

Figure 2. Threshold values for mathematical modeling

The correlation analysis of computer technology in mathematical modeling is mainly used in the modeling carrier to find the association and correspondence of the collection of objects existing in the data, which actually refers to some relations between different mathematical data in the mathematical resource database. If one value changes in the modeling process, then another correlation will change accordingly. Association analysis has a strong ability to mine data and can mine many computational methods and skills in mathematical modeling. Computer technology to establish analysis model construction steps, is the mathematical analysis platform from the abstract goal to the final visual presentation, from discovery to solutions so need to have overall thinking, can grasp the data analysis method, also need the ultimate pursuit of mathematical modeling, optimize every mathematical drill process, accurate every data results. People are in practical mathematical modeling, because of the need of model establishment, so it requires the use of computer technology to assist mathematical modeling in the process, to make it accurate data results. Through the application of computer technology, it will comprehensively improve the practical problem solutions in mathematical modeling, help the relevant problems needed to solve in time, then clarify the specific process, and timely apply relevant data algebra to practical problems, so the integration of mathematical modeling and computer technology can further improve the quality and level of mathematical modeling.

$$\sigma_p = \sqrt{\sum_{j=1}^{m}\sum_{k=1}^{m} A_j A_k \sigma_{jk}}$$

$$\frac{\sum_{i=1}^{n}\left[(X_i - \overline{X}) \times (y_i - \overline{y})\right]}{\sqrt{\sum_{i=1}^{n}(X_i - \overline{X})^2} \times \sqrt{\sum_{i=1}^{N}(y_i - \overline{y})^2}}$$

Figure 3. Application formula of computer technology

3. Specific application of computer technology in mathematical modeling

3.1. Computer technology application method

The application methods of computer technology in mathematical modeling mainly include the following categories: first, mechanism analysis, which means to have an objective and correct understanding of the characteristics of things, and then to deduce the model from the basic physical laws and system data information.

The second is the data analysis method, mainly through the statistical analysis of the data measurements, to find out the best mathematical model matching it. The third is simulation, simulation method is actually a common statistical budget method, equivalent to sampling experimental investigation. The law of factor experiment is to start the experiment in the mathematical system locally, and to constantly adjust and perfect according to the actual situation, and to manage the whole system organically. But the law of artificial experiment is based on the understanding and cognition of mathematical systems, considering the future development process of sensitive constituent mathematical models. The steps of mathematical modeling are generally divided into four steps:

1) Model preparation: Understand the actual background of the problem, clarify the modeling purpose, collect the necessary information, and try to understand the characteristics of the object.
2) Model assumption: According to the characteristics of the object and the modeling purpose, make the necessary and reasonable simplification, and make assumptions in accurate language.
3) Model composition: Analyze the causality of the object, use the internal rules and appropriate mathematical tools to construct the equation relationship or other mathematical structure.
4) Model solution: We can solve various traditional and modern mathematical methods, including

solution equations, drawing figures, proof theorem, logic operation, numerical operation and so on. V. Model analysis: to conduct a mathematical analysis of the model answers.

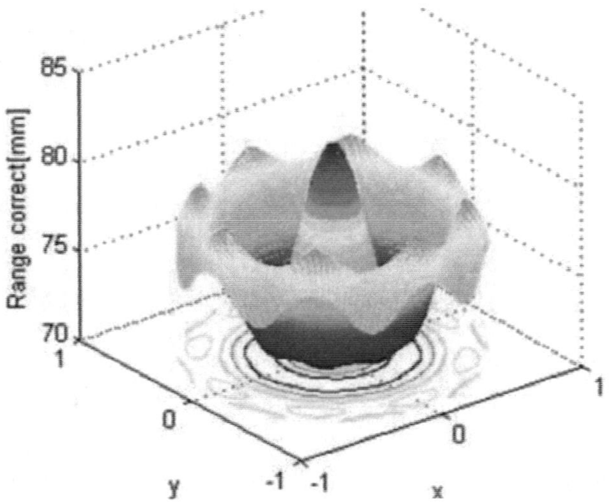

Figure 4. Schematic diagram of the mathematical modeling model

3.2. Mathematical software, supported by computer technology

Usually when computer technology is applied to practical mathematical problems, it needs to use computer help, and computers have developed and designed several mathematical software for mathematical modeling to help in the efficient operation of mathematical modeling. These mathematical software is the same way as the actual operation, but their data processing and analysis operations are capabilities that students cannot reach in the modeling process. Moreover, the reasoning and operation ability of mathematical software are very powerful, so in the application of computer technology for mathematical modeling, mathematical software with low difficulty coefficient will be used. Lingo is often used to solve secondary linear mathematics problems. In its practical application process, it can timely and clearly solve the problems in mathematical modeling. Because in the development of modern science and technology, Lingo is often used to solve the corresponding problems, so that the problem processing and solution can get the basic processing requirements.

Mathematical modeling in addition to using a variety of equations will use a variety of mathematical statistics knowledge, so in the process of learning mathematical modeling, also need to use statistical knowledge to deal with various problems in mathematical learning, so as in practical problems to solve related problems, improve the ability of mathematics to solve problems. Some more cumbersome and complex data are often encountered in mathematical modeling, and the processing process is very complex. But computer technology can help translate these complex mathematical knowledge back to my problem. Through the application of computer technology, the received data information can be systematically drawn in time, and the data processed tables and data can be drawn in time. The composition drawing of mathematical models also occupies a considerable part of the current mathematical model establishment. The application of these drawing software can express the mathematical graphic drawing in time. The drawing is more obviously intuitive under the application of computer technology.

4. Conclusion

Computer technology has great practical significance in mathematical modeling, because the computational model of a computer has a certain relevance in mathematical modeling, which can better provide strong technical support for the construction and calculation of a mathematical model on the basis of computer technology analysis. The establishment of data model of computer technology can improve the efficiency of mathematical modeling, so that the professionalism, accuracy and operational efficiency of mathematical modeling are quickly improved. The mutual integration of computer technology and mathematical models promotes the development and improvement of computer generation, accelerates the structural transformation in the field of mathematical modeling, and strengthens the application of mathematical models to technology.

References

[1] Shi S 2021 The Application of Computer Technology in Mathematical Modeling *Electronics Technology* **50**(08) 242-3
[2] Zhu Q, Zhou P 2021 Study on GDP Growth of National Economy Based on LSTM Model *Guide to Economic Research* **19** 5-9+37
[3] Wang J 2021 The Prediction Model and Simulation of Scientific and Technological Achievements Transfer Based on the Induction Learning Network *Electronic Design Engineering* **29**(07) 161-5
[4] Zhang S 2020 The Effective Application of Computer Technology in Mathematical Modeling *Information and Computer (Theory Edition)* **32**(22) 17-8
[5] Li C 2020 Organic Integration Strategy of Computer Technology and University Mathematics Teaching —— Review of The Application of Computer in Chemical Industry *Plastic Industry* **48**(10) 182
[6] Huang J, Li C, Lin W 2020 Application of Computer Technology in Advanced Mathematics Curriculum in New Engineering Background *Computer Knowledge and Technology* **16**(18) 111+117
[7] Zhang C, Zang W, Zhang J 2020 Construction of Higher Vocational Mathematics Curriculum System under the Certification of Engineering Technology Education *Heilongjiang Science*, **11**(01) 88-9
[8] Qiu Y 2016 Promote the higher mathematics teaching reform with the idea of mathematical modeling and the application of computer technology *Journal of Higher Education* **19** 139-40
[9] Gao J, Lin G 2016 Important Application of Computer Technology in Mathematical modeling *Journal of Shenzhen Information Vocational and Technical College* **14**(03) 54-7
[10] Zhao J, Wan F, Ju J, Cui D 2015 Overview of the Application of Computer Technology and Mathematical algorithm in Certificate Standardization *TCM Guide* **21**(18) 85-8
[11] Liu L, Zhou Q 2015 On the application of computer technology and mathematical methods to urban planning *Electronic Technology and Software Engineering* **17** 162-3
[12] Li R, Xu H 2015 Research on Advanced Mathematics Curriculum Reform Based on Computer Technology in Medical Colleges *Science and Education* **23** 56-7
[13] Li Y 2010 The Application of Computer Technology in Higher Mathematics Teaching *Technology Information* **35** 96
[14] Feng Y 2009 Application of Computer Technology in "Mathematical experiments" and "Mathematical Modeling" *Tangshan College Journal* **22**(03) 91-3
[15] Gao H, Meina Wu 2001 Analysis of the Teaching Mode with Computer Technology Integretedwith Math Teaching *Distance Education* **03** 69-71+79

Flutter Suppression of Wind Turbine Blade based on Adaptive Fuzzy Sliding Mode Control

Yingbo Wang, Changle Sun, Haipeng Sun, Xiaolin Zhang and Tingrui Liu*

College of Mechanical & Electronic Engineering, Shandong University of Science & Technology, Qingdao 266590, China

*Corresponding author's e-mail: liutingrui9999@163.com

Abstract. Aiming at the divergent unstable vibration of wind turbine blades, an adaptive fuzzy sliding mode control algorithm based on sliding mode control is proposed to solve the classic flutter fracture failure problem of blades. The structural model is a typical blade section model based on spring-mass-damper, and the aeroelastic equation is obtained by combining the steady aerodynamic model suitable for pure pitch motion. The adaptive fuzzy sliding mode controller gives the pitch signal, and the second-order pitch actuator performs the pitch action to suppress flutter. The saturation function with variable boundary layer thickness is used instead of symbolic function, and the approach rate parameter is adjusted appropriately by fuzzy rules, so as to realize variable-parameter adaptive fuzzy sliding mode control. Simulation results show that, compared with the traditional sliding mode controller, the proposed control algorithm can not only suppress the classical flutter, but also effectively reduce the amplitude and frequency of chattering, and has good robustness.

1. Introduction

With the increasing size of wind turbines, the length of blades increases. Under the incoming wind speed, the loads on blades are coupled with each other under the action of aerodynamic force, elastic force and inertial force, which easily leads to aeroelastic problems. Aeroelastic divergent unstable vibration is an important cause of blade fracture failure, among which the classical flutter caused by the coupled movement in flap/twist direction is one of the manifestations. The recent fracture failures of wind turbine blades all over the world have also verified this problem, so it is of positive significance to explore the method of suppressing the coupled flutter of blades [1-2].

According to different blade structure models, many classical flutter research methods of coupled flap/twist motion of wind turbine blades have been developed, among which the typical section model has been widely used because of its efficiency, convenience, and ability to focus on the analysis of bending-torsion divergent instability vibration problems. Chaviaropoulos et al. carried out aeroelastic analysis of blades based on typical section model, and studied different aeroelastic instability phenomena by analyzing coupled displacement of flap/lag and flap/twist [3]. Liu had studied in detail the classical flutter and stall flutter of thin-walled composite blades of wind turbines under undamped and damped conditions [4]. SAYED M A et al. studied the classical flutter instability phenomenon of large displacement deformation, and the blade tip flap/lag/twist displacement was studied through time domain response and frequency structure [5]. The members of the research group researched the active control of wind turbine blades based on piezoelectric feedback in the early stage, and suppressed the classical flutter of wind turbine blades through piezoelectric action [6].

Sliding mode control has complete robustness and can resist external uncertain disturbances, which is often used to control the divergence instability of rotating bodies. Wind turbine blade system is a nonlinear, multi-disturbance, time-varying and strongly coupled complex system, which is very suitable for sliding mode control. In this paper, from the angle of variable pitch, the variable pitch signal is given by the designed adaptive fuzzy sliding mode controller, and the variable pitch actuator performs the action according to the variable pitch signal to suppress blade flutter. Aiming at the chattering problem of traditional sliding mode, the control algorithm in this paper can not only suppress chattering, but also effectively reduce the amplitude and frequency of chattering, accelerate the convergence speed of the system, and have good robustness.

2. Structure model and aeroelastic motion equation

According to NACA 0015 airfoil, a typical section model with large aspect ratio is adopted, and the section is located at r (the radius of the section) from the blade root, and the bending-torsion coupling stability analysis and vibration control method based on time domain response are studied. The sectional model is shown in Figure 1. In which: The z direction is the waving movement direction; y direction is the shimmy movement direction; x direction is the direction of blade extension; θ is the direction of torsional motion; α is the angle of attack; β is the pitch angle; U is wind speed; V_0 is relative wind speed; ψ is the relative wind speed angle; c is chord length; F_z and M_θ represent aerodynamic lift and moment respectively.

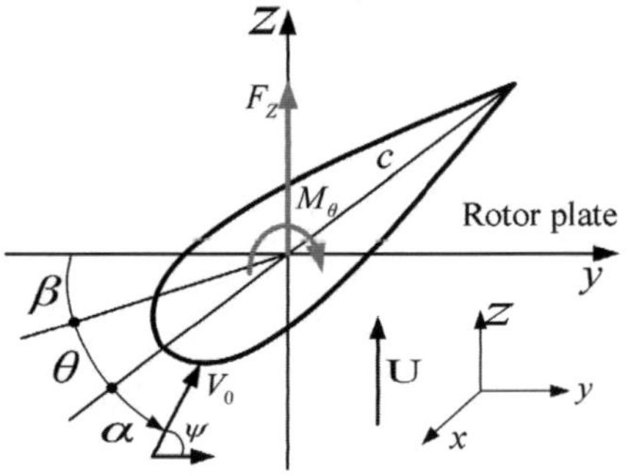

Figure 1. Displacement coordinates and aerodynamic forces of 2D airfoil section.

According to the blade section model in reference [3], under the condition of constant speed, the lag displacement is ignored and the pitch motion is not considered, the 2-DOF bending-torsion motion equation describing the flap/twist displacement can be described as follows:

$$\begin{bmatrix} 1 & r \\ r & r^2 \end{bmatrix}\begin{bmatrix} \ddot{z} \\ \ddot{\theta} \end{bmatrix} + \kappa \begin{bmatrix} \xi_z \omega_z & 0 \\ 0 & r^2 \xi_\theta \omega_\theta \end{bmatrix}\begin{bmatrix} \dot{z} \\ \dot{\theta} \end{bmatrix} + \kappa^2 \begin{bmatrix} \omega_z^2 & 0 \\ 0 & r^2 \omega_\theta^2 \end{bmatrix}\begin{bmatrix} z \\ \theta \end{bmatrix} = \frac{1}{2}\begin{bmatrix} R_f c_L \\ R_f c_M \end{bmatrix} \tag{1}$$

where ω_z is the natural frequency of flap direction; ξ_z is damping ratio in waving direction; ω_θ is the natural frequency of torsion direction; ξ_θ is the torsional damping ratio; Ω is angular velocity; $\kappa=\Omega c/v$ is expressed as reduced frequency; $R_f=c^2\rho_a/\rho$ is the ratio of air density and cross-sectional line density; ()'indicates differentiations about the reduction time $\tau=(V/c)t$. Because the classical flutter acts in the linear flow region, the time domain analysis under aerodynamic force can be studied [7]. The lift and moment coefficients can be defined as:

$$c_L = \frac{\partial C_L}{\partial(\theta+\beta)}\left(\dot{z}+\theta+\beta\right), c_M = \frac{\partial C_L}{\partial(\theta+\beta)}\left(\dot{z}+\theta+\beta\right)\ell \tag{2}$$

where $\partial c_L / \partial(\theta+\beta)$ is the partial differential equation factor of lift coefficient versus angle of attack, representing the slope of lift coefficient curve, and its constant value is -2π, ℓ is the chord fraction, and the value is -0.15.

In order to suppress flutter, a hydraulic pitch actuator is added, and the flutter gradually converges by executing the pitch motion. Introducing the second-order pitch exciter designed earlier [8]:

$$I_\beta\ddot{\beta} + C_\beta\dot{\beta} + K_\beta\beta=\beta_u \tag{3}$$

where β_u is the requested pitch angle of the pitch controller, that is, the input in the hydraulic actuator; I_β value is 0.2; C_β value is 1.1; K_β value is 1. The second-order differential equation of formula (1) and formula (3) are combined and converted into dimensionless form, and the 3-DOF bending-torsional coupling aeroelastic equation of the blade is obtained as follows:

$$\begin{bmatrix} 1 & r & r \\ r & r^2 & r^2 \\ 0 & 0 & I_\beta \end{bmatrix}\begin{bmatrix} \ddot{z} \\ \ddot{\theta} \\ \ddot{\beta} \end{bmatrix} + \kappa\begin{bmatrix} \xi_z\omega_z & 0 & 0 \\ 0 & r^2\xi_\theta\omega_\theta & r^2\xi_\beta\omega_\beta \\ 0 & 0 & C_\beta/\kappa \end{bmatrix}\begin{bmatrix} \dot{z} \\ \dot{\theta} \\ \dot{\beta} \end{bmatrix} + \kappa^2\begin{bmatrix} \bar{\omega}_z^2 & 0 & 0 \\ 0 & r^2\omega_\theta^2 & r^2\omega_\beta^2 \\ 0 & 0 & K_\beta/\kappa^2 \end{bmatrix}\begin{bmatrix} z \\ \theta \\ \beta \end{bmatrix} = \frac{1}{2}\begin{bmatrix} R_f c_L \\ R_f c_M \\ -2\beta_u \end{bmatrix} \tag{4}$$

For subsequent control analysis, the above equation needs to be transformed into first-order aeroelastic equations, defined $\boldsymbol{x}=[x_1,x_2,x_3,x_4,x_5,x_6]^T=[z,\dot{z},\theta,\dot{\theta},\beta,\dot{\beta}]^T$, then the above equation can be reduced to:

$$\dot{\boldsymbol{x}}=\boldsymbol{f}(\boldsymbol{x})+\boldsymbol{g}(\boldsymbol{x})\beta_u \tag{5}$$

where $\boldsymbol{f}(\boldsymbol{x})=[f_1, f_2, f_3, f_4, f_5, f_6]^T$, $\boldsymbol{g}(\boldsymbol{x})=[0, 0, 0, 0, 0, 1]^T$.

3. Adaptive fuzzy sliding mode control

The actual wind turbine system has the characteristics of time-varying, hysteresis, and nonlinearity, which is very suitable for sliding mode control [9]. In this paper, the traditional sliding mode controller is designed to suppress the classical flutter of blades. At the same time, in order to reduce the amplitude and frequency of chattering in the traditional sliding mode, the saturation function with variable boundary layer thickness is used instead of the symbol function, and the approach rate parameter is adjusted appropriately by fuzzy rules, to realize the variable parameter adaptive fuzzy sliding mode control. The control system block diagram is shown in Figure 2.

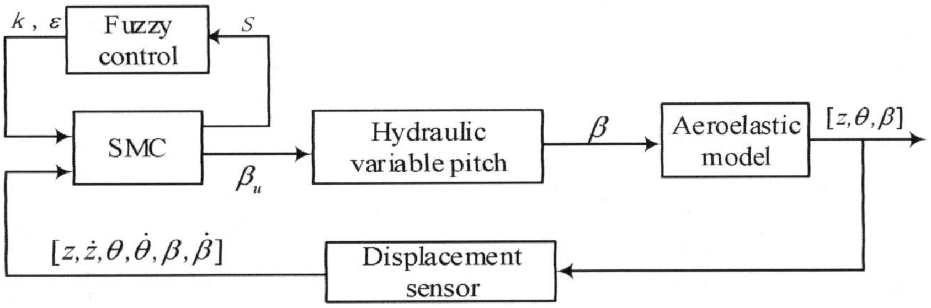

Figure 2. Block diagram of control system

3.1. Classic Sliding Mode Controller

According to the first-order aeroelastic equations of equation (5), the sliding mode pitch controller is designed. The sliding mode surface function is defined as:

$$S=c_1x_1+x_2+c_2x_3+x_4+c_3x_5+x_6 \tag{6}$$

where c_1, c_2, c_3 are all positive real numbers. Differentiate the above equations to get:

$$\dot{S}=c_1\dot{x}_1+\dot{x}_2+c_2\dot{x}_3+\dot{x}_4+c_3\dot{x}_5+\dot{x}_6=c_1x_2+f_2+c_2x_4+f_4+c_3x_6+f_6+\beta_u \tag{7}$$

The sliding mode surface adopts exponential reaching law as follows:

$$\dot{S}=-\varepsilon sgn(s)-ks \quad \varepsilon > 0, k > 0 \tag{8}$$

where ε is the switching gain and k is the exponential convergence rate.

Combining equation (7) with equation (8), the control rate can be deduced as follows:

$$\beta_u=-c_1x_2-f_2-c_2x_4-f_4-c_3x_6-f_6-\varepsilon sgn(s)-ks \tag{9}$$

To verify the stability of the controller, the Lyapunov equation is defined as:

$$V=\frac{1}{2}S^2 \tag{10}$$

Derivation of equation (10) is obtained:

$$\dot{V}=S\dot{S} \leq -\varepsilon|S|-ks^2=-2kV-\varepsilon|S| \leq -2kV \leq 0 \tag{11}$$

According to Lyapunov stability criterion, the designed sliding mode controller can make the system converge gradually from flutter divergence state.

3.2. *Adaptive fuzzy controller*

Although sliding mode controller can suppress classical flutter, it will bring severe chattering problem. In order to reduce system chattering, fuzzy rules are used to adaptively adjust the value of ε and k in exponential reaching law according to the value of sliding surface function S. At the same time, the saturation function with variable boundary layer thickness is used instead of symbol function, which can effectively reduce the amplitude and frequency of chattering.

The fuzzy controller adopts one-dimensional Mamdani fuzzy controller with single input and two outputs, with the value of sliding surface function S as the input variable and the output variable as the ε and k in the reaching law. The fuzzy universe of the input variable S is [-1,1], the fuzzy subset is {NB, NM, NS, ZO, PS, PM, PB}, and the elements in the subset represent negative big, negative middle, negative small, zero, positive small, positive middle and positive big respectively. The fuzzy universe of output variables ε and k is [0,1], the fuzzy subset is {ZO, PS, PM, PB}, and the elements in the subset represent zero, positive small, positive middle and positive big. For convenience of calculation, the values of basic universe are set equal to fuzzy universe. The membership functions of input and output variables choose triangular membership functions, and the barycenter method is used to solve the ambiguity. Fuzzy control rules are shown in Table 1.

Table 1. Fuzzy rules

S	NB	NM	NS	ZO	PS	PM	PB
ε	PM	PS	PS	ZO	PS	PS	PM
k	PB	PB	PM	ZO	PM	PB	PB

To further reduce chattering, the saturation function *sat(S)* with variable boundary layer thickness is used instead of the sign function *sgn(S)* in the control rate, and the saturation function is shown in equation (12):

$$sat(S)=\begin{cases}1 & S > \delta \\ kS & |S| \leq \delta, \quad k=\frac{1}{\delta} \\ -1 & S < -\delta\end{cases} \tag{12}$$

where δ is the thickness of boundary layer. In order to improve the robustness of the controller, an adaptive function is designed in this paper, which can automatically adjust the boundary layer thickness according to the value of sliding surface S. The adaptive function is shown in equation (13):

$$\delta=S^2+\varphi,\ \ \varphi > 0 \tag{13}$$

where φ is a very small positive number, which can be adjusted freely according to actual needs. This paper takes $\varphi = 0.05$.

4. Adaptive fuzzy sliding mode control

According to the above control scheme, the system simulation model is built in simulink to simulate the forced vibration response of the blade. The related parameters of the sectional system, including the normalized structural parameters and external motion parameters are as follows: r=1 m, c=0.5 m, ω_z=4, ω_θ=7, ω_β=5, ξ_z=0.02, ξ_θ =0.01, ξ_β=0.02, R_f=0.02, Ω=2.2rad/s, κ=0.04, 0.06 and 0.08 respectively to represent different wind speed conditions.

The initial flap displacement and twist displacement were set as 0.1. According to the system parameters, the time-domain response of blade flutter was obtained without the action of variable blades, as shown in Figure 3.

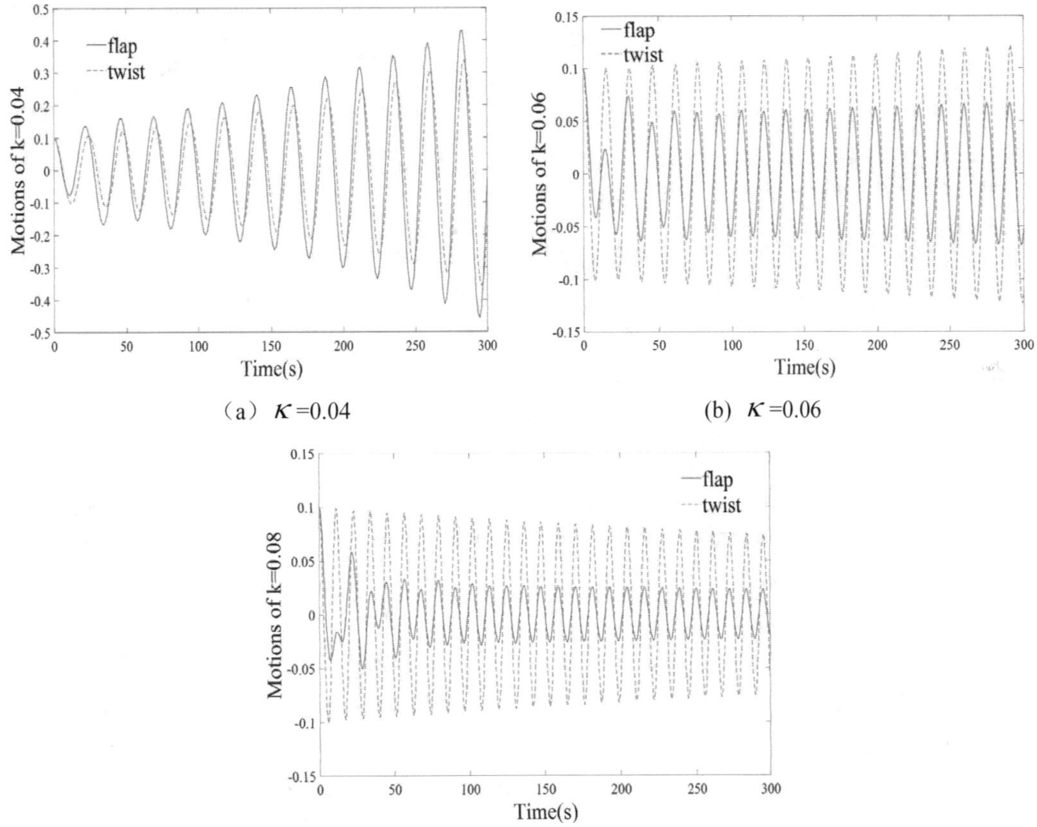

(a) K=0.04 (b) K=0.06

(c) K=0.08

Figure 3. Flap and twist displacements of blades without control

It can be seen from figure. 3 that with the increase of wind speed, the flap and twist displacements of the blade gradually changes from convergent state to divergent unstable state. If it is not controlled, the blade is likely to have fracture effect, so it is necessary to apply pitch control to suppress flutter. Using the classic sliding mode controller designed above, let c_1=1, c_2=3, c_3=1, ε=0.1, k=1, and get the flap and twist displacements at two wind speeds of κ=0.04 and κ=0.06. The control result is shown in Figure 4. It can be seen from the results that the two conditions of blade divergence and critical flutter are well suppressed under the action of variable blade.

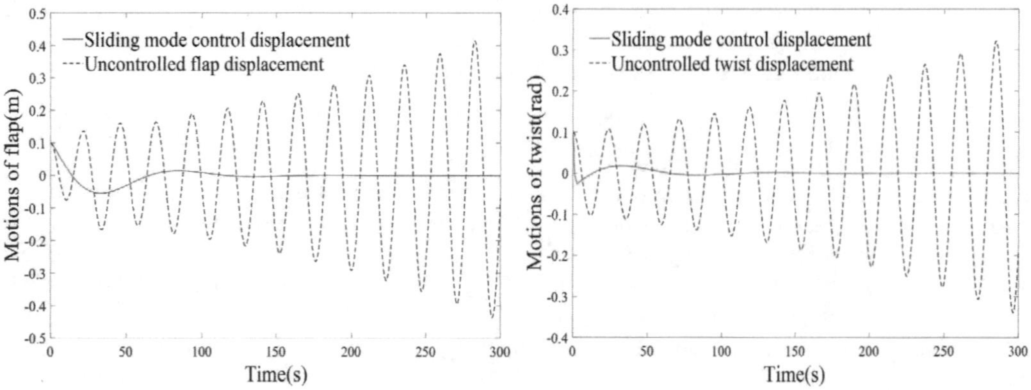

(a) Flap and twist displacements of κ=0.04

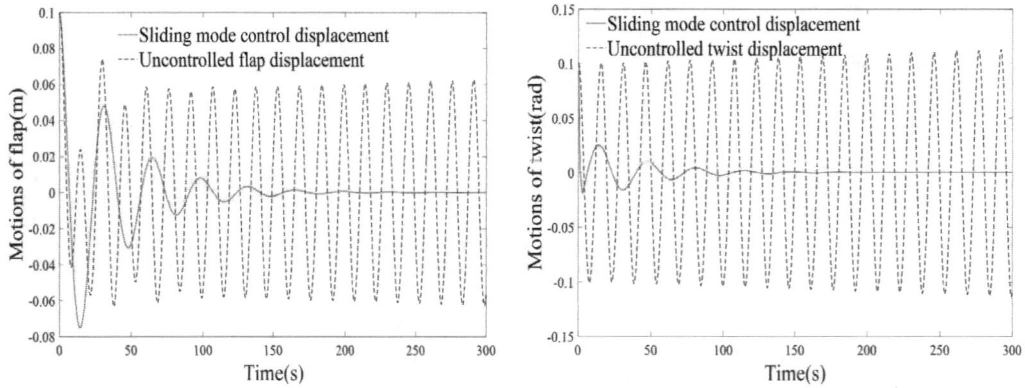

(b) Flap and twist displacements of κ=0.06

Figure 4. The flap and twist control effect of classical sliding mode controller

Although the sliding mode controller can restrain the flutter divergence of blades, high-frequency chattering occurs in both the control variable and the pitch motion. Take κ=0.04 as an example, as shown in Figures 5 and 6 below.

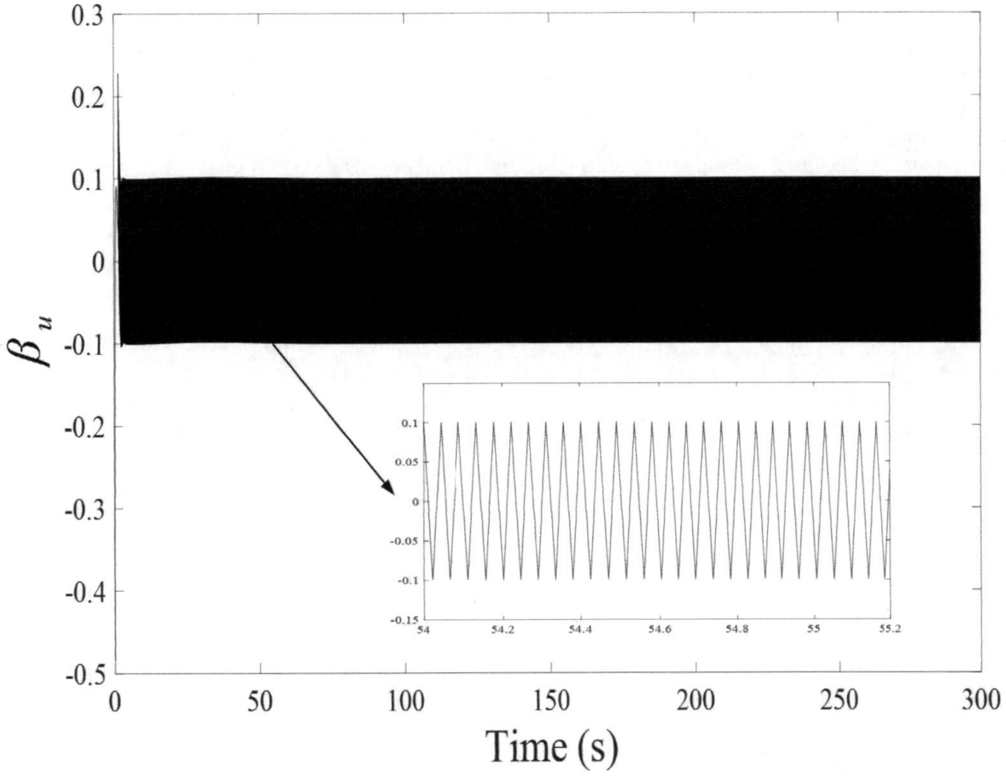

Figure 5. Chattering phenomenon of control

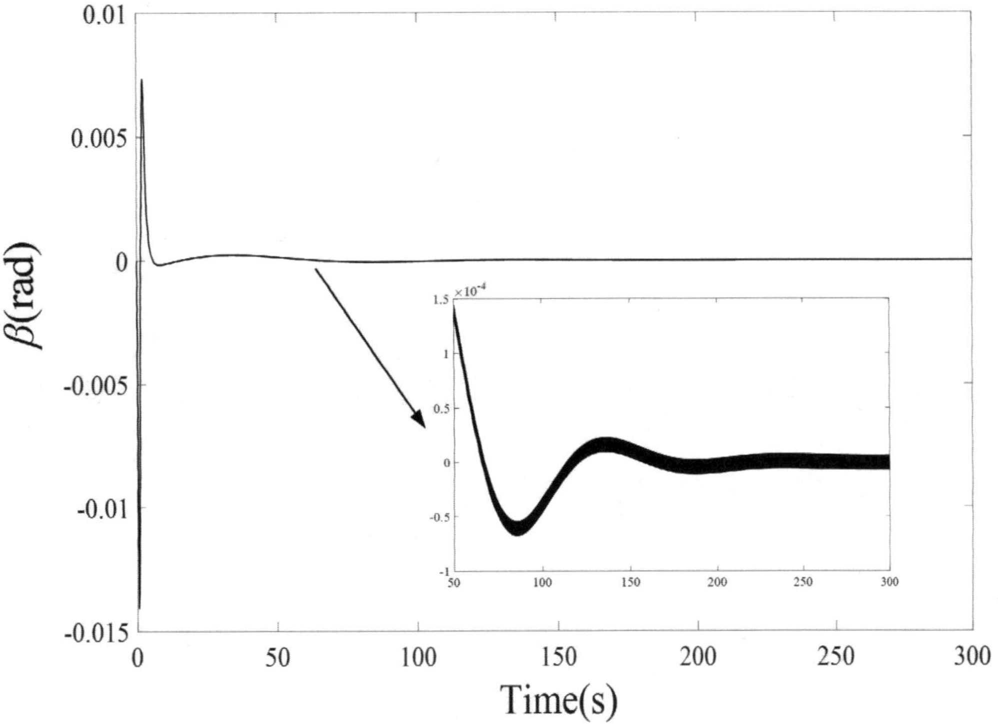

Figure 6. Chattering phenomenon of variable pitch

In order to reduce the amplitude and frequency of chattering, the adaptive fuzzy sliding mode control algorithm designed in this paper is adopted, and fuzzy rules are used to adaptively adjust the value of ε and k in exponential reaching law according to the value of sliding surface function S. At the same time, the saturation function with variable boundary layer thickness is used instead of symbolic function, which can effectively reduce the chattering phenomenon of sliding mode control. The control quantity and pitch action are shown in Figure 7 and Figure 8.

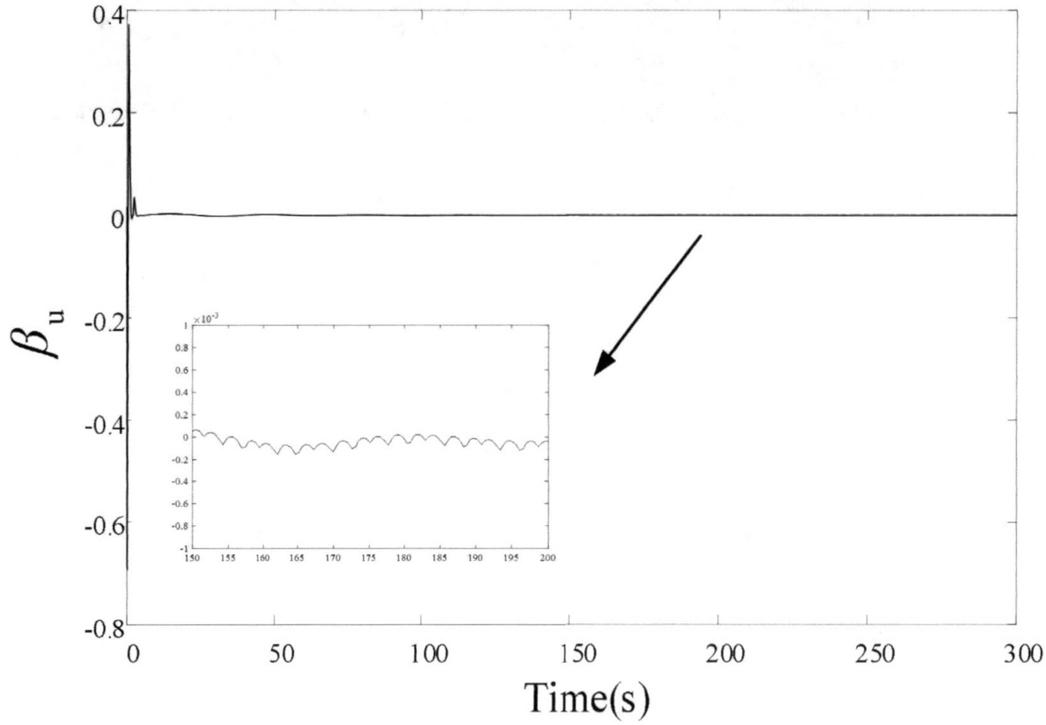

Figure 7. Chattering decrease of control

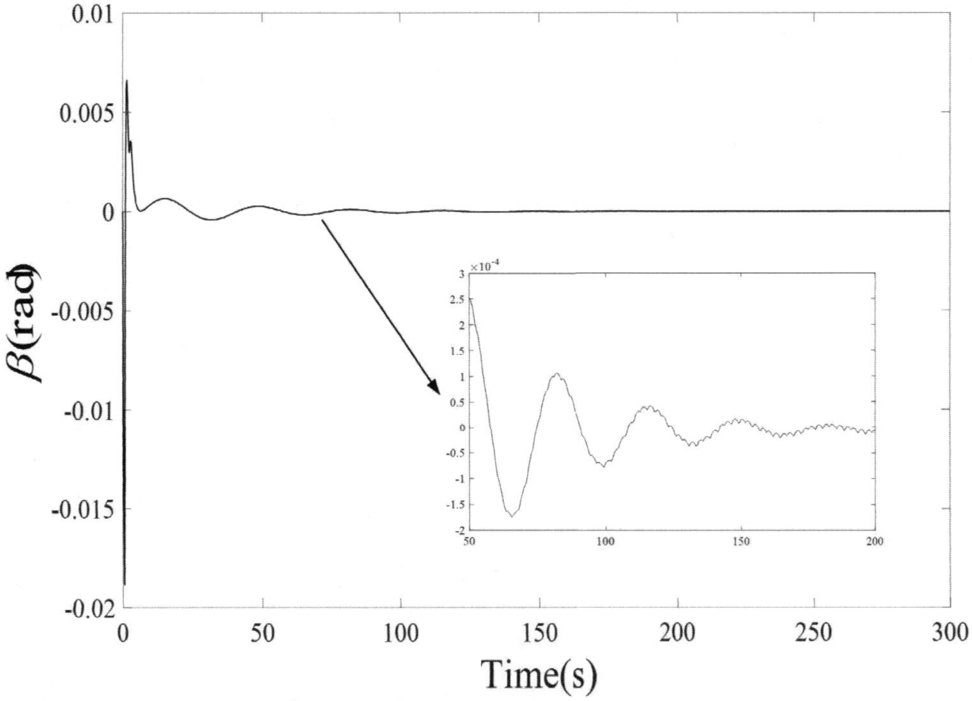

Figure 8. Chattering decrease of variable pitch

From the above simulation results, the classical sliding mode controller can suppress the divergent unstable vibration of blades, but both the control output and the pitch actuator will produce small amplitude high frequency chattering, which will not only affect the control accuracy, but also damage the controller components in serious cases. Adaptive fuzzy sliding mode controller can effectively weaken the amplitude and frequency of chattering, and can automatically adjust the value of ε and k in exponential reaching law, which has excellent robustness.

5. Conclusion

In this paper, an aeroelastic model of flap/twist motion is established, and a 3-DOF coupled aeroelastic equation is obtained by combining the hydraulic pitch actuator. Aiming at the critical flutter and divergence of blades at high wind speed, the flutter suppression is accomplished by the designed sliding mode controller. At the same time, for the chattering problem caused by sliding mode controller, the adaptive function and fuzzy control can greatly weaken the chattering amplitude and frequency, and have good control effect.

Acknowledgement

The authors gratefully acknowledge the support of the National Natural Science Foundation of China (no. 51675315).

References

[1] Liu T, Zhao K, Sun C, Jia J and Liu G 2020 *Energies* **13** 5029.
[2] Kallesoe B S 2006 *Wind Energy* **9** 421−36.
[3] Chaviaropouls P K, Srensen N N, Hansen M O L 2003 *Wind Energy* **6** 387−403.
[4] Liu T 2014 *Journal of Chemical and Pharmaceutical Research* **6**(5) 1544−1550.
[5] Sayed M A, Lutz T 2016 *Progress in Renewable Energies Offshore* (London: Taylor and Francis) chapter 4 pp 105–110.
[6] Liu T 2015 *Shock and Vibration* https://doi.org/10.1155/2015/292368.
[7] Yi C, Sun C, Sun Q, Liu T 2021 *Journal of Shandong University of Science and*

Technology(Natural Science) **40**(4) 111−119.
[8] Liu T, Yu Z 2016 *Central. South. University. (Science Technology)* **47** 3562.
[9] Xia Y P, Liu D, Li R Y 2020 *Control Theory & Applications* **37**(1) 129 – 136.

Flutter Suppression of Wind Turbine Blade Based on RBF Neural Network Compensation Backstepping Control

Changle Sun, Zhiyuan Wang, Yingbo Wang, Haipeng Sun and Tingrui Liu*

College of Mechanical & Electronic Engineering, Shandong University of Science & Technology, Qingdao 266590, China

*Corresponding author's E-mail: liutingrui9999@163.com

Abstract. Aiming at the critical flutter problem of wind turbine blades, which is between classical flutter and stall flutter, a flutter suppression scheme based on radial basis function (RBF) neural network friction compensation backstepping is presented. The structure model is based on the typical 2D section of bending and twist model of spring-mass-damper, and the rotor variable exciter second-order model with friction disturbance is incorporated to control the rotor variable blade. A modified quasi - steady - state aerodynamic model was used for aerodynamics actuation. RBF compensation backstepping control scheme is a block-controlled backstepping controller designed based on the stability theorem of Lyapunov function, which approximates the frictional interference with nonlinear characteristics through RBF network, and cancels the friction existing in the actuator of variable rotor. Four wind speed environments were selected to analyze the response of blades under different wind speeds, and the flutter suppression effects under two wind speeds were selected to verify the ability of RBF network to approach the nonlinear function. The results show that the RBF backstepping control scheme can improve the robustness to suppress the critical flutter problem of wind turbine blades.

1. Introduction

The flutter of wind turbine blade can be timely and effectively suppressed, which is helpful to improve the utilization rate of wind energy and protect the whole structure of wind turbine. Blade flutter can be divided into linear classical flutter and nonlinear stall flutter, which is mainly manifested in the vibration process of blade flap-wise direction coupled with twist direction. In recent years, scholars have discovered a critical flutter in two critical states, where blades are in a quasi-steady state response. Under extreme wind speed, the flutter can lead to the destruction of blades and even the tower body. In view of this, a quasi-steady state aerodynamic model extracted from the helicopter blade stall aerodynamic model is introduced in this paper and modified to study the critical flutter. In order to achieve the accuracy of the modeling of the controlled object, a second-order model of the variable rotor actuator with nonlinear friction interference was introduced to simulate the friction on the variable rotor actuator [1], so as to improve the modeling accuracy and the robustness of the control scheme.

Wind turbine blade flutter suppression problem, in our previous work, such as Liu Tingrui[2] analysis based on piezoelectric driven single chamber thin-walled wind turbine blades composite beams is classic flutter, studied on the transverse shear deformation, warping restraint effect and secondary buckling under the joint action, piezoelectric drive of wind turbine blade beam's classic flutter suppression effect. Chang Lin[3] described the establishment process of the nonlinear aeroelastic model, and proposed the optimal PID control method and the linear quadratic active control process, as well as the comparative analysis of the control results. Aiming at the analysis of the quasi-steady state response of the blade, Liu Tingrui[4] introduced a quasi-steady state aerodynamic model extracted from the helicopter blade stall aerodynamic model and used it to study the critical flutter after modification. The above research did not introduce the friction interference problem, on this basis, this paper introduced the frictional interference with nonlinear characteristics to simulate the internal friction of the propeller variable actuator, so that the control scheme has more practical significance. For friction compensation, many scholars put forward different control schemes. Ruan Wei[5] used the radial basis function neural network (RBFNN) control algorithm to compensate the frictional disturbance torque in the nonlinear electromechanical actuator system based on the sliding mode controller. Guo Xiwen[6] proposed a control scheme based on the nominal model for the friction interference errors existing in the system modeling, and adopted an improved fuzzy compensation adaptive controller to compensate the friction terms and system uncertainties respectively, thus ensuring the stability of the closed-loop system. Lin [7], since the traditional backstepping control method cannot guarantee the robustness, the synovial term is introduced to overcome the interference and ensure the robustness of the controller.

The synovial membrane controller has good nonlinear control ability, but the chattering phenomenon cannot be eliminated. Backstepping control can avoid this problem. Its basic principle is to decompose a complex nonlinear system into many subsystems that do not exceed the order of the system. Then, according to Lyapunov stability theorem, the virtual control quantity in the middle is designed for each subsystem and backward deduced to the whole system, so as to design the whole control rate [8]. Traditional backstepping control has shortcomings, requiring accurate modeling information of the controlled object, and unable to control interference.

In this study, the combination of RBF nerve and backstepping control can avoid the chattering phenomenon in synovial control and improve the robustness of backstepping control. The backstepping control rate was designed based on Lyapunov function, and the frictional disturbance with nonlinear characteristics was approximated by RBF neural network. The weight was adjusted adaptively based on Lyapunov stability theory to ensure the stability of the closed-loop system. The RBF backstepping controller can effectively suppress the critical flutter of wind turbine blades. By analyzing the approximation ability of RBF neural network under different wind speed environments, the robustness of the control scheme is guaranteed.

2. Blade Structure Model and Aeroelastic Equation

The analysis section of a classical symmetrical airfoil with large aspect ratio is considered in Figure 1. The mass of the blade structure is composed of springs and dampers on the y axis and Z axis. b is half chord length of blade; O is the center of the section; G is the aero-elastic center; U is the input wind speed; Z axis is the flap-waving bending direction; α is blade twist angle; β is variable propeller angle; F and M are divided into pneumatic lift force and torque.

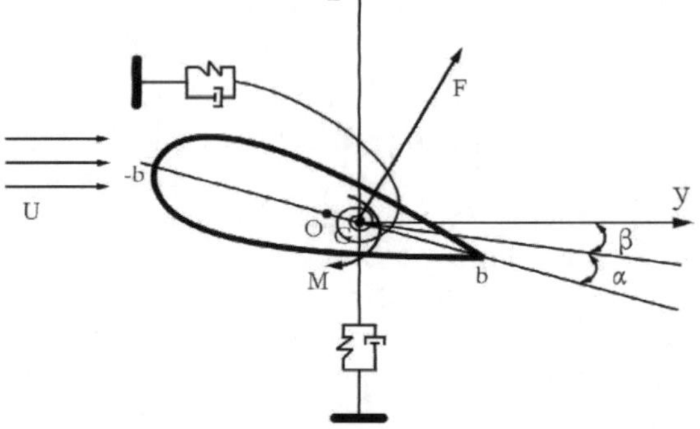

Figure 1. 2D classical symmetrical airfoil cross section with large aspect ratio.

By summing the mass, stiffness and damping of the system, the balance equation of force and moment is derived, which is set as the total force and moment generated by pneumatic lift force, and the motion equation of flap-wise and twist direction can be obtained. The formula is defined as [9]:

$$F = \ddot{z} + 2\xi_z\omega_z\dot{z} + \omega_z^2 z \text{ , } M = \ddot{\alpha} + I_\beta\ddot{\beta} + 2\xi_\alpha\omega_\alpha\dot{\alpha} + C_\beta\dot{\beta} + \omega_\alpha^2\alpha + K_\beta\beta \tag{1}$$

where, Z is flap-waving bending displacement; α is the twist displacement; β is variable paddle displacement; ξ is the damping ratio; ω is the natural frequency; K and C is the stiffness coefficient and damping coefficient; I is the mass moment of inertia of variable propeller direction.

The quasi-steady state aerodynamic model was extracted from the helicopter blade stall aerodynamic model. According to W.P.Jones' approximation of Wagner's function, the aerodynamic load was modeled under the assumption of unsteady aerodynamics, and the strip theory was applied to the system [10]. Pneumatic lift and torque are expressed as:

$$F = 2\pi\rho bU\left[\frac{b}{2U}\ddot{z} - \frac{ab^2}{2U}\left(\ddot{\alpha}+\ddot{\beta}\right) + \dot{z} + b(1-a)\left(\dot{\alpha}+\dot{\beta}\right) + U(\alpha+\beta) + \lambda_1 + \lambda_2\right] \tag{2}$$

$$\begin{aligned} M = 2\pi\rho bU\Bigg[&\frac{ab^2}{2U}\ddot{z} - \frac{b^3}{2U}\left(\frac{1}{8}+a^2\right)\left(\ddot{\alpha}+\ddot{\beta}\right) + b\left(\frac{1}{2}+a\right)\dot{z} + ab^2\left(\frac{1}{2}-a\right)\left(\dot{\alpha}+\dot{\beta}\right) \\ &bU\left(\frac{1}{2}+a\right)(\alpha+\beta) + b\left(\frac{1}{2}+a\right)(\lambda_1+\lambda_2)\Bigg] \end{aligned} \tag{3}$$

$$\dot{\lambda}_1 = -0.165\ddot{z} - 0.165b\left(\frac{1}{2}-a\right)\left(\ddot{\alpha}+\ddot{\beta}\right) - 0.165U\left(\dot{\alpha}+\dot{\beta}\right) - 0.041\frac{U}{b}\lambda_1 \tag{4}$$

$$\dot{\lambda}_2 = -0.335\ddot{z} - 0.335b\left(\frac{1}{2}-a\right)\left(\ddot{\alpha}+\ddot{\beta}\right) - 0.335U\left(\dot{\alpha}+\dot{\beta}\right) - 0.32\frac{U}{b}\lambda_2 \tag{5}$$

where, ρ is air density; a is the percentage of half chord length of the distance between the center of section and the center of aeroelasticity; λ_1 and λ_2 are aerodynamic variables.

In this paper, a second-order variable paddle exciter model is designed as [11] :

$$I_\beta\ddot{\beta} + C_\beta\dot{\beta} + K_\beta\beta = K_\beta\beta_{ref} \tag{6}$$

where, β_{ref} is the requested variable blade Angle of the controller, namely the control quantity u in the control system; According to the actual size requirement of the controlled object, the model parameter of the actuator I_β is 1; C_β is 11.6118; K_β is 495.1565;

Let $X = \begin{bmatrix} z & \alpha & \lambda_1 & \lambda_2 & \beta \end{bmatrix}^T$,and combing Eqs. (1) - (6), and the system aeroelastic equation can be obtained as follows:

$$I_M\ddot{X} + C_M\dot{X} + K_MX = Q_M\beta_{ref} \tag{7}$$

where, I_M, C_M, K_M are mass, damping and stiffness matrices respectively.

3. Backstepping Control

3.1. State Space Equation

According to the system's aeroelastic equation (7), let $X_1 = \begin{bmatrix} z & \alpha & \lambda_1 & \lambda_2 & \beta \end{bmatrix}^T$, $X_2 = \begin{bmatrix} \dot{z} & \dot{\alpha} & \dot{\lambda}_1 & \dot{\lambda}_2 & \dot{\beta} \end{bmatrix}^T$, Convert it to state space form:

$$\begin{cases} \dot{X}_1 = X_2 \\ \dot{X}_2 = -P\dot{X}_1 - RX_1 + Nu \end{cases} \tag{8}$$

where, $P = I_M^{-1}C_M, R = I_M^{-1}K_M, N = I_M^{-1}Q_M$, $u = \beta_{ref}$.

3.2. Block Controlled Backstepping Controller Design

The purpose of the block controlled backstepping controller is to introduce the error state vector so that the flutter divergence trend of the wind turbine blade can track the desired state trajectory stably.

1) The error state vector of the first subsystem is defined as:

$$E = D - X_1 = \begin{bmatrix} e_1 & e_2 & e_3 & e_4 & e_5 \end{bmatrix}^T = \begin{bmatrix} x_{1d} - z \\ x_{2d} - \alpha \\ x_{3d} - \lambda_1 \\ x_{4d} - \lambda_2 \\ x_{5d} - \beta \end{bmatrix} \tag{9}$$

where, $D = \begin{bmatrix} x_{1d} & x_{2d} & x_{3d} & x_{4d} & x_{5d} \end{bmatrix}^T$ is the trajectory of the expected state, and the derivative of the error state vector is:

$$\dot{E} = \dot{D} - \dot{X}_1 = \dot{D} - X_2 \tag{10}$$

Building the Lyapunov function:

$$V_1 = \frac{1}{2} E^T E \tag{11}$$

The derivative with respect to time:

$$\dot{V}_1 = E^T \dot{E} = E^T \left(\dot{D} - X_2 \right) \tag{12}$$

According to the Lyapunov stability theorem, given that V_1 is positive definite, $\dot{V}_1$ must be negative definite. Let $\dot{D} - X_2 = -k_1 E,$ and $k_1 > 0$. Available:

$$\dot{V}_1 = -k_1 E^T E < 0 \tag{13}$$

2) The error state vector F of the second subsystem is defined as:

$$F = H - X_2 \tag{14}$$

Let the second subsystem expect to track the trajectory $H = X_2$ Into the formula $\dot{D} - X_2 = -k_1 E$ .Available:

$$H = \dot{D} + k_1 E \tag{15}$$

Building the Lyapunov function:

$$V_2 = V_1 + \frac{1}{2} F^T F \tag{16}$$

The derivative with respect to time:

$$\dot{V}_2 = E^T \dot{E} + F^T \dot{F} \tag{17}$$

Substitute Eq. (14) and (15) into Eq. (17) to obtain:

$$\dot{V}_2 = -k_1 E^T E + F^T \left(E + \dot{F} \right) \tag{18}$$

According to the Lyapunov stability theorem, given that V_2 is positive definite, $\dot{V}_2$ must be negative definite. Let $E + \dot{F} = -k_2 F$,and $k_2 > 0$ .Available:

$$\dot{V}_2 = -k_1 E^T E - k_2 F^T F < 0 \tag{19}$$

According to Lyapunov stability theorem, when the trajectory D of the expected state is set as the origin, when F is asymptotically stable at the origin, X_2 can be guaranteed to be asymptotically stable at the trajectory H of the expected state, so that X_1 is asymptotically stable at the origin.

According to the formula $E + \dot{F} = -k_2 F$,and combing Eq. (8) and (14). We can obtain:

$$\dot{F} = \dot{H} - \dot{X}_2 = \dot{H} - \left(-P\dot{X}_1 - RX_1 + Nu \right) = -E - k_2 F \tag{20}$$

Block controlled backstepping control rate:

$$u = k_3 N^T \left(E + \dot{H} + k_2 F + P\dot{X}_1 + RX_1 \right) \tag{21}$$

where, k_3 is satisfied $k_3 N^T N = 1$.

3.3. RBF-Block Controlled Backstepping Controller Design

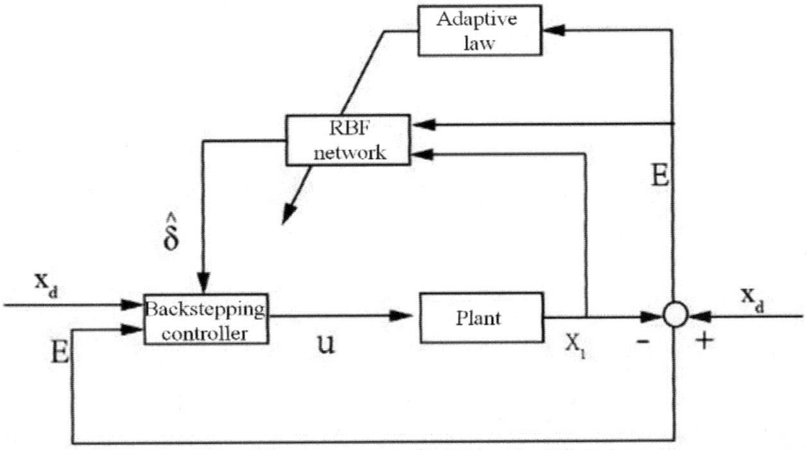

Figure 2. RBF-Block controlled backstepping controlling flow chart.

The second-order variable propeller actuator is designed based on the hydraulic actuator. Due to the internal friction in the actual mechanism, the original backstepping control system cannot deal with such nonlinear friction interference. The interference quantity is designed as $\delta = \beta + 0.01\,\text{sgn}(\beta)$, and the second-order variable propeller actuator is redefined as:

$$I_\beta \ddot{\beta} + C_\beta \dot{\beta} + K_\beta \beta = K_\beta \left(\beta_{ref} - \delta \right) \tag{22}$$

The input and output algorithm of RBF network is :

$$h_j = \exp\left(\frac{\left\|x - c_j\right\|^2}{2b_j^2}\right), \quad j = 1 \sim 10 \tag{23}$$

$$\delta = W^{*T} h(x) + \varepsilon \tag{24}$$

where, b_j is the width of the Gaussian basis function, and the value is 1.75; c_j is the node center vector of hidden layer, and the value is $c_i = [-4 \quad -2 \quad 0 \quad 2 \quad 4]$; h_j is the Gaussian function of RBF neural network; W^* is the ideal weight of the network; $\hat{W}$ is the weight estimation of the neural network; $\hat{\delta}$ is the output of the network.

According to the formula $\delta = \beta + 0.01\operatorname{sgn}(\beta)$ .Taking network input $x = [E \quad X_1]^T$, $E = A - X_1 = [e_1 \quad e_2 \quad e_3 \quad e_4 \quad e_5]^T, X_1 = [y \quad \alpha \quad \lambda_1 \quad \lambda_2 \quad \beta]^T$. Then the network output is:

$$\hat{\delta} = \hat{W}^T h \tag{25}$$

then,

$$\hat{\delta} - \delta = \hat{W}^T h + \varepsilon - W^{*T} h = \left(\hat{W}^T - W^{*T}\right) h + \varepsilon \tag{26}$$

Backstepping control rate based on RBF neural network approximating δ is expressed as:

$$\hat{u} = k_3 N^T \left(E + \dot{H} + k_2 F + P\dot{X}_1 + RX_1\right) + \hat{\delta} \tag{27}$$

Stability analysis of the control scheme. In the second step of backstepping control, combing Ep. (8) and (22), the first derivative of the error state vector F is:

$$\dot{F} = \dot{H} - \dot{X}_2 = \dot{H} - \left[-P\dot{X}_1 - RX_1 + N(u - \delta)\right] \tag{28}$$

Substituting Eq. (27) into Eq. (28). Available:

$$\dot{F} = -E - k_2 F - N\left(\hat{\delta} - \delta\right) = -E - k_2 F - Nh\left(\hat{W}^T - W^{*T}\right) \tag{29}$$

Building the Lyapunov function:

$$V_3 = \frac{1}{2} E^T E + \frac{1}{2} F^T F + \frac{1}{2}\left(\hat{W} - W^*\right)^T \left(\hat{W} - W^*\right) \tag{30}$$

Combing Ep. (29), the derivative with respect to time:

$$\begin{aligned} \dot{V}_3 &= E^T \dot{E} + F^T \dot{F} + \left(\hat{W} - W^*\right)^T \dot{\hat{W}} \\ &= -k_1 E^T E - k_2 F^T F + \left(\hat{W} - W^*\right)^T \left(\dot{\hat{W}} - F^T Nh\right) \end{aligned} \tag{31}$$

According to the Lyapunov stability theorem, given that V_3 is positive definite, $\dot{V}_3$ must be negative definite. When $\dot{\hat{W}} = F^T Nh, \dot{V}_3 < 0$ and the system is asymptotically stable.

The adaptive rate is:

$$\dot{\hat{W}} = F^T Nh \tag{32}$$

4. Simulation and Analysis

Take the basic structural parameters and aerodynamic parameters: $\omega_z = 8 rad/s$; $\omega_\alpha = 6 rad/s; \xi_z = \xi_\alpha = 0.02$; $\rho = 1.225 kg/m^3; b = 0.55m; a = 0.2$; $k_1 = 20$; $k_2 = 35$; $k_3 = 1/495.1565^2$; $D = [0 \quad 0 \quad 0 \quad 0 \quad 0]^T$.

In Figure 3, analysis of wind turbine blade flutter at different wind speeds under the condition of no propeller. The wind turbine structure parameters mentioned above, and aerodynamic parameters are certain, selecting four groups of wind speed: $U_1 = 5m/s, U_2 = 10m/s, U_3 = 15m/s, U_4 = 20m/s$. Wind speeds of Case1 and Case2 are $5 m/s$ and $10 m/s$, respectively. The results show that when the wind speed is $U < 10 m/s$, the self-damping characteristics of the blade structure will inhibit flutter until it gradually converges. The wind speeds of Case3 and Case4 are $15 m/s$ and $20 m/s$ respectively. The results show that at $U > 10 m/s$, the wind turbine blades respond to flutter divergence, which is mainly in the flutter direction of blades. It is demonstrated that in the bending-twist coupling process, the divergence instability caused by flutter direction is an important reason for blade fracture failure [12].

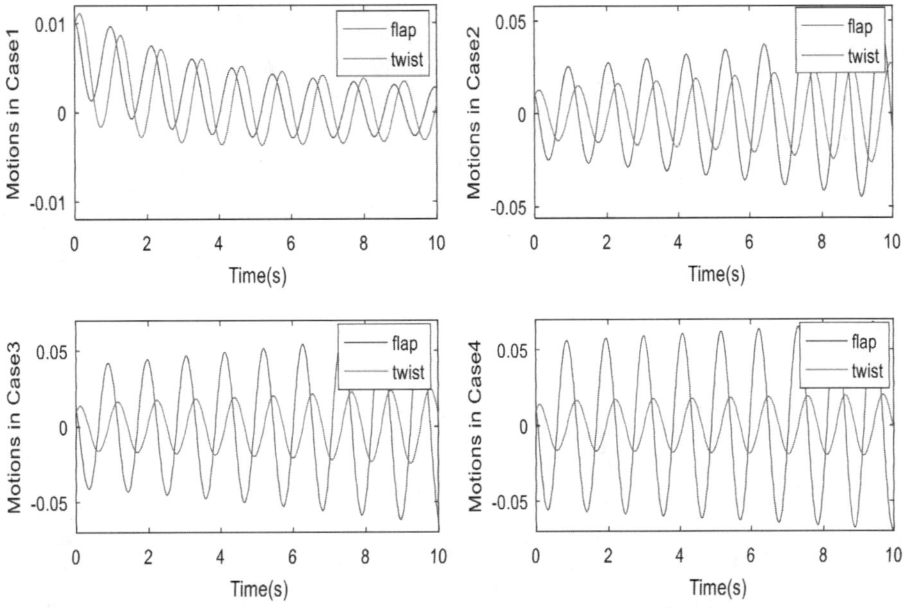

Figure. 3 The response of wind turbine blades under four wind speeds.

In Figure 4, based on the above situation of flutter divergence of wind turbines at extreme wind speeds, the response of blades at wind speeds $U_2 = 10 m/s$ and $U_3 = 15 m/s$ were selected as control objects. In the case of frictional interference or not, the inhibition effect of the backstepping controller on blade flutter is compared. In the case of wind speed $U_2 = 10 m/s$, Case1 is the backstepping of the controller's effect on blade flutter suppression when the system has no frictional interference. Case2 is the inhibition effect of the backstepping controller on blade flutter when the system has frictional interference. Case3 represents the inhibition effect of RBF- backstepping control on blade flutter. Case4, Case5 and Case6 are the corresponding flutter suppression effects of blades when the wind speed is $U_3 = 15\,m/s$. By comparing the flutter suppression effects of the two groups of wind speed blades, it can be seen that as the wind speed increases, the flutter suppression effect will weaken. However, under extreme wind speed, the flutter can still converge to 0 m within 5 seconds, which is of practical significance. Compared with the situation before and after the introduction of friction interference, the traditional backstepping controller failed to suppress the flutter divergence, while the RBF- backstepping control can effectively suppress the flutter divergence, indicating that the RBF neural network improves the robustness of the backstepping control scheme.

In Figure 5, compared to two set of neural network approximation error curve with the wind blowing, as shown in the previous 0.5 seconds approximation effect is the same, then when the wind speed of 15m/s approximation curve mutation, due to the presence of friction, actuators nonlinear characteristics of its positive and negative jump, RBF neural network approximation of friction, will be affected, cause approximation error is bigger. But the approximation error still goes to zero within 5s. It is shown that with the increase of wind speed, the flutter amplitude becomes larger, and the convergence time of RBF backstepping controller to suppress blade flutter becomes longer. In terms of its control effect, it can still converge to 0 m within 5 seconds, which has practical significance.

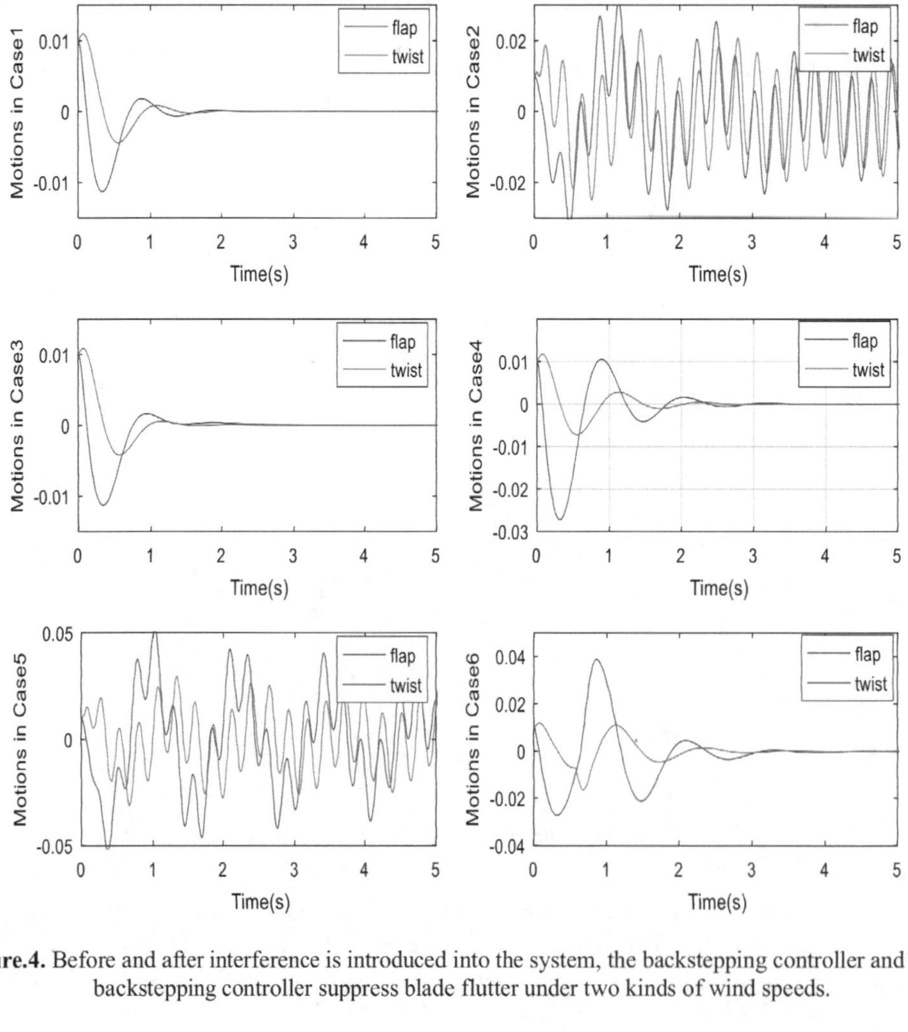

Figure.4. Before and after interference is introduced into the system, the backstepping controller and RBF - backstepping controller suppress blade flutter under two kinds of wind speeds.

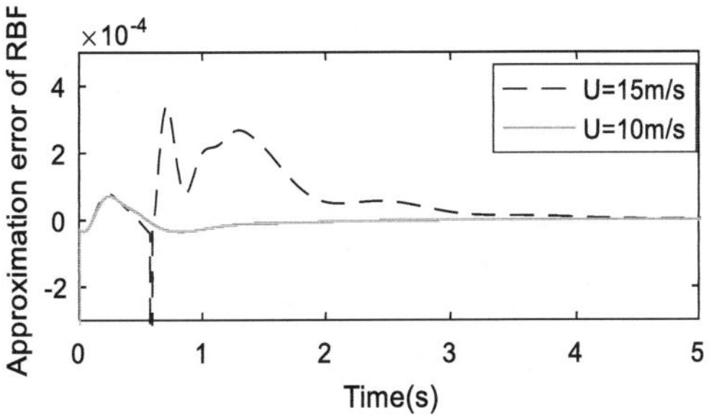

Figure.5. RBF neural network approximation error under the wind speed is 10m/s and 15m/s.

5. Conclusions

- Divergent instability analysis is based on 2D classical symmetrical airfoil section of blades with large aspect ratio, and a quasi-steady aerodynamic model extracted from the helicopter blade stall aerodynamic model is introduced to study the critical flutter after modification.
- A second-order model of the variable rotor actuator with nonlinear friction interference is introduced to simulate the friction interference on the actuator. The backstepping controller can effectively suppress blade flutter without frictional interference, and for blade flutter with frictional interference, the RBF- backstepping controller has a better effect.
- Comparing the approximation effect of RBF neural network at different wind speeds, the approximation object is nonlinear, resulting in a large mutation of the approximation error, which affects the blade flutter suppression effect. Since the approximation error converges to 0 within 5 seconds, the RBF backstepping controller can suppress blade flutter in time and effectively.

Acknowledgement

The authors gratefully acknowledge the support of the National Natural Science Foundation of China (no. 51675315).

References

[1] Liu J 2014 Matlab Simulation of Synovial Variable Structure Control (Tsinghua University Press) chapter 9 pp 338–377.
[2] Liu T 2015 Shock & Vibration 1 1-13.
[3] Chang L, Liu T, Xu W and Wu H 2018 Mechanical design and Research 34(2): 199-204.
[4] Liu T, Chang L 2018 Vibration and Shock 37(13) 123-129.
[5] Wei R, Dong Q and Zhang X 2021 Sensors 21(4) 1508.
[6] Guo X, Wang Q and Li Get 2011 Proceedings of the Csee 31(015) 75-81.
[7] Lin F ,Shen P and Hsu S 2002 Electric Power Applications(IEE Proceedings) 149(3) 184-194.
[8] Madani T,Benallegue A 2007 IEEE/RSJ International Conference on Intelligent Robots & Systems.
[9] Yi C 2020 Earth and Environmental Science(IOP Conference Series) 602(1) 012003.
[10]Thiago S , Dqt B and Mvd A 2020 Mechanical Systems and Signal Processing 1 143.
[11]Liu T 2017 Journal of Vibroengineering 19(7) 5114-5136.
[12]Liu T 2011 Theoretical Modeling and Numerical Study on Aeroelasticity of Large Wind Turbine Blade Power Stall(Shandong University of Science and Technology).

Prestressed Modal Analysis of Wind Turbine Blade Considering Fluid-solid Coupling

Shanyao Li, Yingbo Wang, Haipeng Sun, Changle Sun, Xiaolin Zhang and Tingrui Liu*

College of Mechanical & Electronic Engineering, Shandong University of Science & Technology, Qingdao 266590, China

*Corresponding author's E-mail: liutingrui9999@163.com

Abstract. Wind turbine blade is a kind of slender elastic body structure, which is prone to vibration and deformation during operation, seriously affecting the smooth operation of wind turbine. Taking the 1.5MW wind turbine blade as the research object, the three-dimensional shell model of the blade is established based on Solidworks, and the composite material layer design of the blade is carried out in ANSYS ACP. The stress and deformation of the blade in the flow field are studied based on the fluid structure coupling method. The first six natural frequencies and modes of the blade are obtained by taking the stress of the structural field as the prestress of the modal analysis. The results show that flapping and shimmy are the main vibration forms of the blade. When resonance occurs, the fluctuation of the blade in the flapping direction is the largest, and the maximum stress appears near the blade root.

1. Introduction

With the development of large-scale wind turbine, the power is increasing, and the blade length and flexibility are also increasing, so the wind turbine blade is very prone to vibration. Severe vibration has a great impact on the fatigue of wind turbine blade material, greatly reducing its effective life, and even causing wind turbine blade fracture, resulting in huge economic losses. Blade is the most expensive component in wind turbine, its quality is less than 16% of wind turbine, but it accounts for 14% - 21% of wind turbine cost [1]. Therefore, it is very important to study the blade vibration of wind turbine. For the structural characteristics of wind turbine blades, scholars at home and abroad have done a lot of research. Alireza maheri [2] takes pre twist distribution, fiber direction and skin thickness as objective optimization variables, establishes aeroelastic blade structure model based on genetic algorithm, and studies its influence on wind energy utilization. Based on aeroelastic tailoring, Bottasso [3] studied the influence of all material coupled blade and part material coupled blade on load shedding capacity. Jundong [4] analyzed the aeroelastic complex mode of large flexible blade under rotating condition by using super element method. In reference [5], based on the Lagrange Hamiltonian system of a rotating beam, the dynamic equation of a wind turbine blade is established and the modal analysis of the blade is carried out. Shan Lijun [6] uses multi-objective genetic algorithm to optimize the design of wind turbine blades, which improves the natural frequency of wind turbine blades. The influence of centrifugal load on natural frequency of small wind turbine blade was analyzed by finite element method [7]. In reference [8] , the blade is discretized by using Timoshenko beam with two nodes. Considering the shear effect and centrifugal hardening effect of blade, the inherent properties of blade vibration are analyzed.

In most of the above studies, the blade element momentum theory is used to solve the aerodynamic load of the blade. The aerodynamic load is simplified as a concentrated force loaded on the aerodynamic center of the blade element, or a constant pressure is directly added on the blade surface. With the blade structure becoming slenderer, this loading method may lead to a large deviation between the dynamic response and the real value of the blade. In this paper, based on the fluid structure coupling method to solve the aerodynamic load of the blade, the mode of the blade with prestress is studied, and the frequency response of the blade is obtained by frequency sweep analysis.

2. Blade Modeling

The blade of horizontal axis wind turbine is a very complex three-dimensional curve. Because the chord length, twist angle and airfoil of each section are different, it is difficult to model directly in ANSYS. In this paper, SolidWorks software is used for 3D modeling of wind turbine blade. The first step is to convert the standard airfoil data from DAT format to TXT format and import it into the 3D drawing software Solidworks. In the second step, according to the ratio of chord length calculation value of blade section airfoil to standard airfoil chord length, the initial airfoil coordinate is enlarged in equal proportion in the 3D drawing software Solidworks. The third step is to take the aerodynamic center of the airfoil as the origin to translate the airfoil along the horizontal direction. The aerodynamic center of the airfoil is taken as the center of the circle, and the twist angle is taken as the rotation angle to rotate the airfoil. The fourth step is to take the vertical direction of chord line as translation direction and the section radius of airfoil on blade as translation distance to translate the airfoil coordinate. In the fifth step, the profile of each airfoil structure is drawn, and then two spline curves are drawn with the leading edge and trailing edge points of each airfoil structure as the connection points, and then the three-dimensional shell model of wind turbine blade is generated by sweeping the connected curves of each airfoil structure by segments. The final blade model is shown in Figure 1.

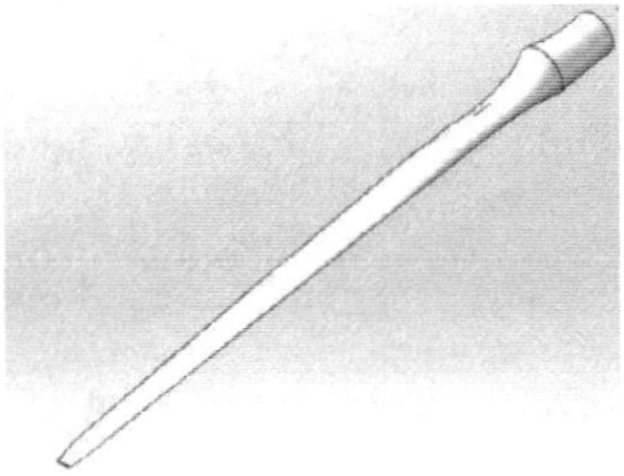

Figure 1. Three dimensional model of blade

3. Fluid Structure Coupling Principle

3.1. Fluid Governing Equations.

The fluid structure interaction follows the basic assumptions and related laws of fluid mechanics, and the fluid domain follows the conservation of mass, momentum and energy:

$$\begin{cases} \dfrac{\partial \rho_{\mathrm{f}}}{\partial t} + \nabla\left(\rho_{\mathrm{f}} v\right) = 0 \\ \dfrac{\partial \rho_{\mathrm{f}}}{\partial t} + \nabla\left(\rho_{\mathrm{f}} v - \tau_{\mathrm{f}}\right) = f_{\mathrm{f}} \\ \dfrac{\partial \rho_{\mathrm{f}} E}{\partial t} + \nabla\left(\rho_{\mathrm{f}} v E - \tau_{\mathrm{f}} v\right) + q_{\mathrm{f}} = f_{\mathrm{f}} + q_{\mathrm{f}} \end{cases} \tag{1}$$

where t is time; FF is the volume force; ρ_{f} is the fluid density; v is the velocity vector of fluid; τ_{f} is the shear force tensor; E is the internal energy per unit mass; q_{f} is the heat loss per unit volume.

3.2. Solid Governing Equations

According to Newton's second law, the governing equation of solid can be deduced as:

$$\rho_s \ddot{d}_s = \nabla \delta_s + f_s \tag{2}$$

where ρ_s is the density of solid; δ_s is the Cauchy stress tensor; f_s is the volume force vector; $\ddot{d}_s$ is the local acceleration vector in solid domain.

3.3. Fluid Structure Coupling Equation

According to the conservation law, the stress and deformation of fluid and solid should be conserved one by one at the contact interface of fluid solid coupling:

$$\begin{cases} \tau_f n_f = \tau_s n_s \\ d_f = d_s \end{cases} \tag{3}$$

where τ_f is the stress of fluid; τ_s is the stress of solid; n_f is the unit direction vector of fluid; n_s is the unit direction vector of solid; d_f is the displacement of fluid; d_s is the displacement of solid.

3.4. Analysis Flow

As shown in Figure 2, different from the ordinary flow field calculation, the fluid structure coupling calculation of composite wind turbine blade first needs to carry out the layer design in ACP module, and then upload the model to the structure field in the form of shell element, and realize the data transmission of flow field and structure field by means of coupling surface. The blade pressure obtained in the fluid structure coupling solid domain is uploaded to the modal analysis module as the prestress of modal analysis, and the vibration response of the blade in the presence of prestress is solved.

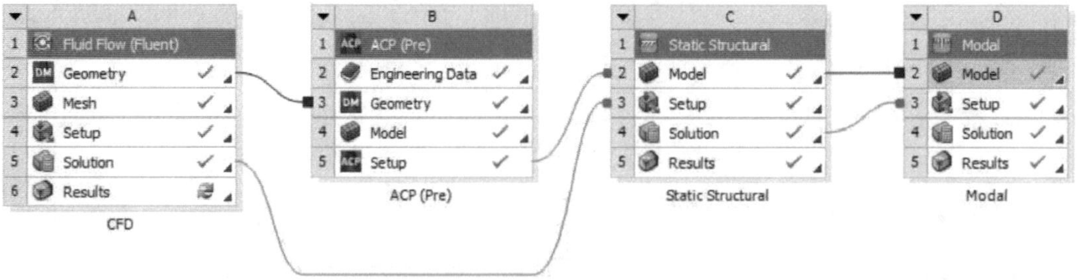

Figure 2. Analysis flow chart

4. Prestressed Modal Analysis Results

If the natural frequency of the blade is close to the external excitation frequency, the blade may be damaged by resonance, so the modal analysis of the blade is necessary. The blade root is set as a fixed end constraint to limit the degree of freedom of the root. A large number of research results show that the low-order natural frequency of wind turbine blade structure is close to the rotating excitation frequency of wind turbine, so the low-order natural frequency is mainly analyzed. In this paper, the sixth order vibration frequency with prestress of blade under wind pressure is extracted. The vibration modes of each order are shown in Figure 3.

The vibration mode of blade tends to be complex from low order to high order. The first mode is the bending vibration perpendicular to the rotation plane, that is the flapping vibration; The second mode is the bending vibration in the rotating plane, that is the shimmy vibration; The fourth-order vibration modes are second-order swing vibration and shimmy vibration; The fifth mode is the third-order swing vibration; The sixth mode is the coupled motion of torsional vibration and flapping vibration of blade around the variable pitch axis. It can be seen that the first five vibration modes are bending vibration with low frequency, so bending vibration is the main vibration mode of the blade, which is the main reason for the fatigue failure of the blade.

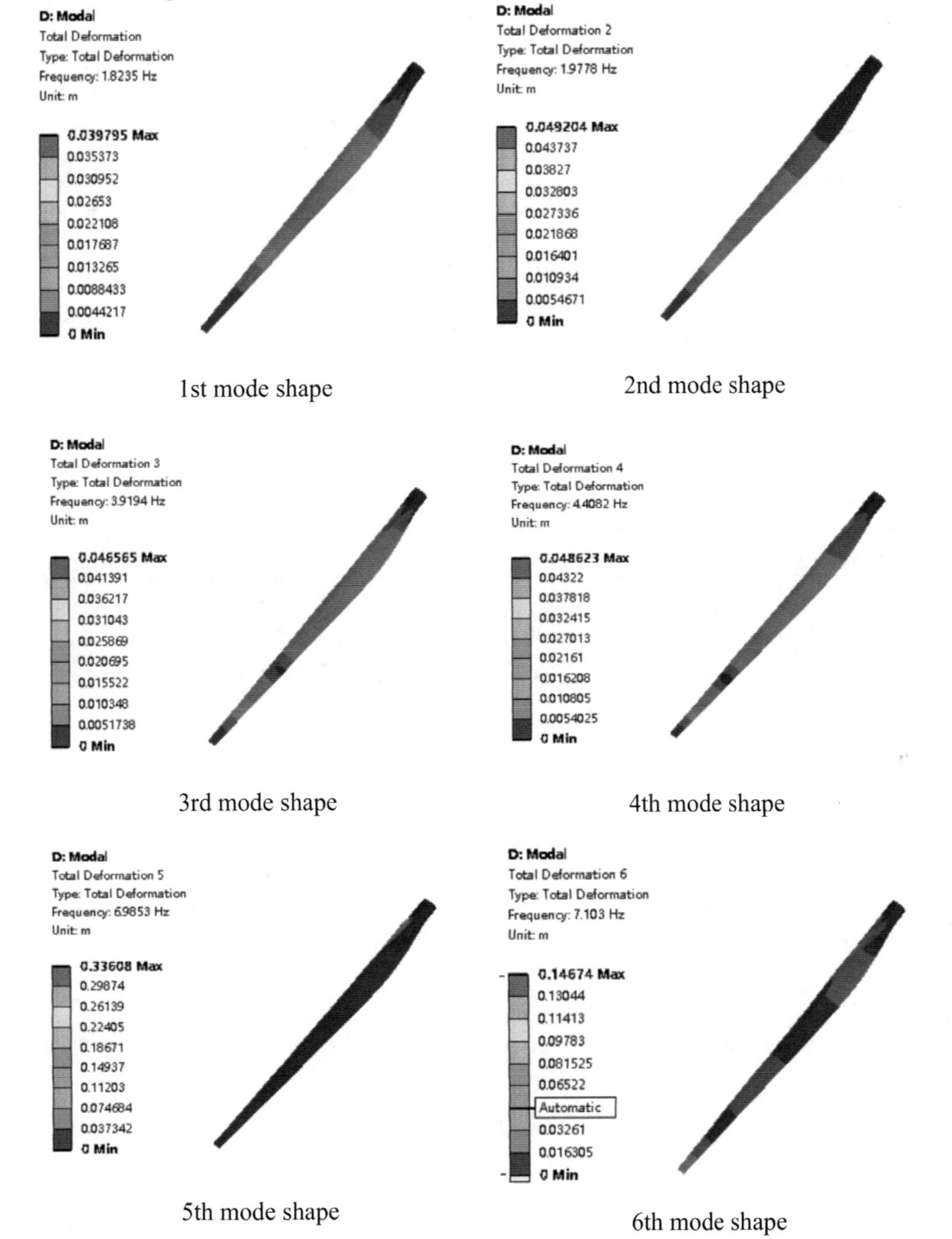

Figure 3. First six vibration modes of the blade

5. Conclusion

In this study, based on the finite element method, the blade solid modeling, fluid structure coupling analysis and dynamic response research of the impeller are carried out. The three-dimensional model of the wind turbine blade is established in Solidworks by using the practical engineering methods. Combined with the basic theory of fluid structure coupling analysis, the unidirectional fluid structure coupling analysis technology is adopted to analyze the flow field and structure field of the wind turbine blade, the flow field distribution, and the deformation of the blade under wind pressure are obtained. According to the analysis results of the wind turbine structure field, the blade root of the impeller bears a large load, so the structural strength of the blade root must be strengthened in the actual manufacturing process of the impeller. Taking the stress of the blade in the process of fluid structure coupling as the initial condition, the first six natural frequencies of the blade are solved. From the results, the vibration forms of the blade are mainly swaying and flapping, and the natural frequencies of the blade increase in turn, but they are far away from the rotation frequency and passing frequency of the blade, so there is no resonance. The design modeling of the surface blade is reasonable and reliable.

Acknowledgement

The authors gratefully acknowledge the support of the National Natural Science Foundation of China (no. 51675315).

References

[1] Maheri A,Noroozi S and Vinney J 2007 Renewable Energy 2011–2018.
[2] Bottasso C,Campagnolo F and Croce A 2013 Wind Energy 2013 1149–1166.
[3] Jundong H,Hongjian X and Deyuan L, 2020 Chinese Journal of Mechanical Engineering 180–187.
[4] Fengqin Z 2020 Harbin Engineering University 19–30.
[5] Lijun S, Jingyuan L 2019 Renewable energy 1067–1072.
[6] Junkai G, Jiaqi Z, Jianwen W, Zhaoguo M, Xiaotian L 2019 Acta Solar Energy 2148–2153.
[7] Yaojie Y, Mingfu L, Pin L 2017 Acta Solar Energy 2552–2558.
[8] Xiong L, Gangqiang L 2010 Journal of Mechanical Engineering 128–134.

Crankshaft-Bearing Assembly Based on Vision-Guided and Attractive Regions in the Environment

Wenjun Zhu, Haida Feng, Mengyi Zhang and Yang Yi*

College of Electrical Engineering and Control Science, Nanjing Tech University, Nanjing, Jiangsu, China.

*corresponding author's e-mail: yiyang@njtech.edu.cn

Abstract. Aiming at the problem of sensor-less high-precision manipulation, this paper designs and constructs a crankshaft-bearing assembly system based on vision-guided and attractive region in environment. A pre-analysis and evaluation of the attitude measurement accuracy method is proposed to describe the trusted region of the current assembly pose through vision. Then its high-dimensional attractive regions of environment constraints is constructed based on the mechanical constraints between the assembly object and the assembly position. Combine the complementary advantages of vision and mechanical information, a robot adjust its pose to a high precision manipulation pose. A platform of an industrial robot with a vision system is built and experiments on the crankshaft and bearing are successfully assembled. It is validated that using the proposed the vision-guided and attraction region in environment system, high precision assembly manipulation is realized.

1. Introduction

Accurate pose and position information of industrial parts is a core technology in the application of robot assembly systems, and it is also one of the bottlenecks that restrict the promotion of robots in the customization and personalization in the industrial production field [1]. Visual object localization [2-3] are mainly used to provide pose and position measurement for grasping and assembling. However, when the part and the assembly object are in contact process, it is difficult to obtain a good assembly effect only relying on vision information. As in crankshaft-bearing assembly, for example, there are also some difficulties: (1)There is a tight fit between the middleware and bearings; (2)It is difficult to predict the contact state of the middleware and the bearing during the assembly process; (3)Because of clamping error and assembly error make it difficult for the crankshaft to be moved to the right above the bearing fit position in the assembly process [4-5].

This paper proposes a crankshaft-bearing assembly method based on the vision-guided and attractive region in the environment (ARIE) , combining the different characteristics of vision features and mechanical information of the assembly process. The attractive region in environment was proposed by Qiao in the process of high precision spindle hole assembly of the robot, and the existence and the stability of the environmental constraint domain were proved in a series of subsequent works[6, 7]. AREA is the constraint domain existing in the configuration space of the system which can be used to eliminate part or all the uncertainties of the system [8]. The vision sensor provides the position and the attitude measurement to guide the process of grasping, approaching, and inserting process. A pre-analysis and evaluation of the attitude measurement accuracy method is proposed to describe the trusted region of the current assembly pose through vision. Then the high-dimensional ARIE is constructed based on the mechanical constraints between the assembly object and the assembly position for the insertion operation. Combine the complementary advantages of vision and mechanical information, a robot adjust its pose to a high precision manipulation pose.Experiments on the crankshaft and bearing are successfully assembled by the guidance of the vision system.

2. Vision-Guided and ARIE Construction system

2.1. A pre-analysis evaluation of attitude measurement accuracy

It finds that when the workpiece is located in certain poses, the error rate of pose detection will also increase greatly. To solve this problem, a pose pre-analysis method was proposed to analyze whether the above measurement errors were likely to occur in the pose measurement process of a specific target in advance and determine the pose measurement accuracy. The method is as follows:

Based on the CAD model of the workpiece and the detection results of the camera, the two-dimensional view set V is generated on the observation space first. The similarity degree between each view and other views is calculated. The calculation method of view similarity measure is shown in Eq. (1), where $p_i(i=1,\mathrm{L}\ ,n)$ and s_i are the gradient magnitude of the corresponding pixel points in two adjacent views, respectively.

$$c=\left|\frac{1}{n}\sum_{i=1}^{n}\frac{\langle p_i,s_i\rangle}{\|p_i\|\cdot\|s_i\|}\right| \tag{1}$$

If the similarity between view v_i and any other views in the non-adjacent range is less than T_{th} in the view set, the corresponding pose of the view is defined as the trusted pose. If the similarity between view v_i and other views in non-adjacent ranges is greater than or equal to T_{th} in the view set, the corresponding pose of the view is defined as the untrusted pose. All of the trusted poses form a trusted region on the Gaussian sphere, which forms the accuracy estimation of the workpiece.

It can be seen the pre-analysis and evaluation of attitude measurement accuracy is used to improve the accuracy of workpiece pose measurement.

2.2. The ARIE construction

The ARIE an effective operation strategy is designed to eliminate the uncertainty of the system state during the operation of the robot, so that the system can reach a certain state. The following is the mathematical definition of the attractive region in environment : the attractive region in environment is the constraint domain existing in the configuration space of the system which can be used to eliminate some or all the uncertainties of the system. For nonlinear systems:

$$\frac{dX}{dt}=f(X,u,t) \tag{2}$$

Where X is the state of the robot in the configuration space. u is the input of the system independent of state. If there is a state potential energy function $g:X\in R^N\to P\in R$ and a state X_0 formed by environmental constraints, it satisfies:

- （1） $g(X)$ is a local minimum if and only if $X=X_0$;
- （2） $g(X)>0$ if and only if $X\neq X_0$;
- （3） $g(X)$ has a continuous partial derivative with respect to X;
- （4） $dg(X)/dt<0$

Then, under the action of u, the nonlinear system is stable in the neighboring region $U^o(X_0,\delta)$ of the state $(\delta\in R,\delta>0)$, which is called the "environmental attraction region", and the system state X_0 is the stable state of this attraction region.

2.3. Complementary combination of vision detention and ARIE

According to the general operating strategies for utilizing the high-dimensional environmental constraints, the following strategies can be used for crankshaft-bearing assembly. The high-dimensional configuration space of the crankshaft-bearing assembly system is decomposed into two subspaces, $\left(g(X),X^a\right)\in \hat{C}^1_{conf}$ and $\left(g(X),X^a\right)\in \hat{C}^2_{conf}$. According to two subspaces, crankshaft-bearing assembly is divided into two stages: eliminating the uncertainty of crankshaft attitude angle and eliminating the uncertainty of crankshaft displacement and attitude angle. Under the constraints of the attractive region in environment, eliminating the uncertainty and completing the assembly of the crankshaft and bearing and completing the assembly of the crankshaft and bearing.

- (1)Select the system state variable X that describes the robot operation process, and select the appropriate state potential energy function $g(X)$ according to the environmental constraints to construct the robot operation configuration space $\left(g(X),X\right)\in C_{conf}$.
- (2)According to the specific operation tasks of the robot, the subspace $\hat{C}_{conf}$ with the attractive region in environment was found in the robot operation configuration space, and the system state was divided into attraction variable X^a and limited variable X^c according to the subspace. So there's a mapping $f: X^c \to X^a$ from the attractive variable to the constrained variable.
- (3)Obtain the initial system state $X_0=\left[X_0^a,X_0^c\right]$ and fix the limited variable $X_1^c=X_0^c$. An independent state-input u^a is found in the configuration subspace $\left(g(X),X^a\right)|_{X^c=X_0^c}$ to eliminate the uncertainty of X^a under X_1^c, and then $X_1^a=X_{0,\min}^c$ and the local minimum $g(X)_{1,\min}$ of $g(X)$ in the attraction domain of the subspace are obtained.
- (4)Under the system state $X_1^s=\left[X_1^a,X_1^c\right]$, adjust the control variable X^c to X_2^c.
- (5)Repeat steps (3) and (4) until $g(X)$ converges to the minimum.

2.4. A Vision-guided and ARIE system for crankshaft-bearing assembly

In the assembly process of crankshaft-bearing, using visual navigation method to complete important assembly steps. The specific implementation method is as follows:

- (1)Bearing pose detection and grasping: According to the calibrated parameters and the calculated pose data, first moves the robot to the middle position and then moves the robot to the target grasping position.
- (2)Bearing placement: The bearing is placed on the fixed seat of the crankshaft-bearing pressing device by grasping the robot hand-mounted at the end of the industrial robot. Since the fixed seat posotion in the world coordinate system is fixed, the bearing placement is completed according to the fixed path manually marked.
- (3)Pose detection and grasping of crankshaft: The position and pose of the crankshaft relative to the camera coordinate system are obtained by computer analysis. If the pose is located in the trusted region of the crankshaft accuracy estimation model, the industrial robot is guided to grab the crankshaft; If the pose is located in the untrusted region, the industrial robot is guided to change the observation pose until the credible detection pose is obtained, and then the crankshaft is grabbed.
- (4)Insertion of crankshaft: The crankshaft insertion method based on the environmental attraction domain is proposed to meet the requirements of high precision assembly.

3. Experimental results and analysis

The experimental platform is composed of a 6-DOF industrial robot, an industrial camera that constitutes the Eye-In-Hand system; the computer is connected with the control circuit of the crankshaft-bearing pressing device through the RS232 interface. The crankshaft-bearing assembly process proposed in this chapter is given, as shown in Figure 1.

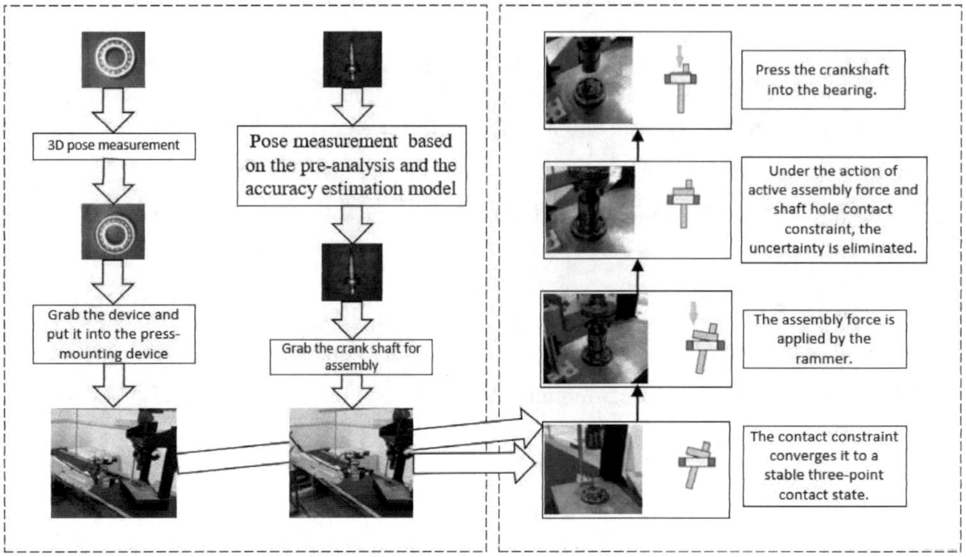

Figure 1. Flowchart of crankshaft-bearing assembly.

Firstly, an industrial camera is used to randomly collect 200 frames of images, respectively, for 3D attitude measurement. The accuracy estimation model of the crankshaft drawn from the obtained data is shown in Figure 2.

The industrial camera will guide the robot pose to the trusted area until it is reach a high accuracy trust region of the attitude measurement . In the experiment, simple background (white background) is used to minimize the impact of the background environment and occlusion on the experiment. The 3Dpose detection is carried out under the natural light environment (weak light) and light source illumination in the experiment. The test results are shown in Table 1 and Table 2.

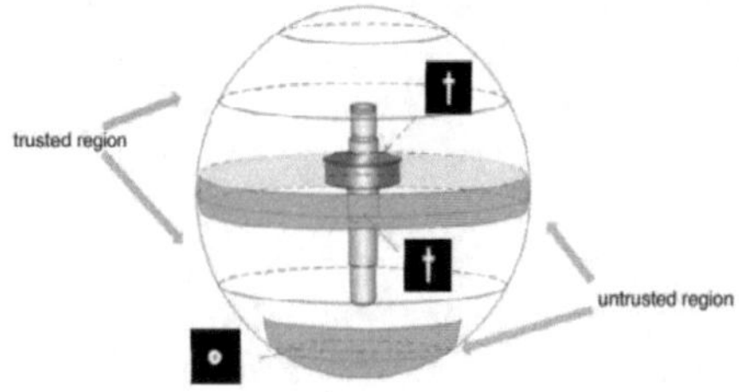

Figure 2. The schematic diagram of the accuracy estimation model of the crankshaft.

Table 1. Success rate of 3D pose detection in different regions.

Area	Illumination	Detection success rate
Untrusted area	Natural light	58.0%
Trusted area	Natural light	95.5%
Untrusted area	Light source illumination	73.5%
Trusted area	Light source illumination	97.5%

Table 2. Pose detection success rate based on the pre-analysis and accuracy evaluation method.

Illumination	Detection success rate
Natural light	94.5%
Light source illumination	98.5%

It can be seen from Table 1 that the success rates of pose detection (view-based monocular vision) are different under different poses for the same workpiece. The success rate of pose detection can be greatly improved under suitable observation pose conditions. Meanwhile, it has a better success rate of pose detection under a good lighting environment. According to Table 2, in the pose detection method based on the pre-analysis and accuracy estimation model, the success rate of the pose detection of the workpiece guided by the robot to the trusted area is equivalent to that of the workpiece located in the trusted area. The method proposed in this paper is compared with the view-based monocular vision pose detection method proposed in the literature[9].The experimental results are shown in Table 3. Experimental results show that the method proposed in this paper has a higher detection success rate.

Table 3. Comparison of success rate of pose detection methods

Approach	Detection success rate
literature	87.5%
The method proposed in this paper	98.5%

The ARIE of crankshaft-bearing assembly is presented in Figure 3 in the 3D configuration space. It can be seen that if the structure state has nothing to do with the input The uncertainty of will be reduced accordingly, so that it can be constructed to use the configuration space environment constraints to carry out the operation of the shaft hole assembly strategy.

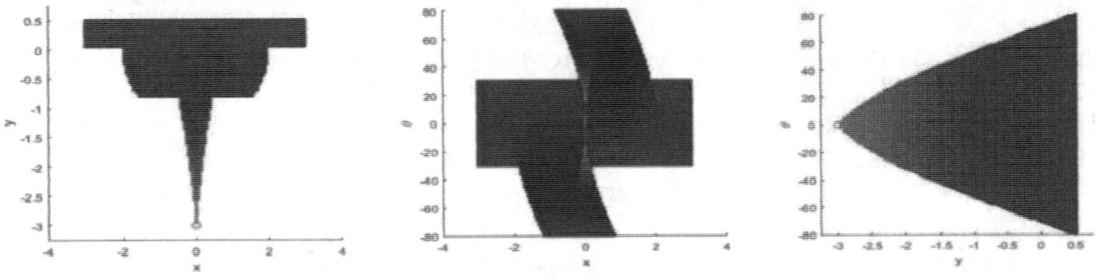

Figure 3. The ARIE of the crankshaft-bearing assembly.

The full assembly of the crankshaft-bearing assembly system in the laboratory is shown in Figure 4:

(a)The industrial camera collects the image of the crankshaft-bearing on the worktable and completes the measurement of the position and posture of the crankshaft-bearing; (b)The industrial robot is guided to grab the bearing and place it in a fixed base of the crankshaft-bearing pressing device; (c) The industrial robot is guided to grab the crankshaft and put it into the hole of the bearing with a fixed pose $(x, y, \alpha) = \left(0, -0.3, \frac{\pi}{9}\right)$; (d) Driving the pressing head of the crankshaft-bearing pressing device to complete the pressing of the crankshaft and bearing.

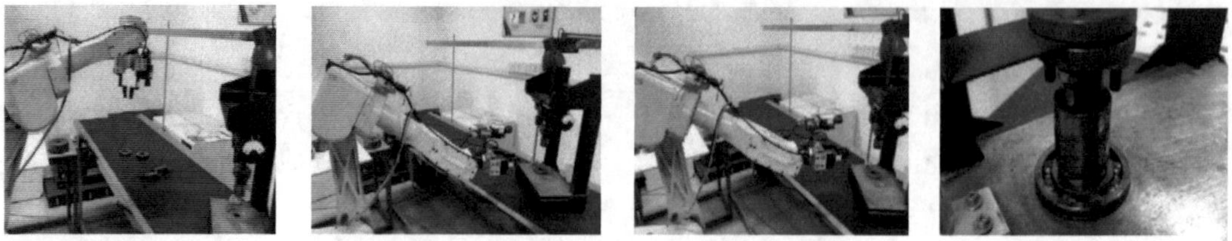

(a)3D pose detection of workpiece (b)Grab and place bearings (c)Grab and insert the crankshaft (d)Assemble bearing and crankshaft

Figure 4. A crankshaft-bearing assembly system assembled in the laboratory.

4. Conclusion

A crankshaft-bearing assembly method based on the vision-guided and attractive region in environment is proposed and implemented in this paper. The pose detection experiments and crankshaft-bearing assembly experiments were carried out by using this method. The experiments show that this method improves the accuracy of workpiece pose measurement and realizes the sensor less high-precision assembly of crankshaft-bearing.

Acknowledgment

The financial supports by the National Natural Science Foundation of China (61803198), Natural Science Foundation of Jiangsu Province of China (BK20180693). Natural Science Foundation of the Higher Education Institutions of Jiangsu Province of China (21KJB520007)

References

[1] Xing W, Ou Y. On the study of workpiece localization with a cost-effective vision based method[C]//2015 IEEE International Conference on Cyber Technology in Automation, Control, and Intelligent Systems (CYBER). IEEE, 2015: 457-461.
[2] Lin H I, Chen Y Y, Chen Y Y. Robot vision to recognize both object and rotation for robot pick-and-place operation[C]//2015 international conference on advanced robotics and intelligent systems (aris). IEEE, 2015: 1-6.
[3] Murtagh F, Contreras P. Algorithms for hierarchical clustering: an overview, II[J]. Wiley Interdisciplinary Reviews: Data Mining and Knowledge Discovery, 2017, 7(6): e1219
[4] Li R, Qiao H. Condition and strategy analysis for assembly based on attractive region in environment[J]. IEEE/ASME Transactions on Mechatronics, 2017, 22(5): 2218-2228.
[5] Qiao H. Attractive regions formed by constraints in configuration space-attractive regions in motion region of a polygonal or a polyhedral part with a flat environment[C]//Proceedings 2001 ICRA. IEEE International Conference on Robotics and Automation (Cat. No. 01CH37164). IEEE, 2001, 2: 1071-1078
[6] Qiao H, Ma C, Li R, et al. Simple coarse sensing to achieve high precision From attractive region in environment to constrained region in environment[J]. Systems Science & Control Engineering, 2017, 5(1): 16-24.
[7] Inoue T, Choudhury S, De Magistris G, et al. Transfer learning from synthetic to real images using variational autoencoders for precise position detection[C]//2018 25th IEEE International Conference on Image Processing (ICIP). IEEE, 2018: 2725-2729.
[8] Yerlikaya U, Balkan R T. Obtaining High-dimensional Configuration Space for Robotic Systems Operating in a Common Environment[J]. International Journal of Electrical and Information Engineering, 2021, 15(4): 158-167.
[9] Ulrich M, Wiedemann C, Steger C. Combining scale-space and similarity-based aspect graphs for fast 3D object recognition[J]. IEEE transactions on pattern analysis and machine intelligence, 2011, 34(10): 1902-1914.

Development of Mobile Robot System Based on Nano-scale Micro-surface Measurement

Zhongwei Ji, Pan Wang, Jiahao Jia,Yan Ding, Kang Zhao and Tingrui Liu*

College of Mechanical & Electronic Engineering, Shandong University of Science & Technology, Qingdao 266590, China

*Corresponding author's E-mail: liutingrui9999@163.com

Absrtact. The system development of the mobile robot mainly includes two parts: one is to build a laser interference system based on piezoelectric ceramic dynamic carrier; The other part is the image processing of the interference fringe pattern collected by the interference system. The laser interference system can be installed on the front-end manipulator of the robot as an integral structure. The shape information of the measured micro-surface can be obtained by processing the interference fringes collected by the image acquisition card on the interference system, so as to achieve the purpose of high-precision surface roughness measurement.

1. Introduction

Industrial robot is widely used in the industrial field of multi joint manipulator or multi degree of freedom machine device, which can rely on its own power and control ability to achieve a variety of industrial processing and manufacturing functions. Industrial robot is widely used in the fields of electronics, logistics, chemical industry and other industries because of its high efficiency and processing accuracy. It has become an indispensable part of modern industrial production [1].

As one of the industrial robots, surface inspection robot is widely used in chip processing equipment and high precision instrument manufacturing. At present, the implementation of 5G and the development of artificial intelligence all need to be applied to semiconductor chips, which not only improves the market scale, but also poses greater challenges to semiconductor technology. At present, the most advanced semiconductor chip processing technology in the world, such as extreme ultraviolet lithography manufacturing system for wafer foundry production, has tens of thousands of high-precision nanoscale micro surfaces on many large parts. In the process of large-scale machining, the accuracy of nano scale micro surface directly determines the performance and success of the lithography machine.

The existing nano scale micro surface measurement equipment is a nano scale laser interferometer, such as Zygo, which converts the mechanical quantity of the measured object into displacement to detect, so the displacement sensor has become the most basic sensor in the mechanical quantity sensor [2]. At present, most of the existing measuring equipment at home and abroad can only measure the horizontal surface of the moving, small "movable element". As for the micro surface, especially the vertical surface, which can't be measured by the existing interferometer in the process of on-line machining. In this paper, we want to design a set of laser interferometry system based on piezoelectric dynamic carrier wave, and combine it with the manipulator to achieve the purpose of moving detection. In addition, for the interference fringes collected by the interference system, the phase distribution of the part surface is obtained by using Fourier transform and space carrier technology, and then the roughness information of the part surface is obtained. In the current industrial robot technology development trend of more integration, intelligence, and full automation [3], the combination of laser detection system and mobile manipulator and the application of interference fringe pattern Fourier transform technology not only provide a strong guarantee for the development of China's high-precision manufacturing industry, but also provide a new choice for the improvement of instrument detection technology in the future.

2. Construction of Laser Interferometry System Based on Piezoelectric Dynamic Carrier

Figure 1 (a) shows the principle diagram of the laser interferometry system, and figures 1 (b)~(c) show the overall structure diagram of the laser interferometry system and the structure diagram of the long flange manipulator respectively.

In physics, interference is the phenomenon that when two or more waves meet in space, they are superimposed or offset to form a new waveform. For the two interference beams, the condition of the same frequency and constant phase difference should be satisfied. Based on the principle of interferometry, the essence of the interferometric scanning measurement for the surface is the superposition of the coherent characteristics of light waves in the transmission process, which makes the light intensity become stronger or weaker in the imaging area. Different interference fringes images are obtained [4]. These images are processed and analyzed.

The working principle of the system is as follows: the divergent laser is refracted into a beam of parallel light by a convex lens, and the parallel light is divided into two parts by a spectroscope: one part refracts to the reference plane, and the other part is reflected by the reference plane and transmitted by the spectroscope; The other part is projected on the measured surface, which is reflected by the measured surface and refracted by the spectroscope. Because the two beams are the same laser and satisfy the same frequency condition, the interference image will be formed on the camera, which represents the optical path difference (OPD) of the beam from the reference plane and the measured plane to the observation plane. After the PZT is electrified, the reference plane will vibrate and form different fringe shapes. It represents different OPD. The OPD reflects the phase information of the surface to be measured. Therefore, in the interferometry, the interference fringe contains the roughness information of the surface to be measured.

In addition, there are two key points in the system:1) PZT is a solid solution of $PbZrO_3$ and $PbTiO_3$, which has positive piezoelectric effect and negative piezoelectric effect, and has the characteristics of high Curie temperature, strong piezoelectric property, and good stability [5]. When PZT is consolidated with the reference surface, the sinusoidal signal will drive the reference surface to vibrate up and down in a small amplitude. When the interference system scans the measured surface, each step of the reference plane is equivalent to scanning the measured surface once to obtain the light intensity values of each point [6]. If the reference plane is still, the interference image we get is Newton's ring instead of fringe pattern. It is impossible to calculate and solve the image.2) The refraction part of the reflected light from the reference plane passing through the spectroscope and the transmission part of the reflected light from the measured plane passing through the spectroscope are very weak and almost have no influence on the interference process, so they can be ignored. It is convenient to form clear fringe pattern and calculate.

The structure is divided into mirror box, laser, mirror tube, PZT actuator and CCD camera. The mirror box is a cube structure, which can ensure the accurate tilt of the splitter lens by 45°. The PZT actuator is located in the center of the whole system, and the lower opening is connected with the PZT actuator by two adjustable bolts; The vertical part of the cross slider is connected with the CCD camera, and the horizontal part of the cross slider is directly sheathed on the front end of the manipulator; The front and rear parts are respectively connected with the lens barrel and the laser. According to the schematic diagram, the structure of the interference system is designed, which includes some structures that can realize the flexible movement of each part, and it is installed on the flange of the front section of the manipulator to realize the structural integration of robot motion and detection.

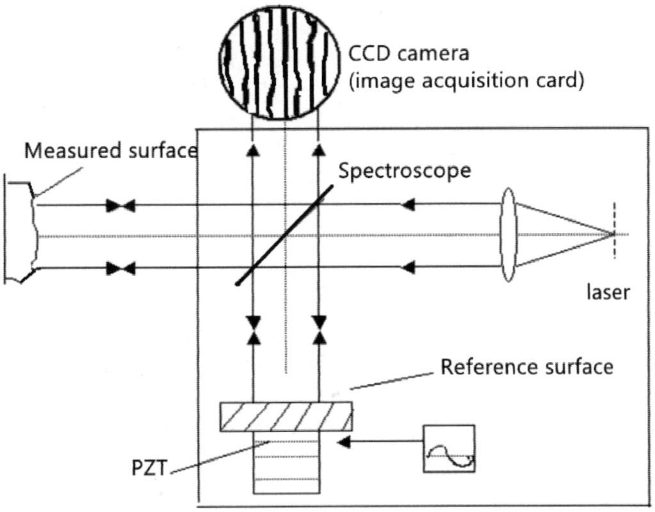

(a) Schematic diagram

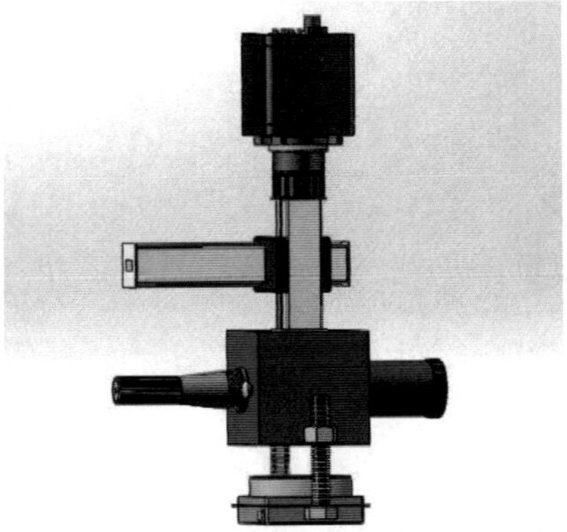

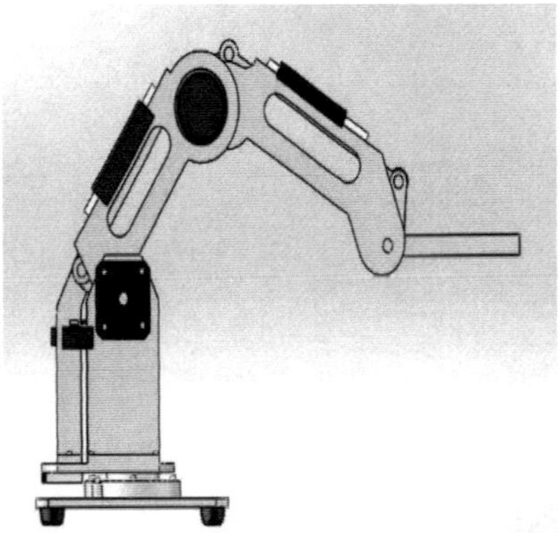

(b) Structure chart (c) Long flange manipulator

Figure 1. Principle and structure of interference system

3. Fringe pattern processing

Generally, an interference fringe pattern can be expressed as follows:

$$I(x,y)=A(x,y)+B(x,y)\cos\left[\phi(x,y)+\phi_0(x,y)\right] \tag{1}$$

In Eq.(1): $A(x,y)$ is the background gray, $B(x,y)$ is the fringe contrast, $\phi(x,y)$ is the phase distribution of the measured surface, $\phi_0(x,y)$ is the initial phase.

The purpose of interferogram analysis is to obtain the phase distribution. There are many analysis methods of interferogram, including fringe tracking technology, phase-shifting interferometry technology and space carrier technology. In the main interferogram processing technology, the speed of fringe tracking detection technology is relatively slow and the accuracy is low. In theory, it can only achieve the measurement accuracy of λ/50 (far lower than this index for the actual system). Although the phase-shifting interferometry technology has high precision (λ/150) and high speed, its equipment is complex and needs a precise phase shifter, and it is not suitable for processing the transient dynamic wavefront with time. Fourier transform method and sine fitting method belong to space carrier method.

They are easy to operate, convenient to calculate, and suitable for dealing with dynamic and static wavefront. However, due to the sampling mismatch, the sinusoidal fitting method leads to sharp error, which affects the accuracy of wave surface recovery; The accuracy of Fourier transform method is relatively high (about λ/100). Although the speed is not as good as sine fitting method, with the continuous updating of computer, the cost of computer is constantly reduced, but the calculation speed is faster and faster, and the calculation speed of FFT is also constantly improved [7].

Different from the above-mentioned phase-shifting technology, space carrier technology can solve the phase information from a single interferogram, and the generation of space carrier does not need special equipment, only needs to tilt the reference plane, so the operation is very simple. After the wave surface is tilted, the fringes become denser and straighter, and the influence caused by non-uniformity of illumination is reduced accordingly [8]. Because it only needs to analyze a single interferogram, it is suitable for information processing of dynamic measurement such as vibration measurement, especially on-line processing. However, due to its small amount of information, its accuracy is lower than that of phase-shifting technology. Space carrier method is divided into Fourier transform method and sine fitting method. Fourier transform transforms interferogram analysis from spatial domain to frequency domain. Sine fitting is an approximate form of Fourier transform.

The processing of single interferogram is widely used in workshop production because of its real-time and simple operation. In this paper, according to the requirements of the subject, we decided to use the Fourier transform processing technology of single interferogram. According to the principle of FFT processing interferogram, the processing block diagram is as follows:

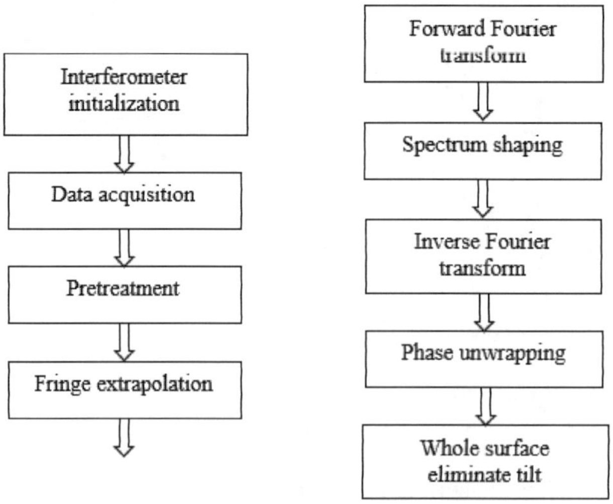

Figure 2. Fringe pattern processing flow

According to the above flow, the spatial carrier can be introduced into the interference diagram by using the inclined reference wave surface, and the image can be expressed as:

$$I(x,y) = A(x,y) + B(x,y)\cos\left[\phi(x,y) + 2\pi f_{0x}\right] \tag{2}$$

The phase is solved by Fourier transform and frequency domain filtering, and the two-dimensional case is transformed, that is, eq(2) is rewritten as:

$$\begin{aligned} I(x, y) &= A(x, y) + C(x, y)\exp\left(2\pi j f_{0x}' x + 2\pi j f_{0y}' y\right) \\ &+ C^*(x, y)\exp\left(-2\pi j f_{0x}' x - 2\pi j f_{0y}' y\right) \end{aligned} \tag{3}$$

where $C(x,y)=\frac{1}{2}B(x,y)\exp[j\varphi(x,y)]$, $C^*(x,y)$ is a complex conjugate function of $C(x,y)$. $\phi(x,y)$ is the required phase position.

By two-dimensional Fourier transform of Eq.(3), it can be concluded that:

$$I(u,v)=A(u,v)+C(u-f_{0x}',\ v-f_{0y}')+C^*(u+f_{0x}',\ v+f_{0y}') \tag{4}$$

After filtering with band-pass filter, the background spectrum $A(u,v)$ is filtered to obtain $C(u-f_{0x}',\ v-f_{0y}')$ or $C^*(u+f_{0x}',\ v+f_{0y}')$, which is moved to the origin to become $C(u,v)$, and $c(x,y)$ is obtained by inverse Fourier transform (IFFT). It can be obtained from the following formula:

$$\phi(x,y)=\tan^{-1}\frac{\text{Im}[c(x,y)]}{\text{Re}[c(x,y)]} \tag{5}$$

where $\text{Im}[\varphi(x,y)]$ and $\text{Re}[\varphi(x,y)]$ are the imaginary part and the real part of $C(x,y)$.

4. Simulation of microstructure wave surface flow

It is known that in MATLAB, the functions fft, fft2 and fftn can implement one-dimensional, two-dimensional, and n-dimensional DFT algorithms respectively; The functions ifft, ifft2 and ifftn are used to calculate DFT; The function ffishift can transfer the zero-frequency component in the result of Fourier operation (fft, fft2, fftn) to the center of the matrix [9], so that the spectrum can be easily observed. Here we use the fringe pattern processing flow combined with the above MATLAB function to simulate wavefront detection to verify the effect of the interference system.

According to the actual situation, the background gray and light intensity can be set to,, and then the spatial carrier analysis can be carried out. First, the original interferogram and spatial carrier interferogram are made through the relevant program, as shown in Fig.3(a)~(b). The original interferogram without dynamic carrier is Newton's ring, which can't be processed further.

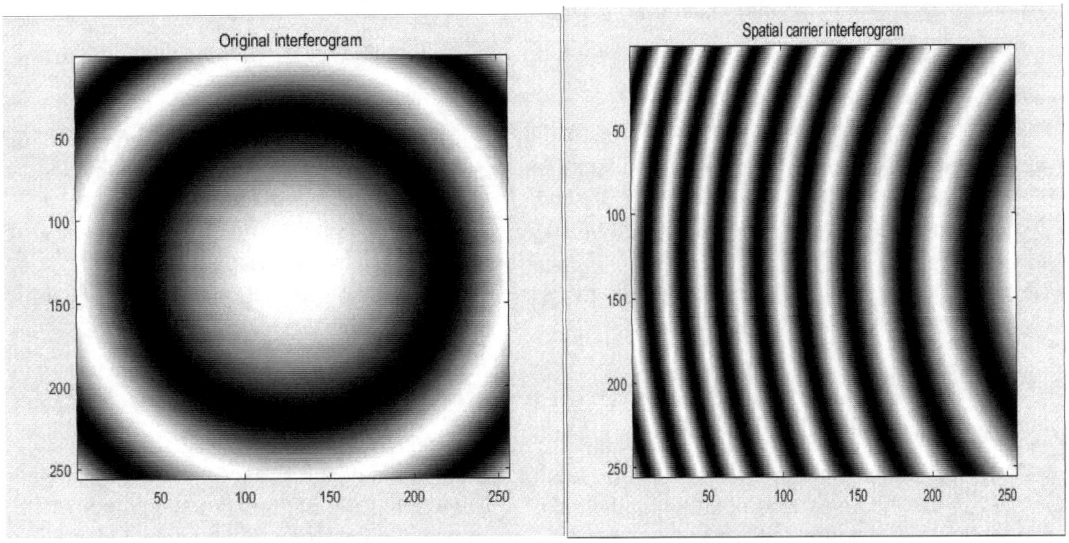

(a) Original interferogram (b) Spatial carrier interferogram

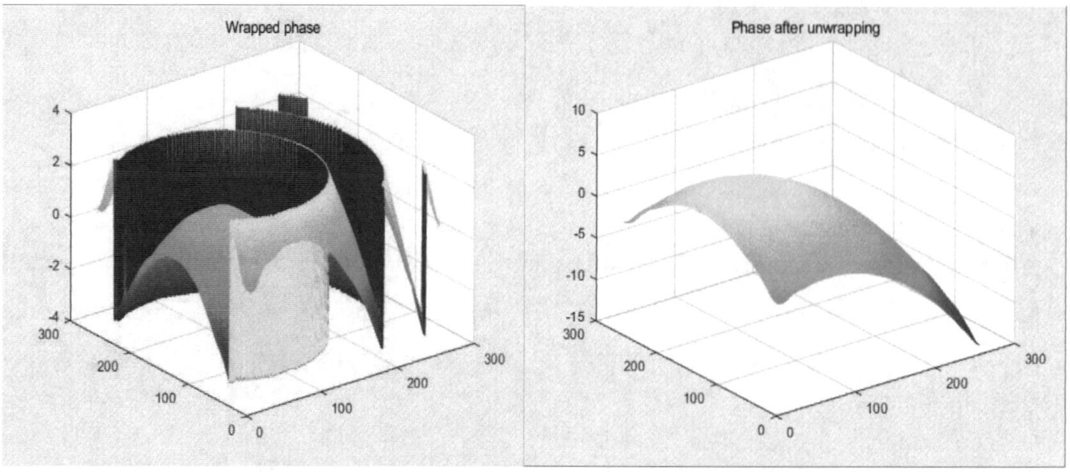

(c) Wrapped phase (d) Phase after unwrapping

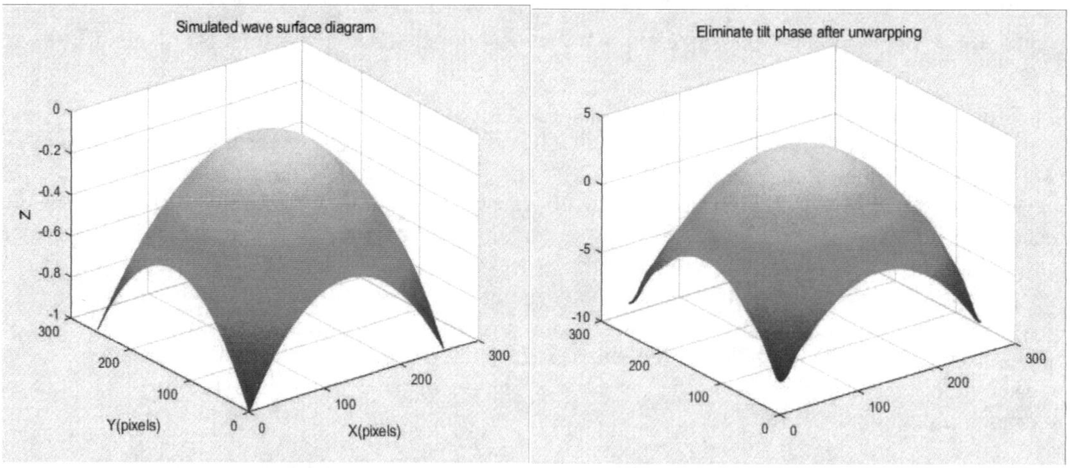

(e) Simulated wave surface diagram (f) Eliminate tilt after unwarpping

Figure 3. Simulation of microstructure wave surface flow

The phase diagram is obtained by Fourier transform, and the zero-frequency center and the fundamental frequency center are found to determine the distance between them. By frequency domain filtering, a rectangular filter is generated, and then the sidelobe is removed and moved to the center.

Finally, inverse Fourier transform is used to calculate the wrapped phase, and unwrapping phase, as shown in Fig.3(c)~(d). Finally, the de skew phase after unwrapping is obtained. As shown in Fig.3(e) ~(f), compared with the simulated original wave surface, they are very similar. It can be seen that the processing effect meets the requirements.

5. Conclusion

Through the research, it is proved that the laser interference system is feasible: the Fourier transform spectrum can only solve the square area, so the fringe pattern can be removed, reduced to a square area, and then solved, but obviously this is a compromise, because part of the surface structure is missing. Or, we can use the image interpolation method to complete the black part around, and get the restored surface shape after solving. Then remove the surrounding part to get the original surface shape. The simulation result of the system is the same as that of the existing high-precision detector. By combining the system with the robot, we can detect the movement of large parts. Of course, there is still room for further improvement in this design, that is, using the high-precision spatial carrier phase-shifting method shown in reference [10], which will be the next work of the author.

Acknowledgement
The authors gratefully acknowledge the support of the National Natural Science Foundation of China (no. 51675315).

References

[1] Hao W 2018 Technology, economy and management 2 109–110
[2] Xiaoyan D, Dongmei R 2012 Measurement technology 32 (6) 1–5
[3] Xuewei Z, Jiawei Z, Dongli W 2021 Research and development of mechanical equipment 127–128
[4] Rongfu Q, Fang B 2016 Machine tools and hydraulics 44 (16) 157–160
[5] Aili Z, Youjun M 2013 Shandong Ceramics 36 (3) 31–32
[6] Yangfeng S, Yunxia F, Lihua L, Yuan L, Xiaoyu C, Dongsheng L 2016 Micro nano electronic technology 53 (10) 679–684
[7] Tingrui L, Haitao H 2005 Optical instruments 25 (5) 10–14
[8] Tingrui L, Yongrui Z,Guangqing C 2004 Optical instruments 26 (5) 26–30
[9] Hongpeng R 2019 Information technology 11–12
[10]Tingrui L, Ailing G, Yongrui Z, et al.2006 SPIE - The International Society for Optical Engineering 6280 1–6.

Incentive and Game of Information Sharing Based on Blockchain Technology

Yuhong Xin, Cheng Ran* and Dehui Liu

School of Computer Science, Guangdong Polytechnic Normal University, Guangzhou, Guangdong Province, China

*Corresponding author's e-mail: ranchengx@foxmail.com

Abstract. With the development of the information age, the efficiency of information resource sharing has not been fully released, and the willingness of participants to share information needs to be improved. It is of great significance to promote the willingness of participants to share information and increase the value of information utilization. Aiming at the problem of low information sharing rate of participants, this paper constructs an evolutionary game model between information sharing and non-sharing participants under the blockchain technology platform and introduces the critical value corresponding to the initial proportion of participants under the condition of a stable strategy as a trigger Conditions, dynamic adjustment of incentive coefficients to maintain a stable strategy of information sharing, simulation with MATLAB. Research shows that the use of blockchain smart contract technology to dynamically adjust the evolutionary stability strategy is shown as: When the initial proportion of participants is lower than the critical value, the incentive parameters are transformed into shared incentive benefits to motivate the participants to actively participate in information sharing; when it is higher than the critical value, the incentive parameters are transformed into sharing costs and dynamically adjusted to reach the initial participants. The proportion is maximized, which provides a strategy for participants to choose information-sharing strategies.

1. Introduction

In the era of rapid development of information technology, the value of the development and utilization of information resources has attracted more and more people's attention, and they have realized the value of information resources. Information resource sharing has evolved into modern information life where people obtain more and more information. The important use of the way [1]. In 2020, my country's "China Blockchain Financial Application and Development Research Report" pointed out that blockchain technology is evolving and mature and has a certain degree of adaptation to the real world. Although there are little researches on information sharing combined with blockchain technology at present, the research found that the features of blockchain technology, such as tamper-proof, decentralized, and distributed, provide information on the use-value, economic utility, and methods of information. New choices and ideas have laid the cornerstone for the reconstruction of the trust model in the era of big information [5]. Information sharing can increase the value of information utilization and reduce the cost of information utilization. The development of modern information technology and the establishment of new information exchange systems provide certain technical support for information sharing. However, under the development of the modern information economy system, the lack of economic benefits, asymmetry of information, high cost of information flow, lack of information security, low transmission efficiency, and insufficient incentives have caused the phenomenon of information "stay" [2].

As an analysis method of modern mathematics, the evolutionary game has been widely used in economics, management, and other fields. Researches on information sharing based on evolutionary game theory include: Wang et al. studied the problem that the disclosure of supply chain demand information reduces the willingness of retailers to share information, and have built a predictive information-sharing model[7-9]. Because of the problems of asymmetry in information sharing, the high cost of information sharing, and lack of economic benefits, in one case, the information sharing system will introduce a third party for supervision. Li et al. established a game mechanism of information sharing among mineral resource enterprises under government supervision [11]. Wei et al. established an evolutionary model of data sharing between enterprises with government participation, the government supervised and encouraged them, and analyzed that the willingness of enterprises to share will change due to data, mutual trust, and liquidated damages [12-13]. Lee et al. found that the integrated digital twin and blockchain framework make all data transactions traceable, and realize the sharing of accountable project-related information among stakeholders [15]. Kumar et al. discussed how blockchain technology guarantees the security of data transmitted by the Internet of Things [16]. At present, there is not much research on the evolutionary game of information sharing based on blockchain technology at home and abroad. Therefore, this paper constructs a game model of information sharing under the blockchain smart contract mechanism, introduces the critical value corresponding to the initial proportion of participants under the condition of a stable strategy as the trigger condition, analyzed the evolution mechanism of the information sharing of the participants, and conducts Numerical simulation analysis.

2. Basic model

2.1. Basic assumptions and parameters of the evolutionary game model

Participants with limited rationality under shared information are more willing to share data through a platform with trust and security. Therefore, establishing an incentive mechanism in the blockchain platform can realize the effective sharing of information.

Hypothesis 1: Participating subjects can choose to share or not share information based on the benefits obtained in participating in information sharing, so the selection set of participating subjects is {information sharing, information not sharing};

Hypothesis 2: For information sharing, the shared benefits obtained by the two parties in the game can be divided into direct benefits and value-added benefits. Direct benefit means that both parties obtain the information shared by the other party so that they can profit from it and record it as direct benefit D. Synergy benefit is the benefit created by the mutual fusion of the participants when sharing information, which is recorded as value-added benefit T.

Hypothesis 3: Shared cost C. In the information-sharing platform, information sharing requires investment in fixed information sharing costs.

Hypothesis 4: Excitation coefficient R. Under the blockchain smart contract incentive mechanism, the incentive coefficient is expressed as reward income or participation cost.

Hypothesis 5: During the game, the probability that participant A chooses to "share" is x ($0<x<1$), on the contrary, the probability of choosing "not to share" is 1-x; the other party to the game, the participant B chooses to "share" The probability is y ($0<y<1$), otherwise, the probability of choosing "not sharing" is 1-y;

Hypothesis 6: When there is no sharing among participating users, the benefits obtained by both are 0.

According to the above assumptions, the income matrix of participating entities' information sharing can be obtained, as shown in Table 1:

Table 1. The profit matrix of the players in the game.

	Participant B	
Participant A	Sharing(y)	Non-sharing(1-y)
Sharing(x)	D+T-C-R，D+T-C-R	-C-R，D
Non-sharing (1-x)	D，-C-R	0，0

The following is an analysis of the income matrix:

Situation 1: In the case where participants A and B do not share information, there is no benefit or cost, that is, the benefit is zero.

Situation 2: Participants A and B share information, not only can obtain the benefits of mutual information sharing, but also collaborate to promote mutual reference and integration of information, and can create new value that is obtained Value-added benefits, enabling participants to obtain more benefits. But information sharing requires a certain investment cost. In the blockchain platform, to share information between users, an incentive mechanism can be created to promote both parties in the game. The parameters in the incentive mechanism include positive promotion and negative promotion. When the excitation parameter C is getting closer and closer to the threshold, the positive promotion will eventually reach the saturation state of the gain. When the excitation parameter exceeds the threshold, the negative promotion will cause a negative gain. Therefore, when mutual information is shared, the benefits of the participants are all D+T-C-R.

Situation 3: In the blockchain information sharing platform where one party chooses to participate, the other party chooses not to participate in information sharing. However, the choice of the sharing party does not benefit from information sharing. On the contrary, the sharing party bears the investment cost by itself. However, because the other party does not share the information, it cannot make additional profits and must compensate for certain incentive costs. Usually in this case, after a game with the other party, the sharing party with bounded rationality will adopt a non-sharing strategy to lose -C-R. For the non-sharing party, the benefit derived from the lack of sharing only comes from the direct benefit D shared by the game party.

In the strategy of encouraging information sharing in the blockchain platform, participants with limited rationality do not share two behavioral strategies. It is difficult to make the best decision in one decision, but the participants will pass long-term imitations, Learn, constantly adjust their strategy and make dynamic adjustments cyclically, and finally reach a state of balance and stability.

2.2. Copy dynamic equations and equilibrium solutions

According to the income matrix and the game process shown in the table, a copy dynamic equation can be obtained, which describes the dynamic evolution process of participant's information sharing, and the analysis shows that the participant A chooses "sharing" or "not sharing" strategy performance The expected return and the average expected return.

Participant A's shared strategy has the expected benefits:

$$U_{\mathrm{i1}}^{1} = y(D+T-C-R)+(1-y)(-C-R) = yD + \mathrm{yT} - C - R \qquad (1)$$

The expected benefits of Participant A's non-sharing strategy are: $\boldsymbol{U_{i2}^{1} = yD}$

The average expected return of participant A is: $\overline{\boldsymbol{U_{i}^{1}}}$=xyT+yD-xD-xR

The expected benefits of Participant B's sharing strategy are:$\boldsymbol{U_{\mathrm{j}1}^{1}}$=x(D+T-C-R)+(1-x)(-C-R)

Participant B does not share the expected benefits of the strategy: $\boldsymbol{U_{\mathrm{j}2}^{1}}$=xD

The average expected return of participant B is:$\overline{\boldsymbol{U_{j2}^{1}}}$=xyT-yC-yD-yR

Synthesize the dynamic equation of the information-sharing behavior of participants A and B in the above structure:

$$\boldsymbol{F_1(x,y) = \frac{dx}{dt} = x(U_{\mathrm{i1}}^{1} - \overline{U_{i}^{1}})}\text{=x(1-x)(yT-C-R)} \qquad (2)$$

$$G_1(x,y)=\frac{dy}{dt}=y(U_{j1}^1-\overline{U_{j2}^1})=y(1-y)(xT-C-R) \tag{3}$$

Nash equilibrium solution, based on the copy dynamic equation under information sharing, find the equilibrium point of the evolutionary game:

$$\begin{cases}F_1(x,y)=0\\G_1(x,y)=0\end{cases} \tag{4}$$

Find the balance point $E_1(0,0)$、$E_2(0,1)$、$E_3(1,0)$、$E_4(1,1)$、$E_5(x^*,y^*)=(\frac{C+R}{T},\frac{C+R}{T})$, using the method of Friedman [3], through the analysis of the Jacobian local stability, the corresponding Jacobian matrix is as follows:

$$J_1=\begin{bmatrix}\frac{\partial F_1}{\partial x} & \frac{\partial F_1}{\partial y}\\ \frac{\partial G_1}{\partial x} & \frac{\partial G_1}{\partial y}\end{bmatrix}=\begin{bmatrix}(1-2x)(yT-C-R) & x(1-x)T\\ y(1-y)T & (1-2y)(xT-C-R)\end{bmatrix} \tag{5}$$

The determinant of J_1: $\det J_1=(1-2x)(1-2y)(yT-C-R)(xT-C-R)-x(1-x)Ty(1-y)T$

The rank of J_1: $trJ_1=(1-2x)(yT-C-R)+(1-2y)(xT-C-R)$

Bring the five equilibrium points into the Jacobian and the value of its trace. When det J_1 >0, tr J_1<0, the conditions of the evolutionary stability strategy are met, and the local stability of the participants' information sharing can be obtained by bringing the points into the equation, such as Table 2 shows:

When the condition $T>C+R>0$ is met, the stability analysis of the system is shown in Table 2.

Table 2. The profit matrix of the players in the game.

	T > C + R > 0			$C+R>T>0$		
Equilibrium point	$\det J_1$	trJ_1	Locally stable	$\det J_1$	trJ_1	Locally stable
$E_1(0,0)$	+	-	stable	+	-	stable
$E_2(0,1)$	+	+	Unstable	-	+	Unstable
$E_3(1,0)$	+	+	Unstable	-	+	Unstable
$E_4(1,1)$	+	-	stable	+	+	Unstable
$E_5(x^*,y^*)$	-	0	Saddle	-	0	Saddle

2.3. Evolutionary game analysis of participants

Participants A and B with bounded rationality, in the blockchain platform incentive model, assume that the probability of selecting information sharing and non-sharing is symmetrical, and the dynamic equation for copying is simplified as $\boldsymbol{F_1(x,y)=G_1(x,y)=F(\omega)}$, $\boldsymbol{F(\omega)=\omega(1-\omega)(\omega T-C-R)}$.Assumption $\boldsymbol{F(\omega)=0}$, It can be seen from the balance point, When the condition $\boldsymbol{\omega_1=0}$、$\boldsymbol{\omega_2=1}$、$\boldsymbol{\omega_3=\frac{C+R}{T}}$ is met. According to the evolutionary stability strategy (ESS) condition, it can be known that the stability condition in the differential equation must be satisfied, must satisfy $\boldsymbol{F'(\omega)}<0$.

Case 1-1: If T>R>0 and T>C+R, it can be seen that and are both the ESS of the participant's information sharing evolutionary game, not the ESS of the participant's information-sharing evolutionary game. The proof is as follows:

$$F(\omega)=\omega(1-\omega)(\omega T-C-R) \tag{6}$$

$$F'(\omega)=(1-2\omega)(\omega T-C-R)+T\omega(1-\omega) \tag{7}$$

Analyze the three stable states in the equilibrium point respectively, when A and B are:

$$F^{'}(0) =-(C+R) < 0 \tag{8}$$

$$F^{'}(1) =-(T-C-R) < 0 \tag{9}$$

$$F^{'}(\tfrac{C+R}{T}) = (C+R)(1-\tfrac{C+R}{T}) > 0 \tag{10}$$

From (8) and (9), the slope of the tangent line is less than zero. Therefore, both are the ESSs of the evolutionary game of information sharing among participants. From (10), the slope of the tangent line at $\omega_3=\frac{C+R}{T}$ is greater than 0, so it is not the ESS of the evolutionary game of information sharing between participants.

From this further analysis, the evolution path is shown in Figure 1. The broken line BEC in the figure divides the area of the figure into two parts, the size of which depends on the specific position of the saddle point E. When $\frac{C+R}{T}>\omega>0$, the initial state of the game between the two parties is below the broken line BEC, then ESS tends to (0,0), that is, not participating in sharing is the only strategy to finally achieve the equilibrium of the evolutionary game; when $1> \omega>\frac{C+R}{T}$, the initial state of the game between the two parties is above the broken line BEC, and the ESS tends to (1,1), that is, participation in sharing can achieve evolutionary game equilibrium. This shows that $\frac{C+R}{T}$ is the critical value between the strategic choice of sharing and non-sharing of information. If the initial participant's participation ratio is greater than the critical value, the rest of the participants will not participate in the information sharing as the number of games evolves. Will participate in information sharing. On the contrary, if the initial participant's participation ratio is less than its critical value, the remaining information sharers of the participants will eventually not participate in information sharing as the number of games increases.

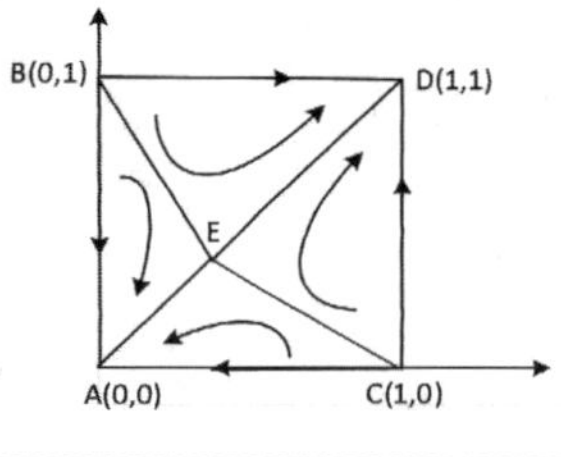

Figure 1

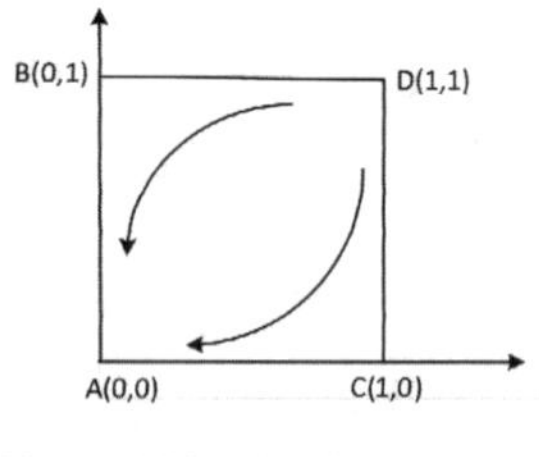

Figure 2

Case 1-2: If T>R>0 and, only is the ESS for the information sharing of the participants. The proof is as follows:

$$F^{'}(0) =-(C+R) < 0 \tag{11}$$

$$F^{'}(1) =-(T-C-R) \geq 0 \tag{12}$$

$$F^{'}(\tfrac{C+R}{T}) = (C+R)(1-\tfrac{C+R}{T}) \leq 0 \tag{13}$$

When verification $\omega 3= (C+R)/T$, since and T+R>T, and$\omega 3 >1$, formula (13) does not hold. From this further analysis, $\omega 1 =0$ is the only ESS under this condition, which means that regardless of the initial proportion of information sharing or non-sharing occupied by participants, they will not participate in information sharing after repeated games, as shown in Figure 2. In this case, the party will choose not to share information, and analyze the reasons: the cost of investment is higher than the benefits of information sharing, resulting in a decrease in the willingness of the participants to share information. Starting from the bounded rationality of the participants' pursuit of maximizing their interests, they will eventually tend to choose the information non-sharing strategy (0,0), that is (no sharing, no sharing).

Case 1-3: If R>0 and ESS is the only ESS under these conditions, similar to the second), it means that due to the high cost of sharing information, the participants will eventually tend not to participate in the selection strategy.

Case 1-4: If R<0 and, it can be obtained as the only ESS, the proof is as follows:

$$F'(0) =-(C+R) > 0 \tag{14}$$

$$F'(1) =-(T-C-R) < 0 \tag{15}$$

$$F'(\frac{C+R}{T}) = (C+R)(1-\frac{C+R}{T}) \leq 0 \tag{16}$$

But since $0 < \omega < 1$ and $C + R \leq 0$, we know that $\omega_3=\frac{C+R}{T}>1$, therefore, (16) does not hold under this condition. From (15), we can get that the slope of the tangent line at $\omega_2 = 0$ is less than 0. Therefore, $\omega_2 = 1$ is the only ESS in the evolutionary game of information sharing among participants, and other ω values are not valid.

Further analysis of this, only to meet the conditions of the evolutionary game stable strategy, which means that the R cost value has become a positive benefit. This cost value becomes the shared incentive cost, which incentivizes the participants to share information, that is, generates reward benefits. Under the evolutionary game, the final ESS evolves into an active information sharing strategy for all participants, as shown in Figure 3. When the benefit of information sharing exceeds the cost of input, the greater the negative effect of non-information sharing, the greater the positive benefit of revenue. At this time, the participants will tend to the information-sharing strategy (1,1), that is, (share, share).

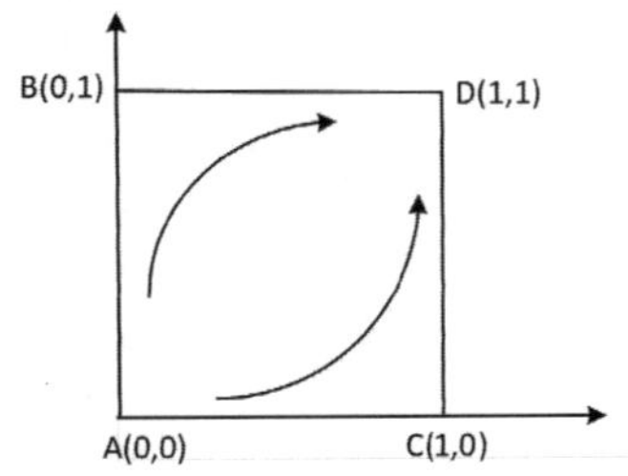

Figure 3. Phase diagram of the evolution path of information sharing

Therefore, there are two evolutionary stable strategies, namely participation and non-participation in information sharing, and the decision-making conditions of the participants need to be analyzed in detail according to the situation. To further understand that the information sharing of the participants is potentially dependent on the ESS in the two participation and non-participation strategies, the evolution trend of the corresponding conditions is further explained, which can clearly analyze the intuitiveness of the selection trend of the participants.

The stable change trend of the evolutionary game under the information sharing incentive mechanism in the blockchain platform is affected by the initial information sharing and the non-sharing ratio of the participants and the incentive parameters. Based on the evolutionary game theory, it is concluded that there are three evolutionary stable strategies and the incentive mechanism is modeled and analyzed. This paper builds a blockchain smart contract incentive system platform to promote the willingness of participants to share information and increase the long-term stability of information utilization value. The development provides strategy.

3. Information sharing model based on the evolutionary game of blockchain smart contracts

By setting different values of C, R, D, and T, it is verified that the ESS in the evolutionary game of the information sharing participants and the role of the incentive mechanism in the case of different setting incentive coefficients. The experiment is divided into three scenarios: For scenario 1,we analyzed the two evolution strategies of the participants in scenario 1-1 and analyzed in detail the changing laws of their evolution images; scenario 2 targets the incentive system and conducts scenarios 1-2 and 1-3 Categorization analysis, select the values of the parameters that meet the conditions, and verify the influence of the incentive mechanism on the evolutionary game; Case 3 also analyzes Cases 1-4.

Scenario 1: Based on the blockchain evolutionary game incentive mechanism, adjust the value of key factors according to the situation. The parameters are as follows: sharing cost C=9, direct benefit D=20, and value-added benefit T=15. Under the conditions of Case 1-1, the excitation coefficient R satisfies 0<R<9. When R takes 2, 4, and 5, the critical values corresponding to $\boldsymbol{\omega} = \frac{C+R}{T}$ are $\boldsymbol{\omega_1} \approx \mathbf{0.73}$、$\boldsymbol{\omega_2} \approx 0.87$、 and $\omega_3 \approx 0.95$ respectively. The probability that the participant chooses to participate in information sharing depends on the critical value of the initial participant. When the initial proportion of participation is less than the critical value $\boldsymbol{\omega}$, the system eventually tends not to participate in the stabilization strategy. Conversely, when the initial proportion of participants is greater than the critical value $\boldsymbol{\omega}$, the system eventually tends to participate in a stable strategy. Use Matlab to simulate the game evolution process of the two parties involved. When C=4, Figure 4(a) and (b) describe the evolution results of $\boldsymbol{\omega} = 0.6$ and $\boldsymbol{\omega} = 0.9$, respectively. Figure 4(a) shows that the initial proportion of participants is less than the critical value, the final stabilization strategy of the participating subjects tends to not participate in information sharing, and Figure 4(b) shows that the initial proportion of participation is greater than the critical value, and the final stabilizing strategy is that all participating subjects choose to participate in information sharing.

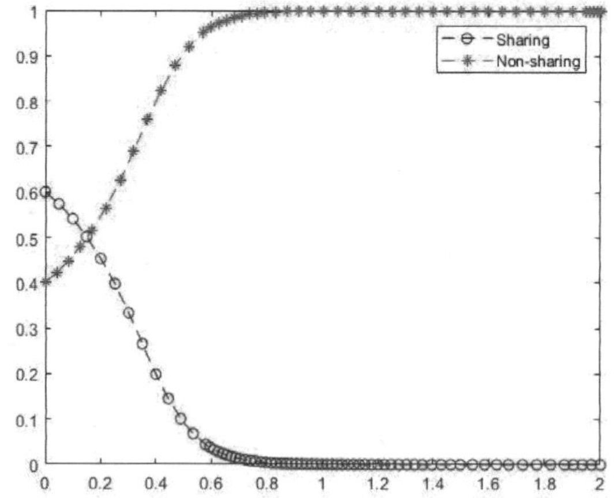

(a)

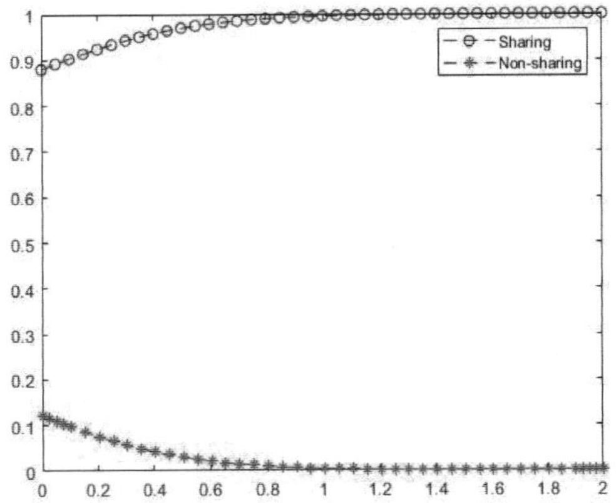

(b)

Figure 4. Evolution process of evolutionary game strategy

Figure 5(a) comprehensively describes the overall trend of the evolutionary stability strategy under the conditions of Case 1-1. The stability points of the participant's evolutionary game are $E_1(0,0)$ and $E_4(1,1)$. The final evolution result depends on the initial probability of the participant's choice to participate in information sharing. Evolutionary stability strategy refers to information sharing or non-sharing.

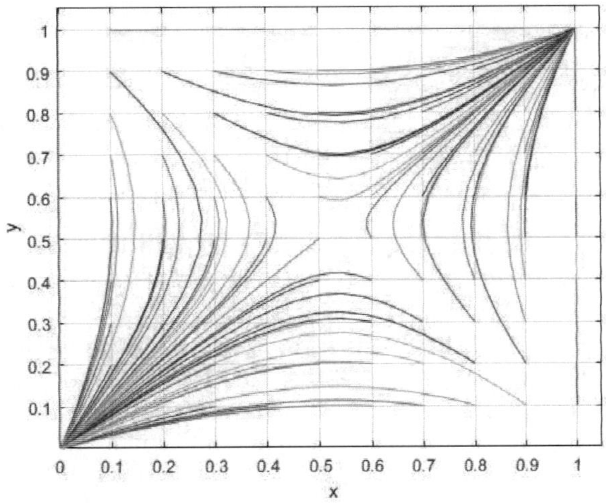

(a)

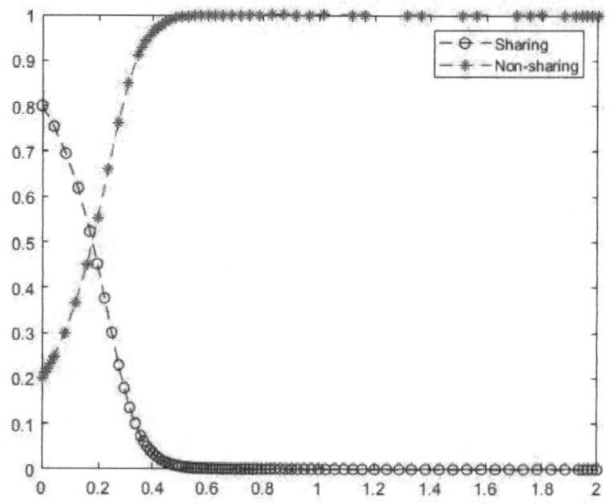

(b)

Figure 5. Evolution process of evolutionary game strategy

Scenario 2: Based on the blockchain evolutionary game incentive mechanism, adjust the key factors and conditions according to the situation. The parameters are as follows: sharing cost C=9, direct benefit D=20, and value-added benefit T=15. Under the conditions of Case 1-2, when the incentive coefficient R satisfies $T > R > 0$ and $T \leq C + R$, regardless of the value of the incentive coefficient R and the sharing cost C under the conditions, the final evolutionary stability strategy of the participants in the information sharing is not to participate in the information Sharing, as shown in Figure 5(b). In the same way, in the case of cases 1-3, the participants will eventually stabilize at the place where they do not participate in information sharing, and the evolution is shown in Figure 6(a). In the above two situations, the entities participating in information sharing will not share information regardless of the initial proportion of participation in the final stabilization strategy.

Scenario 3: Based on the blockchain evolutionary game incentive mechanism, adjust the key factors and conditions according to the situation. The parameters are as follows: direct income D=20, value-added income T=15. Under the conditions of Case 1-4, when the incentive coefficient R satisfies $R < 0$ and $C + R \leq 0$, regardless of the value of the incentive coefficient R and the sharing cost C under the conditions and the initial proportion of participants, the final evolutionary stability strategy is the participation information Sharing, as shown in Figure 6(b), even when the proportion of participating users is $\omega = 0.1$, the part that does not participate in the information subject will eventually stably participate in the information-sharing place after many dynamic game adjustments. Analyzing the situation, when the incentive coefficient R becomes a negative benefit, the participant will be rewarded. The limited participating users on the platform will choose to participate in information sharing to maximize their interests.

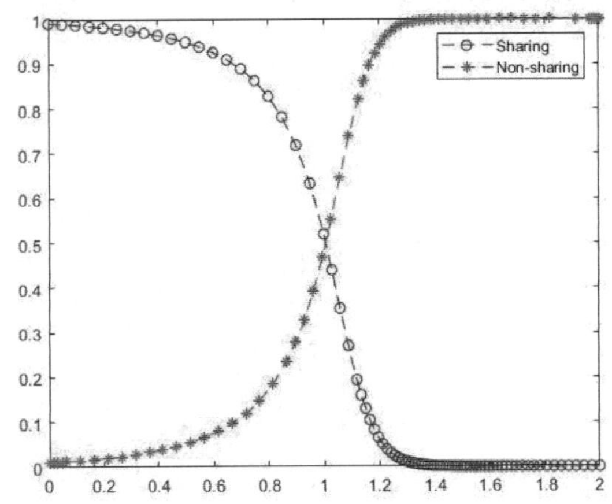

(a)

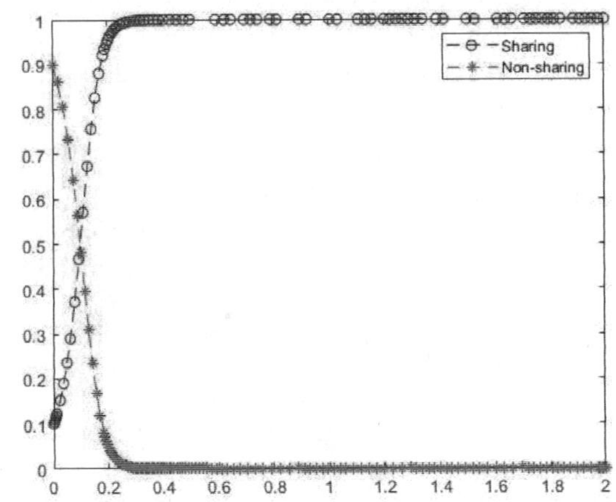

(b)

Figure 6. Evolution process of evolutionary game strategy

The design of the incentive mechanism of the blockchain smart contract will affect the choices of the participants, and the selection strategy of the shared participants is affected by the incentive coefficient and the initial state. Through the automatic generation and execution of contracts triggered by external events, its distributed ledger technology can achieve information sharing, information transparency and credibility and authenticity [4][6]. Use blockchain smart contract technology to dynamically adjust the evolutionary stability strategy. When the initial proportion of information sharing participants is lower than the critical value, refer to cases 1-4. Under this incentive mechanism, incentives are given according to the proportion, that is, the incentive parameters are transformed into Positive benefits are rewards to motivate participants to actively participate in information sharing. According to the maximization of their benefits, non-participants with bounded rationality will tend to participate in the information sharing strategy, thereby promoting the evolution of the non-participating direction to participate in the information sharing trend. When the participant's participation ratio exceeds its corresponding threshold, referring to the incentive mechanism of Scenario 1-1, information sharing participants still tend to share- information strategies. At this time, there is no need for incentives, that is, the conversion of incentive coefficients into negative benefits is ensuring Satisfy the shared stability strategy to dynamically increase the cost of participation to maximize the proportion of initial participants. When the participation strategy in the blockchain smart contract incentive system platform cannot be maintained, the incentive mechanism in Scenario 3-4 is cycled, and so on, until the participating information-sharing party reaches the maximum participation in the blockchain smart contract incentive system platform The ideal state.

4. Conclusions and prospects

This paper addresses the problems of low information sharing rate and insufficient incentives of participants in information sharing, and constructs an evolutionary game model between information sharing and non-sharing participants based on the blockchain smart contract incentive mechanism. Get the following conclusions:

1) The selection strategy of the shared participant is affected by the incentive coefficient and the proportion of the initial state of the participant;
2) Under the conditions of meeting the stable strategy of participating in information sharing, the threshold corresponding to the initial proportion of participants is introduced as the trigger condition, and the automatic generation and execution of the contract are triggered by the blockchain smart contract technology to achieve a state of dynamic adjustment of the incentive coefficient.
3) When $\boldsymbol{\omega} \in (\mathbf{0}, \frac{C+R}{T})$, the overall stabilization strategy for participating in information sharing tends to be non-sharing of information. It encourages participants to share information, and the incentive coefficient is transformed into shared incentive benefits under trigger conditions to encourage participants to actively participate in information sharing; When $\boldsymbol{\omega} \in (\frac{C+R}{T}, \mathbf{1})$, The overall stability strategy of participating in information sharing tends to share information. To maximize the proportion of initial participants, the incentive parameters are transformed into sharing costs, that is, to dynamically adjust the cost while ensuring that the sharing stability strategy is met to maximize the proportion of initial participants.

Therefore, different from the traditional information sharing method, this paper uses the information sharing in the blockchain platform to adjust the parameters of the incentive coefficient according to the conditions under the incentive mechanism of the blockchain smart contract, so that the information sharing participants can achieve Based on the state of maximum participation in the evolutionary game mechanism, it promotes more participants to tend to information-sharing strategies and gain benefits, maximize the interests of the entire participants, and help increase the information sharing rate of participants and encourage more participants to conduct information Sharing long-term and stable development.

Acknowledgement

Projects are supported by the National Natural Science Foundation of China (71571072), supported by the natural science foundation of Guangdong Province (2016a030313725)

References

[1] Chen Baoguo, Li Xiangrong. Decision analysis of information resource sharing based on utility scalability [J]. China Soft Science, 2004, (5): 74-77.
[2] Li L. Cournot oligopoly with information sharing[J]. The RAND Journal of Economics, 1985: 521-536.
[3] Friedman D. Evolutionary games in economics[J]. Econometrica: Journal of the Econometric Society, 1991: 637-666.
[4] Zhang Lu. Research on blockchain-driven supply chain financial innovation from the perspective of game [J]. Economic Issues, 2019, 476(4): 48-54.
[5] Yang Dong, Xu Xinyu. Blockchain and court work innovation-building a judicial credit system for information sharing[J]. Application of Law, 2020, (No.442): 12-22.
[6] Liang X, Chen W, Li J, et al. Incentive Mechanism of Medical Data Sharing Based on Information Entropy in Blockchain Environment[C]//Journal of Physics: Conference Series. IOP Publishing,

2019, 1302(2): 022056.

[7] Wang Wenlong, Wang Chengjun, Hu Haihua. Research on Retailer Demand Forecast Information Sharing Considering Information Leakage[J]. Operations Research and Management, 2020, 29(7): 80-88.

[8] Zhou Jianheng, Ran Yun. Supply chain information sharing strategy based on strategic competition game[J]. Chinese Management Science, 2019, (v.27;No.176): 88-102.

[9] Ye Y, Ju Z, Li Y. An Evolutionary Game Analysis of Information Sharing Behavior for Cluster Supply Chain[C]//Fourth International Conference on Economic and Business Management (FEBM 2019). Atlantis Press, 2019.

[10] Li Zhigang, Zhou Wei. Research on the information sharing game mechanism of mineral resource enterprises under government supervision [J]. China Management Science, 2015, (v.23): 199-203.

[11] Wei Yihua, Chen Xulin, Zou Xiaofeng. Data Sharing, Corporate Strategy and Government Supervision Incentives——Based on Evolutionary Game Analysis [J]. Science of Finance and Economics, 2020, (No.385): 107-120.

[12] Yang Yangyang, Xie Xuemei. The evolutionary game study of government information resource sharing supervision from the collaborative perspective[J]. Information and Documentation Services, 2020, (v.41;No.236): 72-78.

[13] Lee D, Lee S H, Masoud N, et al. Integrated digital twin and blockchain framework to support accountable information sharing in construction projects[J]. Automation in construction, 2021, 127: 103688.

[14] Ramamoorthi S, Kumar B M, Sithik M M, et al. Enhanced security in IOT environment using Blockchain: A survey[J]. Materials Today: Proceedings, 2021.

Combating Covid19 Using a Mobile App

A H Wijaya[1], T M W Andhini[1], Y P Rahmadi[1] and H L H S Warnars[2,*]

[1]Computer Science Department, School of Computer Science, Bina Nusantara University, Jakarta, Indonesia
[2]Computer Science Department, BINUS Graduate Program, Bina Nusantara University, Jakarta, Indonesia

[*]Corresponding author's e-mail: spits.hendric@binus.ac.id

Abstract. Through this pandemic, the world has experienced two major crises, health crisis, and economic crisis. It would be dangerous for us to continue with our "normal" daily lives. We have been forced to stay at home, and some have been dismissed from their jobs; even street vendors are forced to close. The primary purpose of this paper is to represent our solution idea on the troubles regarding COVID-19 to help and support people by providing them all the information and services needed about the COVID-19, such as the high-risk COVID-19 area, statistical growth, safety tips, online swab test booking, etc. through a mobile-based application. In this paper, we designed a model with the use case and class diagrams and implemented using Android Studio.

1. Introduction

The COVID-19 pandemic in Indonesia has been going on since 2019 has captured public attention and affected many aspects of human life. The human life aspects include how to socialize, how to travel, how to communicate, how to be cultured, how to dress, how to maintain cleanliness, and many other human activities[1]. Others have changed and adapted since being affected by the conditions of this covid-19 pandemic. This pandemic has affected the way we live our daily lives. Nowadays, we should wear masks, wash our hands more often than usual, keep our distance, or even minimize activities outside the home[2]. It is easy to be afraid of this situation because we may worry about whether COVID-19 patients have also visited the places we are going to and the places we are visiting, whether sick people have visited those near us. COVID-19 patient, or is a COVID-19 quarantine place[3]. We want to use technology to help people, especially in Indonesia, deal with this challenging situation.

2. Previous related research

We have conducted several research methods to obtain the data and information needed to implement this COVID-19 application. First, we conducted survey research using Google Forms. Second, we conduct research and reviews based on mobile applications available on the Google Play Store and App Store. With this survey research method, we ask the public to answer a total of 10 questions to get all the information and data needed to make an application that is suitable and meets all the needs that the community needs. We send survey forms to people who work outside the home because we believe this app will be helpful while keeping other potential users in mind. The survey contains questions to explore respondents' opinions about the COVID-19 applications offered and comments and expectations on our application, especially on the features we plan to make.

In addition, we also conduct research and reviews on relevant applications. After we tend to browse and search for all the COVID-19 applications that are offered based in Indonesia that is available on every Google play store and App Store, we decided to take a look at the application with the name "PeduliLindung," which is similar to our application idea where the application itself is in the form of a mobile application[4]. This app brings features like map view, statistics growth, doctor consultation, etc. This application was designed by the Ministry of Information Communications and the Indonesian Ministry of SOEs. This application was developed to support people who are forced to leave their homes for activities to get knowledge, news, and info about the COVID-19 zone in the nearest location[5].

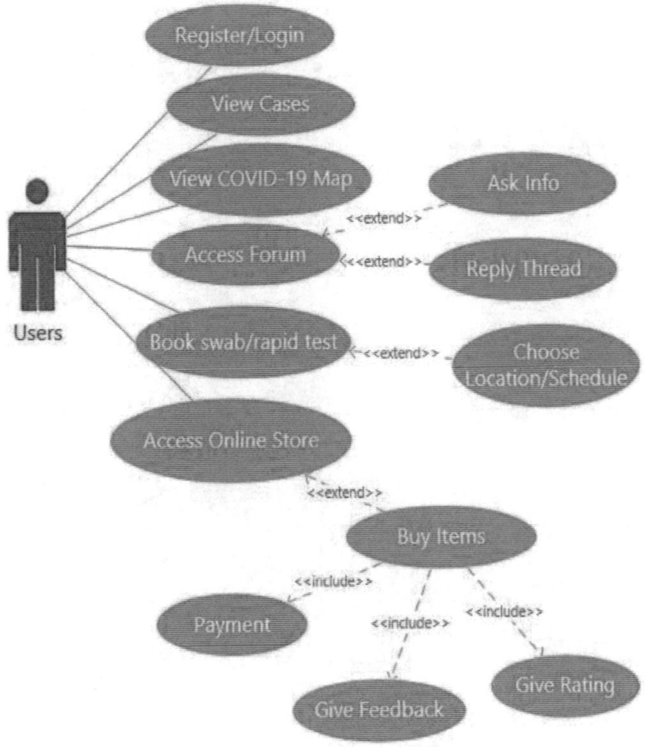

Figure 1. Use case diagram

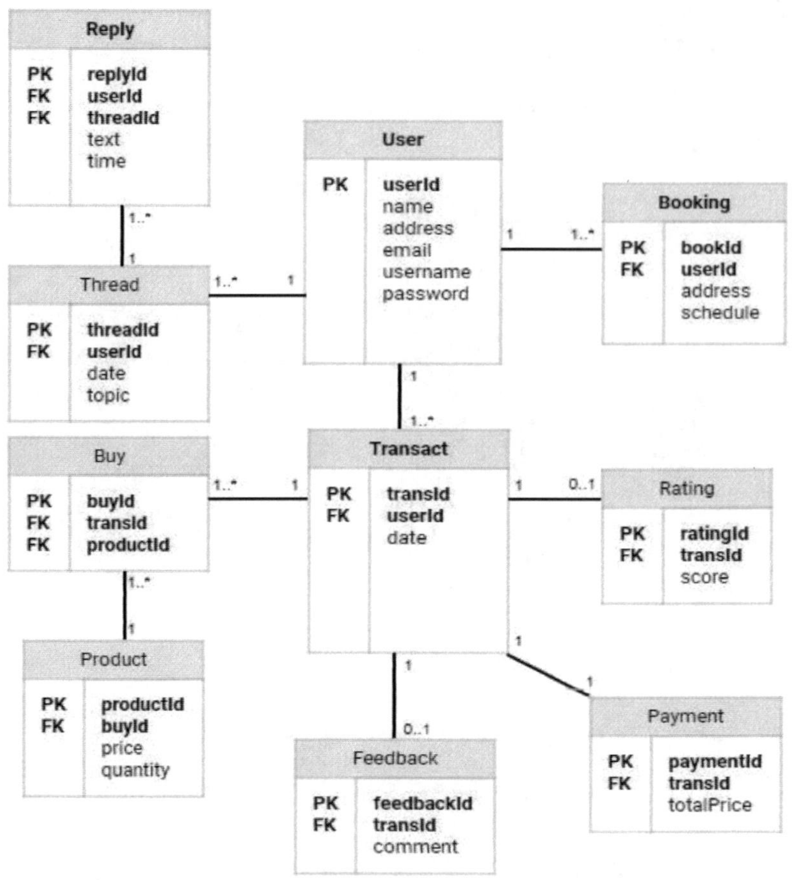

Figure 2. Class diagram

The use of mobile devices such as smartphones, tablets, etc., has shown an enormous increase from year to year. The main reason is its advanced technology, where various applications can be developed on the platform[6]. Apart from that, other factors like portability and speed also play a significant role in increasing the usage of mobile devices. Today, people spend more time on their mobile devices than on their desktops, so developing apps on mobile platforms is a must in the current trend[7].

There are several mobile applications created to assist in providing information related to COVID-19. This is an example of applications such as Care Protect, Health Lynked COVID COVID-19 tracker, HSE COVID-19, SM Covid-19 Support, and much more[8]. From those applications, we tend to learn that the most effective applications to help deal with COVID-19 are free, available to all age groups, available to the public (including those affected by COVID-19). Moreover, those applications have notification features, provide up-to-date COVID-19 data, and correct in real-time, with information about COVID-19 (when to self-quarantine or go to the hospital, ways to avoid contracting the virus, etc.), and which has real-time forums and/or chats[9]. We will try and apply these characteristics to our application.

3. Implementation result

Figure 1 is a use case diagram for the proposed idea, which consists of several use case activities such as register/login, view case, view covid-19 Map, forum access, book swap/rapid test, and online store access. To register/login is a process where new users must register the data stored in the user class, as shown in Figure 2. After having a username and password, the user can enter the system, and the User

Journal of Physics: Conference Series **2173** (2022) 012035 doi:10.1088/1742-6596/2173/1/012035

interface (UI) can be seen in Figure 2. 3, where the username and password will be checked on the user class to check their validity. Meanwhile, the case display will refer to the covid19 data from the Ministry of Health of the Republic of Indonesia, and the data will be updated using the Application Program Interface (API). The Covid19 Map display contains information on each zone in Indonesia's status, marked with dots that have three colors, namely red as danger, yellow as alert, and green as safe area, and Figure 5 shows the UI for the Covid19 Map. As shown in Figure 6, forum access will be used by people who want to discuss anything about covid19, for example, the safest place to do something that has social distancing or other things that users want to know. In addition, the use case book swap/rapid test, as shown in Figure 7, shows that the user can perform a book swap or rapid test by entering the location or schedule for the swab or rapid test. The following use case is accessing an online store, as shown in Figure 8, to show how users can buy COVID-19 items such as masks, hand sanitizers, and other things that can help us fight COVID-19.

Figure 2 shows a database model that has ten classes where each class is represented as a table in the database, namely User, Thread, Reply, Book, Transaction, Rating, Feedback, Buy, Payment, and Product which is designed in the form of a class diagram where we can represent attributes of each data table and their relationship to other tables. In the class diagram above, the relationship between the tables in the database is shown. This will help us develop the database on how data from a particular table can be used as a reference for other data in different tables, which will help us manage database size and prevent excessive storage space. The database that we will use uses MySQL, an open-source database management system (DBMS).

User class stores user data in the registration process while reply and thread classes are used for forum processes where users can create a thread as a question and users can make comments reply from a thread, which is the communication used in the forum process. Meanwhile, the booking class is used to record the book swap test or rapid test activity, which records the schedule and location for the swap test. The product class is a database table that stores goods sold to users related to Covid19, one of which is to prevent covid, such as masks or hand sanitizers. Meanwhile, the transact class records transactions made by the user where the details of the items purchased are stored in the buy class. Furthermore, the payment class is used to record payments made by the user, and after the user obtains the goods, the user can provide a rating in the form of scores and feedback comments stored in the rating and feedback class.

Figure 3. UI login

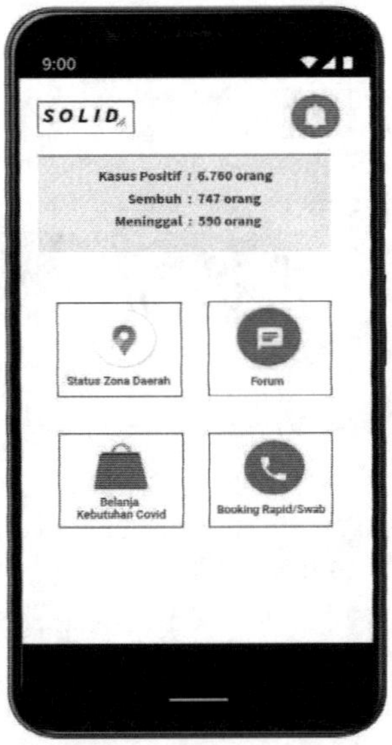

Figure 4. UI main page

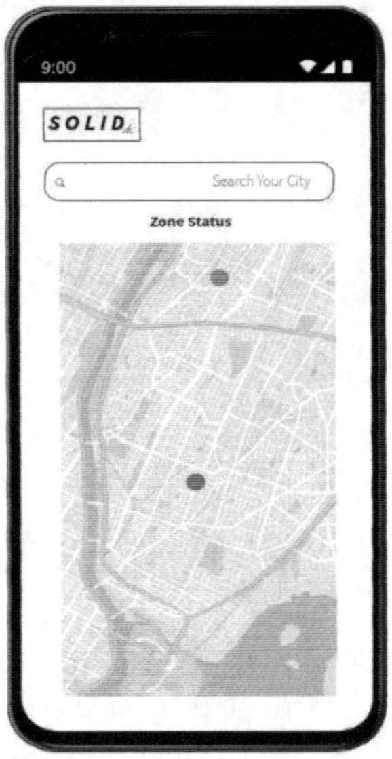

Figure 5. UI view map

The following in Figure 3 to Figure 8 is the User interface (UI) which describes the communication between the user and the application created. When users open the app for the first time, they will be asked to fill out a registration form. Users have to fill in their identity like name, address, phone number, etc. If the user has opened the application previously, they only need to fill in the login form as shown in the UI figure 3. After the username and password are entered according to the user class in Figure 2, the user is given access to the main menu application screen. They are seen in Figure 4, where several menus will direct the user to certain features provided in the application. Users can see the number of COVID-19 cases in Indonesia, view high-risk areas/locations on the map, ask questions and provide information in the forum section, buy antiseptic kits, and order swab/rapid tests online. As seen in figure 5, the map feature will show the user's current location and the surrounding high-risk areas of COVID-19 and notify them if they are in a specific location. In addition, as seen in Figure 6, the forum section can be used by users to obtain or provide information about COVID-19 to other users and communicate things related to covid19. This forum has multiple topics where each different topic will have its chat room, so any information related to the thread topic will not be ignored by other irrelevant information. It is also essential to know that users can ban/suspend certain users if they violate forum chat rules or share false information (hoax).

Furthermore, the user can also order a swab test schedule or a quick test at the test location closest to him, and figure 7 shows the UI. The list of locations provided to the user is limited to the nearest hospital or clinic within a radius of about 5 km from the user's current location. In addition, another feature provided by this application is an online store, as shown in Figure 8, where users can buy all their equipment needs for the pandemic period without having to leave the house.

Figure 6. UI forum

Figure 7. UI book swap

Figure 8. UI buy items

4. Conclusion

This paper discusses mobile-based applications that can help others survive this pandemic era and make their lives easier. We call this application SOLID which contains some information such as cases, zone status, store recommendations and helps users order quick tests or swabs, so users do not have to leave the house. In the future, its development needs to apply Artificial Intelligence (AI) technology to run automatically so that technological intelligence can be seen.

References

[1] Almalki, M. and Giannicchi, A., 2021. Health Apps for Combating COVID-19: Descriptive Review and Taxonomy. *JMIR mHealth and uHealth*, *9*(3), p.e24322.
[2] Lin, L. and Hou, Z., 2020. Combat COVID-19 with artificial intelligence and big data. Journal of

travel medicine, 27(5), p.taaa080.
[3] Ullah, A., Pinglu, C., Ullah, S., Abbas, H.S.M. and Khan, S., 2021. The role of e-governance in combating COVID-19 and promoting sustainable development: a comparative study of China and Pakistan. *Chinese Political Science Review*, *6*(1), pp.86-118.
[4] Teixeira, R. and Doetsch, J., 2020. The multifaceted role of mobile technologies as a strategy to combat COVID-19 pandemic. *Epidemiology & Infection*, *148*.
[5] Waheed, A. and Shafi, J., 2020, October. Successful role of smart technology to combat COVID-19. In *2020 Fourth International Conference on I-SMAC (IoT in Social, Mobile, Analytics, and Cloud)(I-SMAC)* (pp. 772-777). IEEE.
[6] Sarkar, S., 2021. Breaking the chain: Governmental frugal innovation in Kerala to combat the COVID-19 pandemic. *Government information quarterly*, *38*(1), p.101549.
[7] Latif, S., Usman, M., Manzoor, S., Iqbal, W., Qadir, J., Tyson, G., Castro, I., Razi, A., Boulos, M.N.K., Weller, A., and Crowcroft, J., 2020. Leveraging data science to combat covid-19: A comprehensive review. *IEEE Transactions on Artificial Intelligence*, *1*(1), pp.85-103.
[8] Shahid, O., Nasajpour, M., Pouriyeh, S., Parizi, R.M., Han, M., Valero, M., Li, F., Aledhari, M., and Sheng, Q.Z., 2021. Machine learning research towards combating COVID-19: Virus detection, spread prevention, and medical assistance. *Journal of Biomedical Informatics*, *117*, p.103751.
[9] Dubey, S. and Parwekar, P., 2021. Combating COVID-19 Through Emerging Digital Technologies. In *Data Engineering and Intelligent Computing* (pp. 141-149). Springer, Singapore.

Design of a Small Sweeping Robot Based on STM32

Xiaosheng Zhang[a], Gaopeng Zhan[b], Xin Xu[c] and Kai Fang[d,*]

School of Electrical & Information Engineering, Hubei University of Automotive Technology, Shiyan Hubei, China

E-mails: [a]915009621@qq.com, [b]2214102951@qq.com, [c]87677759@qq.com, [d]Fangkai027@sina.com

Abstract. With the continuous improvement of people's living standards and the emergence of various intelligent appliances, a household floor sweeping robot is designed in this subject, which can automatically clean the ground in the room. The robot takes **STM32F103ZET6** single chip microcomputer as the controller, uses two DC motors and L298N driving chips to drive the robot to walk, and uses ultrasonic sensors to detect indoor obstacles, The infrared sensor is used to detect the indoor edge position, the straight travel control of the robot is realized through the coding disk combined with PID algorithm, and the turning control is realized through differential control and PWM speed regulation. The robot can realize the garbage cleaning in the regular room, and has a certain application prospect.
Keywords. Sweeping robot; Path Identification; Single chip microcomputer control; PWM technology

1. Introduction

With the continuous development of science and technology, smart appliances are also changing with each passing day. Floor sweeping robot is a robot product that can liberate people's housework and improve the quality of life, which is loved by young people. The first floor sweeping robot was born in 1997. Since then, it has developed continuously. It has experienced five generations of technological renewal, and now its technology is becoming more and more mature. The sweeping robot has also developed from random type to planning type to continuously meet people's various needs.

This topic aims to design a small and convenient household sweeping robot suitable for ordinary families, and realize the functions of walking, path planning, obstacle avoidance, cleaning and dust collection of the sweeping robot through various modules. The robot takes STM32F103ZET6 single chip microcomputer as the controller, uses two DC motors and L298N drive chip to drive the robot to walk, uses ultrasonic sensor to detect indoor obstacles, uses infrared sensor to detect indoor edge position, realizes robot human direct travel control through coding disk combined with PID algorithm, and realizes turning control through differential control and PWM speed regulation.

This paper is organized as follows. The overall design tasks of the system is reviewed in Sec. II. The specific and detailed system design is presented in Sec. III, and the conclusions are given in in Sec. IV.

2. Design tasks

This paper designs a small sweeping robot, which takes the 32-bit micro-controller STM32F103ZET6 as the core control unit, uses infrared sensor and ultrasonic sensor as the detection unit, designs PID speed controller and PID direction controller to adjust the automatic guidance and angle of the robot in real time, and controls the sweeping and dust collection unit by motor. Different from references [2]and[4], this paper adopts a simpler structure to clean the space with a small number of obstacles, and the price of the robot is also cheaper, only about 100 yuan. The system structure is shown in Figure 1. The system includes sensor module, core controller module, drive module, motor module and power module.

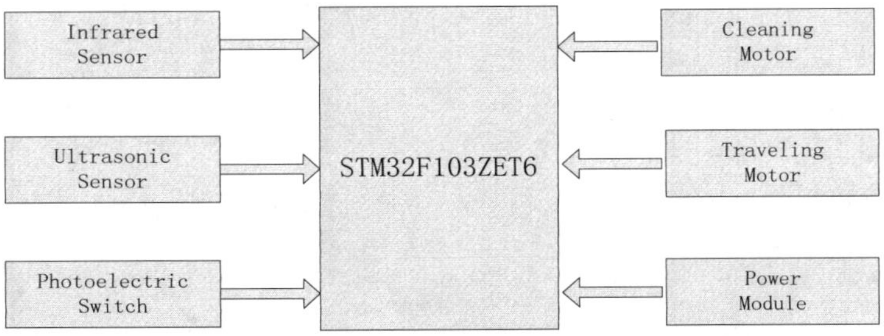

Figure 1. Plane of intelligent material handling robot field setting.

3. System Design of sweeping robot

A. Control System

Using STM32F103ZET6 single chip microcomputer as the controller, the minimum system of single chip microcomputer is built, and the implementation circuit is shown in Figure 2.

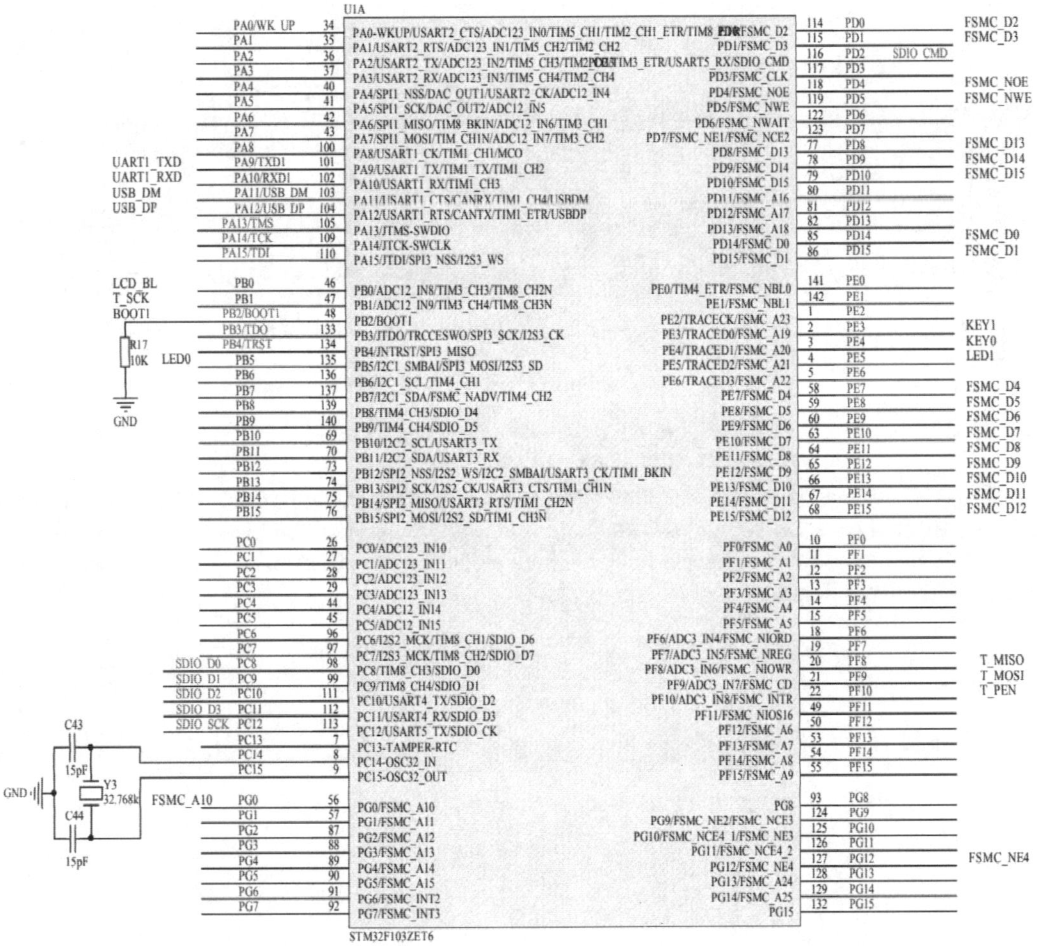

Figure 2. Circuit diagram of minimum system of single chip microcomputer

Specifically include:

1) Sensor module

Select the infrared sensor E18-D80NK, which can detect a distance of up to 80cm. HC-SR04 ultrasonic sensor is selected to measure the obstacles in front of the sweeping robot. The circuit diagram of the obstacles measurement module by ultrasonic sensor is shown in Figure 3.

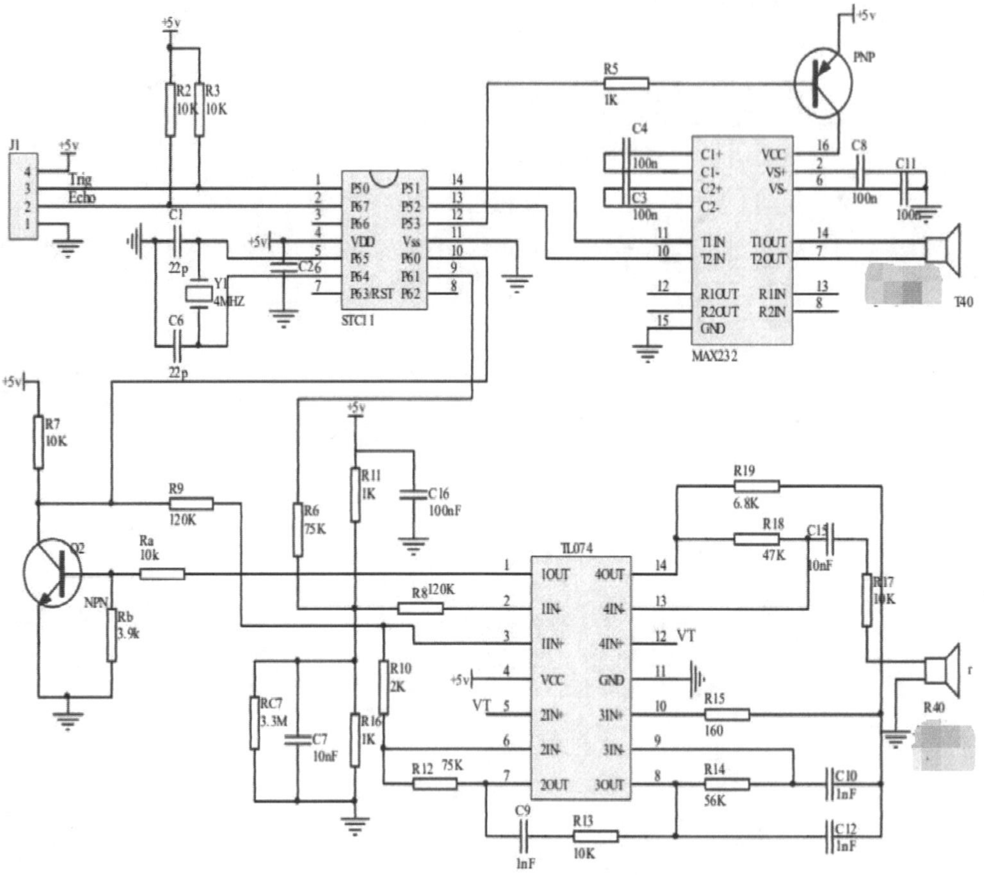

Figure 3. Circuit diagram of obstacles measurement module

The speed measurement module adopts slot type opposed photoelectric sensor, which is composed of an infrared light emitting diode and an NPN light spot triode. Schmidt trigger is used to remove the jitter pulse. The speed measurement performance is stable. The circuit diagram of the speed measurement module is shown in Figure 4.

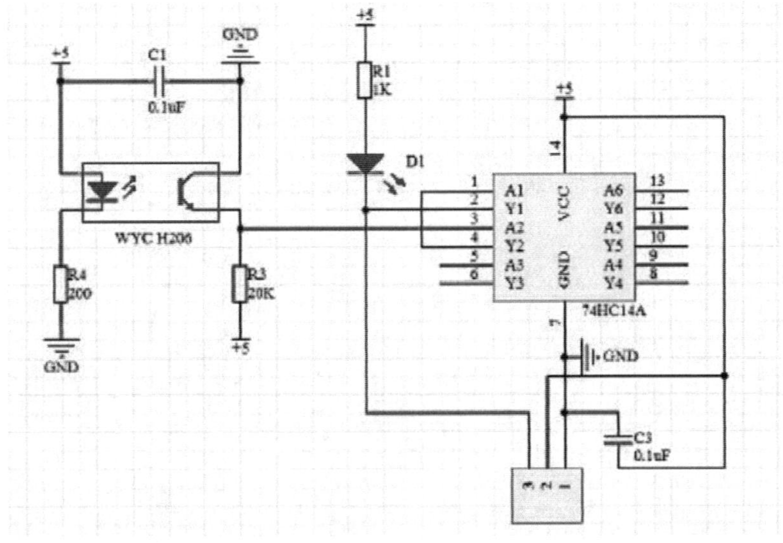

Figure 4. Circuit diagram of speed measurement module

2) Core control module

STM32F103ZET6 single chip microcomputer is the core part of the system. It is responsible for receiving sensor measurement information, walking control of sweeping robot, garbage cleaning and other actions.

3) Drive module

This module includes the drive motor and steering gear. When receiving the command of the single chip microcomputer, it performs the corresponding operation. At the same time, the information acquisition module collects the state information of the motor and steering gear and feeds back to the single chip microcomputer. Thus, the whole system can form a closed loop to make the sweeping robot complete each action.

4) Cleaning module

The speed control cleaning module is mainly composed of a small motor fan and a motor drive module. The motor is installed in the cut plastic bottle and separated by cardboard to form a small vacuum cleaner that can absorb small garbage. In the L298N module, the enable terminal is directly connected to the high level, and the small fan is connected to in1 and in2 terminals, Among them, in1 is terminated on channel 1 of TIM2 in the single chip microcomputer. Channel 1 outputs PWM wave and controls the duty cycle. The fan speed can be adjusted to make the fan speed reach the appropriate speed. When the single chip microcomputer is connected to the power supply, the program burned in the single chip microcomputer starts to initialize. After TIM2 is initialized, the fan starts to rotate, and then the garbage will be absorbed and collected through the plastic bottle mouth. It will not stop until the single chip microcomputer is powered off. It is used to simulate the cleaning module of the floor sweeping robot. The cleaning module will start to work after the single chip microcomputer is initialized until the single chip microcomputer stops running.

5) Power module

Provides power for all parts of the sweeping robot during operation. The robot uses external batteries for power supply. The power supply is composed of two 18650 rechargeable lithium batteries in series. The normal power supply voltage can reach 7.4V, the capacity of each lithium battery is 2600mAH, and the sustainable current can reach 2.6A. The battery is placed in a special 18650 battery box and connected directly to the motor drive module.

B. Software Design of the System

The software design needs to configure the register value of each module of MCU to make each module of MCU work normally. Initialization includes: MCU clock configuration, I /O port configuration, PWM module configuration, A/D module configuration and real-time interrupt configuration. After initialization, start the sensor and clean the motor, and start sampling the sensor input signal. After one sampling, input the sampling value into the control algorithm. The control algorithm is adjusted by PID to obtain the actual controlled motor speed. By changing the value of the internal register of the PWM module, the square wave signals with different duty cycles can be obtained to realize the speed regulation control of the motor, The robot steering is realized by differential control.

1) System Overall operation

The overall operation flow chart of the system is shown in Figure 5 below.

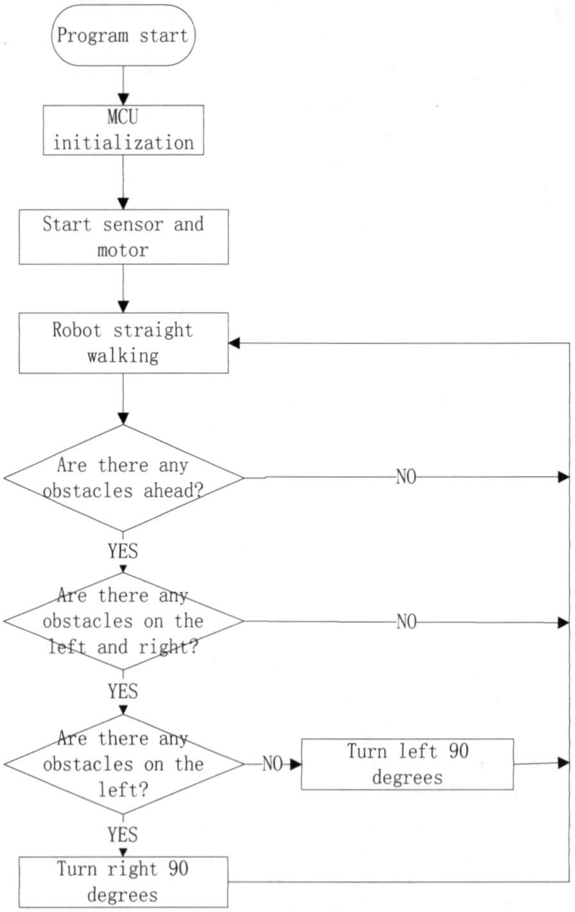

Figure 5. The overall operation flow chart

2) Robot walking program

The flow chart of robot walking program is shown in Figure 6. The classical PID control is adopted for the control of the steering gear, and the specific parameters of each link should be adjusted repeatedly to achieve the adaptability to various types of logistics lanes.

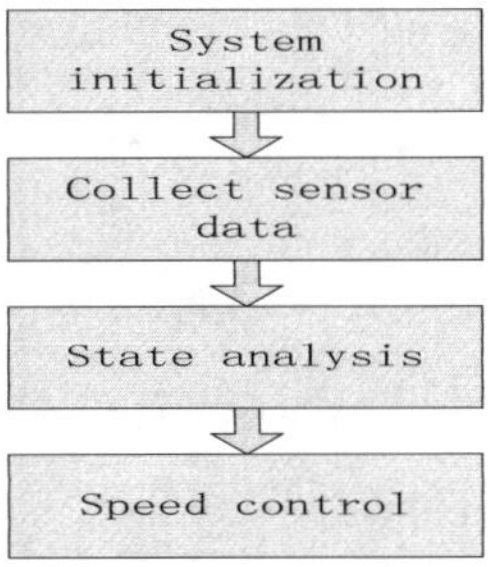

Figure 6. Robot walking program flow chart

C. Physical debugging of the System

Firstly, the sub modules of ultrasonic sensor module, infrared sensor module, photoelectric switch module and drive module are combined into the whole sweeping robot according to the hardware structure, then the independent debugging of each module is completed, and then the comprehensive effect is completed. Finally, the whole system is debugged. We test the sweeping robot in a rectangular room, and turn on the power of the robot to give a one second delay time. This delay time is to allow the program to have enough time to initialize each module while the power is turned on. When the robot starts working, it will turn left and back 90 degrees when it meets the wall in front of it without detecting obstacles on both sides, and then turn right and back 90 degrees when it meets the wall for the second time to realize turning and tracking. At the same time, it will complete the task of cleaning and collecting garbage.

4. Conclusion

In this topic, a small sweeping robot is designed, which takes the micro-controller as the control core, the battery as the power, equipped with ultrasonic sensor, infrared sensor and photoelectric sensor as the sensing mechanism of the robot, and uses the motor control to complete the garbage cleaning. The basic function of its automatic tracking is the roaming mode, with a small number of obstacles in the regular space as the background, Achieve predetermined functions. The small sweeping robot can be widely used as a sweeping module in smart home. Due to proper motor selection, the sweeping robot runs faster, and compared with the sweeping robot on the market, our robot has simpler structure and cheaper price.

Acknowledgment

This work was supported by Foundation of Innovation and Entrepreneurship Training Program for College Students of Hubei University of Automotive Technology under Grant DC2020028, DC2019036, and Key projects of science and technology plan of Hubei Provincial Department of Education under Grant D20201802.

Reference

[1] Xue Huixin. Research and Analysis on design principle of intelligent home sweeping robot. Information technology, 2014 (11): 196-198.
[2] Wang Xin, Zhao Xiangrui, Wang Kangqiang, et al. Intelligent bicycle cleaning robot based on STM32 single chip microcomputer:, CN209776397U, 2019.
[3] Kong Deping. Design of intelligent sweeping robot based on AVR MCU. Scientific and technological innovation and application, 2017 (16): 81-81.
[4] Xu Shenghua. Research and design of intelligent sweeping robot based on STM32. Guangxi Normal University, 2016.
[5] Jian Yi, Gao Bin, Zhang Yue, et al. Research on a full traversal path planning method for indoor sweeping robot. Sensors and Microsystems, 2018, 01 (v.37; No.311):37-39.
[6] He xiuhao, Zhou Wei, Fang Kai. Design of a modular deformable search and rescue robot [J]. File, 2015, (5).

Research on Multi-modal Medical Image Fusion Based on Multi-Scale Attention Mechanism

Baiqing He

Nanchang Institute of Technology, Nanchang, Jiangxi, China

*corresponding author's e-mail: 86035262@qq.com

Abstract. Image fusion technology has been widely used and researched in the fields of medicine, aviation, military and civil. Compared with the traditional image fusion technology, the fusion image technology based on intelligent algorithm generates more realistic, clear and reliable images, and has a lot of detailed information. Deep neural networks have rapidly developed into a research hotspot in medical image analysis, which can automatically extract hidden pathological information from medical image data. Multi-scale and attention mechanism are two important modules in neural networks, which can significantly enhance the features extracted by the network. The problems that need to be solved for medical image fusion based on multi-scale attention mechanism include: how to build a multi-scale module, how to build an attention module, and how to combine multi-scale and attention mechanisms. Construct a deep learning network through a multi-scale attention mechanism, and then adjust the parameters and train the network to achieve multi-modal medical image fusion. In the process of building neural networks, multi-scale and attention mechanisms are incorporated, and multi-scale features of multi-modal medical images are extracted and enhanced for fusion. Through a large number of experiments: (1) The edge strength of the fused image is expected to increase by 10%-20% based on the average value of the existing algorithm; (2) The fused image can achieve high color fidelity and rich detailed information; (3)) The time required for the fusion algorithm is expected to be reduced by 1%-10% based on the average value of the existing algorithm.
Keywords. Multi-scale; Attention; Construction of deep learning network; Multi-modal medical image fusion

1. Introduction

Multi-modal image fusion [1-2] refers to obtaining images with complementary information based on different sensors. Nowadays, image fusion methods can be divided into three types: image fusion acquired by multiple sensors at different times, image fusion acquired by multiple sensors at the same time, and image fusion acquired by a single sensor at different times or under different environmental conditions. At present, most of the images are collected by image sensors. A single sensor can only obtain a single-modal image, and the information is single, which cannot well reflect the multiple original information of the image. Therefore, through the fusion technology, the important information of the multi-modal image can be fused, and the salient features of the target in the image can be improved, thereby overcoming the problem of limited information in the single-modal image. In addition, the fused image can better conform to the human visual perception mechanism.

In recent years, Deep Learning [3] has developed rapidly and has been widely used in the image field. Currently, deep convolutional neural networks (Convolutional Neural Networks) [4] has rapidly developed into a research hotspot in medical image analysis, which can automatically extract hidden pathological information from medical image data. Multi-scale and attention mechanism are two important modules in neural networks, which can significantly enhance the features extracted by the network. Therefore, how to use the multi-scale attention mechanism is the focus of this research. The problems to be solved for medical image fusion based on multi-scale attention mechanism include: (1) How to build a multi-scale module: In deep learning, a multi-scale convolution kernel [5] is used to convolve the input image to obtain multi-scale features . The difficulty in the construction of multi-scale modules lies in the choice of convolution kernel, step size and padding size. The convolution feature obtained by the large convolution kernel is fuzzy, but the feeling is wild; the convolution characteristic obtained by the small convolution kernel is clear, but the feeling is small. In addition, if the step size is large, the calculated feature is relatively fuzzy, but the time-consuming amount is small; the step size is small, the calculated feature is clearer, but the time-consuming amount is large. The size of Padding needs to be adapted to the size of the convolution kernel and the step size; (2) How to construct the attention module: attention mechanism [6] is one of the hotspots of current deep learning research. Attention comes from the human visual system. When humans observe external things, they generally do not see things as a whole, and tend to selectively obtain certain important content of the observed things according to their needs. For example, when you see a person, you usually pay attention to the face first, and then other content such as height, clothing, weight, and so on. Finally, these different types of information are combined to form an overall impression of the observed object. The attention mechanism in deep learning is similar to the human visual system. It determines the feature importance of certain regions by assigning feature weight values. The difficulty in constructing the attention module lies in what type of attention mechanism is used and in which areas of the network the attention module is built. The commonly used attention mechanisms include spatial attention and channel attention mechanisms. The spatial attention mechanism is to assign the corresponding weight value to the feature space, and the channel attention mechanism is to assign the corresponding weight value to the feature channel. In addition, the number of attention modules and their positions in the network are also difficult to study; (3) How to combine multi-scale and attention mechanisms: multi-scale mechanisms can extract features of different scales, and attention mechanisms can significantly enhance the extraction feature. How to combine multi-scale and attention mechanism is the focus and difficulty of this research. Commonly used methods of combining are to directly enhance the extracted multi-scale features with the attention mechanism, or first use the attention mechanism to enhance the extracted features, and then extract the enhanced multi-scale features. In addition, multi-scale and attention mechanisms can be alternately used to continuously enhance features. However, the final combination method should be continuously improved in conjunction with the fusion effect. (4) How to improve the quality of medical image fusion [7-8]: Traditional medical image fusion algorithms are mostly based on image decomposition based on different principles. In the process of image decomposition and fusion, it is inevitable that details are lost, edges are weakened, and content is blurred. The shortcomings of the fusion process are complex and time-consuming. If these shortcomings of current medical image fusion are overcome, improving the quality of medical image fusion is a problem that needs to be addressed in this research.

2. Related work

2.1. Attention mechanism

Attention models have been widely used in various fields of deep learning in recent years. It is easy to encounter attention models in various tasks such as image processing, speech recognition or natural language processing. The attention mechanism is a mechanism that focuses on local information, such as a certain image area in an image. As the task changes, the attention area often changes. The essence of the attention mechanism is to locate the information of interest and suppress useless information. The results are usually displayed in the form of probability graphs or probability feature vectors. In principle, , It is mainly divided into three types: spatial attention model, channel attention model, spatial and channel hybrid attention model.

(1) Spatial attention model

Not all regions in the image are equally important for the contribution of the task. Only the task-related regions need to be concerned, such as the main body of the classification task. The spatial attention model is to find the most important part of the network for processing.

Two representative models, the first is the STN network proposed by Google DeepMind [9]. It completes preprocessing operations suitable for the task by learning the deformation of the input, and is a space-based attention model.

Compared with Spatial Transformer Networks, which completes target positioning and affine transformation adjustment in one step, Dynamic Capacity Networks [10] uses two sub-networks, namely a low-performance sub-network (coarse model) and a high-performance sub-network (fine model).). A low-performance sub-network (coarse model) is used to process the entire image and locate the region of interest. The high-performance sub-network (fine model) refines the region of interest. When the two are used together, lower calculation cost and higher accuracy can be obtained.

(2) Channel attention mechanism

For a CNN that inputs a 2-dimensional image, one dimension is the scale space of the image, that is, the length and width, and the other dimension is the channel. Therefore, channel-based attention is also a very common mechanism.

SENet (Sequeeze and Excitation Net) [11] is the champion network of the 2017 ImageNet classification competition. It is essentially a channel-based attention model. It models the importance of each feature channel and then enhances or suppresses it for different tasks. Different channels are divided into a bypass branch after the normal convolution operation. First, the Squeeze operation (that is, Fsq(·) in the figure) is performed, which compresses the spatial dimensions, that is, each two-dimensional feature map changes As a real number, it is equivalent to a pooling operation with a global receptive field, and the number of characteristic channels remains unchanged. Then there is the Excitation operation (ie Fex(·) in the figure), which generates weights for each feature channel through the parameter w, and w is learned to explicitly model the correlation between the feature channels. In the article, a fully connected layer + Sigmoid function with a two-layer bottleneck structure (first dimension reduction and then dimension increase) is used to achieve this. After the weight of each feature channel is obtained, the weight is applied to each original feature channel. Based on a specific task, the importance of different channels can be learned.

Applying its mechanism to several benchmark models, with a small increase in the amount of calculation, a more obvious performance improvement has been obtained. As a general design idea, it can be used in any existing network and has strong practical significance. Then SKNet and other methods combine such channel weighting ideas with the multi-branch network structure in Inception, and also achieve performance improvements. The essence of the channel attention mechanism is to model the importance of each feature. For different tasks, the feature can be assigned according to the input, which is simple and effective.

(3) Integration of spatial and channel attention mechanisms

The aforementioned Dynamic Capacity Network is to pay attention from the spatial dimension, and SEnet is to pay attention from the channel dimension. Naturally, spatial attention and channel attention mechanisms can also be used at the same time.

CBAM (Convolutional Block Attention Module) [12] is one of the representative networks, and the attention of channel direction models the importance of features. At the same time, the maximum pooling and mean pooling algorithms are used, and then the transformation results are obtained through several MLP layers, and finally applied to the two channels respectively, and the sigmoid function is used to obtain the attention results of the channels. The spatial orientation of attention models the importance of spatial location. First, the channel itself is reduced in dimensionality, and the maximum pooling and average pooling results are obtained respectively, and then stitched into a feature map, and then a convolutional layer is used for learning.

These two mechanisms have learned the importance of the channel and the importance of space respectively, and can also be easily embedded in any known framework. In addition, there are many researches related to attention mechanism, such as residual attention mechanism, multi-scale attention mechanism, recursive attention mechanism, etc.

The attention mechanism integrates multiple scales and has a wide range of usage scenarios in various application fields of deep learning. Mainly used in medical image retrieval, semantic segmentation and other scenarios.

This method adaptively balances the relationship between shallow image texture features and deep image semantic features by fusing multi-scale features and setting learnable weight coefficients, and improves the network's ability to extract pathological features at different scales. At the same time, the attention module is introduced to perform channel weighted summation on the feature maps output by the network to improve the feature expression ability of key feature channels, so that the network can pay more attention to the recognizable pathological feature areas in the image [13].

This method is a method based on attention mechanism combined with multi-scale prediction. Research has found that predictions on specific scales can better solve specific failure modes, and the network learns various scales in this case in order to produce better predictions [14].

2.2. Research on traditional medical image fusion algorithm

Since the 1980s, image fusion technology based on transform domain has attracted special attention from scholars at home and abroad. Multi-scale (MST) [15-18] is one of the most commonly used transform domains. MST can obtain features of different scales by transforming the image to form a feature pyramid. Since features of different scales contain a wealth of different details, flexible algorithms can be used to complete complex image tasks. Common image fusion algorithms based on MST transform include wavelets [19-22], pyramids [23-25], NSCT (non-subsampled contourlet transform) [26-29], NSST (non-subsampled Shearlet transform) [30-32] Wait. The principle of the wavelets-based fusion method is to transform the image domain into wavelet domain coefficients and process the coefficients, and then obtain the fused image through inverse wavelet transform. The advantage of this method is that the transform domain has a large degree of freedom to choose, and different wavelet basis functions can be selected to realize the transformation; the disadvantage is that the processing of wavelet coefficients is difficult, and the fusion rules for wavelet coefficients are simple, and it is difficult to achieve complex image fusion effects. The fusion method based on pyramids is to build a multi-resolution pyramid feature map, and use specific fusion rules to achieve fusion on features of different resolutions. Although the pyramid-based fusion method can keep the edge information of the fused image better, it is easy to lose a lot of detailed information in the process of building pyramid feature maps of different resolutions, which leads to the blurred image of the fusion. The fusion method based on NSCT and NSST inherits the advantages of multi-resolution and multi-directional MST transformation, overcomes the Gibbs effect, and reduces the loss of detailed information to a certain extent, so that the fused image retains more detailed information. However, NSCT and NSST transform and inverse transform have a large amount of calculation, consume a lot of time in the process of transform, and need to design suitable fusion rules, and the algorithm complexity is high. Although MST transform has been widely used in the image field due to its unique advantages, they still have their own shortcomings, resulting in poor fusion image quality. In addition, SR (Sparse Representation) [33-36] is also one of the commonly used fusion algorithms for medical images. The SR-based fusion method realizes fusion by establishing a complete dictionary and optimizing the sparse coefficient, which can achieve better fusion effects than other traditional algorithms. However, the SR algorithm is based on the optimization process of the dictionary and the sparse coefficient, so the amount of calculation and time-consuming is large. In addition, the optimization of dictionaries and sparse coefficients is quite complicated and difficult.

2.3. Research on image fusion algorithm based on deep learning

Compared with traditional algorithms, deep learning is a method that allows computers to automatically learn pattern features, and integrates feature learning into the process of model building, thereby reducing the incompleteness caused by artificial feature extraction. In addition, deep learning has a simple implementation process, flexible model building, and a wide range of applicable data, so it has been widely used in the field of image fusion. In recent years, many DL-based image fusion algorithms have been proposed at home and abroad. Medical image fusion based on deep network does not require complex algorithms, and can adapt to multiple tasks only by changing the network structure or loss function, and its flexibility and adaptability have been greatly improved. [37] proposed an unsupervised multi-exposure image fusion algorithm based on DL. This method uses a deep convolutional neural network to achieve the fusion of multi-exposure images, which improves the quality and efficiency of multi-exposure image fusion to a certain extent. [38] proposed a Densenet-based infrared and visible image fusion algorithm. It extracts the deep dense features of the input image through the encoder, then fuses in the feature layer, and finally obtains the fused image through the decoder. The algorithm can automatically fuse multi-focus images, infrared and visible light images, and achieve better fusion results. [39] proposed an image fusion algorithm based on CNN, which can be used for the fusion of multi-focus images, infrared and visible images, and medical images. This method uses CNN to extract the weight map of the image, and then uses the Gaussian pyramid for feature fusion. Although it achieves a better fusion effect, it still belongs to the traditional algorithm in essence.

At present, deep learning has been widely used in the field of image fusion. However, the fusion algorithm based on deep learning depends on the network architecture and the choice of loss function. The quality of the network architecture directly affects the result of image fusion. In addition, the loss function has different responses to different network structures, which in turn has different effects on the image fusion effect. Therefore, the design of the network architecture and the choice of the loss function must adapt to each other. The current mainstream method is to select a suitable network architecture and loss function through a large number of experiments to improve the image fusion effect.

3. Method

The overall research plan of this study is shown in Figure 1, which is mainly composed of five parts: The first part (the yellow part in Figure 1) is the construction of multi-modal medical image data, including image acquisition of MRI, CT, PET and SPECT, as well as the training set and The construction of the test set; the second part (the purple part in Figure 1) is the network construction based on the multi-scale attention mechanism. The main research content includes the use of multi-scale and attention mechanisms, the construction of network models, the research and use of convolutional layers, activation layers, and pooling layers; the third part (the green part in Figure 1) is network training. The research content includes the determination of training times and training error analysis; the fourth part (the orange part in Figure 1) is the adjustment of deep learning parameters. The research content includes the exploration of the learning rate, the determination of the batch size, the selection of the optimization algorithm and the construction and research of the loss function; the fifth part (the red part in Figure 1) is the network test, and the medical fusion image is obtained through the network test; the sixth part (The blue part in Figure 1) is the quality evaluation, which comprehensively evaluates the quality of the fusion image by selecting multiple index evaluations.

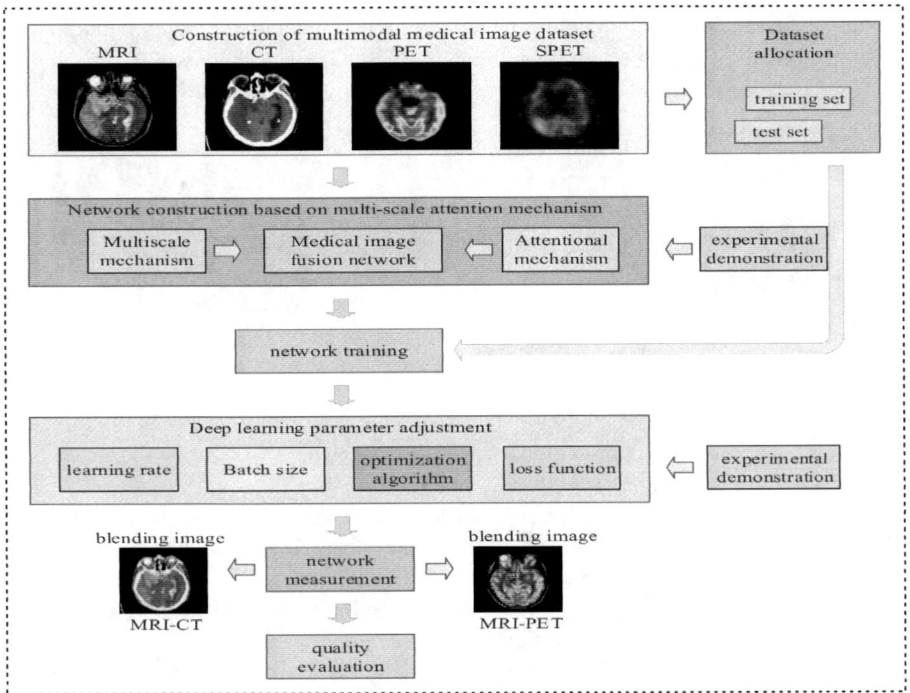

Figure 1. The overall design of the project

3.1. Construction of multi-modal medical image data set

As an important source of medical information, multi-modal medical images have been widely used in medical techniques such as surgical positioning, lesion positioning, radiotherapy plan formulation, organ function analysis, and neurosurgery guidance. Due to the different imaging principles, different medical imaging equipment provides medical images of different modalities. Different modal medical images reflect different information of organs and tissues. For example, MRI (magnetic resonance) images can well reflect the anatomical structure of soft tissues. CT images can clearly reflect bone structure, while PET (positron emission computed tomography) images mainly provide functional information such as metabolism. For example, the white matter of the brain is very bright in T1-weighted MRI (nuclear magnetic resonance) images, while in PET (Positron emission computed tomography) The image is very dark. Multimodal medical image fusion is the organic combination of image information from different imaging equipment to provide more detailed and accurate data information for clinical diagnosis.

The primary task of medical image fusion is to construct a multi-modal medical image data set, including MRI, CT, PET and SPECT images. In addition, medical images that can be fused must have been registered. Since domestic hospitals or university research institutions have not published the corresponding data sets so far, we use the medical image database of Harvard Medical School to construct the multi-modal medical image data set. The website is http://www.med.harvard.edu/AANLIB/ home.html. We select 100-300 registered MRI-CT, MRI-PET and MRI-SPECT image pairs from the Harvard image library as a multi-modal medical image data set, as shown in Figure 2.

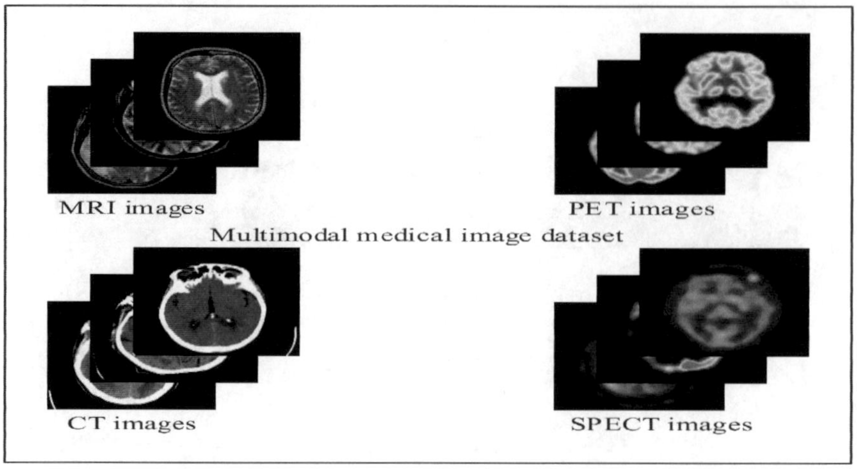

Figure 2. Multimodal medical image data set

3.2. Network construction based on multi-scale attention mechanism

The construction of neural networks is very flexible. Any number of convolutional layers, activation layers and pooling layers can be combined at will to build different networks to adapt to different image tasks. However, in the process of network design, if the number of network layers is deeper, the parameters are larger, and if the training data set is too small, it is easy to fall into over-fitting. Secondly, the deeper the network, the greater the computational complexity and time-consuming, and it is difficult to apply in real time. In addition, the deeper the network, the easier it is for the gradient to disappear, making it more difficult to optimize and converge. Therefore, the balance between network performance and complexity must be fully considered in the network design process to design a reasonable network architecture.

At present, the network architecture based on deep learning needs common convolutional layer, activation layer and pooling layer, and it needs a variety of deep learning mechanisms to enhance the quality of extracted features. Commonly used deep learning mechanisms include multi-scale mechanisms and attention mechanisms. The multi-scale mechanism is widely used in deep learning. The extraction and fusion of features of different scales through convolution kernels of different scales greatly improves the performance of the entire network. However, due to the extraction and fusion of multi-scale features, the amount of calculation is doubled and a large number of gradient vectors are retained, which consumes a large amount of memory, which results in high hardware requirements. The attention mechanism is one of the most widely used neural network structures in recent years. It simulates the human brain's mechanism of observing the environment to extract features. When humans observe the environment, the brain often only pays attention to a few particularly important local data, obtains the required information, and constructs a certain description of the environment. Similar to the brain observation environment, the attention mechanism gives the importance of local features, selectively enhances the local features with high importance, reduces the attention of low-importance information, and selectively filters the extracted features according to the importance to improve information processing Efficiency and accuracy. In this study, the combination of multi-scale and attention mechanism can enhance the feature extraction ability and improve the quality of medical image fusion.

4. Experimental verification scheme

This research will complete the deep learning experimental verification of multi-scale attention mechanism by constructing a multi-modal medical image data set. The experimental plan is shown in Figure 3. First, build a deep neural network through a multi-scale mechanism, and observe the effect of image fusion after training. Then build a deep neural network through the attention mechanism, and observe the effect of image fusion after training. Finally, a neural network is built through the multi-scale attention mechanism, which combines the multi-scale and attention mechanism, and observes the experimental results after training. The experimental verification scheme of this research mainly includes the following contents:

(1) Establish a multi-modal medical image data set (the green part of Figure 3). Construct a multi-modal data set containing MRI, CT, PET and SPECT images, and distribute the training set and test set according to the ratio range of 2:1-4:1. Download pre-registered multi-modal images from the Harvard Medical School Brain Image Database website as the experimental data set.

(2) Use a multi-scale mechanism to build a multi-scale neural network (orange part in Figure 3). The research uses 3-5 different scale convolution kernels to extract the features of multimodal medical images, and uses 3×3 small convolution kernels to increase the depth of the network during the feature learning process. In order to prevent network overfitting and gradient disappearance, BN (Batch Normalization) processing is appropriately added to the network. The purpose of this experiment is to verify the influence of multi-scale mechanism on multi-modal medical image fusion.

(3) Use the attention mechanism to build an attention neural network (the blue part in Figure 3). The research uses spatial attention and channel attention mechanisms to extract and enhance the features of multi-modal medical images. In the process of research, the spatial attention network and the channel attention network were constructed separately, and the influence of the two attention mechanism networks on the fusion effect of medical images was observed after training. Then, a neural network is constructed by mixing spatial attention and channel attention mechanisms, and after training, observe the difference between the fusion effect of a separate spatial attention and channel attention network. The purpose of this experiment is to verify the influence of different attention mechanisms on the fusion of multimodal medical images.

(4) Use multi-scale and attention mechanism to build a multi-scale attention mechanism neural network (Figure 3 yellow part). The research intends to use four different scale convolution kernels of 1×1, 3×3, 5×5 and 7×7 to construct a multi-scale feature extraction and fusion module, and integrate spatial attention and channel attention in the feature extraction and fusion module Force mechanism to enhance features. In addition, in order to reduce the risk of over-fitting or gradient disappearance explosion, the Relu or Sigmoid nonlinear activation layer, batch normalization BN layer and pooling layer are appropriately added to the feature extraction fusion module. When the network training is over, observe the difference between the image fusion quality and the multi-scale network and attention mechanism network. The purpose of this experiment is to verify the advantages of a multi-scale attention mechanism network over a separate multi-scale network and attention mechanism network.

(5) Develop a set of image fusion demonstration system based on python platform and matlab platform. The system also includes neural network modules such as feature extraction, feature fusion, feature reconstruction, multi-scale mechanism, attention mechanism, etc., and can realize the functions of MRI-CT, MRI-PET and MRI-SPECT image display and fusion.

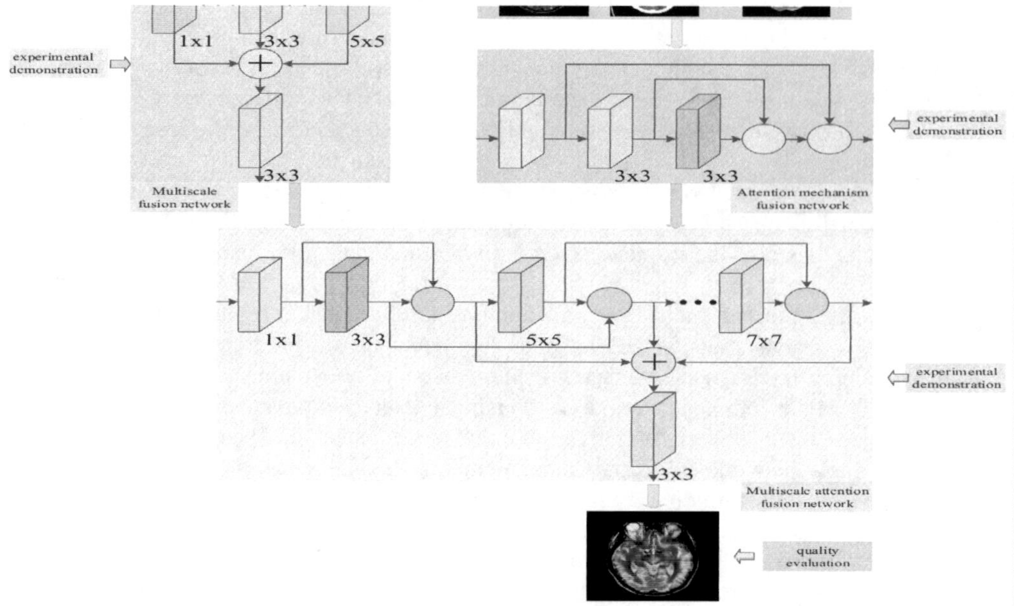

Figure 3. Experimental verification scheme

5. Conclusion

Based on the reference to a large number of related literatures, the principle of existing image fusion algorithms, multi-modal image features and related technical principles of image processing, this research studies the multi-scale attention mechanism of medical images and proposes the research content of this research And research plans, in-depth demonstration of network construction, the use of multi-scale attention mechanisms, optimization algorithms, loss functions, and evaluation indicators to form scientific and reasonable research plans and ideas.

According to the difference of multi-modal medical images, this study extracts the features of multi-modal medical images by constructing a suitable neural network, and through in-depth analysis of multi-modal feature data, explores the influence of different network architectures on feature extraction. The research can be adapted to many Image fusion algorithm of different modalities. The project team refers to the existing image fusion algorithm, uses the advantages of multi-scale and attention mechanism, uses multiple attention mechanism to improve the expression ability of extracted features, explores the adaptive fusion method of multi-modal medical image data, and conducts multi-modal medicine. Image data fusion research is feasible.

This research will conduct in-depth research on residual and attention mechanism models, and propose a feature extraction method for multiple residual attention mechanisms by constructing multiple compound residual attention models. At present, the advantages of residual error and attention mechanism have been deeply studied at home and abroad, and they have been widely used in the image field. These research structures prove that the residual attention model can be well applied in image feature extraction, important information expression and image fusion. Therefore, the medical image fusion algorithm based on the multi-scale attention mechanism proposed in this study is feasible.

Acknowledgement

Fund Project: The Science and Technology Project of Jiangxi Provincial Department of Education "Research on Experimental Equipment of Digital Circuit "Pocket Machine" Based on "Pocket Laboratory"", No. GJJ191008.

Fund Project: Humanities and Social Sciences of Jiangxi Provincial Department of Education "Research on the Path and Strategy of Incorporating Ideological and Political Education in the New Era into the Collaborative Education of Private Colleges and Universities", No. JC20113.

Fund Project: Nanchang Key Laboratory of Intelligent Manufacturing Electronic Assembly Innovation and Development, 2021-NCZDSY-018

References

[1] S. Li, X. Kang, L. Fang, J. Hu, and H. Yin 2017 Pixel-level image fusion: A survey of the state of the art Information Fusion **33** 100–12

[2] Du J, Li W, Lu K and Xiao B 2016 An overview of multi-modal medical image fusion Neurocomputing **215** 3–20

[3] Zuo Y, Huang G, Nie S 2021 The application and challenges of deep learning in the intelligent processing of medical images Journal of Image and Graphics **26**(02) 305-15

[4] Huang J, Zhang G 2020 Overview of target detection algorithms for deep convolutional neural networks Computer Engineering and Applications **56**(17) 12-23

[5] Zhou Y 2021 Research on Bearing Fault Diagnosis Based on Multi-scale Convolutional Neural Network Southern Agricultural Machinery **52**(01) 116-7

[6] Chen H, Wu G, Li J, Wang J, Tao H 2021 Research progress of deep learning recommendation based on attention mechanism Computer Engineering and Science **43**(02) 370-80

[7] Yuan Y, Wan R 2020 Multi-source image fusion algorithm review and fusion quality analysis Information Technology and Information Technology **08** 80-2

[8] Wei X 2020 Research on the Quality Evaluation System of Image Fusion Processing Guide Journal of Science & Technology Economy **28**(24) 29+31

[9] Sønderby S K, Sønderby C K, Maaløe L and Winther O 2015 Recurrent Spatial Transformer Networks. arXiv preprint arXiv **1509** 05329

[10] Almahairi A, Ballas N, Cooijmans T, Zheng Y, Larochelle H, Courville A 2016 Proceedings of The 33rd International Conference on Machine Learning (Proceedings of Machine Learning Research) **48** (25) 49-58

[11] Hu J, Shen L, & Sun G 2018 Squeeze-and-excitation networks. In Proceedings of the IEEE conference on computer vision and pattern recognition 7132-41

[12] Woo S, Park J, Lee J Y, & Kweon I S 2018 Cbam: Convolutional block attention module. In Proceedings of the European conference on computer vision (ECCV) 3-19

[13] Zhou L, Yao J, Yan Q, Lin Z 2021 Medical image retrieval combining multi-scale features and attention mechanism Liquid Crystal Display **36**(08) 1174-85

[14] Tao, A., Sapra, K., & Catanzaro, B 2020 Hierarchical multi-scale attention for semantic segmentation. arXiv preprint arXiv **2005** 10821

[15] Liu Y, Liu S, and Wang Z 2015 A general framework for image fusion based on multi-scale transform and sparse representation *Information Fusion* **24** 147–64

[16] Chen J, Li X, Luo L, Mei X, Ma J 2020 Infrared and visible image fusion based on target-enhanced multiscale transform decomposition Information Sciences **508** 64–78

[17] Zhao J, Chen Y, Feng H, Xu Z and Li Q 2014 Infrared image enhancement through saliency feature analysis based on multi-scale decomposition Infrared Physics & Technology **62** 86–93

[18] Hamza A B, He Y, Krim H and Willsky A 2005 A multiscale approach to pixel-level image fusion Integrated Computer-Aided Engineering **12**(2) 135-46

[19] Wady S M A, Bentoutou Y, Bengermikh A, Bounoua A and Taleb N 2020 A new IHS and wavelet based pansharpening algorithm for high spatial resolution satellite imagery Advances in Space Research **66**(7) 1507-21

[20] Singh R and Khare A 2014 Fusion of multimodal medical images using Daubechies complex wavelet transform – A multiresolution approach Information Fusion **19** 49–60

[21] Hill P, Al-Mualla M E and Bull D 2017 Perceptual image fusion using wavelets IEEE Transactions on Image Processing **26**(3) 1076–88
[22] Li S and Yang B 2008 Multifocus image fusion by combining curvelet and wavelet transform Pattern Recognition Letters **29**(9) 1295-301
[23] Zhang Q, Wang L, Li H, Ma Z 2011 Similarity-based multimodality image fusion with shiftable complex directional pyramid Pattern Recognition Letters **32** 1544–53
[24] Du J, Li W, Xiao B and Nawaz Q 2016 Union Laplacian pyramid with multiple features for medical image fusion Neurocomputing **194**(19) 326–39
[25] Du J, Li W and Xiao B 2017 'Anatomical-functional image fusion by information of interest in local laplacian filtering domain IEEE Transactions on Image Processing **26**(12) 5855–66
[26] Zhang Q and Guo B 2009 Multifocus image fusion using the nonsubsampled contourlet transform Signal Processing **89**(7) 1334–46
[27] Bhatnagar G, Wu Q M J and Liu Z 2013 Directive contrast based multimodal medical image fusion in NSCT domain IEEE Transactions on Multimedia **15**(5) 1014–24
[28] Yang Y, Que Y, Huang S and Lin P 2016 Multimodal sensor medical image fusion based on type-2 fuzzy logic in NSCT domain IEEE Sensors Journal **16**(10) 3735–45
[29] Ganasala P and Kumar V 2014 CT and MR image fusion scheme in nonsubsampled contourlet transform domain Journal of Digital Imaging **27**(3) 407–18
[30] Singh S, Gupta D, Anand R S and Kumar V 2015 Nonsubsampled shearlet based CT and MR medical image fusion using biologically inspired spiking neural network Biomedical Signal Processing and Control **18** 91–101
[31] Luo X, Zhang Z, Zhang B, Wu X 2017 Image Fusion With Contextual Statistical Similarity and Nonsubsampled Shearlet Transform IEEE Sensors Journal **17**(6) 1760–71
[32] Gao G, Xu L, Feng D 2013 Multi-focus image fusion based on non-subsampled shearlet transform IET Image Processing **7**(6) 633-9
[33] Liu Y and Wang Z 2015 Simultaneous image fusion and denosing with adaptive sparse representation IET Image Processing **9**(5) 347–57
[34] Liu Y, Chen X, Ward R and Wang Z 2016 Image fusion with convolutional sparse representation IEEE Signal Processing Letters **23**(12) 1882–6
[35] Yang B and Li S 2010 Multifocus image fusion and restoration with sparse representation IEEE Transactions on Instrumentation and Measurement **59**(4) 884–92
[36] Kim M, Han D K and Ko H 2016 Joint patch clustering-based dictionary learning for multimodal image fusion Information Fusion **27** 198–214
[37] Ram Prabhakar, K., Sai Srikar, V., & Venkatesh Babu, R 2017 Deepfuse: A deep unsupervised approach for exposure fusion with extreme exposure image pairs. In Proceedings of the IEEE international conference on computer vision 4714-22
[38] Li H and Wu X 2019 "DenseFuse: A Fusion Approach to Infrared and Visible Images," IEEE Transactions on Image Processing **28**(5) 2614-23
[39] Liu Y, Chen X, Cheng J, Peng H and Wang Z 2017 "Multi-focus image fusion with a deep convolutional neural network," Information Fusion **36** 191–207

Design and Implementation of Virtual Campus Roaming System Based on Unity3d

Yuxuan Li[1,*], Hua Luo[2] and Yiren Zhou[3]

[1]School of computer information engineering, Nanchang Institute of Technology, Nanchang, Jiangxi, China
[2] College of architectural engineering, Nanchang Institute of Technology, Nanchang, Jiangxi, China
[3]Asset Division, Nanchang Institute of Technology, Nanchang, Jiangxi, China

*corresponding author's e-mail:124705508@qq.com

Abstract. With the rapid development of Internet technology and the maturity of 5g technology, virtual reality gradually appears in the public vision, and its involved fields are also expanding. Using virtual reality technology and head mounted display, users' immersion and authenticity can be improved to the greatest extent. In this paper, an immersive virtual campus roaming system is realized by using 3ds Max tool to create a model, unity 3D tool to build a scene, c# language to write human-computer interaction script, and action one headset device to take Nanchang Institute of technology as an example.

1. Introduction

With the rapid development of information technology and Internet technology, virtual reality has gradually stepped into the vision of more and more people. Using virtual reality technology, people can browse the landscapes all over the world without leaving home; Technically, modeling tools can be used to reproduce the buildings and environment of the real world, then unity 3D combined with 3D model can be used to build a virtual world, and human-computer interaction can be realized through script code. Finally, users can load the virtual environment into the device through the virtual reality device to realize virtual reality roaming[1].

2. Design of virtual roaming system

This project uses unity 3D and 3ds Max as development tools, c# as the main programming language, and develops a campus roaming system based on Nanchang Institute of technology based on action one hybrid reality equipment developed by shadow creator. The goal of the system design is to hope that users can have a real feeling of walking in the campus. Therefore, in the modeling of buildings and the editing of terrain in the campus, we must strive to restore the real appearance of the whole school. In the system, players can choose autonomous roaming and automatic roaming; When the player approaches the building, a bubble box will pop up automatically to show the detailed explanation of the building; You can check the campus map, get your position in the campus at any time, and click the map to transmit it[2].

3. Realization of campus roaming system

3.1. 3D modeling

In the development process of this system, whether the scene is realistic mainly depends on the fineness of modeling. Campus modeling is carried out with 3ds Max tools. Firstly, the whole campus is divided into teaching area in the West and living area in the north according to the terrain characteristics of the campus, and the buildings in each area are modeled respectively. First, collect photos of buildings from different angles and distances and photos of the surrounding environment, and then refer to the scale of Baidu real map to create the corresponding three-dimensional model.The real map of Baidu on campus is shown in Figure 1.

In the process of modeling, considering that the library and teaching building provide indoor browsing function, the models of the two buildings are analyzed

Relatively high model shall be adopted, while other building models use low model. The fidelity of the model shall be improved through mapping to improve resource utilization. For the vegetation in the campus, some representative species are selected for modeling. When the model is created, export the model to a. FBX format file[3].The modeling of some buildings on campus and the layout effect of the scene are shown in Figure 2.

3.2. Scene construction

Restore the campus terrain in the terrain system in unity 3D, and build lakes, steep slopes, flat lands and lawns according to the actual scene of the campus; Then import the FBX format file exported from 3ds Max into unity 3D, and then place the models of all buildings in the corresponding positions in proportion according to the school layout. After placement, start to arrange the surrounding vegetation, street lights and other public facilities.

Figure 1. Campus plan (screenshot from Baidu map)

Figure 2. Modeling and scene layout of some buildings on campus

3.3. Roaming

Roaming can be divided into autonomous roaming and automatic roaming. In the action one headset, the device will recognize the user's hands, and the user can interact with the virtual world directly through gestures. In the process of autonomous roaming, the user only needs to click the forward and backward button to move, and the direction can be changed by turning the head, so as to realize autonomous roaming[4].

The automatic roaming mode cruises the whole campus at a fixed speed according to the set roaming route. In unity 3D, multiple navigation points need to be created. Each navigation point is equipped with a trigger. The navigation points are stored in the array. The navigation coordinates are used to guide the moving target of the viewing angle. The viewing angle moves along the navigation point. In the process of moving, the user can observe the surrounding environment, And select pause roaming and continue roaming. Every time a navigation point is touched, the navigation coordinate points to the next navigation point, and the viewing angle continues to move towards the next navigation point until the last navigation point. At this point, the navigation ends, and the user can choose whether to return to the starting point. The key codes are as follows:

```
void Update()
    {
        if (current<=point.Length-1)
        {
            transform.LookAt(point[current]);//转向导航点方向
            transform.Translate(-transform.forward*Time.deltaTime*speed);//移动
        }
    }
private void OnTriggerEnter(Collider other)
    {
        //每当触发一次触发器时，导航指标指向下一个导航点
    other.GetComponent<Navigator>().current=other.GetComponent<Navigator>().current+1;
    }
```

3.4. Interaction

3.4.1. Common interactive systems. In the user view, the user can directly operate the menu system by hand, and different functions can be realized by clicking the buttons in the menu. In order to enable users to understand the situation of the school more quickly, text panels are appropriately set around important buildings, which are hidden by default, and triggers are added to the buildings. When the user is close, the triggers of the buildings are triggered, and the text panel appears to introduce the main functions of the buildings to users. The key codes are as follows:

```
public GameObject TipsFace;//信息面板

    //当用户触发碰撞器

    private void OnTriggerEnter(Collider other)

    {

        TipsFace.SetActive(true);//显示

    }

    //当用户离开碰撞器

    private void OnTriggerExit(Collider other)

    {

        TipsFace.SetActive(false); //隐藏

    }
```

3.4.2. Small map interactive system. The function of small map in the system can make users more convenient to understand their location in the campus, and can also realize the transmission between different buildings through the interactive system of small map. Firstly, it is necessary to select an appropriate transfer point for each building, and store the corresponding relationship between the transfer point and the building through the dictionary in unity 3D. When the user clicks the building in the small map, the system queries the corresponding transfer point position from the dictionary and transfers the user to the corresponding transfer point, so as to improve the degree of freedom of the system. The key codes are shown as follows:

```
public static void Register(string bulidName,Transform transferPoint)

    {

        instance.bulidList.Add(bulidName, transferPoint);

    }

    public static Transform GetPointByName(string bulidName)

    {

        return instance.bulidList[bulidName];//获得该建筑的传送点

    }
```

```
    public void transfer(Transform transferPoint)
{
    this.transform.position = transferPoint.position;
}
```

3.4.3. Lighting interaction system. In order to simulate a more real campus environment, the lighting system in unity 3D is used to set natural lighting. By changing the lighting angle, the light intensity changes in different time periods can be realized, so that users can experience a more real and richer campus environment. To realize this interactive function, the x-axis of natural light rotation needs to be controlled by the script, and the user can adjust the time period on the right side of the interface[5].

3.5. Optimization

In the campus scene, there are a large number of buildings, trees and other entities. If all entities in the scene are rendered in each frame, the rendering efficiency will be reduced, which will affect the performance of the system and cause the system to get stuck. In order to solve this problem, occlusion elimination technology can be used to eliminate those entities that can not be seen by the user. Even if the object is not displayed, it will not affect the user's experience, reduce the amount of data rendered per frame, and improve the system performance and rendering efficiency.

4. Conclusion

With the development of science and technology, virtual reality technology continues to mature, and there are more and more application scenarios of virtual reality technology, and its advantages are slowly displayed. Compared with the traditional picture campus roaming system, the use of virtual reality technology can not only improve the user's sense of immersion to a great extent, but also promote the construction of digital campus in Colleges and universities.

Acknowledgment

This paper belongs to the scientific and technological research project of Jiangxi Education Department: The research of virtual campus roaming system based on unity 3D and MR technology. Topic number:GJJ202101, one of the results of the conclusion.

At the same time, I would like to thank the science and technology research fund of Jiangxi Provincial Department of Education (GJJ191006) for its support and take it as one of its concluding achievements.

Reference

[1] Cao min. research on Design of multi-user interactive system based on unity3d [J]. Journal of Xi'an University of Arts and Sciences: Natural Science Edition, 24 (2): 5
[2] Pan Shujie. Design and implementation of virtual reality interactive system based on unity3d [J]. Digital world, 2020, No. 182 (12): 256-257
[3] Ma Cheng, Yan Junjun, Xu Jian. Research and construction of virtual campus roaming system based on VR [J]. Software, 2019, 40 (03): 112-120
[4] Liu Xia, Li Xiaohua. Virtual campus roaming system based on unity3d [J]. Science and technology horizon, 2020, No. 300 (06): 237-238
[5] Zang Hui, Yuan Yong. Design and implementation of Virtual Campus Based on unity3d [J]. Computer knowledge and technology: Academic Edition, 2020 (24): 23-25

Research on Virtual Campus Roaming System Based on unity 3d Technology

Hua Luo[1,*], Yiren Zhou[2] and Yuxuan Li[3]

[1]College of architectural engineering, Nanchang Institute of Technology, Nanchang, Jiangxi, China
[2]Asset Division, Nanchang Institute of Technology, Nanchang, Jiangxi, China
[3]School of computer information engineering, Nanchang Institute of Technology, Nanchang, Jiangxi, China

*corresponding author's e-mail:705413069@qq.com

Abstract. with the development of virtual simulation technology and network technology, the application of virtual technology in education, games and so on is deepening. Use 3ds Max and unity 3D software to organically combine 3D technology with virtual reality. Taking the hero campus of Nanchang Institute of technology as the design case, this paper aims to develop a multi-user virtual campus roaming system, so that users can truly experience the campus environment anytime and anywhere through the network and roaming equipment, and play a positive role in the information construction of colleges and universities in the future.

1. Significance of system development

1.1. Strengthen and improve the information construction of the school
The implementation of virtual campus roaming system provides a three-dimensional information exchange platform for visiting users. By accessing the architecture and landscape model in the virtual roaming system, visitors can have an immersive access experience. Through this system, we can map the real campus environment in real life to the computer, and display it in front of visitors in many ways and flexibly through plane images, three-dimensional graphics and three-dimensional animation. Therefore, the design and research of this system is of great significance for the construction of campus informatization of Nanchang Institute of Technology[1].

1.2. It is conducive to improving the image and popularity of the school
In addition to teaching software and hardware resources, the school's image and campus environment are also favorable conditions to attract students. The research and application of the virtual campus roaming system can play a very positive role in the publicity work of the school. Compared with the traditional oral and poster publicity methods, the application of this system enables students and their parents to walk around the campus through the virtual campus roaming system through computers and virtual wearable devices, Simulate and feel the same campus environment as reality. The realization of virtual campus roaming system not only improves the publicity effect in the process of publicizing the school image, but also can be used as an auxiliary tool to strengthen the campus planning ability and improve the level of campus modern management. At the same time, it can also improve the popularity of the school and bring positive impact and objective social benefits to the school[2].

1.3. It is conducive to the construction of resource-saving and environment-friendly campus
In the past, most schools used paper brochures as the main publicity means for enrollment publicity. This traditional publicity method requires a large number of printed materials printed on paper, which is not conducive to resource conservation. Moreover, the brochures are still plane images, which can not bring visual and sensory impact to people. The design and development of virtual campus roaming system based on unity 3D is digital, visual, intelligent and networked. It can be reused many times after one development, and can give visitors a very real sensory experience. It not only saves resources, but also is novel. Lay a technical foundation for building a resource-saving and environment-friendly campus.

2. Development process of virtual campus roaming system based on unity 3D
According to the life cycle of software development, the system can be divided into three stages: preliminary work, system implementation, system testing and application. The specific contents of each stage are as follows: the preliminary work mainly includes the collection, sorting and analysis of relevant research literature on virtual campus development, virtual scene construction and roaming technology. On the basis of comparison and induction, combined with the application of unity 3D technology and its technical points, determine the model classification of Virtual Campus roaming system Theoretical framework and implementation path. At the same time, investigate the successful cases of unity 3D technology in other fields, especially in game applications, and enhance the technical feasibility of virtual roaming system[3].

Secondly, after analyzing the layout of the real campus environment, design and plan the development process of 3ds Max and unity 3D technology combination. The development of virtual campus environment first needs to build the model, including the model of terrain, buildings, campus scenic spots and other accessories. The virtual campus scene is built through 3dsmax, and then imported into unity 3D software for interactive design, including scene roaming, collision detection and voice interaction.

Finally, the virtual campus roaming system of the production number is imported into windows, Android, IOS and other system platforms. Through practical application, the prototype is constantly modified in the process of use, and finally the system is improved.

3. Design and production of virtual campus roaming system

3.1. Data collection and sorting

3.1.1. Application of Baidu map satellite map. As a virtual globe software, baidu map satellite map can arrange satellite photos, aerial images and GIS on a three-dimensional model of the earth. The production of this roaming system is to guide the location of roads and buildings through Baidu map satellite map software, and collect the proportion with the help of Baidu map satellite map. Locate the search location to the hero campus of Nanchang Institute of technology and take a screenshot, import it into 3dsmax as a reference map, and outline the overall distribution map of the campus with the road as a reference[4].

3.1.2. Combination of design drawings and field measurement. By consulting the design drawings and referring to the sand table model, record the size of campus office building, teaching building, graphic information center and playground, and measure the surrounding ancillary objects, such as information display screen, by using measurement tools. With the help of camera and mobile phone, collect the map texture of the building, collect the required map material and texture, and reduce the map by using the "map scaler" modifier in 3dsmax to avoid distortion caused by image reduction.

3.2. 3D model construction of virtual roaming system
The models in virtual campus scene can be divided into terrain model, building model and campus landscape model. Taking the road as a reference, combined with the actual landform of the campus, it is first constructed, and then each building complex, teaching building, playground, graphic

information center, college characteristic decoration, flowers, trees and other objects are constructed and placed in the corresponding position. Combined with the structural characteristics of the real object, the hybrid modeling method is analyzed and selected to build the model, and the optimal scheme of establishing the virtual campus scene model is analyzed and given.

3.2.1. Create terrain and trees. unity 3D software has built-in powerful terrain editor, which can quickly create trees, rocks and grass. It has the characteristics of detail mapping, integrated lightmaps and automatic filling optimization. Therefore, unity 3D is used to create terrain. The specific methods are as follows: click terrain || createterrain, and different terrain, flowers and trees can be established by operating the inspector panel. The tree model is established by first importing the material into the assets folder and then dragging it into the scene. Its parameters can be modified and edited in the inspector panel. You can also use 3dsmax and Photoshop software to supplement and edit some models such as trees, rocks and grass that are not in the built-in editor of unity 3D software, and then import them into the unity 3D scene editor[5].

3.2.2. Building model. The building model of the virtual campus roaming system is mainly created by 3dsmax. Taking the graphic information center building as an example, at the initial stage of modeling, the UAV is used to take multi angle photos of the graphic information center building with a fisheye lens, and combined with the relevant data of the design drawings to build the initial building model of the graphic information center, and then take photos by mobile phone to view the details of the scene, Get the data of the detail part, and then improve each detail part of the model through this data. In order to ensure that the built 3D model is proportional to the objects in the real world, "cm" is set as the system unit. After the building model is made, export it as a *. FBX format file that unity 3D can recognize and read, and place it in the assets folder of the project rollout.

3.3. Interactive design of virtual campus roaming system

In the virtual campus roaming system, the design of interaction is very important. It can not only simulate the motion law of the real world and give the visiting users a real user experience, but also explain some specific scenes through interaction. Therefore, the virtual campus roaming system should not only pay attention to the construction of campus model, but also strive to be true. We also need to design and implement a large number of campus interaction functions based on c# language and real scenes on the basis of the built campus model. For example, fast switching between two scenes, introduction of main scenes, and automatic roaming function of main scenes. Through these interactive designs, users can finally roam freely in the unity 3D environment to achieve the immersive effect.

4. Conclusion

Through the research and practical application of unity 3D virtual campus roaming system, the roaming scene model and map model are constructed in 3DMAX, saved as *. FBX format file and imported into unity 3D software. The free walking in the scene is realized through the keyboard and mouse, the interactive function of handle and gesture is realized through the code, and finally transplanted to multiple platforms for testing and application. The research of virtual campus roaming system not only builds a good platform for the digital construction of the college, but also makes a positive significance for the application in this field. Of course, there are still many problems in the system, such as artificial intelligence and scene interaction. In the future research, we will further strengthen the functional construction of these aspects, and finally complete a virtual campus roaming system which is very good in scene, model, interaction and intelligence.

Acknowledgment

This paper belongs to the scientific and technological research project of Jiangxi Education Department: The research of virtual campus roaming system based on unity 3D and MR technology. Topic number:GJJ202101, one of the results of the conclusion.

At the same time, I would like to thank the science and technology research fund of Jiangxi

Provincial Department of Education (GJJ191006) for its support and take it as one of its concluding achievements.

Reference

[1] Ge Y, Wu F, Wang Z H et al. 2019 Design and development of virtual campus roaming system based on unity 3D *[J]. Digital technology and application* **V37 No 348(06)** 177-179.

[2] Liu X and Li X H 2020 Virtual campus roaming system based on unity 3D *[J]. Science and technology horizon* **No. 300 (06)** 237-238

[3] Yin D W, Zhang X R, Li H H, et al. 2019 Research on Virtual Campus Roaming System Based on unity 3D and its key technologies *[J]. Software engineering* **(9)**

[4] Zhang F 2020 Development and application of roaming interactive system of virtual school history museum based on unity 3D -- Taking the school history museum of Liaoning Transportation College as an example *[J]. Wireless Internet technology* **V.17; No.187(15)** 62-63

[5] Sun Y N, Zhao H D, Sun D Q 2020 Design of MTS virtual simulation experiment system based on unity 3D *[J]. Laboratory research and exploration* **v.39; No.293(07)** 103-105

Underwater Sonar Image Segmentation by a Novel Joint Level Set Model

Yue Wang[1], Kefa Zhou[2,3], Wei Tian[1], Zhe Chen[1,*], Dewei Yang[2,3]

[1]Computer and Information college, Hohai University, Nanjing 211100, China
[2]Nanjing Hydraulic Research Institute, Nanjing 210029, China
[3]Country Dam Safety Management Center of the Ministry of Water Resources, Nanjing 210029, China.

[*]corresponding author's e-mail: chenzhe@hhu.edu.cn

Abstract. This paper proposes a novel joint level set model for underwater sonar image segmentation. Combining features in points and regions in our novel joint level set (LS), it can achieve excellent performance for underwater sonar image segmentation. Regional information guides the model to locate the object of interest, whereas the point information accurately delineates contours. In addition, the unified Markov random field (UMRF) is taken to measure the neighboring relation between points and regions, which can overcome the problems of the high speckle noise, strong bias and low resolution of underwater sonar images. Our novel model can segment underwater sonar images into three partitions, such as the objects of interest, shadow and backgrounds. In contrast to current segmentation methods, outstanding results are demonstrated by our model. Moreover, another advantage of our model lies in its high efficiency.
Keywords. Side-scan sonar; Image segmentation; Joint level set; Unified Markov random field

1. Introduction

Due to the low attenuation of the acoustic beam, side-scan sonar imaging technologies have been well developed for underwater exploration, such as underwater navigation, salvage and resource exploitation [1-3]. However, the intrinsic problem for underwater sonar imaging technology lies in the high speckle noise, strong bias and low resolution, which pose challenges for image segmentation [4]. In specific, the noise and intensity bias will make the segmentation results deviate from objects of interest, the low resolution of sonar images seriously reduce the accuracy of segmentation results. In general, the objectives of various strategies for underwater sonar image segmentation are to remove the influence of image noises and improve the accuracy [5]. For example, the regional models can locate the global saliency that indicates the existence of objects. However, a smooth effect can be found at the regional edges for these regional models, presenting fuzzy segmentation results [6,7]. Edge or pixel feature-based models are much sensitive to the region contours, but likely fall into local convergence. Recently, joint models have demonstrated their outstanding performance in image segmentation.

Underwater sonar images contain drawbacks in multiple aspects, and diverse performance has been achieved by different strategies. Level set (LS) models can find the best optimization, which can delineate the details of region contours. Alternatively, the unified Markov random field (UMRF) has a better performance in noise removal, since it considers the neighboring relation in local regions. Integrating these two models can provide a tradeoff between model accuracy and robustness. This idea motivates the work in this paper. In specific, UMRF is combined with LS to establish a novel model type. In practice, our model has a capability to jointly exploit the pixel-level, region-level and neighboring correlation information for comprehensively delineating contours and removing noises.

2. Related works

2.1. Sonar image segmentation

In contrast to the optical imaging technology, sonar imaging has its excellent advantage under water. For example, the sonar imaging has a long-range visibility, whereas the images contain a large amount of noises and illumination bias which are two main reasons for performance degeneration of underwater image segmentation [8, 9]. Aiming to solve this problem, there are two typical methods: one is based on the supervised strategy and the other one introduces the learning strategy. In general, previous methods are independent of the model training process and achieve segmentation by evidence in the image for segmentation. Typical samples include LS and MRF models [8-12]. The LS models have a good accuracy but they are much susceptible to the noise points which may make model convergence fall into the local optimization. Moreover, if the LS initialization should be close to the objects of interest, whereas an improper initialization will cause a large time cost and degenerated results. Aiming to solve this problem, Bayes theory-based statistical models have been introduced for sonar image segmentation, such as the MRFs. This type of model has a better ability in noise removal, whereas they have large model scales and low efficiency in image segmentation. There are also clustering-based segmentation methods, such as the fuzzy C-means based sonar image segmentation methods.

Later, the development of the machine learning methods has motivated the supervised strategy for underwater sonar image segmentation [13-15]. Various shallow and deep architectures have been evaluated in this field. However, for these learning-based method a main drawback lies in the requirement of the model training effort where a large training dataset is required. Although samples on ground are sufficient for model training, few underwater samples are available for underwater models, especially underwater sonar images. Hence it is somewhat questionable for applying deep learning architectures in underwater sonar image segmentation. They likely fall into an overfitting state that the model generalization seriously degenerates. Another challenge for the deep learning-based underwater sonar image segmentation methods lies in their high computational cost. Different from the devices on ground, the long-term power supply is difficult for underwater devices, so that few underwater devices can support large computational loads.

2.2. Level set

Historically, various level set model-based methods have won successes in many tasks. Existing level set models generally fall into two categories: edge-based and region-based models. A typical edge-based model is the Local Binary Fitting (LBF) model [16,17] which develops a regularization term to prevent the deviation of level set function (LSF) by using the signed distance:

$$E(\varphi)=\lambda\int_{\Omega} g\delta(\varphi)|\nabla\varphi|\,dxdy+v\int_{\Omega} gH(\varphi)\,dxdy+\mu\int_{\Omega}\frac{1}{2}\left(|\nabla\varphi-1|^2\right)dxdy\ , \tag{1}$$

where g is the edge stop function, $H()$ is the Heaviside function and $\delta()$ is the Dirac function. There are three terms in the LBF model. The first is the penalization of the contour length, the area inside the level set is penalized by the second term and the third term is the regularization function which maintains the regularity of LSF and therefore eliminates the requirement of periodical re-initialization. These edge-based models are much sensitive to the image deviation which strongly indicates the tumor contour. However, the drawback of the edge-based models lies in their sensitivity to the noisy points and textures, which may result in the boundary leakage.

Above issue motivates the development of the region-based methods. In contrast to the local fitting, this type of models is sensitive to the regional features, thus they have more opportunities to locate the globally most salient regions:

$$E(c_1,c_2,\varphi)=\varepsilon\cdot Len(\varphi)+\beta_1\cdot\int_{inside(\varphi)}|u_0(x,y)-c_1|^2dxdy+\beta_2\cdot\int_{outside(\varphi)}|u_0(x,y)-c_2|^2dxdy, \quad (2)$$

where c_1 and c_2 are the averages of the intensity features $u_0(x,y)$ inside and outside the zero level set φ , ε, β_1,β_2 are the positive moderation parameters. Assuming that pixels within a region are largely homogeneous, the region-based model, such as the Chan-Vese (CV) model, is robust against point-like noises, but maybe degenerated if hue transitions in complicated scenes are more obvious than object appearances..

3. Proposed model

A novel joint level set model is proposed in this paper. The joint form can be demonstrated in two aspects: the first is the integration of the LS and MRF models and their merits can be synergistically exploited, and the second is the combination of the edge and region features to correctly locate and accurately delineate partitions. In specific, UMRF is calculated as the energy term in LS.

$$E=E_{UMRF}+E_{reg} \quad (3)$$

where the first term is calculated by UMRFs and the second term maintains the regularization of LS evolution.

UMRF term E_{UMRF}

Recall that both the pixel-level information and region-level information are combined in our novel joint model, and UMRF is taken advantage to generate energy terms:

$$E_{UMRF}=E(X_{object}|Y)H(\varphi_1)(1-H(\varphi_2))+E(X_{shadow}|Y)H(\varphi_2)(1-H(\varphi_1))+E(X_{background}|Y)(1-H(\varphi_1))(1-H(\varphi_2)), \quad (4)$$

where $E(X_{object}|Y)$, $E(X_{shadow}|Y)$ and $E(X_{background}|Y)$ measure the energy of the object, shadow and background regions, φ_1 and φ_2 are LS functions (LSFs) for object and shadow regions, and

$$E(X|Y)=-\log P(X|Y)=-\log\left[P(Y^P|X)P(Y^R|X)P(X)\right], \quad (5)$$

As aforementioned in Section 2, $P(X)$ is of Gibbs distribution which is to estimate the correlation between points x_c in a local region *C*:

$$P(X=x)=(1/Z)\exp(-\sum_{c=C}U(x_c,x_{N_c})), \quad (6)$$

where the clique energy function of the Gibbs density $U(x_c,x_{N_c})$ is defined:

$$U(x_c,x_{N_c})=\begin{cases}-\beta\exp\left(-\left|f(x_c)-f(x_{N_c})\right|\right), & if\ f(x_c)=f(x_{N_c})\\ -\beta\left(\dfrac{2}{1+\exp\left(-\left|f(x_c)-f(x_{N_c})\right|\right)}-1\right), & if\ f(x_c)\neq f(x_{N_c})\end{cases}, \quad (7)$$

where x_{N_c} is the neighboring points of the point x_c, $f(x_c)$ and $f(x_{N_c})$ are the features in the points. For the pixel-level information $f(\bullet)$ denotes the intensity, and it denotes the textual density for the regional feature.

$P(Y|X)$ is a Gaussian distribution:

$$P(Y|X)=\prod_{c\in C}\frac{1}{\sqrt{2\pi}\sigma_{X_c}}\exp\left(-\frac{(I_c-\mu_{X_c})^2}{2\sigma_{X_c}^2}\right), \quad (8)$$

where μ_{X_c} and σ_{X_c} are the means and standard deviation of the features.

Regularization term
The second term in Eq.1 is to regularize the level set evolution by combining the length and potential factors:

$$E_{reg} = \nu L(\varphi) + \tau P(\varphi), \tag{9}$$

where v and τ are the moderation parameters, $L(\varphi)$ and $P(\varphi)$ are the length and potential regularization terms:

$$L(\varphi)=\int \left|\nabla H(\varphi_1)\right| dx + \int \left|\nabla H(\varphi_2)\right|, \tag{10}$$

$$P(\varphi)=\int p(\left|\nabla \varphi_1\right|)dx + \int p(\left|\nabla \varphi_2\right|)dx, \tag{11}$$

where ∇ is the gradient operator and $p(x)=0.5(x-1)^2$, such that $P(\varphi)$ reach the minimum when $\left|\nabla \varphi_1\right|=1$ and $\left|\nabla \varphi_2\right|=1$ which is the characteristic of a signed distance function.

Energy minimization
As shown in Eq.2, two LSFs are included in our novel model to generate three partitions over the whole image. A two-phase optimization strategy is proposed in this paper. First, the gradient flow calculation is operated with the LSF φ_1 and the second-phase optimization is achieved with the LSF φ_2.

$$\frac{\partial \varphi_1}{\partial t} = \delta(\varphi_1)\left\{\left[E\left(X_{\text{object}}\left|Y\right.\right) - E\left(X_{background}\left|Y\right.\right)\right]\left[1 - H(\varphi_2)\right] + \left[-E\left(X_{shadow}\left|Y\right.\right)\right] H(\varphi_2)\right\} + \nu\delta(\varphi_1)\, div\left(\frac{\nabla \varphi_1}{\left|\nabla \varphi_1\right|}\right) + \tau div\left(d_p\left(\left|\nabla \varphi_1\right|\right)\nabla \varphi_1\right), \tag{12}$$

$$\frac{\partial \varphi_2}{\partial t} = \delta(\varphi_1)\left\{\left[-E\left(X_{\text{object}}\left|Y\right.\right)\right]H(\varphi_1) + \left[E\left(X_{\text{shadow}}\left|Y\right.\right) - E\left(X_{\text{background}}\left|Y\right.\right)\right]\left[1 - H(\varphi_1)\right]\right\} + \nu\delta(\varphi_2)\, div\left(\frac{\nabla \varphi_2}{\left|\nabla \varphi_2\right|}\right) + \tau div\left(d_p\left|\nabla \varphi_2\right|\nabla \varphi_2\right), \tag{13}$$

where *div*( ) is the divergence operator and ∇ is a Laplacian operator.

Then, the LSF can be stepwise updated by

$$\varphi^{t+1} = \varphi^t + \Delta\varphi^t * \Delta t. \tag{14}$$

4. Experimental analysis

In this section, experiments are conducted on excessive underwater side sonar samples to analyse the performance of our proposed model. The performance of our proposed model is demonstrated by comparing to the state-of-the-art models for underwater sonar image segmentation. At the same time, the time cost of different models is recorded to present the computational efficiency. Quantitative results are evaluated by the Dice coefficient that measures the overlaps of the segmented region to the ground truth:

$$Dice(P_{object}, P_{shadow}, T_{object}, T_{shadow}) = \left(\frac{\left|P_{object} \wedge T_{object}\right|}{\left(\left|P_{object}\right| + \left|T_{object}\right|\right)/2} + \frac{\left|P_{shadow} \wedge T_{shadow}\right|}{\left(\left|P_{shadow}\right| + \left|T_{shadow}\right|\right)/2}\right) \Big/ 2, \tag{15}$$

where P is the segmented area, T is the ground truth. For the Dice coefficient, a large value means a better performance.

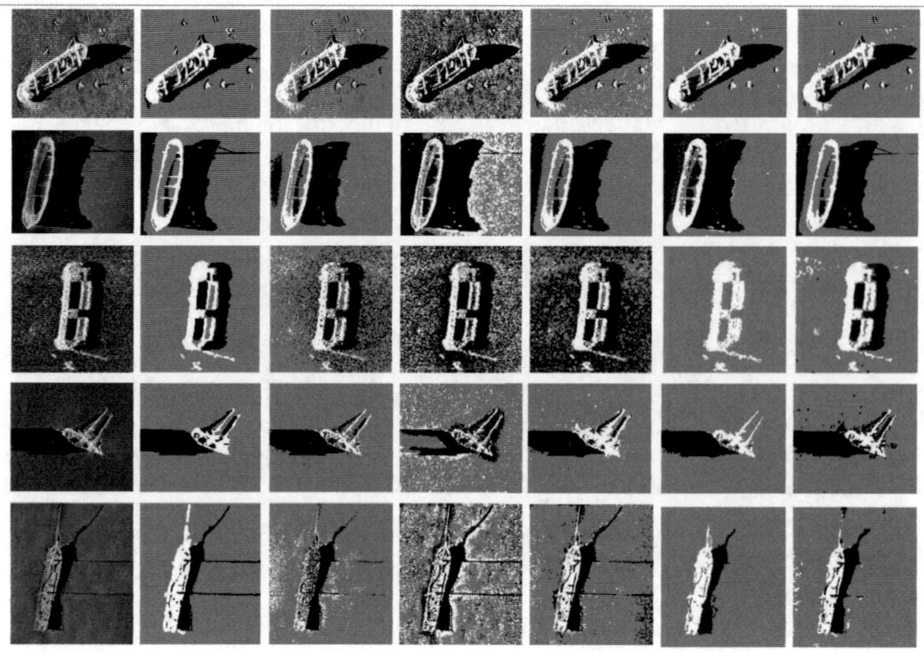

(a) (b) (c) (d) (e) (f) (g)

Figure 1. Experimental comparisons, (a) original images, (b) ground truth, (c) edge-level LS(MS model), (d) region-level LS(LIF), (e)pixel-level MRF(ICM-MRF), (f) region-level MRF(MS-MRF), (g) ours.

In this subsection, our novel model is compared with the state-of-the-art models in the field of underwater sonar image segmentation, such as the edge-based LS (MS model) [20], region-based LS (LIF) [21], pixel-level MRF (ICM-MRF) [18], and region-level MRF (MS-MRF) [22]. The experimental comparison results are shown in Fig.1. From this result, various segmentation performances can be obtained by different methods. In general, the edge-level LS is much sensitive to the texture features, so that it performs well in the case with few noises. However, edge-level and pixel-level models degenerate when there are heavy image noises. Alternatively, region-level models have better robustness against imaging noises, while they can not classify backgrounds and shadows. The high-density noises will be mistaken as the object appearances, causing a large part of false segmentation, as shown in Fig.1(d). The best performance is provided by our novel model which exploits the advantage in both LS and MRF. In specific, the accuracy of LS and robustness of MRF for image segmentation are comprehensively developed. The quantitative evaluations are presented in Tab.1, from which the outstanding performance of our novel model is clearly described.

Table 1. Performance comparisons

Sample	1	2	3	4	5
MS model	0.8596	0.8803	0.8516	0.9044	0.7278
LIF	0.8064	0.6392	0.6734	0.7244	0.5708
ICM-MRF	0.8753	0.8202	0.8629	0.9331	0.4628
MS-MRF	0.8740	0.8244	0.8801	0.8237	0.9097
Ours	0.9531	0.9358	0.9756	0.9614	0.9575

5. Conclusion

A joint LS model is proposed in this paper, which combines the regional and edge features in a unified form, so that our model can achieve a good performance in segmentation accuracy and robustness. Moreover, the local neighboring relation is modeled by the UMRF model which can largely prevent the influence of heavy noises. Different from the image segmentation tasks on ground, samples under water are limited and can not provide enough prior for segmenting underwater sonar images. This problem impedes the development of machine learning-based strategies for underwater sonar image segmentation. To solve this problem, the optical imaging provides complementary information which can be extracted as the guidance for underwater sonar image segmentation.

Acknowledge

This work was supported in part by the National Key Research and Development Program of China (Grant No. 2018YFC0407106), National Natural Science Foundation of China(51909174 and 61801169).

References

[1] Ye X F, Yang H B, Li C L, et al. 2019 A Gray Scale Correction Method for Side-Scan Sonar Images Based on Retinex *Remote Sensing* **11(11)**
[2] Lu Z, Chen Y, Zhang T, et al. A Sonar Image Segmentation Algorithm based on Two-Dimensional Spatio-Temporal Fuzzy Entropy. *2018 IEEE 8th Int. Conf. on Underwater System Technology: Theory and Applications (USYS)*
[3] Marco R and Ovidio S 2016 Seafloor analysis and understanding for underwater archeology. *Journal of Cultural Heritage* **24** 147-156
[4] Kaeser A, Litts T and Tracy T 2013 Using low-cost side-scan sonar for benthic mapping throughout the lower flint river. *River Research and Application* **29(5)** 634-644
[5] Ran S, King N, Li S and Li H 2018 Interactive object segmentation in two phases *Signal Processing: Image Communication* **65** 107-114
[6] Liu G, Zhao Z and Zhang Y I2015 mage Fuzzy Clustering Based on the Region-Level Markov Random Field Model. *IEEE Geoscience and Remote Sensing Letters* **12(8)** 1770-1774
[7] Gao R L, Wang C J and Wang S 2016 Color Image Segmentation Based on Local Region MRF model *2016 Int. Conf. on Management Science and Engineering(ICMSE 2016)* 89-94
[8] Liu J, Jiang P and Zhu H 2021 A Local Region-Based Level Set Method With Markov Random Field for Side-Scan Sonar Image Multi-Level Segmentation. *IEEE Sensors Journal* **21(1)** 510-519
[9] Estellers V, Zosso D, Lai R, et al. 2012 Efficient algorithm for level set method preserving distance function *IEEE transactions on image processing* **21(12)** 4722-4730
[10] Shao X, Lin X and Shang T 2017 Liver CT image segmentation algorithm research based on CV model *2017 3rd IEEE Int. Conf. on Computer and Communications (ICCC)* 1889-1892
[11] Wang H, Huang T Z, Xu Z, et al. 2014 An active contour model and its algorithms with local and global Gaussian distribution fitting energies *Information Sciences* **263** 43-59.
[12] Gao M, Chen H, Zhang S, et al. 2016 Texture image segmentation using fused features and active contour *2016 23rd Int. Conf. on Pattern Recognition (ICPR)*; 2036-2041.
[13] Bai J W, Li P A and Wang K H 2016 Automatic whole heart segmentation based on watershed and active contour model in CT images. *2016 5th Int. Conf. on Computer Science and Network Technology (ICCSNT)* 741-744.
[14] Ding G, Song Y, Guo J, et al. Side-Scan Sonar Image Segmentation using Kernel-based Extreme Learning Machine. *2017 IEEE UNDERWATER TECHNOLOGY (UT)*
[15] Zhang L and ,Ji Q 2010 Image segmentation with a unified graphical model. *IEEE transactions on pattern analysis and machine intelligence* **32(8)** 1406-1425.
[16] Chan T F and Vese L A 2001 Active contours without edges. *IEEE Transaction on Image Processing* **10(2)** 266-277
[17] Li C, Kao C Y, Gore J C, et al. 2008 Minimization of region-scalable fitting energy for image segmentation. IEEE Transaction on Image Processing **17(10)** 1940-1949
[18] Peng Z, Wee W and Lee J H. 2005 MR brain imaging segmentation based on spatial Gaussian mixture model and Markov random field. *IEEE Int. Conf. on Image Processing* 1-313
[19] Li S Z 2009 *Markov random field modeling in image analysis* (London: Springer)
[20] Vese L A and Chan T F 2002 A multiphase level set framework for image segmentation using the Mumford and Shah model *Int. J. of Computer Vision* **50(3)** 271-293

[21] Li C, Huang R, Ding Z, et al. 2011 A Level Set Method for Image Segmentation in the Presence of Intensity Inhomogeneities With Application to MRI. *IEEE Transactions on Image Processing* **20(7)** 2007-2016

[22] Zhang B, Ma G R, Zhang Z, et al. 2013 Region-based classification by combining MS segmentation and MRF for POLSAR images. *J. of Systems Engineering and Electronics* **24(3)** 400-409.

Application Research on Key Points of Software Security Development Technology

He Zhou

School of Computer and Information Engineering, Nanchang Institute of Technology, Nanchang, Jiangxi 330044, China

*corresponding author's e-mail: 51001202@qq.com

Abstract. With the rapid development of network technology, software security has attracted more and more attention. Based on the software security lifecycle presents an improved suitable for small and medium-sized enterprise software security development process, make the software safety to cover the whole software life cycle, improve the level of safety of encoding software. The model used in the software development phase, test phase and release phase three software security key technologies are studied and discussed in detail. In addition, it also designed a software security development management platform, the software development process so that the model can better apply to the actual enterprise to the platform. After a number of units for a period of time of trial, get a better evaluation of the households.
Keywords. Software security; key technology; development; network

1. Introduction

The software product with everything else, but also through birth, birth, growth, maturity, decline stage, in software engineering we commonly known as the software life cycle [1]. Through the whole life cycle of software is divided into several stages, each stage has a clear mandate, the scale is big, easy to control and management of complex software the development and management of the complex [2]. Usually, the software life cycle including feasibility analysis and development plan, requirement analysis, design, encoding, testing, maintenance and other activities, these activities can be in a proper way assigned to different stages to complete [3]. This method divided the process is a principle, in software engineering, which follow the prescribed order step by step, each stage, have defined work, review, document for communication or for reference, in order to improve the quality of software.

This stage is to discuss the software development and demand, mainly to determine the development goals and the feasibility of the software [4]. The feasibility study is usually the feasibility of the project report to top management for software development funds, usually from the economic, business or organization, and technology in project feasibility assessment. In addition, feasibility to continue the project tracking and research progress of information management system of the evaluation team also can be used as a reference [5]

In the software development is feasible; the detailed analysis of the various functions of the software needs to realize [6]. The demand analysis stage is a very important stage, this stage is well done, will be a good foundation for the software development project success lay. "The only constant is change itself, the same." demand is constantly changing during the whole process of software development and in-depth, so we must make changes in demand plan to deal with this change, in order to protect the smooth progress of the project.

This stage is mainly based on the results of requirement analysis, the design of the function and operation of the entire software system, such as system framework design, database design, user interface layout, business rules, and system flow and so on. At the end of this period, we will make the whole system design described as a series of modules or subsystems of software design [7]. Generally divided into the overall design and detailed design for each demand system, will be on or make a prototype system to produce one or more key points of the design. The software design phase is a crucial stage in the whole software life cycle. In the design stage if there is some design errors later, in the software you want to restore the price is very expensive. And good software design for software programming and lays a good foundation.

To go through rigorous testing in the software design is completed, to found in the whole design process of software problems and correct. The whole test process is divided into three stages of unit testing, assembly test and system test. The test methods include white box testing and black box testing in two. In the course of the testing a detailed test plan and test strictly according to the test plan, test to reduce the randomness.

2. Overview of community Gymnastics Teaching

Traditional software development model mainly includes the waterfall model, spiral model, incremental model. They mainly focus on the development of the quality of the software process, the development efficiency, and not too much attention to software security. Due to the lack of emphasis on security model, to bring a variety of security risks to the software products in the development process of software project, the security of the software cannot be improved, may eventually be due to some security vulnerabilities in the software to software developers and end users have suffered varying degrees of loss.

2.1. Disadvantages of traditional gymnastic teaching methods

In order to make the software to deal with current security threats, it is necessary to improve the safety awareness in the process of software development, the software in the whole life cycle of the safety of the software, improve software security products. In this paper, the software security development processes are introduced, and makes comparison and analysis on them at present the software security development process mainly needs more resources based on small and medium enterprises, barriers to the status quo in the application process, the software security development lifecycle is proposed based on an improved suitable for small and medium-sized enterprise software security development process, make the software safety to cover the whole software life cycle, safety and cost saving enterprises, improve the level of software security.

At present, the security threats faced by software products, which mainly includes two aspects: the people's attention. On the one hand is a software product of the crack, illegal dissemination and use, and security vulnerabilities on the other hand from the attacker for the software product itself for the attack. The two, whether it is from which aspects of the threat, will bring huge losses to the enterprise.

The problem of copyright protection on software, while software developers tried various ways to use disk protection, hardware protection, software encryption shell and other technical means, hackers can always find ways to solve the corresponding, to completely prevent piracy is not too realistic. And there are some security vulnerabilities in software is often break and software are the relationship between hackers can obtain information related to the logical flow of software through software security vulnerabilities, which makes it possible to break the software.

The problem of copyright protection of software are mainly on the illegal copying and illegal use of software, namely software piracy software. As a product often will bring huge economic and social benefits, and its value has also been more attention. And because of the characteristics of digital software, the software has the cost of reproduction low copy, high efficiency characteristics, so often become the object of copyright infringement. From the development of Internet technology, but also to the spread of digital products including software has brought great convenience, so that the issue of intellectual property protection has become increasingly prominent as shown in figure 1.

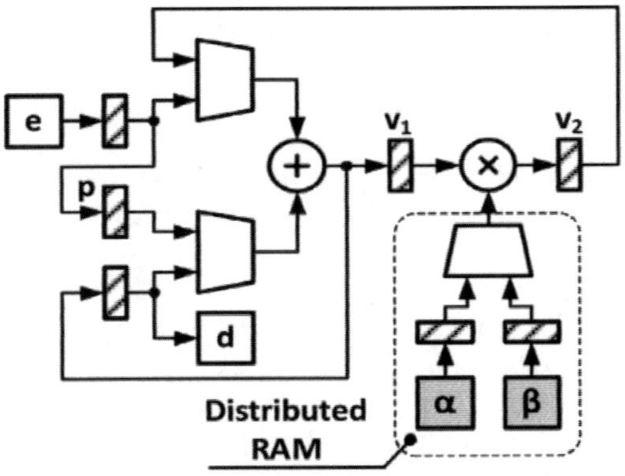

Figure 1. Software internal control flow chart

2.2. Feasibility analysis of application of multimedia technology in the teaching of Gymnastics

With the development of the concept of cloud computing, more and more software available to users through the network service, which undoubtedly brings users great convenience, users need only a browser can be used for shopping, entertainment and communication services of all kinds. For enterprise users, enterprises can also be through the release of information, network services, network marketing, customer service, and business partners as well as online contract or commodity trading business. However, due to security vulnerabilities of software services and browser vulnerabilities, steal information, data tampering and other illegal attacks it is often seen. Often cause serious harm, and losses to the use of software users and software network the service of software developers.

The Internet enables people to use the software more convenient, but also give the attacker the opportunity to bring more risk to the software. The open nature of the Internet makes the hacker software system of enterprise development at the same time because whenever and wherever possible access to the platform is open to the enterprise is very difficult to distinguish between legitimate users and hackers. This makes hackers than in the past the more time and space to attack the software system, found in the software security defects and use it.

At present, the software application system scalability are relatively high, often through the installation and update the plug-in to the system functions are expanded to meet the changing needs of users. For example, support plug-ins and browsers on its official website provides a large number of various functions by third party development plugin for the user to choose the use of the operating system can support a variety of hardware devices by dynamically loading device driver according to their own needs. The scalability of the software brings users great freedom and can be customized, but to the security of software has brought great challenges. Because the third party development level encoding expansion function and attention the degree of safety of the uneven, often lead to a security vulnerability in the extended function cause the entire software security defense collapsed as shown in formula (1).

$$\frac{-x \pm \sqrt{x^2 - 4yz}}{2y} = \sqrt{x^2 + z^2}\,\sqrt{x^2 - 4yz} \tag{1}$$

Most programmers think in the preparation of the program the program is run in a way. In fact, the modern computer is very fast in task switching, in multi-core, multi or distributed system, often have a time to happen at the same time the exact time. Between the programming ideas and the real situation of existence a lot of program weaknesses. These security vulnerabilities by thread, process, time, and state between the non expected interaction caused.

3. The development and application of multimedia technology in the teaching of community sports

Error and error handling on behalf of a class, but with the error handling security issues is very common, and as a separate category. There are two kinds of error handling cases will have security vulnerabilities. One is the lack of error or no error handling, a situation this is the most common. The other one is in error when leaked too much information, or processing error is safe.

3.1. The necessity of making gymnastic courseware

At present, the Internet has become a very important platform, and with the concept of cloud computing, many enterprises are based on the development of software system, to provide customers with more convenient and efficient service support. These applications in function and performance, are constantly improving and improving, but in safety very important, but did not get enough attention. Most of the enterprises will be a large number of investment spending in the network and server security, there is no guarantee that the application of its own security in the real sense, to hackers. According to the annual report, in the application security vulnerabilities list of the ten most serious were the top three cross site scripting, injection vulnerabilities and malicious files. The cause of the implementation [three vulnerabilities is direct and not to do enough to check user input caused. Seven vulnerabilities are also and there is more or less user input.

XSS vulnerability is due to a system without a good filter out user submitted content in the executable script caused. Using XSS vulnerabilities, an attacker can remotely submit malicious scripts into the system to access the system; users will perform these attacks from malicious script by ". A XSS attack usually to the attacker, and attack the client site out of the three groups.

The user must suffer by their own initiative to complete the submission to make malicious scripts, malicious scripts can access to the user's privacy information, so the attacker will usually constructs a malicious script link, using a variety of methods to lure victims of active users to click on the link in order to complete the attack as shown in figure 2.

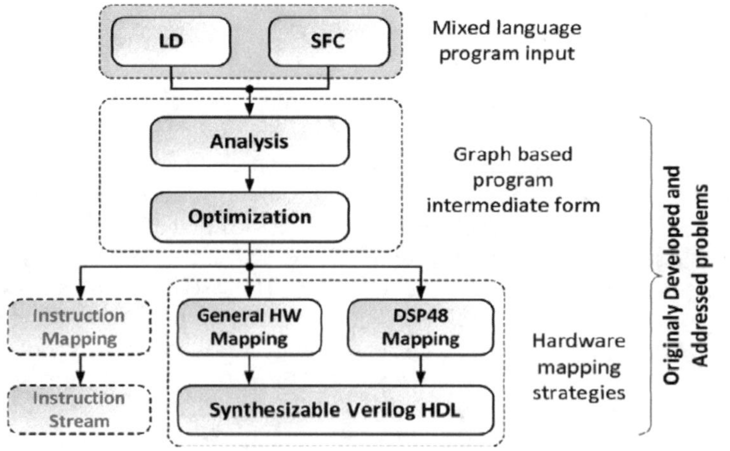

Figure 2. The software flow chart analysis

The above example is the use of a non-persistent cross site scripting attacks. This kind of vulnerability is usually due to the usual user supplied data is parameters in the request or submit the page from the data in the form; the server did not filter directly used to generate returns to the user page results formed. Submit data is not stored in the server. This case is as shown in figure 3.

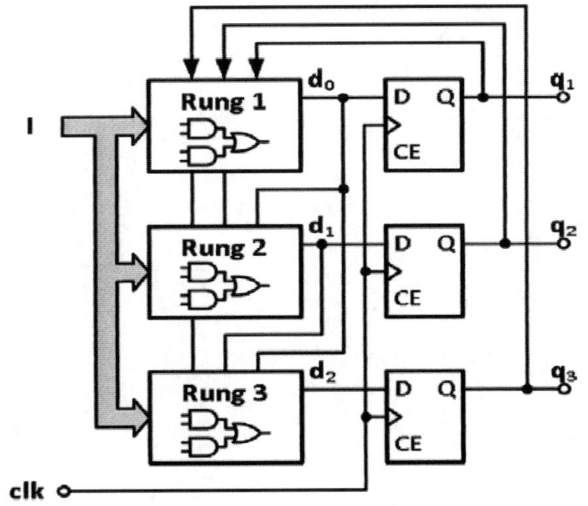

Figure 3. Control gate program

3.2. The function, principle, method and condition of developing multimedia technology

Persistent XSS vulnerability is usually more serious. When kept in the server stored in the database such as persistent attacker submitted data, then the data is not verified displayed on other users usually may access the page, will be stored in persistent XSS vulnerabilities. A typical example is that if an online forum allows users to use directly in the code at the post, it will cause persistent XSS vulnerabilities, because online forum posting is usually stored in the database, and posting content can be all users to see. A big problem existing in software security is that many software systems do not have a good sense of safety in the construction process. Software developers mostly take a better attitude; will focus more on the repair of software security issues, rather than an effort to create safe software from the beginning.

Even some software developers have a misunderstanding on software security issues, often the safety investment used in the procurement of equipment such as firewall, old for protection from the network layer attacks, leaving only the security of the software itself for the rest. In fact, the statistical data presented show that only the attack is realized by network protocols or hardware vulnerabilities, and the hacker attacks on the application layer, the use of software security vulnerabilities exist in the attack. For these attacks using software vulnerabilities, security devices in the network layer can play a protective role. Only the safety investment against a possible attack this is also the reason that some software security is poor as shown in figure 4.

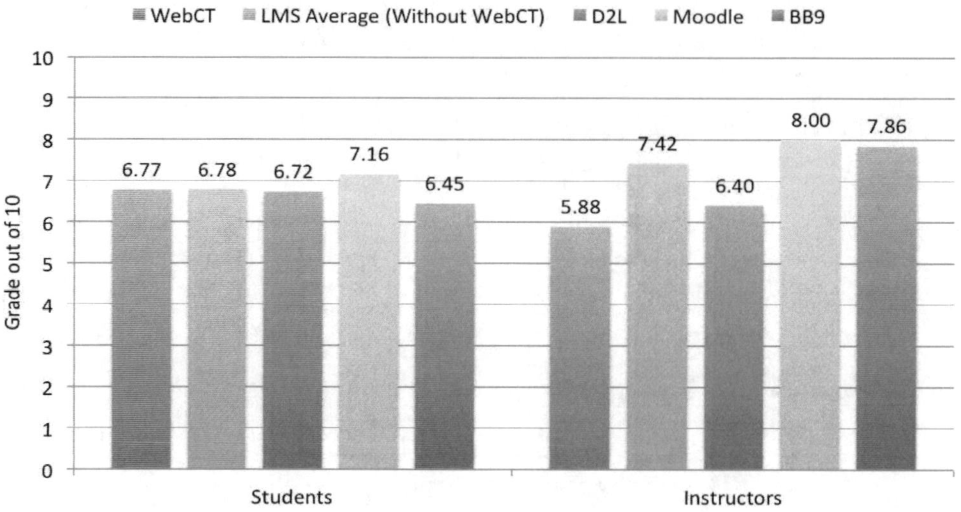

Figure 4. User praise

4. Discussions

Although the enterprise in the software development process for software security problems have been more and more attention, but the application of the software security development process is still not ideal. The survey report pointed out that the scale of the enterprise is one of the important factors that influence the enterprise whether the security development process in the software development process. At present, several major software security development process needs large resources of large enterprises, they can afford more safety tools and safety activities overhead for funds, the lack of human resources of small and medium enterprises to say, application software security development process, there is still a significant obstacle. Therefore, it is necessary for the lack of resources of small and medium sized enterprises some consider must optimize the software security development process.

On the basis, the process is simplified, a better applied to the lightweight software security development process of small and medium enterprises, and the first two chapters put forward dynamic watermarking based on dynamic data dependence graph, software protection technology of dynamic security analysis of dynamic data dependence and some other existing software security technology based on the combination of. Has the same security as the core, to software security coverage to the characteristics of each stage of the software life cycle.

And different, involving safety software security development process proposed by this paper, is not fully in accordance with the order, but there is overlap in time, in order to save time and cost. We also emphasize the use of free security tools, and security tools for automatic configuration, a part of the regular automatic safety activities the work, save money, time and manpower. Software security development process is the goal of software security issues in the design and implementation of the reduction in the software development life cycle as early as possible during the macro grasp of software safety in the whole process of software development, improve software security, severity reduce the impact of any security issues not found in the software.

5. Conclusion

The work of this paper is the first from the two aspects of copyright protection and security vulnerabilities summarizes the security threats facing the software. In order to make the software to deal with current security threats, it is necessary to improve the safety awareness in the process of software development, software in the whole life cycle of software safety concerns, improve safety software products. Based on the software security development processes are introduced, and they are compared and analyzed. Then the current software security development process mainly needs more resources based on small and medium enterprises, barriers to the status quo in the application process, based on software security development life cycle forward improved suitable for small and medium enterprises software development process, software security is able to cover the entire software life cycle, save enterprise security Total cost, improve the level of software security. The software security development process based on the improved design and implementation of a software development management platform, the purpose is to enable enterprises to better in the software development process of application software security development process, the software development process of effective management, as far as possible early discovery and repair security vulnerabilities in the software, security the final software was significantly improved. The platform after a number of units for a period of time after the trial, get a better evaluation.

Acknowledgement
Project Fund: Collaborative Education by the Ministry of Education, Fund No. 202101196010, Application of Network Crawber in Big Data Teaching.

References
[1] Evren G 2016 Preparation for Central Common Examination is not a Torment but Fun *Procedia Social and Behavioral Sciences* **232**(24) 47-56
[2] Hu Y N 2017 Research on the Application of Fault Tree Analysis for Building Fire Safety of Hotels *Procedia Engineering* **135**(1) 45-6
[3] Flaherty G T 2016 Research on the move: the potential applications of mobile health technology in travel medicine *Journal of Travel Medicine* **23**(6) 49-55
[4] Maja S J, Radmila B, Bojan L 2016 A Proposal for Research Framework Based on The Theoretical Analysis and Practical Application of MLQ Questionnaire *Conomic Themes* **4**(4) 5-6
[5] Petros T, ikolaos S 2016 Editorial to the IFSET Special Issue "Advances in research and applications of nonthermal technologies for food processing and preservation" based on the 2015 International Nonthermal Processing Workshop *Innovative Food Science and Emerging Technologies* **38**(1) 220-7
[6] Pattanapong W 2016 Errors in translation made by English major students: A study on types and causes *Kasetsart Journal of Social Sciences* **783**(26) 89-102
[7] Agnes A O 2017 The Abolishment of Colleges and its Implications on ELT in Nigeria *Procedia - Social and Behavioral Sciences* **232**(24) 567-9

Research on the Development and Application of Parallel Programming Technology in Heterogeneous Systems

Linghong Wu

School of Computer and Information Engineering, Nanchang Institute of Technology, Nanchang, Jiangxi 330044, China

*corresponding author's e-mail: 103294693@qq.com

Abstract. In recent years, with the continuous improvement of high performance computing hardware platform,the application of high performance for a wide range of high-performance, traditional parallel programming environment tools cannot meet the growing demand for high-performance applications and the face of parallel software development environment in a relatively backward state. The heterogeneous system of the new coprocessor with a high performance and power ratio can have high performance computing needs to provide the task to accelerate. As a result, heterogeneous systems are becoming a new trend in high performance computing. However, with the introduction of the new coprocessor and the expansion of the system scale, heterogeneous systems are difficult to program the problem is also increasingly prominent. In view of this problem, this paper adopts the abstract technology to enhance the granularity of parallel programming, and designs a structured parallel programming prototype system which can achieve different levels of implicit parallelism
Keywords. High performance computing; Heterogeneous systems; Programmability; Collaborative programming model

1. Introduction

In the international high-tech increasingly competitive today, high-performance parallel computing technology and its application level has become a show of national science and technology strength of one of the signs. In the process of developing a new generation of computers, a prominent feature is the use of parallel technology to create high-performance computers; parallel technology has become an effective means to achieve high-performance computer computing power [1]. In the new application context, high performance computing has a broader application environment compared with traditional parallel computing, and the application field of parallel computing is gradually tilted from the field of scientific research to commercial applications.

With the continuous expansion of parallel computing applications, parallel computing software, the scale and complexity of the growing, the existing parallel software development technology continue to be new challenges, the development of a certain size and complexity of the software system and the traditional serial program development Compared to the large differences, so that programmers directly to the details of the parallel programming implementation is no doubt similar to the use of assembly language for large-scale software development [2]. These problems, whether from the programming model and language abstract point of view to re-understand the construction of parallel programming environment, or from the software engineering point of view of parallel software development in all aspects of the parallel software developers, especially parallel software development environment, Many questions.

Heterogeneous systems are computing systems that are connected by functions or different performance through a certain interconnection structure [3]. Currently, heterogeneous systems are commonly used by general purpose processors (such as general purpose multicore CPUs) and coprocessors (eg, GPU, Intel XeonPhi, etc.). Based on this structure, heterogeneous systems can rely on general-purpose processors to provide general computing power, to deal with a variety of types of applications, on the other hand can use coprocessor to accelerate the application of specific areas, with low power consumption the performance. In June 2013 announced the TOP500 supercomputer rankings, China's "Tianhe -2" supercomputer through the integration of universal multi-core CPU and Intel Xeon Phi coprocessor, measured performance 33.86Pflops ranked first [4]. At the same time, there are four top-level supercomputers based on heterogeneous architecture design. However, in the heterogeneous system to obtain high computing performance at the same time, how to play in the practical application of this performance is still facing a huge challenge. On the one hand, the increase in the size of the heterogeneous system on the application itself put forward higher requirements, in order to use heterogeneous systems hundreds of millions of millions or even more calculation of the number of nuclear, the user must face the previous never tried parallel degree [5]. On the other hand, heterogeneous systems have complex structures, and their internal computing devices each have different system structures, instruction sets, and programming models, which add to the burden of user development applications.

In this paper, we focus on the development of parallel programming technology for heterogeneous systems, and propose a collaborative programming model, which divides the main body of application development into application experts and computer experts to meet the challenges of application and architecture.

2. Parallel programming techniques

2.1. Messaging interface

MPI is a factual specification that describes a set of functions used to manage the migration of data between processes. It has established a binding relationship with languages such as C, FORTRAN, and C ++. MPI defines functions for point-to-point communication between two processes [6]; aggregate functions for multiple interposes communication, and functions for parallel I / O and process management. In addition, MPI support for communicators facilitates the development of modular programming and reusable libraries. The communicator in the MPI specifies the type and layout of the communication data, allowing the MPI to proceed with the reference operation of the discontinuous data in the content and provide support for the heterogeneous cluster. MPI programs are usually implemented using the SPMD model, where all processes perform the same program logic. MPI has an excellent advantage in portability, and people have developed a number of software libraries based on MPI, which efficiently implement some common algorithms [7]. But explicit messaging programming will give developers an additional burden. Some of the other techniques may be more useful when parallelizing existing programs.

2.2. Parallel virtual machine

Parallel Virtual Machine (PVM) represents another common implementation of the messaging model, which was generated before MPI and was the first implementation standard for developing portable messaging parallel programs. Although PVM has been replaced by subsequent MPIs on tightly coupled multiprocessors and multi-core processors, it is still widely used in workstation cluster environments [8]. The primary goal of PVM is to make portability portable, and even to provide portability for multiple combinations of heterogeneous nodes, although this will degrade program performance. On the other hand, MPI provides high-performance communications. MPI - 1 not only provides an inflexible static processing model, but MPI - 2 adds an extensible dynamic processing model. MPI provides a rich set of communication functions, when the application needs to use a special communication mode, MPI advantage to show up, and PVM cannot provide such a function. If the application run on cluster, especially when the cluster is composed of heterogeneous nodes, the use of PVM more suitable. PVM also has advantages in fault tolerance.

2.3. Parallel compiler

Because parallel programming is difficult, people want to use the compiler to do all the work. So automatic parallelization, that is, by the compiler from the serial program to extract parallel information, has become a parallel computing software to achieve the goal of everyone dream, especially in the vector machine has been made after the success of the automation method is even more so. But automatic parallelism did not achieve success as automatic victimization [9]. Compiler analysis and parallel machine hardware features are highly complex, resulting in automatic parallel failures. Because of these difficulties, automatic parallelization only succeeds in shared storage systems and small-scale processors. But the performance is usually not very good. In some cases, programmers often use instructions to communicate to improve parallel performance. In these cases, Open MP can provide this information in a more portable way. On an extensible parallel machine, the HPF can provide automatic parallelism for regular data parallel programs written in FORTRAN90.

2.4. High performance FORTRAN

High performance FORTRAN (HPF) is similar to Open MP and is extended on the basis of serial languages, adding some instructions and libraries, and some new language constructs that provide data parallel, implicit parallel programming models. The place at Open MP is that it emphasizes support for user-managed data distribution to support the migration of programs on all scalable parallel machines, especially in distributed storage environments, and efficient execution of programs [10]. HPF can express complex parallel algorithms and describe the distribution of data in a machine-independent manner. But the current HPF is primarily effective for algorithms that are defined on regular grids and sparse linear algebraic algorithms. Although the extended HPF - 2 standards defines extensions that can handle irregular calculations, few compilers can implement these extensions.

3. Programming technology for heterogeneous systems

3.1. Research on architecture

This type of research usually provides hardware-related details of the programming elements, the user in parallel development to understand the characteristics of the underlying hardware, so the use of parallel components for the calculation method to optimize the program. Such research usually only provides a mechanism for the use of parallel components, rather than how to use the strategy.

NVIDIA CUDA programming model is the GPU manufacturer NVIDIA introduced CPU-GPU heterogeneous system programming model [11]. It provides a language mechanism for class C that exposes parallel details within the GPU to a user in a single instruction flow multithreading (SIMT). Users can perform hardware performance by designing a large number of threads that can be executed in parallel. Figure 1 shows the user see the CUDA thread model diagram. It contains the way threads are managed: Each section is mapped to a function on the GPU (called a computational kernel function in CUDA) that corresponds to a thread grid, which is in the first layer of the threading model, and each thread grid By a number of thread blocks (thread block), each thread block by the number of threads [12]. It also contains the thread of the scheduling implementation of the way: each thread is organized into the form of warp, each warp fixed with 32 threads, any time there is always a warp on the implementation of the war and warp threads are active The Users can understand the thread management and scheduling details, you can control the thread to complete the calculation of work, you can share data through high-speed shared data effectively, improve the effective bandwidth, you can synchronize the way through the coordination of the behavior of the thread.

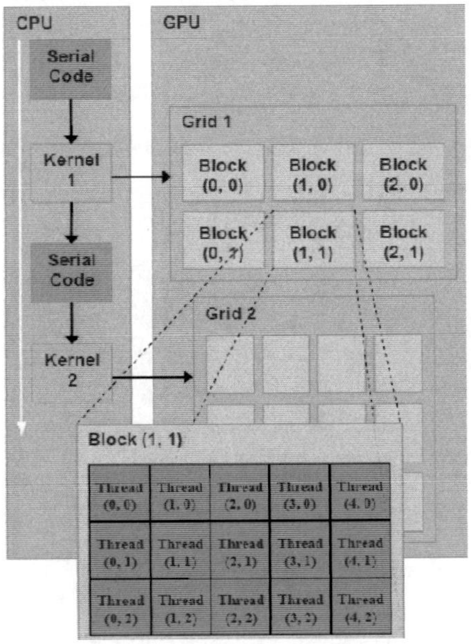

Figure 1. CUDA's threading model

3.2. Collaborative programming framework for heterogeneous systems

Aiming at the challenge of heterogeneous system programming, this paper proposes a collaborative programming model, which divides the main body of application development into application experts and computer experts. In order to cooperate with both, this paper designs a collaborative programming framework for heterogeneous systems. The framework provides high-performance applications for heterogeneous systems by combining application-oriented code provided by application specialists and code for hardware architectures provided by computer experts. How to make the code developed between the two naturally together, and can effectively use the heterogeneous system of computing potential is difficult to study the problem.

Based on the study of heterogeneous system typical development model, this paper abstracts the common task parallel operation set of co-author of cooperative programming. On this basis, we design a collaborative programming framework including application layer, system structure layer and support layer module And then implement the framework of the application layer module to the programming interface to the application of experts to provide these parallel operations to shield the details of the underlying hardware system; then to achieve the framework of the system structure layer module for the computer experts to dig the specific parallel operation As well as the serial core algorithm in the specific heterogeneous system to provide support for parallelism to filter the application-related details; framework also reserved optimization methods related to the support layer module.

4. Research on collaborative programming framework for heterogeneous systems

It is often not possible to understand the heterogeneous parallel details when applying heterogeneous system application development. For example, for CPU-GPU heterogeneous systems, you need to use special programming models such as CUDA, Open CL and other programming, but also need to have a more profound understanding of the architecture in order to develop efficient procedures, and this understanding is the application The lack of experts. Likewise, it is unrealistic for system designers or computer experts to fully understand the algorithms and models of the application, and for large-scale application development. Therefore, we believe that the future large-scale heterogeneous systems on the development of large-scale applications should be applied by experts and computer experts to cooperate more reasonable.

Based on this assumption, this paper proposes a collaborative programming model on heterogeneous systems, which divides the main body of heterogeneous system application development into application specialists and computer specialists, and defines two tasks: the former is responsible for describing the application itself Models, features, parameters, etc., while the latter is responsible for the full use of various computing resources based on architectural features. As shown in figure 2.

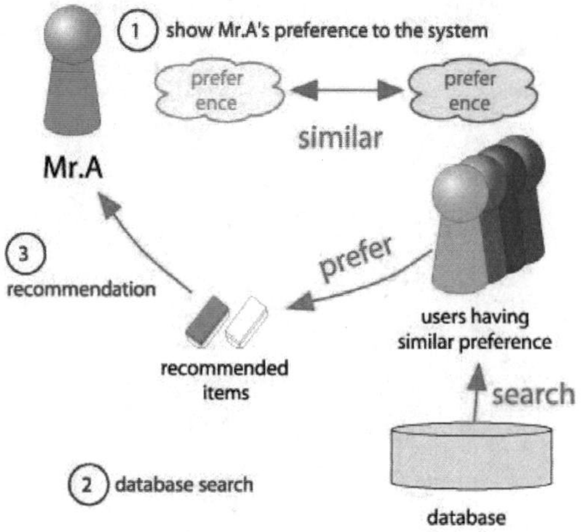

Figure 2. Application of expert and computer experts' collaborative programming model diagram

4.1. Collaborative programming framework design

With the study of application development model on typical heterogeneous systems, we get the enabler of collaborative programming: hierarchical task parallel operation set. Based on these parallel operations, we designed the overall structure of the collaborative programming framework CPF4H, as shown in Figure 3, which provides three modules: application layer, system structure layer and support layer. The application layer mainly provides the parallel operation set of the encapsulation in the way of programming interface, so as to hide the details of the hardware structure of the heterogeneous system so that the application specialists can focus on the implementation of the application; and the architecture layer provides the parallel operation and string of the computer expert Core algorithm to hide the application-specific details, so that it is committed to the parallel operation and serial core algorithm in the specific heterogeneous system to achieve; support layer to provide heterogeneous programs in the performance optimization related methods, including how to load balancing, How to carry out access optimization.

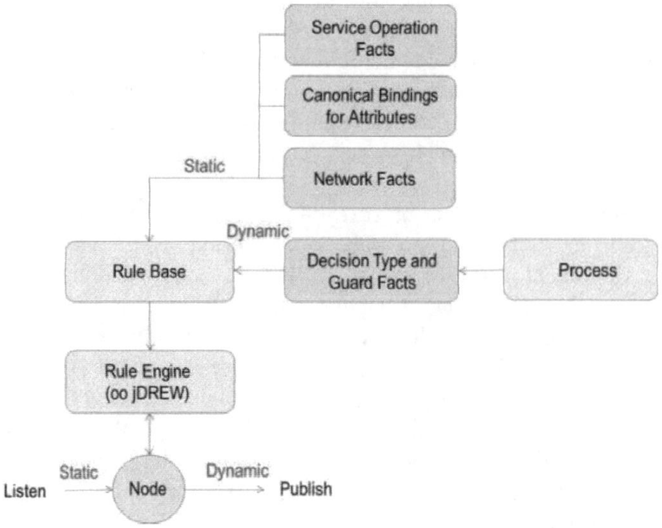

Figure 3. The overall structure of the collaborative programming framework

4.2. Programming interface and typical task flow

Because the CPF4H framework uses an external function library to provide a parallel operation of the collection, so it's programming interface mainly for the function call. Table 1 shows the list of functions and parameters provided by the framework.

Table 1. CPF4H provides a list of functions and their parameters

Function name	Scope of action	parameter list
Coarse Farm	Task	Data slice, function fragment
Mid Farm	Task	Data slice, unit, function fragment
Fine Farm	Task	Data slice, element, element, function fragment
Pipe	Task	Data slice, unit, function fragment
Map	Data	Element set
Reduce	Data	Element set

If the computer experts to complete the application of expert call to the parallel operation of the manual implementation, the efficiency will be very low. In fact, most of the parallel operations have their common implementation. Only when the core algorithm related to the code, it is necessary for computer experts to carry out special analysis and implementation. Thus, the CPF4H framework provides a variety of implementations of common parallel operations at the system fabric layer. This section first describes these implementations and then examines how these implementations are managed.

5. Conclusion

Aiming at the problem that the heterogeneous system is difficult to be programmed at present, this paper studies the parallel programming technology for heterogeneous systems. This paper first analyzes the factors that affect the programmability of heterogeneous systems. On this basis, a new programming model for heterogeneous systems - collaborative programming model is proposed. The main body of application development on heterogeneous systems is divided into application experts and Computer experts and the development of the two tasks. Based on the collaborative programming model, this paper designs a collaborative programming framework for heterogeneous systems, which supports the development and application of two kinds of development subjects and application performance optimization. Finally, the article also validates the validity of the collaborative programming model and the cooperative programming framework by implementing the molecular dynamics application based on the cooperative programming framework.

Although the research has made some research results, but still a lot of work still need to be further improved. At present, the collaborative programming model mainly relies on the parallel task set of abstract task to realize the cooperation between the application expert and the computer expert. Although the collaborative development of the two is realized, there is still insufficient in the development efficiency and application performance. The next step can further study the parallel operation of the data in the application to enhance the ability of the collaborative programming model to describe the parallel operation of the data level.

Acknowledgement

Project fund: Science and Technology Research Project of Jiangxi Provincial Department of Education, Project No. 202106, Research and application of VR action capture algorithm.

References

[1] Dong J 2014 An Empirical Study of the Teaching Mode of Scenarios Simulation in Business English under Verbal Behavior Theory *Theory and Practicein Language Studies* **4**(5) 21-45

[2] Tariq Elyas M P 2010 Saudi Arabian educational history: impacts on English language teaching Education *Business and Society: Contemporary Middle Eastern* **3**(2) 78-104

[3] Ma X 2011 Expansion Dynamic Game Analysis on Education Quality Control for Distance Education System *Advanced Materials Research* **127**(30) 45-90
[4] Zheng M, Lin Y, Kang Y, Yu Q 2013 A Parallel Programming Pattern Based on Directed Acyclic Graph *Applied Mechanics and Materials* **8**(33) 12-37
[5] Chi S 2014 Application of MATLAB Parallel Programming Technology *Applied Mechanics and Materials* **65**(6) 74-98
[6] George T 2016 The Features of Parallel Program for 3-D Modeling of Dynamic Processes in Porous Objets with Heat-Evolution *Key Engineering Materials* **40**(6) 48-85
[7] Anonymous 2011 Research and Markets; Professional Parallel Programming with C#: Master Parallel Extensions with NET4 *ComputerWeeklyNews* **5**(7) 47-90
[8] Weltzin C 2010 Reduce parallel programming pain with data flow languages *Electronic Engineering Times* **10**(15) 84-95
[9] Can Ö, Balara S, Boleslaw K S 2014 Compiler Technology for Parallel Scientific Computation *Scientific Programming* **3**(3) 72-87
[10] Liu D, Sohl J, Wang J 2010 Parallel Programming and Its Architectures Based on Data Access Separated Algorithm Kernels *International Journal of Embedded and Real-Time Communication Systems (IJERTCS)* **1**(1) 35-64
[11] Rodolfo O, Benoît M, José R, Didier M 2013 Nonlin earsystem identification using heterogeneous multiple models *International Journal of Applied Mathematics and Computer Science* **23**(1) 12-43
[12] VikasK S, Manju A, Kanwar S 2010 Optimal Structurein Heterogeneous Multi-state Series-parallel Reliability Systems *Economic Quality Control* **25**(1) 24-63

Research on Software Defect Prediction and Analysis Based on Machine Learning

Xuemei Peng

School of Computer and Information Engineering, Nanchang Institute of Technology, Nanchang, Jiangxi, 330044, China

*corresponding author's e-mail: 37666461@qq.com

Abstract. The defects of machine learning prediction technology can be more comprehensive and automatic learning model to find the defects in software has become the main method of defect prediction, selection and study of algorithm is the key to improve the accuracy and efficiency of machine learning. Comparing different machine learning defect prediction methods reveals that the algorithms have different advantages in different evaluation indicators, the use of these advantages and combining the stacking ensemble learning method in machine learning is put forward different prediction algorithm of prediction results. As software metrics and again the prediction model of software defect prediction combined machine learning algorithm is based on the experiment with the model of Eclipse, the data sets show the effectiveness of the model.
Keywords. Software defect prediction; machine learning; ensemble learning; combination; Eclipse prediction data set

1. Introduction

Software systems being made today are complex hence the difficulty of producing software without defects is increasing. If producing software without defects is nearly impossible to achieve, then minimization of defects needs to be done (Shepperd M and Bowes D, 2014). This is not easily achievable. It is important to detect the defects early in the development lifecycle of a software in order to avoid the cost losses (Shepperd M, 2017). Furthermore, early detection will ensure that reliable, good quality and optimum cost software are being delivered to the customers (Tantithamthavorn C, 2016). Various Object Oriented (OO) software metrics which quantify structural characteristics of a software which include inheritance, cohesion etc. can be used for developing prediction models for detecting defective classes in a software (Czibula G and Marian Z, 2014). Past releases of a software can be used for collecting these metrics and further these metrics are used for developing the prediction models for later releases (Shepperd M and Song Q, 2013).

A number of ML techniques such as Bayesian learning, neural networks, Support Vector Machine (SVM) and so on have been analysed by a number of studies in literature for developing prediction models which identify classes which may contain defects. However, what is important to understand is that all of these techniques perform differently for different data sets according to data sets characteristics (Rathore S S and Kumar S, 2017). Moreover, certain other important factors which needs to be considered for empirical comparison are the size and number of the data sets used for empirical validation in the study, the performance measure selected for performance assessment and the statistical tests (Peters F and Menzies T, 2013). In the past, researchers had limited data sets available for empirical validation, which is an important constraint. Also, the usage of statistical tests has been meagre in literature which develops and validates software defect prediction models. Therefore, it is essential to perform more studies which statistically compare a number of ML techniques using different data sets to reach a commonly acceptable and well-formed conclusion (Bhardwaj C and Kaswan K, 2015). In order to adhere to the above stated issues, we compare the performances of 17 ML techniques on a dataset obtained from the Tera-Promise data repository named Xerces. This enables us to find out if a specific technique performs better than the other techniques and we statistically rank the techniques on the basis of their performance using Friedman test (Jing X Y and Wu F, 2017). Also, the best technique is pairwise compared with other ML techniques to evaluate whether it is better than the other used ML techniques.

In practical application, code metrics can easily get from the current version of the source code of a software system, and the measurement process measurement and historical defects collecting expensive, from the version control system and defect tracking system (Rana Z A and Mian M A, 2015). Therefore, this paper focuses on the problem of extraction method to construct the model presents an improved prediction model of defective combination machine the learning algorithm based on different machine learning combined prediction algorithm, the first prediction results as predicted second new metrics to join data set, at the same time using feature selection method, improve the efficiency of defect prediction. In order to show the practicability of the model, based on Eclipse standard data sets for the defect prediction and verify the validity of this method.

2. Software defect prediction model based on machine learning

This section will introduce the related concept of defect prediction and machine learning. The defect prediction is from the measure metadata from the software module, classification, and machine learning by regression and clustering method to find out the relationship between the defect metrics, establish the appropriate model to predict the distribution of whether new software modules contain defects or the prediction model of defects, as shown in Figure 1.

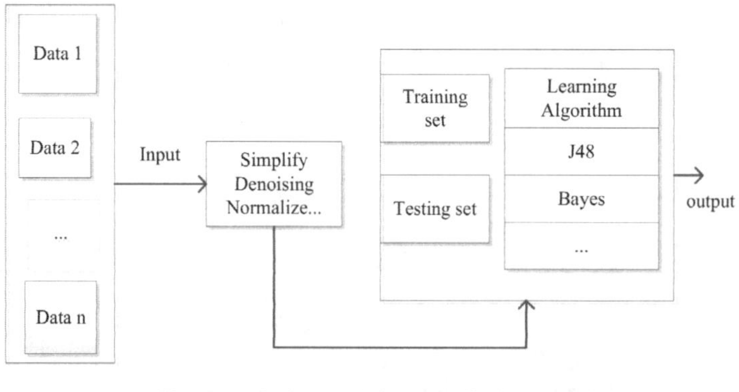

Figure 1. Software defects prediction model

According to the purpose of prediction can be defect prediction model is divided into 2 categories:

1) prediction software defect classification problem, namely the prediction software module is defective;

2) software defect prediction scheduling problem, the number of software defect prediction module, and according to the number of defects sorting module are given. In this paper according to different machine learning the combination of algorithm classification and scheduling problem of defect prediction are studied respectively. Based on the prediction of software defect classification problem is to predict whether the software module contains a defect, guide software developers will test the resources on the prediction for the defective module According to this kind of defect prediction model has a lot of research results, the prediction effect is good. It is impossible to completely correct prediction of all software modules. When the error will be non-defect prediction module for defects. Test the waste of resources; when the error will be defect prediction module without defects, there are defects not found to be associated with the risk in time. Repair the two kinds of mistakes are not the same, how to balance the two, find out more suitable for software defect prediction model has been the focus of research. Based on the defect prediction model for scheduling test resource unknown - the

software modules are sorted according to the number of defects, preferred test modules containing a large number of defects, when tested with abundant resources, can continue to contain less defect module for testing. For this kind of defect prediction model, it is difficult to achieve accurate prediction the number of defects contained in the software module, the key is to sort the high quality software module are given. The software defect prediction scheduling algorithm model is mainly constructed by regression algorithm in machine learning. The decision tree algorithm is a common modeling method for classification, by constructing a decision tree classification rule discovery the data contained in the application of decision tree is established. When the defect prediction model, need treatment, pruning technology, derived values in predicting variables missing The corresponding improvement of rules and other aspects, it should also avoid over fitting problems, try to reduce the length of the path, in order to improve the performance of the prediction model. The linear regression is the optimal combination of multiple variable to predict or estimate the dependent variable, compared with a variable in the estimation is more effective, more practical linear. Regression is used for regression algorithm scheduling problem is the most basic, through the establishment of software metric linear equation between the yuan and the number of defects to calculate the number of defects may contain unknown module. Logistic regression is a modified linear regression technique, the logarithm transform linear regression results for 0-1 mapping between the value of the two classification problems therefore to predict whether the defective software can be easily established model. Bias classifier based on Bias theorem is established in simple statistical feature independence based learning classifier, it is ideal for a given class variables, a given feature does not depend on other features. Bias classifier only needs little training to estimate the parameters of the prediction model, performance a good correlation between the independent feature' attributes or attribute in less obvious situations. Bias network is a kind of uncertain causality model with uncertain problem solving ability strong, can effectively carry out multi information expression and fusion, which can be widely used in fault prediction. Artificial neural network is a kind of imitation animal behavior characteristics, neural network distributed computing model of information processing in parallel. In most cases, artificial neural network can change the internal structure of the external network based on information, is a kind of adaptive system. Multilayer perceptron neural network is used to use the standard BP algorithm, is a kind of nonlinear statistical data modeling tools commonly used to model the complex relationship between input and output, or to explore the data model, is a relatively simple and effective method to solve the complexity of the problem. In addition, there is support vector machine and clustering analysis method can be used for software defect prediction, need to use historical data to construct and validate the model predictions, the technology depends on the data itself and their accuracy, stability and generalization degree comparison and analysis.

3. Performance evaluation index analysis

In the process of prediction, it is necessary to select reasonable indicators to evaluate the performance of each prediction model. In this experiment, we use the area -AUC index below the ROC curve to evaluate, and first introduce the most basic performance evaluation index. Software defect prediction is a two classification problem, and its prediction process will produce four different results, as shown in Table 1, where the defect is positive, and no defect is negative. Rows represent actual classes, and lists indicate categories of forecasts.

Table 1. Predicted results

Predicted class			
		defective	There is no defect
Actual	defective	Real cases	False negative cases
category	There is no defect	False positive cases	True negative cases

The true instance rate (TPR) is the ratio that is correctly predicted as the positive number and the actual positive number, which is correctly predicted as the ratio of the defective sample number to the actual defective sample number, such as formula (1):

$$TPR = \frac{TP}{TP + FN} \quad (1)$$

False positive rate (FPR) is the ratio of the number of false positive cases to the actual negative ones, that is, the ratio of the number of defective samples to the actual number of defective samples, as shown in formula (2):

$$FPR = \frac{FP}{FP + TN} \quad (2)$$

Based on this, the ROC curve is a graphical method to describe the relationship between the true rate TPR and the false positive rate FPR of the classification model. As shown in Figure 2, the abscissa represents the false positive rate, and the ordinate represents the true instance rate.

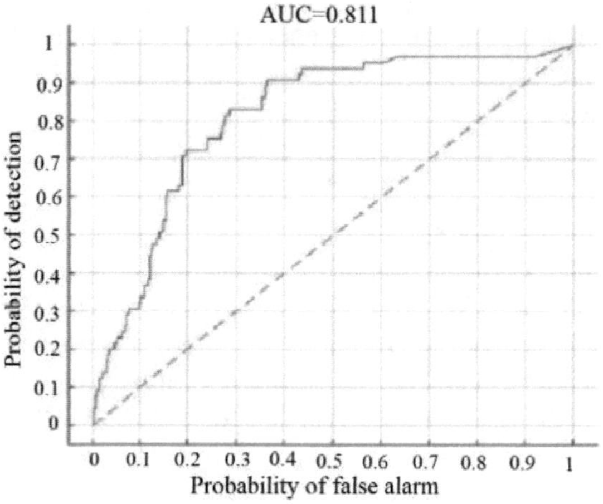

Figure 2. The abscissa represents the false positive

4. Experimental results and analysis

The first set a simple prediction algorithm for data, the experimental results due to excessive and similar effects, so only a partial list of the experimental results. The algorithm used a random forest, linear regression and neural network. It can be seen from Figure 3 the establishment of the 3 algorithms in the model the effect of the same sort, simple linear regression and random forest model structure in the practical test, the operation speed is faster, and the neural network time overhead is very large, the number of hours required to complete 18 experiments. Based on the combination of machine learning algorithm to obtain a new prediction model, and according to the prediction results of values from large to young module sort, make CLC curve and CE curve, as shown in Figure 4. In the CLC evaluation system, a neural network prediction model of linear regression metric is improved, CE index and single prediction algorithm. However, the improvement of the space is not large, the reason was due to the effect of 3 machine learning algorithms used for similar correct prediction combination model can provide more useful information, and each kind of machine learning algorithm of the CE curve and the results are close to the linear slope of 1, which is close to a random sequence of results, suggesting that the data set in order to prediction can't get the good effect.

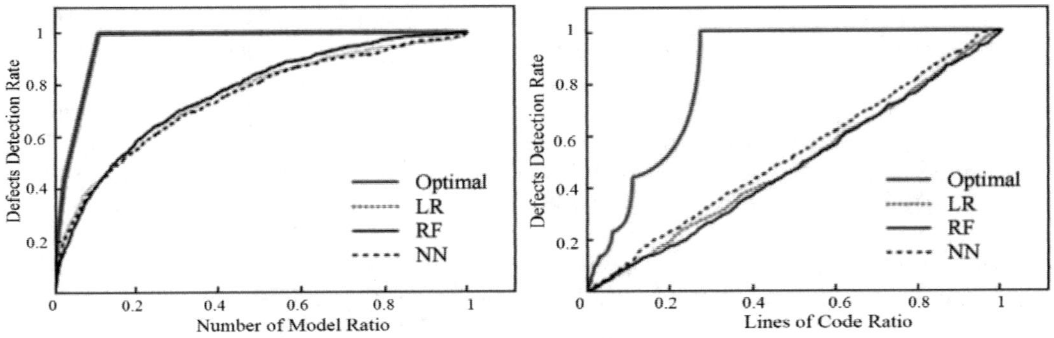

(a) CLC (b) CE

Figure 3. Results of different machine learning model (File 2.0 as training set, File 2.1 as testing set)

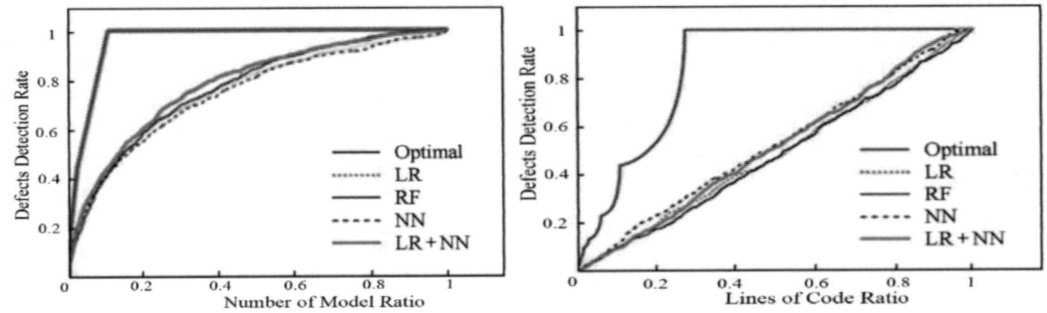

(a) CLC (b) CE

Figure 4. Result of combined machine learning model (File 2.0 as training set, File 2.1 as testing set)

5. Conclusion

In this paper, we have addressed the question of what factors influence the predictive performance of software defect classifiers measured by the Matthews Correlation Coefficient. The question is important and the reasons are following: (1) being able to effectively predict the location of defects would be useful for software engineers, (2) there is no real congruence between the many empirical results that have been generated such that no classifier technique—amongst the many proposed—dominates. We conclude from the meta-analysis that there is strong evidence that current experiments in software defect prediction are extremely limited in their effectiveness and that we must take into account the dominant factor which is researcher group. Although our focus has been on dichotomous defect classifiers we anticipate that these kind of problems will be manifest in other areas where prediction systems are being researched. Given problems of bias have been reported across many branches of science, it is difficult to imagine that other areas of software engineering will be immune. Research to confirm, otherwise, this conjecture is urgently required.

Acknowledgement

Project fund: Ministry of Education, Fund No.: 202101196012,Curriculum construction and research based on big data storage technology

References

[1] Shepperd M, Bowes D, Hall T 2014 Researcher Bias: The Use of Machine Learning in Software Defect Prediction *IEEE Transactions on Software Engineering* **42**(11) 1092-4

[2] Shepperd M, Hall T, Bowes D 2017 Authors' Reply to "Comments on 'Researcher Bias: The Use of Machine Learning in Software Defect Prediction' " *IEEE Transactions on Software Engineering* **99** 1

[3] Tantithamthavorn C, Mcintosh S, Hassan A E, et al 2016 Comments on "Researcher Bias: The Use of Machine Learning in Software Defect Prediction" *IEEE Transactions on Software Engineering* **42**(11) 1092-4
[4] Czibula G, Marian Z, Czibula I G 2014 Software defect prediction using relational association rule mining *Information Sciences* **264**(183) 260-78
[5] Shepperd M, Song Q, Sun Z, et al 2013 Data Quality: Some Comments on the NASA Software Defect Datasets *IEEE Transactions on Software Engineering* **39**(9) 1208-15
[6] Rathore S S, Kumar S 2017 A decision tree logic based recommendation system to select software fault prediction techniques *Computing* **99**(3) 255-85
[7] Peters F, Menzies T, Gong L, et al 2013 Balancing Privacy and Utility in Cross-Company Defect Prediction *IEEE Transactions on Software Engineering* **39**(8) 1054-68
[8] Bhardwaj C, Kaswan K 2015 Classification Rule Mining Based Software Defect Prediction *Nucleic Acids Research* **18**(17) 5293
[9] Jing X Y, Wu F, Dong X, et al 2017 An Improved SDA Based Defect Prediction Framework for Both Within-Project and Cross-Project Class-Imbalance Problems *IEEE Transactions on Software Engineering* **43**(4) 321-39
[10] Rana Z A, Mian M A, Shamail S 2015 Improving Recall of software defect prediction models using association mining *Knowledge-Based Systems* **90**(C) 1-13

Research on Factors Influencing Tensile Property in CGHAZ of DP780 Steel

Dongqin Xu[1], Yonghuan Guo[1,*], Tian Yan[1], Qin Yang[1], Lulu Li[2] and Xiying Fan[1]

[1]School of Mechanical and Electrical Engineering, Jiangsu Normal University, Xuzhou 221116, Jiangsu, China
[2]CNPC Engineering Technology R&D Company Limited, Beijing 102206, China

[*]Corresponding author's email: gyh8421@126.com

Abstract. Tensile property of coarse grain heat affected zone (CGHAZ) is one of the important mechanical property of DP780 cold-rolled dual phase steel, and factors influencing tensile property of CGHAZ include peak temperature, heating rate and cooling rate. Based on response surface method (RSM), two response surfaces models were established. By using Design-Expert software, three factors were combined in pairs. Analysis of variance (ANOVA) was performed on the response surface models. The results showed that tensile property of DP780 was influenced by peak temperature most, then cooling rate, and heating rate made the least influence.
Keywords: DP780; CGHAZ; RSM; ANOVA

1. Introduction

The proportion of dual-phase high-strength steel used has been increasing in recent years, and energy saving, environmental protection and safety are the development direction of the future automobile industry, Therefore, it has become a trend to use high-strength steel instead of traditional steel to achieve lightweight and improve safety in automobile frames [1-3]. Studies have shown that every 10% reduction in body weight can reduce its fuel consumption per 100 kilometers by 5% [4]. But after this kind of steel is welded, the quality of the weld will be severely reduced, its coarse grain heat affected zone (CGHAZ), which is a component of the heat affected zone (HAZ) of welded joints and is close to the melting zone, is indirectly affected by the welding heat source, thus make coarse grains generate in this area, eventually lead to a serious decline in performance in this area. Therefore, CGHAZ has become the focus of research on the performance of welded joints. In actual welding, the width of HAZ is very limited, only about a few millimeters, the width of CGHAZ is even smaller, making it very difficult to study the tiny area of CGHAZ through actual welding. Thermal simulation of tensile specimens is used in this study, Response surface method (RSM) and Analysis of variance (ANOVA) were performed to determine the degree of effect of each influencing factor on the tensile strength of dual-phase steel, and then a reference was provided for formulating a reasonable welding plan and improving the quality of welded joints.

2. Materials, methods and results

2.1. Materials

DP780 cold-rolled high strength steel(HSS), produced by the Shanghai Baosteel Group, was used in this work, and chemical composition and mechanical properties of it are shown in Tables 1and 2.

Table 1. Chemical composition of DP780 HSS (wt. %)

C	Si	P	S	Al	Mn	Cr	Fe
0.15	0.35	0.012	0.013	0.01	1.65	0.54	Balance

Table 2. Mechanical properties of DP780 HSS

Yield Strength (YS)(MPa)	Ultimate Tensile Strength (UTS) (MPa)	YS/UTS (%)	Elongation (%)
435	780	56	16

The influence of factors such as peak temperature, heating rate and cooling rate on the tensile strength of the thermally simulated coarse grain heat affected zone (CGHAZ) were discussed. The size of tensile sample is shown in Figure 1.

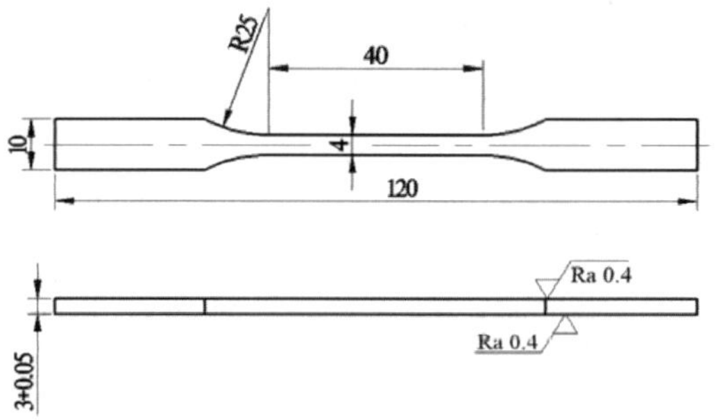

Figure 1. Tensile sample

2.2. Methods

The DST1000-PC dynamic thermal simulation tester was used to simulate the CGHAZ of the tensile sample. It was pointed out by Indacochea J. E. et al. [5] that the peak temperature of CGHAZ is 1100~1500°C, therefore, the range of peak temperature was set at 1100~1250°C in this study, and the range of heating rate was set at 55~80°C/min. The cooling rate also has a very important effect on the transformation of the microstructure, which is mainly reflected on the precipitation of α-ferrite and other phases. The range of cooling rate was set at 1000~1500°C/min in this study. The schematic of the thermal simulation process is shown in Figure 2.

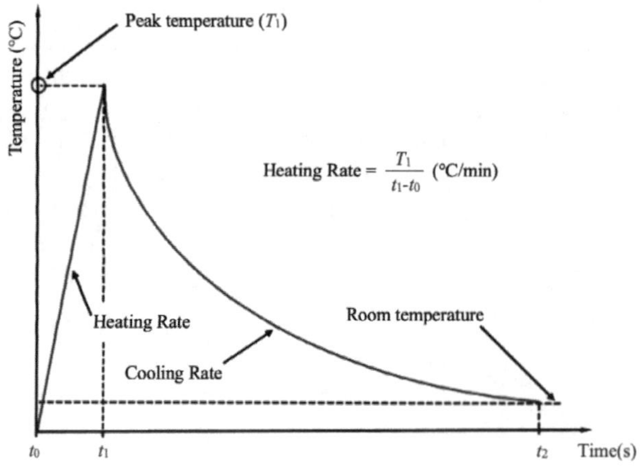

Figure 2. Schematic of the thermal simulation process

The response surface model is generally a three-dimensional surface, thus there can only be two influencing factors in the established response surface model. The number of influencing factors that need to be studied is three, and they are combined in pairs. The test parameter table is shown in Table 3.

Table 3. Schedule of test parameters

No.	Peak Temperature (℃)	Heating Rate (℃/min)	Cooling Rate (℃/min)	No.	Peak Temperature (℃)	Heating Temperature (℃/min)	Cooling Rate (℃/min)
1	1250	65	1300	7	1160	80	1400
2	1220	70	1200	8	1100	60	1300
3	1250	80	1100	9	1220	55	1400
4	1190	60	1000	10	1160	55	1100
5	1190	75	1500	11	1130	70	1000
6	1100	75	1200	12	1130	65	1500

2.3. Results

The LNK-150-1000 electronic tensile torsion testing machine to statically stretch the samples in Tab.3, and the results are shown in Tab.4.

Table 4. Results

No.	Tensile Property (MPa)	No.	Tensile Property (MPa)
1	579.0	7	723.3
2	671.1	8	732.4
3	563.9	9	693.7
4	719.1	10	742.3
5	747.1	11	746.7
6	730.6	12	743.2

3. Establishment of response surface models

3.1. Response surface model of peak temperature, heating rate, and tensile property

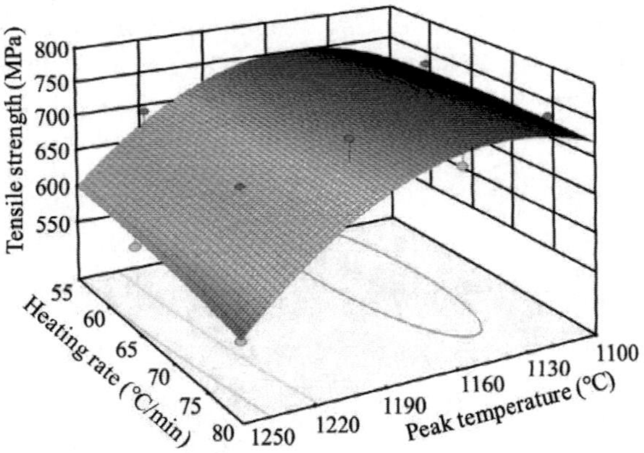

Figure 3. Model 1: Response surface model of peak temperature, heating rate, and tensile property

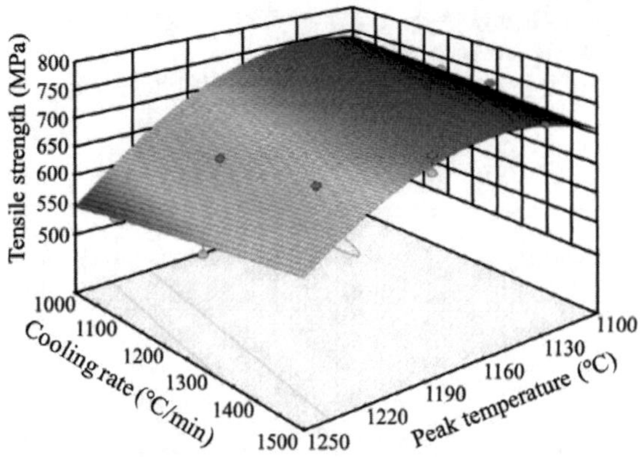

Figure 4. Model 2: Response surface model of peak temperature, cooling rate, and tensile property

Combining the test parameters in Tab.3 and the test results in Tab.4, the influence of the peak temperature and heating rate of the thermal simulated CGHAZ on its tensile strength can be analyzed by using Design-Expert, then the response surface model of peak temperature, heating rate, and tensile property (model1) is established as shown in Fig.3. The following conclusion can be drawn from Fig.3: the peak temperature has a greater effect on the tensile strength of thermally simulated CGHAZ, as the peak temperature decreases, the tensile strength and the toughness both increase; when the range of peak temperature is 1100~1160°C, the tensile strength of thermal simulated CGHAZ almost maintain at a constant value, and the maximum tensile strength is about 730MPa. As the peak temperature increases, the tensile strength of the thermal simulated CGHAZ will drop significantly.

3.2. Response surface model of peak temperature, cooling rate, and tensile property

In the same way, the response surface model of peak temperature, cooling rate, and tensile property (model 2) is established as shown in Fig.4. The following conclusions can be drawn from Fig.4: as the peak temperature increases, the tensile strength of the thermal simulation CGHAZ decreases extremely significantly. At low temperatures, that is, around 1100°C, the tensile strength is affected by the cooling rate to a certain extent. When the peak temperature of the thermal simulated CGHAZ is determined, as the cooling rate gradually increases, the tensile strength will gradually decrease. Therefore, in actual welding, if the cooling rate of thermal simulated CGHAZ is faster, its performance will be worse.

4. ANOVA on response surface models

In addition to building response surface models, Design-Expert also has the function of performing ANOVA on the test results. This function can be used to analyze the significance of the model while analyzing the significant factors of the tensile strength. The results of model 1 are shown in Tab.5. Among them, the peak temperature (°C) is represented by A, and the heating rate (°C/min) is represented by B.

Table 5. ANOVA of peak temperature, heating rate and tensile property

Source of variation	Mean square	DOF	Mean square error	F value	P value	Significance
Mode 1	42916.08	5	8583.22	24.85	0.0006	Significant
A-A	24875.38	1	24875.38	72.03	0.0001	Significant
B-B	318.58	1	318.58	0.92	0.3739	
AB	196.06	1	196.06	0.57	0.4797	
A^2	11777.74	1	11777.74	34.10	0.0011	Significant
B^2	60.00	1	60.00	0.17	0.6913	
Residual	2072.08	6	345.35			
Total	44988.16	11				

It can be seen from Table5 that the P value of model 1 is 0.0006, which is less than 0.05, indicating that the model term is significant. In model 1, the influence of A-A and A^2 on the test results is significant. Therefore, the peak temperature of the thermal simulated CGHAZ has a significantly higher impact on the tensile strength than the heating rate. That is, the main influencing factor of the tensile strength of thermal simulated CGHAZ is the peak temperature in model 1.

Using the same method to perform ANOVA on model 2, the P value of model 2 is 0.0003, which is less than 0.05, also indicating that the model term is significant. In model 2, the influence of A-A and A^2 on the test results is significant. Thus, the peak temperature of the thermal simulated CGHAZ has a significantly higher impact on the tensile strength than the cooling rate. That is, the main influencing factor of the tensile strength of thermal simulated CGHAZ is also the peak temperature in model 2.

5. Conclusion

(1) The most significant factor affecting the tensile strength of thermal simulated CGHAZ in DP780 cold-rolled dual-phase HSS is the peak temperature.

(2) The tensile strength of the thermal simulated CGHAZ of DP780 cold-rolled dual-phase HSS is lower than that of the original DP780 base metal.

(3) In actual welding, the performance of CGHAZ will become worse as the cooling rate gets faster. A combination of low heat input and slow cooling rate can be adopted in order to ensure that the welded joints have good property.

Acknowledgments

This research was financially supported by the Xuzhou City Science and Technology Plan Projects (grant KC18239), and Jiangsu Province Innovation and Entrepreneurship Program for Undergraduates (grant 202010320183H). The authors would express their sincere thanks.

References

[1] Yan T, Guo Y H, Fan X Y, et al. Numerical simulation and welding parameters optimization for minimum deformation of AHSS based on RSM & QGA[J]. Materials Research Express, 2019(6):1165e9.

[2] Wang C Y, Yang J, Chang Y, et al. Development trend and challenge of advanced high strength automobile steels. Iron and Steel. 2019, 54(2):1-6.

[3] Moeini G, Ramazani A, Sundararaghavan V, et al. Micromechanical modeling of fatigue behavior of DP steels[J]. Materials Science and Engineering: A. 2017, 689:89-95.

[4] Kastensson Å. Developing lightweight concepts in the automotive industry: taking on the environmental challenge with the SåNätt project[J]. Journal of Cleaner Production, 2014, 66:337-346.

[5] Indacochea J E, Wang G, Seshadri R, et al. Creep rupture properties of high-temperature bainitic steels after weld repair[J]. Journal of Engineering Materials and Technology, 2000, 122(3):259-263.

Nonlinear Stall Flutter Suppression of Wind Turbine Blade Based on LMI Method

Xiaolin Zhang, Haipeng Sun, Yingbo Wang, Shanyao Li, Changle Sun and Tingrui Liu*

College of Mechanical & Electronic Engineering, Shandong University of Science & Technology, Qingdao 266590, China

*Corresponding author's E-mail: liutingrui9999@163.com

Abstract. Aiming at the failure of stall flutter in aeroelastic system of wind turbine blade, the active control process of stall flutter based on linear matrix inequality (LMI) design is described. The structural model is a typical blade section model based on spring-mass-damper, and the aerodynamic force is the ONERA stall aerodynamic model suitable for pure pitch motion. Based on the state variables, the nonlinear aeroelastic equations are expanded by Taylor series and linearized by low order approximation. The state feedback gain is calculated through LMI, and the time domain response stability analysis and stall flutter suppression method based on linearization are studied. The simulation results show that the maximum amplitude is greatly reduced after flutter control, and the system can be stabilized in a short time.

1. Introduction

Aeroelasticity is a coupling problem among aerodynamic force, elastic force and inertial force. Stall flutter is a typical aeroelastic instability phenomenon, which is an important reason for the fracture failure of wind turbine blades. Stall flutter usually occurs at high angle of attack, accompanied by periodic partial or complete separation of the airflow from the airfoil. The airflow injects energy into the blade structure of the wind turbine and induces self-excited vibration of the blade. In order to study the aeroelastic stability of dynamic stall of wind turbine blades, the widely used dynamic stall models are the ONERA model and the Beddoes Leishman (B-L) model. Based on the semi-empirical ONERA unsteady and nonlinear aerodynamic model, Ren and Lin [1] studied the stability of blade dynamic stall-induced flutter by directly integrating the nonlinear aeroelastic equations in the time domain. Li et al. [2] calculated the periodic time-varying dynamic stall aerodynamic load based on the B-L nonlinear aerodynamic model according to the periodic variation characteristics of attack angle of rotating blade, and analyzed the aeroelastic stability of the system based on the time-varying aeroelastic model and the Flqouet theory. With the trend of large-scale blade, the flutter caused by aeroelastic instability will become a more important factor in blade design.

In recent years, many scholars have explored different methods to suppress stall flutter of wind turbine blades. Liu et al. [3] studied the stability analysis and stall flutter suppression method based on time domain response based on piezoelectric feedback actuator theory and three-weight hybrid sensitivity robust control. At the same time, based on the servo aeroelastic pitch control technology, the optimal PID, fuzzy PID and neural network PID controllers were used to suppress the stall flutter of two-dimensional blade section [4]. Chang et al. [5] studied the flutter suppression of coupled motion of wind turbine blades by linear quadratic (LQR) control. With the continuous development of control technology, in the design of feedback control system, many constraints of the system often need to be considered. In dealing with robust control problems and other control theory problems, the control problem can be transformed into a linear matrix inequality or optimization problem with linear matrix inequality constraints. In this study, aiming at the flap-lag stall nonlinear flutter failure problem of wind turbine blade, based on the large angle of attack nonlinear aerodynamic model suitable for pure pitch motion, based on linearization, the feedback controller is designed based on LMI, and the simulation is carried out in SIMULINK.

2. Structural Model and Aeroelastic Equation

Using the typical section model of the blade with a large aspect ratio (the section rotation radius is *r*), the distance between the design aeroelastic center and the center of gravity is small enough, so the elastic torsion of the blade profile can be ignored, and the section mass body is respectively suspended by the spring and damper in the direction of *z* and *y*, as shown in Figure 1, where the direction of *z* and *y* represents the motion in the direction of flap and lag respectively, the pitch angle is denoted by ϕ; the angle of attack is denoted by α; *U* is the wind speed; *c* is the chord length; V_0 is the relative wind speed; ψ is the relative wind speed angle; Ω is the impeller speed; F_C and F_{NC} are the cyclic and non cyclic aerodynamic lift, respectively; F_D is the aerodynamic drag.

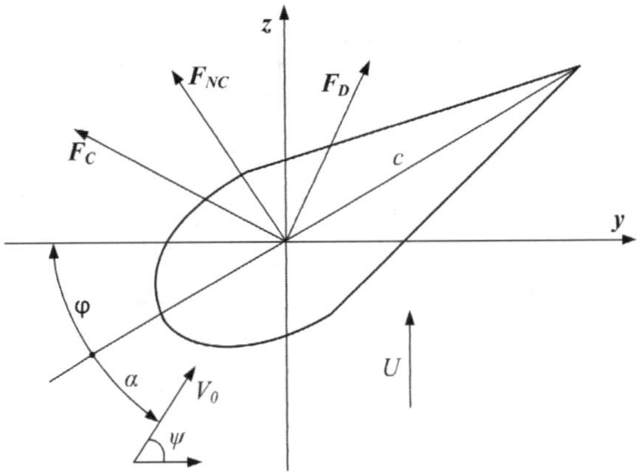

Figure 1. Coordinate systems and aerodynamic forces

Based on the calculation of kinetic energy and potential energy of the system, using the Lagrange equation and ignoring the high-order nonlinear term [6], the equations of motion in the direction of flap and lag can be expressed as follows:

$$\ddot{z}+2\xi_z\omega_z\dot{z}+\omega_z^2 z+\ddot{\varphi}y=\rho_b^{-1}(F_C\cos\psi+F_{NC}\cos\varphi+F_D\sin\psi) \tag{1}$$

$$\ddot{y}+2\xi_y\omega_y\dot{y}+\omega_y^2 z-\ddot{\varphi}z=\rho_b^{-1}(F_D\cos\psi-F_{NC}\sin\varphi-F_C\sin\psi) \tag{2}$$

where ω_z and ω_y are the first natural frequencies of the flap and lag directions respectively ; ξ_z and ξ_y are the damping ratios in the flap and lag directions respectively; ρ_b is the section density; On the right side of the equation is the aerodynamic expression.

According to the previous research results of the research group [7], the harmonic term of nonlinear aerodynamic force is extracted by using the Fourier transform, and the nonlinear ONERA aerodynamic model at high angle of attack is simplified, and the nonlinear stall aerodynamic force expression neglecting elastic torsion can be obtained.

The expressions of aerodynamic lift are:

$$F_{NC}=\frac{1}{2}\rho S_L(q_L b v_0\dot{\varphi}+l_L b v_0\dot{\varphi}+k_L b^2\ddot{\varphi}),\quad F_C=\frac{1}{2}\rho S_L(v_0^2 C_{1L}+v_0^2 C_{2L}) \tag{3}$$

The expression of aerodynamic drag is:

$$F_D = \frac{1}{2}\rho c(0.014v_0^2 + v_0^2 C_{D2}) \tag{4}$$

where C_{1L} and C_{2L} are the lift structural variables of the cycle term and C_{D2} is the drag structural variables, their expressions are respectively:

$$\begin{aligned}
&\dot{C}_{1L} + \lambda_L \frac{v_0}{b} C_{1L} = \lambda_L a_{0L} \frac{v_0}{b}\varphi + \lambda_L \sigma_L \dot{\varphi} + \gamma_L a_{0L}\dot{\varphi} + \gamma_L \sigma_L \frac{b}{v_0}\ddot{\varphi} \\
&\ddot{C}_{2L} + a_L \frac{v_0}{b}\dot{C}_{2L} + r_L \frac{v_0^2}{b^2} C_{2L} = -r_L[\frac{v_0^2}{b^2}\Delta C_L + e_L \frac{v_0}{b}\frac{\partial \Delta C_L}{\partial t}] \\
&\ddot{C}_{D2} + a_D \frac{v_0}{b}\dot{C}_{D2} + r_D \frac{v_0^2}{b^2} C_{D2} = -[r_D \frac{v_0^2}{b^2}\Delta C_D + e_D \frac{v_0}{b}\dot{\varphi}]
\end{aligned} \tag{5}$$

Where v_0 is the relative wind speed, which is related to the input wind speed and the impeller speed, $v_0=\sqrt{U^2+(r\Omega)^2}$; b is the half chord length of the airfoil section; ρ is the air density; S_L, q_L, l_L, k_L are the aerodynamic term coefficients respectively; ΔC_L and ΔC_D are the difference between the linear extension line and the nonlinear part of the static aerodynamic lift curve and the static aerodynamic resistance curve respectively.

According to reference [8], the values of constant aerodynamic parameters are shown in Table 1.

Table 1. Constant Aerodynamic Parameters

Constant aerodynamic parameters	Values	Constant aerodynamic parameters	Values
SL	c	qL	π
lL	0	kL	0.5π
λL	0.15	a0L	5.9
σL	5.9	γL	0.55

The values of other parameters in Eq. (5) are as follows:

$$\Delta C_L = (a_{L1} + a_{L2})\alpha - 0.1396 a_{L1} - 0.3142 a_{L2} \qquad \Delta C_D = -(a_{D1}\varphi + a_{D2}\varphi^2 + a_{D3}\varphi^2)$$

$$\begin{cases} a_{L1} = a_{L2} = 0, & \alpha \le 0.1396 \\ a_{L1} = 6.32284, a_{L2} = 0, & 0.1396 < \alpha \le 0.3142 \\ a_{L1} = 6.32284, a_{L2} = -0.42284, & \alpha > 0.3142 \end{cases} \qquad \alpha = \arctan[U/(\Omega r)],$$

$$a_L = 0.25 + 0.4(\Delta C_L)^2,\ r_L = [0.2 + 0.23(\Delta C_L)^2]^2,\ e_L = -2.7(\Delta C_L)^2,$$

$$a_D = 0.25 + 0.4(\Delta C_D)^2,\ r_D = [0.2 + 0.1(\Delta C_D)]^2,\ e_D = -0.015(\Delta C_D)^2.$$

3. Equation Linearization and Solution

In conjunction with Eqs. (1)~(5), let the state variable $X=[z\ y\ C_{1L}\ C_{2L}\ C_{D2}\ \phi]$, then a second-order nonlinear aeroelastic equation system can be obtained:

$$M\ddot{X} + C\dot{X} + KX = Q \tag{6}$$

Because it is difficult to solve the nonlinear system and analyze the stability of the system, it is necessary to linearize the nonlinear equations. Suppose there is an equilibrium point: $X_0=[z_0\ y_0\ C_{1L0}\ C_{2L0}\ C_{D20}\ \phi_0]$, each order derivative of the system exists at the equilibrium point. The system state variables are expanded by Taylor series at the assumed equilibrium point, and the high-order term ε of the system can be ignored because it has little influence on the system state. Take the first two low-order models for approximate analysis, and bring them into Eq. (6) to obtain the equations of the system at the equilibrium point:

$$K_0 X_0 = Q_0 \tag{7}$$

The term containing ε is extracted, and the perturbation equations including state variables are obtained, that is the system of linear approximate equations:

$$M_M \ddot{X}_1 + C_M \dot{X}_1 + K_M X_1 = 0 \tag{8}$$

Because the coefficient of Eq. (8) contains unknown equilibrium point variables, it is necessary to solve Eq. (7) by Newton Raphson iterative method to obtain the equilibrium point under specific simulation parameters. By bringing the obtained equilibrium point back to Eq. (8), the blade aeroelastic equation after linearization can be obtained. In order to analyze the time-domain response of the linearized Eq. (8), we need to transform the second-order Eq. (8) into the first-order equations.

Defining $Y = [X_1^T \ \dot{X}_1^T]^T$, and the simultaneous state equations $\dot{X}_1 = \dot{X}_1$, can be obtained:

$$\dot{Y}(t) = AY(t) + Bu \tag{9}$$

where the coefficient matrix A can be expressed as:

$$A = \begin{bmatrix} 0_{6\times6} & I_{6\times6} \\ -M_M^{-1}K_M & -M_M^{-1}C_M \end{bmatrix}$$

4. Linear System Control and Simulation Based on LMI

At present, the technology of linear matrix inequality (LMI) has become an effective tool in control engineering and other fields. Using LMI to solve control problems is an important direction in the development of control theory [9].

The control goal is to make Y track the desired state trajectory Y_d asymptotically and stably, and define the tracking error $e = Y - Y_d$, then

$$\dot{e} = \dot{Y} - \dot{Y}_d = AY + Bu - \dot{Y}_d \tag{10}$$

The control rate is designed as:

$$u = KY + u_r \tag{11}$$

where K is the state feedback gain, which can be obtained by LMI design. Taking the feedforward control term $u_r = -KY_d - B^{-1}AY_d + B^{-1}\dot{Y}_d$, the control input is obtained as:

$$u = KY - KY_d - B^{-1}AY_d + B^{-1}\dot{Y}_d = Ke - B^{-1}AY_d + B^{-1}\dot{Y}_d \tag{12}$$

Substitute Eq. (12) into Eq. (10) to obtain:

$$\begin{aligned} \dot{e} &= AY + B(Ke - B^{-1}AY_d + B^{-1}\dot{Y}_d) - \dot{Y}_d \\ &= AY + BKe - AY_d + \dot{Y}_d - \dot{Y}_d \\ &= Ae + BKe \end{aligned} \tag{13}$$

The Lyapunov function is constructed as:

$$V = e^T P e, \quad P = P^T > 0 \tag{14}$$

The derivative with respect to time:

$$\begin{aligned}\dot{V} &= (e^T P)'e + e^T P\dot{e} = \dot{e}^T Pe + e^T P\dot{e} = (Ae + BKe)^T Pe + e^T P(Ae + BKe) \\ &= e^T A^T Pe + e^T K^T B^T Pe + e^T PAe + e^T PBKe \\ &= e^T (A^T P + K^T B^T P + PA + PBK)e = e^T \Gamma e\end{aligned} \quad (15)$$

where $\Gamma = A^T P + K^T B^T P + PA + PBK$

In order to ensure $\dot{V} < 0$, we need to satisfy $\Gamma < 0$, as:

$$A^T P + K^T B^T P + PA + PBK < 0 \quad (16)$$

In the LMI Eq. (16), since both K and P are unknown, in order to solve LMI, we need to linearize it, let $N = PBK$, then the LMI can be expressed as:

$$A^T P + N^T + PA + N < 0 \quad (17)$$

The solution of LMI inequality (17) is a complex problem, which is an optimization problem constrained by linear matrix inequality. It can be simplified by using YALMIP toolbox of MATLAB. Through LMI, N and K can be obtained, thus $K=(PB)^{-1}M$, then the closed-loop system composed of controlled object Eq. (9) and control law Eq. (12) is asymptotically stable.

Take the basic parameters: c=0.3m, b=c/2, ρ_b=15 kg/m, ω_z =8 rad/m, ω_y =12 rad/m, ξ_z=0.01, ξ_y =0.02, λ=10, U=15 m/s, r=1 m, In order to simplify the analysis process during the simulation, the constant pitch angle is set to be $\pi/6$. Combined with the above controller scheme, the module is built in Simulink for simulation [10].

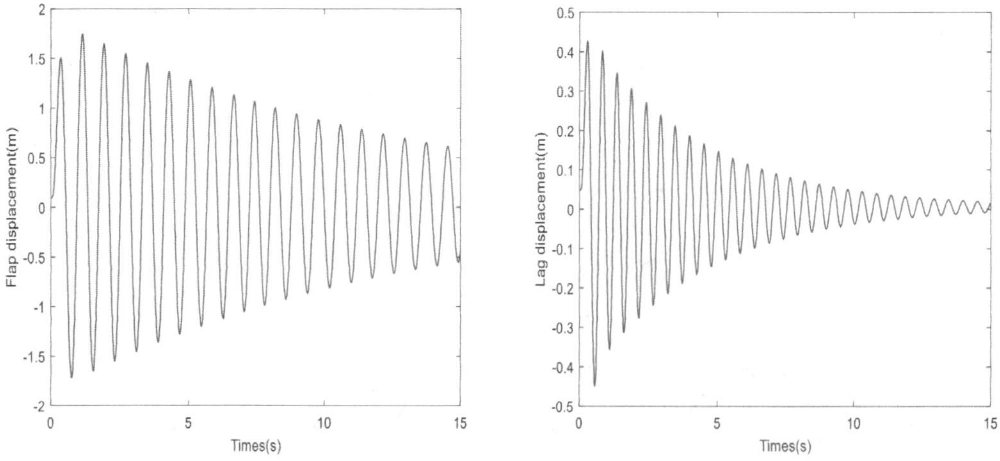

(a) (b)

Figure 2. Vibration displacement in flap and lag direction without control

It can be seen from Figure 2 that the maximum amplitude in the flap direction has reached 3.4 meters, the maximum amplitude in the lag direction has reached 0.9 meter, and the chord length of the section is 0.3 meter. Therefore, under this condition, if the vibration condition is as shown in the simulation results, the blade will inevitably be damaged, and the control should be applied to reduce the amplitude.

It can be seen from the observation of Figure 3 that after the control is applied, the maximum amplitude of the flap and lag direction of the system decreases greatly, the system response is finally stable, and the stability time is greatly shortened, and the control effect is good.

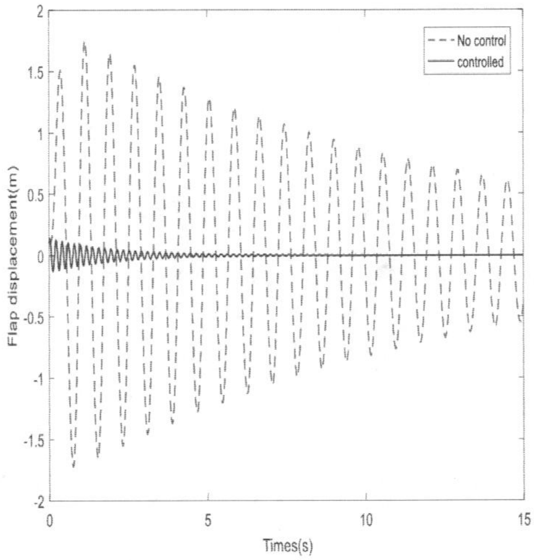

(a)

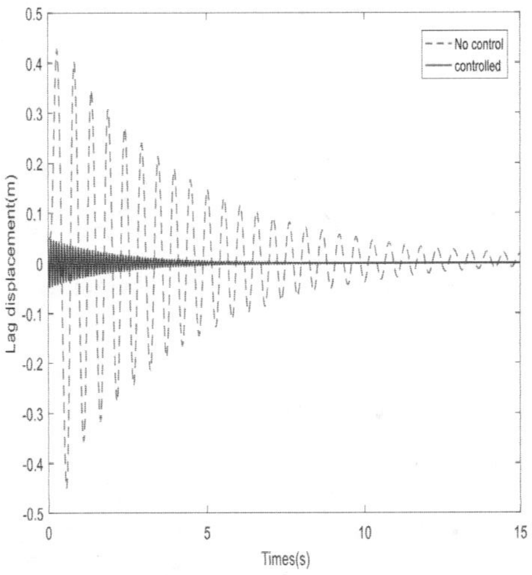

(b)

Figure 3. Comparison of flap and lag direction vibration after control

5. Conclusion

The stall flutter analysis in this study is based on the premise of neglecting the elastic torsion. Based on the modified nonlinear ONERA aerodynamic model, the aeroelastic equations of wind turbine blade flap/lag coupling are established, and the nonlinear aeroelastic equations are linearized by expanding to Taylor series at the equilibrium point. By constructing the Lyapunov function, a linear matrix inequality satisfying the system stability is designed. The state feedback gain is obtained by using the YALMIP toolbox in MATLAB, and the control input which makes the system stable is obtained. The simulation results show that: before the control is applied, the maximum amplitude of the flap/lag can lead to the blade fracture failure. By adding the control input designed in this study, the maximum amplitude of the flap/lag is greatly reduced, and the time to reach the steady state is also greatly shortened, so as to achieve the purpose of stall flutter suppression.

Acknowledgement

The authors gratefully acknowledge the support of the National Natural Science Foundation of China (no. 51675315).

References

[1] Ren Y and Lin X 2019 Shock and Vibration 29 121-124.
[2] Li N, Mu A and Balas M 2015 Shock and Vibration 34 82-88.
[3] Liu T and Wu H 2017 Shock and Vibration 36 265-272.
[4] Liu T 2015 Shock and Vibration 5 1-20.
[5] Chang L, Liu T, Xu W and Wu H 2018 Machine Design & Research 34 199-204.
[6] Liu T, Ren Y and Yang X 2013 Journal of Fluids and Structures 42 488–502.
[7] Liu T and Yu Z 2016 Journal of Central South University (Science Technology) 47 3562-3569.
[8] Liu T and Ren Y 2012 Acta Energiae Solaris Sinica 33 105-112.
[9] Liu J 2015 Sliding mode control design and MATLAB Simulation-the Basic Design Method (Beijing: Tsinghua University Press) chapter 11 pp 421–431.
[10]Xue D 2012 Computer Aided Control Systems Design Using MATLAB Language (Beijing: Tsinghua University Press) chapter 9 pp 379–387.

Variable Pitch Control of Wind Turbine Based on Fuzzy Feedforward and Feedback Linearization Sliding Mode

Haipeng Sun, Xiaolin Zhang, Yingbo Wang, Shanyao Li, Changle Sun and Tingrui Liu*

College of Mechanical & Electronic Engineering, Shandong University of Science & Technology, Qingdao 266590, China

*Corresponding author's E-mail: liutingrui9999@163.com

Abstract. When the wind turbine operates above the rated wind speed, the pitch angle is usually adjusted to make the output power stable at the rated value. As the traditional linear controller is difficult to achieve better control effect for the large inertia and strong nonlinear wind power generation system, a variable pitch control method combining fuzzy feedforward and feedback linearization sliding mode is proposed in this study. Firstly, the nonlinear wind power model is globally linearized by input/output feedback linearization, and then the feedback controller is designed by combining with sliding mode control. The fuzzy feedforward controller can give the appropriate pitch angle according to the change of wind speed and realize dynamic feedforward compensation. The simulation results show that the proposed control method can make the output power of the wind turbine stable, and has better control accuracy and robustness.

1. Introduction

Wind energy is one of the most promising renewable energy sources and has made great progress in the past decade [1]. When large-scale wind turbine operates above the rated wind speed, the variable pitch control method is usually used to ensure the safe operation and stable output power [2]. Due to the randomness of wind speed variation and the strong nonlinearity and time delay of wind turbine, it is difficult for traditional linear controller to achieve better control effect.

In recent years, many researches on pitch control methods have been carried out at home and abroad. In [3], sliding mode control is applied to variable pitch control, which can meet the requirements of constant power output of wind turbine by using the characteristics of fast response and strong robustness of sliding mode control. In [4], the method of wind speed feedforward is used to estimate the wind speed through Kalman filter and Newton Raphson algorithm, and then the appropriate pitch angle feedforward value is given to solve the problem of system control lag. In [5], the mixed H_2 / H_∞ pitch control method based on global exact linearization is proposed, which can ensure good control performance for the large-scale variety of operation points, and has a better interference suppression effect.

In this study, a variable pitch control method based on fuzzy feedforward and feedback linearization sliding mode is proposed. The input-output feedback linearization is used to linearize the nonlinear wind power model, and the sliding mode control is used to improve the robustness of the feedback controller. According to the change of wind speed, the fuzzy feedforward controller can infer the appropriate pitch angle feedforward value, which is added with the output value of the feedback controller as the reference pitch angle input to the wind power generation system. Simulation results show that compared with the traditional control method, this method can reduce the fluctuation of system output, improve the response speed of the system, and has good control effect.

2. Wind Turbine Model

2.1. Aerodynamic Model

The aerodynamic torque captured by wind turbine from wind energy can be expressed as:

$$T_r = \frac{P_r}{\omega_r} = \frac{1}{2\omega_r}\rho\pi R^2 v^3 C_p(\lambda,\beta) \tag{1}$$

$$\lambda = \frac{\omega_r R}{v} \tag{2}$$

where P_r the aerodynamic power of wind turbine, R is the rotor radius of wind turbine, ρ is the air density, v is the wind speed, ω_r is the rotor speed of wind turbine, λ is the tip speed ratio, β is the pitch angle, $C_p(\lambda,\beta)$ is the wind energy utilization coefficient, which is a nonlinear function of the tip speed ratio and the pitch angle [6].

$$\begin{cases} C_p(\lambda,\beta) = 0.22(\frac{116}{\lambda_i} - 0.4\beta - 5)e^{-12.5/\lambda_i} \\ \frac{1}{\lambda_i} = \frac{1}{\lambda + 0.08\beta} - \frac{0.035}{\beta^3 + 1} \end{cases} \tag{3}$$

2.2. Transmission System Model

The transmission system model of wind turbine is composed of wind wheel, low speed shaft, gearbox, high speed shaft and generator [4]. In order to simplify the calculation, it is assumed that transmission shaft is rigid and has no friction loss. The dynamic equation of the low speed shaft near the wind wheel can be expressed as:

$$J_r\dot{\omega}_r = T_r - nT_m \tag{4}$$

where J_r is the rotor inertia, T_m is the high-speed shaft torque, n is the gear ratio, which can be expressed as $n = \omega_g / \omega_r$.

The dynamic equation of the high-speed shaft near the generator can be expressed as:

$$J_g\dot{\omega}_g = T_m - T_e \tag{5}$$

where J_g is the generator inertia, ω_g is the generator speed, T_e is the electromagnetic torque of generator.

Substituting Eq. (5) into Eq. (4), the transmission system equation of wind turbine is expressed as:

$$(J_r + n^2 J_g)\dot{\omega}_r = T_r - nT_e \tag{6}$$

As this study mainly studies the performance of pitch controller for wind turbine operating above the rated wind speed, the electromagnetic torque of generator can be set as the rated value.

2.3. Pitch Actuator Model

The pitch actuator of wind turbine can be simplified as a first-order inertial system:

$$\dot{\beta} = \frac{1}{\tau}(\beta_r - \beta) \tag{7}$$

where τ is the time constant, β_r is the reference pitch angle of the pitch actuator.

3. Design of Pitch Controller

When the wind speed is higher than the rated wind speed, the pitch angle is changed to limit the aerodynamic power absorbed by the wind turbine, so as to achieve the purpose of constant output power. Due to the strong nonlinearity and time delay of wind turbine, the traditional feedback controller can only pitch when there is deviation. To solve this problem, this study proposes a pitch control method combining fuzzy feedforward and feedback linearization sliding mode. The control structure is shown in Figure 1.

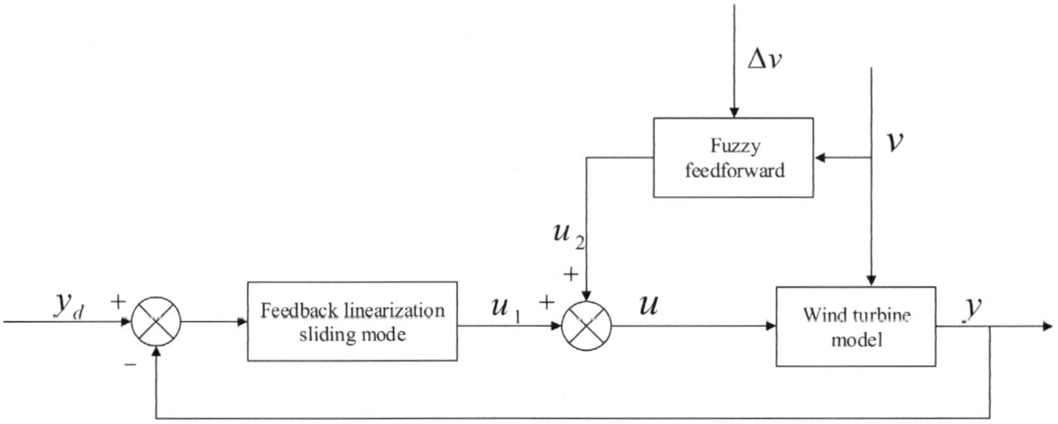

Figure 1. The structure diagram of control system

3.1. Model Linearization

As the wind power system is a strong nonlinear system, it needs to be linearized. Different from the linearization method of Taylor expansion at the equilibrium point, this study uses input-output feedback linearization to globally linearize the nonlinear wind power model. In fact, the basic idea of the feedback linearization theory is to repeatedly derive the output function y until the input variable u appears in the expression of the output function, and then design a reasonable control law u to transform the nonlinear system into a linear system [7].

According to Eqs. (6)~(7), the state variable of wind power system can be defined as $x = [x_1, x_2]^T = [\omega_r, \beta]^T$, the control input can be defined as $u = \beta_r$. Then the single input single output affine nonlinear model of wind power generation system can be expressed as:

$$\begin{cases} \dot{x} = f(x) + g(x)u \\ y = h(x) \end{cases} \tag{8}$$

where $f(x) = \begin{bmatrix} \frac{1}{J_r + n^2 J_g}\left(\frac{\rho \pi R^2 v^3 Cp(\lambda, x_2)}{2x_1} - nTe \right) \\ -\frac{1}{\tau} x_2 \end{bmatrix}$, $g(x) = \begin{pmatrix} 0 \\ \frac{1}{\tau} \end{pmatrix}$, $y = h(x) = x_1$.

According to Eq. (8), there is no direct relationship between the output function y and the control input u, so the derivative of the output function y can be expressed as:

$$\dot{y} = L_f h(x) + L_g h(x) u \tag{9}$$

Due to $L_g h(x)=0$, we need to continue to derive the output function y.

$$\ddot{y}=L_f{}^2 h(x)+L_g L_f h(x)u \tag{10}$$

where $L_f{}^2 h(x)=\frac{\partial f_1(x)}{\partial x_1}f_1(x)+\frac{\partial f_1(x)}{\partial x_2}f_2(x)$, $L_g L_f h(x)=\frac{\partial f_1(x)}{\partial x_2}\frac{1}{\tau}$.

Due to $L_g L_f h(x)\neq 0$, the input variable u appears in the second derivative of the output function y. And the relative order of the system is equal to the order of the system, which satisfies the condition of feedback linearization. So the control law to realize linear relationship $\ddot{y}=v$ can be expressed as:

$$u=\frac{1}{L_g L_f h(x)}(v-L^2{}_f h(x)) \tag{11}$$

3.2. Feedback Linearization Sliding Mode Controller

Sliding mode control algorithm has the characteristics of fast response and strong robustness. In order to better track the speed of the wind turbine and make it stable at the rated value, this study combines the input-output feedback linearization with the sliding mode control algorithm to design the feedback controller [8].

The tracking error of the system can be defined as $e=y_d-y$, y_d is the rated speed of the wind turbine. Then the sliding mode function can be defined as:

$$s(x,t)=ce+\dot{e} \tag{12}$$

where $c>0$. According to the result of model feedback linearization, the control law can be set as:

$$u_1=\frac{1}{L_g L_f h(x)}(v-L^2{}_f h(x)+\eta\,\mathrm{sgn}(s)) \tag{13}$$

where $\mathrm{sgn}(s)$ is the sign function, v is the auxiliary term of the control law, $\eta>0$.

The Lyapunov function can be defined as:

$$V{=}s^2/2 \tag{14}$$

Then

$$\dot{V}=s\dot{s}=s(\ddot{y}_d-\ddot{y}+c\dot{e}) \tag{15}$$

Substituting Eq. (13) into Eq. (15), and $v=\ddot{y}_d+c\dot{e}$.Then we can get $\dot{V}\leq 0$. According to the stability criterion of Lyapunov function, it can be proved that the designed controller can make the system stable.

In this study, saturation function is used instead of sign function to reduce the chattering caused by sliding mode control. Saturation function can be expressed as:

$$sat(s)=\begin{cases}1 & s>\Delta\\ s/\Delta & |s|\leq\Delta\\ -1 & s<-\Delta\end{cases} \tag{16}$$

where Δ is the thickness of the boundary layer.

3.3. Fuzzy Feedforward Controller

As the wind speed is a measurable external disturbance, and the wind power generation system has a certain time delay [9]. Therefore, the fuzzy feedforward control method is used to give the appropriate pitch angle for dynamic feedforward compensation, which effectively solves the problem of feedback controller delay.

In the fuzzy feedforward controller designed in this study, the wind speed at the last moment $v(k-1)$ and the difference Δv between the wind speed at the current moment and the wind speed at the last moment are taken as the two inputs of the controller. The feedforward pitch angle u_2 is used as the output of the controller.

The basic domain of wind speed is defined as [12m/s, 25m/s], fuzzy domain is defined as {1, 2, 3, 4, 5}. The corresponding fuzzy subset is defined as {LH, RH, H, VH, EH}. The basic domain of wind speed increment is defined as [-3m/s, 3m/s], fuzzy domain is defined as {-3, -2, -1, 0, 1, 2, 3}. The corresponding fuzzy subset is defined as {NB, NM, NS, ZO, PS, PM, PB}.

The basic domain of wind speed is defined as [-8°, 8°]. The fuzzy domain of output is defined as {-6, -5, -4, -3, -2, -1, 0, 1, 2, 3, 4, 5, 6}, The corresponding fuzzy subset is defined as {NB2, NB1, NM2, NM1, NS2, NS1, ZO, PS1, PS2, PM1, PM2, PB1, PB2}.

The input and output membership functions of the fuzzy feedforward controller designed in this study are all triangular membership functions and the center of gravity method is used for defuzzification. According to the above, the control rules of fuzzy feedforward controller are shown in Table 1.

Table 1. Fuzzy feedforward controller rules

Δv	v				
	LH	RH	H	VH	EH
NB	NB2	NB1	NM2	NM1	NS2
NM	NB1	NM2	NM1	NS2	NS1
NS	NM1	NS2	NS1	NS1	NS1
ZO	ZO	ZO	ZO	ZO	ZO
PS	PM1	PS2	PS1	PS1	PS1
PM	PB1	PM1	PS2	PS2	PS1
PB	PB2	PM2	PM1	PS2	PS2

4. Simulation and Discussion

In order to verify the effectiveness of the control method proposed in this study, a 3MW wind turbine model is established by using MATLAB / Simulink, and the control method proposed in this study is compared with the feedback linearization control. The main parameters of the wind turbine used in the simulation are shown in Table 2. The parameters of the feedback linearized sliding mode controller can be set as: $c=15$, $\eta=3$, $\Delta=0.1$.

Table 2. The parameters of 3 MW wind turbine

Wind turbine parameters	Value
Rated power	3MW
Rotor radius	47.5m
Air density	1.20kg/m3
Rated wind speed	12m/s
Cut out wind speed	25m/s
Rotor inertia	6250000kg·m2
Generator inertia	15kg·m2
Gearbox ratio	80
Time constant	0.2s

This study mainly analyzes the performance of variable pitch control of the wind turbine when the wind speed is higher than the rated wind speed, so the random wind speed as shown in Figure. 2 is used for simulation. The simulation results of pitch angle are shown in Figure. 3, and the simulation results of speed and power of wind turbine are shown in Figure. 4 and Figure. 5.

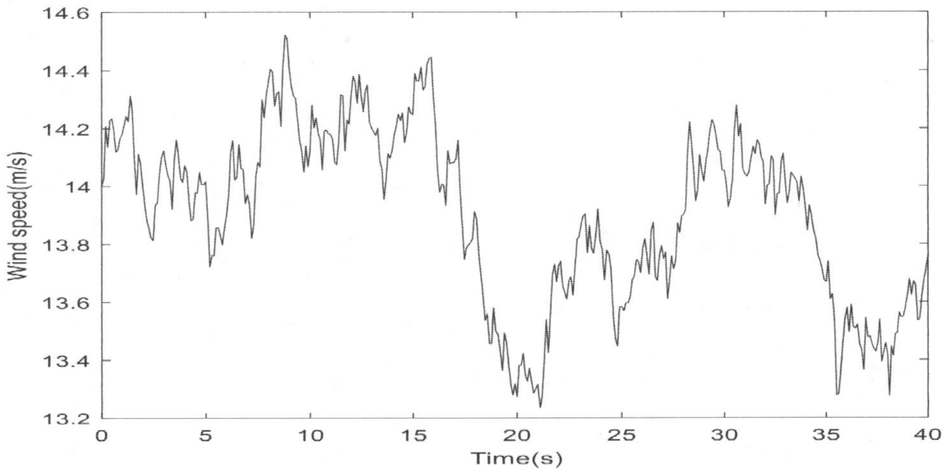

Figure 2. Random wind speed

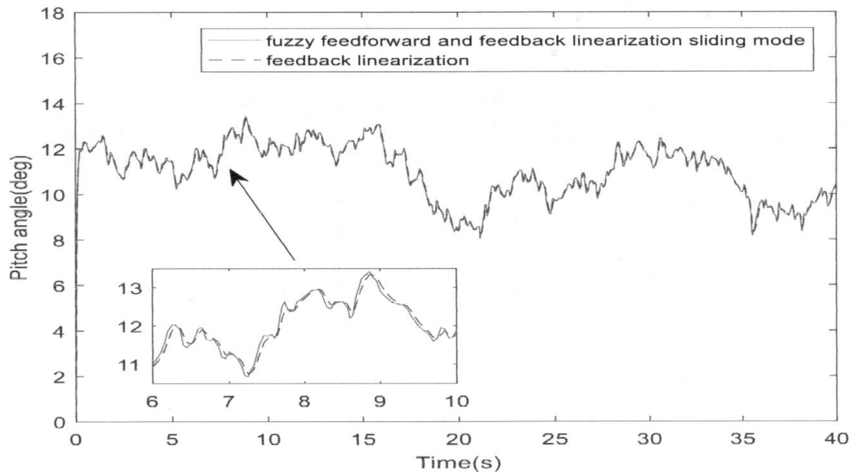

Figure 3. Pitch angle of different controllers

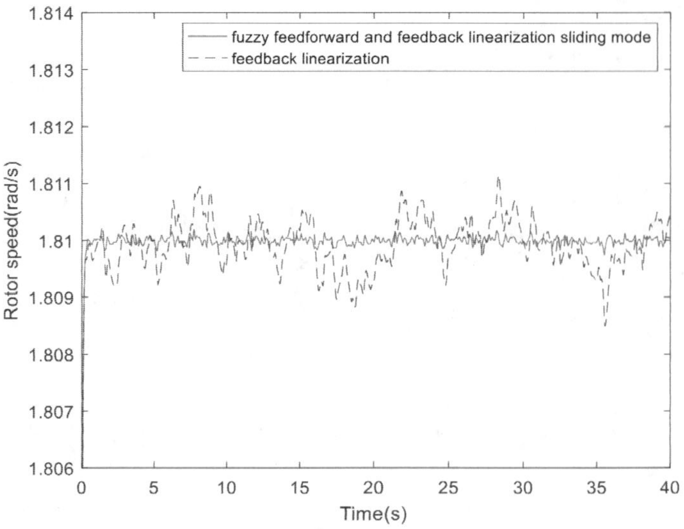

Figure 4. Rotor speed of different controllers

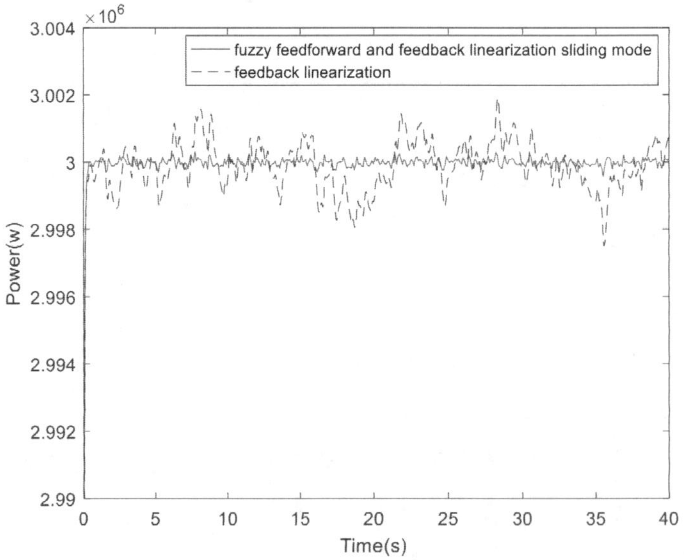

Figure 5. Output power of different controllers

It can be seen from Figure 3, with the constant change of wind speed, the pitch angle can be changed under the effect of two pitch control methods. However, compared with the feedback linearization control, the pitch angle of the output of the pitch control method proposed in this study changes greatly, so that the response speed of the system is faster. It can be seen from Figure. 4 and Figure. 5 that compared with the feedback linearization control, the pitch controller combined with fuzzy feedforward and feedback linearization sliding mode can obtain smaller fluctuation amplitude of speed and output power of wind turbine, which is more stable, so it has better control effect.

5. Conclusion

For the wind speed above the rated wind speed, the traditional pitch controller is difficult to achieve satisfactory control effect for wind turbine. In this study, a variable pitch control method based on fuzzy feedforward and feedback linearization sliding mode is proposed. The feedback linearized sliding mode controller can eliminate the nonlinear problem of the wind power system and enhance the robustness of the system. The fuzzy feedforward controller can give the appropriate pitch angle according to the different wind speed to achieve fast compensation. The simulation results show that when the wind speed changes randomly, the pitch control method in this study can make the speed and output power of the wind turbine stable near the rated value, reduce the output fluctuation, and improve the system response speed.

Acknowledgement

The authors gratefully acknowledge the support of the National Natural Science Foundation of China (no. 51675315).

References

[1] Ren Y, Li L, Brindley J and Jiang L 2016 Control Engineering Practice 50 84–94.
[2] Wei Z, Chen R, Chen J, Chen J, Gong C and Chen Z 2011 Proceedings of the CSEE 31 121–126.
[3] Wang D, Jia Z, Sun J and Xu D 2008 Journal of North China Electric Power University 35 1–3.
[4] He Y, Huang S, Du J, Su D and Li J 2012 Power System Protection and Control 40 15–20.
[5] Zhu M, Liu J, Lin Z and Meng H 2016 International Journal of Machine Learning & Cybernetics 7 921–930.
[6] Zhao Y, Xia C, Song Z, Wang H and Chen W 2008 Proceedings of the CSEE 28 133-138.
[7] Yang J, Zheng J, Yang M and Wu J 2012 Control Theory & Applications 29 1365–1370.
[8] Liu J 2015 Sliding Mode Control System Design and MATLAB Simulation-the Basic Design Method (Beijing: Tsinghua University Press) chapter 3 pp 101–109.
[9] Guo P 2010 Proceedings of the CSEE 30 123-128.

DOA Estimation of Underwater Acoustic Targets Based on Bayesian Compressed Sensing

Fei Liu

School of Electronic Engineering, Department of Information Science and Engineering, Ocean University of China, Qingdao, Shandong, China

*corresponding author's e-mail: 1093091591@qq.com

Abstract. Aiming at the performance degradation or even failure of existing underwater acoustic target direction of arrial (DOA) estimation methods under the conditions of low signal-to-noise ratio and low snapshot, a underwater acoustic target DOA estimation method based on sparse Bayesian compressed sensing theory is studied, and simulation experiments are carried out by MATLAB to verify the feasibility and effectiveness of the proposed method. Simulation results show that compared with CBF , MUSIC and OMP algorithms, the proposed method can improve the estimation accuracy in the case of lower snapshot number, lower number of array elements and smaller array element spacing, and can achieve high resolution DOA estimation in the case of low signal-to-noise ratio and close position of multiple target signals.

1. Introduction

At present, the direction of arrival estimation algorithms of underwater acoustic targets mainly include traditional beamforming algorithm (CBF) and high resolution algorithm of spatial spectrum (such as MUSIC, MVDR) and so on. Compared with the CBF algorithm, the MUSIC algorithm breaks the Rayleigh limit and can achieve high-resolution DOA estimation. However, it is necessary to preset the number of incident azimuths of the target signal in advance, and the estimation performance will decrease due to the correlation between the target signals. Therefore, it is not suitable for practical application; In addition, both types of algorithms require large enough snapshot data to realize the position estimation. In the case of a small number of snapshots or a low signal-to-noise ratio, the estimation performance will deteriorate and the impact will be greater.

Compressed sensing (CS) theory [1] was proposed in 2016. When the target signal satisfies the condition of sparsity in the spatial domain, the underwater acoustic signal can be accurately reconstructed with sampling data much smaller than the Nyquist sampling theorem. According to the sparseness of the signal in the airspace, compressed sensing is applied to DOA estimation, and certain results have been achieved in recent years [2][3][4]. From these research results, it can be seen that the CS-based DOA estimation method has a higher angular resolution than the traditional DOA estimation algorithm [5], and it can be estimated normally under the condition of a single snapshot. According to the theory of compressed sensing, in the sparse reconstruction algorithm, the greedy algorithm cannot obtain the global optimal solution [6]; while the convex optimization algorithm has relatively high complexity and the regularization parameters are difficult to determine, which is not suitable for practical applications [7]. Therefore, the sparse reconstruction of the target signal can be combined with the sparse Bayesian theory based on statistical probability. For example, Zhang et al. make full use of the correlation of the incident signal in the time domain to reconstruct the multi-frame data into a vector, which improves the performance of the sparse signal orientation estimation [8]. Wang et al. use the Dirichlet process to construct a prior probability model and then based on the sparse The mode grouped each frequency point [9], Yang et al. proposed an off-grid (Off-Grid) azimuth estimation technology, which solved the basis mismatch problem by adding an angle offset variable [10].

Aiming at the problem of poor accuracy or even failure of traditional DOA estimation methods under low signal-to-noise ratio and low snapshot, and the problem of grid mismatch in compressed sensing DOA estimation, this paper is based on Bayesian theory, combined with underwater target signals in space Based on the above sparseness, a DOA estimation method for underwater targets based on sparse Bayesian compressed sensing theory is proposed, which realizes high-resolution DOA estimation of underwater target azimuth under the conditions of low signal-to-noise ratio and low snapshot.

2. Sparse representation of underwater acoustic signal

2.1. Basic principles of compressed sensing

The compressed sensing theory can be expressed as:

$$y = \Phi x + n, \tag{1}$$

In the formula, y is the observation data, Φ is the measurement matrix of $M \times N$ dimension, this matrix satisfies the property of restricted isovolume (RIP criterion),and $M < N$. $x = [x_1, x_2, \cdots, x_N]^T$ is sparse underwater acoustic target signal in the airspace.

The spatially sparse underwater acoustic signal x , base matrix φ and sparse vector s satisfy:

$$x = \varphi s. \tag{2}$$

equation (1) can be rewritten as:

$$y = \Phi x + n = \Phi \varphi s + n = \Theta s + n, \tag{3}$$

Among them, Θ is the over-complete dictionary of $M \times N$ dimension. The process of DOA estimation is to recover the sparse underwater acoustic target signal s through the observation signal y received by the array and Θ , which is the process of signal reconstruction.

2.2. DOA estimation model of underwater acoustic target based on compressed sensing

Suppose that there are K incident signals in the space of [0°,180°], denoted as $x = [x_1, x_2, \cdots, x_K]^T$, and their incident azimuth angles are denoted as $\beta_1, \beta_2, \cdots, \beta_K$ respectively. The underwater acoustic target space of [0°,180°] is divided into N grids, so that each divided space grid corresponds to the potential signal incident angle one by one, denoted as $\alpha_1, \alpha_2, \cdots, \alpha_N$. Because in order to combine compressed sensing theory with DOA estimation, a prerequisite requirement is that the target signal must be sparse. Therefore, the number of underwater acoustic target signals that actually exist in the space must be much smaller than the number of potential signals to meet this requirement. That is satisfied $K << N$. The sparse representation signal is denoted as $s = [s_1, s_2, \cdots, s_N]^T$. When there is a real target incident signal on a certain spatial grid, sparseness means that the corresponding position of the signal is not zero, otherwise it is zero.

This completes the sparse representation of the underwater acoustic target signal. The general mathematical model of DOA estimation based on CS, such as formula (5):

$$Y = A(\alpha)S + N, \tag{4}$$

In formula (4), the $M \times L$ observation matrix is Y , the $M \times L$ Gaussian noise matrix is N , $S = [s_1, s_2, \ldots, s_L]^T$ is the $N \times L$ matrix, N is the number of spatial grids, L is the number of snapshots, and M is the number of array elements. $A(\alpha)$ is the array flow matrix, that is, the over-complete dictionary in Section 1.1, denoted as:

$$A(\alpha)=\left[a(\alpha_1),a(\alpha_2),\ldots,a(\alpha_N)\right], \tag{5}$$

It can be seen from equation (6) that the array flow matrix $A(\alpha)$ contains the angle information of all potential signals in the spatial domain. Where $a(\alpha_i)$ is the steering vector at the grid α_i , the steering vector can be written as:

$$a(\alpha_i)=\left[1,e^{j\psi(\alpha_i)},\ldots,e^{j(M-1)\psi(\alpha_i)}\right]^T, \tag{6}$$

In formula (6), $\psi(\alpha_i)=-2\pi\frac{d}{\lambda}\sin(\alpha_i)$, d is the distance between the array elements and λ is the source wavelength.

3. DOA estimation based on Bayesian compressed sensing

3.1. Bayesian compressed sensing theory

The mathematical model of compressed sensing is:

$$y=\Phi x+n, \tag{7}$$

In formula (7), y is the observation data received by the array, Φ is the measurement matrix, x is the target signal, n is Gaussian noise, the mean is 0, and the variance is σ^2. According to the Bayesian estimation theory, the observation data obeys gaussian distribution, the mean is Φx and variance is σ^2, which can be expressed as:

$$p(y\mid x,\sigma^2)=N(y,\mid \Phi x,\sigma^2). \tag{8}$$

The likelihood function of the sum of the parameters x and σ^2 to be estimated is:

$$\begin{aligned} p(y\mid x,\sigma^2) &= \prod_{n=1}^{N} p(y_n,\mid \Phi_n x,\sigma^2) \\ &= (2\pi\sigma^2)^{\frac{N}{2}} \exp\left(-\frac{1}{2\sigma^2}\|y-\Phi x\|\right)^2. \end{aligned} \tag{9}$$

Next, the maximum a posteriori estimation of the parameters x and σ^2 is performed. Under the vector machine theory, the sparse signal x is designated as a Gaussian prior distribution with a mean value of 0, namely:

$$p(x\mid a)=\prod_{i=1}^{N} N(x_i\mid 0,a_i^{-1}), \tag{10}$$

Among them, $a=(a_1,a_2,\ldots,a_N)$ is the hyperparameter, and the reciprocal of the variance of the Gaussian prior distribution is the hyperparameter. Specify the Gamma prior distribution for the hyperparameters a and noise parameters α_0, namely:

$$p(a)=\prod_{i=1}^{N}\Gamma(a_i\mid a,b). \tag{11}$$

$$p(a_0)=\Gamma(a_0\mid c,d), \tag{12}$$

$$a_0 = 1/\sigma^2.$$

By marginalizing the hyperparameters by integration, the probability distribution function of the signal x to be estimated can be obtained:

$$p(x \mid a,b) = \prod_{i=1}^{N} \int_{0}^{\infty} N\left(x_i \mid 0, a_i^{-1}\right) \Gamma\left(a_i \mid a,b\right) da_i. \tag{13}$$

Under Bayesian theory, the posterior probability distribution of the signal x to be estimated is:

$$p(x,\alpha,a_0 \mid y) = \frac{p(y \mid x,\alpha,\alpha_0)\, p(x,\alpha,\alpha_0)}{p(y)}. \tag{14}$$

Solve by means of posterior distribution decomposition, namely:

$$p(x,\alpha,a_0 \mid y) = p(x \mid y,\alpha,\alpha_0)\, p(\alpha,\alpha_0 \mid y), \tag{15}$$

Equation (15) is the form of Gaussian convolution. From the previous analysis, it can be concluded that the equation (15) $p(x,\alpha,a_0 \mid y)$ obeys the Gaussian distribution, namely:

$$\begin{aligned} p(x,\alpha,a_0 \mid y) &= \frac{p(y \mid x,\alpha)\, p(x \mid a)}{p(y \mid a,\alpha_0)} \\ &= (2\pi)^{-(N-1)/2} |\Sigma|^{-1/2} \exp\left\{-\frac{1}{2}(x-u)^T \Sigma^{-1} (x-u)\right\}, \end{aligned} \tag{16}$$

Among them, u represents the posterior mean, Σ represents the posterior variance, and the specific expression is written as:

$$u = a_0 \sum \Phi^T y, \tag{17}$$

$$\Sigma = \left(a_0 \Phi^T \Phi + \Lambda\right)^{-1}, \tag{18}$$

$$\Lambda = diag\left(a_1, a_2, \ldots, a_N\right).$$

$$p(a,a_0 \mid y) \propto p(y \mid a,a_0) \cdot p(a)\, p(a_0). \tag{19}$$

The sum of $p(a,a_0 \mid y)$ and $p(y \mid a,a_0)$ in the equation (20) has the same solution, and the sum of hyperparameters can be calculated by $p(y \mid a,a_0)$. Knowing the sum of hyperparameters, using equation (17) to calculate the posterior mean u and posterior variance Σ, the posterior matrix u obtained is the maximum likelihood estimate of the target signal x.

Therefore, in order to calculate the posterior mean u and variance Σ, it is necessary to estimate the sum of hyperparameters a and α_0 firstly, so next, it is necessary to use $p(y \mid a,a_0)$ in equation (19) to estimate the sum of hyperparameters a and α_0.

3.2. DOA estimation of fast bayesian compressed sensing

In order to improve the convergence speed of the algorithm and reduce the running time of the algorithm, a fast Bayesian compressed sensing DOA estimation algorithm is studied here. The marginal likelihood function of x can be obtained by performing marginal integration, and the expression is:

$$p(y|\alpha,\alpha_0)=\int p(y|x,\alpha_0)p(x|\alpha)$$
$$=(2\pi)^{-N/2}\left|\alpha_0 I+\Phi\Lambda^{-1}\Phi^T\right|^{-1/2}\exp\left\{-\frac{1}{2}y^T\left(\alpha_0 I+\Phi\Lambda^{-1}\Phi^T\right)^{-1}y\right\}, \tag{20}$$

Take the logarithm of equation (20) to get:

$$L(\alpha,\alpha_0)=-\frac{1}{2}\left[N\log 2\pi+\log\left|\alpha_0^{-1}I+\Phi\Lambda^{-1}\Phi^T\right|+y^T\left(\alpha_0 I+\Phi\Lambda^{-1}\Phi^T\right)^{-1}y\right], \tag{21}$$

L is the cost function of the sparse signal x with respect to the sum of hyperparameters a and α_0.

Rewrite formula (20) as:

$$p(y|\alpha,\alpha_0)=(2\pi)^{-N/2}|C|^{-N/2}\exp\left\{-\frac{1}{2}y^T C^{-1}y\right\}, \tag{22}$$

among them:

$$C=\alpha_0^{-1}I+\Phi\Lambda^{-1}\Phi^T. \tag{23}$$

Remove the constant term of equation (22) to get:

$$L(\alpha,\alpha_0)=-\frac{1}{2}\left[\log|C|+y^T(C)^{-1}y\right]. \tag{24}$$

At this time, the problem of solving unknown parameters is transformed into a problem of solving the maximum value of the cost function.

decompose the formula (23):

$$C=\sigma^2 I+\Phi diag(\alpha)^{-1}\Phi^T=C_{-i}+\alpha_i^{-1}\Phi_i\varphi_i^T, \tag{25}$$

Among them, C_{-i} is the matrix formed by the remaining columns after removing the column i of the matrix C, and φ_i is the row i of the matrix Φ.

Therefore, the marginal likelihood function is expressed as $L(\alpha)=L(\alpha_{-i})+l(\alpha_i), L(\alpha_{-i})$ is the removed likelihood function after removing φ_i, and $l(\alpha_i)$ and α_i is related, namely:

$$l(\alpha_i)=1/2\left[\log\alpha_i-\log(\alpha_i+m_i)+n_i^2/(\alpha_i+m_i)\right], \tag{26}$$

$m_i=\phi_i^T C_{-i}^{-1}\phi_i$, $n_i=\phi_i^T C_{-i}^{-1}\phi_i$, m_i is Coefficient factor, n_i is Quality factor.

Find the first-order partial derivative of α_i by $l(\alpha_i)$:

$$\frac{d^2(l(\alpha_i))}{d\alpha_i^2}=\frac{\alpha_i(m_i-q_i^2)+m_i}{2\alpha_i(\alpha_i+m_i)^2}. \tag{27}$$

Through the analysis of equation (26), $l(\alpha_i)$ has a unique maximum point.

While $n_i^2>m_i$,

$$\alpha_i=m_i^2/(n_i^2-m_i). \tag{28}$$

While $n_i^2 < m_i$,

$$\alpha_i = \infty. \tag{29}$$

The algorithm steps are as follows:

- 1) Initialize σ^2 and base vectors ϕ_1, assign σ^2 reasonable initial values; update the hyperparameters $\alpha_1 = \frac{\|\varphi_1\|^2}{\|\varphi_1^T g\| / \|\varphi_1\|^2 - \sigma^2}$, and make other hyperparameters infinite at the same time;
- 2) Calculate the mean and variance, and calculate m_i and n_i;
- 3) Reselect a new basis vector ϕ_1;
- 4) Calculate $\theta_i = n_i^2 - m_i$;
- 5) If $\theta_i > 0, \alpha_i < \infty$, update the hyperparameters according to equation (28); if $\theta_i > 0$, $\alpha_i = \infty$, add the basis vector ϕ_1 to the model, and update the hyperparameters according to equation (28); if $\theta_i \le 0$, delete the basis vector ϕ_1, and assign the hyperparameter to infinity;
- 6) Calculated σ^2 and update α_0;
- 7) Update the mean and variance, m_i and n_i;
- 8) At this time, if the convergence condition is met, stop the iteration, otherwise go to (4) for the next iteration.

4. Simulation analysis and verification

Simulation conditions: The array is a uniform linear array (ULA), the number of array elements is 20, and the array element interval is d, $d = \lambda / 2$. λ is.the target signal wavelength. Assuming that two far-field signals are incident on the hydrophone array, the target incident angle is [40°,80°], the signal-to-noise ratio is 0dB, the number of snapshots is 50, and the division range of the spatial discrete grid for [0°,180°], the space is divided into 180 grids, and the distance between two adjacent grids is 1°.

4.1. Method feasibility analysis

In order to verify the effectiveness of the proposed method, the CBF algorithm, MUSIC algorithm, OMP algorithm and the method in this paper are used to perform DOA estimation simulation experiments. The comparison results are shown in Figure 1. It can be seen from Figure 1 that under this condition, the four algorithms can estimate the orientation information of the target, but in comparison, the OMP algorithm and the method in this paper have better estimation performance and sharper spectral peaks, while the CBF and MUSIC algorithm spectrum The peaks are relatively broad and the resolution is poor. Therefore, it is feasible to use the method proposed in this paper to estimate the DOA of underwater acoustic targets.

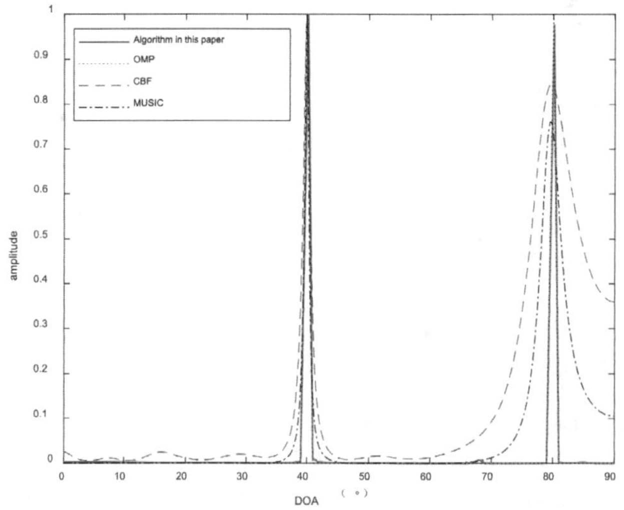

Figure 1. DOA estimation of four algorithms

4.2. Comparison and analysis of DOA estimation performance of various algorithms

4.2.1. Analysis of the resolution ability of each algorithm

- (1) Change the incident angle of the two target signals to [40°, 42°], and other simulation conditions are the same. The CBF algorithm, MUSIC algorithm, OMP algorithm and the method in this article are used to estimate the DOA of the target signal source. The simulation results are shown in the figure. 2 shown. Compared with figure 1, it can be seen that when the two target source angles [40°, 42°] are relatively concentrated in space, the traditional CBF algorithm and MUSIC algorithm basically fail, and only one angle can be estimated. Although as for the OMP algorithm, two angles can be distinguished, but there is still a certain error, but the method in this paper can still accurately estimate the DOA, distinguish two very close target signals, and the estimation performance is remarkable.

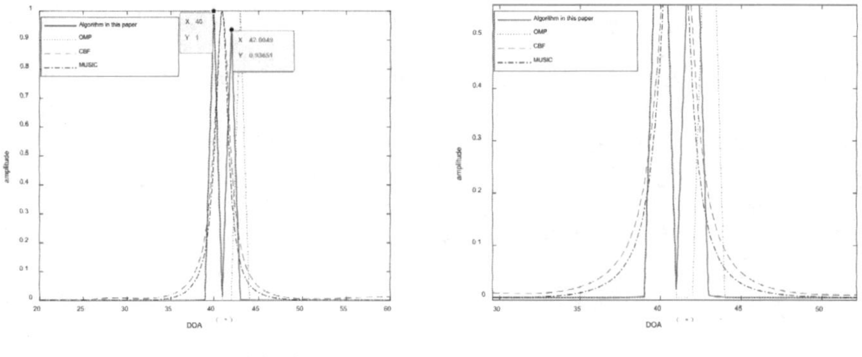

Figure 2. The DOA estimation of the four algorithms when the angle of the target signal source is relatively concentrated

- (2)Change the incident angle of the two target signals again to [46.3°, 78.6°], the division range of the spatial discrete grid is [0°, 180°]. At this time, there is a grid mismatch. problem. Other simulation conditions are the same. Four algorithms are used for DOA estimation. The simulation results are shown in figure 3. When the signal incident angle does not fall on the divided discrete grid, the grid mismatch problem occurs. It can be seen from the simulation results that the CBF, MUSIC, and OMP algorithms are unable to accurately estimate at this time, and can only estimate the approximate angle of the incident angle on the divided grid. The method in this paper carries out DOA estimation on the incident direction of the source, which can obtain a more accurate incident angle of the source, which greatly improves the estimation accuracy.

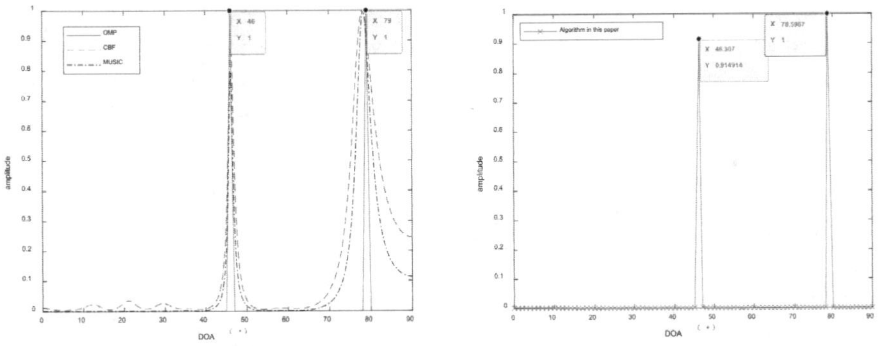

Figure 3. DOA estimation of four algorithms under grid mismatch

4.2.2. Analysis of the impact of the number of snapshots on the DOA estimation performance of each algorithm

Change the number of snapshots in the simulation conditions to 4 for simulation experiments, and other conditions remain unchanged. The simulation results are shown in figure 4. Compared with figure 1, it can be seen that when the number of snapshots is reduced from 50 to 4, the CBF algorithm fails, the performance of the MUSIC algorithm also becomes very poor, the sidelobes increase, and the spectral peaks are broadened. This is because the MUSIC algorithm implements high-resolution DOA estimation based on a large amount of snapshot data. Once the number of snapshots drops sharply, the algorithm performance will be greatly affected. The OMP algorithm also has certain errors, but the method in this paper has no obvious change, and its estimation performance is better than the other three algorithms.

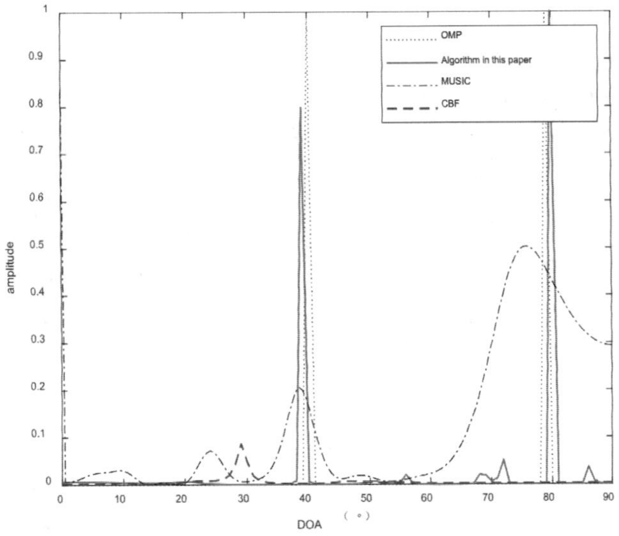

Figure 4. DOA estimation of four algorithms with a snap number of 4

4.2.3. Analysis of the influence of the number of array elements on the DOA estimation performance of each algorithm

Change the number of elements in the simulation conditions to 6 for simulation experiments, and other conditions remain unchanged. The simulation results are shown in Figure 5. Comparing with Figure 1, it can be seen that when the number of array elements is reduced, the spectral peaks of the four algorithms are widened. Among them, the resolution of the CBF and MUSIC algorithms is particularly reduced. At this time, the CBF and MUSIC algorithms have large errors. Even up to 5°, the performance of the OMP algorithm is relatively poor. The spectral peak of the method in this paper has been widened, and the estimated angle has a slight deviation, but it still has significant estimation performance in comparison.

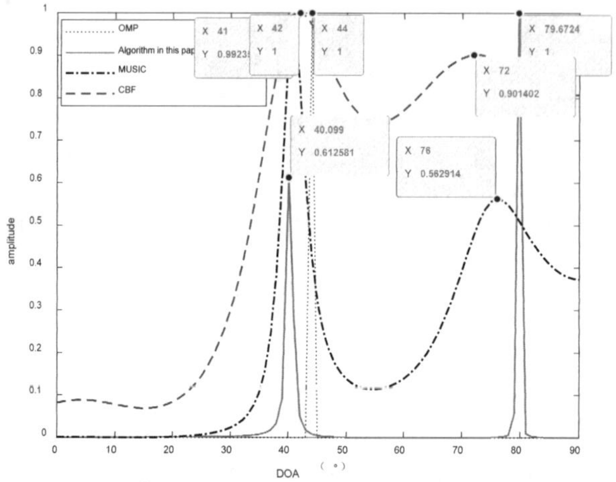

Figure 5. DOA estimation of four algorithms with element number 6

4.2.4. Analysis of the influence of array element interval on the DOA estimation performance of each algorithm

Change the element interval in the simulation condition to $\lambda / 10$ for simulation experiment, other conditions remain unchanged, the simulation result is shown in figure 6. Compared with fig. 1, it can be seen that when the array element interval becomes smaller, the CBF algorithm is limited by the array aperture, and effective DOA estimation cannot be performed, and the estimation effect is very poor. The performance of the MUSIC algorithm is also relatively poor. The performance of the method in this paper has relatively small changes, and effective DOA estimation can still be carried out.

 2173 (2022) 012047 doi:10.1088/1742-6596/2173/1/012047

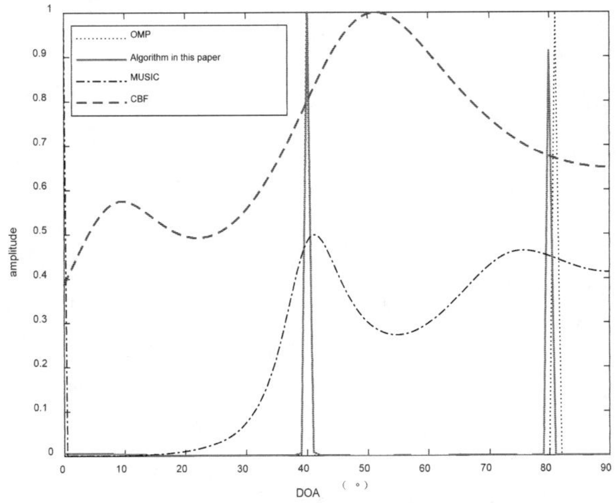

Figure 6. DOA estimation of four algorithms with element interval $\lambda/10$

4.2.5. Root Mean Square Error Analysis

Calculate the root-mean-square error of the CBF algorithm, MUSIC algorithm, OMP algorithm and the method in this paper when the signal-to-noise ratio, the number of samples, and the number of snapshots change through simulation experiments. The calculation formula of the root mean square error is:

$$RMSE=\sqrt{\frac{1}{K*Q}\sum_{q=1}^{Q}\sum_{k=1}^{K}\left(\hat{\theta}-\theta\right)^{2}}$$

Among them, K is the number of target signals, Q is the number of Monte Carlo experiments, θ is the real target wave direction, and $\hat{\theta}$ represents the estimated value of the real DOA parameter in the kth experiment.

- 1) DOA estimation performance analysis under different signal-to-noise ratios

The number of array elements is set to 20, the number of snapshots is set to 50, other simulation conditions remain unchanged, the signal-to-noise ratio changes from -5dB to 15dB in 2dB steps, and the CBF algorithm, MUSIC algorithm, OMP algorithm and the method in this article are used to perform DOA respectively. It is estimated that the root mean square error of each algorithm for DOA estimation is calculated through 100 Monte Carlo simulation experiments, and the simulation results are shown in figure 7. It can be seen from figure 7 that as the signal-to-noise ratio increases, the root-mean-square errors of the four algorithms gradually decrease. When the signal-to-noise ratio is high, the root-mean-square errors of the four algorithms are not very different, but When the signal-to-noise ratio is low (SNR<-10dB), the error of this method can be within 0.5°. It can be seen that in the case of low signal-to-noise ratio, the method in this paper has high accuracy, good estimation performance, and good anti-noise ability.

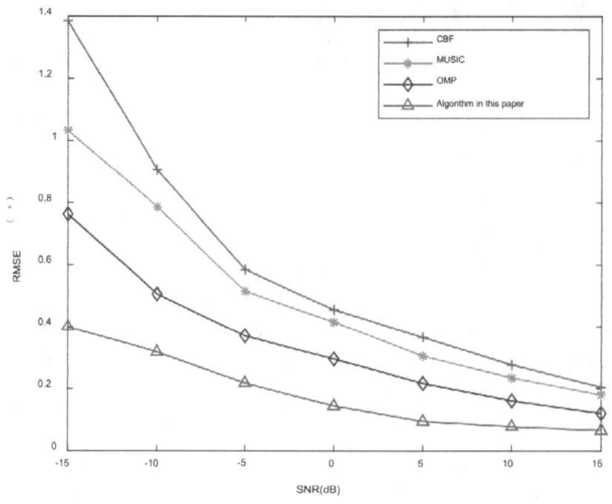

Figure 7. Root mean square error under different signal-to-noise ratio

- 2) DOA estimation performance analysis under different number of snapshots

The signal-to-noise ratio is set to 0dB, the number of array elements is set to 20, and other simulation conditions remain unchanged. The number of snapshots changes from 10 to 50 in steps of 5. The four algorithms are used for DOA estimation, and 100 Monte Carlo simulations are used. The experiment calculates the root mean square error of each algorithm to achieve DOA estimation, and the simulation result is shown in Figure 8. It can be seen from Figure 8 that as the number of snapshots increases, the root mean square error of each algorithm will always decrease, and in the case of high snapshots, the estimation accuracy of the four algorithms is not very different. However, under low snapshots, the root mean square error of the method in this paper is small, less than 0.6°, the estimation performance is better than the other three algorithms, and the estimation accuracy is higher.

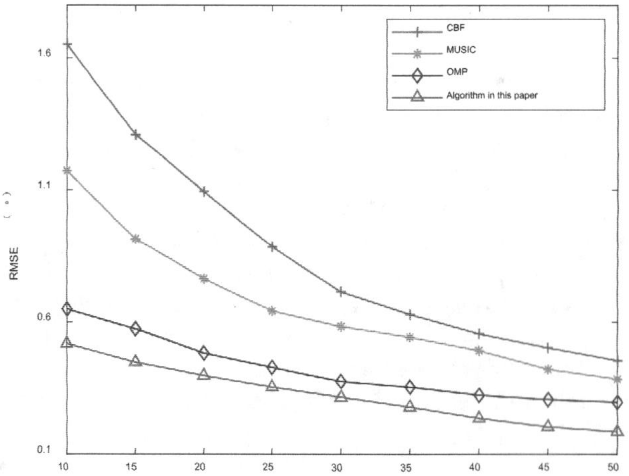

Figure 8. Root mean square error under different snapshot numbers

- 3) DOA estimation performance analysis under different number of array elements

The signal-to-noise ratio is set to 0dB, the number of snapshots is set to 50, and the other simulation conditions remain unchanged. The number of array elements is changed from 6 to 20 in a step of 2. The four algorithms are used for DOA estimation respectively, and the Monte Carlo simulation is passed 100 times The experiment calculates the root mean square error of each algorithm for DOA estimation. The simulation result is shown in Figure 9. It can be seen from Figure 9 that with the increase of the number of array elements, the DOA estimation accuracy of the four algorithms will always improve, and the accuracy of the DOA estimation of the four algorithms will always increase. In terms of yuan, the estimation accuracy of the four algorithms is not much different. But under the low number of elements, the root mean square error of the method in this paper is smaller, which is obviously better than the other three algorithms, and the estimation accuracy is higher.

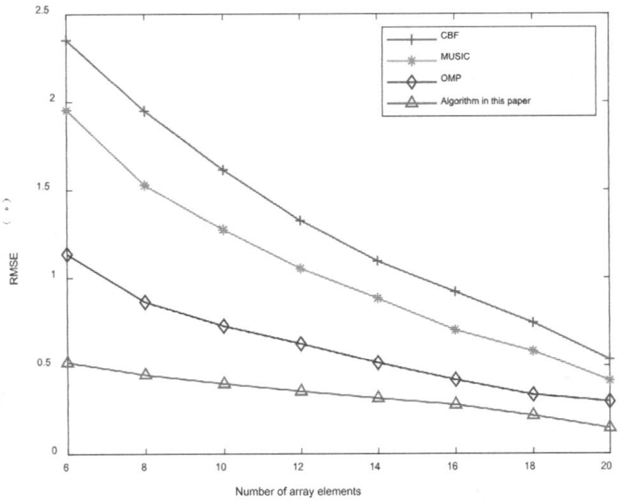

Figure 9. Root mean square errors under different array elements

Through the above simulation experiments, it can be seen that when the incident angles of multiple target signals do not fall on the divided grids and cause grid mismatch problems, the CBF, MUSIC, and OMP algorithms cannot be accurately estimated, and can only be estimated the incident angle which falls on the approximate angle of the adjacent grid, and the method proposed in this paper uses the posterior probability distribution of the signal to be estimated, and the azimuth angle can be accurately estimated without mesh refinement, and the estimation performance is better. When the number of array elements and the number of snapshots decrease sharply, the CBF algorithm basically fails, and the estimation performance of the MUSIC algorithm becomes poor. The reduction of the number of snapshots has a particularly obvious impact on the algorithm performance. This is because these two estimation algorithms are implemented with a large amount of snapshot sampling data and a large amount of calculations. At the same time, when the signal-to-noise ratio is relatively low, the estimation accuracy of the other three algorithms is significantly reduced. In summary, compared with the other three estimation algorithms, the method in this paper still has relatively good estimation performance under the conditions of fewer array elements, lower snapshots, and smaller receiving array size, and it also has relatively good estimation performance at low signal-to-noise ratios. In the case of, the estimation accuracy is high, and the estimation performance is remarkable.

5. Conclusion

On the basis of compressed sensing theory, combined with sparse Bayesian theory and the characteristics of narrow-band underwater acoustic target DOA estimation, a DOA estimation method of underwater acoustic target is proposed. Through simulation comparison experiments with commonly used CBF algorithm, MUSIC algorithm and OMP algorithm, it is confirmed that the method proposed in this paper can still achieve high-resolution DOA estimation even when the signal-to-noise ratio and the target signal source position are relatively close. In addition, this method has better estimation performance in the case of fewer array elements and fewer snapshots, and has low requirements on the array size. It is suitable for the estimation of the azimuth of underwater targets on small platforms such as buoys and AUVs. The above alleviates the problems of low accuracy or failure that may occur in the actual application of traditional DOA estimation methods. In the next step, the measured data can be used in the method proposed in this article to further verify the effectiveness of the proposed method.

References

[1] EJ Candès. Compressive sampling[J]. Marta Sanz Solé, 2006, 17(2):págs. 1433-1452.
[2] Bilik I . Spatial Compressive Sensing for Direction-of-Arrival Estimation of Multiple Sources using Dynamic Sensor Arrays[J]. IEEE Transactions on Aerospace & Electronic Systems, 2011, 47(3):1754-1769.
[3] Ying W , Leus G , Pandharipande A . Direction estimation using compressive sampling array processing[C]// Statistical Signal Processing, 2009. SSP '09. IEEE/SP 15th Workshop on. IEEE, 2009.
[4] Gretsistas A , MD Plumbley. A Multichannel Spatial Compressed Sensing Approach for Direction of Arrival Estimation[C]// International Conference on Latent Variable Analysis & Signal Separation. Springer Berlin Heidelberg, 2010.
[5] Lu Yi, Wentao W U , Ping L I . The application of compressed sensing in synthetic transmit aperture medical ultrasound imaging[J]. Shengxue Xuebao/Acta Acustica, 2013(03):426-432.
[6] Jia Yunfei. Research and Application of Matching Pursuit Algorithm in Compressed Sensing [D]. Xidian University,2020.
[7] Hao Dongbin. Research and Application of Convex Optimization Algorithm Based on Compressed Sensing [D]. Harbin Engineering University,2019.
[8] Zhang Z , Rao B D . Sparse Signal Recovery With Temporally Correlated Source Vectors Using Sparse Bayesian Learning[J]. IEEE Journal of Selected Topics in Signal Processing, 2011, 5(5):912-926.
[9] Wang L , Zhao L , Bi G , et al. Novel Wideband DOA Estimation Based on Sparse Bayesian Learning With Dirichlet Process Priors[J]. IEEE Transactions on Signal Processing, 2015, 64(2):275-289.
[10]Yang Z , Xie L , Zhang C . Off-Grid Direction of Arrival Estimation Using Sparse Bayesian Inference[J]. IEEE Transactions on Signal Processing, 2013, 61(1):38-43.

Strong $\gamma_c - \gamma_{cl}$ H_∞ Stabilization for Networked Control System Based on Lyapunov Function and DoS Attacks

Ying Zheng[1,*] and Ying Zhou[1]

[1]College of Automation & College of Artificial Intelligence, Nanjing University of Posts and Telecommunications, Nanjing, Jiangsu, China

[*]corresponding author's e-mail: 237717077@qq.com

Abstract. This paper is concerned with the strong $\gamma_c - \gamma_{cl}$ H_∞ stabilization for networked control system under DoS attacks with limited communication bandwidths. The quantizer is employed to the network transmission channel from sensor-to-controller in order to reduce the unnecessary data transmissions. In the network environment, it is assumed that there exist DoS attacks in the transmission channel from controller-to-actuator. By describing DoS attacks as the probability models subject to the Bernoulli distribution, the network control system is modeled as the system with random parameters. With this model, a dynamic output feedback controller is designed, which satisfies the given $H\infty$ bound norm γ_c. Under the action of the controller, the closed-loop system is exponential stable in the mean-square sense and satisfies $H\infty$ bound norm γ_{cl}. Especially, a Lyapunov function based on quantization error and attack probability is proposed in the stability analysis. Finally, the effectiveness of the proposed method is verified by a simulation example.

1. Introduction

With rapid developments and widespread applications of computer science and communication network, networked control system (NCS) is playing an increasing role in the fields of smart grid, gas distribution, and modern public transportation [1]. However, symbiotic with the advantages, at the same time, the insertion of communication networks also results in some challenging problems such as limited network bandwidth, communication delay, packet loss and attacks.

Aiming to transmit less data and save energy efficiently, the signal can be quantized before it be transmitted. However, the introduction of the quantizer will also produce the corresponding quantization error, so that the performance of the control system decreases. For the problem of limited bandwidth resources, reference [2] proposed a new event-triggered quantizer for NCS with transmission delay, which avoided the transmission of repetitive data. In addition to the problem of limited communication resources, because the signal is transmitted through the network, it may be subjected to malicious attacks, resulting in data loss, which makes the NCS unable to work properly. Therefore, it is necessary to carry on research about the attacked NCS. Attacks are divided into two categories, deception attacks [3] and DoS (Denial-of-Service) attacks [4,5,6]. Reference [3] studied the security problem of the NCS under deception attacks, a new estimation method was proposed and the sufficient conditions for the stability of the system were obtained. In order to avoid the Zeno phenomenon, an event-triggered scheme based on fixed lower bound is proposed in reference [4], the $L\infty$ control performance of Markov closed-loop jump system under DoS attacks is studied.

On the other hand, when there exist attacks in the network system, the unstable controller will also cause the system performance to degrade, which influence the stable operation of the system. Strong robust stability, that is, adding a stable controller with $H\infty$ performance bound to make the control system operate stably under attacks. Reference [7] studied the strong stabilization problem for continuous-time linear systems with an unknown input delay via dynamic output feedback controllers and a new criterion for dynamic output feedback stability is proposed in terms of matrix equation with the separation of the controller gain with the Lyapunov matrix and the system matrices.

Motivated by the above discussions, this paper focuses on the strong $\gamma_c - \gamma_{cl}$ $H\infty$ stabilization for NCS under DoS attacks. For each sensor, the quantizer is employed to quantify signals to save transmission resources. By describing DoS attacks as the probability models subject to Bernoulli distribution, the original NCS model is modeled as a system with random parameters. A dynamic output feedback controller is designed and the Lyapunov function which depends on quantization error and attack probability is selected to analyze the stability of the system. Finally, the effectiveness of the proposed method is verified by simulation example.

2. Problem description

The structure diagram of NCS under DoS attacks and quantization is given in figure 1, where the sensor-to-controller transmission signal is quantized, assuming that DoS attacks in controller-to-actuator transmission channel.

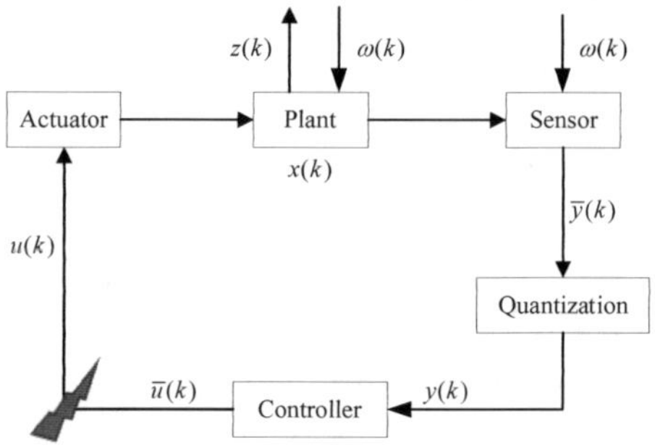

Figure 1. Structure diagram of NCS under DoS attacks and quantization.

Considering the system in the figure 1 is described as the following discrete equations:

$$\begin{cases} x(k+1) = Ax(k) + Bu(k) + G\omega(k) \\ \bar{y}(k) = Cx(k) + F\omega(k) \\ z(k) = Dx(k) \end{cases} \tag{1}$$

where $x(k) \in R^n$, $\bar{y}(k) \in R^p$, $z(k) \in R^n$ are system state, sensor output and control output, respectively. $u(k) \in R^p$ and $\omega(k) \in R^n$ are system input and external interference, respectively. A, B, C, D, F, G are known constant matrices with appropriate dimensions.

In this paper, the logarithmic quantizer is used to quantify the signal. Its definition and characteristics are as follows:

Definition 1: A quantizer is called a logarithmic quantizer if it has the following quantization level set [8]:

$$\lambda = \{\pm\lambda_i, \lambda_i = \rho^i \lambda_0, i = 0, \pm1, \pm2 \ldots\} \cup \{\pm\lambda_0\} \cup \{0\}$$

and has the following mapping functions:

$$f(\xi) = \begin{cases} \lambda_i, & \frac{1}{1+\delta}\lambda_i \le \xi \le \frac{1}{1-\delta}\lambda_i \\ 0, & \xi = 0 \\ -f(-\xi), & \xi < 0 \end{cases}$$

where ξ is the input of the quantizer, $f(\xi)$ is the output of the quantizer, $\delta = \frac{1-\rho}{1+\rho}$, ρ is the quantizer density, and $0 < \rho < 1$, $\lambda_0 > 0$. The quantization strategy of logarithmic quantizer is shown in figure 2:

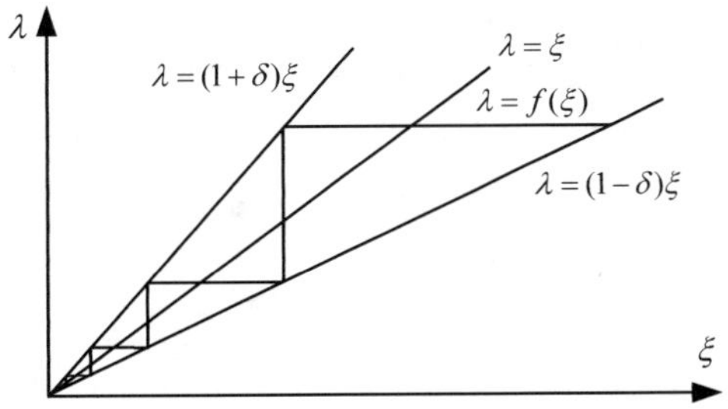

Figure 2. The quantization strategy for the logarithmic quantizer.

In this paper, the sensor output $\bar{y}(k)$ is quantified:

$$y(k) = f\left(\bar{y}(k)\right) \tag{2}$$

where $y(k)$ is the quantized signal transmitted to the controller through the network.

According to the sector bounded method [9]:

$f(\bar{y}(k)) = (I + \Delta(k))\bar{y}(k)$, $\Delta(k) = diag\left\{\Delta_1(k), \Delta_2(k), ..., \Delta_p(k)\right\}$, $\Delta_i(k) \in [-\delta, \delta]$, $i = 1,2,...,p$.

Now assume that the communication network is vulnerable to an attacker who can prevent the transmission of signals. Suppose DoS attacks occur in the transmission channel from controller to actuator. The output signal of the controller is $\bar{u}(k)$, and the input signal of the actuator is $u(k)$. DoS attacks are molded as random variables obeying Bernoulli distribution has following form:

$$u(k) = \alpha(k)\bar{u}(k) + \left(1 - \alpha(k)\right)u(k-1) \tag{3}$$

where $\alpha(k) \in \{0,1\}$, obeys the Bernoulli distribution, and satisfies the following probability:

$$prob\left\{\alpha(k) = 1\right\} = E\left\{\alpha(k)\right\} = \alpha, \; prob\left\{\alpha(k) = 0\right\} = 1 - E\left\{\alpha(k)\right\} = 1 - \alpha$$

where α is the probability value, satisfies $0 < \alpha < 1$. When $\alpha(k) = 1$, it means that the communication network is smooth, and the data are successfully transmitted; when $\alpha(k) = 0$, it indicates that the network is attacked during the transmission, the current data is lost, and the controller still uses the data received at the previous moment.

When DoS attacks exist, the controller should be designed to make the system stable. Therefore, a dynamic output feedback controller is designed as follow:

$$\begin{cases} \varepsilon(k+1) = K_a\varepsilon(k) + K_b y(k) \\ \bar{u}(k) = K_c\varepsilon(k) \end{cases} \tag{4}$$

where $\varepsilon(k)$ is the state of the dynamic output feedback controller, K_a, K_b, K_c are controller parameters to be solved.

Combining equations (1) and (4), the following closed-loop stochastic system is obtained:

$$\begin{cases} \eta(k+1) = A_1\eta(k) + \Delta(k)C_1\eta(k) + [\alpha(k) - \alpha]B_1K\eta(k) + A_2\omega(k) + \Delta(k)A_3\omega(k) \\ z(k) = \bar{D}\eta(k) \end{cases} \tag{5}$$

where $\eta(k) = \left[x^T(k) \quad \varepsilon^T(k) \quad u^T(k-1)\right]^T$, $A_1 = \begin{bmatrix} A & \alpha BK_c & B(1-\alpha) \\ K_bC & K_a & 0 \\ 0 & \alpha K_c & 1-\alpha \end{bmatrix}$, $C_1 = \begin{bmatrix} 0 & 0 & 0 \\ K_bC & 0 & 0 \\ 0 & 0 & 0 \end{bmatrix}$,

$$B_1 = \begin{bmatrix} 0 & B & -B \\ 0 & 0 & 0 \\ 0 & I & -I \end{bmatrix},\ K = \begin{bmatrix} 0 & K_a & K_b \\ 0 & K_c & 0 \\ 0 & 0 & I \end{bmatrix},\ A_2 = \begin{bmatrix} G \\ K_b F \\ 0 \end{bmatrix},\ A_3 = \begin{bmatrix} 0 \\ K_b F \\ 0 \end{bmatrix},\ \bar{D} = \begin{bmatrix} D & 0 & 0 \end{bmatrix}.$$

The purpose of this paper is to study the robust control problem of NCS under DoS attacks and design a dynamic output feedback controller so that the system satisfies the following conditions:

(1)When $\omega(k) = 0$, the closed-loop stochastic system (5) is exponential mean square stability.

When $\omega(k) \neq 0$, for $\gamma_{cl} > 0$, under the zero initial condition, the control output $z(k)$ satisfies:

$$\sum_{k=0}^{\infty} E\left\{\|z(k)\|^2\right\} < \gamma_{cl}^2 \sum_{k=0}^{\infty} E\left\{\|\omega(k)\|^2\right\} \tag{6}$$

(2)When $y(k) = 0$, the dynamic controller satisfies asymptotic stability.

When $y(k) \neq 0$, that is, when $\bar{y}(k) \neq 0$, for $\gamma_c > 0$, under the zero initial condition, the controller output $\bar{u}(k)$ satisfies:

$$\sum_{k=0}^{\infty} E\left\{\|\bar{u}(k)\|^2\right\} < \gamma_c^2 \sum_{k=0}^{\infty} E\left\{\|\bar{y}(k)\|^2\right\} \tag{7}$$

3. Main results

Based on the contents in last section, in this section, the sufficient conditions for the closed-loop system (5) to satisfy strong $\gamma_c - \gamma_{cl}\ H_\infty$ stabilization under the control of dynamic controller (4) is obtained.

Lemma 1[10] For a given symmetric matrix $S = \begin{bmatrix} S_{11} & S_{12} \\ S_{21} & S_{22} \end{bmatrix}$, where S_{11} is a square matrix, the following conditions are equivalent:

(1) $S < 0$; (2) $S_{11} < 0, S_{22} - S_{12}^T S_{11}^{-1} S_{12} < 0$; (3) $S_{22} < 0, S_{11} - S_{12} S_{22}^{-1} S_{12}^T < 0$.

Lemma 2[10] For matrix Σ_1, Σ_2, Σ_3, where $\Sigma_1 = \Sigma_1^T$, if there exists a symmetric positive definite matrix Γ, satisfies: $\Sigma_1 + \Sigma_3 \Gamma^{-1} \Sigma_3^T + \Sigma_2^T \Gamma \Sigma_2 < 0$. Then, $\Sigma_1 + \Sigma_3 \Sigma_2 + \Sigma_2^T \Sigma_3^T < 0$.

Lemma 3[10] For $\kappa > 0$, $0 < \omega < 1$, the following equation holds: $E\left\{\|x(k)^2\|\right\} < \kappa\omega E\left\{\|x(0)^2\|\right\}$, where $\forall x(k) \neq 0$, $x(k) = \left[x_1^T(k), x_1^T(k), \ldots, x_n^T(k)\right]^T$, the closed-loop system is exponential stability in the mean-square sense.

Lemma 4[11] $V(x(k))$ is a Lyapunov function, if there are positive real numbers $\lambda \geq 0$, $\mu > 0$, $\nu > 0$, $0 < \psi < 1$, satisfies:

$$\mu\|x(k)\|^2 \leq V(x(k)) \leq \nu\|x(k)\|^2,\ E\{V(x(k+1))|x(k)\} - V(x(k)) \leq \lambda - \psi V(x(k))$$

then $E\left\{\|x(k)^2\|\right\} \leq \dfrac{\nu}{\mu}\|x(0)\|^2 (1-\psi)^k + \dfrac{\lambda}{\mu\psi}$.

Theorem 1: For given scalars $\gamma_c > 0$, $\gamma_{cl} > 0$, the closed-loop stochastic system (5) and the dynamic controller (4) satisfy strong $\gamma_c - \gamma_{cl}\ H_\infty$ stabilization conditions if there exist symmetric positive definite matrices P_1, P_2, P_3 and control gain K such that the following conditions hold:

$$T_1(k) = \begin{bmatrix} T_{11} & T_{12} \\ T_{12}^T & T_{13} \end{bmatrix} < 0 \tag{8}$$

$$\Lambda = \begin{bmatrix} \Lambda_{11} & \Lambda_{12} \\ \Lambda_{12}^T & \Lambda_{13} \end{bmatrix} < 0 \tag{9}$$

where $T_{11} = (A_1^T + \delta C_1^T)P(k+1)(A_1 + \delta C_1) + \beta K^T B_1^T P(k+1)\beta BK - P(k) + \bar{D}^T \bar{D}$,
$T_{12} = (A_1^T + \delta C_1^T)P(k+1)(A_2 + \delta A_3)$, $T_{13} = (A_2^T + \delta A_3^T)P(k+1)(A_2 + \delta A_3) - (\gamma_{cl})^2 I$,
$\Lambda_{11} = K_a^T P_3 K_a - P_3 + K_c^T K_c$, $\Lambda_{12} = K_a^T P_3 K_b[I + \Delta(k)]$, $\Lambda_{13} = [I + \Delta(k)]^T K_b^T P_3 K_b [I + \Delta(k)] - (\gamma_c)^2 I$.

Proof: For the system (5) and dynamic controller (4), the following Lyapunov functions are selected:

$$V_1[\eta(k)] = \eta^T(k)P(k)\eta(k),\quad V_2[\varepsilon(k)] = \varepsilon^T(k)P_3\varepsilon(k)$$

where $P(k) = [\delta I + \Delta(k) + \alpha I]P_1 + [\delta I - \Delta(k) + \alpha I]P_2$, P_1, P_2, P_3 are symmetric positive definite matrices.

When $\omega(k) = 0$, due to $E\{\alpha(k) - \alpha\} = 0$, $E\{(\alpha(k) - \alpha)^2\} = \alpha(1-\alpha) = \beta^2$, differentiating function $V_1(\eta(k))$ along the system (5) yields:

$$E\{V_1(\eta(k+1))|V_1(\eta(k))\} - E\{V_1(\eta(k))\} \le \eta^T(k)\mathrm{T}(k)\eta(k) \tag{10}$$

where $T(k) = (A_1^T + \delta C_1^T)P(k+1)(A_1 + \delta C_1) + \beta K^T B_1^T P(k+1)\beta BK - P(k)$. It can be seen from equation (8) that $T(k)$ is its principal sub-matrix. When equation (8) holds, it implies $T(k) < 0$, therefore:

$$E\{V_1(\eta(k+1))|V_1(\eta(k))\} - E\{V_1(\eta(k))\} \le \eta^T(k)T(k)\eta(k) \le -\lambda_{\min}(-T(k))\eta^T(k)\eta(k) < -a\eta^T(k)\eta(k)$$

where $0 < a < \lambda_{\min}(-\mathrm{T}(k))$, by defining $\bar{\lambda} = \lambda_{\max}(P(k))$, $\underline{\lambda} = \lambda_{\min}(P(k))$, we can get:

$$\underline{\lambda}\|\eta(k)\|^2 \le V_1(\eta(k)) \le \bar{\lambda}\|\eta(k)\|^2$$

Therefore, $E\{V_1(\eta(k+1))|V_1(\eta(k))\} - E\{V_1(\eta(k))\} < -a\eta^T(k)\eta(k) < -\frac{a}{\bar{\lambda}}V_1(\eta(k)) = -\psi V_1(\eta(k))$

From **Lemma 3**, **Lemma 4**, the system (5) satisfies the exponential stability in the sense of mean-square.

When $\bar{y}(k) = 0$, for dynamic controller (4), differentiating function along the system (4) yields:

$$V_2[\varepsilon(k+1)] - V_2[\varepsilon(k)] = \varepsilon^T(k+1)P_3\varepsilon(k+1) - \varepsilon^T(k)P_3\varepsilon(k) = \varepsilon^T(k)[K_a^T P_3 K_a - P_3]\varepsilon(k) < 0$$

If (9) holds, then $K_a^T P_3 K_a - P_3 < 0$ holds, and the dynamic controller is asymptotically stable.

Next, when $\omega(k) \ne 0$, $\bar{y}(k) \ne 0$, it is proved that the system (5) and dynamic controller (4) satisfy $H\infty$ performance bounds. For the system (5):

$$E\{V_1(\eta(k+1))|V_1(\eta(k))\} - E\{V_1(\eta(k))\} + E\{z^T(k)z(k)\} - (\gamma_{cl})^2 E\{\omega^T(k)\omega(k)\} \le \chi^T(k)T_1(k)\chi(k)$$

where $\chi(k) = [\eta^T(k) \quad \omega^T(k)]^T$. When $\mathrm{T}_1(k) < 0$, the system satisfies:

$$\sum_{k=0}^{\infty} E\{\|z(k)\|^2\} < (\gamma_{cl})^2 \sum_{k=0}^{\infty} E\{\|\omega(k)\|^2\}$$

Therefore, the closed-loop system (5) satisfies $H\infty$ performance index γ_{cl}.

For controller (4):

$$E\{V_2(\varepsilon(k+1))|V_2(\varepsilon(k))\} - E\{V_2(\varepsilon(k))\} + E\{\bar{u}^T(k)\bar{u}(k)\} - (\gamma_c)^2 E\{\bar{y}^T(k)\bar{y}(k)\} = \begin{bmatrix} \varepsilon(k) \\ \bar{y}(k) \end{bmatrix}^T \Lambda \begin{bmatrix} \varepsilon(k) \\ \bar{y}(k) \end{bmatrix}$$

When $\Lambda < 0$, $\sum_{k=0}^{\infty} E\{\|\bar{u}(k)\|^2\} < (\gamma_c)^2 \sum_{k=0}^{\infty} E\{\|\bar{y}(k)\|^2\}$ holds, therefore, system (4) satisfies $H\infty$ performance γ_c and it completes the proof.

Next, equation (8) and equation (9) are transformed to provide an effective method for solving the controller gains and Lyapunov matrices.

Theorem 2: For given scalars $\gamma_c > 0, \gamma_{cl} > 0$, the closed-loop stochastic system (5) and the dynamic controller (4) satisfy strong $\gamma_c - \gamma_{cl}$ $H\infty$ stabilization conditions (6), (7) if there are symmetric positive definite matrixes $P_1, P_2, P_4, W_1, W_2, W_3, Y_1, Y_2$ and control gain K such that the following conditions hold:

$$T_2 = \begin{bmatrix} T_{21} & T_{22} \\ T_{22}^T & T_{23} \end{bmatrix} < 0 \tag{11}$$

$$\begin{bmatrix} -P_4 & 0 & 0 & K_c^T & K_a^T & 0 & K_a^T P_4 & 0 & 0 \\ * & -\gamma_c & K_b^T & 0 & 0 & K_b^T & 0 & 0 & I \\ * & * & -P_4^{-1} & 0 & 0 & 0 & 0 & \delta K_b & 0 \\ * & * & * & -\gamma_c & 0 & 0 & 0 & 0 & 0 \\ * & * & * & * & -P_4^{-1} & 0 & 0 & 0 & 0 \\ * & * & * & * & * & -Y_1 & 0 & \delta K_b & 0 \\ * & * & * & * & * & * & -Y_1^{-1} & 0 & 0 \\ * & * & * & * & * & * & * & -Y_2 & 0 \\ * & * & * & * & * & * & * & * & -Y_2^{-1} \end{bmatrix} < 0 \tag{12}$$

where $T_{21} = \begin{bmatrix} T_{31} & T_{32} \\ T_{32}^T & T_{33} \end{bmatrix}$, $T_{22} = \begin{bmatrix} \frac{P_2 - P_1}{2}\delta & I & 0 & A_1^T + \delta C_1^T & 0 & \beta K^T B_1^T \\ 0 & 0 & 0 & A_2^T + \delta A_3^T & 0 & 0 \\ 0 & 0 & (P_1 - P_2)\delta & -\frac{I}{2} & 0 & 0 \\ 0 & 0 & 0 & 0 & (P_1 - P_2)\delta & -\frac{I}{2} \end{bmatrix}$,

$T_{23} = diag\{-W_1, -W_1^{-1}, -W_2, -W_2^{-1}, -W_3, -W_3^{-1}\}$, $T_{31} = diag\{-\delta(P_1 + P_2) - \alpha(P_1 + P_2) + \bar{D}^T\bar{D}, -(\gamma_{cl})^2 I\}$,

$$T_{32} = \begin{bmatrix} (A_1^T + \delta C_1^T)[\delta(P_1 + P_2) + \alpha(P_1 + P_2)] & \beta K^T B_1^T[\delta(P_1 + P_2) + \alpha(P_1 + P_2)] \\ (A_2^T + \delta A_3^T)[\delta(P_1 + P_2) + \alpha(P_1 + P_2)] & 0 \end{bmatrix},$$

$T_{33} = diag\{-\delta(P_1 + P_2) - \alpha(P_1 + P_2), -\delta(P_1 + P_2) - \alpha(P_1 + P_2)\}$.

Proof: For equation (8), by using **Lemma 1**, is equivalent to:

$$\begin{bmatrix} -P(k) + \bar{D}^T\bar{D} & 0 & A_1^T + \delta C_1^T & \beta K^T B_1^T \\ 0 & -(\gamma_{cl})^2 I & A_2^T + \delta A_3^T & 0 \\ A_1 + \delta C_1 & A_2 + \delta A_3 & -P^{-1}(k+1) & 0 \\ \beta B_1 K & 0 & 0 & -P^{-1}(k+1) \end{bmatrix} < 0$$

Multiplying the above formula to the left and right by $diag\{I, I, P(k+1), P(k+1)\}$ respectively, splitting, scaling and merging the formula, it can be obtained:

$$\begin{bmatrix} -\delta(P_1+P_2)-\alpha(P_1+P_2)+\bar{D}^T\bar{D} & 0 & (A_1^T+\delta C_1^T)P(k+1) & \beta K^T B_1^T P(k+1) \\ 0 & -(\gamma_{cl})^2 I & (A_2^T+\delta A_3^T)P(k+1) & 0 \\ P(k+1)(A_1+\delta C_1) & P(k+1)(A_2+\delta A_3) & -P(k+1) & 0 \\ \beta P(k+1)B_1K & 0 & 0 & -P(k+1) \end{bmatrix}+$$

$$\begin{bmatrix} \frac{P_2-P_1}{2} \\ 0 \\ 0 \\ 0 \end{bmatrix}\delta\begin{bmatrix} I & 0 & 0 & 0 \end{bmatrix}+\begin{bmatrix} I \\ 0 \\ 0 \\ 0 \end{bmatrix}\delta\begin{bmatrix} \frac{P_2-P_1}{2} & 0 & 0 & 0 \end{bmatrix}<0$$

According to **Lemma 2** and **Lemma 1**, the above equation can be transformed into the following form:

$$\begin{bmatrix} -\delta(P_1+P_2)-\alpha(P_1+P_2)+\bar{D}^T\bar{D} & 0 & (A_1^T+\delta C_1^T)P(k+1) & \beta K^T B_1^T P(k+1) & \frac{P_2-P_1}{2}\delta & I \\ 0 & -(\gamma_{cl})^2 I & (A_2^T+\delta A_3^T)P(k+1) & 0 & 0 & 0 \\ P(k+1)(A_1+\delta C_1) & P(k+1)(A_2+\delta A_3) & -P(k+1) & 0 & 0 & 0 \\ \beta P(k+1)B_1K & 0 & 0 & -P(k+1) & 0 & 0 \\ \frac{P_2-P_1}{2}\delta & 0 & 0 & 0 & -W_1 & 0 \\ I & 0 & 0 & 0 & 0 & -W_1^{-1} \end{bmatrix}<0$$

The same method is used to deal with $P(k+1)$ in the above equation, and finally it is obtained that equation (11) holds. Similar to the above proof, let $P_4=\frac{P_3}{\gamma_c}$, from to **Lemma 1**, **Lemma 2**, finally it is clear to prove equation (9) is equivalent to the equation (12).

4. Simulation example

In order to illustrate the effectiveness of the method proposed in this paper, consider the system described by equation (1) with arbitrary parameters as follows:

$A=\begin{bmatrix} -0.5 & -1.1 \\ 0 & 0.16 \end{bmatrix}$, $B=\begin{bmatrix} 0.2 \\ 0.1 \end{bmatrix}$, $C=\begin{bmatrix} 0.5 & 0 \end{bmatrix}$, $D=\begin{bmatrix} -1 & 0.2 \end{bmatrix}$, $G=\begin{bmatrix} -0.2 & -0.1 \\ 0 & 0.01 \end{bmatrix}$, $F=\begin{bmatrix} 0.5 & 1 \end{bmatrix}$,

$x(0)=\begin{bmatrix} 1 & 0.8 \end{bmatrix}^T$, $\varepsilon(0)=\begin{bmatrix} 0 & 0 \end{bmatrix}^T$, quantization density $\rho=0.6$, $\delta=0.25$, probability of attack $\alpha=0.6$, $\lambda_0=0.1$, $\gamma_c=1$, $\gamma_{cl}=2$.

The equations (11) and (12) can be solved by the MATLAB and Yalmip toolboxes. Since equation (11) is a bilinear matrix equation, analysis of the matrix (11) shows that the controller gains need to be separated from the positive definite symmetric matrixes P_1, P_2, W_2, W_3. In the simulation process, use MATLAB software to randomly generate positive definite symmetric matrices with appropriate dimensions. By this method, the bilinear matrix inequation (11) is transformed into linear matrix inequation when solving the controller gains. By solving the linear matrix equation with MATLAB, we can get:

$$K_a = \begin{bmatrix} -0.0953 & 0.0624 \\ 0.0624 & -0.0370 \end{bmatrix},\ K_b = \begin{bmatrix} -0.2405 \\ 0.0963 \end{bmatrix},\ K_c = \begin{bmatrix} 0.8520 & -0.7693 \end{bmatrix}.$$

The simulation results are shown in the following figures.

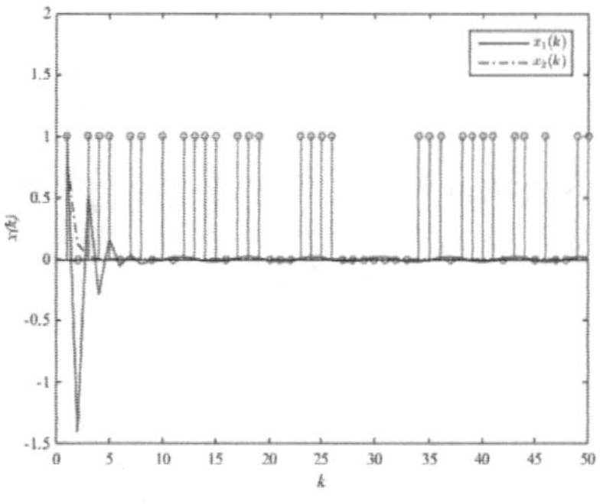

Figure 3. State response curve with $\omega(k) \neq 0$.

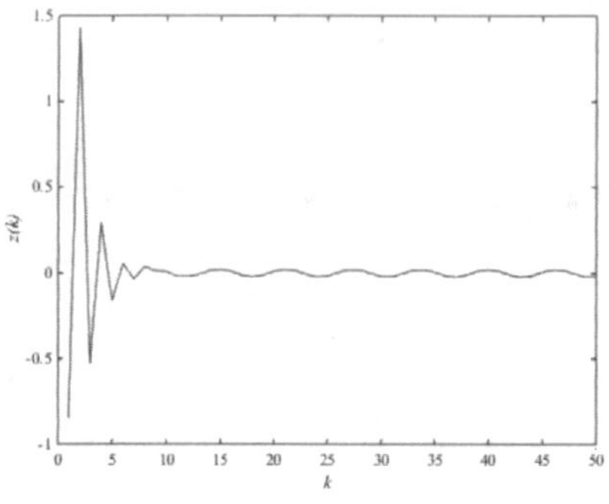

Figure 4. Control output curve with $\omega(k) \neq 0$.

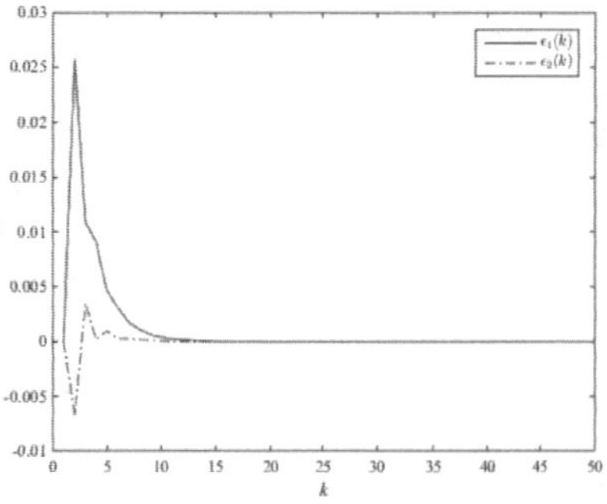

Figure 5. State curve of dynamic controller.

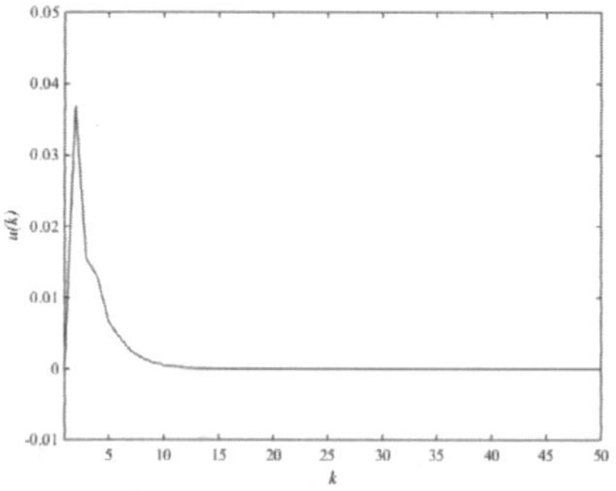

Figure 6. Actuator input signal $u(k)$.

In figure 3, the gray needle-shaped graph represents the specific moment when the DoS attacks occurred. From figure 3 and figure 4 that under the action of the dynamic controller, the system state gradually tends to be stable and satisfies $H\infty$ performance, indicating that the designed stable dynamic controller ensures the stability of the closed-loop system under DoS attacks. It can be seen from the figure 5 and figure 6, the designed dynamic controller is also stable. Simulation results show the effectiveness of the method.

5. Conclusion

This paper studies the problem of strong $\gamma_c - \gamma_{cl}$ $H\infty$ stabilization for networked control system under DoS attacks. In the signal transmission, the quantizer is added to achieve the purpose of saving bandwidth resources. Based on the stability principle, a dynamic output feedback controller is designed, and sufficient conditions for the closed-loop system satisfies strong $\gamma_c - \gamma_{cl}$ $H\infty$ stabilization are given by selecting the Lyapunov function which depends on the quantization error and attack probability. In particular, when solving the bilinear matrix equation, the stochastic matrices satisfying the conditions are generated by MATLAB due to the need to separate the controller gain from matrixes, and the controller gains of the dynamic controller are obtained. Finally, the effectiveness of the control method is verified by simulation example.

Acknowledgments
Authors wishing to acknowledge assistance or encouragement from colleagues. This work was supported by the National Natural Science Foundation of China under Grant 62073172.

6. References

[1] Mahmoud M S and Hamdan M M 2018 Fundamental issues in networked control systems IEEE/CAA Journal of Automatica Sinica 5 902–22

[2] Chen Z R, Zhang B Y, Li Y M, Ma Q and Zhang Y J 2020 Design of event-triggered networked control systems via specific triggering quantizer Int J Robust Nonlinear Control 30 6947–60

[3] Zhang Y L, Zhu Y F and Fan Q Q 2020 A novel set-membership estimation approach for preserving security in networked control systems under deception attacks Neurocomputing 400 440–9

[4] Zeng P Y, Deng F Q, Liu X H and Gao X B 2021 Event-Triggered Resilient $L\infty$ Control for Markov Jump Systems Subject to Denial-of-Service Jamming Attacks IEEE transactions on cybernetics (Preprint DOI: 10.1109/TCYB.2021.3063244)

[5] Cao E Z, Zhang B L, Cai Z H, Han Q L and Wang B R 2021 Observer-based state feedback H∞ control for offshore steel jacket structures under denial-of-service attacks ISA Transactions (Preprint DOI: 10.1016/J.ISATRA.2021.01.006)

[6] Liu F C, Wang C C and Geng Q 2021 Observer-based MPC for NCS with actuator saturation and DoS attacks via interval type-2 T-S fuzzy model IET Control Theory & Applications 14 3537-46

[7] Du B, Lam J and Shu Z 2012 Strong stabilisation by output feedback controller for linear systems with delayed input IET Control Theory & Applications 6 1329–40

[8] Elia N and Mitter S K 2001 Stabilization of linear systems with limited information IEEE Transactions on Automatic Control 46 1384-1400

[9] Fu M Y and Xie L H 2005 The sector bound approach to quantized feedback control IEEE Transactions on Automatic Control 50 1698–1711

[10] Stephen B, Laurent E G, Eric F and Venkataramanan B 1994 Linear Matrix Equation in System and Control Theory (Philadelphia: SIAM)

[11] Zhu Y F, Yang F W, Li C J, Zhang Y L and Han Q L 2019 Strong $\gamma_c - \gamma_{cl}$ $H\infty$ stabilization for networked control systems under denial of service attacks Journal of the Franklin Institute 365 2723–41

Strength Analysis on the Frame of Off-way Wide-body Mining Dump Truck Based on Finite Element Analysis

Weiwei Liu[1], **Jiang Liu**[2,*]

[1]Mechanical Engineering College, Beihua University, Jilin, Jilin Province, China
[2]Transmission and Transformation Technology College, Northeast Electric Power Jilin, Jilin, China

[*]corresponding author's e-mail: 1510462047@qq.com

Abstract. The frame is an important component for the off-way wide-body mining dump truck. The carrying capacity of the truck is greatly affected by the strength of the frame. The finite element analysis was used to simulate the strength of the frame under the work condition of full load and lifting. The maximum stress of the frame is 214MPa under the full load condition, and the maximum stress of the frame is 459MPa under lifting condition. The upper surface of the frame near the rear axle is the more dangerous section for the frame under the two working conditions. The stiffness of the front leaf spring has very impact to the maximum stress of the frame for both full load condition and lifting condition, but has some impact to the maximum stress near the support of the front leaf spring.

1. Introduction

The off-way wide-body mining dump truck is a kind of vehicle appeared in recent years. It is obtained on the basis of combining the high-load capacity of electric mining dump truck and the layout form of heavy truck, which is widely used for the transportation of small and medium-sized coal mines and metal mines. At present, some research on the finite element analysis and fatigue life calculation of the frame to the electric mining dump truck has been implemented. But the analysis of the frame for the off-way wide-body mining dump track is not deep enough because it came later. Yancheng Liu etc. analysed the dynamic stress distribution of the vehicle under full load condition and bump obstacle condition according the established finite element and multi-body dynamic models respectively. Then the fatigue life of the frame was analysed by the obtained dynamic load history, which provided a method of fatigue life prediction for the frame[1]. Chengji Mi etc. provided a means to predicting the fatigue life of the frame for an electric mining dump truck. The dangerous points of the frame were obtained through the finite element analysis, and then the dynamic experiments were conducted to verify the results of finite element model. What's more, the load spectrum of the frame driving on class B random uneven road was obtained by the experiments. Furthermore, the fatigue life of the frame was predicted by fatigue software, which provided a new method for the fatigue life prediction of the frame[2-4]. Weiwei Liu etc. analysed the driving performance of an off-way wide-body mining dump truck, and the experiments were conducted to obtain the dynamic stress of the frame[5-7].

In conclusion, the finite element analysis of the frame for the off-way wide-body mining dump truck is not enough. In this article, the 3D model was imported into Hypermesh to mesh after some simplification of the model, and then the external loads and constraints were applied in Ansys to analyse the strength of the frame under the working conditions of full load and lifting respectively, which provided a method of finite element analysis of the frame for the off-way wide-body mining dump truck.

2. Finite element model of the frame

2.1. Geometric model processing

The 3D model of the off-way wide-body mining dump truck was provided by the manufacturer. However, the model was not suitable for the finite element analysis since the 3D model was established according to the sequence of manufacturing, processing and assembly. The 3D model was simplified on the premise of ensuring the accuracy of calculation results. The off-way wide-body mining dump truck and simplified frame models were shown in Figure 1. The material of the frame Q460CFD, the elastic modulus 200GPa, the poison's ratio 0.27 and the density 7800kg / m^3.

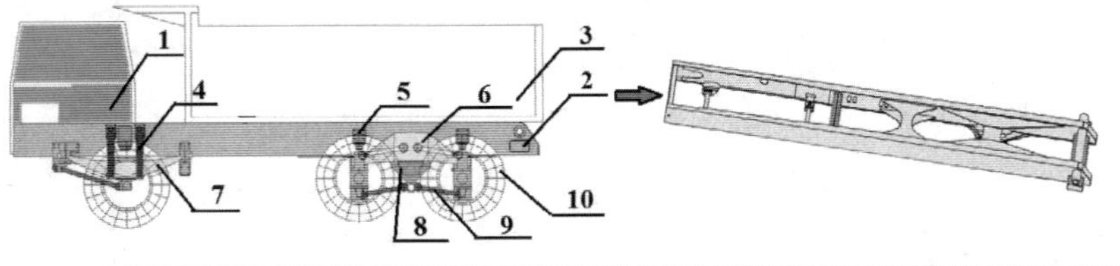

Figure 1. Vehicle and frame 3D model. (1-cab;2-frame;3- carriage;4-shock absorber;5-limit block;6-rear leaf spring;7-front leaf spring;8- balance shaft;9-draw bar;10-tire.

2.2. Meshing of the frame

The frame plate and shell element finite element model (shell181) were established in HyperMesh in order to obtain the accurate stress distribution of the frame under various working conditions, which can more concretely reflect the stress distribution of the frame than the beam element. What's more, the front and rear axles connected to the frame were simulated by the beam element beam188, and the draw bar was simulated by bar element link180. The bar beam element combination and node coupling were adopted at the balance axis. The final finite element model had 226018 nodes and 234097 elements, shown in Figure 2.

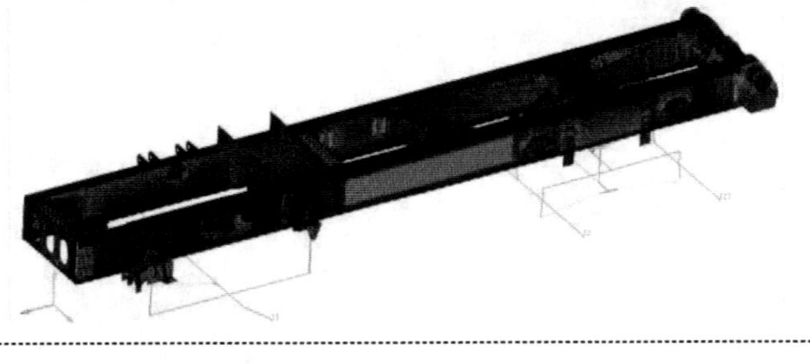

Figure 2. Frame meshing model

2.3. External loads and constraints

The limit blocks were added on the basis of the original model in order to simulate the actual working conditions. What's more, the spring elements were used to simulate the actual loading characteristics of the limit blocks. The external loads included material 64t, carriage weight 8.3t, dead weight of the frame 3.5t and cab 2.9t. The front, middle and roar axles were imposed constraints on the X, Y, Z directions, as shown in Figure 3.

Figure 3. External loads and constraints of full load condition

3. Analysis of typical conditions

The full load and lifting conditions were selected to analyse according to the actual working environment of the off-way wide-body mining dump truck. The strength analysis of the frame was conducted by the finite element model established above.

3.1. Full loading condition

It can be seen from Figure 4 that the maximum stress was 214 MPa, which was located at the upper surface of the frame near the middle and rear axles. The maximum stress at the bottom of the frame near the rear axle was 58MPa, which was located at the connection between the lower surface of the frame and the balance shaft. The maximum stress at the bottom of the frame near the middle axle was 65MPa, which was located at the lower surface. The maximum stress at the rear support of the front leaf spring was 90MPa, which was greater than the front support of the front leaf spring.

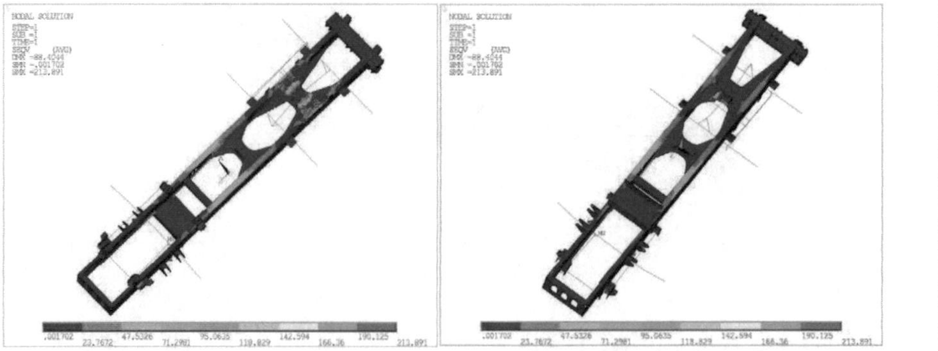

Figure 4. Mises stress nephograms of full load condition

3.2. Lifting condition

In lifting condition, the lifting angle 0 degree was the most dangerous condition, as shown in Figure 5. The forces could be formatted as

$$\begin{aligned} m'g - F_1 - F_2 = 0 \\ F_1 h = m'gh' \end{aligned} \tag{1}$$

Where m' is the total weight of the material and carriage, F_1 is the force of rear hinge point, F_2 is the force of front hinge point, h is the distance from the front hinge to the rear hinge, 5.2m, h' is the distance from the center of gravity to the front hinge point, 2.185m. So F_1 and F_2 can be calculated, which were $5 \times 10^5 N$ and $3.62 \times 10^5 N$ respectively.

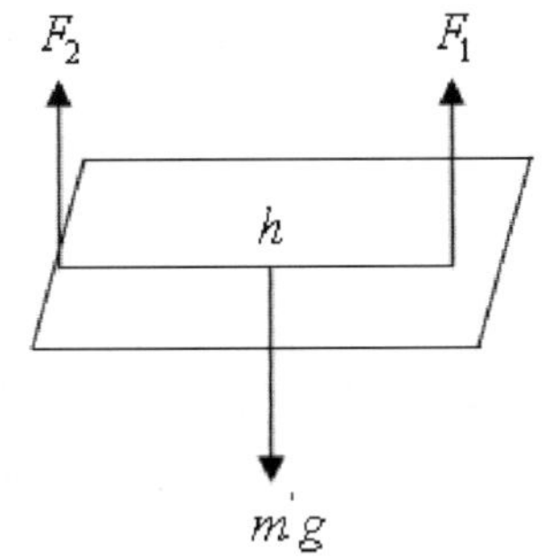

Figure 5. Fore nephograms of lifting condition

The concentrated loads were applied at the front-point of the hydraulic cylinder and the rear hinge point of the frame respectively, and other loads and constraints were same as the full load condition, as shown in Figure 6. It can be seen from the Figure 7 that the maximum stress was 459 MPa, which was located at the upper surface of the frame near the rear axle. The maximum stress at the bottom of the frame near the rear axle was 360MPa, which was located at the connection between the lower surface of the frame and balance shaft. The maximum stress at the rear support of the front leaf spring was 150MPa, and the front support of the front leaf spring was 100MPa. What's more, the maximum stress at the bottom of the frame near middle axle was 280MPa, which was located at the lower surface.

Figure 6. External loads and constraints of lifting condition

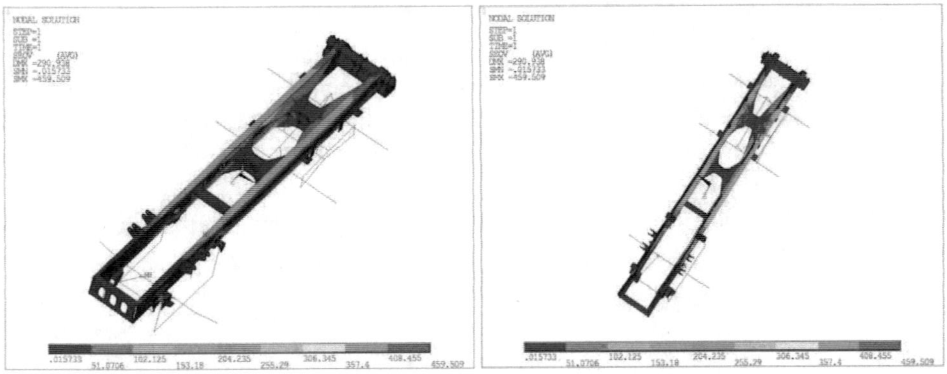

Figure 7. Mises stress nephograms of lifting condition

4. Influence of the stiffness of front leaf spring

The effect of stiffness of front leaf spring to the frame was conducted by comparing the stiffness of 1000MPa and 1500MPa, as shown in Figure 8. It can be seen that the stiffness of the front leaf spring had very little impact to the maximum stress of the frame for both full load condition and lifting condition. However, the stiffness of the front leaf spring had some impact to the maximum stress near the front and rear support of the front leaf spring. The maximum stress near the rear support was 95MPa and 115MPa respectively in the stiffness of 1000MPa and 1500MPa in full load condition, which increased by 21.05%. What's more, the maximum stress near the rear support was 150MPa and 187MPa respectively in the stiffness of 1000MPa and 1500MPa in full load condition, which increased by 24.67%.

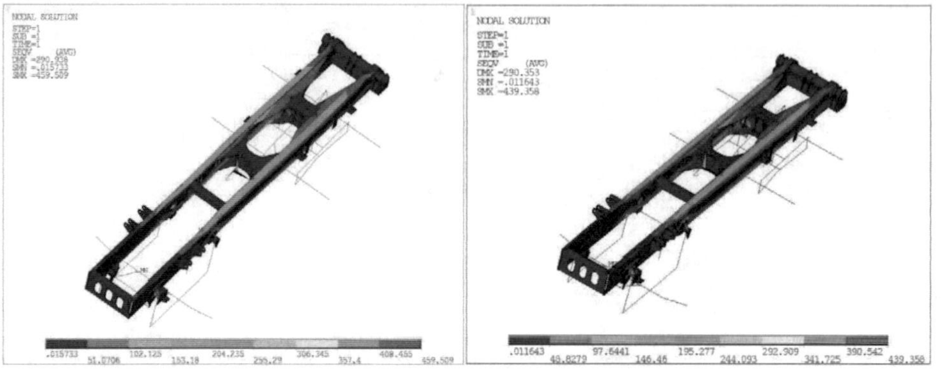

Figure 8. Stess nephograms of full load condition and lifting condition for 1500MPa

5. Conclusion

Based on the finite element model of the frame, the strength of the frame under two working conditions was analysed. The following conclusions were obtained:

- The maximum stress of the frame is 214MPa under the full load condition, which is located at the upper surface between the middle and rear axles.
- The maximum stress of the frame is 459Mpa under lifting condition, which is located at the upper surface near the rear axle.
- The maximum stress of the frame under lifting condition is greater than the full load condition. Despite this, the strength of the frame is still enough since the yield limit of the frame is 890MPa.
- The upper surface of the frame near the rear axle is the dangerous section according to the analysis results of the full loading condition and lifting condition.
- The stiffness of the front leaf spring has very impact to the maximum stress of the frame for both full load condition and lifting condition, but has some impact to the maximum stress near the support of the front leaf spring.

Acknowledgments

The research was financially supported by the Technological Research Projects of Department of Education of Jilin Province (NO. JJKH20210033KJ).

References

[1] LIU Y C.Structural Strength Analysis and Test Study of Dump Truck[D].Chongqing：Chongqing University，2014.

[2] Mi, CJ.Gu, ZQ.Zhang, Y. et al. Frame weight and anti-fatigue co-optimization of a mining dump

truck based on Kriging approximation model [J]. Engineering Failure Analysis, 2016, 66: 99-109.
[3] Chengji Mi, Zhengqi Gu, Qingquan Yang. Frame fatigue life assessment of a mining dump truck based on finite element method and multibody dynamic analysis[J]. Engineering Failure Analysis, 2012, 23: 18-26.
[4] MI C J, LI W T, NI Z S,et al. Static and Dynamic Characteristics Analysis and Optimization of Electric Wheel Dump Track Frame[J].Journal of Mechanical Strength,2020,42(03):639-647.
[5] CHENG K,LIU W W,LIAO Q D,et al. ADAMS-based dy namic stress analysis on mining dumper frames[J]. CHINESE JOURNAL OF CONSTRUCTION MACHINERY ,2014,12(2):158-162.
[6] LIU W W. Simulation Analysis on Driving Performance of Off-way Wide-body Mining Dump Truck[D].Changchun: Jilin University,2015.
[7] LIU W W,CHENG K. Simulation Analysis on Steady Turning Performance of Off-way Wide-body Mining Dump Truck[J]. JOURNAL OF JILIN UNSTITUTE OF CHEMICAL TECHNOLOGY,2017,34(05):42-47.

High-Frequency Vibration Control of Rotating Beam Based on Trailing-edge Flap

Jian Wang, Shanyao Li, Changle Sun and Tingrui Liu*

College of Mechanical & Electronic Engineering, Shandong University of Science & Technology, Qingdao 266590, China

*Corresponding author's e-mail: liutingrui9999@163.com

Abstract. This study is to investigate an active control method, based on trailing-edge flap, in high-frequency vibration control of rotating blade beam. The structure is modeled as composite blade beam with circumferentially asymmetric stiffness configuration, which is based on analysis of the elastic flap-wise/twist displacements and incorporates the angle control of trailing-edge flap driven by a stepping motor. Aerodynamic expressions are based on a novel quasi-steady model suitable for trailing-edge flap. Based on a strip theory, the multi-body dynamic aeroelastic system is analyzed and solved. The active control is realized by H∞ algorithm using linear matrix inequality (LMI) design. This study is to investigate time-domain stability analysis and robust control method, realize displacement response analysis and robust performance analysis, and input signal display of trailing-edge flap angle. The optimization mechanism of LMI is to optimize uncertain robust performance parameters based on the selection of robust control parameters, so that the controlled displacement and control input are kept within reasonable ranges. The reliability and robustness of H∞ control algorithm based on LMI are verified by comparisons of the results of high-frequency vibration control using different robust performance parameters.

1. Introduction

The high frequency vibration of flapping / torsion (bending torsion) coupling displacement based on classical flutter is an important reason for the fracture failure of rotating blade structure [1-2]. In recent years, trailing edge flap structure [3-7] has been widely used in blade aeroelastic flutter control. Haselbach [3] proposed an advanced blade modeling method based on shell element model, in which the bond line in the trailing edge area is discretized by solid brick elements, which improves the reliability of structural prediction of wind turbine blades. Lorente[4] and others focused on the change of aerodynamic performance of wind turbine blades with serrated tail edge, as well as the specific design process and prediction methods. Qin [5] and others designed a trailing edge flow control device and discussed its performance. Micro insert was installed on the wind wheel and micro jet unit was designed on the trailing edge to achieve the purpose of load reduction. Mu Anle et al. [6] simplified the wind turbine blade as a composite cantilever beam laminated plate with spanwise distributed trailing edge flap, realized the structural modeling by Rayleigh Ritz method, established the aeroelastic model combined with Theodorsen aerodynamic force, and finally realized the main dynamic vibration control by using the model predictive control algorithm; The detailed analysis process and flutter suppression conclusion, including vibration frequency analysis, can be referred to reference [7].

Under normal working conditions, a kind of blade under the condition of "low amplitude high frequency" vibration [8-9] for a long time will form a hidden crack defect, which makes the blade more prone to fracture failure under extreme flutter conditions. The high-frequency vibration is easy to

IOP Publishing
2173 (2022) 012050 doi:10.1088/1742-6596/2173/1/012050

generate at the trailing edge of turbulent flow [9], or at the stall state when the aerodynamic torque of flexible blade shows low amplitude and high-frequency oscillation [10]. Based on the finite element analysis, reference [11] also confirmed the low amplitude and high frequency vibration effects at the trailing edge.

Due to the high frequency characteristics of vibration, many intelligent control algorithms fail due to the large number of iterations or the difficulty of optimization. This paper presents a kind of tail flap structure distributed along the blade length, driven by stepping motor, and realizes "low amplitude high frequency" vibration control by H_∞ control method based on LMI design. At the same time, based on the simulation environment of PLC and MATLAB, a "real-time communication technology" is developed. A process control experiment is adopted to verify the feasibility of the control algorithm in engineering application. It provides a feasible scheme for real-time engineering application for intelligent control algorithm which cannot be run in the conventional controller hardware due to the complexity of the algorithm.

2. Aeroelastic System Equation

The blade beam adopts solid composite structure with fixed chord length, and the airfoil adopts NACA64418 airfoil with quadratic fitting [12], as shown in Figure 1.

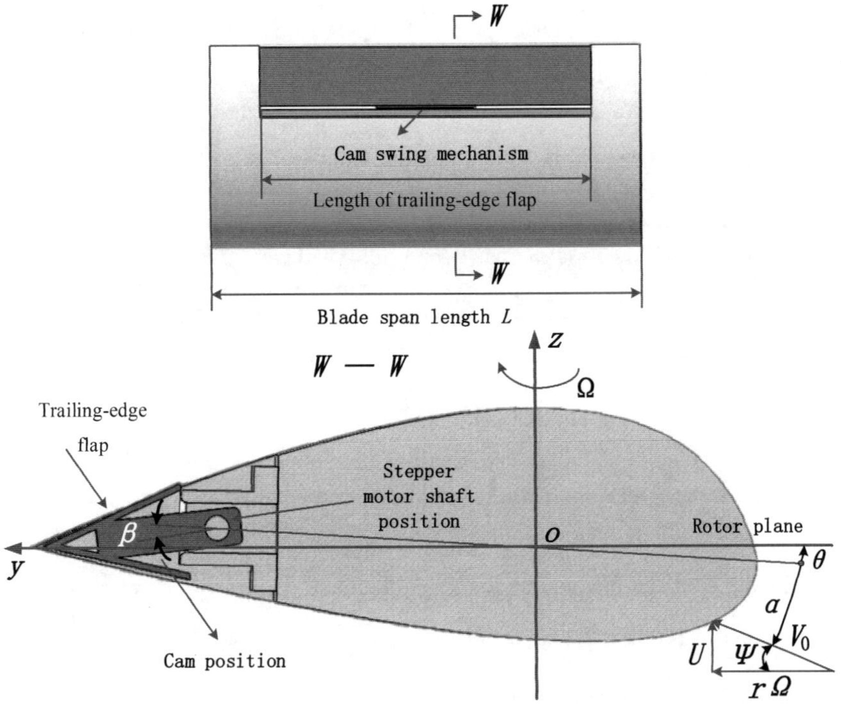

Figure 1. Structure and parameters of blade beam.

L is blade span, r is displacement from airfoil section to rotation center; y、z is the motion (displacement) of swing and swing direction, and x is assumed to be the direction of length; θ is torsional displacement; α is the angle of attack; U is wind speed; c is blade chord length; β is flap swing angle, L_0 is flap 2D section length (along chord direction); V_0 is the inflow wind speed; λ_0 is the speed ratio coefficient; Ω is the constant speed. The wing is made of light space aluminum material, and the two sides of the flap are connected with the blade beam matrix in a loose fit hinged way. The flap is driven by the stepper motor through the flat key, and the middle part of the hollow flap is connected with the blade beam by the light smooth cam swing structure, which plays the role of smooth transmission (the mass of all light structures is not included, and the stepper motor is fixed at the trailing edge of the middle part of the solid blade).θ

Composite laminates are symmetrical in chord direction and designed based on circumferential antisymmetric stiffness (CAS). Under CAS ply, the blade coupling shows a single bending torsion coupling displacement, that is, the flapping/torsion coupling displacement [13]. Then the swing z or twist θ coupling motion of a single section can be described as:

$$\left[GJ+\frac{1}{2}m_c\Omega^2K_A^2\left(L^2-x^2\right)\right]\theta''+K_cz'''-m_c\Omega^2K_A^2x\theta'-m_cK_m^2\ddot{\theta}-m_c\Omega^2\left(K_{m2}^2-K_{m1}^2\right)$$
$$-m_c\Omega^2\left(K_{m2}^2-K_{m1}^2\right)\theta=M \tag{1}$$

$$K_c\theta'''+EIz''''+m_c\ddot{z}-\frac{1}{2}m_c\Omega^2\left[z''\left(L^2-x^2\right)+z'\left(-2x\right)\right]=F \tag{2}$$

where, GJ, EI and K_c are the corresponding torsional stiffness,1 bending stiffness and bending torsion coupling stiffness respectively; m_c is the aerofoil mass per unit length; K_A、K_m is the correlation coefficient of moment of inertia ; M、F corresponding quasi steady aerodynamic torque and aerodynamic lift.

Reference [14] shows the quasi-steady aerodynamic force of a cantilever vibration 2D airfoil based on trailing edge flap. In this paper, the absolute wind speed in the flap excitation aerodynamic equation in reference [14] is changed to the relative wind speed of the wind turbine, the pitch angle is replaced by the angle of attack α of the wind turbine blade, and the flap excitation aerodynamic term caused by relative rotation is added. Based on the specific length of the flap 2D section (along the chord direction) is $L_0=c/6$, the quasi-steady aerodynamic expression of the rotating airfoil suitable for the flap structure is obtained, it can be described as follows:

$$F=\frac{1}{2}\rho_a\left\{\pi b[V_0(\ddot{z}/V_0-\dot{\theta})-b\ddot{\theta}/2]+2\pi\dot{z}+2\pi[V_0^2\frac{C_{1\beta}}{C_{1\alpha}}(\psi+1)\beta-V_0^2\frac{C_{1\beta}}{C_{1\alpha}}\theta-bV_0\dot{\theta}]\right\} \tag{3}$$

$$M=\rho_ab[\frac{\pi bV_0(\dot{\theta}-\ddot{z}/V_0)}{4}+\frac{\pi bV_0\dot{\theta}}{4}+\frac{3\pi b^2\ddot{\theta}}{16}]+\rho_a\pi V_0^2\frac{C_{m\beta}}{C_{m\alpha}}\beta \tag{4}$$

where, ρ_a is the air density; $b=c/2$; $C_{l\alpha}$, $C_{m\alpha}$, $C_{l\beta}$ and $C_{m\beta}$ are the corresponding chord flap matching coefficients, And satisfy the following formula: $C_{l\alpha}$=6. 28, $C_{m\alpha}=\left(0.5+c/6\right)C_{l\alpha}$, $C_{l\beta}$=3. 358, $C_{m\beta}=-0.635$.

3. Solution of Aeroelastic System

3.1. Multibody Dynamics Analysis Based on Strip Theory

In order to avoid the integral problem in the conventional discrete integral algorithm (such as Galerkin method) of equation (1-2) and facilitate the implementation of mathematical algorithm in the hardware of controller, the multi-body dynamic analysis based on strip theory [15] is adopted to divide the blade into $\bar{N}$ segments along the full length of flap (the essence of which is to replace the integral operation with summation), When $\bar{N}$ is large enough (satisfactory results can be obtained when $\bar{N}$=6 in this design), each sub segment can be regarded as a 2D airfoil, on which the aerodynamic forces of the segments (3-4) are distributed. Assuming that the coupling displacement of bending and torsion is described as $z(x,t)$ and $\theta(x,t)$ respectively, the space element $z_j(x)$ and $\theta_j(x)$ of $z(x,t)$ and $\theta(x,t)$ can be set as the mode function satisfying the boundary problem of cantilever beam [16]:

$$z_j(x) = (x/L)^{1+j}\left\{6 + i^2(1-x/L)^2 + j\left[5 - 6x/L + (x/L)^2\right]\right\} / \left[j(1+j)(2+j)(3+j)\right] \quad (5)$$

$$\theta_j(x) = \sin(j\pi x/L) \quad (6)$$

The full length of the flap is $L_f = 0.92L$, and the aerodynamic force of the $0.08L$ part on both sides of the blade beam is ignored by applying the strip processing, then the structure of $2\bar{N}$ sub equations can be obtained:

$$M_0\ddot{X} + C_0\dot{X} + K_0X = Q_0\beta \quad (7)$$

where, M_0, C_0 and K_0 are $2\bar{N} \times 2\bar{N}$ mass, damping and stiffness matrices respectively; Q_0 is a $2\bar{N} \times 1$ matrix. And:

$$M_0 = \begin{bmatrix} \sum_{m=1}^{\bar{N}}[(M_{011}(x)\theta_j)|_{x=m\Delta x}\,\Delta x & \sum_{m=1}^{\bar{N}}[(M_{012}(x)\theta_j)|_{x=m\Delta x}\,\Delta x \\ \sum_{m=1}^{\bar{N}}[(M_{021}(x)z_j)|_{x=m\Delta x}\,\Delta x & \sum_{m=1}^{\bar{N}}[(M_{022}(x)z_j)|_{x=m\Delta x}\,\Delta x \end{bmatrix} \quad (8)$$

$$Q_0 = \begin{bmatrix} \sum_{m=1}^{\bar{N}}[(f_{\theta\bar{N}}(x)\theta_j)|_{x=m\Delta x}\,\Delta x \\ \sum_{m=1}^{\bar{N}}[(f_{z\bar{N}}(x)z_j)|_{x=m\Delta x}\,\Delta x \end{bmatrix}_{\Delta x=\frac{L}{\bar{N}},\, j=1,2,\ldots,\bar{N}} \quad (9)$$

Among them, M_{011}, M_{012}, M_{021} and M_{022} are mass elements similar to those described in [15], $f_{\theta\bar{N}}$ and $f_{z\bar{N}}$ are force elements similar to those described in [15]; Similarly, C_0 and K_0 have similar expression.

By transforming equation (7) into a first-order system, we can obtain:

$$\dot{Y} = A_0Y + B_0\beta \quad (10)$$

where, $Y = \begin{bmatrix} X & \dot{X} \end{bmatrix}^T$ is the state variable, and:

$$A_0 = \begin{bmatrix} 0_{2\bar{N}\times 2\bar{N}} & I_{E(2\bar{N}\times 2\bar{N})} \\ -M_0^{-1}K_0 & -M_0^{-1}C_0 \end{bmatrix}, B_0 = \begin{bmatrix} 0_{2\bar{N}\times 1} \\ M_0^{-1}Q_0 \end{bmatrix} \quad (11)$$

Compared with the discrete integral Galerkin method, the multi-body dynamic analysis based on strip theory may reduce the accuracy of the solution, but it improves the speed of operation and is more suitable for practical engineering application.

3.2. System Improvement Based on Engineering Application

The mass of light structure in system figure 1 can be ignored, but the mass and dynamic driving effect of stepper motor cannot be ignored. Assuming that the mass of the stepping motor is $m_s = 2\text{kg}$, it can be regarded as a kind of elastic rotating body's payload [17], which can suppress the vibration of the coupled motion. However, for the control input, it is also equivalent to a disturbance signal. The disturbance model based on the simulation time can be approximately expressed as

$\omega = 2m_s(L/2)\sin(2\pi f_0 t)/(t+1)$, where t is the simulation time [17]; $f_0 = 5/(2\pi)$ is the pulse frequency of the change of the driving angle β of the stepper motor, which affects the input of the control system with a sinusoidal signal effect [18]. Therefore, considering the actual driving effect of the stepping motor, the system equation (11) can be improved as follows:

$$\begin{cases} \dot{Y} = A_0 Y + B_1\omega + B_0\beta \\ Z = C_1 Y + D_1\beta \end{cases} \tag{12}$$

where, $B_1 = \begin{bmatrix} 0_{1\times 2\bar{N}} & 1_{1\times 2\bar{N}} \end{bmatrix}^{\mathrm{T}}$, The component relative to the displacement part is 0, while the component relative to the velocity part is 1. This is because the measurement of state variables in the full feedback process is obtained by using the acceleration sensor to measure the swing / torsion acceleration [19]; From the measured acceleration signal, after filtering the low-frequency disturbance, the integral operation obtains the velocity signal, and then after the second filtering, the integral operation again obtains the displacement signal, so the disturbance signal from the input first affects the acceleration state variable, and then the velocity state variable. C_1 and D_1 are the coefficient matrix of H_∞ control.

The improved system equation (12) cannot be solved by the conventional intelligent control method. In this design, an H_∞ control theory is adopted to realize the optimal control.

4. H_∞ Control

For the system equation (12), given the robust performance coefficient $\gamma > 0$, there are matrices $P_1 = P_1^{\mathrm{T}} > 0$ and P_2, according to Schur co lemma, if the linear matrix inequality (LMI) [17]:

$$\begin{bmatrix} A_0P_1 + P_1A_0^T + B_0P_2 + P_2^TB_0^T + \gamma^{-2}B_1B_1^T & (C_1P_1 + D_1P_2)^T \\ C_1P_1 + D_1P_2 & -I \end{bmatrix} < 0 \tag{13}$$

then the state feedback controller of H_∞ control can be designed as follows:

$$\beta = KY = P_2P_1^{-1}Y \tag{14}$$

where $K = \begin{bmatrix} k_1 & k_2 & \cdots & k_{4\bar{N}} \end{bmatrix}$.

The control objective is designed as follows: for any disturbance $\omega > 0$, the robustness of the closed-loop system can be expressed as:

$$\int_0^\infty \left[\sum_{j=1}^{4\bar{N}} Y_i^2 + \rho^2\beta^2 \right] dt < \gamma^2 \int_0^\infty \omega^2 dt \tag{15}$$

where, the robust control parameter ρ is related to the control input signal, which is the factor influencing the control input; γ is the robust performance parameter to be optimized.

In the control output expression $Z = C_1Y + D_1\beta$, $C_1 = \begin{bmatrix} I_{4\bar{N}\times 4\bar{N}} & 0_{4\bar{N}\times 1} \end{bmatrix}^{\mathrm{T}}$, $D_1 = \begin{bmatrix} 0_{1\times 4\bar{N}} & \rho \end{bmatrix}^{\mathrm{T}}$, then there are:

$$Z = C_1Y + D_1\beta = \begin{bmatrix} Y_1 & Y_2 & \cdots & Y_{4\bar{N}} & \rho\beta \end{bmatrix}^{\mathrm{T}} \tag{16}$$

therefore:

$$\|Z\|_2^2 = \int_0^{\infty} \left[\sum_{j=1}^{4\bar{N}} Y_i^2 + \rho^2 \beta^2 \right] dt < \gamma \|\omega\|_2 \tag{17}$$

therefore, the H_∞ controller satisfying equation (17) which makes the closed-loop system internally stable is exactly the controller K satisfying the control objective design (15).

The solution of LMI inequality (13) is a complex problem, which is an optimization problem constrained by linear matrix inequality. The YALMIP toolbox of MATLAB can be used to simplify the solution process. Based on the selection of robust control parameter ρ ,the uncertain robust performance parameter γ is optimized to keep the controlled displacement and control input β in a reasonable range. The above is the complete design process of H_∞ control based on LMI.

4.1. Response without Control

The structure parameters of composite material, blade structure parameters and motion parameters are as follows:

$$GJ = 3.3935\times10^7\,\mathrm{Nm}^2, K_c = -4.8869\times10^5\,\mathrm{Nm}^2,\ EI = 1.8822\times10^7\,Nm^6, K_A = 1.418\times10^{-1},$$
$$m_c = 1.703\times10^2\,Kg, K_m = 1.239\times10^{-1}, K_{m1} = 3.023\times10^{-2}, K_{m2} = 1.201\times10^{-1}.$$
$$L=5\,\mathrm{m};\ U=15\,\mathrm{m\cdot s^{-1}};\ c=0.6\mathrm{m};\ \Omega=\lambda_0 U/L = 1\times U/L.$$

Figure 2 shows the high frequency vibration displacement of swing / torsion without control. It can be seen that the two displacements are in the "low amplitude high frequency" response state: Although the vibration frequency of torsional displacement is higher, its amplitude order of magnitude is very small (10^{-18}), so that any existing industrial sensor cannot measure its value, which can be completely ignored, so there is no potential failure of fracture failure; But for the swing displacement, its vibration amplitude can cause the potential failure of fracture at high frequency, which is the purpose of "high frequency vibration" control in this paper. In the follow-up active control, the control effect of swing vibration will be mainly discussed. In addition, because the torsional displacement is very small, it will not affect the torsional motion of the trailing edge flap along the span, which provides the conditions for the realization of the swing control of the trailing edge flap.

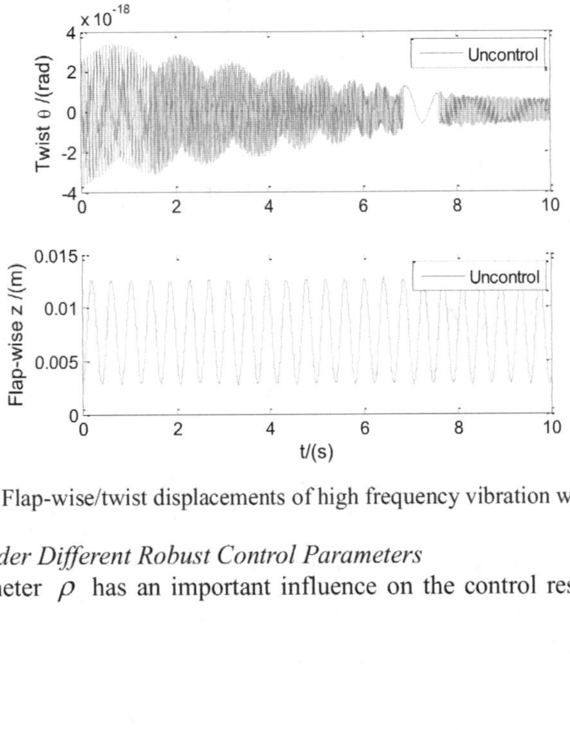

Figure 2. Flap-wise/twist displacements of high frequency vibration without control.

4.2. Control Effect under Different Robust Control Parameters

Robust control parameter ρ has an important influence on the control results. The purpose of this design is to control the high frequency vibration of the swing motion. For the high frequency vibration case in Figure 2, figure 3 shows the flap swing angle (a) and the controlled swing displacement (b) under different parameters ρ. From the actual controlled swing displacement, the influence of different parameters ρ is not significant, and the control effect can be maintained well. However, from the angle of flap swing controlled by the actual control, when $\rho = 3$, control input β is maintained in the range of $\pm 0.2 rad$, i.e., $\pm 12^{\circ}$, which is a moderate range. Because too large input swing angle β not only means too much power consumption, but also easy to bring physical interference of different components in physical implementation, and too small input swing angle β means that the cumulative error of pulse drive of stepper motor is relatively large, which affects the control accuracy. Therefore, the value of robust control parameter ρ can directly affect the physical input. But from the time-domain fluctuation of the controlled swing displacement, compared with Figure 2, the system is in a stable convergence state, and no matter what ρ value is, the vibration frequency is greatly reduced, which achieves the purpose of high-frequency vibration control.

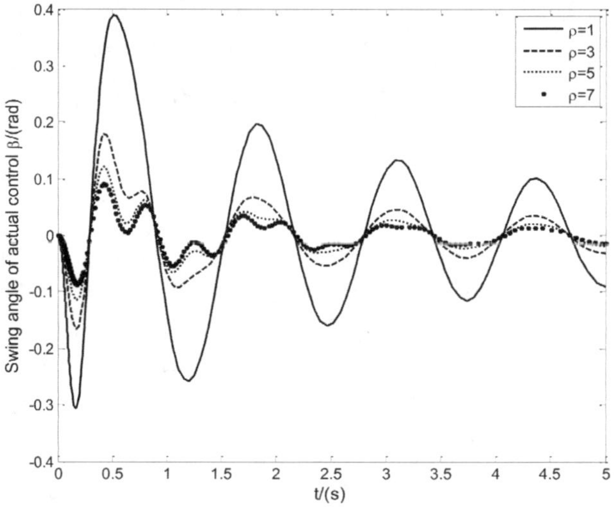

(a) Swing angle of actual control under different parameters ρ

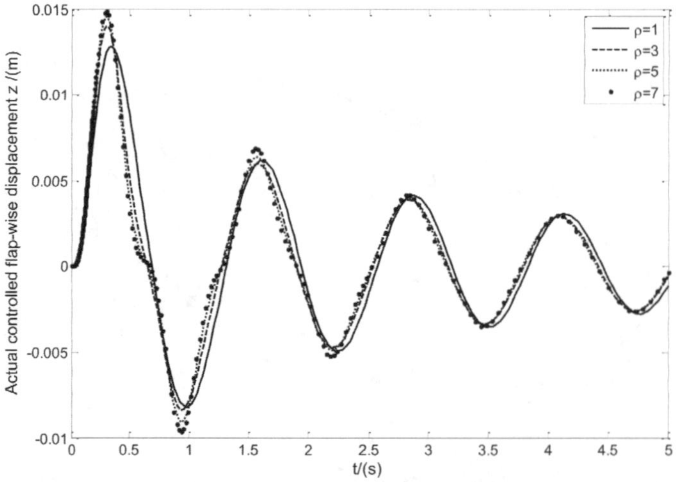

(b) Actual controlled swing displacement under different parameters ρ

Figure 3. The trailing-edge flap angles (a) and the controlled flap-wise displacements (b) of practical control under different parameters ρ

5. Conclusions

(1) The high frequency vibration analysis is based on the rotating composite blade beam with solid CAS layer driven by trailing edge flap with 92% span distribution; Aerodynamic force is based on a quasi-steady aerodynamic model suitable for flap structure. The aeroelastic equation is solved by the multi-body dynamic analysis method based on strip theory, and the mathematical integration is replaced by the summation based on strip number, which is more suitable for engineering application in controller hardware than the conventional integral discrete algorithm.

(2) The H_∞ control is based on LMI design, the LMI solution based on YALMIP toolbox can simplify the calculation, and the actual H_∞ improvement based on stepping motor effect can realize the control of low amplitude high frequency vibration.

(3) The influence of robust control parameters in H_∞ control algorithm is discussed, which provides a basis for similar H_∞ control; And the robust performance is stable, which provides a feasible scheme for the physical realization of the blade flap structure and the system control drive.

Acknowledgement

The authors gratefully acknowledge the support of the National Natural Science Foundation of China (no. 51675315), and the Graduate Science and Technology Innovation Project in SDUST.

References

[1] Yi C,Sun C and Sun Q 2021 *Journal of Shandong University of Science and Technology(Natural Science)* in press.
[2] Hayat K, Alvaro G M L, Moriones C D, et al 2016 *Journal of Sound and Vibration* **370** 149-162.
[3] Haselbach P U 2017 *Composite Structures* **180** 521-530.
[4] Llorente E, Ragni D 2020 *Renewable Energy* **147** 437-446..
[5] Chen H, Qin N 2017 *Renewable Energy* **105** 419-435.
[6] Mu A, Zhang G and Li N 2018 *Journal of Vibration and Shock* **37**(14):79-85.
[7] Zhang G 2017 *The active flow control of high-aspect-ratio wind turbine blade based on trailing-edge flaps* (Xi'an University of Technology).
[8] BolinK, Bluhm G and Eriksson G 2011 *Environmental Research Letters* **6**(3) 035103.
[9] Jianu O, Rosen M A and Naterer G 2012 *Sustainability* **4** 1104-1117.
[10]Xiong L, Cheng L and Godbole A R 2017 *Applied Energy* **185** (2) 1109-1119.
[11]Zuheir S, Abdullah O I and Mustafa A M 2019 *Journal of Mechanical Engineering Research and Developments* **42**(4) 14-19.
[12]Wang X, Chen J and Wenzhong S 2010 *China Mechanical Engineering* **20**(2) 211-213.
[13]Liu T, Ren Y 2012 *Acta Energiae Solaris Sinica* **33**(1) 105-112.
[14]Sahjendra N S, Yoosoon Y 2003 *Aerospace Science and Technology* **7** 23-31.
[15]Ren Y, Liu T and Yang S 2011 *Journal of Mechanical Engineering* **47**(12) 113-125.
[16]Hodges D H, Pierce G A 2002 *Introduction to Structural Dynamics and Aeroelasticity* (Cambrige: Cambrige University Press) pp 55-75.
[17]Liu J 2017 *Robot Control System Design and MATLAB Simulation* (Beijing: Tsinghua University Press).
[18]Gong Z 2012 *The Applied Technology Based on S7-200/300/400 PLC, the Improved Version* (Posts and Telecom Press Co. LTD).
[19]Yu Z 2016 *Numerical simulation of free vibration of piezoelectric composite cantilever blade* (Shandong University of Science & Technology).
[20]Yuan Q, Li C and Yang Y 2018 *Energiae Solaris Sinica* **39**(7) 2027-2035.

Mechanical Analysis of Structural Damage Detection Device for Cylinder Tower

Zhiyuan Wang, Changle Sun, Jian Wang, Kang Zhao and Tingrui Liu*

College of Mechanical & Electronic Engineering, Shandong University of Science & Technology, Qingdao 266590, China

*Corresponding author's e-mail: liutingrui9999@163.com

Abstract. Wind turbines suffer from severe weather and complex environment for a long time, so wind turbine towers can cause structural damage. As an important part of wind turbine, tower damage detection is particularly important. In this study, a structural damage detection device for cylinder tower is introduced and its mechanical analysis is carried out. Due to the special structure design of the device, it has the characteristics of comprehensive detection, accuracy and efficiency.

1. Introduction

The wind turbine tower is one of the important components of wind turbine, among which cylinder tower is widely used because of its simple structure, simple shape and convenient manufacture. The weight of the tower accounted for about one-half of the total weight and the manufacturing cost accounted for about one-fourth of the total weight. Therefore, the tower is particularly important in the manufacturing process of the whole unit [1].

Wind turbines are usually built in high-altitude windy mountains, vast plains, Gobi Desert or near the sea without reefs. They are often affected by bad weather and complex environment [2]. For example, in the wind sand environment, due to long-term erosion by wind sand, wind turbine tower surface will appear scratches. For the offshore wind turbine, due to the influence of high humidity and high salt spray environment, the tower made of metal materials is more likely to be corroded [3]. If it is not found and repaired in time, the original slight damage of the tower surface is likely to aggravate the damage, produce cracks or more serious damage, which will lead to the tower breaking and the collapse of the wind turbine, causing huge losses of life and property.

Damage detection device for wind turbine tower is produced by installing a camera, an ultrasonic flaw detector, etc. detection machine on a device that can move on the tower. By checking the wind turbine tower, various kinds of damage can be found, which can enable workers to repair damage on the tower in time, avoid aggravating damage and provide guarantee for safe operation of wind turbine. However, as the wind turbine tower body is three-dimensional structure, it is not convenient to detect. The existing detection devices based on tower climbing robots and unmanned aerial vehicles are powered by electric power and cannot utilize new energy. In addition, when the climbing robot and unmanned aerial vehicle are inspecting the tower, they need to adjust the running angle several times in order to inspect the tower as completely as possible.

The advantage of the structural damage detection device for cylinder tower introduced in this paper is that the wind wheel is mounted on the body, the renewable energy of wind energy is used to provide driving force for the whole device, and the positive and negative stroke motion of the detection device around the surface of the tower body is realized. During the movement, the surface and interior of cylinder tower can be scanned in all directions by digital camera and ultrasonic flaw detector installed on the body. Compared with existing detection devices, the advantages are comprehensive detection, high accuracy, wind energy utilization and environmental protection.

2. Model Introduction of Detection Device

2.1. Structure of Detection Device

The structural damage detection device for cylinder tower mainly consists of frame, balance wind wheel, power wind wheel, digital camera, ultrasonic flaw detector and universal ball. Figure 1 shows the front view of the detection device model. Figure 2 shows the front view of the reverse perspective of the detection device model.

There is a support frame in the middle of the device frame and a balancing wind wheel is fixed on the support frame at an angle of 45 degrees. The balancing wind wheel consists of a wind wheel frame, paddle blade and blade hub. A pair of airfoil paddle blades are fixed and mounted on the blade hub of the wind turbine frame, which forms a rotating pair with clearance between the hub and the wind turbine frame. The interior of the wind wheel frame fits with the clearance of the blade hub to form a rotating pair.

Two power wind wheels are welded on the left and right sides of the device frame, which are composed of small wind wheel frame, small paddle blade and small blade hub. The volume of the power wind wheel is half of that of the balance wind wheel. The structure and matching mode of the components of the power wind wheel are the same as those of the balancing wind wheel.

A digital camera and an ultrasonic flaw detection are installed in front of the frame. The digital camera consists of a digital camera probe and a camera information memory. The ultrasonic flaw detector device consists of an ultrasonic flaw detection probe and a flaw detection information memory. Install two rows of universal balls at a 45-degree angle at the bottom of the frame.

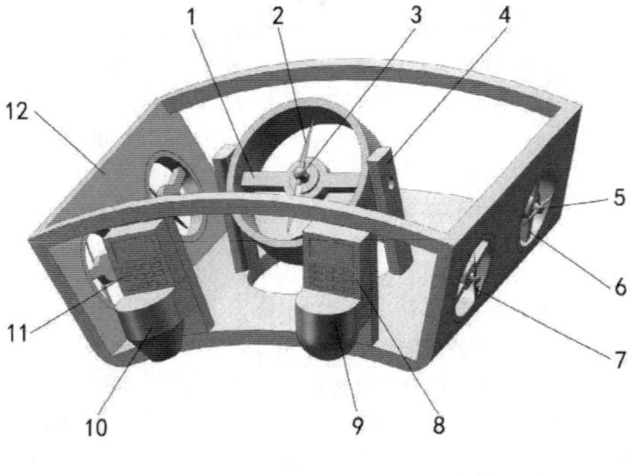

Figure 1. Detection device model (front).

Figure 2. Detection device model (reverse).

where 1 is wind wheel frame, 2 is blade, 3 is blade hub, 4 is support frame, 5 is small blade hub, 6 is small wind wheel frame, 7 is small blade, 8 is camera information memory, 9 is digital camera probe, 10 is ultrasonic flaw detection probe, 11 is flaw detection information memory, and 12 is device frame and 13 is universal ball.

2.2. Operation Mode of Detection Device

Turn on the digital camera and ultrasonic flaw detector before working, contact the tower surface with one side of the universal ball installed on the device, and point the side of the digital camera and ultrasonic flaw detector downwards to the ground. Open the right two power wind wheels, close the left two power wind wheels, adjust the device according to the wind direction. The blade and two pairs of paddles on the right are rotated by the wind to generate power. The power wind wheel generates the driving force and the driving device moves. Balancing wind wheels mainly produce balancing forces, which can be broken down into two parts, one part to counteract part of the gravity of the device itself and the other to attach the device to the tower surface.

Since the bottom of the frame is curved and the universal ball is arranged at an angle of 45 degrees, the device will spiral upward around the surface of the tower body. When the device is traveling to the top of the tower, two power wind wheels on the right side are closed and two power wind wheels on the left side are opened by the control of PLC switch. The two pairs of small paddles on the left side rotates under the action of wind, driving the device to move in reverse direction, reaching the bottom of the tower and completing the whole working process.

Digital camera and ultrasonic flaw detector can scan the tower body in all aspects during the spiral forward and reverse travel movement of the device, and store the test results in each confidence information memory, which can achieve a one-time comprehensive detection without adjusting the angle, multiple tests and high efficiency. When the unit returns to the bottom of the tower, the worker can diagnose the structural damage caused by the tower by viewing the scanned detected data from two information memories.

3. Mechanical Analysis

3.1. Force Analysis of Universal Ball

For the detection device in working state, two universal balls are selected as the research objects for force analysis.

The device works in positive stroke: open the two power wind wheels on the right and close the two power wind wheels on the left. Figures 3-5 show the analysis diagrams of device positive stroke motion model with z-y coordinates, z-x coordinates, and y-x coordinates, respectively. The balancing wind wheel generates a balancing force F_b, The power wind wheel generates the driving force F_d.

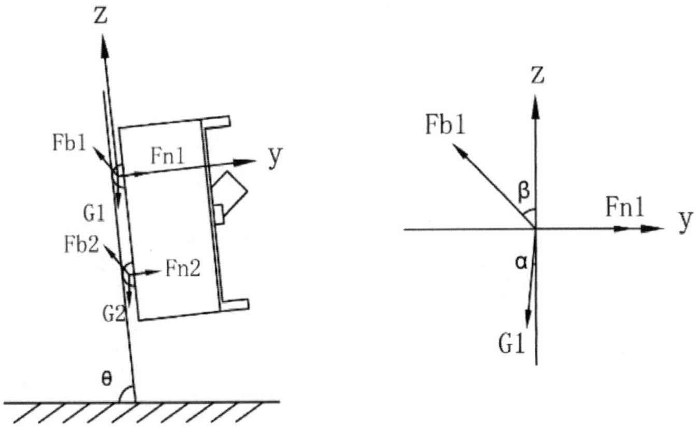

Figure 3. Analysis diagram of device positive stroke motion model (z-y coordinates).

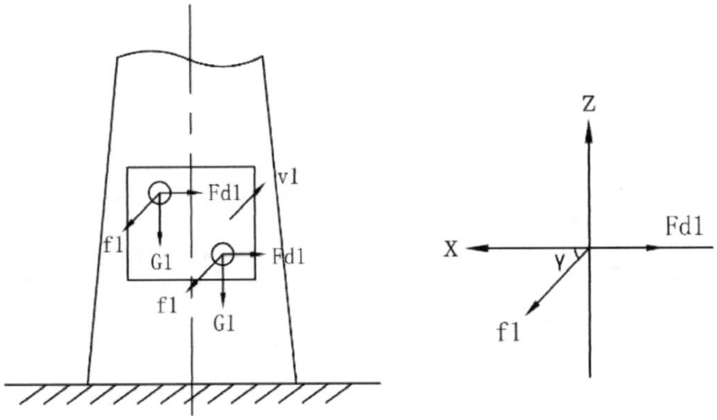

Figure 4. Analysis diagram of device positive stroke motion model (z-x coordinates).

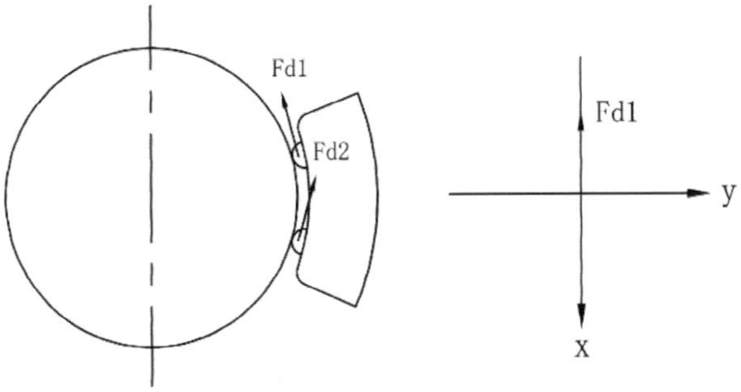

Figure 5. Analysis diagram of device positive stroke motion model (y-x coordinates).

where G_1 is the gravity of the No.1 universal ball, F_{n1} is the positive pressure on the No.1 universal Ball, F_{b1} is the component of F_b on the No.1 universal ball, v_1 is the direction of movement of the device, F_{d1} is the component of F_d on the No.1 universal ball, and f_1 is the rolling friction on the No.1 universal ball. The forces acting on No.2 and No.1 are the same.

The device works in reverse stroke: open the two power wind wheels on the left and close the two power wind wheels on the right. Figures 6-8 show the analysis diagrams of device reverse stroke motion model with z-y coordinates, z-x coordinates, and y-x coordinates, respectively.

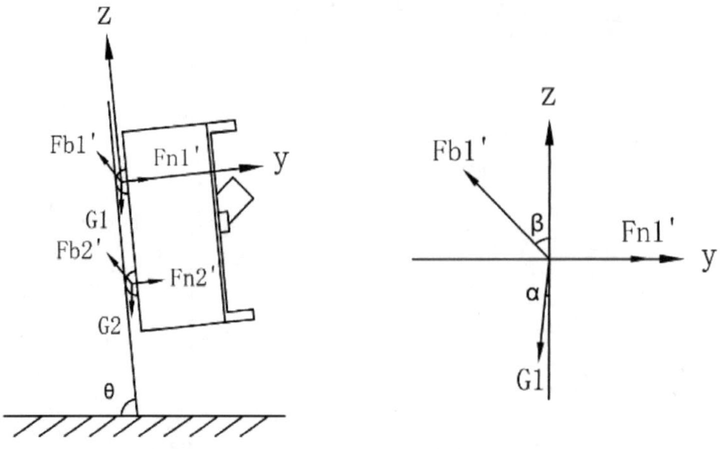

Figure 6. Analysis diagram of device reverse stroke motion model (z-y coordinates).

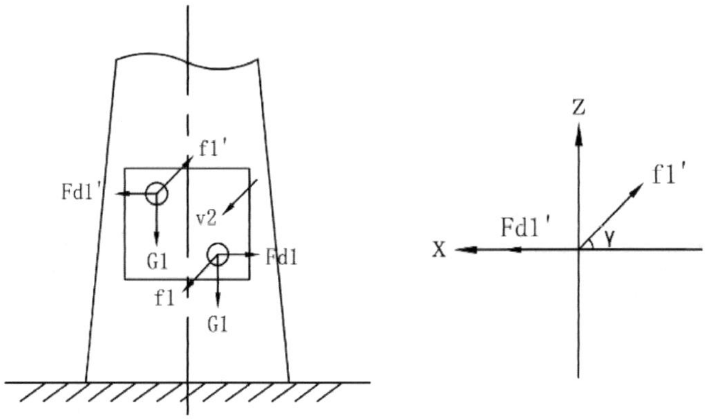

Figure 7. Analysis diagram of device reverse stroke motion model (z-x coordinates).

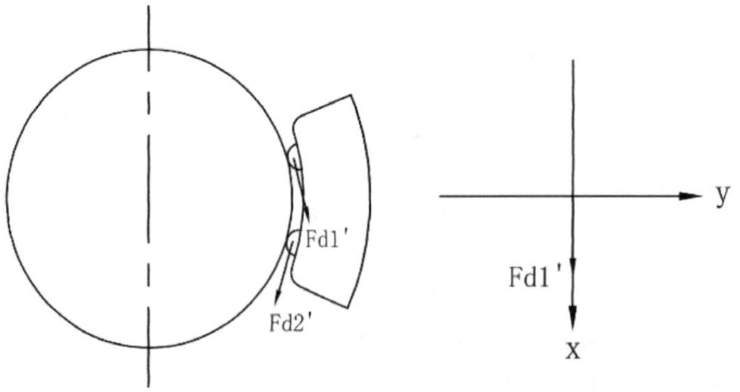

Figure 8. Analysis diagram of device reverse stroke motion model (y-x coordinates).

where F_{n1}' is the positive pressure on the No.1 universal ball, F_{b1}' is the component of F_b' on the No.1 universal ball, v_2 is the direction of movement of the device, F_{d1}' is the component of F_d' on the No.1 universal ball, and f_1' is the rolling friction on the No.1 universal ball. The forces acting on No.2 and No.1 are the same.

The force analysis of the No.1 universal ball is carried out and a space rectangular coordinate system is established. The angle between tower and ground is θ, so α is $90° - \theta$. Since the balancing wind wheel is fixed on the support frame at an angle of 45 degrees, so β is $45°$. Since the universal ball is arranged at an angle of 45 degrees at the bottom of the frame, the trajectory of the device on the surface of the tower can be regarded as a conical helix with an elevation of 45 degrees, so γ is $45°$.

The following three special states are analyzed when the device is in operation.

When the balancing wind wheel rotates and the power wind wheel closes, if the device is stationary on the surface of the tower body:

$$\begin{cases} \Sigma F_z = 0, & F_{b1}\cos\beta + f_s - G_1\cos\alpha = 0 \\ \Sigma F_y = 0, & F_{n1} - F_{b1}\sin\beta - G_1\sin\alpha = 0 \end{cases} \tag{1}$$

where f_s is the static friction on the No.1 universal ball.

When the device is operating with positive stroke, if at some point the device is in balance:

$$\begin{cases} \Sigma F_z = 0, & F_{b1}\cos\beta - f_1\sin\gamma - G_1\cos\alpha = 0 \\ \Sigma F_y = 0, & F_{n1} - F_{b1}\sin\beta - G_1\sin\alpha = 0 \\ \Sigma F_x = 0, & f_1\cos\gamma - F_{d1} = 0 \end{cases} \tag{2}$$

When the device is operating with reverse stroke, if at some point the device is in balance:

$$\begin{cases} \Sigma F_z = 0, & F_{b1}'\cos\beta + f_1'\sin\gamma - G_1\cos\alpha = 0 \\ \Sigma F_y = 0, & F_{n1}' - F_{b1}'\sin\beta - G_1'\sin\alpha = 0 \\ \Sigma F_x = 0, & F_{d1}' - f_1'\cos\gamma = 0 \end{cases} \tag{3}$$

3.2. Analysis of Contact Stress

CY-15A model is selected for the universal ball as depicted in Figure 9. The central ball and auxiliary ball are made of carbon steel and the housing is made of galvanized steel.

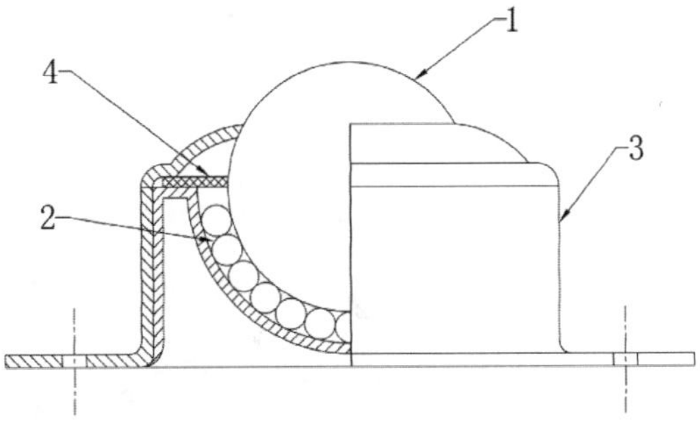

Figure 9. CY-15A universal ball.

where 1 is central ball, 2 is auxiliary ball, 3 is housing, and 4 is seal ring.

Figure 10 shows the schematic diagram of contact between two balls.

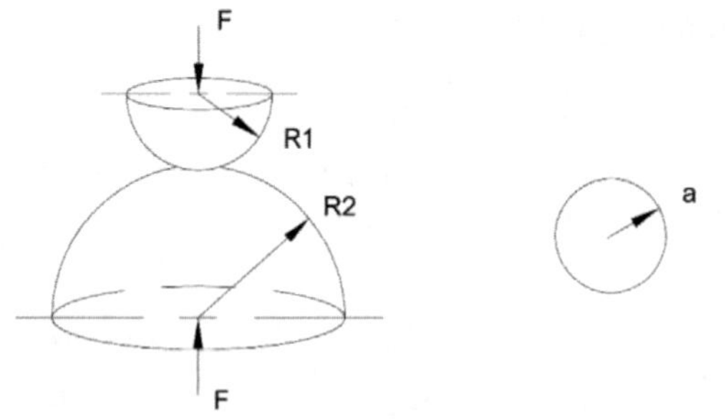

Figure 10. Schematic diagram of contact between two balls.

where R_1 is the radius of the ball 1, R_2 is the radius of the ball 2, F is the external load, and a is the radius of the contact area. The maximum contact stress is:

$$\sigma_{\max}=0.578\left[\frac{F\left(\frac{R_1+R_2}{R_1R_2}\right)^2}{\left(\frac{1-\upsilon_1^2}{E_1}+\frac{1-\upsilon_2^2}{E_2}\right)^2}\right]^{\frac{1}{3}} \tag{4}$$

Equivalent elastic modulus is:

$$\tilde{E}=\frac{1}{\frac{1-\upsilon_1^2}{E_1}+\frac{1-\upsilon_2^2}{E_2}} \tag{5}$$

then:

$$\sigma_{\max}=0.578\left[F\tilde{E}^2\left(\frac{R_1+R_2}{R_1R_2}\right)^2\right]^{\frac{1}{3}} \tag{6}$$

Set the radius of the cross section at a point in the wind turbine tower to be R_t, the external load between the tower and the universal ball is Q_{tc}, and equivalent elastic modulus is $\tilde{E}_{tc}$. The radius of the central ball is R_c, and the radius of the auxiliary ball is R_a. The external load between the central ball and the auxiliary ball is Q_{ca}, equivalent elastic modulus is $\tilde{E}_{ca}$. Then the maximum contact stress is:

$$\sigma_{\max tc}=0.578\left[Q_{tc}\tilde{E}_{tc}{}^2\left(\frac{R_t+R_c}{R_tR_c}\right)^2\right]^{\frac{1}{3}} \tag{7}$$

$$\sigma_{\max ca}=0.578\left[Q_{ca}\tilde{E}_{ca}{}^2\left(\frac{R_c+R_a}{R_cR_a}\right)^2\right]^{\frac{1}{3}} \tag{8}$$

where $\sigma_{\max tc}$ is the maximum contact stress between the tower and the universal ball, $\sigma_{\max ca}$ is the maximum contact stress between the central ball and the auxiliary ball.

4. Summary
Due to special structures such as the curved bottom of the device and the universal ball mounted at 45 degrees, the device can spiral around the surface of the tower body, thus realizing one-off comprehensive detection without manual adjustment of the device position. The mechanical analysis of the universal ball shows that the device is theoretically realized.

Acknowledgement
The authors gratefully acknowledge the support of the National Natural Science Foundation of China (no. 51675315), and the Graduate Science and Technology Innovation Project in SDUST.

References
[1] Shan L 2009 Structural Performance and Design of Wind Turbine Towers, Ph.D. Dissertation. Harbin: Harbin Institute of Technology.
[2] Zhang Y 2019 Research on Image Recognition Method of Wind Turbine Blade Surface Damage, Ph.D. Dissertation. Kunming: Kunming University of Science and Technology.
[3] Shi S, Xu Q, Yun H and Pan H 2010 Corrosion & Protection 31(11) 875-877+885.

Online Performance Monitoring of Processes Subject to Non-stationary Stochastic Disturbance

Yichao Li and Chunqing Huang*

Department of Automation, Xiamen University, Xiamen, Fujian, China

*corresponding author's e-mail: chqhuang@xmu.edu.cn

Abstract. Since non-stationary disturbances are common in practical situation due to the varying stochastic disturbance or some deterministic disturbance injected, the traditional algorithm based on FCOR is not valid anymore. This paper develop a new approach, which combines EMD method with FCOR algorithm, to handle this problem such that the minimum variance (MV) performance of the closed-loop system can be monitored online. In the presence of varying stochastic disturbance and step/sine disturbance, the numerical example demonstrates the effectiveness of the proposed algorithm.

1. Introduction

The earliest research work of performance assessment is developed by Astrom [1] and Harris [2], in which the output variance under MV control is used as a benchmark to evaluate the controller performance. Due to the minimal requirement of a priori knowledge regarding the process, MV control has been widely used and is regarded as a reference bound on achievable performance [3]. So far, various benchmarks such as user-specified benchmark, optimal PID benchmark (OPID), historical data benchmark (HIS), generalized minimum variance (GMV) benchmark, generalized Hurst exponent based benchmark [4] and integrated absolute errors (IAEs) [5] etc., have been developed (see the survey paper [6, 7, 8]).

Although FCOR algorithm seems to be effective for on-line estimate of MV benchmark [10, 13], it should be noted that this technique involves applying time series analysis to estimate the source white noise under an assumption of stationarity for output data [10, 12, 13]. However, such assumption is usually problematic in industrial process due to the facts: either the periodic disturbances, such as sine wave and rectangular signal, are common in industrial environment [15]-[19]; or deterministic disturbances such as sudden step changes may occur randomly in time [21]-[23]; or changes of disturbances dynamics are often encountered in chemical processes [14], [19].

In the presence of non-stationary disturbances or output data, the time series analysis techniques becomes invalid in application to CPA problem [13, 14]. It may yield an unmeaningful performance evaluation if simple stochastic disturbance form is taken into account for benchmark calculation [23]. Actually, the practical application of MV-based CPA technique is restricted significantly by this factor though MV-based CPA technique is mature by the plentiful attainment in the academic control community [9]-[11].

Motivated by the aforementioned facts, this paper is to deal with the performance assessment online for the industrial processes that is subject to non-stationary disturbances. The stochastic component is to separate from the composite output signal by using EMD method such that the MV benchmark can be estimated from the non-stationary signal data and the performance monitoring can be conducted online.

2. Problem description

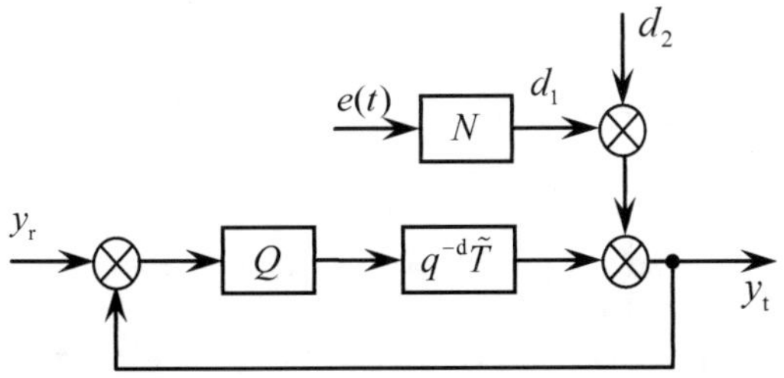

Figure 1. Schematic of closed loop system

Consider a SISO industrial closed-loop process as illustrated in Fig.1. The output $y(t)$ is described by the following model:

$$y(t)=\frac{N(t)}{1+q^{-d}\tilde{T}(q^{-1})Q(q^{-1})}e(t) \tag{1}$$

Where q^{-1} is the one-step backward shift operator; y_r is the reference input, Q is the controller transfer function, d is the time delay, $\tilde{T}$ is the process transfer function without delay. d_1 / d_2 denotes the stochastic/deterministic disturbance; the N is the disturbance transfer function; $e(t)$ is a white noise with zero mean and variance σ_e^2.

For a stable closed-loop system depicted in Fig.1, the output $y(t)$ can be rewritten in the form of MA model with infinite order as

$$y(t)=(g_0+g_1q^{-1}+\cdots+g_{d-1}q^{-d+1}+g_dq^{-d}+\cdots)\cdot e(t) \tag{2}$$

According to minimum variance control, the MV benchmark can be represented by the first d feedback control invariant terms

$$\sigma_{mv}^2=(g_0{}^2+g_1{}^2+\cdots+g_{d-1}{}^2)\cdot\sigma_e^2 \tag{3}$$

The performance indices of the control system is defined as

$$\eta=\frac{\sigma_{mv}^2}{\sigma_y^2} \tag{4}$$

Where σ_y^2 is the system output variance.

3. Online estimate of MV benchmark

For stable SISO closed-loop system, Huang [24] proposed a relatively simple method to calculate the closed-loop impulse response coefficients from $e(t)$ to $y(t)$. As mentioned before, $y(t)$ can be represented by an MA model with infinite order. Multiply both the left and the right by $e(t)$, $e(t-1)$, $e(t-2)$, $\cdots$ $e(t-n+1)$ respectively and calculate the expectation on both sides of the above equation simultaneously to yield

$$\begin{cases} r_{ye}(0) = E\{y(t)e(t)\} = g_0\sigma_e^2 \\ r_{ye}(1) = E\{y(t)e(t-1)\} = g_1\sigma_e^2 \\ \quad\vdots \\ r_{ye}(d-1) = E\{y(t)e(t-d+1)\} = g_{d-1}\sigma_e^2 \\ r_{ye}(d) = E\{y(t)e(t-d)\} = g_d\sigma_e^2 \\ \quad\vdots \end{cases} \tag{5}$$

Where $E\{\cdot\}$ is the expectation. By using the above FCOR algorithm, the first d impulse response coefficients of the closed-loop system can be estimated online from the measurement data under routine operating condition.

Note that noise sequence $e(t)$ is unknown, it can be estimated by the residual sequence of the fitting AR or ARMA model for the measurement data $y(t)$, where $y(t)$ is required to be stationary. However, when the disturbance dynamics is changed or step disturbance is involved in the stochastic disturbance, the output disturbance is non-stationary and hence it can't be modeled by AR or ARMA model. In order to make this signal stationary, EMD method is introduced to separate the deterministic signals from the closed-loop output. Note that EMD method, which is provided in MATLAB, can be invoked directly to deal with the non-stationary output signal. As a result, the composite disturbance can be separated into two different components, i.e. stochastic $\hat{d}_1$ and deterministic disturbance $\hat{d}_2$ respectively. Based on the data $\hat{d}_1$, the residual $\hat{e}(t)$ of the fitting model for $\hat{d}_1$ can be estimated by using the corresponding function in MATLAB™. Subsequently, FCOR algorithm in (5) is utilized to estimate the impulse response coefficients of the closed-loop system $\hat{g}_i$, $i = 1, 2, \cdots, d-1, d, \cdots$. The MV benchmark can be estimated as follows

$$\hat{\sigma}_{mv}^2 = (\hat{g}_0{}^2 + \hat{g}_1{}^2 + \cdots + \hat{g}_{d-1}{}^2) \cdot \hat{\sigma}_e^2 \tag{6}$$

Where $\hat{\sigma}_e^2$ is normalized as $\hat{\sigma}_e^2 = 1$. The output variance can also estimated from the data $\hat{d}_1$. Thus the performance monitoring is conducted online as the data is collected constantly as time goes by.

4. MV-based performance monitoring online

Based on the combination of the FCOR algorithm and EMD method, the performance monitoring can be conducted online and the standard procedure is presented as follows,

- Step-1 Collect the closed-loop output data of the system. Use EMD algorithm to preprocess the data within each data window and separate the stochastic component from the original signal.
- Step-2 Use a time series model (e.g. an ARMA model) to fit the statistical data and check stationarity of the residual;
- Step-3 Take the residual of the established model as an estimate of stochastic white noise, and the FCOR algorithm is used to calculate the minimum variance benchmark at the current moment. The output variance of the system is not difficult to find, and then the Harris index is obtained.
- Step-4 Update the data at each interval, then go to Step-1 and use EMD method to process the window data.
- Step-5 Repeat step 2, 3 and 4 utilizing the updated data window, and obtain the real-time performance index of the system according to the real-time output data of the system.

In the course of performance monitoring, the residual of the AR or ARMA model fitting should be check by sightseeing whether it is stationary at each calculation time in Step-2. If it is not, the order of the fitting model should be adjusted to be larger in general. This is an effective way to make the residual stationary.

5. Simulation

Consider a time-variant control system given in [20]

$$q^{-d}\tilde{T}(q^{-1})=\frac{0.33}{1-0.67q^{-1}}q^{-4} \tag{7}$$

The controller Q is given by

$$Q(q^{-1})=\frac{0.7-0.47q^{-1}}{0.33-0.1q^{-1}-0.23q^{-2}} \tag{8}$$

The disturbance model varies from N_1 to N_2 and N_3 with $e(t)$ that is a white noise with mean of zero and variance of $\sigma_e^2=1$, where

$$\begin{cases} N_1=\dfrac{1-0.4q^{-1}}{1-0.67q^{-1}} \\ N_2=\dfrac{1-0.4q^{-1}}{1-q^{-1}} \\ N_3=\dfrac{1-0.4q^{-1}}{1-0.87q^{-1}} \end{cases} \tag{9}$$

As time goes by, the disturbance transfer function changes from N_1 to N_2, and then to N_3, respectively at the 10001st sampling time and 20001st sampling time.

Using the algorithm proposed in this article, the performance index over time are calculated online for each interval $k=200$ sampling times and is listed in Fig.3, in which the data window is 8000 sampling times.

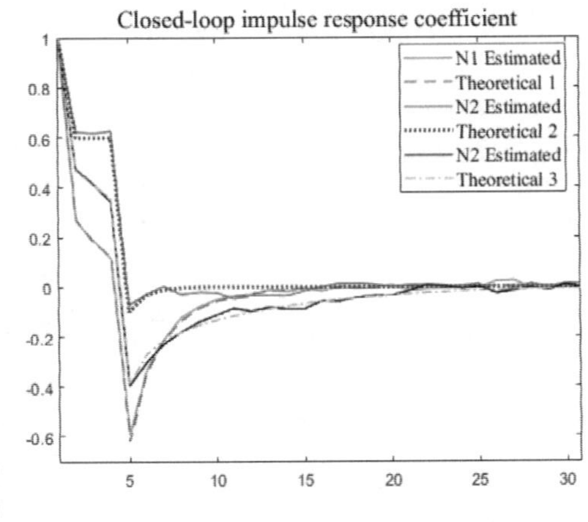

Figure 2. Estimated impulse response coefficient under the varying disturbance dynamics

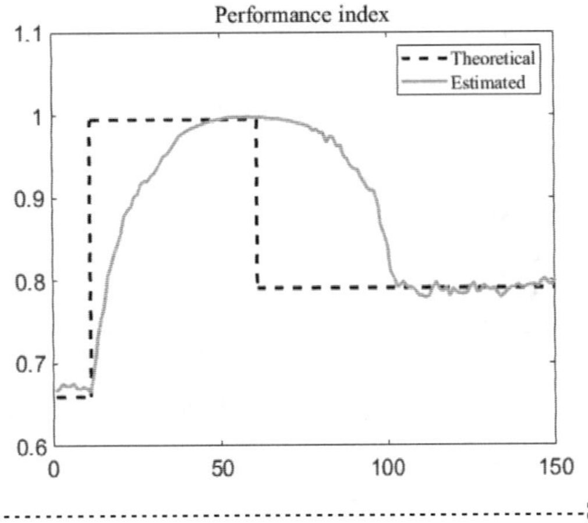

Figure 3. Performance monitoring under stationary disturbance.

Figure 2 illustrates the closed-loop impulse response coefficient estimated at 6th, 54th, 126th interval respectively. It shows that the estimated impulse response coefficients are closed to the real ones. As a result, the performance index as shown in Figure 3 is also close to the theoretical value during their responding steady operating stage.

In Figure 3, before the moment of the first disturbance model changed into the second one abruptly, the process data used to build the time series model is only affected by the first noise model N_1. After this moment, part of the data involved in the modeling is affected by N_2. As time goes on, the amount of this part of the data will increase gradually until the data window is all filled with statistics affected by N_2. The same situation applies to the next downward step point. The figure shows that even if the noise model changes abruptly, the online monitoring algorithm can track the performance indicators of the system well.

Consider a sine signal injected at d_2 with amplitude of 20 and period of 1000 sampling times. The simulation results are as follows.

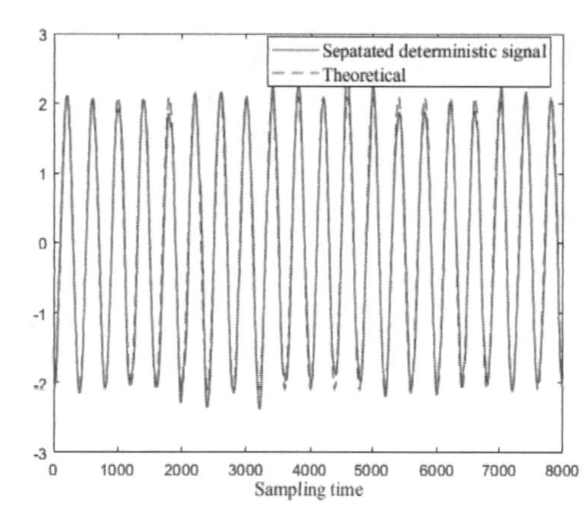

Figure 4. Deterministic signal separated from the data window

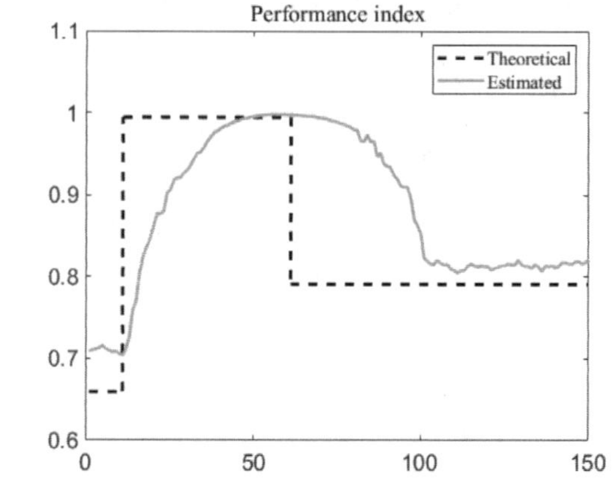

Figure 5. Performance monitoring subject to non-stationary disturbance with sine signal.

In Figure 4, the solid line represents the deterministic part separated from the original data of a data window by EMD algorithm. The dashed line represents the output of the control loop when only sine disturbance exists. After preprocessing by the EMD algorithm, most of the deterministic signal can be removed. Compared with the situation where the system is not disturbed by any deterministic dynamics, Figure 5 shows that, the proposed online monitoring algorithm can track the performance indices of the control loop affected by a sine signal. Although the accuracy of the performance indices corresponding to N_1 and N_3 is not particularly good, the trend of performance index can be well estimated by this algorithm.

6. Conclusions

In the presence of non-stationary disturbances, a new approach for performance monitoring online is proposed in this paper, in which EMD method is utilized to separate the stochastic disturbance from the non-stationary output data. The impulse response coefficient of the closed-loop system is estimated online by using FCOR algorithm that is conducted on the separated stochastic data. Meanwhile, the output variance is also estimated by this data and subsequently the performance index is calculated constantly. The effectiveness of the proposed approach is demonstrated by numerical example that is subject to varying stochastic disturbance or along with step/sine output disturbance.

Reference

[1] K.J. Astrom, Introduction to Stochastic Control Theory, Academic Press, San Diego, CA, 1970.
[2] T.J. Harris, Assessment of control loop performance, Can. J. Chem. Eng. 67(10) (1989) 856–861.
[3] B.S. Ko, T.F. Edgar, Performance assessment of multivariate feedback control systems, Automatica

37 (2001) 899-905.
[4] B. Srinivasan, T. Spinner, R. Rengaswamy, Control loop performance assessment using detrended fluctuation analysis (DFA), Automatica 48(2012) 1359-1363.
[5] Z. Yu, J. Wang, B. Huang, Performance assessment of PID control loops subject to setpoint changes, 21(2011) 1164-1171.
[6] M. Jelali, An overview of control performance assessment technology and industrial applications, Control Engineering Practice 14 (2006) 441–466.
[7] Y. Shardt, Y. Zhao, F. Qi, K. Lee, X. Yu, B. Huang and S. Shah, Determining the state of a process control system Current trends and future challenges, Can. J. Chem. Eng. 90 (2012) 217-245
[8] M. Bauer, A. Horch, L. Xie, M. Jelali, N. Thornhill, The current state of control loop performance monitoring – A survey of application in industry, J. Process Control 38 (2016) 1-10
[9] S. X. Ding and L. L. Li, "Control performance monitoring and degradation recovery in automatic control systems: A review, some new results, and future perspectives," Control Engineering Practice, vol. 111, no. 104790, 2021.
[10]M. Bauer, A. Horch, L. Xie, M. Jelali, and N. Thornhill, "The current state of control loop performance monitoring–A survey of application in industry," J. Process Control, vol. 38, pp. 1-10, 2016.
[11]Y. Shardt, Y. Zhao, F. Qi, K. Lee, X. Yu, B. Huang, and S. Shah, "Determining the state of a process control system Current trends and future challenges," Can. J. Chem. Eng., vol. 90, no. 2, pp. 217-245, 2012.
[12][12] L. Das, B. Srinivasan, and R. Rengaswamy, "Multivariate control loop performance assessment with Hurst exponent and Mahalanobis distance," IEEE Trans. Control Systems Technology, vol. 24, no. 3, pp. 1067-1074, 2016a.
[13]L. Das, R. Rengaswamy, and B. Srinivasan, "Data mining and control loop performance assessment: The multivariate case," AIChE J., vol. 63, no. 8, pp. 3311-3328, 2017.
[14]L. Das, B. Srinivasan, and R. Rengaswamy, "A novel framework for integrating data mining with control loop performance assessment," AIChE J., vol. 62, no. 1, pp. 146-165, 2016b.
[15]P. Shen, H. X. Li, "A multiple periodic disturbance rejection control for process with long dead-time," J. Process Control, vol. 24, no. 9, pp. 1394-1401, 2014.
[16]Q. G. Wang, Y. Zhang, and X. G. Huang, "Virtual feedforward control for asymptotic rejection of periodic disturbance," IEEE Trans. Ind. Electron., vol. 49, no. 3, pp. 566-573, 2002.
[17]M. A. A. S. Choudhury, S. L. Shah, and N. F. Thornhill, Diagnosis of Process Nonlinearities and Valve Stiction-Data Driven Approaches. Springer-Verlag, 2008.
[18]A. W. Ordys, D. Uduehi, and M. A. Johnson, Process Control Performance Assessment-From Theory to Implementation. Springer-Verlag, 2006.
[19][19] S. J. Dong, T. Liu, W. Wang, J. Bao, and Y. Cao, "Identification of discrete-time output error model for industrial processes with time delay subject to load disturbance," J. Process Control, vol. 50, pp. 40-55, 2017.
[20]F. Olaleye, B. Huang, and E. Tamayo, "Performance assessment of control loops with time-variant disturbance dynamics," J. Process Control, vol. 14, no. 8, pp. 867-877, 2004.
[21]Q. Yuan and B. Lennox, "Control performance assessment for multivariable systems based on a modified relative variance technique," J. Process Control, vol. 19, no. 3, pp. 489-497, 2009.
[22]J. F. MacGregor, T. J. Harris, and J. D. Wright, "Duality between the control of processes subject to randomly occurring deterministic disturbances and ARIMA stochastic disturbances," Technometrics, vol. 26, no. 4, pp. 389-397, 1984.
[23]B. Huang, "Performance assessment of processes with abrupt changes of disturbances," Can. J. Chem. Eng., vol. 77, no. 5, pp. 1044-1054, 1999.
[24]Huang B, 1997. Multivariate statistical methods for control loop performance assessment. University of Alberta.

Theoretical Investigation of Particles Manipulation in Fluids Using Acoustic Technique in a Microfluidic Channel

Yi Yang, Min Liu*

Department of Automation, Tsinghua University, Beijing, China

*corresponding author e-mail: lium@tsinghua.edu.cn

Abstract. The present work is focused on discussing standing acoustic wave-based particles separation in PMMA microfluidic channel at tilted angle. A theoretical model is developed to describe the interaction of the particles at different sizes under an acoustic pressure with tilted angles. Tilted angle, acoustic pressure, acoustic wavelength, flow velocity and buffer solution viscosity are as key parameters to be investigated. It is demonstrated that these parameters can affect the particle motion in microfluidic channel, which play the important roles in particle separation. It's shown the vital criteria of the microfluidic chip design for specific applications.
Keywords: microfluidics, acoustic wave, particle separation

1. Introduction

The separation of particles in complex fluid module, which contain different sizes of microparticles, cells, droplets and other contaminative particles, has become a critical issue in many applications, including industrial procedures [1], environmental assessment [2], biochemical analyses [3], and clinical diagnosis [4]. Recently, several groups have investigated separation approaches with microfluidic channels [5-8]. Low efficient isolation methods and low throughput restrict the development of these applications [9].

One of the limitations of standing surface acoustic wave (SSAW) applied in PDMS (polydimethylsiloxane) channels has the drawbacks of high attenuation, and difficulty in fabrication for large scale applications, in addition, the SSAW microfluidic channel has limits on channel height which reduces throughput and size range of the particles [10, 11].

To address the limitations of SSAW tilted angle particle separation devices, a PMMA tilted angle acoustofluidic-based microfluidic theoretical analysis has been studied. Bulk acoustic waves (BAW) are excited by PZT (lead zirconate titanate) transducers through the PMMA into a tilted angle channel, particles within microfluidic flow in this acoustic field are separated by size. The PMMA tilted angle channel design allows higher throughput, easier to fabricate in large scale and more mechanically robust than a tilted angle SSAW device.

2. Theoretical Analysis

Excited transducers propagate the bulk waves through the PMMA to form standing waves and acoustic node lines in the channel. Different particle sizes are separated based on the acoustic and drag force in the flow.

In a tilted angle microfluidic acoustic environment, particles experience forces from both the acoustic field and the drag force induced by the fluid. With the acoustic force itself, the particles travel towards the node lines, providing a visual tracing of the locations of the node lines in the system. With the drag force alone, the particles follow the streamlines of the water flow. Neither configuration alone induces particle separation. The velocity flow profile is approximated by applying Poiseuille Flow in rectangular section.

$$v_{\hat{x}}(y,z) = \frac{\Delta p w^2}{8\mu l}\left(1 - \frac{4\hat{y}^2}{w^2} - \sum_{n=0}^{\infty} \frac{32(-1)^n \cosh(\frac{(2n+1)\pi\hat{z}}{w}) \cos(\frac{(2n+1)\pi\hat{y}}{w})}{(2n+1)^3\pi^3 \cosh(\frac{(2n+1)\pi h}{w})}\right) \quad (1)$$

In Which, h, w, l are the height, width, and length of the channel, respectively. $\triangle p$ is the pressure gradient across the length of the channel, μ is the viscosity of the buffer solution.

In a standing acoustic wave field, the force on particles towards the nodes or anti-nodes of the acoustic field depends on the acoustic contrast of the particle material with respect to the medium. The acoustic contrast factor as presented by Lenshof et al [12], is shown in equation (2),

$$\phi = \frac{\rho_p + \frac{2}{3}(\rho_p - \rho_f)}{2\rho_p + \rho_f} - \frac{\rho_f c_f^2}{3\rho_p c_p^2} \quad (2)$$

Where ρ_p, ρ_f are the density of particles and fluid. β_p and β_f denote the compressibility of the particle and fluid, respectively. When the acoustic contrast factor is positive, the particle will move towards a node. The force applied on the particles in an acoustic field is,

$$F_r = -\left(\frac{4\pi^2 p_0^2 r^3 \beta_f}{6\lambda}\right)\left(\frac{5\rho_p - 2\rho_f}{2\rho_p + \rho_f} - \frac{\beta_p}{\beta_f}\right)\sin\left(\frac{4\pi}{\lambda}x\right) \quad (3)$$

Where p_0 is pressure amplitude, r is particle radius, λ is acoustic wavelength, x is particle position. Also, in microfluidic channels, the flow is laminar. Particles suspended in water in a microfluidic channel will move along the channel at approximately the velocity of the water. The drag force given by flow is,

$$F_v = -6\pi\eta r v \quad (4)$$

Where, η is the viscosity of water and v is the relative velocity between particles and water. Combining the equation (2) and (3), particles are described by fundamental acoustic and flow fields. The given force in $\hat{x}$ direction,

$$F_{\hat{x}} = -\left(\frac{4\pi^2 p_0^2 r^3 \beta_f}{6\lambda}\right)\left(\frac{5\rho_p - 2\rho_f}{2\rho_p + \rho_f} - \frac{\beta_p}{\beta_f}\right)\sin\left(\frac{4\pi(\hat{x}\sin(\theta) - \hat{y}\cos(\theta))}{\lambda}\right)\sin(\theta) + 6\pi\eta r(v_w - \dot{\hat{x}}) \quad (5)$$

The force in y direction,

$$F_{\hat{y}} = -\left(\frac{4\pi^2 p_0^2 r^3 \beta_f}{6\lambda}\right)\left(\frac{5\rho_p - 2\rho_f}{2\rho_p + \rho_f} - \frac{\beta_p}{\beta_f}\right)\sin\left(\frac{4\pi(\hat{x}\sin(\theta) - \hat{y}\cos(\theta))}{\lambda}\right)\cos(\theta) + 6\pi\eta r\dot{\hat{y}} \quad (6)$$

To simplify the equations, axes are rotated, where x axe is parallel to acoustic node line, and y axe is perpendicular to x axe, which gives,

$$F_x = 6\pi\eta r(v_w\cos(\theta) - \dot{x}) \quad (7)$$

$$F_y = -\left(\frac{4\pi^2 p_0^2 r^3 \beta_f}{6\lambda}\right)\left(\frac{5\rho_p - 2\rho_f}{2\rho_p + \rho_f} - \frac{\beta_p}{\beta_f}\right)\sin\left(\frac{4\pi y}{\lambda}\right) + 6\pi\eta r(v_w\sin(\theta) - \dot{y}) \quad (8)$$

Where ρ_p and ρ_f denote the density of the particle and fluid respectively. c is the speed of sound in the material, and β_p and β_f are the compressibility of the particle and fluid. θ is the tilted angle of the device and vw as the flow velocity of the water. From equation (8), the drag force equation is,

$$F_{yd} = 6\pi\eta r(v_w\sin(\theta) - \dot{y}) = 6\pi\eta r v_w \sin(\theta)\left(1 - \frac{\dot{y}}{v_w\sin(\theta)}\right) \quad (9)$$

The acoustic force is,

$$F_{ya} = -\left(\frac{4\pi^2 p_0^2 r^3 \beta_f}{6\lambda}\right)\left(\frac{5\rho_p - 2\rho_f}{2\rho_p + \rho_f} - \frac{\beta_p}{\beta_f}\right)\sin\left(\frac{4\pi y}{\lambda}\right) \quad (10)$$

If $F_{ya} \gg F_{yd}$, the acoustic force dominate the particle motion, the particles follow the acoustic node lines, thus, in which, if the acoustic pressure $p_0 \gg \sqrt{\frac{36\pi\eta v_w \sin(\theta)}{4\pi^2 r^2 \beta_f(\frac{5\rho_p-2\rho_f}{2\rho_p+\rho_f}-\frac{\beta_p}{\beta_f})}}$, the particles follow node lines, if the acoustic pressure $p_0 \ll \sqrt{\frac{36\pi\eta v_w \sin(\theta)}{4\pi^2 r^2 \beta_f(\frac{5\rho_p-2\rho_f}{2\rho_p+\rho_f}-\frac{\beta_p}{\beta_f})}}$, particles follow streamlines. So the acoustic pressure needs to be at the same order of magnitude, which is $p_0 \approx \sqrt{\frac{36\pi\eta v_w \sin(\theta)}{4\pi^2 r^2 \beta_f(\frac{5\rho_p-2\rho_f}{2\rho_p+\rho_f}-\frac{\beta_p}{\beta_f})}}$. The undulated path behavior of particles occurs.

3. Results and discussion

The polystyrene spheres are used as particles in the simulations. The microfluidic channel is simulated at dimension of 1.5mm in width, 6mm in length and 0.5mm in depth with 2 mm/s average flow rate and 0.2 MPa acoustic pressure rate at different tilted angle. The particles are selected from 1um to 10um at radius. The graph of the particle deflection at different tilted angle is shown in the Figure 1. As seen in the figure, the tilted angle that generates the deflection varies with respect to particle sizes. The simulation results indicate that as particle sizes and tilted angle increase, the maximum deflection is induced, which reveals that it's necessary to consider the relationship between tilted angle and particles size in specific application.

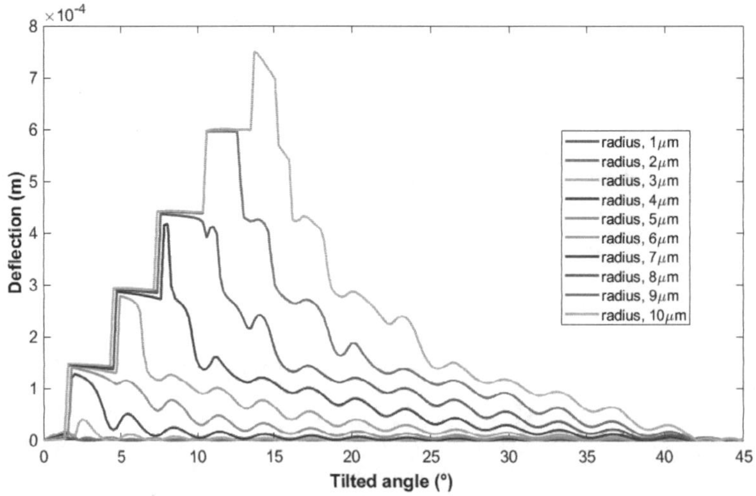

Figure 1. Particle deflection varies with respect to the change of tilted angle of the channel for different particles size.

The further investigation is to explore the acoustic pressure effect to particles deflection. At tilted angle of 20°, the results of particles deflection vary with respect to the change of acoustic pressure is shown in the Figure 2, in which, the deflection of particles increase when higher acoustic pressure applied, the lager particle has more intense reaction to the acoustic pressure.

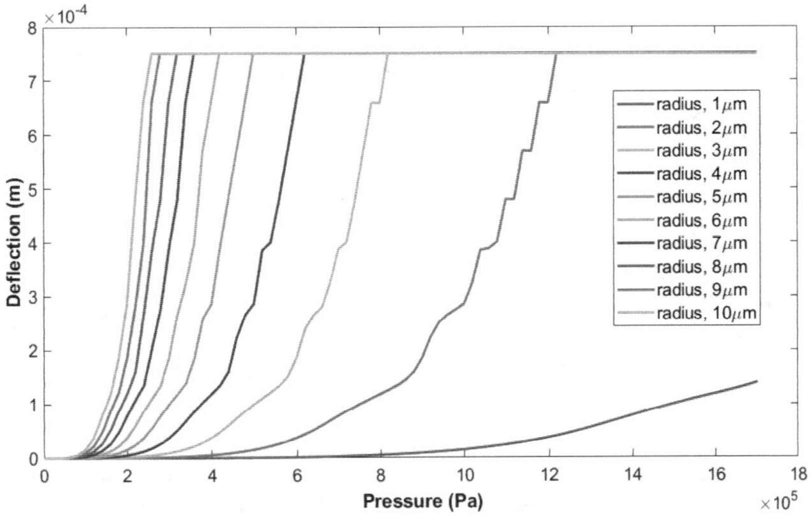

Figure 2. Particle deflection varies with respect to the change of acoustic pressure for different particles size.

Acoustic wavelength is one of the key parameters in the design. The particle deflection versus the applied acoustic wavelength is also simulated, as shown in **Figure 3**. The smaller wavelength requires the higher acoustic frequency which will generate the more node lines, it will also provide the larger deflection of particles. With the decreasing of the node lines, the particles motion will be smoother due to the particles will have enough time to travel between node lines.

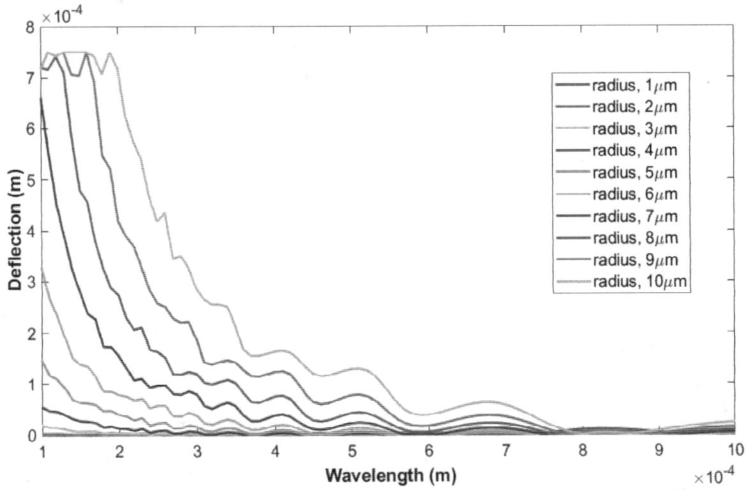

Figure 3. Particle deflection varies with respect to the change of acoustic wavelength in the channel for different particles size.

The flow velocity also plays an important role in the separation procedure. With increasing of the flow velocity, the particle deflection will decrease, which is demonstrated in **Figure 4**. In order to achieve the high throughput of particles separation, high flow velocity is desired. However, it has to be balanced between throughput and particle sizes to be separated in specific applications.

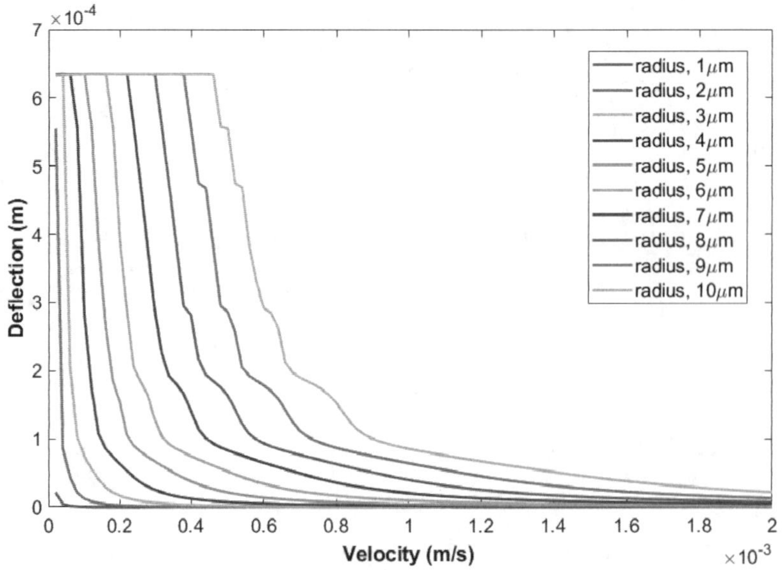

Figure 4. Particle deflection varies with respect to the change of flow velocity in the channel for different particles size.

In the last, the effect of the buffer solution viscosity is evaluated, by increasing the viscosity of the buffer solution, the particles deflection become larger, as shown in the **Figure 5**. In certain applications, under the same dimension of the microfluidic chips, similar sizes of particles could also be separated through adjusting the buffer solution viscosity.

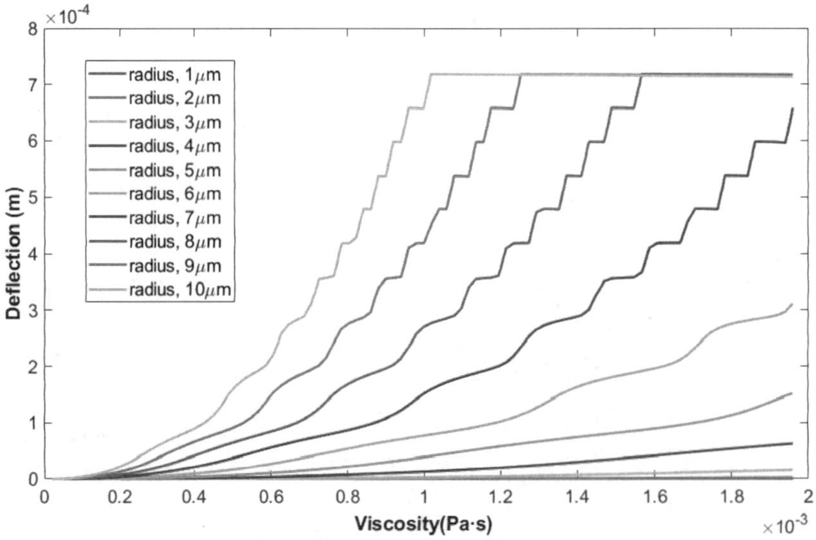

Figure 5. Particle deflection varies with respect to the change of buffer solution viscosity in the channel for different particles size.

4. Conclusion

In order to study the effect of multiple parameters to different size of particles deflection, the theoretical analysis is conducted. By increasing the tilted angle and acoustic pressure within a proper range, the particles will deflect further. At the same time, applying the lower acoustic wavelength and flow velocity will also provide the larger deflection of particles. To be noted that, increasing viscosity of the buffer solution would help to separate the particles at the small difference. The results are useful for the further design of particles separation devices.

References

[1] Cao X and Bian J 2019 Supersonic separation technology for natural gas processing: A review. *Chemical Engineering and Processing-Process Intensification* **136** 138-51

[2] Felsing, S, Kochleus C, Buchinger S, Brennholt N, Stock F and Reifferscheid G 2018 A new approach in separating microplastics from environmental samples based on their electrostatic behavior. *Environmental Pollution* **234** 20-8

[3] Zhang H and Lyden D 2019 Asymmetric-flow field-flow fractionation technology for exomere and small extracellular vesicle separation and characterization *Nature protocols* **14**(4) 1027-53

[4] Zhou, B, Xu K, Zheng X, Chen T, Wang J, Song Y, Shao Y, and Zheng S 2020 Application of exosomes as liquid biopsy in clinical diagnosis *Signal transduction and targeted therapy* **5**(1) 1-14

[5] Chen, X, Ren Y, Liu W, Feng X, Jia Y, Tao Y and Jiang H 2017 A Simplified Microfluidic Device for Particle Separation with Two Consecutive Steps: Induced Charge Electro-osmotic Prefocusing and Dielectrophoretic Separation *Analytical Chemistry* **89**(17) 9583-92

[6] Simon G, Pailhas Y, Andrade M A B, Reboud J, Marques-Hueso J, Desmulliez M P Y, Cooper J M, Riehle M O and Bernassau A L 2018 Particle separation in surface acoustic wave microfluidic devices using reprogrammable, pseudo-standing waves *Appl. Phys. Lett.* **113**(4) 044101

[7] Ghadami S, Kowsari-Esfahan R, Saidi M S, and Firoozbakhsh K 2017 Spiral microchannel with stair-like cross section for size-based particle separation *Microfluidics and Nanofluidics* **21**(7) 115

[8] Li Y, Zhang H, Li Y, Li X, Wu J, Qian S, and Li F 2018 Dynamic control of particle separation in deterministic lateral displacement separator with viscoelastic fluids *Scientific reports* **8**(1) 1-9

[9] Bayareh M 2020 An updated review on particle separation in passive microfluidic devices *Chemical Engineering and Processing-Process Intensification* **153** 107984

[10]Ding X, Peng Z, Lin S-C S, Geri M, Li S, Li P, Chen Y, Dao M, Suresh S and Huang T J 2014 Cell separation using tilted-angle standing surface acoustic waves *Proc. of the National Academy of Sciences* **111**(36) 12992

[11]Ni Z, Yin C, Xu G, Xie L, Huang J, Liu S, Tu J, Guo X and Zhang D 2019 Modelling of SAW-PDMS acoustofluidics: physical fields and particle motions influenced by different descriptions of the PDMS domain *Lab on a Chip* **19**(16) 2728-40

[12]Lenshof A and Laurell T 2010 Continuous separation of cells and particles in microfluidic systems. *Chemical Society Reviews* **39**(3) 1203-17

An Evaluation Method of Vehicle Functional Safety Controllability Based on Driver-in-the-Loop Platform

Ronghuang Liu[a], Ke Lu, Yuan Zhu and Zhihong Wu[b,*]

College of Automotive Studies, Tongji University, Shanghai 201800, P.R.China

E-mails: [a]liu_rh@foxmail.com; [*,b]zhihong.wu@tongji.edu.cn

Abstract. Existing vehicle functional safety controllability evaluation methods are difficult to implement in practice and lack quantitative evaluation basis. In this paper, we propose a simulation-based scenario evaluation method, which can achieve verifiable, quantitative, and generalized functional safety controllability evaluation. The method focuses on the driver's reaction model. By analyzing the driver's behavior in emergency braking scenarios, we propose the "remaining collision distance" metrics as the controllability criterion. To investigate the reaction time of drivers in different emergencies, we experimented in a driver-in-the-loop platform and classified the scenario factors that might affect controllability. The results show that our method is feasible to evaluate the controllability of vehicle-functional safety in emergency braking scenarios.

1. Introduction

In recent years, ADAS(Advanced Driving Assistance Syste) has been developed rapidly. Conducive to the advancement of autonomous driving technology, excellent object perception and DPC(Decision, Planning & Control) technology can better avoid traffic accidents. ADAS sensors(camera, lidar, radar, etc.) monitor road situation and detect dangerous targets for DPC module to response. In most cases, the rationality of countermeasure depends on the current state of the driver. For example, if the driver is distracted, advance warning may be appropriate. If the driver has been focusing on the driving situation, this kind of advance warning may be too early. This phenomenon is often referred to as the "warning dilemma" [1]. To ensure sufficient reliability, the system needs to be consistent with the driver on the perceptual level in corner cases. Even so, the driver's decision and the decision of DPC may diverge, resulting in unexpected reactions from the driver. In order to detailed evaluate the vehicle functional safety controllability of ADAS, we propose a simulation-based scenario evaluation method, which can achieve verifiable, quantifiable, and generalized functional safety controllability evaluation. Our studies have established driver behavior model and a controllable criterion to make the evaluation more reliable and precise.

2. Relate work

2.1. Functional safety standards

ISO 26262 international standard was established in 2011 to set requirements for the safety process of electronic and electrical systems in the automotive industry. In the "hazard analysis and risk assessment ", the potential failures, the resulting hazards and their causes must be determined. In order to assess the risk, it is necessary to evaluate the exposure (E), controllability (C) and severity (S) of the hazard, and divide it into different stages (for example, C1—simply controllable to C4—difficult to control or uncontrollable) [2]. Based on these levels, ASIL(Automotive Safety Integrity Level) can be derived according to Table 1.

Table 1. ASIL Determination Matrix[1]

Severity	Exposure	Controllability		
		C1	C2	C3
S1	E1	QM	QM	QM
	E2	QM	QM	QM
	E3	QM	QM	A
	E4	QM	A	B
S2	E1	QM	QM	QM
	E2	QM	QM	A
	E3	QM	A	B
	E4	A	B	C
S3	E1	QM	QM	A
	E2	QM	A	B
	E3	A	B	C
	E4	B	C	D

ASIL defines the safety requirements of functional software and hardware components. Take PMHF(Probabilistic Metric for random Hardware Failure) for example. ASIL-B requires the PMHF index to be 100FIT, which is equivalent to a safety goal failure of about 1142 years of continuous work. ISO 26262 applies to the functional safety of electronic and electrical systems in vehicles weighing less than 3.5 t. However, its applicability to unexpected reactions is questionable [3]. Strictly speaking, the method described in ISO 26262 is not necessarily suitable for the unexpected response of ADAS. In most cases, the system works within its specification, and the failure is caused by the difference between the driver's condition assessment and the system's condition assessment. For example, underestimation of driving emergencies due to driver distraction. Such unexpected reactions are difficult to detect.

2.2. Existing methods

According to the ISO 26262, the system's exposure to specific situations and the controllability of events caused by unexpected reactions are two factors that affect system development.

Considering that the driver's awareness of the system's functional safety and controllability will directly affect the judgment of takeover behavior, code of practice of ISO 26262 [4] recommends experiments with naive drivers. However, the kind of experiment is very time-consuming. To reach the C2 level of assessment, at least 20 valid trials are required, and no single uncontrollable event occurs. Detailed statistical analysis shows that in 20 trials, the probability of success of this method is quite low [3].

[5] describes the controllability research based on the dynamic driving simulator, using the collision/no-collision criteria recommended in code of practice. [6] use subjective and objective value evaluation methods to evaluate the controllability of adaptive cruise control based on public transportation experiments. In order to supplement these technologies and improve the verifiability and adaptability of the evaluation methods , we propose a verifiable, quantitative and generalized controllability evaluation method.

3. Controllability evaluation method

In order to evaluate the controllability of emergency scenarios, a quantitative controllability criterion is needed. Code of practice[4] nominally defines the standard of controllability. But it depends on whether the reaction of the drivers can avoid the collision. For the actual controllability test experiments, a more detailed and quantifiable criterion is needed.

Generally speaking, the risk level of driving scenarios is from low to high, and the occurrence of emergencies will increase a certain risk. If the driver does not take countermeasures, the risk factor will increase accordingly until the collision event occurs.

The controllability of the evaluation scenario needs to meet at least two experimental requirements. First of all, there are unpredictable emergencies in the scene, and the risk of collision will greatly increase when the event occurs. Secondly, the driver has the ability to respond to emergencies and has a certain margin to take over the vehicle, otherwise the controllability will be zero in any case.

According to the second requirement, the driver's limit takeover margin should be measurable. Therefore, we propose the "remaining collision distance" metrics, which is defined by the distance between the position where the vehicle is bound to collide and the position where the driver starts to take over, and the relative speed of the vehicle. It describes how much distance the driver has left to avoid a collision after the driver responds to the emergency. If the value is negative, it means that the driver will not be able to prevent the collision; if the value is positive, the corresponding remaining collision distance can be calculated by bringing in the response time data of different naive drivers, so as to further get the proportion of drivers can avoid collision behavior in this scene.

3.1. Driver behavior model

For the analysis and modeling of driver response behavior, a time-based view is needed. Manual processing and action parameters are usually time-based. For example, take over time, reaction time, foot movement time, etc. [6].

When there is an accident such as jaywalking, if ADAS cannot handle it, the system must automatically issue a warning until the driver's attention returns to the driving task, hands on the steering wheel, and be prepared to take over the control of the vehicle. The time consumed in this stage is called TOR-time(take over request time). It depends on the driver's reaction performance, the suddenness of the event and the complexity of the scene

[7], [8] use the advance warning time as the independent variable to study the TOR-time distribution of subjects. Therefore, we take the in-vehicle scene where the driver is operating the intelligent vehicle system as an example, and take the average TOR-time under the 5s advance warning time in [7] as our precondition.

Then it still needs some PR-time(perception reaction time)for the driver to comprehend the current scene and put the foots on the brake pedal. PR-time reflects the driver's reaction performance, and there are some quantitative methods to evaluate reaction performance. [9] estimate the relative speed and distance between the vehicle and the target based on the change in the size of the object in the field of view, and further derive the driver's estimated limit safe distance, i.e. the probability of vehicle controllability at this distance is greater than 50%. [10], [11] studied the influence of different scene factors on the driver's emergency braking reaction time. The driver reaction time collected by these studies can be used in most scenarios, but the paper is based on experiments in a specific scenario, i.e. jaywalking. The data provided by [10], [11] is not completely applicable, so we use DYNA4, a latest virtual driving simulation platform, to build a DIL(driver-in-the-loop) platform collects the data of naive driver's reaction time.

After the driver depresses the brake pedal, the vehicle will not decelerate immediately. The brake signal is transmitted to the brake system, and the hydraulic brake starts to work. The time required at this stage is defined as BSR-time (brake system reaction time).

Then the vehicle gradually slow down until it stops and this the time required is called BK-time(braking time). The time view of the driver behavior model is shown in Fig. 1:

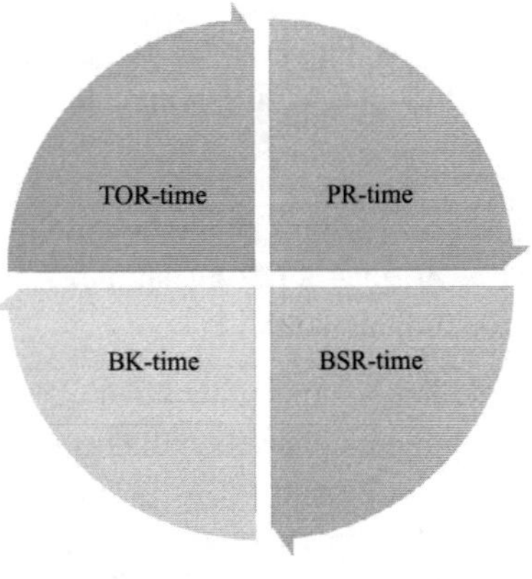

Figure 1. Time view of driver behavior model.

3.2. Controllability criterion

Our simulation driving scene is that VUT(Vehicle Under Test) keeps driving on a straight road, and a jaywalker suddenly appears in the blind area of the front visual field. In such emergency situations, most drivers will take emergency braking operations, but when the driver does not react in time, a collision will occur. By analyzing the driver's behavior model, the driving distance of the vehicle in each time period can be calculated.

Before the driver takes over the vehicle, VUT maintains a uniform linear motion during the TOR-time period, and the driving distance as follows:

$$S_{TOR} = v_{init} * t_{TOR} \tag{1}$$

where v_{init} is the initial velocity before the car is taken over. According to the TOR-time distribution of the subjects under the 5s advance warning time group in [7], we set t_{TOR} to 1.45s as mean TOR-time.

Then in the PR-time and BSR-time phases, VUT cannot slow down immediately until the brake signal is sent to the system and executed. The driving distances generated in the corresponding time phase are S_{PR} and S_{BSR}:

$$S_{PR} = v_{init} * t_{PR} \tag{2}$$

$$S_{BSR} = v_{init} * t_{BSR} \tag{3}$$

where t_{PR} is the data we need to collect by experiment on subjects through our DIL platform, t_{BSR} means the delay time of the braking system and $t_{BSR} = 0.1s$. So before the brake system is activated, VUT maintains a uniform linear motion and the distance travelled by VUT is S_{ULM}:

$$S_{ULM} = S_{TOR} + S_{PR} + S_{BSR} \tag{4}$$

After the hydraulic brake is activated, the vehicle starts to decelerate to a stop. Consider that the pressure output response of the hydraulic brakes is very fast, ignoring the time required to lift the pressure. The braking distance S_{BK} is calculated by the following formula:

$$S_{Bk} = \frac{v_{init}^2}{2*a_{dec}} \tag{5}$$

where a_{dec} is the deceleration under maximum braking pressure. In our simulation system we set a_{dec} to 8 m/s^2.

The sum of driving distances produced at each time stage of the driver's behavior model is S_{TOTAL}:

$$S_{TOTAL} = S_{ULM} + S_{BK} \tag{6}$$

And the remaining collision distance S_{RCD} can be calculated by the following formula:

$$S_{RCD} = S_{TARGET} - S_{TOTAL} \tag{7}$$

where S_{TARGET} is the distance between the location of VUT and the location where target(jaywalker) suddenly appears. If the value of S_{RCD} is negative, the collision will occur and the current scene is not controllable. On the contrary, if the value of S_{RCD} is positive, it means the scene is controllable.

4. DIL simulation test platform

In order to examine the controllability of vehicle functional safety in jaywalking scenario and solve the problems such as the limitation of weather and the difficulty of road test on real car and so on, we developed a DIL simulation test platform based on software such as DYNA4, Matlab, Simulink and hardware such as virtual cockpit.

4.1. Framework of DIL simulation test platform

Our DIL simulation test platform is composed of a software platform and a hardware platform. The software platform with DYNA4, Matlab and Simulink can help us quickly build virtual driving scenarios and configure vehicle and sensor models flexibly. The hardware platform consists of a driver control system constructed of Logitech equipments such as the cockpit, steering wheel, accelerator pedal, brake pedal, shifter and clutch. Fig. 2 show the framework of the platform.

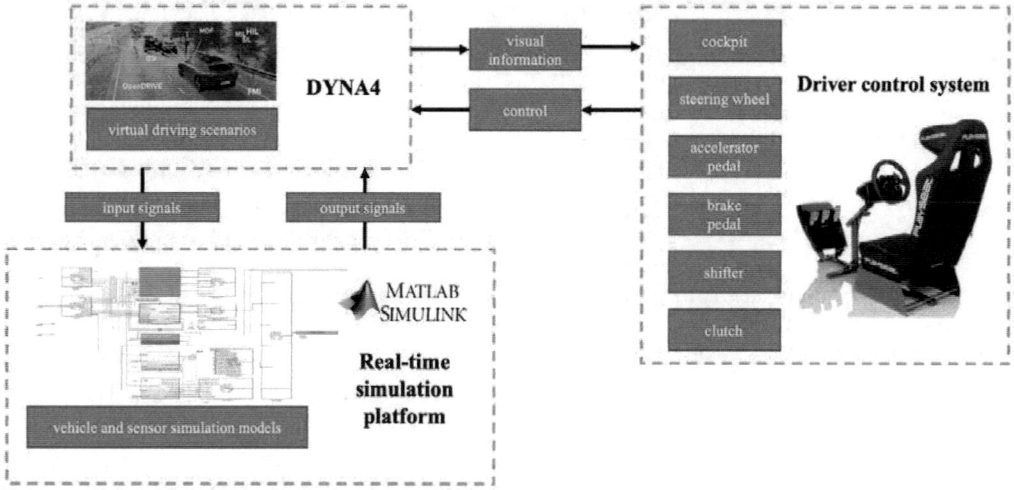

Figure 2. Driver-in-the-Loop system.

The system has the functions of simulated driving, real physics engine to build virtual driving scene, multi-angle playback scene, real-time simulation, data acquisition and so on. The rich scenario model library and first-person perspective provide the subjects with an immersive driving experience close to reality, which makes the driver's PR-time data collected by the experiment more reliable. For the virtual driving scenarios, see Fig. 3.

Figure 3. Virtual driving scenarios.

4.2. Virtual driving scenarios design

There are many scenario parameters to consider when designing a virtual driving scenario, and they can be classified into three categories[12]:

- environment: e.g. weather, light intensity, road visibility, road friction coefficient, traffic congestion, etc.
- driver: e.g. standards of driving, driving experience, instant reflexes, attention level, fatigue level, etc.
- vehicle: e.g. car dimension, blind zone range, braking performance, system signal transmission rate, velocity, etc.

Table 2 shows the influence of different scenario categories on controllability:

Table 2. Scenario Parameters

Category	Parameter	Influence
Environment	Weather	Road friction coefficient and road visibility
	Lighting conditions	Comprehension of current scene
	Road visibility	Drvier's perception reaction
Driver	Fatigue level	Driver's attetion, concentration and operation
	Driving experience	Sight, hearing, speed judgment
	Instant reflexes	The ability to control the vehicle
Vehicle	Car dimension	Blind area range
	Braking performance	Braking time and braking distance
	velocity	Judgment and decision making

Journal of Physics: Conference Series **2173** (2022) 012054 doi:10.1088/1742-6596/2173/1/012054

Jaywalking is one of the most frequent and dangerous traffic accidents. When pedestrian crosses a roadway where regulations do not permit doing so, the driver cannot see until he or she runs out of the blind spot. Therefore, this is a suitable scenario to collect different subjects' PR-time data and to verify the accuracy and feasibility of the controllability assessment method proposed in this paper. We designed eight simulation scenarios with velocity, lighting conditions and S_{TARGET} as variables while keeping the other parameters consistent. The specific settings for the simulation scenarios are shown in Table 3.

Table 3. Simulation scenarios

Scenario	Velocity(km/h)	Lighting conditions	S_{TARGET}(m)
1	20	daytime	18
2	40	daytime	36
3	60	daytime	54
4	80	daytime	72
5	20	night	18
6	40	night	36
7	60	night	54
8	80	night	72

S_{TARGET} directly affects the driver's understanding of the current scene and the execution of the decision(e.g. brake or swerve). When the S_{TARGET} setting is too small, the driver is too late to react and hits the pedestrian, and when the S_{TARGET} setting is too large, the driver does not necessarily brake even if the pedestrian appears in the middle of the road. So we need to set the S_{TARGET} value reasonably to ensure that the driver's PR-time data can be reliably collected.[13] stipulates that the TTC minimum value of the vehicle collision warning system should be greater than 2.1s, and when the TTC is greater than 4.0s, the vehicle collision warning system can suppress or delay the alarm. We set the TTC warning value to 3.2s and set the value of S_{TARGET} by formula (8):

$$S_{TARGET} = v_{init} * TTC \tag{8}$$

The 54 subjects were from Tongji University and between 20 and 40 years old. 13 subjects were female(24%) and 41 male(76%). Only a small percentage of the subjects had experienced driving simulators before, and the vast majority were unfamiliar with the characteristics of simulated driving. Before the simulation test begins, we'll do some demonstrations so that the subjects understand the purpose of the experiment. In order to ensure the authenticity and reliability of the experimental data, we used random test method, each subject will be randomly tested in 8 different virtual driving scenarios, and finally collected a total of 432 sets of PR-time data in 8 sets of scenarios.

5. Analysis

We analyzed the subject's PR-time and found that the data obeys the logarithmic normal distribution. Fig. 4 shows the Q–Q plots(Quantile-Quantile plot) of per set of scenario's data(54), and the Q-Q plot of all data(432) is shown in Fig. 5. The percentiles(50%, 90%, 99%) of PR-time are also described in Table 4.

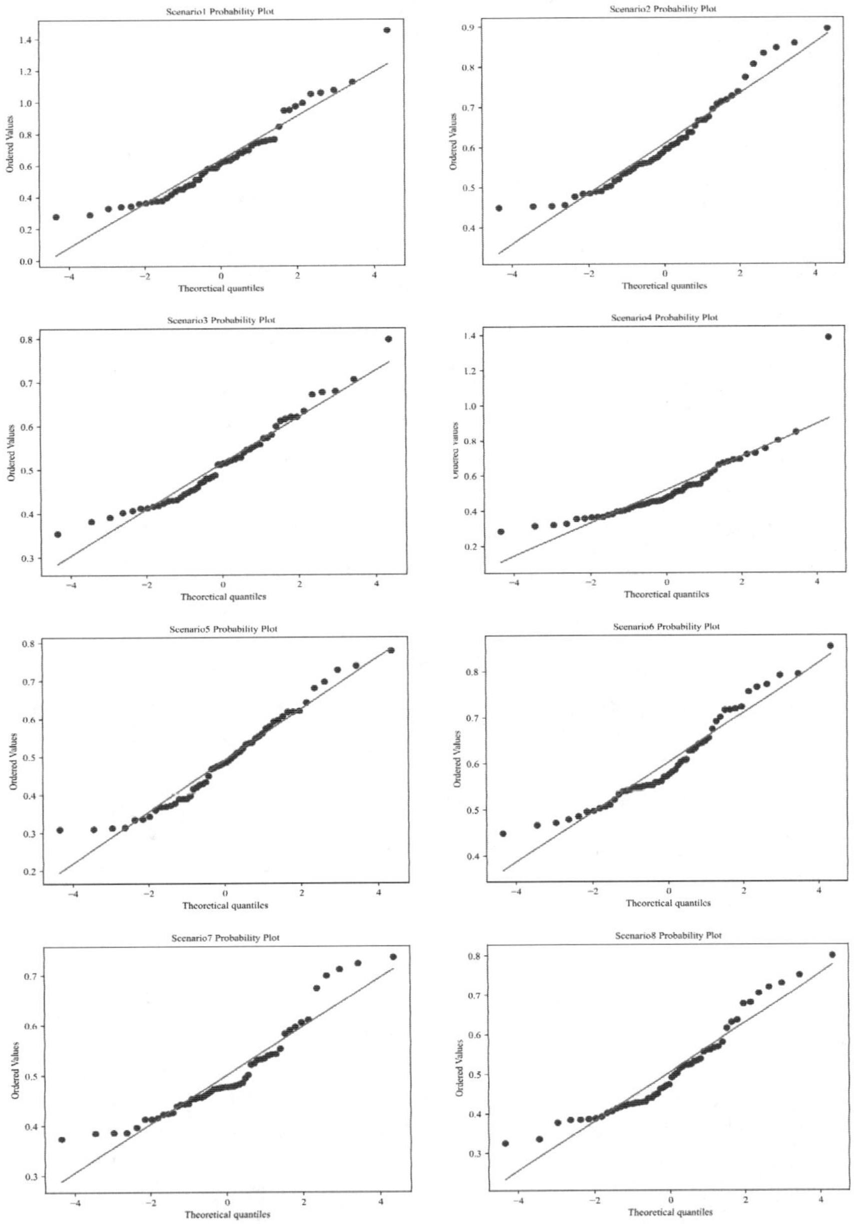

Figure 4. Q-Q plots of per set of scenario's data.

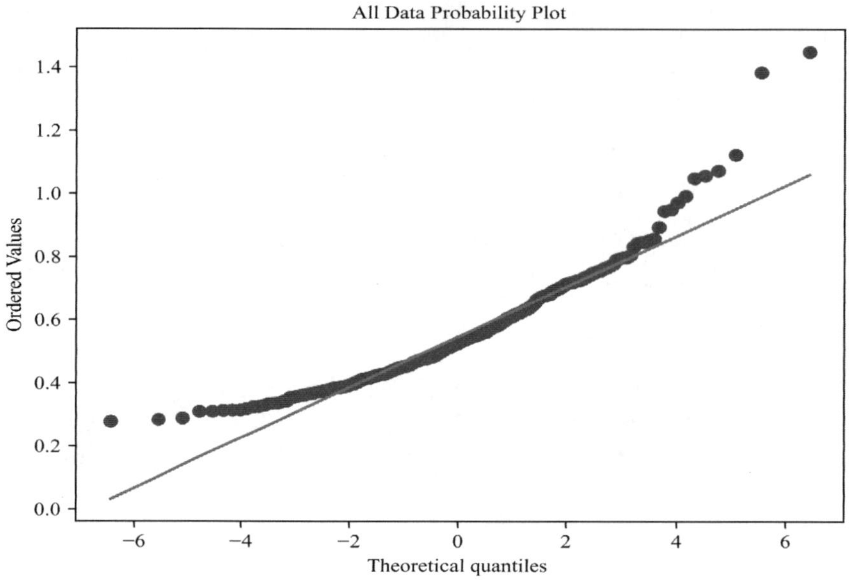

Figure 5. Q-Q plot of all data.

Table 4. The percentiles of PR-time

Scenario	99%(s)	90%(s)	50%(s)
1	1.277	0.987	0.616
2	0.875	0.763	0.592
3	0.748	0.630	0.513
4	1.100	0.710	0.473
5	0.757	0.635	0.483
6	0.822	0.746	0.574
7	0.728	0.609	0.477
8	0.773	0.680	0.482
all data	1.055	0.725	0.527

S_{RCD} can be calculated according to the driver behavior model and the controllability criteria formula, results are shown in Table 5. In order to obtain the controllability assessment results, we first define the controllability level: when more than 99% of the subjects can avoid collisions in the experiments, the controllability level is C1, when more than 90% of the subjects can avoid collisions, the controllability level is C2, and when 90% of subjects are unable to avoid collisions, the controllability level is C3. The specific controllability assessment results are shown in Table 6.

Table 5. Remaining collision distance

Scenario		S_{ULM}(m)	S_{BK}(m)	S_{TOTAL}(m)	S_{TARGET}(m)	S_{RCD}(m)
1	99%	15.71	1.93	17.64	18	0.36
	90%	14.09		16.02		1.98
2	99%	26.94	7.72	34.66	36	2.34
	90%	25.70		33.42		2.58
3	99%	38.30	17.36	55.66	54	-1.66
	90%	36.33		53.69		0.31
4	99%	58.89	30.86	89.75	72	-16.25
	90%	50.22		81.08		-9.08
5	99%	12.82	1.93	14.75	18	3.25
	90%	12.14		14.07		3.97
6	99%	26.36	7.72	34.08	36	1.92
	90%	25.51		33.23		2.77
7	99%	37.97	17.36	55.33	54	-1.33
	90%	35.98		53.34		-0.66
8	99%	51.62	30.86	82.48	72	-10.48
	90%	49.56		80.42		-8.42

Table 6. Controllability assessment results

Scenario		S_{RCD}(m)	Collision	Controllability level
1	99%	0.36	No	C1
	90%	1.98	No	
2	99%	2.34	No	C1
	90%	2.58	No	
3	99%	-1.66	YES	C2
	90%	0.31	No	
4	99%	-16.25	YES	C3
	90%	-9.08	YES	
5	99%	3.25	No	C1
	90%	3.97	No	
6	99%	1.92	No	C1
	90%	2.77	No	
7	99%	-1.33	YES	C3
	90%	-0.66	YES	
8	99%	-10.48	YES	C3
	90%	-8.42	YES	

From Table 5, 6, it can be seen that the velocity is the key factor influencing driver's reaction performance. When the velocity up to a certain threshold, the controllability level rises sharply and the vehicle becomes difficult to control. At the meantimethe lighting conditions directly influences driver's perceptual ability and attention, subjects were more focused at night, as can be seen from the PR-time results of Table 4.

6. Conclusion

(1)We proposed an evalution method of vehicle functional safety controllability baed on DIL platform. The method consists of driver behavior model, controllability criterion and DIL simulation test platform. We applied the method to the virtual jaywalking scenarios set which we built in DYNA4, and got controllability assessment results as shown in Table 5 and 6.

(2)The quantitative controllability criterion "remaining collision distance" achieve collision prediction and assessment of controllability. For ADAS, it helps optimize the functional safety of the vehicle. For drivers, it helps ADAS dynamically adjust advance warning time based on its reaction performance. And because the criterion is established in based of the driver behavior model of open road and the vehicle environment, it has better adaptability.

(3)The DIL simulation test platform solved the problems of weather limitation and difficulty of road test on real car and so on. The virtual driving scenarios and simulator can give subjects a fully immersive driving experience. The perception reaction time collected through the platform obeys the logarithmic normal distribution.

References

[1] Abe, G., & Richardson, J. (2006). Alarm timing, trust and driver expectation for forward collision warning systems. Applied ergonomics, 37(5), 577-586.
[2] ISO. 26262:2018 Road vehicles – Functional safety. 2018
[3] Weitzel, A. (2013). Objective controllability assessment for unintended ADAS reactions. In Automotive Systems Engineering (pp. 135-145). Springer, Berlin, Heidelberg.
[4] Knapp, A., Neumann, M., Brockmann, M., Walz, R., & Winkle, T. (2009). Code of practice for the design and evaluation of ADAS. Preventive and Active Safety Applications, eSafety for road and air transport, European Commission Project, Brüssel.
[5] DeGuzman, C. A., Hopkins, S. A., & Donmez, B. (2020). Driver takeover performance and monitoring behavior with driving automation at system-limit versus system-malfunction failures. Transportation research record, 2674(4), 140-151.
[6] Green, M. (2000). " How long does it take to stop?" Methodological analysis of driver perception-brake times. Transportation human factors, 2(3), 195-216.
[7] Gold, C., Damböck, D., Lorenz, L., & Bengler, K. (2013, September). "Take over!" How long does it take to get the driver back into the loop?. In Proceedings of the human factors and ergonomics society annual meeting (Vol. 57, No. 1, pp. 1938-1942). Sage CA: Los Angeles, CA: Sage Publications.
[8] Radlmayr, J., Gold, C., Lorenz, L., Farid, M., & Bengler, K. (2014, September). How traffic situations and non-driving related tasks affect the take-over quality in highly automated driving. In Proceedings of the human factors and ergonomics society annual meeting (Vol. 58, No. 1, pp. 2063-2067). Sage CA: Los Angeles, CA: Sage Publications.
[9] Krueger, G. P. (2011). Book Review: Forensic Aspects of Driver Perception and Response.
[10]Pala, S. (2009). Book review: Automotive Lighting and Human Vision.
[11]Zheng, G., Zu, Z., & Kong, Z. (2021). The Collision Avoidance Strategy of Automatic Emergency Braking System Considering the Response Time of the Driver. Journal of Chongqing University of Technology(Natural Science), 34(12), 45-52.
[12]Shu, H., Yuan, K., Xiu, H., Xia, Q., & He, S. (2019). Construction of Basic Test Scenarios of Automated Vehicles. China Journal of Highway and Transport(11), 245-254. doi:10.19721/j.cnki.1001-7372.2019.11.025.
[13]GB/T 33577-2017, Intelligent Transportation Systems-Forward Vehicle Collision Warning System-Performance Requirements and Test Procedures[S].

Research on Position Sensorless Vector Control of Synchronous Reluctance Motor Based on High Frequency Injection Method

Dengke Li*, Yukun Liu

Shanghai University, School of Mechatronic Engineering and Automation, Shanghai, China

*corresponding author's e-mail: 1627140925@qq.com

Abstract. In order to achieve the high-performance control position sensorless vector control of synchronous reluctance motor (SynRM), this paper investigated its running-up performance and capability at zero or low speed range. The contribution of this paper mainly includes two parts. Firstly, it takes cross-saturation effect into consideration, Secondly, the paper improves the control stiffness and reliability under low-speed sensorless operation condition by adopting the pulsating sinusoidal high frequency voltage injection snesorless method. By injecting high-frequency voltage excitation signal, the rotor position is estimated from the high-frequency response current. The simulation results verified the feasibility of the control algorithm, and fulfilled the running-up of the SynRM at zero speed, the capability of stability is evaluated via simulation results.

Keywords. Synchronous reluctance motor; Sensorless control; Low speed start; High frequency voltage injection method

1. Introduction

SynRM have the advantages of simple structure, low material cost, high efficiency, and wide speed range, and its research is becoming more and more popular [1]. SynRM rely on stator windings for excitation. There are no permanent magnets on the rotor. They are only laminated with special-shaped high-permeability materials. With its specific rotor structure, it generates high saliency and generates high reluctance torque [2]. SynRM are expected to be applied in the field of electric vehicles [3].

The vector control strategy of SynRM is relatively mature. The research on high-performance control strategies of SynRM began in the 1990s [4]. By adjusting the currents of the d and q axis, a variety of control methods can be easily realized, such as Maximum Torque Per-Ampere (MTPA) control, and maximum power factor. Control (Maximum Power Factor Control, MPFC), Maximum Efficiency Control (MEC), etc. [5-7]. The vector control strategy has higher control accuracy, but the torque response and speed response are slower. In the vector control of synchronous reluctance motors, mechanical sensors are usually used to detect the rotor position, but the use of mechanical sensors will increase costs and reduce reliability [8]. Position sensorless control methods can be roughly divided into two categories. One is to use the fundamental wave model of the motor to directly or indirectly extract the rotor position information from the extended back EMF [9-11]. Because the fundamental wave electrical signal of the motor is very weak at low speeds, this type of method is suitable Applied in the range of medium and high speed. The second type is to use the saliency of the motor's physical structure to inject high-frequency voltage excitation, estimate the rotor position from the high-frequency response current, and apply position identification at zero speed and low speed [12].

According to the different injection methods, the high frequency signal injection method can be divided into rotating high frequency injection, pulse high frequency injection, square wave signal injection and so on. The basic principle of the rotating high-frequency voltage injection method is to superimpose a three-phase balanced high-frequency voltage excitation on the fundamental excitation, then detect the corresponding current response generated in the motor, and obtain the rotor position information through a specific signal processing process. The vibration high-frequency voltage signal injection method only injects high-frequency sinusoidal voltage signals on the d axis in the estimated synchronous rotating d-q coordinate system.

This paper establishes the SynRM mathematical model under high-frequency excitation, studies the position tracking technology based on the pulsating high-frequency voltage signal injection method, and realizes the zero-speed and low-speed start and stable operation of the SynRM. Finally, the feasibility of the scheme is verified by simulation.

2. Mathematical Model of SynRM

The mathematical model of the SynRM is similar to that of the permanent magnet synchronous motor. Because there is no permanent magnet on the rotor, there is no permanent magnet flux in the d-axis flux linkage of the SynRM. The mathematical model of the SynRM is established in the d-q axis coordinate system, and its voltage equation is:

$$\begin{cases} u_d = R_s i_d + \dfrac{d\psi_d}{dt} - \omega_e \psi_q \\ u_q = R_s i_q + \dfrac{d\psi_q}{dt} + \omega_e \psi_d \end{cases} \tag{1}$$

Where, u_d, u_q are the components of the stator voltage vector on the d-q axis; i_d, i_q are the components of the stator current vector on the d-q axis; ψ_d, ψ_q are the components of the stator flux vector on the d-q axis; ω_e is the electrical angular velocity, R_s is the stator resistance.

The motor stator flux linkage can be expressed as:

$$\begin{cases} \psi_d = L_d i_d \\ \psi_q = L_q i_q \end{cases} \tag{2}$$

SynRM runs on reluctance torque, and its electromagnetic torque can be expressed as:

$$T_e = \frac{3}{2} p(\psi_d i_q - \psi_q i_d) \tag{3}$$

After substituting equation (2) into equation (3), the electromagnetic torque can be expressed as:

$$T_e = \frac{3}{2} p(L_d - L_q) i_d i_q \tag{4}$$

Where, p is the number of pole pairs of the motor, and L_d and L_q are the d and q axis inductances respectively.

The motion equation of the synchronous reluctance motor is:

$$T_e - T_L - B\omega_m = J\frac{d\omega_m}{dt} \tag{5}$$

Where, T_L is the load torque, B is the damping coefficient, J is the moment of inertia, and ω_m is the mechanical angular velocity of the motor.

3. Rotor position detection based on pulsating high frequency injection method

Due to the salient pole characteristics of SynRM, high-frequency signal injection method is used to observe the position angle at static and low speed. Its central idea is to inject high-frequency signals above the fundamental frequency of the motor, and analyze the response based on the saliency of the motor after injection. The purpose of tracking the position of the salient pole is achieved by adopting some control algorithms, so as to extract the motor speed and position information at the current moment.

The pulsating high-frequency voltage injection method is mainly to inject a sinusoidal high-frequency voltage signal into the direct axis to saturate the d-axis inductance, and accurately estimate the rotor position by analyzing the high-frequency model of the motor. This method is only suitable for stationary or low speeds.

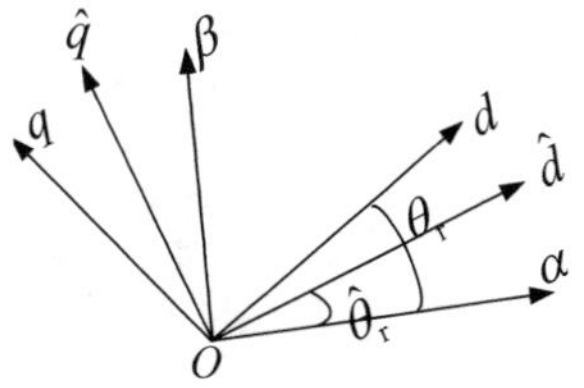

Figure 1. The relationship between the actual rotor and the estimated rotor synchronous rotation coordinate system

Figure 1 is the actual synchronous rotating coordinate system and the estimated position relationship diagram. The high-frequency signal $u_{in}\cos\omega_{in}t$ is injected into the d axis, where u_{in} represents the amplitude of the injection voltage, and ω_{in} is the injection frequency. Generally, 10-40 times the fundamental frequency of the motor is selected.

$$\begin{cases} \hat{u}^{r}_{din} = u_{in}\cos\omega_{in}t \\ \hat{u}^{r}_{qin} = 0 \end{cases} \tag{6}$$

When injecting high frequencies, because the injected voltage frequency is much higher than the fundamental frequency of the SynRM, the resistance voltage drop can be ignored. At this time, the voltage equation of SynRM can be simplified as:

$$\begin{cases} u_{din} \approx L_{d}\dfrac{di_{din}}{dt} \\ u_{qin} \approx L_{q}\dfrac{di_{qin}}{dt} \end{cases} \tag{7}$$

The current response corresponding to the high-frequency injection voltage can be written as:

$$\begin{bmatrix} \dfrac{d\hat{i}^{r}_{din}}{dt} \\ \dfrac{d\hat{i}^{r}_{qin}}{dt} \end{bmatrix} = \begin{bmatrix} \cos(\Delta\theta_r) & \sin(\Delta\theta_r) \\ -\sin(\Delta\theta_r) & \cos(\Delta\theta_r) \end{bmatrix}^{-1} \begin{bmatrix} \dfrac{1}{L_d} & 0 \\ 0 & \dfrac{1}{L_q} \end{bmatrix} \begin{bmatrix} \cos(\Delta\theta_r) & \sin(\Delta\theta_r) \\ -\sin(\Delta\theta_r) & \cos(\Delta\theta_r) \end{bmatrix} \begin{bmatrix} \hat{u}^{r}_{din} \\ \hat{u}^{r}_{qin} \end{bmatrix} \tag{8}$$

Rewrite the above equation :

$$\begin{cases} \dfrac{d\hat{i}^{r}_{din}}{dt} = \dfrac{1}{L^2-\Delta L^2}\left[(L+\Delta L\cos(2\Delta\theta_r))\hat{u}^{r}_{din} + (\Delta L\sin(2\Delta\theta_r))\hat{u}^{r}_{qin}\right] \\ \dfrac{d\hat{i}^{r}_{qin}}{dt} = \dfrac{1}{L^2-\Delta L^2}\left[(\Delta L\sin(2\Delta\theta_r))\hat{u}^{r}_{din} + (L-\Delta L\cos(2\Delta\theta_r))\hat{u}^{r}_{qin}\right] \end{cases} \tag{9}$$

Where:

$$\begin{cases} L = (L_d + L_q)/2 \\ \Delta L = (L_d - L_q)/2 \end{cases} \tag{10}$$

Substituting formula (6) into formula (9) :

$$\begin{cases} \hat{i}_{\mathrm{din}}^{\mathrm{r}} = \dfrac{u_{\mathrm{in}} \sin \omega_{\mathrm{in}} t}{\omega_{\mathrm{in}} (L^2 - \Delta L^2)} (L + \Delta L \cos(2\Delta\theta_{\mathrm{r}})) \\ \hat{i}_{\mathrm{qin}}^{\mathrm{r}} = \dfrac{u_{\mathrm{in}} \sin \omega_{\mathrm{in}} t}{\omega_{\mathrm{in}} (L^2 - \Delta L^2)} (\Delta L \sin(2\Delta\theta_{\mathrm{r}})) \end{cases} \tag{11}$$

Further calculation:

$$\hat{i}_{\mathrm{qin}}^{\mathrm{r}} \sin \omega_{\mathrm{in}} t = \frac{u_{\mathrm{in}} \Delta L \sin(2\Delta\theta_{\mathrm{r}})}{2\omega_{\mathrm{in}} (L^2 - \Delta L^2)} (1 - \cos 2\omega_{\mathrm{in}} t) \tag{12}$$

Through the first-order Low pass filter (LPF):

$$LPF\left\{\hat{i}_{\mathrm{qin}}^{\mathrm{r}} \sin \omega_{\mathrm{in}} t\right\} = \frac{u_{\mathrm{in}} \Delta L \sin(2\Delta\theta_{\mathrm{r}})}{2\omega_{\mathrm{in}} (L^2 - \Delta L^2)} = K_{\mathrm{si}} \sin(2\Delta\theta_{\mathrm{r}}) \tag{13}$$

When $\Delta\theta_r$ relatively small, the error is:

$$\varepsilon = K_{\mathrm{si}} \sin(2\Delta\theta_{\mathrm{r}}) = 2K_{\mathrm{si}} \Delta\theta_{\mathrm{r}} \tag{14}$$

Therefore, the speed estimated is:

$$\hat{\omega}_{\mathrm{r}} = K_{\mathrm{p}} \varepsilon + K_{\mathrm{i}} \int \varepsilon \mathrm{d}t \tag{15}$$

The electrical angle of the rotor is:

$$\hat{\theta}_{\mathrm{r}} = \int \hat{\omega}_{\mathrm{r}} \mathrm{d}t \tag{16}$$

The rotor position angle can be extracted from the shaft high-frequency current component, as shown in Figure 2.

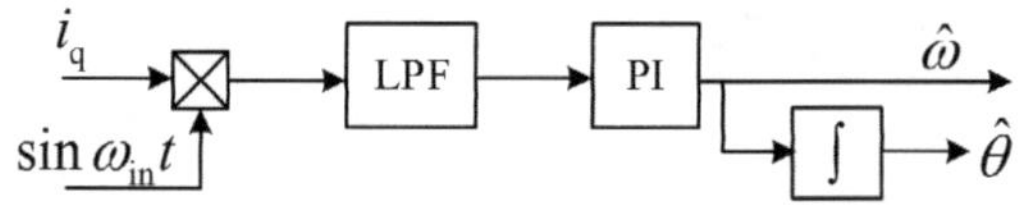

Figure 2. Rotor position extraction module of pulsating high frequency injection method

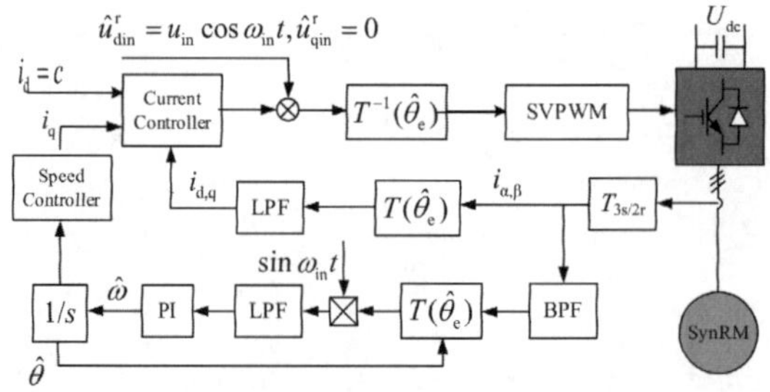

Figure 3. Control block diagram of pulsating high frequency voltage signal injection

Figure 3 shows the control block diagram of pulsed high frequency voltage signal injection. It extracts the position angle information by processing the q-axis current. The key is the choice of the filter, the interference signal is removed through the filter, and the useful signal is extracted.

4. Simulation results and analysis

In order to verify the research method of this article, the method is simulated and analyzed in the low-speed range under the MATLAB/Simulink environment. The parameters of the synchronous reluctance motor are shown in the following table. A 5.5kw two-pole motor is selected, and the d-axis and q-axis inductances change with the current.

Table 1. Parameters of SynRM

Parameters	Value
Rated speed(rpm)	1500
Rated voltage(V)	350
Rated power(kw)	5.5
Rated current(A)	12.8
Stator resistance(Ω)	0.50
Inertia(kg· m^2)	0.019
L_d (mH)	116.16-149.07
L_q (mH)	22.06-56.09
Pole pairs	2

The block diagram of the simulation analysis is shown in the figure below. The system is a typical double closed-loop control system with an outer speed loop and an inner current loop. The current PI control module is mainly to ensure that the control system can track the current values of the d-axis and q-axis stably, accurately and quickly, enhance the dynamic response of the system, limit the upper and lower boundaries, and suppress the noise of the current inner loop; The speed PI control module is mainly to ensure that the control system can track the given speed stably, accurately and quickly, and restrain the outer loop speed fluctuation and load disturbance. In the simulation, the frequency of the injected high-frequency voltage is set to 1000Hz, and the amplitude is 20V.

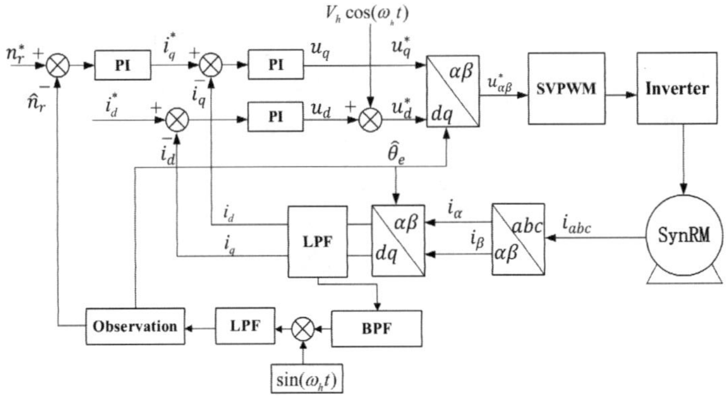

Figure 4. The block diagram of the position sensorless control system of the high-frequency pulsation injection method

Figures 5 and Figures 6 are the speed and its error waveforms from 0 r/min to 50r/min. The red color is the real speed and the other is the estimated speed. It can be seen from the figure that the real speed and the estimated speed waveform basically coincide. The speed estimation error is very small.

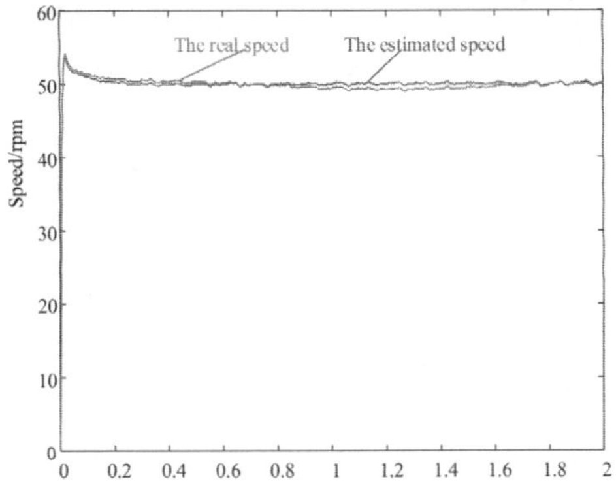

Figure 5. Actual speed and Estimated speed

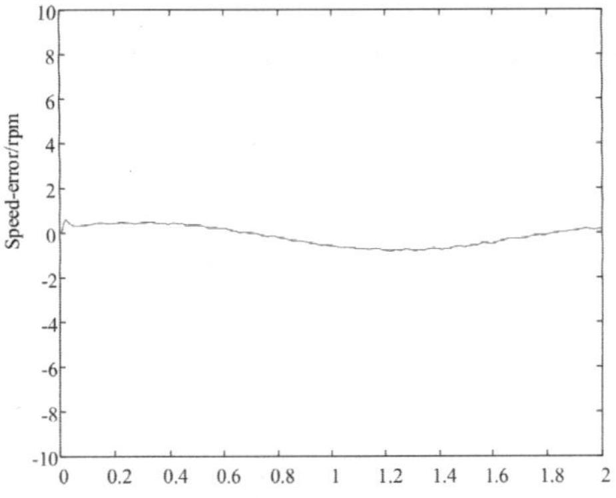

Figure 6. Speed error

Figure 7 and Figures 8 is the position and its error waveform. The red represents the real angle of the rotor, the other represents the estimated angle. It can be seen from the figure that the real angle and the estimated angle waveform basically coincide, and the angle estimation error is very small.

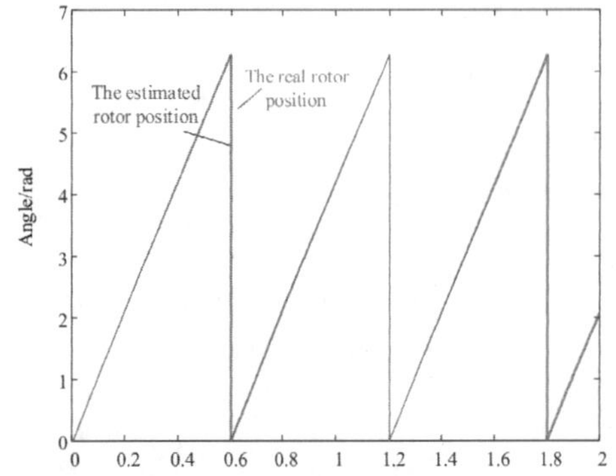

Figure 7. Rotor position

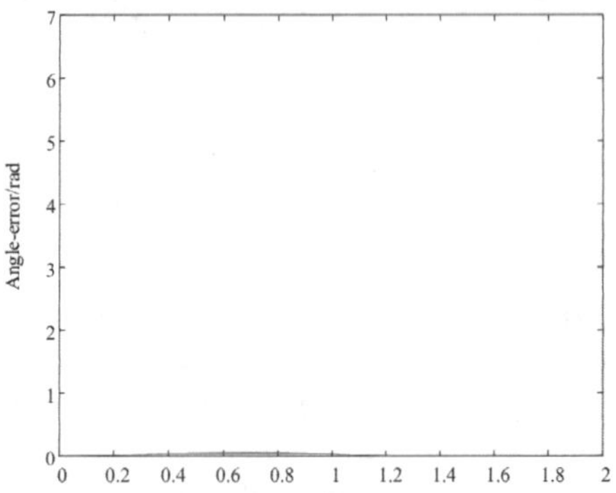

Figure 8. Position error

Figures 9 and Figures 10 are the speed and its error waveforms from 0 r/min to 50r/min to 100r/min. The red color is the real speed and the other is the estimated speed. It can be seen from the figure that the real speed and the estimated speed waveform basically coincide. The speed estimation error is very small.

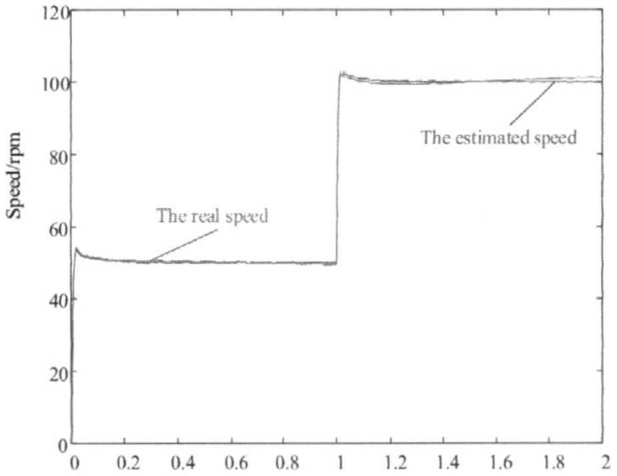

Figure 9. Abrupt change of speed

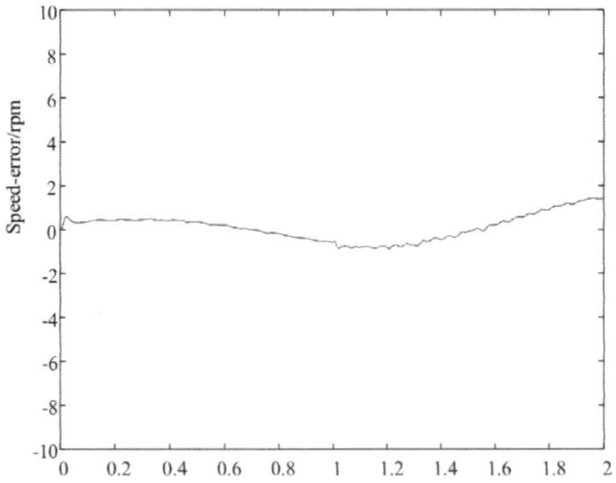

Figure 10. Speed error

Figure 11 and Figures 12 is the position and its error waveform. The red represents the real angle of the rotor, the other represents the estimated angle. It can be seen from the figure that the real angle and the estimated angle waveform basically coincide, and the angle estimation error is very small.

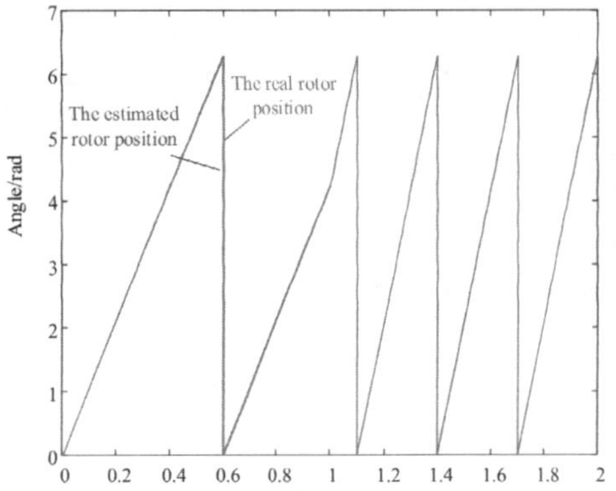

Figure 11. Rotor position

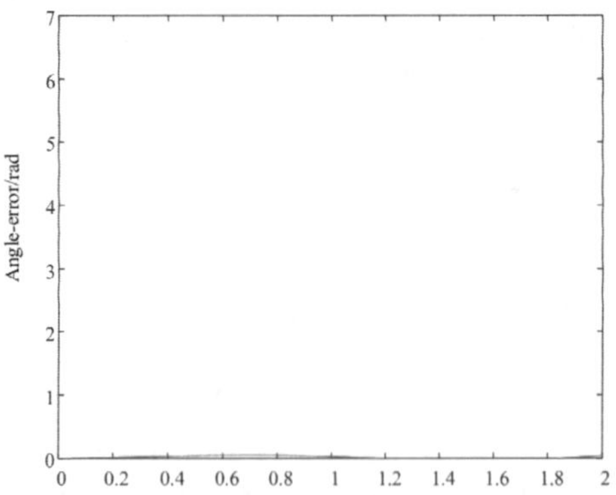

Figure 12. Position error

5. Conclusion

Firstly, the mathematical model of the SynRM is established according to the control principle of the SynRM, and the simulation model of the double closed-loop vector control system is set up in the Matlab/Simulink environment, which includes the current inner loop and the speed outer loop, the saturation effect of the SynRM is taken into account. In addition, a sensorless control method is used to realize zero-speed start-up and sensorless control at low speed. The simulation waveform shows that the speed estimation error and the angle estimation error are all within the acceptable error range, which ensures sufficient precision and control performance, to achieve zero-speed start and stable operation.

References

[1] R. R. Moghaddam, F. Magnussen and C. Sadarangani, "Theoretical and Experimental Reevaluation of Synchronous Reluctance Machine," in IEEE Transactions on Industrial Electronics, vol. 57, no. 1, pp. 6-13, Jan. 2010, doi: 10.1109/TIE.2009.2025286.

[2] L. Lavrinovicha and J. Dirba, "Comparison of permanent magnet synchronous motor and synchronous reluctance motor based on their torque per unit volume," 2014 Electric Power Quality and Supply Reliability Conference (PQ), 2014,pp. 233-236, doi: 10.1109/PQ.2014.6866817.

[3] N. Bianchi, S. Bolognani, E. Carraro, M. Castiello and E. Fornasiero, "Electric Vehicle Traction Based on Synchronous Reluctance Motors," in IEEE Transactions on Industry Applications, vol. 52, no. 6, pp. 4762-4769, Nov.-Dec. 2016, doi: 10.1109/TIA.2016.2599850.

[4] Betz R E. Theoretical aspects of control of synchronous reluctance machines[J]. IEE Proceedings B-Electric Power Applications, 1992, 139(4): 355-364.

[5] Kang S J, Sul S K. Efficiency optimized vector control of synchronous reluctance motor[C]//Industry Applications Conference, San Diego, USA, 1996: 117-121.

[6] Singh B, Varshney A. Maximum power factor based vector control approach for synchronous reluctance motor driven solar PV array fed water pumping system[C]//IEEE Power India International Confer ence, Delhi, India, 2016: 1-6

[7] Betz R E, Lagerquist R, Jovanovic M, et al. Control of synchronous reluctance machines[J]. IEEE Transa ctions on Industry Applications, 1993, 29(6): 1110- 1122.

[8] Jansen P L,Lorenz R D.Transducerless field orientation concepts employing saturation-induced saliencies in induction machines. Industry Applications IEEE Transactions on,1996,36(6):1380-1393.

[9] R.Wu,G.R.Slemon.A Permanent Magnet Motor Drive Without a Shaft Sensors. IEEE Transactions on Industry App lications,1991,27(5):1005-1011.

[10] Ichikawa S,Tomita M,Doki S,et al. Sensorless Control of Synchronous Reluctance Motors Based on Extended EMF Models Considering Magnetic Saturation With Online Parameter Identification.IEEE Transactions on Industry Applications,2006,42(5):1264-1274.

[11] Ichikawa S,Tomita M, Doki S,et al.Extended EMF Models of Synchronous Reluctance Motors and Selection of Main Flux Direction. Ieej Transactions on Industry Applicatins,2004,123(12):1507-1515.

[12] A.Ravikumar Setty,S.Wekhande, and K. Chatterjee. Comparison of high frequency signal injection techniques for rotor position estimation at low speed to standstill of pmsm. In Power Electronics（IICPE）,2012 IEEE 5th India International Conference on, pages 1-6,2012.

Aerostat Airbag Damage Detection Based on VMD Decomposition and GenSVM

Mei Wang[1], Zhipeng Chen[2], Pingan Wang[1,*], Haiping Zhu[2] and Liangzhi Fan[3]

[1]NO.38 Research Institute of CETC, Hefei, China
[2]School of Mechanical Science and Engineering, Huazhong University of Science and Technology, Wuhan, China
[3]School of Mechanical engineering and automation, Wuhan Textile University, Wuhan, China

[*]corresponding author's e-mail: 9764777@qq.com

Abstract. The capsule of the buoy is usually woven with composite materials. The existence of material anisotropy makes it difficult to establish the degradation mechanism model of the capsule, which affects the defect detection and health management of the capsule. To this end, this paper proposes a data-driven airbag defect detection method based on stress wave measurement. Firstly, strain signals under acoustic excitation are collected by sticking strain gauges on the surface of the capsule. Then, VMD is used to decompose the signal and extract the energy characteristics of each modal signal component to construct the feature vector. Finally, the feature vector is input into the GenSVM model for defect type recognition. The example analysis shows the effectiveness and applicability of the proposed method.

1. Introduction

Due to the characteristics of long-stay time, large load capacity, and good safety, the aerostat is widely used in climate monitoring, geographical mapping, and military reconnaissance [1]. As the key component of the buoy, the buoyancy capsule often works in high temperatures, high ultraviolet intensity, and a strong wind environment, which will greatly aggravate the aging damage of the capsule. Some small structural cracks will also suddenly increase during the task of the buoy, affecting the operation of the equipment and causing economic losses. Therefore, carrying out accurate defect detection for the capsule of the floater can timely and effectively discover the early defects of the capsule and avoid the occurrence of the accident of the floater.

The buoyancy capsule is generally made of floating composite material. The anisotropy of the material and the complexity of the capsule structure make the mechanism of capsule deterioration difficult and apply to fault prediction [2]. The existing film defect detection methods are mainly based on optical detection. The principle is mainly to use parallel light to irradiate the film to be detected and to obtain the defect signal by using the linear array charge-coupled (CCD) technology of the photoelectric conversion device. Finally, the defect is identified by the image recognition algorithm [3]. Zhong et al. [4] improved the image segmentation algorithm of optical film defect detection and proposed an image segmentation algorithm based on error correction theory to enlarge the difference of pixel gray value between defect area and normal area, to improve the accuracy of defect area segmentation. Wang et al. [5] improved the spectral residual saliency model applied to the defect detection of the TFT-LCD panel and improved the detection ability of the model for fibers, stains, and scratches. Wei et al. [6] designed an automatic detection system for film surface defects based on computer vision, which realized the rapid detection of film defects. Although these thin-film defect detection algorithms based on optical and computer vision have made some achievements in thin-film defect detection, the capsule film of the floater capsule is formed by the bonding of multi-layer composite materials. Therefore, whether the existing CCD technology can realize the defect detection of the floater capsule needs further research.

Similar to CCD detection technology, defect detection technology based on stress waves are also widely used in the field of nondestructive testing of equipment. Wang et.al [7] explored the application of Lamb wave tomography in the field of structural crack damage imaging monitoring and realized the evaluation of crack direction and length. Liu et al. [8] studied the health monitoring of composite cylindrical structures and used stress wave technology combined with probabilistic imaging methods to monitor and identify damage. Xu et al. [9] proposed the theory and method of non-contact vibration measurement based on the fringed film, realized the measurement of stress wave on the surface of capsule film, and realized the modal analysis of beam structure. Qiu et al. [10] further analyzed and tested the modal of the plane membrane structure of spacecraft, and studied the modal influencing factors and internal laws of membrane structure. These studies not only analyzed the possibility of stress wave technology in the detection of capsule defects from the application level but also verified the influence factors and action law of the vibration mode of the capsule from the theoretical analysis.

In summary, this paper will study the defect detection method of the airbag based on the stress wave with the crack broke yarn defect type that has a great influence on the airbag [11]. The authors first determined the influence of yarn breakage defects on the vibration mode of capsule membrane from the aspect of mechanism analysis, and then designed a capsule stress wave signal acquisition system based on acoustic excitation to collect the capsule stress signal. Finally, a defect-recognition model based on VMD decomposition and GenSVM classification model is proposed to realize the detection of the capsule crack defect of the floater.

2. Stress-wave based damage detection

The aerostat capsule is mostly an elliptical sphere woven by composite materials. Due to the special anisotropic properties of composite materials, it is difficult to obtain good diagnostic results on the aerostat capsule by conventional defect methods [12]. The damage identification technology based on stress waves can judge the structural health state by exciting a series of wave signals on the surface of the measured object and detecting the propagation and loss of the wave. This method has the advantages of long propagation distance, fast detection speed, and many application scenarios. It has been widely used in industry [8]. Similar to the microphone film, the capsule will also undergo forced vibration after inflation under external excitation. Because the diaphragm is mainly the first-order vibration mode, the principle formula of the first-order resonant frequency of the capsule is analyzed.

For a membrane vesicle in a free state with a long axis and a short axis of R and S, respectively, it is assumed that the tension is T_0 (N · m^{-1}) in the static state. When subjected to acoustic excitation, the membrane will vibrate. Assuming that the sound pressure is, the differential equation of vibration can be expressed as

$$\pi R^2 \sigma \frac{\mathrm{d}^2 s}{\mathrm{d}t^2} = -2\pi T_0 s - \pi R^2 \eta \frac{ds}{dt} + \pi R^2 P_0 \sin \omega t \tag{1}$$

where σ represents per unit area of the thylakoid membrane, s is amplitude and η indicates air viscosity coefficient.

The formula can be simplified as

$$\frac{\mathrm{d}^2 s}{\mathrm{d}t^2} + 2k\frac{ds}{dt} + \omega_0^2 s = \frac{P_0}{\sigma} \sin \omega t \tag{2}$$

where $\omega_0 = \frac{k_0}{R}\sqrt{\frac{T_0}{\sigma}}$ is the natural frequency of capsule vibration, k_0 is determined by the root of the first Bessel function. If the k_0 is set as 2.4048, then

$$f_0 = \frac{\omega_0}{2\pi} = \frac{k_0}{2\pi R}\sqrt{\frac{T_0}{h\rho}} \tag{3}$$

where h is the thickness of the thylakoid membrane, ρ represents the density of the thylakoid membrane. It can be seen from the above equation that when defects such as broken lines occur in the capsule, the tension T_0 in the local area will be affected, and the vibration frequency in the capsule film area will be further changed.

This change in vibration frequency can be recorded using strain sensors for further analysis. In this paper, the strain gauge adheres to the buoy capsule, and then the acoustic excitation signal is played to promote the vibration of the capsule, and the signal of the strain gauge is recorded by the strain signal acquisition instrument. When defects occur on the capsule surface, the vibration frequency will inevitably change. These changes are reflected by strain gauges and recorded for the next defect diagnosis model.

3. VMD and GenSVM for damage feature extraction and identification

Based on the stress signal of the capsule collection by the strain gauge, this paper proposes a defect-recognition algorithm combining VMD and GenSVM. The specific recognition process is shown in figure 1. The original signal collected first needs to be denoised to improve the signal-to-noise ratio. Then, the VMD decomposition algorithm is used to decompose the original signal to obtain different modal components, and the energy entropy eigenvalues of each modal component signal are extracted. Using these feature vectors, the GenSVM model can be trained to recognize the type of cyst defect.

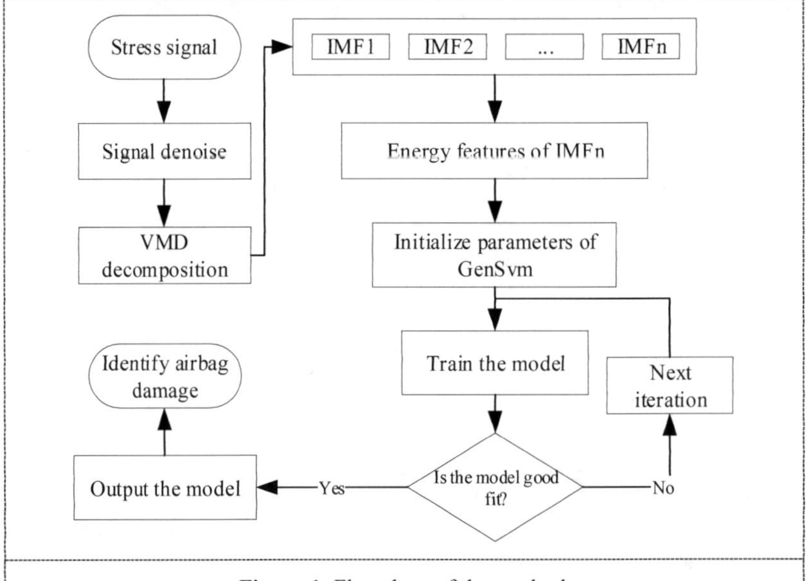

Figure 1. Flowchart of the method

3.1. Feature extraction by VMD

VMD algorithm has great advantages in the decomposition of the unstable noise signal, which is often used in the field of equipment fault diagnosis [13-14]. The principle is that the original signal f (t) is adaptively decomposed into a series of component signals IMF_k containing different frequency centers and bandwidths through Wiener filtering and Hilbert transform, and each IMF signal can be expressed as

$$u_k(t) = A_k(t)\cos(\phi_k(t)), k = 1, 2, ..., K \tag{4}$$

where $u_k(t)$ is the kth IMF signal, $A_k(t)$ represents the instantaneous amplitude of the signal, and $\phi_k(t)$ indicates the signal phase. Where K is a preset number of modal signals that need to be decomposed, it usually needs to be optimized according to experience. The objective of VMD is to minimize the sum of the bandwidth of each IMF signal, and the optimization objective function is as follows

$$\begin{cases} \min_{|n_k/|, w_i|} \{\sum_k \left\| \partial_t [(\delta(t) + \frac{\mathrm{j}}{\pi t}) * u_k(t)] \mathrm{e}^{-j\omega_k t} \right\|_2^2 \} \\ \mathrm{s.t} \sum_K u_k(t) = f(t) \end{cases} \quad (5)$$

where $\delta(t)$ is the unit pulse function, and ∂_t represents the derivation of t, $\{w_k\} = \{w_1, w_2, ...w_k\}$ is the center frequency for signal IMF.

After transforming the above equation into the unconstrained variational problem through the Lagrange operator, the augmented Lagrange operator formula is shown as follows

$$L(\{u_k\}, \{\omega_k\}, \lambda) = \alpha \sum_k \left\| \partial_t [(\delta(t) + \frac{j}{\pi t}) u_k(t)] e^{-j\omega_k t} \right\|_2^2 + \left\| f(t) - \sum_k u_k(t) \right\|_2^2 + \left\langle \lambda(t), f(t) - \sum_k u_k(t) \right\rangle \quad (6)$$

where α is quadratic penalty factor and $\lambda(t)$ is Lagrange multiplier.

In the process of solving the problem, we first initialize and search the saddle point of Equation (6) through continuous iterative updating to obtain the optimal decomposition signal. After obtaining the decomposed modal signal, Hilbert transform is performed on each IMF component, and the obtained Hilbert spectrum can be expressed as

$$H(t, \omega) = \sum_n^{i=1} b_i A_i(t) \mathrm{e}^{\mathrm{j}\omega(i)t}, and \begin{cases} b_i = 1, \omega_i = \omega \\ b_i = 0, \omega_i \neq \omega \end{cases} \quad (7)$$

By Hilbert spectrum, the time spectrum energy characteristic of each IMF can be calculated as

$$E_i = \int_{\omega 1}^{\omega 2} \int_0^T H_i^2(\omega, t) \mathrm{d}t \mathrm{d}\omega \quad (8)$$

where w1 and w2 represent the range of the spectrum, T represents time. The Hilbert energy spectrum can be obtained by integrating the square of the amplitude in the time dimension, indicating the energy of the signal in time T. Finally, solving the energy of each IMF component can construct a multimodal eigenvalue vector for training and diagnosing the GenSVM defect-recognition model.

3.2. GenSVM training

SVM solves the multi-classification problem usually by using heuristic rules or transforming the original problem into a large dual optimization problem, which often greatly increases the time complexity of solving. To solve this problem, Gerrit [15] proposed a new generalized multi-classification support vector machine (GenSVM). The algorithm first maps the input M-dimensional feature data into the new space of the $M-1$ dimension by using the RBF kernel function and then optimizes the weighted summation of each classification error as the objective function. Finally, the determined classification boundary is mapped to the original space to classify the input data. Assuming that the original input data are, it also represents the category label. The mapping formula from raw data to feature space can be expressed as

$$\mathbf{s}_i' = \mathbf{x}_i'\mathbf{W} + \mathbf{t}' \quad (9)$$

where $\mathbf{s}_i'$ is the mapping data of feature space, $\mathbf{W}$ is the transform matrix, and t' represents bias.

The classification error can be calculated by the projection distance from sample i to boundary k and j, and the calculation formula is

$$q_i^{(k,j)} = \left(\mathbf{\Phi}_i' \boldsymbol{W} + \boldsymbol{t}'\right)\left(\boldsymbol{u}_k - \boldsymbol{u}_j\right) \quad (10)$$

where and are the vertices of boundary k and j respectively.

To ensure the flexibility of the model, GenSVM uses the Huber loss function to weighted sum the classification errors of each subclass as the error loss function of the model. For the L classification problem containing n data, the loss function is calculated as follows

$$L_{MSVM}(\boldsymbol{W},\boldsymbol{t}) = \frac{1}{n}\sum_{L}^{l=1}\sum_{i\in G_l}\rho_i\left(\sum_{j\neq k}h^p\left(q_i^{(k,j)}\right)\right)^{\frac{1}{p}} + \lambda tr\boldsymbol{W}'\boldsymbol{W} \tag{11}$$

where $\rho_i = n/(|G_k|L)$ is the classification error weight of each class, h is the Huber loss function, and p is the parameter of Huber loss. $G_l = \{i : y_i = l\}$ is the sample set that belongs to class l, $\lambda tr\boldsymbol{W}'\boldsymbol{W}$ is used to prevent overfitting penalty terms, and is a regularization parameter greater than zero.

After the model construction is completed, the parameters **W** and t need to be optimized. Let $V = [tW']'$ and $L = L_{MSVM}(W,t)$, the optimization objective function be

$$\begin{cases} L(\bar{V}) = g(\bar{V},\bar{V}) \\ L(V_t) \leq g(V_t,\bar{V}) \end{cases} \tag{12}$$

where $\bar{V}$ is a parameter set that needs to be optimized and $L(\bar{V})$ is a derivable function.

3.3. Airbag damage identification

When the online identification of capsule defects is carried out, the strain signal of the capsule is collected first, and the noise brought by the acquisition process is removed by noise reduction. Then, the VMD model is used to decompose the modal components. The energy features are extracted from each obtained modal component to form a feature vector and set to. Then use formula (9) to map the original data to the feature space, and finally calculate the Euclidean distance to each type of boundary optimized in the GenSVM model. The calculation formula is

$$\hat{y}_{n+1} = \underset{l}{\arg\min} \| \mathbf{s}_{n+1}' - \boldsymbol{u}_k' \|^2, \text{for } k = 1,\ldots,L \tag{13}$$

According to the calculation results, the category of the newly collected samples is determined as the category of the points closest to the vertices of each class in the mapping space.

4. Experiment study

To verify the effectiveness of the proposed algorithm, the capsule defect experiment of the floater is designed, and the strain data under different size defects are collected to identify the algorithm. The experiment was carried out at Huazhong University of Science and Technology, Wuhan. The details are as follows.

4.1. Experiment setup

The capsule body used in the experiment is a cylinder with a length of 264 mm and a diameter of 149 mm. The thickness of the capsule membrane is 0.2 mm and the tensile strength is 800 N. The experimental equipment is arranged as shown in figure 2, mainly including the buoy capsule, strain gauge, data acquisition device, air pump, and excitation soundbox. During the experiment, the strain gauge with the model of BF120 - 10AA (11) -P150D was attached to the surface of the capsule, and the pressure of the capsule was maintained at 5 KPa. The sweep signal excitation was played by the speaker. The sensing data of the strain gauge were collected by Donghua DH5981 dynamic strain tester.

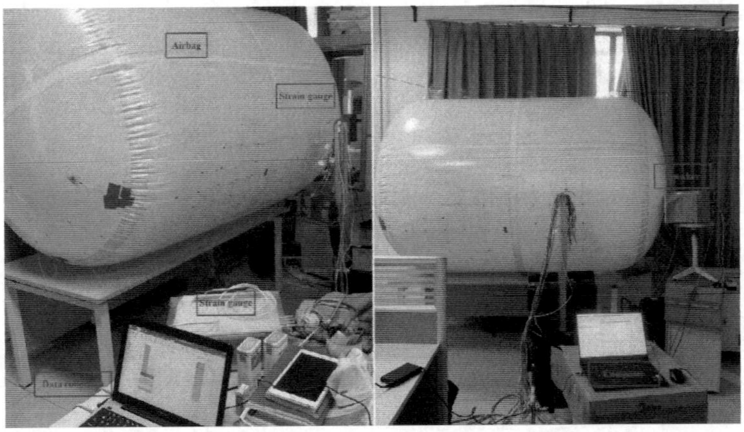

Figure 2. The figure of the platform.

Two kinds of slit defects, 3mm and 5mm, were set up in the experiment. Ten groups of strain data were collected under each state of the capsule, a total of 30 groups. Set the data acquisition frequency of 2000HZ, the acquisition time is 3 minutes, therefore, each group of data has 360000 strain values.

4.2. Result and discussion

The original signal is intercepted, and VMD decomposition is performed with every 200 points as a signal segment to construct a data set. The specific process is shown in figure 3. A section of data was randomly intercepted from the health status detection signal of the capsule, and VMD decomposition was performed. The number of modal components to be decomposed was set to be 6. By analogy, through repeated interception processing, training and test data sets can be obtained.

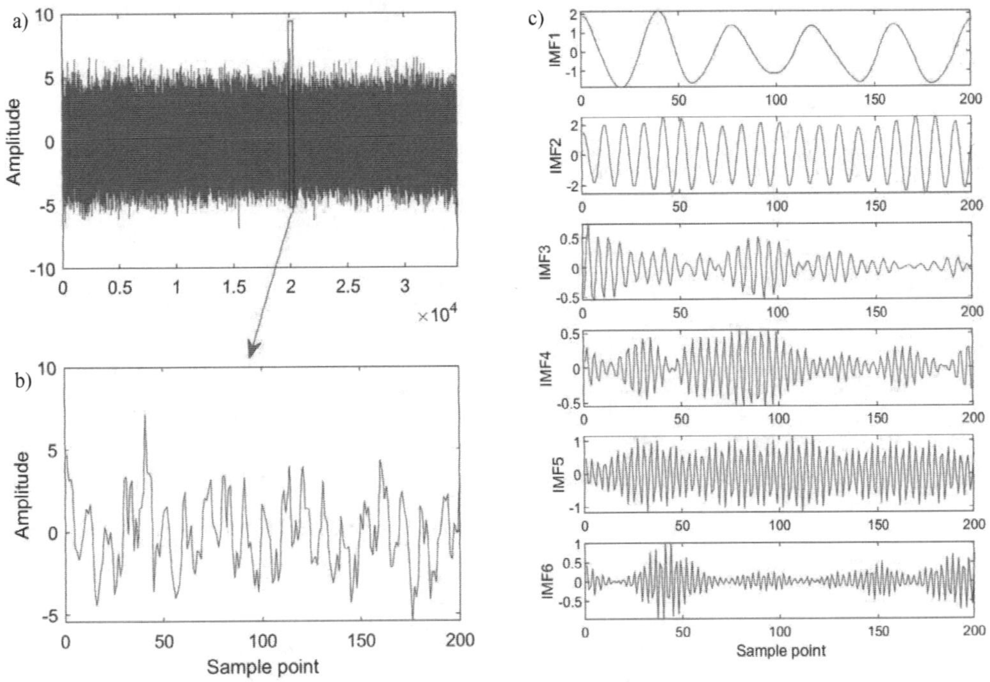

Figure 3. Feature extraction, a)origin signal, b)signal Segment, c) IMF signal.

In the training process of the GenSVM model, the parameter p is set to be 1, λ is 10-6, and the maximum number of iterations for model training is 1000. The training is stopped when the relative change of the inter-generational loss function is set to be less than 0.001 in advance. Referring to other literature, the kernel function uses the most commonly used RBF. To improve the accuracy of the model, a 5 fold cross-validation training method is used. After training, the recognition effect is shown in Figure 4. From the results, it can be seen that the classification effect of the model is better, and each defect type can be accurately judged. Figure 4 shows the classification effect of the trained model on the validation set. In a total of 150 validation set samples, only four samples are classified wrong, indicating the effectiveness of the GenSVM algorithm.

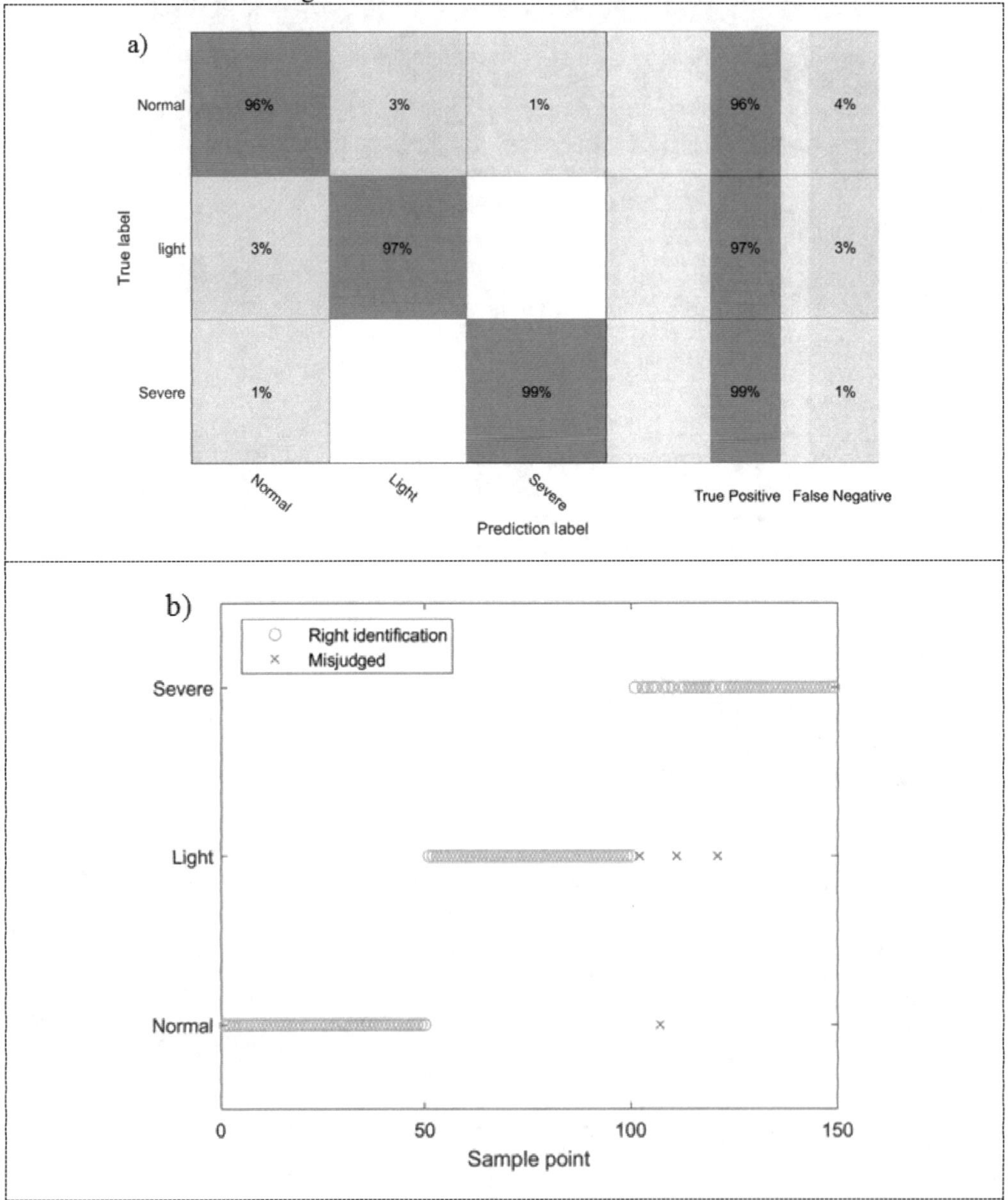

Figure 4. Result of GenSVM, a) training accuracy, b) test accuracy.

To further verify the superiority of the GenSVM model over other classification models, this paper selects the classification tree algorithm (CTree), nearest neighbor (KNN), naive Bayes (NB), multi-layer perceptron (MLP), and ensemble learning (AdaBoost) for comparative experiments. For these algorithms, the maximum splitting number of CTree is set to 20 the number of neighbors of KNN is set to 5, and the distance measurement method is Euclidean distance; NB selects the naive Bayesian algorithm with Gaussian prior distribution; the MLP is set to a neural network consisting of the input layer, hidden layer, and output layer. The hidden layer has 10 hidden nodes. Adaboost uses 30 decision trees as weak classifiers, and the maximum splitting number of the decision tree is 20. At the same time, to reduce the impact of accidental errors, all test algorithms were run 10 times to compare the average.

Table 1. Algorithm comparison

Algorithm	GenSVM	MLP	Adaboost	KNN	CTree	NB
Training accuracy	97.6	98.67	94	89.3	80.4	52.7
Test accuracy	97.32	92.97	91.95	89.93	79.19	48.32
Average	97.46	95.82	92.975	89.615	79.795	50.51

Table 1 shows the running results of all algorithms. GenSVM achieves the best average classification accuracy. Although the training accuracy of MLP is higher than that of GenSVM, its test accuracy is low, indicating that the neural network model may have an over-fitting problem. At the same time, the Adaboost algorithm performs better than CTree, which proves the effectiveness of ensemble learning and improves the accuracy of a single classifier. NB algorithm has the worst effect, training and test accuracy are not high, and the test accuracy is less than 50 %. From the above comparison, it can be seen that the superiority of the GenSVM algorithm can quickly learn the classification criteria on a limited number of data sets, and achieve good results. It has good applicability for the identification of cystic defects.

5. Conclusion

To realize the defect identification of the airbag based on stress wave, this paper first analyzes the influence mechanism of the slit on the vibration mode of the capsule membrane from the theoretical level and verifies it by finite element simulation. After that, the stress-strain data acquisition scheme of the capsule under audio excitation based on strain gauge measurement was designed. Combined with VMD signal decomposition and GenSVM classification algorithm, the defect types of cysts were identified and analyzed. Finally, the GenSVM model is compared with other classification algorithms to verify the effectiveness of the GenSVM algorithm in the identification of cystic defects.

Acknowledgments

This work is supported by the National Natural Science Foundation of China under Grant No. 52075202.

References

[1] Zhang Yunhao, Alimu·Anwaier, Mi Xiang, et. al 2020 Effects of Short-Term Aging and Creasing Damage on Air Leakage Properties of Airship Envelope Materials. Journal of Shanghai Jiaotong University. **54** 1189-1199.

[2] Zhang Zhifu 2019 Experimental Study on Rapid Aging of Tethered Balloon Envelope Materials[J]. Equipment Environmental Engineering. **16** 23-26.

[3] Xu Hengying, The CCD-based Inspecting System of Defects in Plastic Film. Insulating Materials, 2008: 60-62, 65.

[4] Zhong Qiusheng, Li Jingrong, Hu Guanghua 2011 Efficient image segmentation algorithm for defect inspection of optical films. Journal of Mechanical & Electrical Engineering. **28** 1170-1175.

[5] Wang Hongshuo, Yang Yongyue 2018 Surface defect inspection of TFT-LCD panels based on improved saliency model. Journal of Electronic Measurement and Instrumentation. **32** 29-35.

[6] Wei Yuke, Zeng Gui, Fan peng 2014 Design of Cutting Film Machine for Surface Defect Detection. Computer Measurement & Control. **192** 2752-2753, 2757.

[7] Wang Qiang, Xu Jing, Wang Mengxin, et. al 2016 Lamb Wave Tomography Technique for Crack Damage Detection. Journal of Mechanical Engineering. **52** 30-36.

[8] Liu Xiang, Yan Gang, Tang Jianfei 2021 Stress wave-based damage probability imaging for

composite cylindrical shell structures. Composites Science and Engineering. **332** 5-11.
[9] Xu Zonghuang, Zhong Shuncong, Lin Yun-lin, et. al 2020 Noncontact vibration measurement and modal analysis system based on fringe-pattern thin films. Journal of Mechanical & Electrical Engineering. **37** 1158-1164.
[10] Qiu Hui, Li Xiao, Fan Junfeng, et. al 2017 Modal Analysis and Experiment of Spacecraft Flat Membrane Structures. Spacecraft Engineering. **26** 43-49.
[11] Chen si, Wu Guojun, Liu shengqian, et.al 2020 Study on Main Defects of Envelope Material and its Effect on Material Performance. Progress in Textile Science & Technology. **229** 1-4
[12] Liu Bingfeng, Dong Mingli, Sun Guangkai, et. al 2021 Strain transfer characteristics of fiber Bragg grating sensorin aerostat flexible composite skin deformation. Infrared and Laser Engineering. **50** 201-209.
[13] Li, H., et al. 2020 An optimized VMD method and its applications in bearing fault diagnosis. Measurement. **166** 108185
[14] Cheng, H., et al. 2019 A Bearing Fault Diagnosis Method Based on VMD-SVD and Fuzzy Clustering. International Journal of Pattern Recognition and Artificial Intelligence. **33** 1950018.
[15] Burg, Gerrit J. J. Van Den , and P. J. F. Groenen 2016 GenSVM: a generalized multiclass support vector machine. JMLR.org.

Fault Diagnosis Method of Rotating Machinery Based on CEEMDAN-LSTM Model

Peng Wang[1,3], Yiren Zhou[2,*] and Lan Zhang[1]

[1]School of Mechanical Engineering, Tianjin University, Tianjin, China
[2]Nanchang University of Technology, Nanchang, Jiangxi, China
[3]International Institute for Innovative Design and Intelligent Manufacturing of Tianjin University in Zhejiang, Shaoxing, Zhejiang, China

[*]corresponding author's e-mail: renyiz@126.com

Abstract. To overcome the shortcomings of traditional feature extraction methods that rely on prior knowledge and expert experience in the field and the problems of shallow machine learning in fault diagnosis, such as inability to extract deep features, low accuracy of model recognition and poor generalization ability, a fault diagnosis method based on complete ensemble empirical mode decomposition with adaptive noise and long short-term memory (CEEMDAN-LSTM) model is proposed. Firstly, the original signal is decomposed by complete ensemble empirical mode decomposition with adaptive noise (CEEMDAN); secondly, the temporal features of selected component signals are extracted by long short-term memory (LSTM); thirdly the temporal features of each component are concatenated into feature vectors according to the order of high and low frequencies; finally, the feature vectors are input into the softmax layer to get the fault classification category. The results of gear box fault diagnosis test show that the model can effectively identify the fault location. Compared with empirical mode decomposition and support vector machine (EMD-SVM) model and LSTM model, CEEMDAN-LSTM model can obtain the highest fault identification accuracy.

1. Introduction

During the operation of mechanical equipment, if the equipment fails, it may cause huge economic losses, even casualties. Therefore, the research on fault diagnosis of mechanical equipment is of great significance [1]. With the rapid development of advanced sensing technology and the improvement of computer storage performance, sensor time series data can be collected on a large scale in the process of equipment operation. Therefore, the method based on data signal driving has been greatly developed and applied in the field of fault diagnosis [2-4].Among the commonly used data signals, the vibration signal is the most commonly used analysis signal in the field of fault diagnosis because the vibration sensor is easy to install and the vibration signal can reflect the running state of the equipment relatively well. Li et al. [5] and Murad et al. [6] creatively combined energy operator demodulation method and empirical mode decomposition (EMD) method to identify rolling bearing fault types. Cong [7] used local mean decomposition to decompose the signal into PF components, and used support vector multi-classifier for fault identification. Xu et al. [8] used harmonic wavelet packet decomposition to extract features, designed a multi-class support vector machine model, and realized the fault diagnosis of rolling bearings.

However, the traditional fault diagnosis methods have some shortcomings:

(1) Manual feature extraction needs to be designed according to different types of monitoring objects, which depends on prior domain knowledge.

(2) Shallow diagnostic models can not extract deep features and have low recognition accuracy and poor generalization ability.

Therefore, in recent years, many deep models based on adaptive feature extraction have been proposed for fault diagnosis. Tang et al. [9] proposed a deep neural network based on sparse auto encoding, which can automatically extract the inherent feature of rolling bearing vibration signals and realize rolling bearing fault diagnosis. Shao et al. [10] used deep belief network (DBN) for intelligent condition monitoring of induction motors, and used DBN to extract the relevant feature of vibration signals adaptively. Jiang et al. [11] proposed a new type of multi-layer deep learning convolutional neural network (CNN) for fault diagnosis of rolling bearings. The vibration signal was preprocessed by normalized pretreatment method, and the processed signal was sent to CNN for pattern recognition.

However, in the DBN and CNN models, time series data is treated as static spatial data, and the time dependence of the sequence is not considered, resulting in information loss [12]. Considering the timing of signals, some scholars have proposed the use of recurrent neural network (RNN) in the field of fault diagnosis. Hassan et al. [13] used RNN to perform non-parametric system identification of foil air bearings. As an advanced RNN variant, LSTM can adaptively capture long-term dependencies and nonlinear dynamics of time series data. Tang [14] proposed a bearing fault identification method based on LSTM, which does not need to extract fault features in advance. Although the advantages of LSTM in processing time series have been proved, it is difficult to extract the time series features of each frequency band in processing high sampling rate vibration signals at the same time.

Because traditional feature extraction methods rely on expert experience, shallow machine learning can not extract deep features and LSTM can not simultaneously extract the time series features of each frequency band in the signal, CEEMDAN-LSTM fault diagnosis method is proposed. This method can not only take advantages of CEEMDAN to decompose signals into different frequency bands adaptively, but also take advantages of LSTM to extract signal characteristics at different times, so as to extract signal characteristics in time domain and realize fault diagnosis.

The remainder of the paper is organized as follows: In Section 2, the structure of CEEMDAN and LSTM are briefly described. In Section 3 illustrates the procedures of the proposed method. In Section 4, a gearbox fault diagnosis experiment is used to validate the effectiveness of the proposed method. Finally, conclusions are drawn in Section 5.

2. Related work

2.1. CEEMDAN algorithm

CEEMDAN is a signal decomposition method. This method not only adaptively divides the signal into different frequency bands, but also avoids mode mixing in EMD and ensemble empirical mode decomposition (EEMD) by adding white noise of specific frequency bands in the decomposition process. The algorithm flow of CEEMDAN is shown in Figure 1.

The specific steps are as follows:

(1) Similar to EEMD, the expression of $\overline{IMF_1}$ is

$$\overline{IMF_1} = \frac{1}{I}\sum_{i=1}^{I} IMF_{i1}. \tag{1}$$

(2) After removing the modal components from the original signal, the first residual signal is obtained as follows:

$$r_1(t) = x(t) - \overline{IMF_1}. \tag{2}$$

(3) The set residual signal is constructed and decomposed to obtain the $\overline{IMF_2}$ as follows:

$$\overline{IMF_2} = \frac{1}{I}\sum_{i=1}^{I} E_1\left(r_1(t) + \varepsilon_1 E_1\left(No_i(t)\right)\right). \tag{3}$$

(4) For$k = 2,3,\cdots K$, the calculation process is similar. First, the $k - th$ residual signal is obtained, and then the $k + 1 - th\, IMF$ modal component is obtained. The formula is as follows:

$$r_k(t) = r_{k-1}(t) - \overline{IMF_k}. \tag{4}$$

$$\overline{IMF_{k+1}} = \frac{1}{I}\sum_{i=1}^{I} E_1\left(r_k(t) + \varepsilon_k E_k\left(No_i(t)\right)\right). \quad (5)$$

(5) When the pole of residual signal $r_k(t)$ is less than 3, it will not be decomposed and the algorithm terminates. The final decomposition results are as follows:

$$\mathrm{x(t)} = \sum_{\mathrm{k=1}}^{\mathrm{K}} \overline{IMF_k} + r_k(t). \quad (6)$$

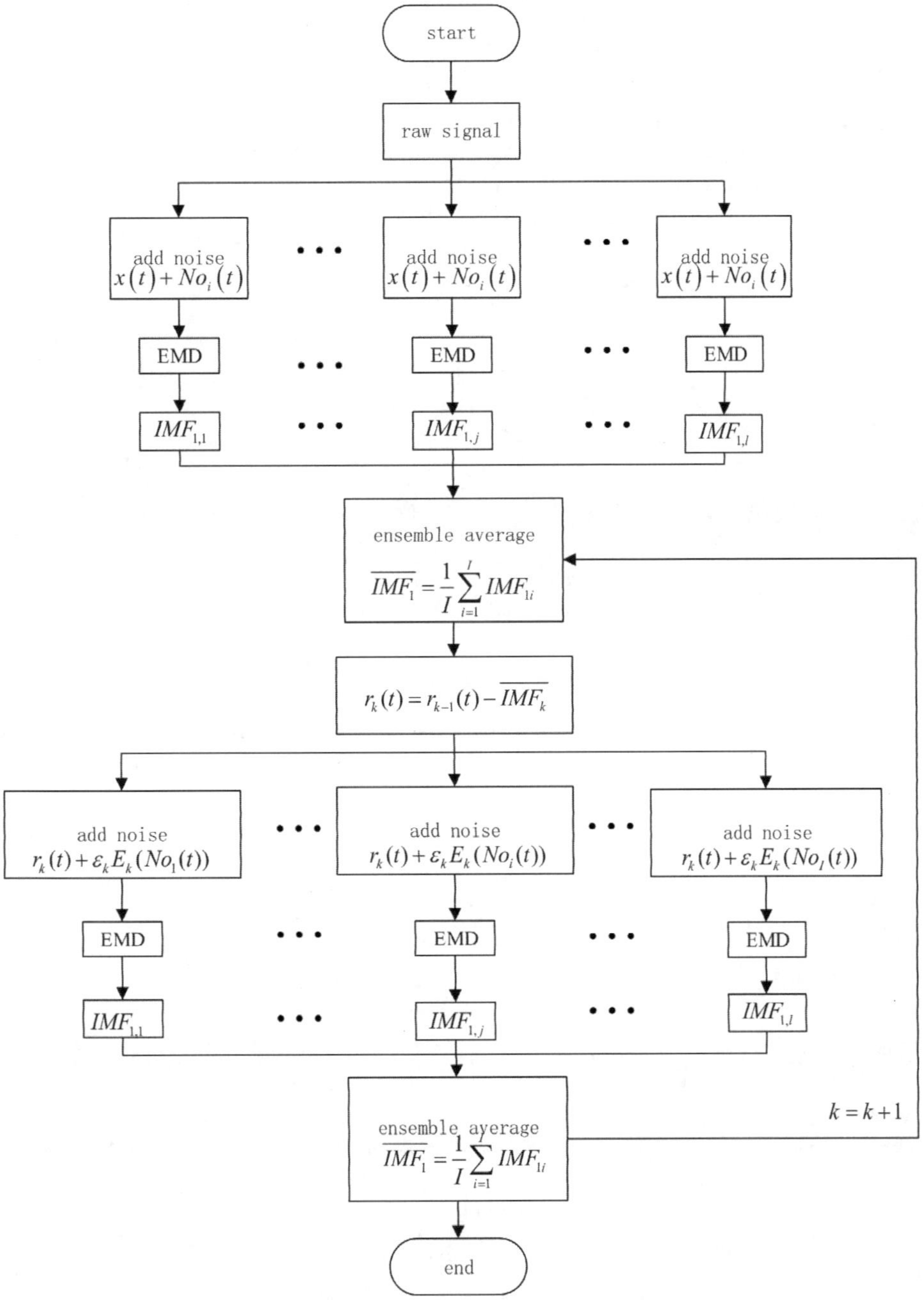

Figure 1. Flow chart of CEEMDAN algorithm

2.2. LSTM model

RNN is a kind of deep learning network model with both feedback and feed-forward connections between network layers. Because of the existence of short-term memory units, it can achieve the purpose of short-term memory and has the ability to adapt to time-varying feature. However, gradient "dissipation" or gradient "explosion" are prone to occur during training. LSTM, as a special structure of RNN, not only avoids the generation of gradient "dissipation" or gradient "explosion", but also has longer time memory than RNN and other models. The hidden layer of RNN has only one state h sensitive to short-term input, and a long-term state memory c is added to the hidden layer of LSTM model. As shown in Figure 2, the LSTM mainly controls the memory unit c through three gate structures. The input gate main controls inputs the temporary status information into the memory unit c; the forget gate mainly controls which information is forgotten at the last moment; the output gate is responsible for controlling whether the status information is output at this time. In the LSTM hidden layer unit structure diagram, x_t is the input vector of the current sample, h_{t-1} is the previous hidden layer output, h_t is the hidden layer output of the current sample, and the input gate includes two parts, i_t and a_t . f_t is the output of the forget gate, o_t is the output of the output gate, c_t is the update of the current state, and c_{t-1} is the last state.

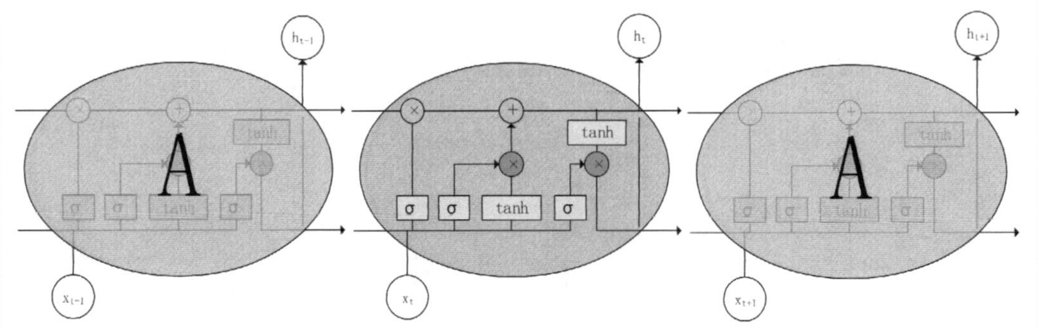

Figure 2. LSTM cell structure

For the LSTM single memory unit c, the algorithm derivation process is as follows:

$$f_t = \sigma\left(W_f h_{t-1} + U_f x_t + b_f\right). \tag{7}$$

$$i_t = \sigma(W_i h_{t-1} + U_i x_t + b_i). \tag{8}$$

$$a_t = tanh(W_a h_{t-1} + U_a x_t + b_a). \tag{9}$$

$$c_t = c_{t-1} \circ f_t + i_t \circ a_t. \tag{10}$$

$$o_t = \sigma(W_o h_{t-1} + U_o x_t + b_o). \tag{11}$$

$$h_t = o_t \circ tanh\,(c_t). \tag{12}$$

In this formula, W and U are weight matrices corresponding to each gate structure, and b are biases.

3. Proposed CEEMDAN-LSTM model

Because the collected signal is non-linearity and non-constant Q, it needs domain knowledge and expert experience to extract features of the signal. CEEMDAN not only can adaptively decompose vibration signals into different frequency bands, but also solve the phenomenon of mode mixing by adding specific white noise. LSTM not only can extract the inherent features of the rolling bearing vibration signal adaptively, but also make use of the memory feature of the network, which can combine the running state of the front and back time to make the classification more accurate. Therefore, CEEMDAN-LSTM model by connecting CEEMDAN and LSTM is proposed to realize fault diagnosis. The overall structure of the model is shown in Figure 3.

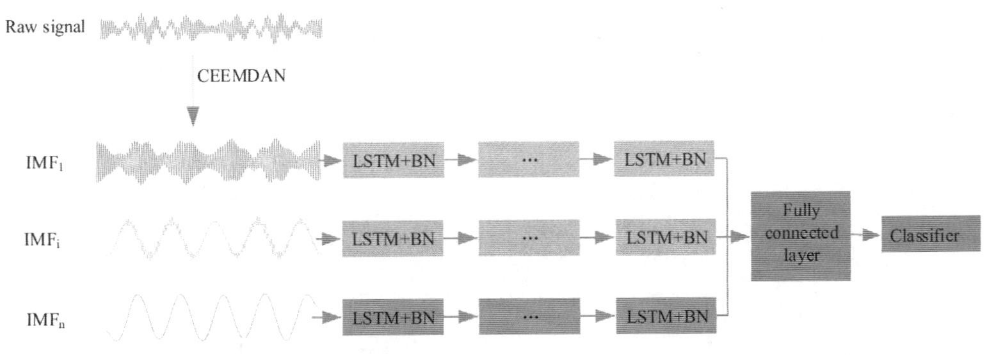

Figure 3. Flow chart of the proposed method

The proposed model first adaptively decomposes the collected fault data into frequency bands using CEEMDAN; secondly, the decomposed signals are filtered using autocorrelation function; thirdly, the selected component signals are fed into the LSTM network model, and the advantages of LSTM network in processing sequential signals are utilized to process the component data with temporal feature and extract the deep features; finally, the extracted features of each frequency band are connected in turn and fed into the classifier to complete the model construction.

The training steps of the model are as follows:

(1) CEEMDAN decomposes M sample signals to obtain IMF_i (i=1,... n);

(2) Screen IMF_i by autocorrelation function, IMF_i' (i=1,..., N) is constituted the samples sets.

(3) The sample sets is divided into training sets and test sets, and the appropriate batch size is selected.

(4) Select the appropriate number of layers and neurons of LSTM network, initialize network parameters, set training times, then, train model. In training stage, the cost function of the current training stage is obtained by forward calculation, and the network parameters are updated by error back propagation until the set number of training times is reached and the cost function converges.

(5) In diagnostic stage, the monitoring signals after the preprocessing of diagnosis will be sent to the trained model. Diagnostic results can be obtained.

4. Verification of results

4.1. Test data

The signals of a gearbox test rig are used in this paper. As shown in Figure 4, the test rig has four kinds of faults: gear broken teeth, gear cracks, big gear cracks + pinion broken teeth and bearing inner ring. The test speed is 863r/min, and the sampling rate is 10.24K. 450 samples were collected for each failure state, and 5120 data points were collected for each sample.

Figure 4. The experimental device. (1) an electric motor; (2) a parallel gear box; (3) a magnetic powder brake.

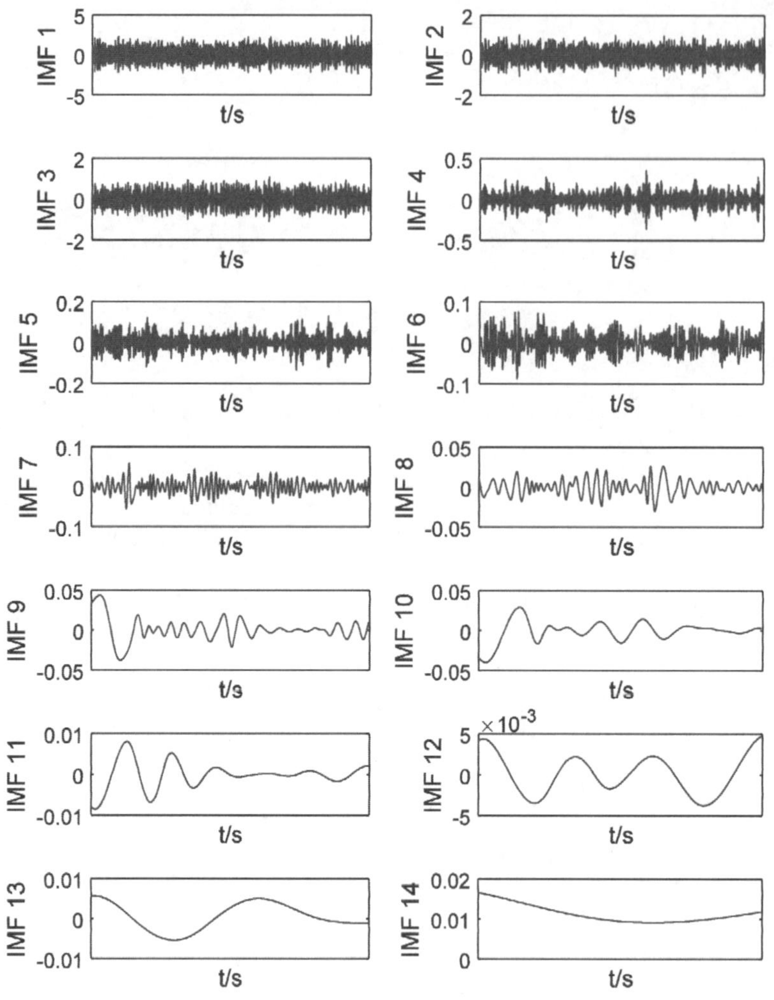

Figure 5. IMF of signal after decomposing by CEEMDAN

4.2. Model training and results

The decomposed IMF is first filtered by autocorrelation function. Ten IMF are selected for each sample, and the length of component sequence is 5120. This constitutes the sample sets. For example, the IMF of a fault sample after decomposing by CEEMDAN are shown in Figure 5. The label in the sample sets is defined by fault types. Subsequently, 1800 samples are randomly divided into training sets and test sets according to the ratio of 1:3. Finally, the initial parameters of LSTM training model are set to 0.2, the number of iterations is set to 10, and the batch size of LSTM training model is set to 10.

The input of LSTM model is (10 * 512). The number of neurons is 100/50/10 in three layers of LSTM model. The average accuracy of the model is 99.5%. Figure 6 shows the relationship between the number of iterations and the accuracy; Figure7 shows the accuracy of the model of 10 experiments.

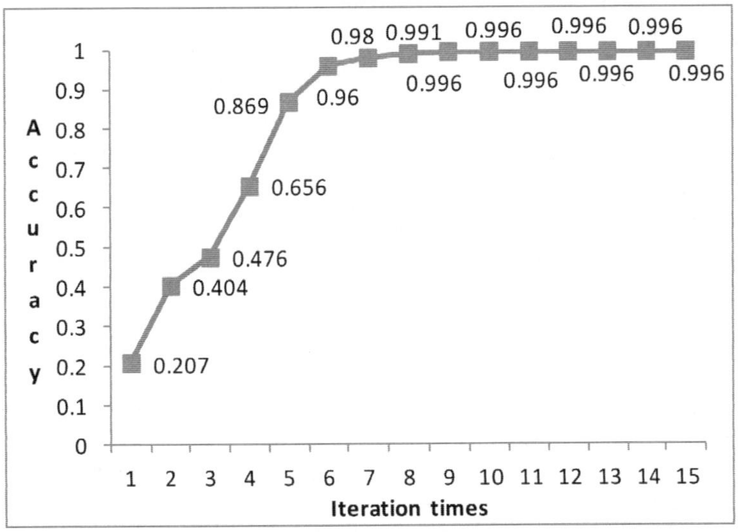

Figure 6. The relationship between iteration times and accuracy

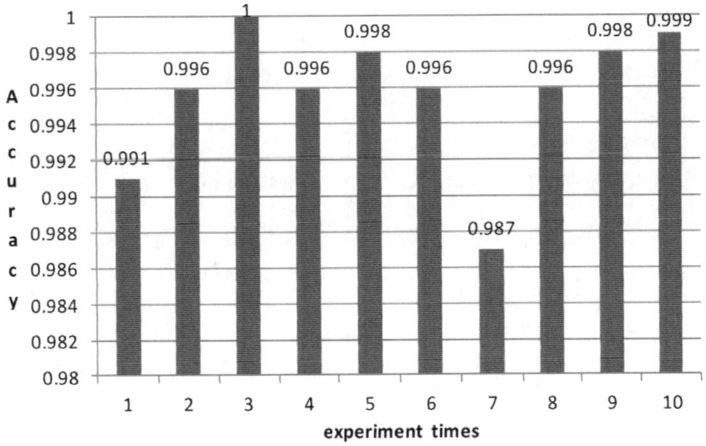

Figure 7. Accuracy of 10 tests

As it shown in Figure6, the accuracy is 0.996 at the 9th iteration of the model, and it does not increase any more. Therefore, the number of iterations selected in this paper is 10, which meets the requirements. As can be seen from Figure 7, the accuracy of the 10 test models is above 0.987, which proves the stability and reliability of the model in this paper.

4.3. Model comparison

To verify the advantage of the model, the model results will be compared with the accuracy of EMD-SVM model and LSTM model. SVM model uses parameters: (C=0.8, kernel='rbf', gamma='auto', decision_function_shape='ovr'), LSTM model uses parameters: learning rate is set to 0.2, iteration number is set to 10, batch size of LSTM training model is selected to be 10, sequence length is 512, and (10*512) is used as input of LSTM model. LSTM model has three layers and the number of neurons is 100/50/10 in turn. The results are as table 1.

Compared with EMD-SVM model and LSTM model, the fault diagnosis model based on our proposed CEEMDAN-LSTM model in this paper has higher accuracy than SVM because of using deep model. Compared with LSTM, because of extracting high and low frequency features of signal in different frequency bands, it has higher accuracy in fault diagnosis of original signal with noise.

Table 1. Accuracy comparison of different models

	Accuracy
EMD-SVM	74.4%
LSTM	98%
CEEMDAN -LSTM	99.5%

5. Summary

A fault diagnosis method based on CEEMDAN-LSTM model has been proposed in this paper. CEEMDAN can adaptively decompose signals into different frequency bands, and LSTM model can adaptively extract the time series feature of signals in different frequency bands. The model does not need manual design and selection of features, and does not rely on manual experience. It can extract the deep features of signals adaptively and improve the generalization ability of the model. In the gearbox fault diagnosis experiment, the results have confirmed CEEMDAN-LSTM model has higher prediction accuracy than EMD-SVM and LSTM model. In future work, we plan to apply the proposed method in machine remaining useful life prediction tasks and continue to optimize our model to get better performance.

Acknowledgments

This work is supported by The Innovation Fund of Industry, education and research of the Science and Technology Development Center of the Ministry of Education(2018 C 01068), Science and Technology Research Fund of Jiangxi Provincial Department of Education(GJJ191006),2020 Nanchang Engineering Research Center and Key Laboratory Foundation(2020-NCZDSY-005).

References

[1] Xu X, Hua Q 2017 Industrial Big Data Analysis in Smart Factory: Current Status and Research Strategies *IEEE Access* 17543-51
[2] Liu R, Meng G, Yang B 2017 Dislocated Time Series Convolutional Neural Architecture: An Intelligent Fault Diagnosis Approach for Electric Machine *IEEE Transactions on Industrial Informatics* 1310-20
[3] Shatnawi Y, Al-Khassaweneh M 2014 Fault diagnosis in internal combustion engines using extension neural network *IEEE Transactions on Industrial Electronics* 1434–43
[4] Qiao H H, Wang T Y, Wang P 2018 A Time-Distributed Spatiotemporal Feature Learning Method for Machine Health Monitoring with Multi-Sensor Time Series *SENSORS* 2932-51
[5] D. Yinfeng, L. Yingmin, X. Mingkui 2008 Analysis of earthquake ground motions using an improved Hilbert–Huang transform *Soil Dynamics and Earthquake Engineering* **28**(1) 7-19
[6] Ni S, Xie W, Pandey M D 2013 Generation of spectrum-compatible earthquake ground motions considering intrinsic spectral variability using Hilbert–Huang transform *Structural Safety* 45-53
[7] Cong J Y 2015 Research on sensor fault diagnosis method based on Lmd and Svm fusion *Liaoning University of Engineering and Technology*
[8] Xu T, Pei A L, Liu Y 2014 Rolling bearing fault diagnosis method based on harmonic wavelet packet and SVM *Journal of Shenyang University of Aeronautics and Astronautics* **31**(4) 50-4
[9] Tang F, Liu Y L, Long H 2018 Sparse auto encoding depth neural network and its application in fault diagnosis of rolling bearings *Mechanical Science and Technology* **37**(3) 352-7
[10] Shao S, Sun W, Wang P 2016 Learning features from vibration signals for induction motor fault diagnosis *2016 Int. Sym. on Flexible Automation (ISFA)* 71-6
[11] Jiang H, Wang F, Shao H 2017 Rolling bearing fault identification using multilayer deep learning convolutional neural network *Journal of Vibroengineering* **19**(1) 138-49
[12] Hassan M F B, Bonello P 2017 A neural network identification technique for a foil-air bearing

under variable speed conditions and its application to unbalance response analysis *Journal of Tribology* **139**(2) 021501
[13] Wielgosz M, Skoczeń A, Mertik M 2017 Using LSTM recurrent neural networks for monitoring the LHC superconducting magnets *Nuclear Instruments and Methods in Physics Research Section A- Accelerators Spectrometers Detectors and Associated Equipment* **867** 40-50
[14] Tang S, He B L, Zhang J Y 2018 Bearing fault recognition based on long-term and short-term memory network *Journal of Automotive Engineering* **8**(4) 297-303

Digital Twin-Based Reconfiguration Time Point Prediction Method for Reconfigurable Manufacturing Systems

Zhihao Liu[1], Taiyong Wang[2], Yiren Zhou[3,*], Wei Zhao[1], Mingliang Zheng[2], Zhisheng Ke[1] and Xiaomeng Zhao[1]

[1]School of Mechanical Engineering, Tianjin University of Technology and Education, Tianjin, China
[2]School of Mechanical Engineering, Tianjin University, Tianjin, China
[3]School of Mechanical Engineering, Nanchang Institute of Technology, Nanchang, Jiangxi, China

* corresponding author's e-mail: renyiz@126.com

Abstract. In the operation cycle of reconfigurable manufacturing system, the system needs to be continuously reconstructed to adapt to the changes of market demand. Aiming at the problem of when to reconstruct the reconfigurable manufacturing system, combined with digital twinning, a prediction method of reconfiguration time point of reconfigurable manufacturing system is proposed. Firstly, the factors affecting the processing capability and function of reconfigurable manufacturing system are analyzed quantitatively, the optimization objective is to minimize the part completion time and reconfiguration cost, and the multi-objective cuckoo algorithm is used to solve it; Then, the digital twin model of reconfigurable machine tool and reconfigurable production line is constructed, so that the state of reconfigurable manufacturing system can be truly mapped in virtual space, and the simulation and prediction framework of reconfigurable manufacturing system is constructed; Secondly, the initial scheme configuration is generated according to the production demand, the set threshold is optimized and simulated, and the simulation sample data is analyzed to obtain the appropriate reconstruction time; Finally, an engineering example is given to verify the feasibility and effectiveness of the method.
Keywords. Reconfigurable manufacturing system; Reconfiguration time point; Digital twin; Cuckoo search algorithm

1. Introduction

At present, the focus on manufacturing competition is constantly shifting, from the amount of product demand to the diversification of product types, the manufacturing industry is facing huge pressure in the market, and customized production has become the trend. Manufacturing companies wants to maintain lasting competitiveness, they need to be able to respond quickly to changes in the market and launch products that meet customer needs in a timely manner [1]. Reconfigurable manufacturing system (RMS) is a highly responsive and highly performance driven manufacturing model developed in recent years and redefined and modified by Professor Koren of the University of Michigan in 1999 [2]. Although the system structure of RMS can respond well to changes in market demand, it is a difficult decision to decide when to refactor the system.

Du et al. resolved the performance of reconfigurable manufacturing systems as signals in the multi-scale direction and used an autoregressive modeling approach to predict the variation of RMS systems at three scales [3]. However, the study was simple in its analysis of system performance and did not investigate the selection of machining cells and machines. Wang et al. proposed to identify factors affecting reconfiguration of reconfigurable manufacturing systems in terms of both processing function and processing capability, information entropy theory to establish a quantitative model of complexity, and cusp mutation theory to identify reconfiguration time points [4]. However, the study simply considered only system-level reconfiguration at the reconfiguration point, and did not consider the multilevel nature of RMS.

Jiewu et al. propose a digitally driven method of rapid reconfiguration of automated manufacturing systems, which consists of two parts, namely semi-physical simulation and optimization, where the semi-physical simulation maps system data and provides input data onto optimization, and the optimized results are fed back to the semi-physical simulation for verification [5]. Hu et al. accurately described the simultaneous mapping of complex product process states from multiple levels in the digital twin space and established associations with individual scale models, and introduced a workflow model for assembly process data collection, which allows prediction and rapid product problem traceability [6].

In this paper, the timing problem of RMS reconfiguration when external factors change is investigated. The method first constructs quantitative mathematical expressions in terms of both machining functions and machining capabilities of RMS, and solves them using the cuckoo algorithm. Secondly, a digital twin mapping model of reconfigurable machine tool and RMS is constructed, and an online simulation prediction mechanism is established. Finally, the scheme is validated using the digital twin model and the sample data are analyzed to demonstrate that the method can effectively guide the timing of RMS reconfiguration.

2. Problem description and model construction

2.1. Problem description

The RMS has a distinctive feature in that its configuration changes in the production cycle to provide the required functionality and capacity for each production cycle [7]. Where problems such as non-conforming products, equipment operational failures, etc. often occur to the initial period of operation. After a period of time, the system runs smoothly and enters a period of efficient production; as internal or external factors affect the system, such as the addition of new parts or earlier order delivery periods, it may cause the system's functionality or capacity to not meeting the demand, which then causes the system to enter a refactoring period. If the reconfiguration time point is delayed, the capacity of the system may not meet the standard, resulting in delayed delivery orders; if the reconfiguration time point is advanced, it may lead to unnecessary cost increase. Therefore, the reconfiguration period needs to be entered at the right time.

In order to meet the variations in the functions and capacities of RMS, the best combination of machines needs to be selected from a set of Reconfigurable machine tool (RMT) that can meet the corresponding functions and capacities. RMT is the core equipment of RMS, which can be combined with different modules to achieve different machining functions and machining capabilities, and its modules can be divided into fixed modules and reconfigurable modules.

2.2. Consider processing capability mathematical model construction

When analyzing the machining capability of RMS, the machining performance of each RMT in different configurations is not the same, that is, the same process is not machined in the same amount of time on different machines. Therefore, the minimum completion time of the part is selected as an indicator, and the following constraints need to be satisfied.

(1) There is no processing sequence required between the processes of different parts, and the processes of the same parts need to be processed in a certain sequence.

(2) If a machine can process multiple processes, the machine can only process one process at the same moment.

(3) The machining completion time for a single part is the sum of the machining time for each process of the part.

$$T_i = \sum_{u=1}^{NOP} \sum_{j=1}^{NM} \Pr Time[u][j] \times NO_i \tag{1}$$

Where: *Ti* indicates the time required to machine part *I*; NO_i indicates number of orders for part *i*; *NOP* indicates number of processes; *NM* indicates number of RMT; *i* indicates Index of parts; *u* indicates Index of process; *j* indicates Index of RMT.

$$\max\{T_1, T_2, ..., T_i\} > OdT \tag{2}$$

Where: *OdT* indicates order delivery time.

When the maximum completion time of RMS is greater than the order delivery time, the system needs to be reconfigured to meet the order requirements.

2.3. Consider the configuration options on processing functions

RMS analyzes their change states at both the system-level and the equipment-levels in order to respond to factors such as order changes. System-level reconfiguration represents the installation and removal of RMT. The reconfigurable modules are connected to the fixed modules through standard interfaces. When the RMT needs to be replaced, only the fixed module needs to be installed and removed. Equipment-level reconfiguration is the installation and disassembly of reconfigurable modules.

Each process operation is considered to have the tool approaching the part in six different directions (TAD) [6], i.e., $\pm x、\pm y、\pm z$. Create process processing matrix $OPTAD = [+x, +y, +z, -x, -y, -z]$, if a process requires three machining directions $+x、-x、+z$, $OPTAD$ = $[1,0,0,1,0,1]$. RMTs offer different degrees of freedom depending on the combination of reconfigurable modules to create the machine tool processing matrix $MTAD$ = $[+x, +y, +z, -x, -y, -z]$, if RMT requires four machining directions $+x、-x、+y、+z$, $MTAD$ = $[1,1,0,1,0,1]$. RMT only be recognized as capable of machining a process if it meets the machining direction of the process being performed.

In the selection process, the minimization of configuration costs was considered. The configuration costs of a RMS consist mainly of the installation of new equipment, the removal of useless equipment and the installation and removal of reconfigurable modules, i.e., they consist of equipment-level configuration costs and system-level configuration costs.

$$\min\{TC_{cost} = C_{\text{mod}} + C_{RMT}\} \tag{3}$$

Where: TC_{cost} indicates total Cost of Configuration; C_{mod} indicates equipment-level configuration costs; C_{RMT} indicates system-level configuration costs.

$$C_{\text{mod}} = \sum_{j=1}^{NM} \sum_{r=1}^{NRMod} \left(RM_{cost}[r] \times |RMod_r[j] - RMod_{r'}[j']| \right) \tag{4}$$

Where: $RM_{cost}[r]$ indicates cost of installing or uninstalling reconfigurable modules *r*; *NRMod* indicates number of reconfigurable modules; *r* indicates index of reconfigurable modules.

$$RMod_r[j] = \begin{cases} 1, \text{ if reconfigurable module } r \text{ is installed on machine } j \\ 0, \text{ otherwise} \end{cases} \tag{5}$$

$$C_{RMT} = \sum_{j=1}^{NM} \sum_{s=1}^{NFMod} \left(FM_{cost}[s] \times |FMod_s[j] - FMod_{s'}[j']| \right) \tag{6}$$

Where: $FM_{cost}[s]$ indicates cost of installing or uninstalling fixed modules *s*; *NFMod* indicates number of fixed modules; *s* indicates index of fixed modules.

$$FMod_s[j] = \begin{cases} 1, \text{ if fixed module } s \text{ is installed on machine } j \\ 0, \text{ otherwise} \end{cases} \tag{7}$$

3. Problem solving

For the above configuration optimization problems of reconfigurable manufacturing systems, the multi-objective genetic algorithm (NSGA-II) is often used to solve such multi-objective optimization problems. However, Wang et al. found that the multi-objective cuckoo algorithm (MOCS) not only improves the operational efficiency over NSGA-II, but also avoids getting trapped in a local optimal solution [8]. Therefore, this paper solves the problem of reconfigurable manufacturing system configuration selection by building a mathematical model of reconfiguration cost and using MOCS.

3.1. Cuckoo algorithm model

The cuckoo algorithm is a nature heuristic inspired by the parasitic nurturing behavior of cuckoos. The algorithm flow is as follows:

(1) Generate a random set of cuckoos in the feasible domain.

(2) The adaptation values of these cuckoos were calculated and the best adaptation values and the location of the birds were recorded.

(3) The location of the nest is then updated by the Levy flight method.

(4) These cuckoo eggs that board into the host nest have a certain chance of being abandoned, at which point the cuckoo needs to find a new nest to lay new cuckoo eggs, keeping the total number of nests in each iteration the same.

(5) After the cuckoo eggs are successfully hatched, the original cuckoo dies and the adaptation value of the new cuckoo is evaluated, and the best adaptation value is updated if there is an adaptation value that is better than the original recorded best adaptation value.

(6) Continue iterating until the required number of iterations is met.

3.1.1. Levy flight way to update the nest. In the algorithm flow, it can be known that updating the nest position is very important, it is related to the convergence speed of the final nest. The nest is updated by randomly generating the length of any direction based on the original cuckoo position, which gives a new nest. And this random generation of steps in any direction has been shown to be effective in finding the global optimal solution by way of Levy flight. The cuckoo's search formula for finding host nest locations and paths is as follows.

$$X_i^{t+1} = X_i^t + \alpha \otimes L(s, \lambda) \tag{8}$$

$$L(s, \lambda) = \frac{\lambda \Gamma(\lambda) \sin \frac{\pi \lambda}{2}}{\pi} \times \frac{1}{s^{1+\lambda}} \tag{9}$$

$$\Gamma(\lambda) = \int_0^\infty \frac{t^{\lambda-1}}{e^t} dt \tag{10}$$

Where: X_i^t indicates the *t-th* generation solution of the *i-th* nest; a indicates step scale factor; $L(s, \lambda)$ indicates random search path; s indicates step length; λ indicates exponential of the gamma function; $\Gamma(\lambda)$ indicates gamma function.

3.2. Solution process

First, enter the processing matrix MTAD that the RMT can provide and the processing matrix OPTAD that the process needs to obtain the candidate list of RMT. Next, the machining time of the candidate machine, the installation and disassembly cost of the reconfigurable and fixed modules and the MOCS parameters are entered. Finally, the MOCS algorithm is applied to solve the mathematical model, and output the solution set. The flow chart is shown in Figure 1.

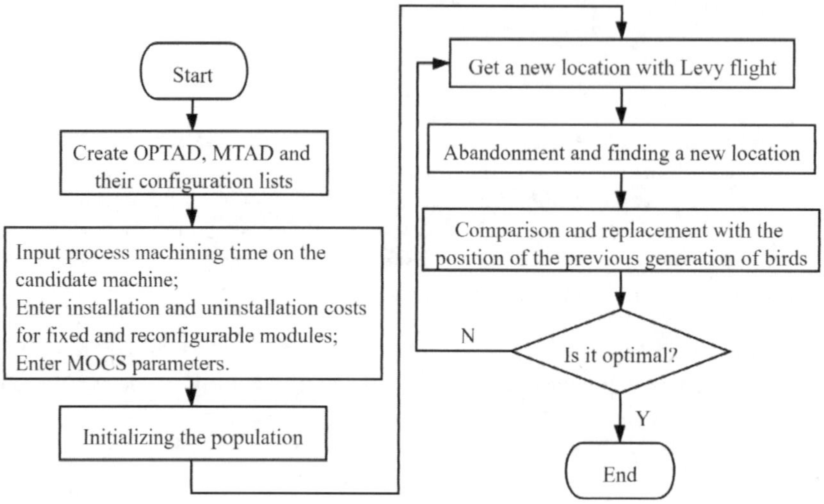

Figure 1. Reconfiguration configuration optimization flow chart.

4. A digital twin-based simulation and prediction framework for RMS

The concept of digital twin, first proposed by Grieves in 2003, is a realistic mapping of physical entities in a virtual space and simulates, monitors, analyzes and computes the physical entities, etc [9]. In addition to reflecting the current state of physical entities in the virtual space, digital twins can also predict the future state and behavior of physical entities. In this section, a digital model of RMS is constructed and an online simulation and prediction mechanism for the operational state of RMS is built to realize real-time, online simulation and prediction.

The digital twin consists of three parts [10], namely the physical layer, the virtual layer and the information layer. The physical layer is a real space where people, machines, and materials are assembled, which is always equipped with various subsystems and deployed sensors, each with different functions that jointly support the operation of the machines and sensors that collect and transmit equipment and environmental data; the virtual layer includes the operation and evolution of the virtual system and consists of multidimensional models, including geometric, information, control, and behavioral models. The information layer is the place where the data onto the physical layer and the virtual layer is fused and analyzed, so as to map the productivity and operation state of the real RMS in real time, and then provide the basis of simulation prediction.

4.1. Digital twin models construction

4.1.1. Rapid definition of RMT. RMT are the basis of realizing RMS. The main difference between RMT and ordinary machine tools is their modular structure and reconfiguration, i.e, RMT can adjust their machining functions and machining capabilities through the reorganization and replacement of machine component modules to meet the various changing needs of machined parts in a timely and efficient manner [11]. As shown in Figure 2.

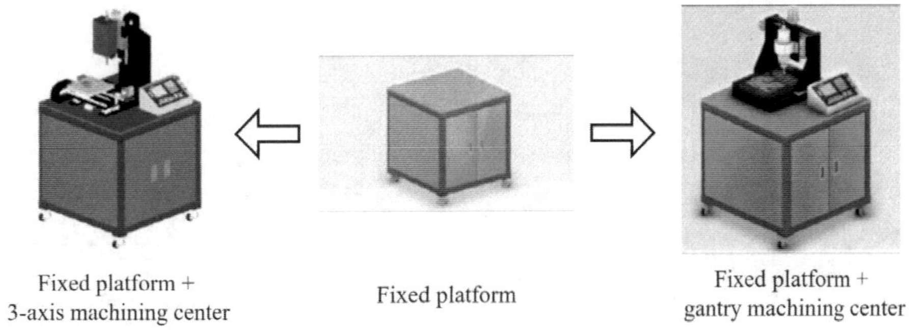

Figure 2. Schematic diagram of the standardized platform for integrating various modules.

Each RMT is functionally different, but they have common features that can be represented in the digital model. In order to be able to quickly construct a virtual model of a RMT for virtual space, define its five-tuple data model.

$$RMT = \{GeoM, InfM, MotM, ConM, ComInt\} \quad (11)$$

Five of these elements are represented as geometry model, information model, motion model, control model and communication interface. Among them *GeoM, InfM* can be further modeled as follows.

$$GeoM = \{ G\_point, G\_line, G\_sur, G\_bod, G\_coo \} \quad (12)$$

The elements in the formula represent points, lines, surfaces, bodies and coordinates for the visual display of the digital twin model in the virtual space.

$$InfM = \{ I\_Macid, I\_Modid, I\_TAD \} \quad (13)$$

The elements in the formula represent the name of the device, the name of the module and the TAD that the RMT can provide, respectively.

The motion model defines the RMT motion nodes by combining modules; the control model parses the real-time data and drives the RMT motion nodes; and the communication interface defines the interaction channel with external data.

4.1.2. RMS modeling. The digital modeling of RMS is mainly for the information interaction between production resources and reconfigurable machine model of. To solve this problem, a Resource Node (RN) models is established, and the control methods of reconfigurable machine model and resource nodes, and resource nodes and resource nodes are established.

The resource node model allows for the identification of passing production resources, and only when the metric meets the requirements will the identified data interact with the corresponding device for information, the resource node model expression is as follows.

$$RN = \left\{PR\_id, PR\_pro, PR\_pos, PR\_op\right\} \quad (14)$$

where: *PR_id* indicates the production resource name; *PR_pro* indicates the processing information of the production resource, i.e., which operation direction is required; *PR_pos* indicates the location information of the production resource; *PR_op* indicates the production resource operation at this resource node, i.e., to determine whether the production resource is shipped to the next resource node or interacts with the device at this resource node.

The logical relationship between reconfigurable machine tools then conforms to the machining process sequence of the parts, only the first process of a part to be processed, the next process can be processed, strictly in accordance with the processing order of execution, the logic control method is shown in Figure 3.

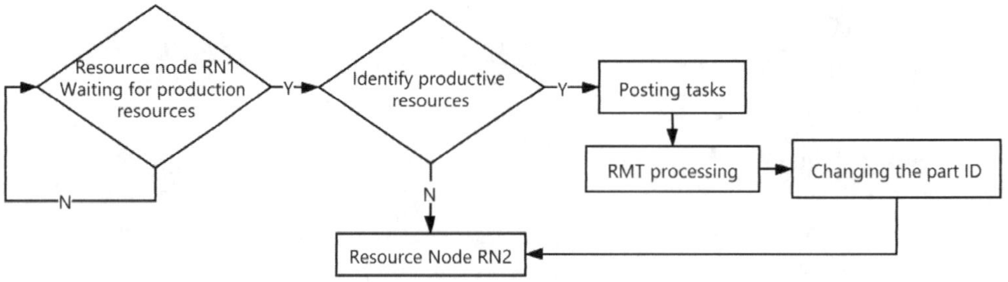

Figure 3. Logic control schematic.

Resource node 1 waits for the inflow of production resources. When the production resources flow into resource node 1, it will judge whether the production resources meet the processing criteria, and if they do, the resource node will issue a task to the RMT for processing and change the part ID to identify the part for primary process processing; if not, it will flow into the next resource node 2.

4.2. Predictive simulation frameworks based on RMS

Reconfiguration of RMS should coordinate multiple variables (including their reconfiguration configurations, process processing, delivery dates, etc.) and their constraints, making the modeling between them more complex and the computational space expanding, leading to huge computational consumption. Digital twin technology provides a powerful approach to support system reconfiguration, optimizing, judging and predicting reconfiguration differences of RMS. The framework of reconfigurable simulation prediction based on digital twin is shown in Figure 4.

The detailed implementation process is as follows.

(1) Data-driven source, real-time collection of real data in physical space (production tasks, equipment status, etc.) through sensing devices such as RFID, sensors and physical systems.

(2) The communication mechanism between the physical layer, information layer and virtual layer is established, and the local area network is set up through the WAN interface, and the TCP/IP communication protocol is used to make the data transmission between each part more stable and with lower latency, and to realize dynamic mapping and real-time updating of data.

(3) The physical layer transmits the collected data to the information layer, which parses the data, and the parsed data is transmitted to the virtual layer as a driver.

(4) The virtual layer performs simulation by driving data, generates sample data (production tasks, equipment status, etc.), and transmits the sample data to the information layer, which analyzes and optimizes the virtual layer transmission, transmits the optimization results to both the virtual layer and the physical layer, and the virtual layer performs optimization verification and the physical layer performs optimization execution, thus realizing online prediction of RMS.

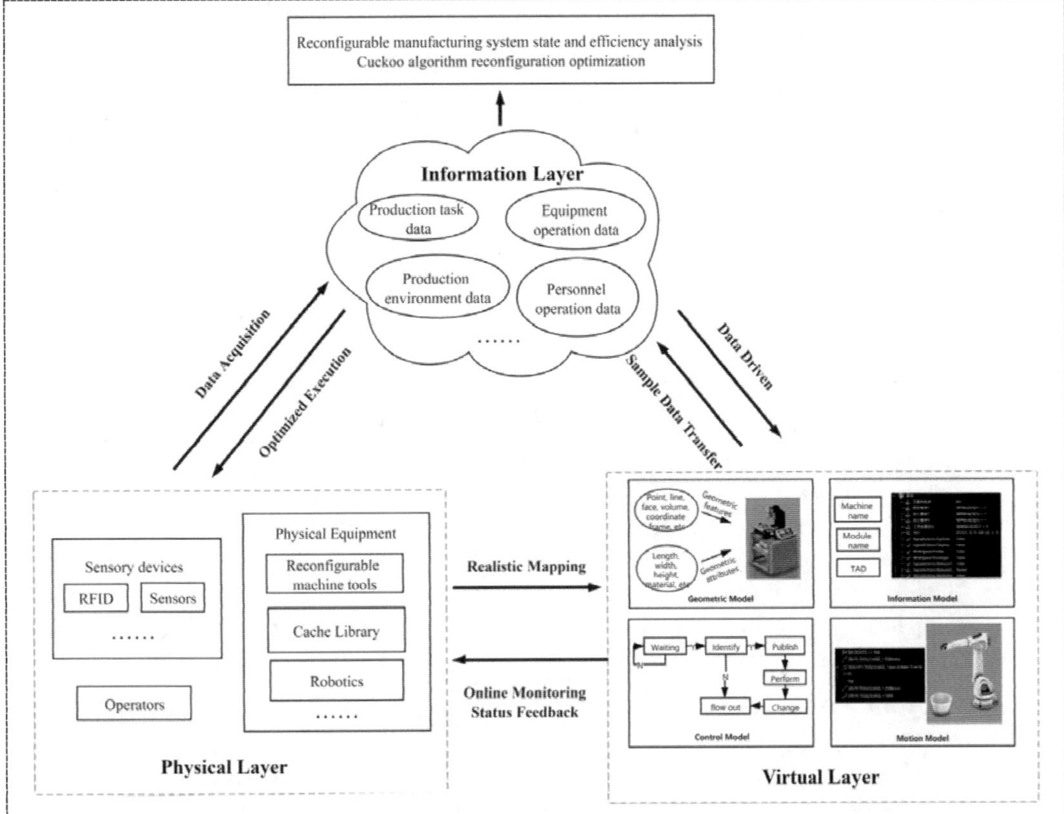

Figure 4. Digital twin-based simulation and prediction framework for RMS.

5.Case study

Suppose there is RMS that produces three parts to form a part family with an order quantity of 200 parts each. Due to the market change, the order delivery period is advanced from 11000min to 9000min. The number of processes of these three parts is 3, 3 and 2, respectively. There are 8 RMT and the module library contains 6 reconfigurable modules and 4 fixed modules.

Table 1 indicates the TAD required for each process of each part. Table 2 indicates the TAD that the RMT can provide in different configurations. Table 3 indicates the machining time of the RMT for the process in different configurations. Table 4 indicates the cost of installation and uninstallation of reconfigurable modules and fixed modules. Table 5 indicates the parameters of MOCS. Table 6 indicates the initial configuration of the RMS. Figure 5 shows the efficiency analysis of the RMS.

Table 1. TAD required for the process.

Parts	Process	TAD					
		+x	+y	+z	-x	-y	-z
P1	OP1	1	1				1
P1	OP2	1			1		
P1	OP3		1	1			1
P2	OP1	1				1	
P2	OP2		1			1	
P2	OP3	1	1		1		
P3	OP1		1		1		
P3	OP2	1		1			

Table 2. TAD provided by RMT.

RMT	Reconfigurable modules	Fixed modules	TAD					
			+x	+y	+z	-x	-y	-z
M1	RMod1、2、4、5	FMod1	1	1		1	1	
M1	RMod1、2、3、4		1	1	1	1		
M1	RMod2、3、4、5	FMod2		1	1	1	1	
M2	RMod1、2、5、6		1	1			1	1
M2	RMod2、4、5、6	FMod3		1		1	1	1
M2	RMod1、2、4、6		1	1		1		1
M3	RMod1、2、3、4	FMod4	1	1	1	1		
M3	RMod1、3、5、6		1		1		1	1

Table 3. RMT processing time.

		M1	M2	M3	M4	M5	M6	M7	M8
P1	OP1					15	16		
P1	OP2	15	12					17	
P1	OP3			15	13	18			
P2	OP1	16					18		14
P2	OP2	12		13			15		
P2	OP3	18						20	
P3	OP1	15			16			14	
P3	OP2		17			20		18	19

Table 4. Handling costs for reconfigurable modules and fixed modules.

modules	RMod1	RMod2	RMod3	RMod4	RMod5	RMod6	FMod1	FMod2	FMod3	FMod4
Loading and unloading costs	5.0	5.1	5.2	5.3	5.4	5.5	24	25	26	27

Table 5. MOCS Parameters.

Parameters	Meaning	Value
MaxGen	Maximum number of iterations	250
n	Number of populations	25
Pa	Probability of discarded solutions	0.2
α	Step size scaling factor	0.1
λ	Exponential of gamma function	1.5

Table 6. RMS initial configuration.

Parts	Process	Initial machine
P1	OP1	M6
P1	OP2	M7
P1	OP3	M3
P2	OP1	M1
P2	OP2	M3
P2	OP3	M7
P3	OP1	M4
P3	OP2	M8

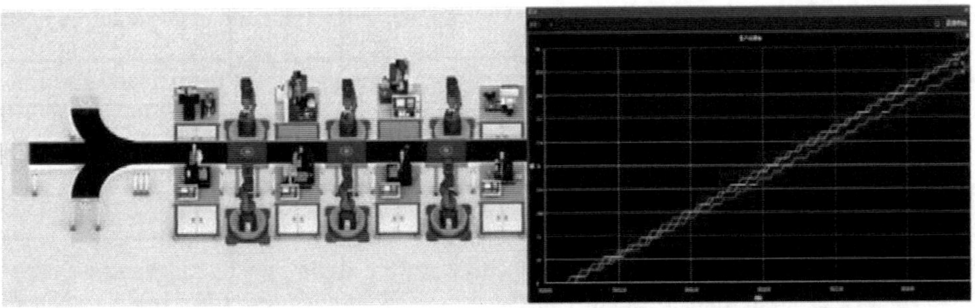

Figure 5. efficiency analysis.

The efficiency analysis of the RMS is shown in Figure 5. When RMTs are selected (M5, M2, M4), (M8, M1, M1), (M7, M2), parts P1, P2, P3 machining the shortest duration of 40, 44, 31, and the completion time of 8800, by the final completion of part P2, configuration cost of 336.4. When RMTs are selected (M5, M7, M4), (M8, M3, M1), (M4, M8), parts P1, P2, P3 machining the shortest duration of 45, 45, 35, and the completion time is 9000, by the final completion of parts P1 and P2, configuration cost is 133.4.

The latter configuration minimizes system-level reconfiguration while meeting order change requirements. Compared with fixed modules, reconfigurable modules are more convenient and economical to install and uninstall, which effectively reducing configuration costs. If the first configuration scheme is used, refactoring within 200 min can meet the order change requirements; if the second configuration scheme is used, immediate refactoring is required to meet the requirements.

6. Conclusion

The problem of when RMS will be reconfigured is combined with digital twins to simulate and predict the reconfiguration time point of RMS. This paper analyzes the machining capability and machining function mapping matrix of RMT, and further constructs a digital twin model of RMT and RMS to map the functions and attributes of RMT in virtual space. Finally, an online simulation prediction mechanism is established to predict the reconfiguration time point by sample data, which is a reference basis for RMS to reconfigure within a reasonable reconfiguration time. However, the article assumes one RMT for one process and does not consider the case of one RMT for multiple processes, which will be further investigated in the future.

Acknowledgements

This work is financially supported in part by University Industry-University-Research Innovation Fund of the Science and Technology Development Center of the Ministry of Education(No. 1: 2018 C 01068); Science and Technology Research Fund of Jiangxi Provincial Department of Education(No. 2: GJJ191006); 2020 Nanchang Engineering Technology Research Center and Key Laboratory Fund(No. 3: 2020-NCZDSY-005).

References

[1] Huang S H, Wang G X, Shang X W, et al 2019 Conceptual model design method for multi-scale configuration of reconfigurable manufacturing system *Computer Integrated Manufacturing Systems* **25**(11) 2803-12

[2] Koren Y 1999 Reconfigurable Manufacturing Systems *Journal of Manufacturing Systems* **29**(4) 130-41

[3] Du J 2015 *Optimal Design of Configuration and Multi Scale Prediction of Reconfiguration Time for Reconfigurable Manufacturing System* (Beijing: Beijing Institute of Technology)

[4] Huang S H, Wang G X, Shang X W, et al 2016 Decision-making of reconfiguration timing for

reconfigurable manufacturing systems considering functionality and capacity *Computer Integrated Manufacturing Systems* **22**(09) 2135-44

[5] Jiewu L et al. Digital twin-driven rapid reconfiguration of the automated manufacturing system via an open architecture model[J]. *Robotics and Computer-Integrated Manufacturing*, 2020, 63

[6] Hu X, Liu J H, Zhuang C B, et al 2021 Digital twin-based management method and application for the complex products assembly process *Computer Integrated Manufacturing Systems* **27**(02) 642-53

[7] Dou J P, Dai X Z and Meng Z D 2010 Optimisation for multi-part flow-line configuration of reconfigurable manufacturing system using GA *International Journal of Production Research* **48**(14) 4071-100

[8] Wang T Y, Sun X R, Tian S L, et al 2021 Research on Multi-objective Optimization Selection Method of Reconfigurable Machine Tool *Journal of Tianjin University: Science and Technology* **54**(9) 9

[9] Liu J, Zhuang C B, Liu J H, et al 2021 Online prediction technology of workshop operating status base on digital twin *Computer Integrated Manufacturing Systems* **27**(02) 467-77

[10] Guo H F et al 2021 A digital twin-based layout optimization method for discrete manufacturing workshop *The International Journal of Advanced Manufacturing Technology* **112**(5-6) 1307-18

[11] Zhao Z M 2010 Research on Modularization Design of Reconstructed Machine Tool *World Manufacturing Engineering & Market* **01** 96-100

Research on Trusted Access of Ubiquitous Power Internet of Things Terminal Based on Hyperledger

Liandong Chen[1], Kai Cheng[1], Conglin Zhao[1], Gang Li[2,*] and Xinying Wang[2]

[1]State Grid Hebei Information & Telecommunication Branch, Shijiazhuang, Hebei, China
[2]North China Electric Power University/School of Control and Computer Engineering, Baoding, Hebei, China

[*]Corresponding author's e-mail: ququ_er2003@126.com

Abstract. The secure access of terminals is an important problem that hinders the construction of ubiquitous power Internet of things. This paper analyzes the requirements and problems of ubiquitous power Internet of things terminal access, introduces the alliance chain framework of blockchain technology, selects super Ledger as the basic framework, and introduces it into the ubiquitous power Internet of things terminal access authentication scheme, and proposes a trusted access scheme of ubiquitous power Internet of things terminal based on Super ledger. Firstly, based on the framework of fabric project in hyperledger, the terminal access framework based on Hyperledger is designed, and the overall framework is mapped into a classic five layer structure of blockchain for functional design. Then, the terminal access process is divided into two stages: terminal access and identity authentication to achieve the authority control and identity authentication of the terminal and ensure the safe access of the terminal. Finally, the feasibility of the proposed scheme is verified by comparing with the traditional scheme through simulation experiments.

1. Introduction

With the development of the Internet of things, many problems are gradually revealed. Wider coverage and more terminal access applications bring higher risks to the ubiquitous power Internet of things, which is more vulnerable to attack. In addition, the traditional power Internet of things adopts a centralized system architecture. Under the same risk, once a malicious node enters the network to attack the central node, or even impersonate as a central component, the impact on the whole system is difficult to estimate [1]. Therefore, in order to meet the needs of the construction and development of the ubiquitous power Internet of things, how to deal with the various risks faced by the power grid under the new energy network model and realize the trusted access of the terminal is imminent.

Blockchain is a distributed ledger model, which has the characteristics of distributed ledger storage, node common maintenance consensus, traceability of transaction information, protection of information from forgery and so on. These characteristics can well adapt to the distributed security authentication requirements of ubiquitous power Internet of things, and effectively promote the development of ubiquitous power Internet of things to a higher form of distributed and intelligent.

This paper proposes a trusted access scheme for ubiquitous power Internet of things terminal based on blockchain. The scheme establishes a trust contract for secure access of terminal based on blockchain, digital signature, PKI system, asymmetric encryption and other technologies, realizes cross domain distributed identity authentication and access control, and solves the problem under the traditional centralized architecture, It is prone to a series of security problems caused by the attack of authentication center node or even being replaced by malicious nodes, so as to ensure the security of access terminals in different types, time and space under the power Internet of things. The simulation experiment is designed to verify the feasibility of the scheme.

2. Related research

Strengthening the access control in the network and system is one of the cornerstone technologies of data protection in the Internet of things. Sandhu proposed RBAC model, and LANM [2] made unremitting efforts to expand the scope of application of RBAC model. They realized the role mining in some complex environments, but there are still great limitations in practical application. Santos DR [3] introduced risk control, and Yuan e et al. The above models have some defects, either the application field is narrow, or they are not suitable to be used alone. In the environment of big data, the access control model is to meet the requirements of distributed, minimum authorization principle, dynamic access control and other new situations. It is necessary to study new access control model to deal with the challenges in the new environment.

Blockchain is a new distributed infrastructure[4], distributed storage and encryption algorithm can maintain the environmental security of power system operation; The consensus mechanism enables all nodes to participate in the maintenance of the account book, ensuring the new credibility of the transaction. These characteristics can well adapt to the distributed security authentication requirements of ubiquitous power Internet of things, and effectively promote the development of ubiquitous power Internet of things to a higher form of distributed and intelligent.

3. Preparatory knowledge

3.1. Blockchain hierarchy

Blockchain [5] is a decentralized account book which is composed of data structures such as blocks and maintained by all users. It is an orderly linked list in which the location of the previous block is confirmed by the hash value of the parent block. Each block is connected in chronological order. At the same time, the consensus mechanism ensures the consistency of the account books maintained by each node, and the smart contract provides the development basis for expanding the application field.

In the research, the architecture of blockchain system is generally divided into six layers, that is, from the bottom data layer to the top application layer. The responsibilities and functions of each layer are different, and there are certain connections and mutual support between them. It is this kind of architecture and the technologies on which each layer of the architecture depends that make the blockchain decentralized and traceable. At the same time, it can also be considered that at the bottom of the six layers architecture, there is the traditional cryptography as the basic support layer. The architecture hierarchy of blockchain technology is shown in Figure 1.

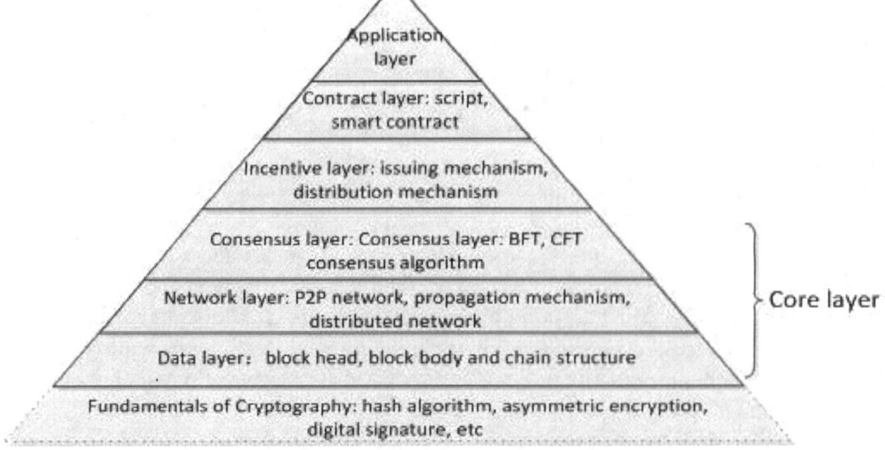

Figure 1. The hierarchy of blockchain

3.2. Hyperledger fabric structure

Hyperledger [6] completes the shared ledger with permission, and it can be divided into several excellent sub projects. Here we use hyperledger fabric as the research framework.

The advantages of hyperledger fabric are as follows:

- The structure of alliance chain meets the demand of incomplete decentralization of power system.
- Identity authentication and authority management are more suitable for the privacy and security needs of the power Internet of things where data cannot be fully disclosed. Through the authentication and management of nodes, the security and concealment of system data are protected.
- Pluggable consensus mechanism can configure more appropriate consensus algorithm to fit the actual application scenarios of power Internet of things. The consensus algorithm can be improved according to the needs of the actual power scene.

4. Terminal access scheme based on hyperledger

This scheme mainly relies on the framework of fabric project in super ledger, combines it with the application scenario of power system terminal equipment access, and designs the overall blockchain architecture. This paper focuses on the design of three core layers. The meanings of the symbols used below are shown in Table 1.

Table 1. comparison of symbols and meanings in the text

Symbol	significance
M	management device
C	controlled end equipment
MP	authentication management network composed of management devices
SKM、PKM	The private key and public key generated by the key generation system for the management device
v	The key in the key chain generated for the controlled end device
r	Merkle root is stored in Merkle tree
ID	Preset identity information for the device
T	Generated timestamp
h()	Secure hash function

4.1. Data layer design

The entity of the whole system is analyzed. Before accessing the network, the devices to be accessed should be divided into management end and controlled end according to their physical properties and functions in the network

Management terminal M: mainly includes the master station and control center equipment as the manager in the system, or the terminal equipment with relatively high hardware facilities and able to complete higher requirements. When the system is initialized or connected to the network, the M nodes that initially exist in the system form the authentication center in the system, which enjoys the accounting right for the block chain in the system and the management right for the node authority.

Controlled end C: basic terminal equipment. Because of its low performance or importance, it doesn't have the right of accounting, so it needs to obtain the right to access the block data indirectly through M.

The whole ubiquitous power Internet of things system should be controlled by： A lot of C equipment; A peer-to-peer authentication network composed of multiple M's; The composition of CA center with preset ID.

4.2. Network layer design

The network layer will be initialized in the initial stage of the whole system construction to build the initial MP network. In the network, the initial M selected by the staff constructs the voting center of the system, which enjoys the rights of accounting and management. More than 50% of the ripple consensus is taken as the consent node to access to the network, and the judgment on C is made: m will join the certification blockchain, and C will only implement device access to the network.

In the next step, the terminal equipment should send the access request to the CA center before the identity authentication of the access system. After receiving the response, the terminal equipment can get the access right of the intranet, and obtain the distributed key and certificate, and then carry out the access identity authentication. Only when the key of CA center is obtained, the whole process of node access can be continued, and then the whole blockchain network can be constructed.

4.2.1. Generate Mi 's key:

Step 1: *Mi* uses *RSA* to generate a random number *n*.

$$SKM_i = n \tag{1}$$

Step 2:

$$PKMi=RSA\ (SKMi\ ) \tag{2}$$

4.2.2. Generate C_{ij} 's key: Because the hardware of Cj device is relatively poor and easy to be intruded, its key should be changed regularly. A key chain of Cij device can be generated and managed by Mi device with better performance.

Step 1: The CA center generates the cipher chain of the C_{ij} device from the key of the M_i device managing the C_{ij}, and v_i is the I item of the chain.

$$vi=hi(PKMi)=h(h(h(...h(PKMi)...))) \tag{3}$$

$$v0= PKMi \tag{4}$$

Step 2: Save as Merkle root ri :

$$ri = h(vi) = vi\text{-}1 \quad i \in\{1 < i \leq n,\ i \bmod 2\neq 0\} \tag{5}$$

4.3. Consensus layer design

Identity authentication and access control is the most practical security technology in the traditional Internet of things security research. We can divide the whole terminal access into two stages: terminal access and identity authentication, so as to better protect the security and credibility of terminal access.

Before access authentication, the terminal needs to check its preset identity information and key by MP to verify its identity, so as to realize the security and credibility of the initial stage of the system. At this time, it will judge the access terminal ID information. If it is an M-type device, it will join the MP to obtain the accounting right and management right of the whole blockchain after it has access permission; If it is a class C device, only access to the intranet. After the new terminal is connected to the network, a new block will be released.

The steps from terminal access to authentication are divided into three stages, as shown in Figure 2.

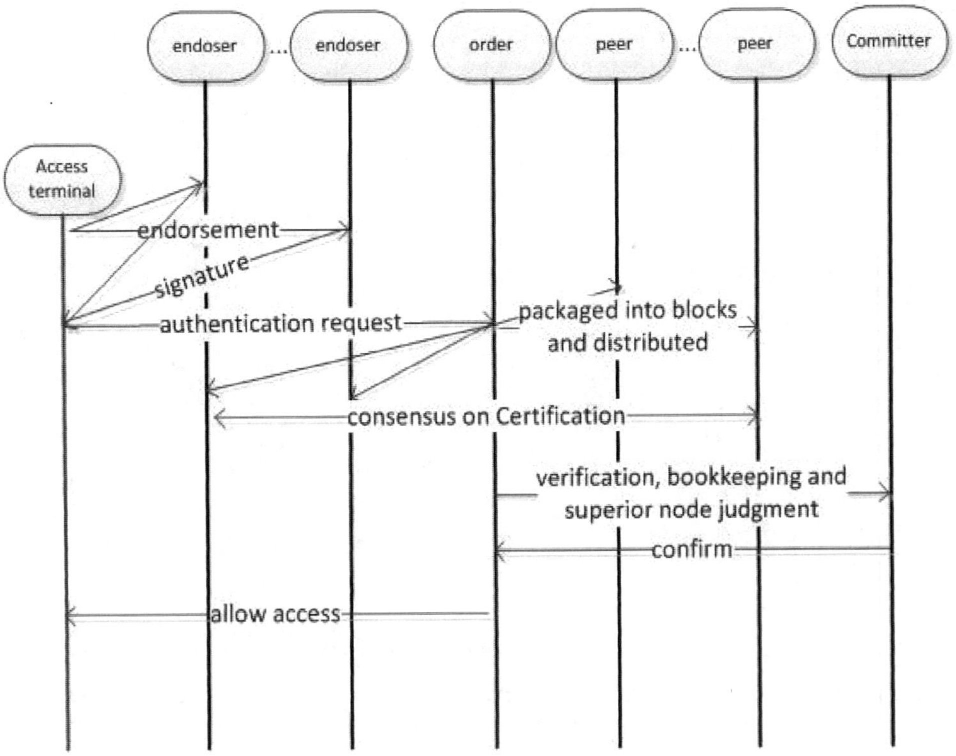

Figure 2. Specific flow chart of terminal access authentication

Endorsement stage. The access terminal applies for endorsement signature from the endoser node. The endorsement policy defines which nodes are required for endorsement transactions to be effective. Access to the internal network can only be allowed after receiving sufficient endorsement signature access response, otherwise access to the network is not allowed.

5. Simulation verification and result analysis

In this section, hyperledger and chain code are deployed in the laboratory to simulate the power Internet of things access authentication process to verify the effectiveness of the proposed algorithm. The experimental environment was Intel Core i5-9300H @ 2.40GHz, 8GB Crucial DDDR 2666MHZ memory, 512GB SSD, NVIDIA gtx1660ti GPU.

The simulation terminal initiates access authentication request, uses fabric network to set authentication node and sorting node on the chain, and deploys PBFT authentication algorithm on the chain code. This paper mainly records the throughput and service delay of the experiment, and tests the terminal trusted access scheme. As an org organization, the certification center composed of each group on the chain builds a network for testing.

In order to reduce the influence of accidental error, the mean value of each experiment was taken five times. Finally, the experimental results are presented in the form of charts. Compare the terminal access authentication time based on blockchain with that under traditional authentication scheme under different node sizes, as shown in Figure 3.

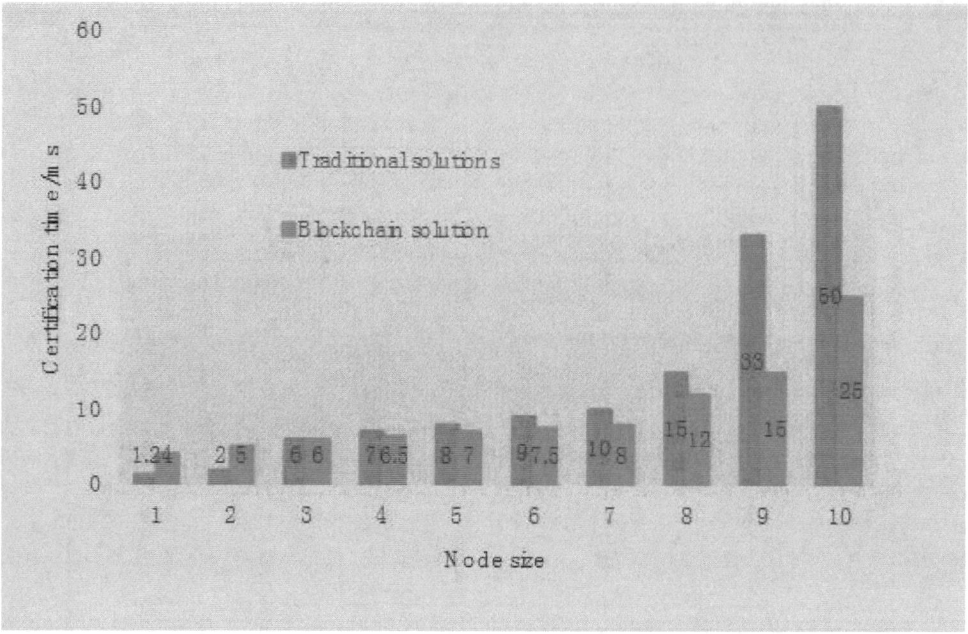

Figure 3. Comparison between traditional scheme and blockchain scheme

Through the comparative analysis of the curve, it can be concluded that when the number of authentication nodes in the traditional authentication scheme gradually increases, the burden of the central node increases, the cost of calculation increases, the corresponding terminal access authentication time increases, the decreasing trend of authentication efficiency is more and more significant. However, as the size of the node increases to a certain extent, compared with the traditional scheme, the computational cost of this scheme increases. Due to the common maintenance of distributed multi nodes, the growth of computational cost is relatively small, and the authentication efficiency is higher.

6. Conclusion

In view of the current situation that the security risk of the terminal is increasing in the development and construction of the ubiquitous power Internet of things, this paper proposes a research scheme of trusted access of the ubiquitous power Internet of things terminal based on blockchain technology, It provides a distributed terminal access security model for the construction and development of new energy network, which is conducive to ensuring the terminal security of the whole power system, allowing the system to process the terminal access requests with higher efficiency, and maintaining the terminal security of the power system under the background of high concurrency.

Acknowledgments

This paper is supported by the Science and Technology Project of State Grid Hebei Information & Telecommunication Branch (No.SGHEXT00WLJS2000144). We would like to thank anonymous reviewers for their helpful comments and suggestions.

References

[1] Fromknecht C and Velicanu D 2014 Authentication System vol. 21, no. 2, pp 857-867.
[2] Lan Molloy, Youngja Park , Suresh and Chari 2012 ACM symposium on access control models and technologies pp 45-55.
[3] Santos, Daniel Ricardo Dos and Westphall 2014 IEEE Network Operations and Management Symposium pp 1-9
[4] Xu D J, Ning H and Shen S 2020 Electronics vol. 9, no. 4, pp 634

[5] Krishna , Mohan and Botcha 2019 Far east journal of electronics and communications vol. 20, no.1, pp 9-19.
[6] Yang L and Blockchain 2018 Journal of Management Analytic vol. 5, no. 4, pp 231-255

Detection of Auxiliary Equipment in Engine Room Based on Improved SSD

Jiahao Qi, Jundong Zhang*, Qingyan Meng, Jiaming Ju and Haotian Jiang

Marine Engineering College, Dalian Maritime University, Dalian, Liaoning, China

*Corresponding author's e-mail: zhjundong@dlmu.edu.cn

Abstract. To improve the intelligent operation and maintenance of ships, the research on the perception of engine room auxiliary equipment (A/E) will realize AUTO-0 and replace the engineers. Due to compact layout of the cabin and the different shapes of equipment, the Single Shot MultiBox Detector (SSD) that uses the traditional convolution method for feature extraction cannot adapt to the complex environment. Therefore, an improved A/E detection algorithm based on SSD is proposed. Firstly, we collected the images of actual ships, annotated them manually, and used the K-Means clustering algorithm to analyze them. Secondly, we replaced the classification loss of SSD with the focal loss, balancing the positive and negative samples in the dataset. Finally, we added repulsion loss for the overlapping targets to improve the detection effect of the model against dense occlusion & overlap. The experiment is based on the dataset containing 18185 training object boxes, and 2044 test object boxes, the mean average precision (mAP) in this paper reaches 78.95%, which is 5.63% higher than the original SSD, and the detection boxes are more suitable.

1. Introduction

The perception of intelligent E-Patrol robots will be way to operate AUTO-0 cabin on ships in the future, and the engine room equipment will be the main research object. To realize the intelligent perception of patrol robot, a series of basic research and scientific problems need to be solved, such as communication [1], positioning [2], information acquisition [3], path planning [4], target perception and detection [5], and target grasping [6]. Among them, target detection is an indispensable part of the environment perception technology in smart ships.

Whether the current equipment system status can be accurately detected is closely related to the operation of the engine room. Based on the detection technology of A/E in the engine room, the E-Patrol robot can use the camera to automatically recognize the front equipment, and provide timely feedback to the crew onshore, which is conducive to preventing accidents and fully guaranteeing the safety of ships. At present, there is no perception model specifically in cabins, but various object detection methods [7-11] based on deep convolutional neural network (DCNN) have achieved significant performance in the field of ocean and marine engineering, such as underwater target detection, surface target detection and port species recognition. Due to the use of a more complex backbone network, the detection accuracy will be reduced [12], we intend to improve the cabin environment of the model while retaining the SSD classical framework in this paper.

This manuscript is ordered as follows. The second part describes the dataset preparation, the original SSD, and its deficiencies in cabins. The improvement of SSD is described in the third part. The fourth part expounds on experimental the methods, provides the training details, analyses the results. The fifth part summarizes main ideas of the paper and points out the inadequacies.

2. Preparation

2.1. Dataset

To collect a large number of A/E images in ships, we realized the technical docking with our 3D virtual research group of ship engine room, a total of 6G images were collected eventually. As shown in Figure 1, the dataset contains many complex scenes, which lay a solid foundation for building the experimental dataset, and have a significant impact on the final performance of the trained model. In the training tasks of deep learning, the most prone to appear is overfitting, and the main reason is the amount of data used in training is not enough so the data cannot contain enough information. To enhance the number of images, this paper adopted data augmentation in Figure 2, which is to perform the following geometric transformations on the original images, such as adding noise, changing brightness, cutout, flip, rotation, shift picture. Finally, we combined one or several augmentations to obtain the final datasets and label them as VOC.

Figure 1. Examples of A/E dataset.

Figure 2. Examples of data augmentation.

2.2. SSD model

2.2.1. Framework. To better combine contextual information, SSD uses a pyramid structure combined with feature maps of different receptive field sizes, which performs position regression and SoftMax classification on multiple scale feature maps simultaneously. The SSD framework is shown in Figure 3, it takes VGG16 as the backbone to extract features, and the fully connected layers FC6 and FC7 are replaced with convolutional layers Conv6 and Conv7, at last, the dropout and classification layer are removed. In addition, 4 convolutional layers are added. Each layer uses a 1×1 kernel to decrease the channel by first and then uses a 3×3 kernel to reduce the scale and increase the channel.

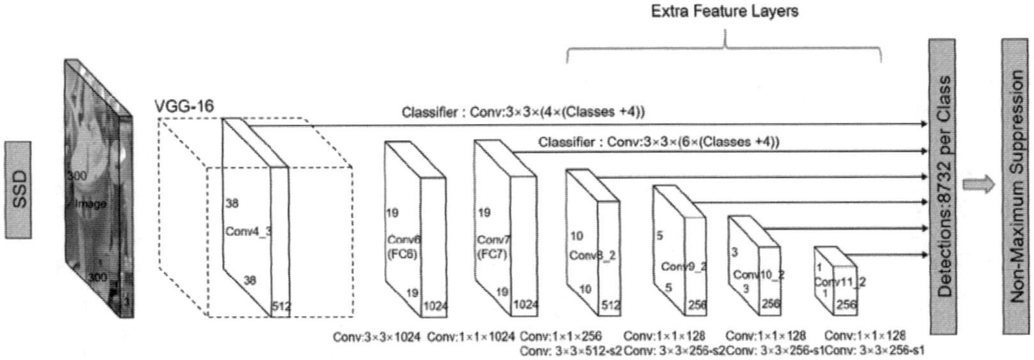

Figure 3. SSD Framework.

2.2.2. Priors. SSD spreads out default boxes of different scales to multiple layers and enforces each layer to focus on predicting objects of specific scale [13], so there will be feature maps of different scales involved in location regression and classification, which will generate the corresponding priors on each feature map. The priors of the SSD have independent detection values, and each prior can be divided into two sections, firstly with the confidence of each category, secondly with the relative location of the bounding box. The scaling factor needs to be calculated according to the sequence of the feature maps, that is, s_k as the scale of the k-th default box:

$$s_k = s_{\min} + \frac{s_{\max} - s_{\min}}{n-1}(k-1), k \in [1,n] \tag{1}$$

$$\begin{cases} w_k^a = s_k \sqrt{a_r} \\ h_k^a = s_k / \sqrt{a_r} \end{cases}, a_r \in \left\{1,2,3,\frac{1}{2},\frac{1}{3}\right\} \tag{2}$$

Where n is the number of feature layers, $s_{\min}$ and $s_{\max}$ represent the minimum and maximum ratio of the default box of the feature map to the input image. It is worth noting that when the default aspect ratio a_r =1, we need to add another ratio:

$$s_k' = \sqrt{s_k s_{k+1}} \tag{3}$$

2.2.3. Loss function. SSD uses a multi-scale training method, every corresponding prior is generated on each feature map, so there will be feature maps of different scales involved in location regression and classification, and its loss function L is the sum of the regression loss L_{loc} and the classification loss L_{conf}:

$$L(x,c,l,g) = \frac{1}{N}(L_{\text{conf}}(x,c) + \alpha L_{\text{loc}}(x,l,g)) \tag{4}$$

Where N is the number of matching between the bounding boxes and ground truth, if $N=0$, L = 0; x is the matching result of the bounding box and the ground truth, if it is a match, x=1, otherwise $x=0$; c is the confidence of the predicted object; l is the position offset information of the bounding box; g is the offset between the ground truth and the bounding boxes; α is the weight parameter of positioning loss.

Regarding location regression, the SSD uses SmoothL1, which is calculated by regressing the offsets of the center coordinates (cx, cy), width (w), and height (h) of the bounding boxes:

$$L_{\text{loc}}(x,l,g) = \sum_{i \in \text{Pos}}^{N} \sum_{m \in \{cx,cy,w,h\}} x_{ij}^k \ smooth_{L1}(l_i^m - \hat{g}_j^m) \tag{5}$$

Regarding classification, the SSD uses the SoftMax:

$$\begin{cases} L_{\text{conf}}(x,c) = -\sum_{i \in \text{Pos}}^{N} x_{ij}^p \log(\hat{c}_i^p) - \sum_{i \in \text{Neg}} \log(\hat{c}_i^0) \\ \hat{c}_i^p = \dfrac{\exp(c_i^p)}{\sum_p \exp(c_i^p)} \end{cases} \tag{6}$$

2.3. Deficiencies of SSD for A/E

2.3.1. Imbalanced samples. When using the original SSD to detect A/E in engine rooms, we found the detection accuracy is far lower than expected. After comparing the dataset carefully, we believed that the imbalance of samples is one of the main reasons. Due to the number of negative samples in image are too large, and negative samples are easy to classify. When detecting valves or meters in the engine room, because of the complex background and numerous equipment, the proportions of these two types of targets in the image are small, a large number of negative samples will be generated. If this phenomenon is not restricted, the model cannot evolve in the right direction during optimization, which will lead to a decrease in accuracy eventually. As shown in Figure 4, the proportion of target meters M1, M2, and M3 in the entire image is small, so many well-classified negative examples will be generated. At the same time, the proportion of difficult-to-positive sample B3 and well-classified negative examples B1, B4 are also much higher than the well-classified positive examples B2, so the proportion of the well-classified positive examples in the entire loss function will also be very small, which is unfavorable for loss function optimization. Therefore, it is necessary to design a new classification loss function to make the model better update the gradient.

2.3.2. Occlusion & Overlap. We found that the original SSD has weaker detection for overlapping targets. Due to the connection characteristics of the pipeline and equipment in engine rooms, the images show dense target occlusion, and the ground truth boxes of equipment are intersected. The bounding boxes generated in the model training will be affected by the interference of characteristics of the non-responsible target object in the regression process, resulting in the detector being unable to accurately locate the target object. As shown in Figure 5, S1, S2 are ground truth boxes, B1, B2 and B3 are bounding boxes. Since the B2 contains a part of S1, it is easily affected by S1 during regression, so that the regression position is not accurate. When B3 does not exist and B2 has the maximum confidence level concerning S1, B1 may be filtered during the NMS and leading to missed detection of S1.

Figure 4. The complexity of default boxes.

Figure 5. The influence of occlusion & overlap.

3. A/E detection model

3.1. Improvement of priors setting

The distribution of the original SSD bounding boxes and the cabin equipment dataset is shown in Figure 6, where the dots represent the width and height distribution of various equipment in the A/E dataset, the lines represent different aspect ratios in the original SSD, and the triangles represent bounding boxes in the original SSD. It can be seen from the figure that most of the bounding box's selection in the original SSD is quite different from the distribution of the A/E dataset. If the dataset is directly applied to the original SSD in this paper, a satisfied detection cannot be obtained. Therefore, it is necessary to redesign the default box, which can better match the real target. In terms of aspect ratio design, we need to perform K-Means clustering on the cabin equipment dataset and set the SSD bounding boxes and aspect ratio around the cluster center. In this paper, our metric is intersection over union (IoU). When the intersection between the two boxes is higher, the IoU is larger, that is, 1- IoU is used as distance judgment. According to the clustering results, we make changes to the original SSD default boxes as shown in Figure 7.

3.2. Improvement of imbalanced samples

For the one-stage detection method, if only the dense sampling window is detected, the number of negative samples will be much larger than the positive, and many of these negative samples are well-classified samples, eventually, the loss of well-classified examples will be greater than the loss of difficult-classified samples. When the loss of the difficult-classified samples does not contribute much to the gradient update of the model, the final trained model cannot achieve good accuracy. Therefore, we use the focal loss [14]:

$$FL(p_t) = -\alpha_t (1-p_t)^{\gamma} \log(p_t) \quad (7)$$

This new loss function is aimed at the imbalance of positive and negative samples, so that in the process of training the model, we can focus more on samples that are more difficult to classify.

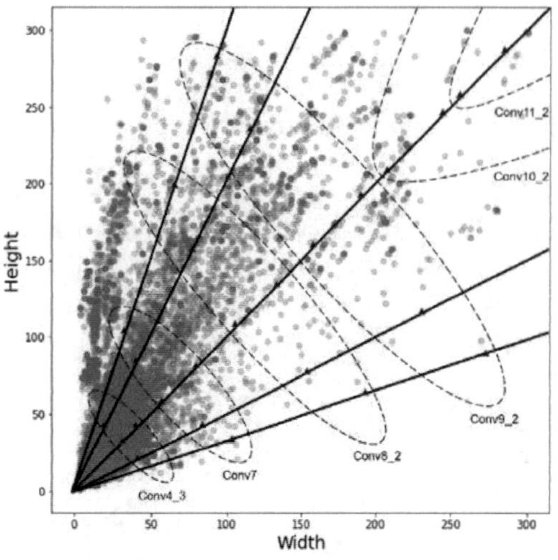

Figure 6. The original scale of default boxes.

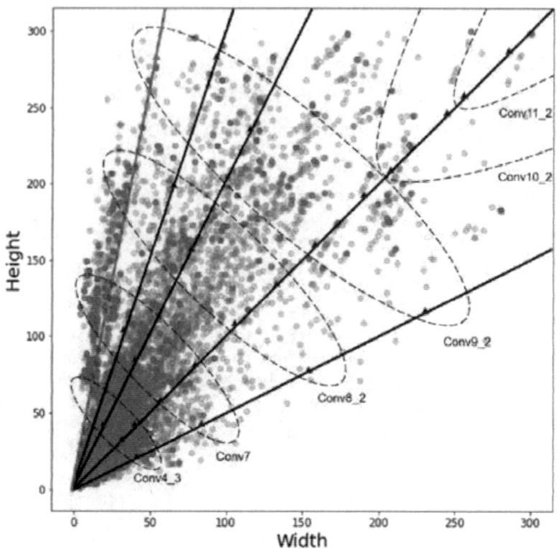

Figure 7. The corrected scale of default boxes.

3.3. Improvement of occlusion & overlap.
Aiming at the poor detection effect on overlapping targets, we add a repulsion loss [15] based on original the loss function of SSD, which can reduce the distance between bounding box and the ground truth responsible for regression and increase the distance between the bounding box and the non-responsible ground truth that intersects with the surrounding and the intersecting bounding box. In other words, the repulsion loss can make the bounding box closer to responsible ground truth, can also be far away from the wrong target, and reduce the mistake rate of the model. This function formula is as follows:

$$L = L_{\text{Attr}} + \alpha L_{\text{RepGT}} + \beta L_{\text{RepBox}} \quad (8)$$

Where L_{Attr} is the regression loss of the original model, which is used to calculate the loss between the bounding box and the responsible ground truth, as shown in formula (9); L_{RepGT} is the loss between the bounding box and the adjacent ground truth, which can far away from the nearby ground truth with the largest intersection ratio, as shown in formula (10); L_{RepBox} is the loss between the bounding box and the adjacent bounding box that is not predicting the same target, it can enlarge the distance between the two bounding boxes, as shown in formula (11); the correlation coefficients α and β are used to balance the two parts of the loss value.

$$L_{\text{Attr}} = \frac{\sum_{p \in P_+} Smooth_{\text{L1}}(B^P, G^P_{\text{Attr}})}{|P_+|} \quad (9)$$

$$L_{\text{RepGT}} = \frac{\sum_{p \in P_+} Smooth_{\text{L1}}(IoG(B^P, G^P_{\text{Rep}}))}{|P_+|} \quad (10)$$

$$L_{\text{RepBox}} = \frac{\sum_{i \neq j} Smooth_{\ln}(IoU(B^{pi}, B^{pj}))}{\sum_{i \neq j} \prod[IoU(B^{pi}, B^{pj}) > 0] + \varepsilon} \tag{11}$$

4. Experimental results

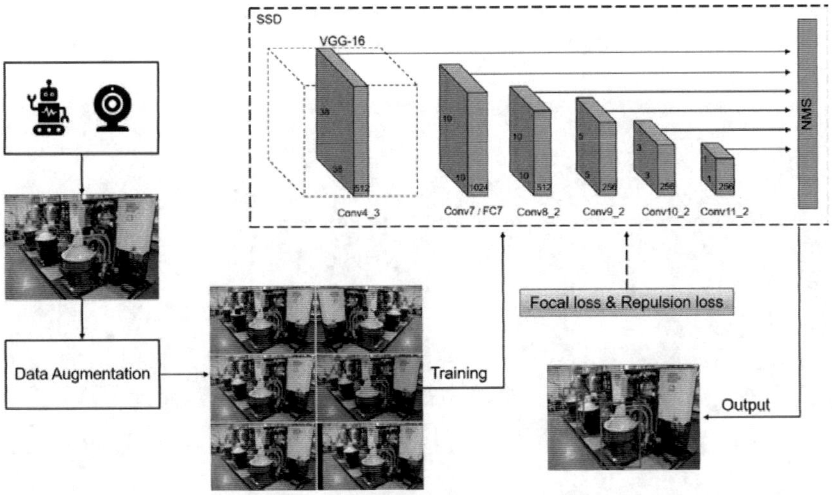

Figure 8. The process of auxiliary equipment detection.

The process of A/E detection is shown in Figure 8. Firstly, we collected a number of original images of A/E through image acquisition devices in ships. Secondly, a large number of images was enhanced through data augmentation, and then the A/E dataset was built. Finally, we used the model in this paper to train the A/E dataset and used the model to recognize and detect the A/E in engine rooms.

4.1. Analysis of dataset

To avoid over-fitting and balance the number of each category, we enhanced the A/E dataset by data augmentation in this experiment, and then, we obtained a self-made dataset containing 2,575 images and 20,229 ground truth object boxes, which include diesel engine, oil separator, reservoir, heat exchanger, meter, pump and valve. The resolution of the captured image was 6000×4000 pixels by Canon digital camera. Finally, we randomly selected 2085 images as the training set, 232 images as the validation set, and 258 images as the test set.

4.2. Model training

Table 1 describes some training details, which includes the configurations of experimental computer, and the setting of learning rate. It's important to note that the weights of VGG were unfrozen when the number of iterations reached the 21st epoch.

Table 1. Details of model training.

	Detail
CPU	Intel i7-9700
GPU	NVIDIA GeForce GTX 1660TI
Memory	32GB+1TB
Lr in 0-20 epoch	1e-3
Lr in 21-60 epoch	1e-4
Lr in 61-80 epoch	1e-5

4.3. Analysis and comparisons.

To optimize the performance of the improved SSD for A/E perception in engine room, we tested 258 images in the trained model, 7 types of equipment, and 2,044 ground truth object boxes. Figure 9 displays the detection results of A/E. Regarding the detection result, we hereby declare the followings: the first is that the motor was rated as a pump when we annotated the dataset, given the structure of the pump body, which explains why the driven motor of oil separator was tipped as pump; the second is the handle of valve was deemed as valve by default when we annotated the dataset, which explains why the valve body is not truly well-detected.

Figure 9. Examples of detection results.

4.3.1. Analysis of transfer learning. We sub-sampled the weights of the SSD300, rewrote the weights of the 6 classification layers in the original SSD, and used them as the pre-training parameters of the improved SSD in this paper. To grasp the significance of transfer learning, the contrast experiment is designed in this part. Compared to non-transfer learning, the mAP of transfer learning method is increased by 19.63%, and the final training loss is reduced by 0.32 in Table 2. Based on the experimental verification and analysis, the method of transfer learning has higher detection accuracy and verifies the performance of the model.

Table 2. Comparison of transfer learning.

	Initial Learning rate	Weight decay	Momentum Factor	Final Learning rate	Training Loss	mAP/%
Transfer	0.001	1e-4	0.9	1e-5	0.19	**78.95**
Non-transfer	0.001	1e-4	0.9	1e-5	0.51	59.32

4.3.2. Balance of focal loss and repulsion loss. When the hyperparameter α and β in the Repulsion Loss are both 0, compared to the original SSD, only the Focal Loss is involved in the optimization at this time. In order to obtain better weight α_t and γ, we designed the contrast experiment in this part and the comparison results are shown in Table 3.

Table 3. Hyperparameter of focal loss.

Weight α_t	Focusing γ	mAP/%
0.20	2.00	72.33
0.20	2.50	73.37
0.25	2.00	74.92
0.25	2.50	**75.51**
0.30	2.00	73.56
0.30	2.50	72.84

Then, we compare the hyperparameters in the Repulsion Loss. According to the best results of mAP in Table 4, which is to ensure that α_t and γ are 0.25 and 2.5 respectively. At this time, we change α, β and find that when they are 0.55 and 0.45 respectively, the perception in this paper is best.

Table 4. Hyperparameter of repulsion loss.

RepGT α	RepBox β	mAP/%
0.30	0.70	76.61
0.40	0.60	77.53
0.50	0.50	78.15
0.55	0.45	**78.95**
0.60	0.40	75.99
0.70	0.30	76.32

4.3.3. Comparison with other models. In order to verify the performance of proposed method, we compare the improved SSD with other methods for A/E perception in this part. We find the mAP is 6.42% higher than Faster R-CNN (72.53%), 9.81% higher than YOLOv1 (69.14%) and 8.09% higher thanYOLOv2 (70.86%), 5.81% higher than original SSD (73.14%), as Table 5 shows.

Table 5. Comparison of model test results.

Model	AP/%							mAP /%
	Engine	Separator	Cooler	Reservoir	Pump	Meter	Valve	
Faster R-CNN	93	80	90	82	68	50	45	72.53
YOLOv1	88	79	86	83	74	38	36	69.14
YOLOv2	91	81	89	76	69	48	42	70.86
Original SSD	100	92	91	83	73	39	34	73.14
Ours SSD	100	100	95	87	79	48	44	78.95

5. Conclusions

Aiming at the original SSD has a poor detection effect in the ship engine room, we propose an improved model based on SSD. We used K-Means to redesign the default box of the SSD to make it distributed in the clustering center of the data, increasing the overlap between the default box and the ground truth; because SSD is weak in detecting overlapping targets, we added the focal loss and repulsion loss based on the original loss function of SSD. The experiment shows that the improved SSD model has the mAP of 78.95% for A/E, which is 5.63% higher than the original SSD model. In addition, we have also noticed that the AP value of small objects such as meter and valve is not very high compared to diesel engine or heat exchanger. for the sake of improving the intelligent maintenance and operation of the ship's engine room, we will not only make small target improvements to the detection model of A/E, but also carry out lightweight processing for the model in this paper to make it suitable for portability, and upgrade the inspection robot's path planning, engine room positioning and other technical upgrades in future, so that it can better replace the engineer.

Finally, we suggest that readers can boldly try advanced technologies such as computer vision and apply them to the field of ship and ocean engineering. For example, UAV assist classification societies to complete remote ship inspections, USV assist rescue and salvage bureaus to complete search and rescue, or make it assist in the diagnosis of ship deformation cracks and other tasks.

Acknowledgments

This work was supported by the National Natural Science Foundation of China, grant number U1905212, and the Fundamental Research Funds for the Central Universities, grant number 3132016316.

References

[1] Xia M, Liu WA, Wang K and Zhang X 2019 Neural Computing & Applications 31(12) 8931-53
[2] Xia M, Liu WA, Shi BC, Wang LG and Liu J 2019 Int. J. Remote Sensing 40(1) 156-70
[3] Qian J, Xia M and Lv P 2015 Int. Conf. on Machine Learning and Cybernetics (Guangzhou) 316-21
[4] Yu Z, Sun FC, Lu X and Song YX 2021 Int. Conf. on Electronics, Circuits and Information Engineering (Zhengzhou) 368-37
[5] Weng LG, Sun XD, Xia M, Liu J and Xu YQ 2020 Neurocomputing 402 171-82
[6] Weng LG, Xu YM, Xia M, Zhang YH, Liu J and Xu YQ 2020 ISPRS Int. Geo-Information 9(4)
[7] Ma KW, Huang B and Yin HS 2019 IEEE Int. Conf. on Power, Intelligent Computing and Systems (Shenyang) 343-47
[8] Chen YT, Zhang XZ, Chen WN, Li YY and Wang JS 2020 IEEE Access 8 102907-19
[9] Hu K, Lu FY, Lu MX, Deng ZL and Liu YP 2020 Complexity
[10] Qiang W, He YY, Guo YJ, Li BQ and He LJ 2020 J. Northwestern Polytechnical University 38(4) 747-54
[11] Zou YY, Zhao LN, Qin SX, Pan MY and Li ZL 2020 IEEE 5th Information Technology and Mechatronics Engineering Conference (Chongqing) 1676-80
[12] Zhao H, Li ZW, Fang LF and Zhang TQ 2020 Neural Processing Letters 51(3) 2789-806
[13] Zheng QY and Chen Y 2021 Multimedia Tools and Applications 80 20283-305
[14] Lin TY, Goyal P, Girshick R, He K and Dollar P 2017 IEEE Int. Conf. on Computer Vision (Venice) 2999-3007
[15] Wang XL, Xiao TT, Jiang YN, Shao S, Sun J and Shen CH 2018 Proc. of the IEEE Computer Society Conf. on Computer Vision and Pattern Recognition (Salt Lake) 7774-83

Journal of Physics: Conference Series **2173** (2022) 012061 doi:10.1088/1742-6596/2173/1/012061

Stability Control of Transport Robot Based on Iterative Learning Control

Tingrui Liu, Yan Ding, Pan Wang, Kang Zhao and Jiahao Jia*

College of Mechanical & Electronic Engineering, Shandong University of Science & Technology, Qingdao 266590, China

*Corresponding author's E-mail: sdustjjh@163.com

Abstract. In this study, stability control for transport process of transport robot subjected to 2R manipulator movement, is investigated based on iterative learning control (ILC). The joint positions, speeds and accelerations are used as variables to establish the expression of driving torques of manipulator joints. According to the experience, the linear interference torque in the process of motion is determined. Three ILC algorithms are applied to achieve stability control, and good trajectory tracking results are obtained. Position tracking, speed tracking, and the maximum position error in the process of tracking are illustrated, and through the absolute value of the maximum error, the optimal iterative algorithm is finally determined.

1. Introduction

In the process of transport, the stability of robot manipulator is always an important control problem in the process of accurate transport, especially for adsorption handling manipulator [1]. The new adsorption manipulator can improve the operation efficiency, greatly reduce the cost of enterprises, and solve the logistics problems of small and medium-sized enterprises in the storage process of light and small composite materials [2]. Therefore, the factors that affect the stability of transport, the estimation of instability parameters and the control method are the problems that must be discussed.

In recent years, the adsorption and handling robots based on planar 2R manipulator are widely used in the product handling process [3]. In order to study the displacement and speed relationship of each connecting rod of planar 2R manipulator, the manipulator is regarded as an open kinematic chain. One end of the open chain is fixed on the base, and the other end is free. It is composed of connecting rods connected in series by rotating joints. The relative motion of joints leads to the movement of connecting rods and makes the end reach the desired state. A coordinate system is fixed on each link, and then the relationship between these coordinate systems is described. The forward and inverse kinematics solutions are solved by homogeneous transformation matrix and Jacobian matrix respectively. The results of many scholars show that the length of the connecting rod and the joint angle determine the position and pose of the end point; The motion state of the end point will also affect the joint angle and angular speed [3].

Most of the handling robots use manipulators controlled by PLC for loading and unloading sheet material [4]. The characteristics of these robots are designed and embodied as ingenuity of structure with small size, simple motion, complete function and high reliability. The source of driving of these robots is often realized by pneumatic transmission and hydraulic transmission that just can match the control performance of PLC perfectly [5-6]. In the present study, a pneumatic manipulator used for loading and unloading sheet material is designed, with its stability analyzed. Due to the stability requirements in the process of handling, some linear or nonlinear disturbances need to be fully considered. Iterative algorithm is often used to solve the mathematical problems of linear and nonlinear interference, and the hardware-in-the-loop simulation technology based on OPC control technology [7] can be perfectly combined with PLC control, so that the iterative algorithm can be applied in the PLC hardware system, so as to realize the perfect control of the robot. Therefore, ILC algorithms based on three different iterative methods are used to realize the stability control of robot.

2. Structure Modeling of 2R Robot Manipulator

Figure 1(a)-(b) show the motion process of 2R manipulator from initial position to final position. Figure 1(c)-(d) show the motion process modeling based on SolidWorks and the geometric dimensions and kinematic coordinates, respectively. The lengths of the two manipulators are l_1 and l_2, the masses are m_1 and m_2, and the angular displacements are θ_1 and θ_2, respectively.

A numerical calculation method can be used to calculate the driving torque of the joint required by the robot to achieve any motion directly by Newton Euler dynamics recursion method when the mass of the connecting rod, the inertia tensor, the mass of the center of mass and the transformation matrix of the coordinate system are known [8]. Firstly, the kinematic and dynamic parameters in the recursive formula are determined by coordinate system. Secondly, the angular speed, angular acceleration, linear acceleration, inertial force and inertial torque of each connecting rod are calculated by extrapolation method. Finally, the torque of each connecting rod is calculated by the method of interpolation. Because the two joints are rotating joints, the corresponding components of the corresponding vectors of each joint are extracted, and the driving torques of the two joints, τ_1 and τ_2, are obtained as follows:

$$\begin{cases} \tau_1 = m_2 l_2^2 \cos\theta_2 (\ddot{\theta}_1 - \ddot{\theta}_2) + m_2 l_1 l_2 (2\ddot{\theta}_1 - \ddot{\theta}_2) + (m_1 + m_2) l_1^2 \ddot{\theta}_1 + m_2 l_1 l_2 \sin\theta_2 \dot{\theta}_2^2 \\ \quad - 2 m_2 l_1 l_2 \sin\theta_2 \dot{\theta}_1 \dot{\theta}_2 + m_2 l_2 g \cos(\theta_1 - \theta_2) + (m_1 + m_2) l_1 g \cos\theta_1 \\ \tau_2 = m_2 l_1 l_2 \cos\theta_2 \ddot{\theta}_1 - m_2 l_1 l_2 \sin\theta_2 \dot{\theta}_1^2 + m_2 l_2 g \cos(\theta_1 - \theta_2) + m_2 l_2^2 (\ddot{\theta}_1 - \ddot{\theta}_2) \end{cases} \tag{1}$$

In view of the requirement of stability control, the disturbances of the system should be estimated. The system output is the angle variable signals mentioned above. The system input is the practical driving torques. System identification is investigated by the least squares algorithm based on the minimized sum of squared residuals (LSR)[9]. Therefore, the interference torques are expatiated as follows.

$$M_i = \pi m_i g l_i v \left(\frac{l_i^2}{16v} \dot{v} - \frac{3 l_i^3}{128 v} \ddot{\theta}_i + \frac{l_i}{2} v + \frac{l_i v}{4} \theta_i + \frac{l_i}{2} \lambda_1 + \frac{l_i}{2} \lambda_2 \right) \tag{2}$$

where v is the linear velocity of the adsorbed object; $i = 1, 2$, represents different variables respectively. λ_1 and λ_2 represent lead-lag vibration items of the two connecting rods, which can be described as:

$$\begin{cases} \dot{\lambda}_1 = -\dfrac{l_1}{l_1 + l_2} \dot{v} - \dfrac{l_1}{l_1 + l_2} v \dot{\theta}_1 - \dfrac{0.5v}{l_1} \lambda_1 \\ \dot{\lambda}_2 = -\dfrac{2 l_2}{l_1 + l_2} \dot{v} - \dfrac{2 l_2}{l_1 + l_2} v \dot{\theta}_2 - \dfrac{v}{l_2} \lambda_2 \end{cases} \tag{3}$$

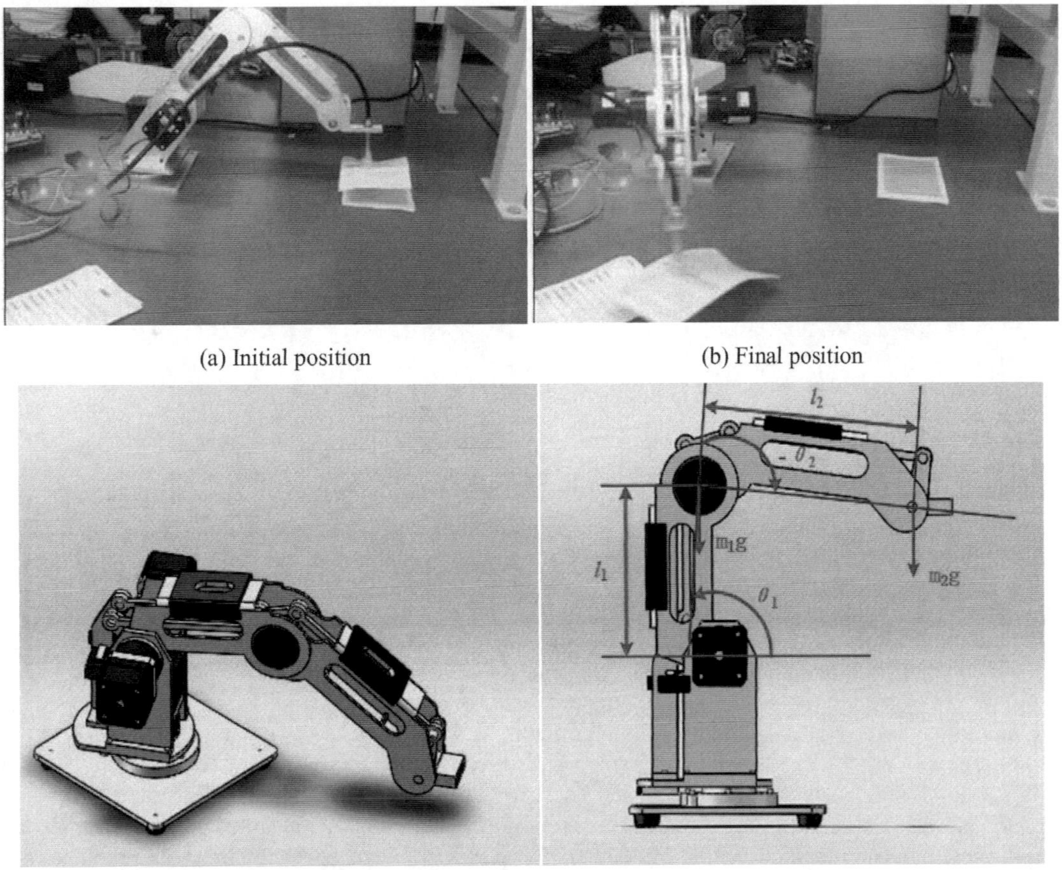

(a) Initial position (b) Final position

(c) Motion process modelling based on SolidWorks (d) Geometric dimensions and kinematic coordinates

Figure 1. Structure, modelling and geometric dimensions of 2R robot manipulator.

3. ILC Based on Trajectory Tracking

Under the action of interference torques, the small amplitude of shimmy will occur in the front and back (lead-lag) direction during the movement of the weight. At this time, it is possible to assume that v is a constant, that is assumed as an average velocity, then formulas (2-3) can be rewritten as follows:

$$M_i = \pi m_i g l_i v\left(-\frac{3l_i^3}{128v}\ddot{\theta}_i + \frac{l_i}{2}v + \frac{l_i v}{4}\theta_i + \frac{l_i}{2}\lambda_1 + \frac{l_i}{2}\lambda_2\right) \tag{4}$$

$$\dot{\lambda}_1 = -\frac{l_1}{l_1+l_2}v\dot{\theta}_1 - \frac{0.5v}{l_1}\lambda_1, \quad \dot{\lambda}_2 = -\frac{2l_2}{l_1+l_2}v\dot{\theta}_2 - \frac{v}{l_2}\lambda_2 \tag{5}$$

Considering Eq. (1) and Eq. (4) in conjunction with the lead-lag expressions interference behavior of Eq. (5), and assuming state variable matrix $q = [\theta_1\ \theta_2\ \lambda_1\ \lambda_2]^{\mathrm{T}}$, the governing system can be obtained as follows.

$$M(q)\ddot{q} + C(q,\dot{q})\dot{q} + G(q) = u - u_d \tag{6}$$

where u is the required control torque vector; u_d is the non-variable disturbance item.

Suppose that the desired trajectory to be tracked by the system is $q_d = [\sin(3t);\cos(3t);0;0]$, and the i-th output in the iterative process of the system is q_i, so that the error $e_i = q_d - q_i$. The task of learning control is to reduce the (i+1)-th motion error e_{i+1} by learning the control law u_{i+1}. The following three feedback-based iterative learning control (ILC) laws are adopted in this design [10].

- 1) The first type is the closed-loop D-type that can be described as:

$$u_{k+1}(t) = u_k(t) + K_d[\dot{q}_d(t) - \dot{q}_{k+1}(t)] \quad (7)$$

- The second type is the closed-loop PD-type that can be described as:

$$u_{k+1}(t) = u_k(t) + K_p[q_d(t) - q_{k+1}(t)] + K_d[\dot{q}_d(t) - \dot{q}_{k+1}(t)] \quad (8)$$

- 3) The third type is exponential variable-gain D-type that can be described as:

$$u_{k+1}(t) = u_k(t) + K_d \exp(K_e t)[\dot{q}_d(t) - \dot{q}_{k+1}(t)] \quad (9)$$

where K_p, K_d and K_e are parameters that needs to be tuned. According to the experience, in the analysis of this study, we can set them as: $K_p = 100$, $K_d = 500$, and $K_e = 0.8$.

4. Simulation and Discussion

According to the above methods in Eqs.(7)~(9), we implement the simulation process of three ILC algorithms. The iterations numbers of the three ILC algorithms are all set as 20. Each iteration includes 301 time points. It should be stated that, in Eq.(4) and Eq.(5), the velocity v is a constant, and formulas (4)~(5) only apply to the case that v is less than or equal to 4 m/s, which is the linear disturbance.

Figure 2(a) shows the results of all tracking processes of 20 iterations under v=2 m/s using the first ILC of closed-loop D-type. The blue curve represents the fluctuation of the given value q_d; All the red curves represent the fluctuation of the two angular displacements in the 20 times tracking process. data i (i=1,2,3,…,40) represents the target value and tracking value of 20 times respectively. It can be seen that with the increase of iteration times, displacement tracking is getting closer and closer, among which, the angular displacement tracking of link 2 is faster.

Figure 2(b)-(c) show the results of the last tracking process of the final iteration under v=2 m/s. It can be seen that the two angular displacement signals both have very good tracking effects in Figure 2(b). Figure 2(c) also shows a good angular speed tracking effect, in which the tracking effect of joint 2 is better.

Figure 2(d) shows the maximum absolute values of error 1 for link 1 and error 2 for link 2 against the iterative times 10, herein, blue data represents the error of the first joint variable, and the red data represents the error of the second joint variable. Both the maximum angular displacement errors and the maximum angular speed errors are shown in 20 iterations. It can be seen that with the increase of the number of iterations, the maximum angular displacement error and the maximum angular speed error both decrease rapidly and converge. The displacement error converges faster and almost converges to zero, which shows a very good tracking effect. Due to the significant performance and convergence of the tracking effects, we only show the maximum error fluctuation of the previous 10 iterations. In fact, when reaching the twentieth iteration, the error is basically close to 0, so we can compare the advantages and disadvantages of the control algorithms even if we only need to detect the errors in the first five iterations. This can be fully reflected in the comparisons of the control effects of the following several algorithms.

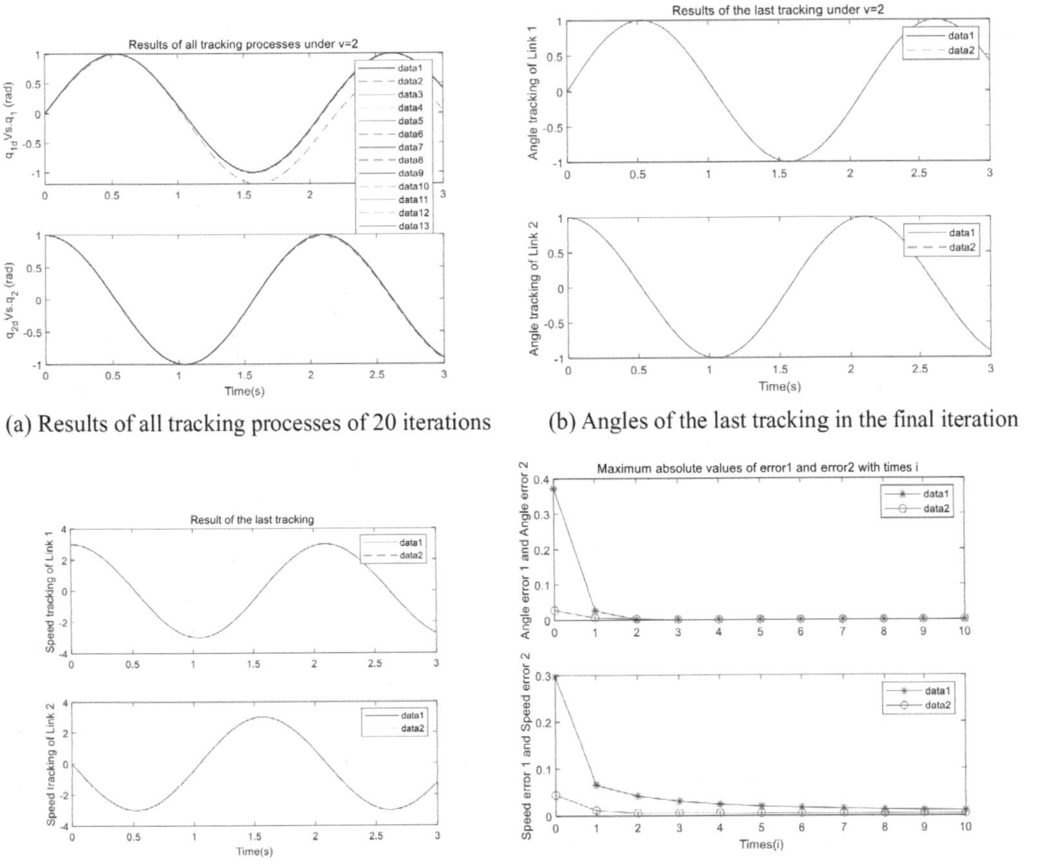

(a) Results of all tracking processes of 20 iterations (b) Angles of the last tracking in the final iteration

(c) Speeds of the last tracking in the final iteration (d) The maximum error comparisons in 10 iterations

Figure 2. Results of all tracking processes of 20 iterations under v=2 m/s using the first ILC of closed-loop D-type, and the maximum error comparisons.

Figure 3(a) illustrates the results of all tracking processes of 20 iterations under v=2 m/s using the second ILC of closed-loop PD-type. Similarly, the blue curve represents the fluctuation of the given value y_d; All the red curves represent the fluctuation of the two angular displacements in the 20 times tracking process. It can be seen that with the increase of iteration times, displacement tracking is also getting closer and closer, among which, the angular displacement tracking of link 2 is also faster. Figure 2(b)-(c) illustrate the results of the last tracking process of the final iteration under v=2 m/s. It can be seen that the two angular displacement signals both have very good tracking effects in Figure 2(b). Figure 2(c) also illustrates a good angular speed tracking effect, in which the tracking effect of joint 2 is better. Compared with the results of (a), (b) and (c) in Figure 2, those in Figure 3 shows almost the same results, reflecting the robustness of the iterative algorithms.

Figure 3(d) also illustrates the maximum absolute values of error 1 for link 1 and error 2 for link 2 with iterative times. Compared with Figure 2(d), it shows almost the same fluctuation trend, However, in terms of amplitude, the maximum error amplitude in Figure 3(d) is reduced, which shows the superiority of the second ILC algorithm over the first ILC algorithm.

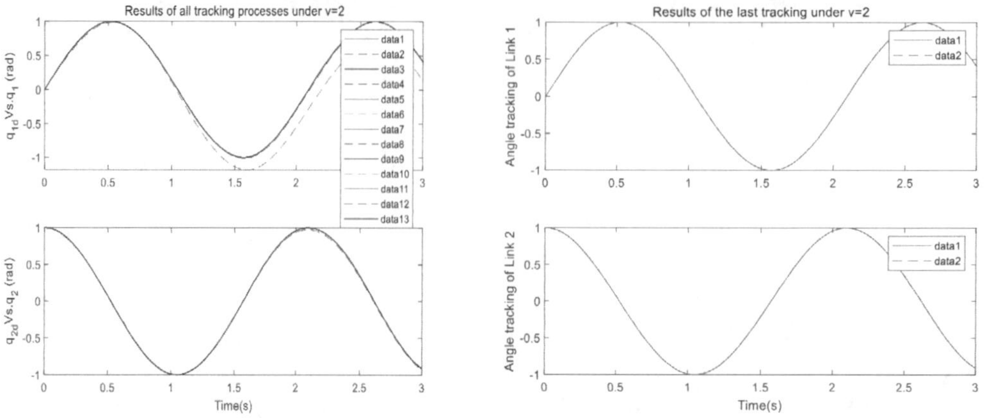

(a) Results of all tracking processes of 20 iterations (b) Angles of the last tracking in the final iteration

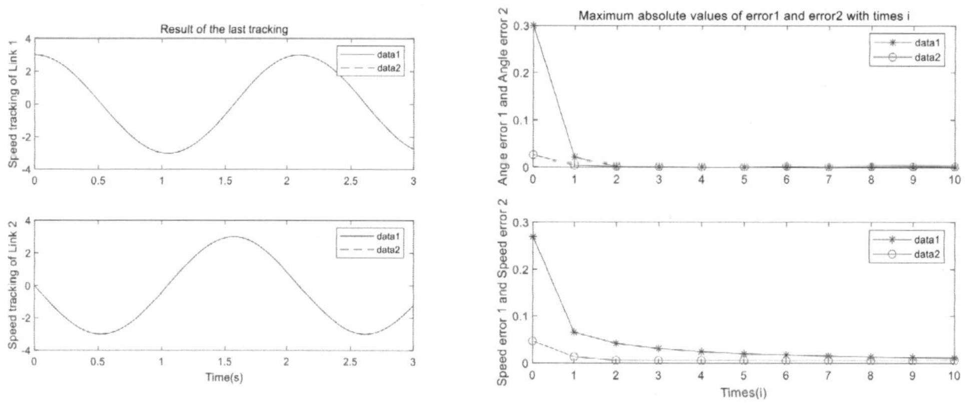

(c) Speeds of the last tracking in the final iteration (d) The maximum error comparisons in 10 iterations

Figure 3. Results of all tracking processes of 20 iterations under v=2 m/s using the second ILC of closed-loop PD-type, and the maximum error comparisons.

Figure 4(a) illustrates the results of all tracking processes of 20 iterations under v=2 m/s using the third ILC of exponential variable-gain D-type. It can be observed from the results of Figure 4(a)-(c) that the third algorithm also obtains excellent simulation results. From Figure 4(d), the third algorithm shows better error results compared with those in Figure 2(d) and Figure 3(d). The third algorithm shows better error results; hence the third algorithm has higher iteration accuracy and is recommended preferentially. At the same time, it can be seen that the best way to judge the superiority of the iterative algorithm is from value fluctuation of the maximum error of the variable in the tracking process. During the tracking processes of the proposed three ILC algorithms, the comparisons of the maximum displacement errors and the maximum speed errors of the two joints are shown in Table 1. The maximum errors of angular displacements and angular speeds of the third algorithm are 0.12 m, 0.005 m, 0.15 rad, and 0.02 rad, respectively. Compared with the results of the former two algorithms, the best advantage is demonstrated.

It should be stated that, in Eq.(4) and Eq.(5), when v is greater than 4 m/s, the disturbance is nonlinear and its formula is exactly expressed as Eqs.(2)~(3). In the present study, we set v=2 m/s, which is exactly the linear vibration mentioned above.

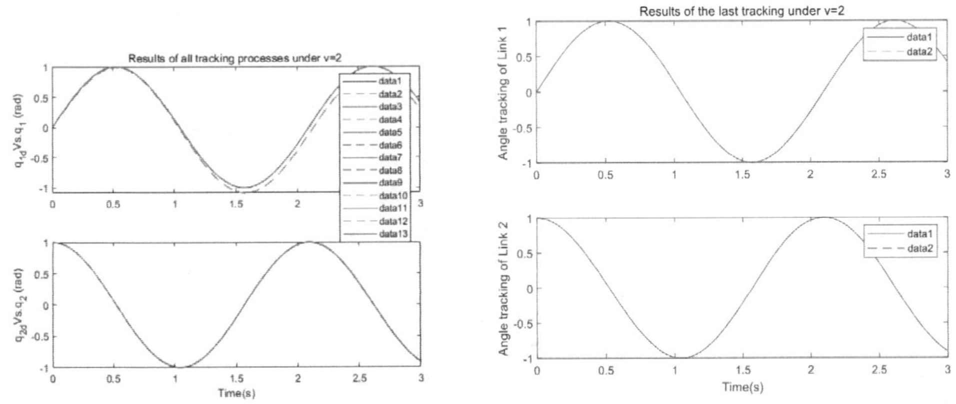

(a) Results of all tracking processes of 20 iterations (b) Angles of the last tracking in the final iteration

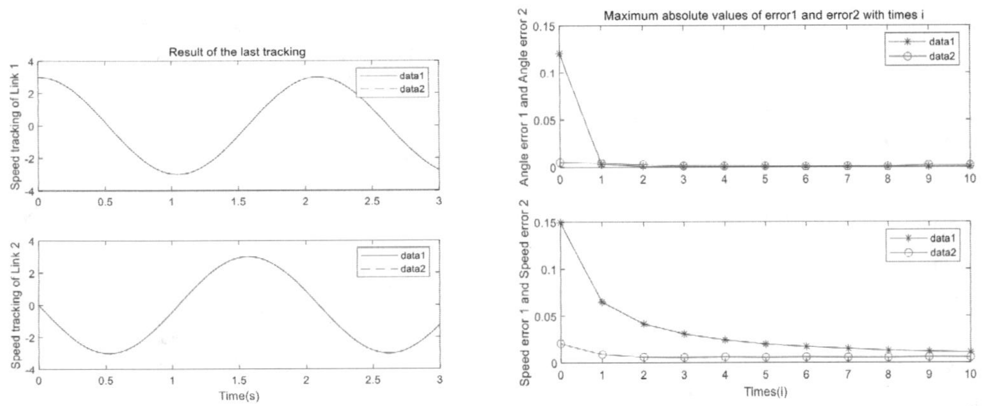

(c) Speeds of the last tracking in the final iteration (d) The maximum error comparisons in 10 iterations

Figure 4. Results of all tracking processes of 20 iterations under v=2 m/s using the third ILC of exponential variable-gain D-type, and the maximum error comparisons.

Table 1. Comparisons of the maximum displacement errors and the maximum speed errors of the two joints

Items	The first ILC	The second ILC	The third ILC
Angular displacement 1	0.38 m	0.3 m	0.12 m
Angular displacement 2	0.03 m	0.02 m	0.005 m
Angular speed 1	0.3 rad	0.28 rad	0.15 rad
Angular speed 2	0.043 rad	0.0.042 rad	0.02 rad

5. Conclusion

In the present study, we present three methods of ILC algorithm for trajectory tracking of transport process, which can be used in any types of transport robots. stability control for robot subjected to 2R manipulator movement is investigated based on ILC. The link joint displacements, speeds, and the maximum errors are displayed, with the obvious superior results demonstrated. According to the results of tracking errors, no matter from the displacement error results or from the speed error results, the third iterative algorithm is the best and should be recommended first. At the same time, it can be seen that the best way to judge the superiority of the iterative algorithm is from value fluctuation of the maximum error of the variable in the tracking process.

Acknowledgement

The authors gratefully acknowledge the support of the National Natural Science Foundation of China (no. 51675315), and the Graduate Science and Technology Innovation Project in SDUST.

References

[1] Si Z, Cao X and Jiang X 2009 Packaging and Food Machinery 6 26–30.
[2] Wang Y, Liu H and Cai Y 2020 Mechanical and Electronical Information 5 85–88.
[3] Huang J and Li Z 2018 Mechanical Engineering & Automation 2(1) 45–47.
[4] Zhu C, Wang D, Li Z and Wang Z 1999 Hydraulics Pneumatics & Seals 6 21–24.
[5] Liu T, Gong A, Song C and Wang Y 2020 Energies 13 1029.
[6] Liu T, Zhao K, Sun C, Jia J and Liu G 2020 Energies 13 5029.
[7] Liu T 2010 Transactions of the Institute of Measurement and Control 3 1–13.
[8] Zhang Q 2019 Robotics Mechanisms, Kinematics, Dynamics and Motion Planning (Beijing: Tsinghua University Press) chapter 6 pp 125–141.
[9] Xue D 2007 MATLAB Solutions to Mathematical Problems in Control (Beijing: Tsinghua University Press) chapter 3 pp 105–110.
[10] Liu J 2016 Robot Control System Design and MATLAB Simulation-the Basic Design Method (Beijing: Tsinghua University Press) chapter 5 pp 125–134.

Operational Effectiveness Evaluation of EW Equipment System Based on Framework Design

Mingqiu Ren[1,*], Yi Leng[1] and Le Zhang[2]

[1]Early Warning Academy, Wuhan, Hubei, China
[2]Army Engineering University, Wuhan, Hubei, China

[*]corresponding author's e-mail: atlas_ren@hust.edu.cn

Abstract. The currently used evaluating method has such problems as complex evaluation index space, complexity of computation, low evaluation efficiency, and low systematicness and intellectualization, which is difficult to meet the requirements of operational effectiveness evaluation for ECM equipment. In order to guarantee the efficiency and accuracy of operational effectiveness evaluation, an integrated and systematic evaluation method was provided based on multi-views system's simulation model architecture, which can show the actual frameworks of equipment system for evaluation, we describe the element, structure, operational process and index system in a standardized manner. Furthermore, a framework of Flexible-intelligent Evaluation towards Combat-capability Space in System-of-systems (FECSS) was proposed. The FECSS method builds a bi-directional mapping relation between the upper and low operational effectiveness indicators by basic indicators-formed space in system-of-systems which combined the complex evaluation index, evaluation process with intelligent evaluation model. The proposed method can effectively analysis the effectiveness results in condition of different types and numbers of equipment. It also makes the electronic countermeasure equipment project demonstration and operational application more feasibly and practically.

1. Introduction

EW equipment system is the synthesis of the system, which is a whole composed of different systems according to a certain interaction relationship. It is the system of the system. However, in the complex electromagnetic environment, the lack or fluctuation of the capability of some ECM equipment will cause great fluctuations in the system capability. How to evaluate the combat effectiveness of ECM equipment system under different combat scenarios, how to scientifically formulate the development plan of ECM equipment system, and how to scientifically apply the theoretical research and evaluation verification means[1-3], finding the weak links of the existing ECM equipment system as soon as possible, which is not only lack of practical test, but also less experience to follow, has become a practical problem to be solved in the development and construction of ECM equipment system.

2. Challenges faced by combat effectiveness evaluation of EW equipment systems

Electronic countermeasure combat effectiveness evaluation has always been the focus of attention. Since the 1960s, the Western military powers represented by the United States have successively established special research institutions to carry out combat effectiveness analysis and evaluation. Combined with MCP missile defense radar, the detection performance of missiles under the background of chaff jamming is deeply analyzed. The theater ballistic missile defense published by the American Aeronautical Society brings together the innovative research results of many missile defense experts in the United States and Israel, and proposes to consider the impact of combat style on the early warning time, identification probability and tracking accuracy of electronic countermeasure equipment from the perspective of system confrontation. After decades of research and exploration, great achievements have been made in the effectiveness evaluation of electronic countermeasures [4,5]. The traditional efficiency evaluation methods mainly include analytic hierarchy process (AHP), network analysis (ANP), fuzzy comprehensive evaluation method, Monte Carlo method, system dynamics method and so on. With the development of computer, network and communication technology, large-scale simulation computing methods such as parallel computing, distributed computing and grid computing have gradually developed and become more and more mature. Some researchers have also introduced the theories of biology, statistics and artificial intelligence into the field of effectiveness evaluation. Exploratory modeling and analysis methods, value centered decision analysis Data mining, machine learning and other new methods.

As a new weapon and equipment system, ECM equipment system has complex composition and many missions and tasks. Compared with conventional operations, ECM has the characteristics of many interference patterns, rapid change of target environment and many uncertain factors in operation process, which puts forward higher requirements for the scientific nature of the system operation efficiency evaluation model. Due to the complexity of the problem and the limitations of related research, the combat effectiveness evaluation of ECM equipment system is still facing great challenges.

2.1. Insufficient ability to deal with uncertain factors

The target environment is becoming more and more complex, and the existing evaluation methods have insufficient ability to deal with uncertain factors. Electromagnetic spectrum warfare is a dynamic game process with incomplete information, which includes not only complex technical factors, but also a large number of fuzzy and uncertain factors. It is very difficult to evaluate the combat effectiveness of the system. Uncertainty mainly comes from three aspects: firstly, due to the limited ability of equipment to distinguish the attributes of incoming targets, there are a series of problems in the discovery and interception of target radiation sources, such as intermittent time, parameter fluctuation, unknown recognition results and so on. Secondly, there is a lack of clear standards for the fusion rules of various sensors within the system, which makes it difficult to unify the results of emitter recognition. These strong fuzzy and uncertain relations make the traditional evaluation methods based on probability theory and mathematical statistics theory no longer applicable.

2.2. Lack of overall framework for system assessment

ECM equipment system involves many factors and complex structural relationships. It is very complex to evaluate its combat effectiveness. At present, due to different research objectives and task requirements, the equipment effectiveness evaluation methods established by various research institutes generally have problems of diverse hierarchical structure, different model granularity and different applicability of evaluation methods, which are not strong guidance and reference for the construction of equipment system. Therefore, we should make full use of mature framework modeling theories such as multi view architecture, comprehensively consider the constituent elements, structure, combat process and index system of the system, design the method framework for the whole process and systematic treatment of evaluation problems, and clarify the model granularity and hierarchy of task model, entity model, structure model and index system, It lays the model foundation and key technical support for the operational effectiveness evaluation of EW equipment system.

2.3. Inadequate applicability of evaluation methods

The core of system combat effectiveness evaluation is to demonstrate the construction requirements of equipment system and relevant combat effectiveness index requirements against the background of combat system. Although some units, experts and scholars have carried out relevant research on this, due to the lack of systematic theories and methods, substantive progress has not been made, especially in the field of electronic countermeasure, which is still in its infancy, and the supporting effect on the construction of electronic countermeasure system and equipment type spectrum demonstration is not obvious. Therefore, it is urgent to establish a complete system effectiveness evaluation method, reveal the problems and shortcomings in the construction and application of EW equipment system, accurately describe the working state and combat effectiveness of EW equipment system under different penetration scenarios, and analyze the impact of typical combat patterns on the combat effectiveness of the system, Provide scientific cognitive means for the construction of electronic countermeasure system.

3. Design of combat effectiveness evaluation framework for EW equipment system

Framework design is the basis of system combat effectiveness evaluation, and the selection of evaluation methods is the key to the effectiveness of evaluation results. How to build a scientific and reasonable equipment system effectiveness evaluation framework and ensure that the evaluation results can effectively support the construction of electronic countermeasure equipment system and equipment combat application is the primary problem to be solved in the research of electronic countermeasure system combat effectiveness evaluation.

3.1. Architecture modeling method for effectiveness evaluation

At present, the research on modeling methods and framework of weapon equipment architecture at home and abroad is more and more in-depth. However, in order to achieve the comprehensiveness and logical rigor of overall description, there are often dozens of related view products. With the further deepening of system research, related products will be more abundant. Therefore, in the system evaluation, we can make use of the modular and pluggable characteristics of the system view products, flexibly select the appropriate view products, and serve the evaluation scientifically and effectively on the basis of accurately describing the system elements. Therefore, on the basis of DoD Architecture Framework dodaf2.0 and mod architecture framework modaf1.5 of the UK, using the idea of architecture modeling for reference, using the mission incentive oriented architecture modeling method, the system product set is cut, and the product modules participating in the evaluation of EW equipment system are selected, Constitute an architecture framework that can meet the requirements of EW equipment system of systems evaluation.

3.2. Design of intelligent Agile Evaluation Framework for system of systems combat capability space

In view of the limitations of the current combat effectiveness evaluation of EW system, a flexible intelligent evaluation towards combat capability space in system of systems (FECSS) method is proposed. FECSS method is a whole that supports each other in theory and closely combines logically. Among them, the last S refers to system of systems, that is, the evaluation object served by the theory. It is precisely because of the complexity, relevance and uncertainty of weapon equipment system that we need to find a fast, efficient and accurate evaluation method CS refers to combat capability space, that is, combat capability space, which is the basis of evaluation. Combat effectiveness value is mapped by combat capability index set through combat process. Therefore, the change of basic index space is the fundamental reason affecting the change of effectiveness value. Where C stands for combat capability, that is, combat capability. It represents the equipment performance index data at the bottom of the evaluation system and is the basic element of the combat effectiveness space; Here S stands for space, that is, the physical form shown by the combat capability index set exists in a nonlinear and high-dimensional data structure, which is jointly determined by the complexity of system elements and the connectivity of structure FE refers to flexible intelligent evaluation, that is, agile and intelligent evaluation methods. Agile means that the evaluation process only studies the change index value, and the previously learned index samples do not need to change, so as to improve the evaluation efficiency. Intelligence means that the statistical learning method is used to predict the effectiveness evaluation value through self-learning weight, so as to reduce the subjectivity of the evaluation results. At the same time, through the optimization of evaluation data samples and evaluation model, the evaluation time complexity is reduced. It can be seen that FECSS method can coordinate and operate the advantages of various evaluation methods through an evaluation production line, and can solve the problem of combat effectiveness evaluation of EW equipment system in a systematic way.

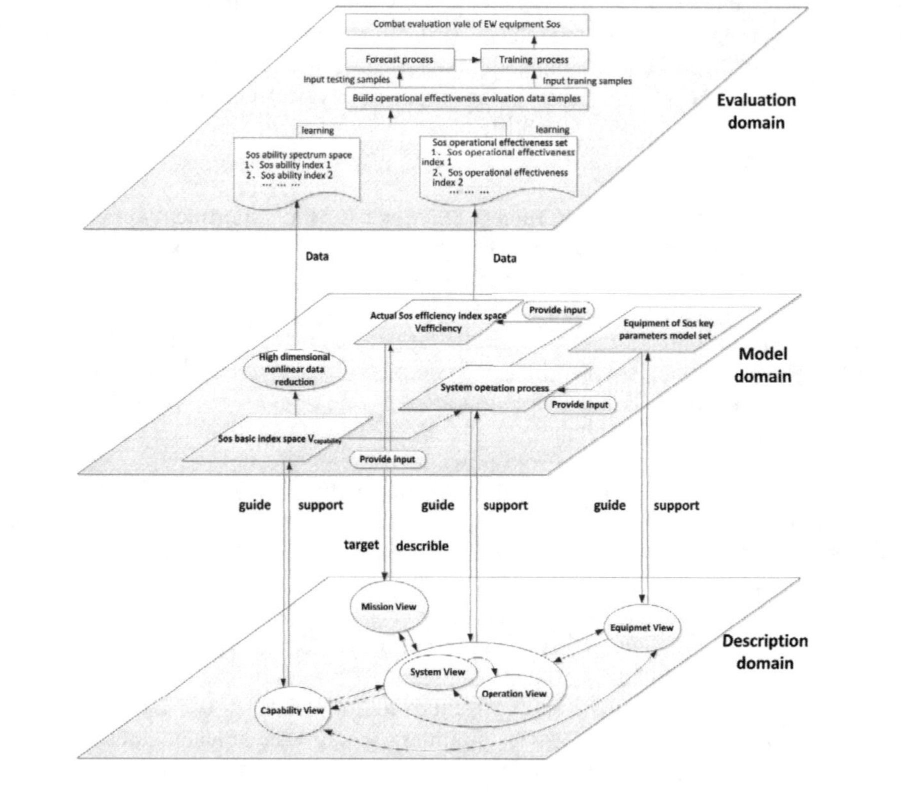

Figure 1. Diagram of FECSS architecture based on architecture products.

3.3. Operational effectiveness evaluation process of system of systems based on FECSS

The operational effectiveness evaluation process of EW equipment system using FECSS method is shown in Figure 1, which mainly consists of the following steps:

- Step 1: establish an operational capability space for operational effectiveness evaluation of EW equipment system. According to the combat mission of EW equipment system, the composition and working principle of equipment system, combined with the characteristics and purpose of combat effectiveness evaluation, the system capability index is decomposed layer by layer by AHP method, and the combat effectiveness evaluation index system of EW equipment system is established. Then, based on the bottom capability index set, the combat capability space is selected by Delphi method.
- Step 2: reduce the combat capability space. Firstly, the basic index space is standardized and the evaluation data sample set is constructed. Then, the mapping relationship between the basic index space and the spectral capability index space is established, and the high-dimensional data reduction technology is used to reduce the basic data space, so as to reduce the computational complexity of the evaluation process.
- Step 3: establish the mapping relationship between basic index space and operational effectiveness index. Firstly, expert scoring, test statistics, combat simulation and other methods are used to collect and evaluate the required combat effectiveness index values. Then, combined with the small sample data reduced in step 2 and the combat effectiveness index value collected in this step, the combat effectiveness spatial data sample of EW equipment system is constructed.
- Step 4: establish the combat effectiveness evaluation model of EW equipment system based on Intelligent Computing, focus on solving the problems of low evaluation efficiency and poor

prediction accuracy, and verify the feasibility and effectiveness of the evaluation method in combination with simulation experiments. Through the analysis of system key capabilities in different scenarios, the weaknesses of EW equipment system operation are found out.

4. Conclusion

Taking the combat capability of the system as the starting point and foothold, this paper systematically combs the research status and existing problems of combat effectiveness evaluation of EW equipment system. On this basis, considering the systematicness, complexity and uncertainty of EW equipment system combat effectiveness evaluation, a design method of intelligent Agile Evaluation (FECSS) framework for system combat capability space is proposed, which focuses on solving the deficiencies of system combat effectiveness evaluation top-level design, evaluation process formulation, evaluation model selection and so on. This method can carry out the simulation analysis of the combat effectiveness of ECM equipment system in combination with different combat tasks, investigate the impact of different types and quantities of ECM equipment on the combat effectiveness of the system, and provide theoretical basis and technical support for the in-depth study of the practical problems of the existing ECM equipment system structure demonstration, equipment system efficiency evaluation, equipment deployment and application.

Acknowledgments

The research is supported by Fundamental Research Funds of Military Theory (No.21KJ3C1-0090R).

References

[1] Qu H, Chang L, Ma L, Xu Y, Yang G 2012 DPEES: DDoS Protection Effectiveness Evaluation System. In: Gaol F. (eds) *Recent Progress in Data Engineering and Internet Technology. Lecture Notes in Electrical Engineering*, vol 157 (Berlin, Heidelberg: Springer) p 155-61

[2] Liu S, Li H 2016 A Human Factors Integrated Methods for Weapon System Effectiveness Evaluation. In: Long S., Dhillon B. (eds) *Man-Machine-Environment System Engineering. MMESE 2016. Lecture Notes in Electrical Engineering* **406** 469-79

[3] Fang Z, Tao Z, Xin W, Xin Jun Z 2013 Research on Simulation-Experimentation-Based Effectiveness Evaluation of Digitized Force's Command and Control System. In: Tan G., Yeo G.K., Turner S J, Teo Y M (eds) AsiaSim 2013. AsiaSim 2013. *Communications in Computer and Information Science* **402** 355-60

[4] Song W, He J, Wu X, Lou S 2016 Effectiveness Evaluation and Operational Application Research of Anti-Radiation Missile Decoy. In: Qi E (eds) *Proc. of the 6th Int. Asia Conf. on Industrial Engineering and Management Innovation* p 869-78

[5] Wang C, Li N, Kong H, Li H 2012 Study on General Effectiveness Evaluation Platform for Combat Aircraft. In: Xiao T, Zhang L, Ma S (eds) *System Simulation and Scientific Computing. ICSC 2012. Communications in Computer and Information Scienc* **326** 128-36

An Improved Dark Channel Prior Method for Dusty Image Restoration

Ya Zhang[1,*] and Weimin Yang[2]

[1]College of Engineering, Zhejiang Normal University, Jinhua, Zhejiang, China
[2]Yiwu Industrial & Commerical College, Yiwu, Zhejiang, China

[*]corresponding author's e-mail: yazhang@zjnu.edu.cn

Abstract. For the images which were obtained in a dusty environment, because of the tiny particles suspended in the atmosphere absorb and scatter the light, there is serious haze in the images. The haze can directly cause colour changes and the loss of image details. As a result, a lot of efficient features in the dusty image were concealed or weakened by the haze, which created difficulties to obtain a natural and clear restored image, and has brought serious impact for aerospace, automatic driving and traffic monitoring in the dusty environment. An improved restoration method based on dark channel prior theory and gradient domain guided filtering was proposed to remove the haze and restore the dusty image. According to the characteristics of the dusty images, a specific guided image was adopted to make gradient domain guided filtering for the dark channel image firstly, so that the transmittance image can be better refined than that in the dark channel prior method, and then the image can be restored according to the image degradation model. Experiments show that the visual effect of the restored image is natural and clear, and can retain most effective features, which lays a foundation for computer vision-based target monitoring and recognition in the dusty environment.
Keywords. Image Restoration, Dark Channel, Guided Filtering, Gradient Domain

1. Introduction

Haze is an atmospheric phenomenon where the tiny particles suspended in the atmosphere absorb and scatter light, which directly causes color changes and the loss of image details. As a result, image information cannot be accurately recognized, which has brought serious impact for aerospace, automatic driving and traffic monitoring [1-3]. Therefore, the research of dusty image dehazing is of great significance, and it is also the research difficulties of dehazing technology.

Many researchers have studied on the image dehazing technology and achieved remarkable results. According to the number of input images, the existed dehazing algorithm can be divided into two kinds: the methods based on several images [4-8] and the methods based on single image [9-13]. The methods based on several images need to obtain the images at different conditions. Because it is difficult to obtain the required images, this kind of methods are rarely used in practice. The dehazing methods based on single image were usually based on atmospheric scattering physical model and the prior information. This kind of methods is popularly used now.

Tan [9] combined the relationship between the wavelength and image contrast quality reducing, and utilized the local contrast to remove the haze. Fatal [10] proposed the hypothesis that the light transmission and the shade of the object surface are locally irrelevant, and removed the haze by estimating the scene illuminance to obtain the propagation image. He et al. [11] found the prior laws of dark channels by counting a large number of haze-free images, and proposed a dehazing method based on dark channel prior theory. This method has a better dehazing effect and has milestone significance in dehazing technology.

In this paper, we integrated gradient information in the dark channel method to make the hazy image in the dusty environment be more natural and clear by guided filtering. Firstly, a specific guided image is proposed. Then, the gradient domain guided filtering is performed on the dark channel image. At last, according to the hazy image degradation model, the haze-free image can be obtained based on the dark channel prior theory. Experiments have proved that the visual effect of the restored image by our method is natural and clear.

2. Image dehazing method based on dark channel prior theory

Image dehazing method based on dark channel prior theory was proposed in [11].

2.1. Hazy image degradation model

The model of the hazy image is:

$$I(x)=J(x)t(x)+A(1-t(x)) \tag{1}$$

Where, I(x) is the hazy image, J(x) is the haze-free image which will be restored, A is the global atmospheric light, and t(x) is the transmittance. Solving the problem of image dehazing is the process of finding t(x) and A from I(x).

2.2. Solving global light A

The global atmospheric light value A can be estimated by using the dark channel of the image. In the dark channel image, select the brightest 0.1% pixel value from the dark prior color channel, and then select the maximum gray value of these pixels in the original input image as the global atmospheric light value A. Among the three RGB channels, an atmospheric light value A is estimated for each channel.

2.3. Solve the transmittance t(x)

(1) Solve the transmittance according to the hazy degradation model

Formula (2) can be obtained from Formula (1):

$$t(x)=\frac{A-I(x)}{A-J(x)} \tag{2}$$

Therefore, an initial rough estimate of transmittance can be obtained from Formula (2):

$$\tilde{t}(x)=1-\frac{I(x)}{A} \tag{3}$$

In real life, even if it is sunny, there are some particles in the air. Therefore, you can still feel the influence of the haze when looking at distant objects. In addition, the existence of haze makes humans feel the existence of depth of field, and it is necessary to retain a certain degree of haze when removing the haze. This can be corrected by introducing a factor between [0,1] in the above formula:

$$\tilde{t}(x)=1-\omega\frac{I(x)}{A} \tag{4}$$

Generally, "ω" can be set to 0.95.

(2) Solve the transmittance and global light according to the dark channel

The mathematical definition of the dark channel is:

$$J^{dark}(x)=\min_{x\in\Omega(x)}(\min_{c\in\Omega\{r,g,b\}}J^{C}(x)) \tag{5}$$

Where, C represents one of the three RGB channels. Formula (8) means that in an input image, we take the minimum value of the gray value in the three channels of each pixel in the image to obtain a gray image.

In this grayscale image, take a rectangular window of a certain size with each pixel as the center, and take the minimum gray value in the rectangular window instead of the gray value of the center pixel (minimum filtering) to obtain the dark channel image of the hazy image.

The dark channel image is a grayscale image. Through a lot of statistics and observation, it is found that the gray value of the dark channel image is very low, so the gray value of all pixels in the entire dark channel image is approximated to 0, namely:

$$J^{dark} \to 0 \tag{6}$$

For the imaging model, divide both sides of Formula (1) by the atmospheric light value of each channel:

$$\frac{I^C(x)}{A^C} = t(x)\frac{J^C(x)}{A^C} + 1 - t(x) \tag{7}$$

Assuming that in a rectangular window of a certain size in the image, the value of the transfer function is a fixed value, and the minimum operators are used to minimize both sides of Formula (7):

$$\min_{x\in\Omega(x)}(\min_c \frac{I^C(x)}{A^C}) = \tilde{t}(x)\min_{x\in\Omega(x)}(\min_c \frac{J^C(x)}{A^C})+1-\tilde{t}(x) \tag{8}$$

Since the scene radiance is a haze-free image, applying the dark channel prior to J, then we can obtain:

$$J^{dark}(x) = \min_{x\in\Omega(x)}(\min_c J^C(x))=0 \tag{9}$$

Since A^C is always positive, we have

$$\min_{x\in\Omega(x)}(\min_c \frac{J^C(x)}{A^C}) = 0 \tag{10}$$

Substituting formula (10) into the minimization formula (8), the estimated value of the transfer function can be obtained:

$$\tilde{t}(x)=1-\min_{x\in\Omega(x)}(\min_c \frac{I^C(x)}{A^C}) \tag{11}$$

Considering the factor of "ω", Formula (12) can be obtained:

$$\tilde{t}(x)=1-\omega\min_{x\in\Omega(x)}(\min_c \frac{I^C(x)}{A^C}) \tag{12}$$

At this point, the haze-free image can be restored. Formula can be obtained:

$$J = (I - A)/t + A \tag{13}$$

In order to prevent calculation errors caused by too small transmittance, set t0 as the minimum transmittance, and substitute the transmittance t(x), atmospheric light A, and hazy image I(x) into Formula (1) , we can obtain:

$$J(x) = \frac{I(x)-A}{\max[t(x),t_0]} + A \tag{14}$$

3. Gradient domain guided filtering

The dehazing method based on dark channel will cause problems such as blocking effects, white borders around the edges, and so on. In order to maintain the edges and restore as many effective features of the image as possible, guided filtering [12] was proposed to refine the transmittance. The guided filter is a non-linear filter that can smooth the image. It can smooth the image while maintaining the image boundary. However, the dehazing image after the guided filtering will have halos, because the local linear model used by the guided filtering cannot represent the image well near some edges. The weighted guided filter [13] can limit the reduction of halo artifacts produced by the guided filter. The weighted guided filter adopts an edge perception factor into the constraints of the guided filter, so that the edges are better preserved in the result image, thereby reducing halos. However, since there are still not sufficiently clear constraints to deal with the edges in the image, the weighted guided filter cannot preserve the edges well in some cases. Kou F [14] et al. combined a clear first-order edge perception constraint into the guided image filter, and thus proposed a gradient domain guided filtering method. This edge perception constraint makes the edge better preserved. In the guided filter, the image itself or the other images can be adopted as the guided image. We should focus on the characteristic of the hazy dusty image to utilize a suitable guided image. In this paper, in order to achieve the best restoring effect of hazy dusty images, we proposed a guided image by performing twice maximum filtering on the dark channel image.

The relationship of between the guided image and the output image can be represented as Formula (15):

$$q_i = a_{p'} G_p + b_{p'}, \forall p \in \Omega_{\zeta_1(p')} \quad (15)$$

Where, q_i is the output and G_p is the guiding image.

$\Omega_{\zeta_1(p')}$ is a window with a radius of ζ_1 and a pixel p as the center.

It can be seen from Formula (15) that

$$\nabla q = a \nabla G \quad (16)$$

Therefore, only where the guided image has an edge, the output image has an edge. By modifying the gradient characteristics of the guided image, the gradient characteristics of the output image can be changed, and the restoring quality of the hazy dusty image will be improved.

Therefore, if a suitable guided image was adopted to make graduated domain filtering in the dark channel prior dehazing method, the visual effect of the dehazing image may be improved in theory.

4. Experimental study

This paper uses Fig.1 for experimental verification. A specific guided image was adopted to make gradient domain guided filtering to overcome the technology difficulties of the dusty image dehazing. Fig.2 is the dark channel image of Fig.1. The guided image was proposed as can be seen in Fig.3. Fig.4 is the transmittance image. Fig.5 is the output dehazing image.

Fig.6 is the output dehazing image based on traditional dark channel prior theory. There are a lot of color blocks in the image in Fig.6, and the dehazing effect is not natural. Compared with the traditional guided filtering method, the restoration effect of the method in this paper is clearer and the distortion is reduced, and the restored effect is significantly improved.

Figure 1. Input image

Figure 2. Dark channel image

Figure 3. Guided image

Figure 4. Transmittance image

Figure 5. output of dehazing image

Figure 6. output of dehazing image by the method in [11]

5. Conclusions

Facing to the image problems caused by the dusty environment, an improved dehazing method is proposed in this paper based on dark channel prior theory. According to the characteristics of the dusty images, a specific guided image was proposed to make guided filtering for the dark channel image so that the transmittance image can be better refined than that in the dark channel prior method, then the image can be restored according to the image degradation model. Compared with the dark channel prior method, the image restored by our method is more natural and real, and can retain more effective features, which lays the foundation for computer vision-based target monitoring and recognition in a hazy environment.

References

[1] Liu F 2016 *The research on image defogging algorithm and its application in traffic surveillance video* (Wuhan: Wuhan University of Science and Technology)
[2] Wu D, Zhu Q S 2015 The latest research progress of image dehazing *Acta Automatica Sinica* **41**(2) 221-39
[3] Guo P, Cai Z X, Xie B, et al 2010 Review and prospect of image dehazing techniques *Journal of Computer Applications* **30**(9) 2417-21
[4] Narasimhan S G, Nayar S K 2000 *Proc. IEEE Comput. Soc. Conf. Comput. Vis. Pattern Recognit* (Silver Spring: IEEE Computer Society) p 598-605
[5] Narasimhan S G, Nayar S K 2003 Contrast restoration ofweather degraded images *IEEE Transactions on Pattern Analysis and Machine Intelligence* **25**(6) 713-24
[6] Schechner Y Y, Narasimhan S G, Nayar S K 2001 *Proc. IEEE Comput. Soc. Conf. Comput. Vis. Pattern Recognit* (Silver Spring: IEEE Computer Society) p 325-32
[7] Shwartz S, Namer E, Schechner Y Y 2006 *Proc. IEEE Comput. Soc. Conf. Comput. Vis. Pattern Recognit* (Silver Spring: IEEE Computer Society) p 1984-91
[8] Fang S, Xia X, Huo X, et al 2014 Image dehazing using polarization effects of objects and airlight *Optics Express* **22**(16) 19523-37
[9] Tan R T 2008 *Proc. IEEE Comput. Soc. Conf. Comput. Vis. Pattern Recognit* (Silver Spring: IEEE Computer Society) p 1-8
[10]Fattal R 2008 Single image dehazing *ACM Transactions on Graphics* **27**(3) 1-9
[11]He K, Sun J, Tang X 2011 Single image haze removal using dark Channel prior *IEEE Transactions on Pattern Analysis and Machine Intelligence* **33**(12) 2341-53
[12]He K, Sun J, Tang X 2013 Guided Image Filtering *IEEE Transactions on Pattern Analysis & Machine Intelligence* **35**(6) 1397-409
[13]Geethu H, Shamna S, Kizhakkethottam J J 2016 Weighted Guided Image Filtering and Haze Removal in Single Image *Procedia Technology* **24** 1475–82
[14]Kou F, Chen W, Wen C, et al 2015 Gradient Domain Guided Image Filtering *IEEE Transactions on Image Processing* **24**(11) 4528-39

The Application of Reprogramming Technology in a Wireless Sensor Network

Jianjun Huang

School of Computer and Information Engineering, Nanchang Institute of Technology, Nanchang, Jiangxi 330044, China

*corresponding author's e-mail: 27458808@qq.com

Abstract. Wireless sensor networks can collect physical world information in a large, long-term and real-time manner in a complex environment. Wireless sensor networks have been successfully applied in many fields. With the development of wireless sensor network, wireless sensor network applications need to update the executable code on the sensor due to the change of application requirements. The researchers call this a wireless sensor network application reprogramming, or simply reprogramming. The efficiency of reprogramming is closely related to the size of the updated code that needs to be distributed and the amount of traffic to be distributed. This paper firstly expands the classical logic language based on the characteristics of wireless sensor network, and implements the language by compiling/interpretation, which can generate the smaller update code. Secondly, the paper also makes full use of the location information of the sensor and the data distribution protocol of low message complexity of telecommunication and channel resources. The research results of this paper can effectively reduce the communication volume of the distribution process.

Keywords. Wireless sensor network; Executable code; Classical logic language; Update code

1. Introduction

New computing models and applications, such as universal computing, the Internet of things, and the physical information fusion system, are thought to lead to the next computing revolution [1]. In these computing models and applications, a common need is to integrate the logical world of information with the real world of physics [2]. Wireless sensor network consists of awareness, treatment, and wireless data transmission ability of multiple sensors in the form of multiple hops, self-organization, can be easily deployed in various complex environment entity, the phenomena of the physical world information such as large-scale, long-term, real-time collection, thus suitable to serve as the bridge of the connection information world and physical world. Because of this, the wireless sensor network has been vigorously developed by academia and industry [3].

In recent years, wireless sensor networks have been used in many fields such as disaster relief, security, health care, home, construction, agriculture, transportation, logistics, and life sciences [4]. As for a single sensor local processing data and multiple sensor coordination ability to deal with data mining, the application of wireless sensor network application is no longer just the data collection, may also need to in situ data processing in a quick response to the change of the physical world. As the abundance of application requirements, the researchers found that wireless sensor network applications often because of application requirements to modify, add, and need to replace the sensor application executable code updates [5]. Therefore, the application of the application has become an important issue in the field of wireless sensor networks. In the academic world, the application of wireless sensor network is referred to as the wireless sensor network application reprogramming (referred to as reprogramming), and a lot of research is carried out.

Aiming at the efficiency of the reprogramming of wireless sensor networks, this paper analyzes the factors that affect the efficiency of the reprogramming process and the relationship between them [6]. On this basis, combining with the application of wireless sensor network platform and demand characteristics, puts forward a set of including the programming language, the language processing system, and the batch data distribution agreement, the application of efficient programming method. And the design and realization of the support platform for multiple wireless sensor network applications can be effectively reprogrammed. This introduction is the first to introduce the background and research motivation of the reprogramming of wireless sensor network and the definition of reprogramming.[7] Then the paper presents the research ideas of reprogramming problems, discusses the challenge of reprogramming and summarizes the current research status [8]. The technical route of this paper is put forward and the main work and contribution of this paper is briefly explained. Finally, the organization structure of this paper is given.

In the academic world, the application of the wireless sensor network is called the wireless sensor network application reprogramming, and a lot of research has been done on it. In these research work, understanding of the wireless sensor network applications reprogramming stays in the executable code level, the main research according to the new executable code update sensor on the original application method of executable code. Yet such understanding is not comprehensive. With the development of wireless sensor network, application demand change has become the main reason for the reprogramming of wireless sensor network. The implementation and efficiency of the application reprogramming of wireless sensor network can be significantly affected by the application of the application of executable code. Therefore, the application of reprogramming of wireless sensor networks requires not only the remote loading process of application executable code, but also the process of generating application executable code based on application requirements.Therefore, the purpose of reprogramming with infinite sensor network is to guarantee the network application and stabilize the system.

2. Material and Methods

According to the general method of software technology research (as shown in figure 1 (a)), software technology research on the problem of reprogramming must take into account software requirements and characteristics of hardware platforms. The essence of reprogramming is to apply update problems. Figure 1(b) shows the process of updating the application based on the classical platform. In a classic platform, application requirements are often complex, so the focus of research is often on how to modify and compile applications efficiently. After you get new executable code, you can easily load it into local memory. In the application update (reprogramming) based on the wireless sensor network platform shown in figure 1 (c), the application requirements are relatively simple, and modifying and compiling applications is usually not too complicated. However, the executable code needs to be mounted remotely to the target sensor. Because of the large scale of executable code and limited communication capability of sensors, remote loading often requires a large number of sensor coordination to carry out a large amount of wireless data transmission (Hu S, Han J, Wei X,2015). The application update process is shown in Figure 1.

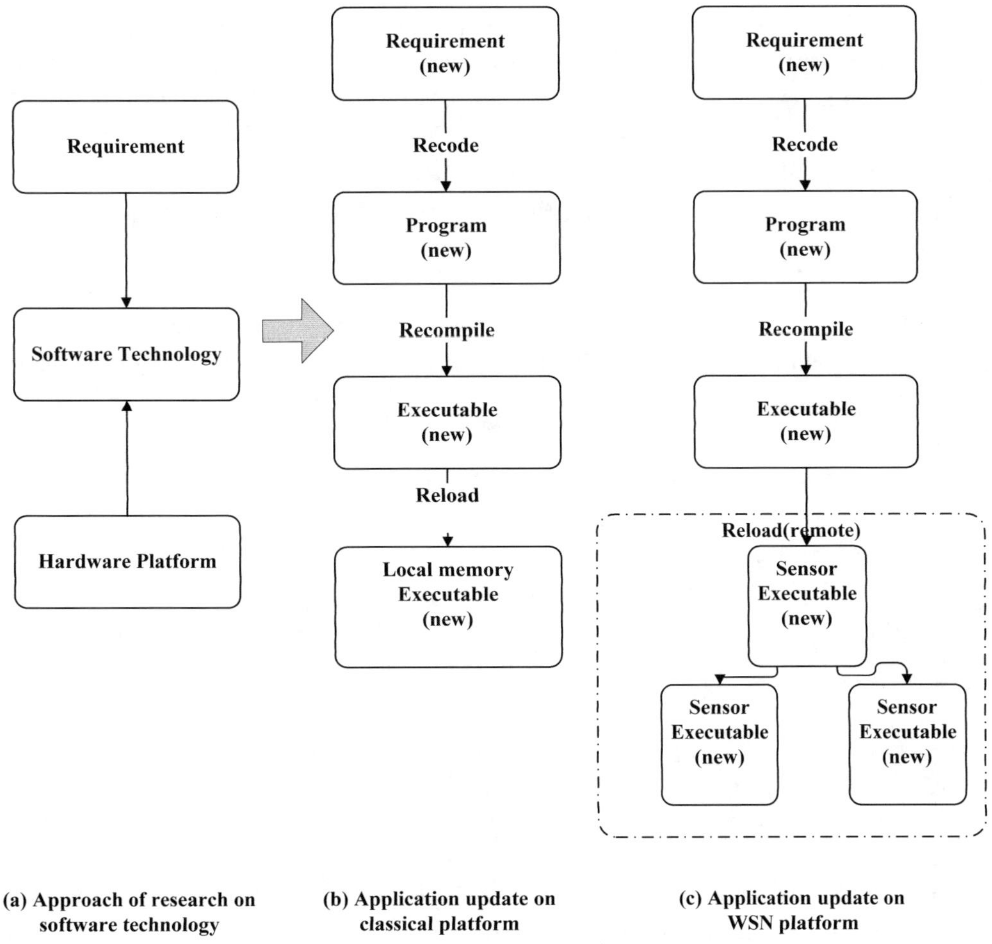

Figure 1. Apply the update process

Reprogramming will have a great impact on the normal operation of wireless sensor network:

First, reprogramming may shorten the lifetime of wireless sensor networks. Sensors are often used batteries, however remote loading need sensors for a large number of wireless data transmission, this can significantly increase the power consumption of the sensor may result in the application of wireless sensor network lifetime shorten. Second, reprogramming reduces the availability of wireless sensor networks. When making remote loading, wireless sensor network need to stop the current running application and all of its resources to complete executable code loading process, which can lead to applications in the process can't provide services.

The above analysis shows that reprogramming can affect the lifetime and availability of wireless sensor networks. The survival period of wireless sensor network is available in formula (1):

$$L_{app} \leq \min\left(\frac{E_{sensor} - E_{repro}}{e_{app}}\right) \tag{1}$$

In the above formula, the application of life cycle (Lapp) won't be longer than the first to run out of energy in wireless sensor network lifetime of sensor and sensor survival and sensors of the total energy (Esensor), and the energy of the sensor used for reprogramming (Erepro). Due to the application to run on a sensor unit (eapp) energy consumption can be predicted based on the experimental data, application developers often according to the desired application of survival for sensor equipment can provide enough battery power. However, when heavy programming consumes too much energy (that is, Erepro is too large), the lifetime of the application will be affected [9].

The availability of wireless sensor network applications can be expressed by formula (2):

$$A_{app} = \frac{T_{app} - T_{repro}}{T_{app}} \tag{2}$$

In the above formula, the availability of the application (App) is limited to the time of reprogramming (Trepro). Given application runtime (Tapp}, the more time you spend reprogramming, the lower the availability of the application.

In order to reduce the impact of reprogramming on wireless sensor networks, it is necessary to reduce the energy consumption and time consumption of reprogramming. Formula (3), (4) indicates that the energy consumption and time consumption of heavy programming are determined by the energy consumption of single reprogramming (erepro) and time consumption (trepro) and the number of reprogramming (n).

$$E_{repro}=\mathrm{n} \bullet \mathrm{e}_{repro} \tag{3}$$

$$T_{repro}=\mathrm{n} \bullet t_{repro} \tag{4}$$

On the basis of summarizing the existing work, this paper makes some explorations on the method of realizing high-efficiency reprogramming, and the technical route is as follows:

In terms of technical framework, this paper proposes a framework of programming language, language processing system, and bulk data distribution protocol to support efficient wireless sensor network application reprogramming. Programming language design considerations can reduce the scale of application modification caused by changes in application requirements. Logical programming style can not only shield the low-level implementation details naturally, but also reduce the disaster degree of business logic and programming. Therefore, it is easy to build the application of loose coupling. At the same time, logical language is suitable for interpretation execution, facilitating the application of executable code that is independent of the sensor platform. Therefore, based on the logic language and based on the characteristics of wireless sensor network, it is feasible to design the programming language.

3. Results

The average number of non-target sensors that are included in the connected dominating set in a randomly generated network of 200, 300 and 400 sensors. We repeated 1000 times experiment, the results show that, compared with MDeluge TZC can have 200, 300, and 400 sensors in the network, respectively, by 31.7% - 60.6%, 45.2% and 69.6%, and 54.3% to 75.3% of the number of the target sensor. These results show that T2C can effectively reduce the number of non-target sensors connected to the centralized control, and the effect is especially evident in the large-scale network. This is due to T2C in the process of establishing a connected dominating set that will converge the target sensor to a smaller path. In large-scale networks, sensors have more opportunities to choose fewer paths for non-target sensors. For this kind of the methods to set up the target sensors connected dominating set concerns is its effect on the depth of the connected dominating set, in the same network configuration of the connected dominating set the average depth. According to the results of the connected dominating set T2C with MDeluge establish an average depth of difference within a jump, the reason is that T2C also using breadth-first way to establish a connected dominating set, at the same time will affect the depth of the connected dominating set in the process of establishing the control node selection operation can only choose the nodes of the tree [10]. As shown in Figure 2 and Figure 3.

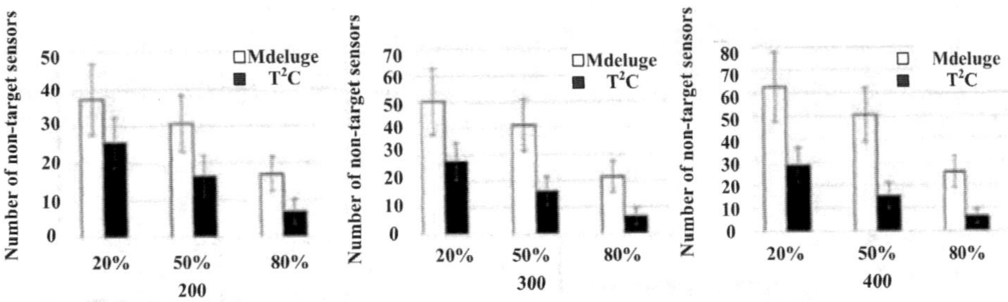

Figure 2. The number of non-target sensors included in the connected dominating set in the random generation network

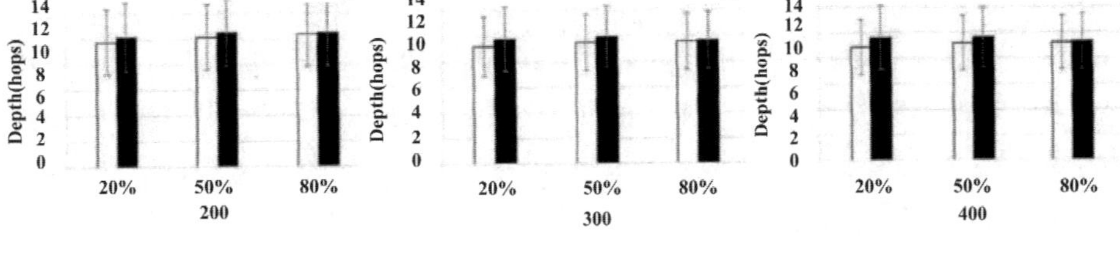

Figure 3. The depth of the connected dominating set in the network is randomly generated

Journal of Physics: Conference Series **2173** (2022) 012064 doi:10.1088/1742-6596/2173/1/012064

When evaluating different types of data distribution protocols, total energy consumption is a fair indicator. Figure 6.12 shows the total energy consumption of CORD, McTorrent, and T2C in the distribution of 5 pages of data in multiple networks. For CORD, total energy consumption consists of the energy consumption and distribution data of the connected dominating set. For T2C, total energy consumption also includes distribution of configuration information. Compared with CORD, TaCi and T2C: the total energy consumption was reduced by 15.5%-43.0% and 33.0%-54.2% respectively. Compared with McTorrent, T2C; The total energy consumption was reduced by 12.8%-45.8% and 30.9%-56.4% respectively. An interesting phenomenon is that when the network size increases, the T2C compared with CORD and McTorrent can reduce more of the total energy consumption, the reason is that T2C in large-scale network requires less data source sensor (control node), so that more sensors (be dominated nodes) can be in the process of data distribution close communication equipment periodically. In addition, the multi-channel communication of T2C can effectively accelerate the distribution process and reduce the number of lost packets, thus the total time of the sensor communication equipment in the working state is shorter. These two factors make the sensor communication equipment used in T2C to monitor and receive operation far lower than CORD and McTorrent, thus the total energy consumption is lower.

Simulation of time consumption. For cord X, time Y,, time consumption includes the time to set up the connected dominating set and the time to distribute the data. For T2C, it also includes the time to distribute configuration information. In the experiment, the length of CORD and T2C was set to 6s. Compared with CORD, TZC1 and C2 reduced 42.1 %-63.2% and 52.7%-79.7% respectively. Compared with McTorrent, T2C1 requires a little more time in the network's smaller size (for example, 50 sensors). This is T2C need to establish a connected dominating set in advance, and the advantages of shortening distribution path length and eliminating communication conflict cannot be fully reflected in the small network during distribution. However, when network scale increases, T2C is compared with McTorrent; Energy consumption can be reduced by 10.2% to 24.9%. For T2C2, compared with McTorrent, it can reduce energy consumption by 1.1 %-39.8 %.

Heavy programming tasks to create interface allows the user to create a new programming tasks, including fill in the creator, the execution time and required application, network, and network in the target sensor (optional). The execution time can be either executed immediately or executed at a specified time. The required application is selected by the user through the specified project name and the program version number. The target sensor can be manually entered into the sensor number, checked in the deployment chart of the wireless sensor network, or selected by specifying the sensor type.

The Server layer is primarily responsible for providing the data and services required to perform a reprogramming task. Server layer includes the application Manager (Program Manager) process Manager (Reprogramming Manager) Network Manager (Network Manager), reweave Reprogramming Server (Reprogramming Server), the Compiler (Compiler), as well as Data distributor (Data Disseminator) six modules, as well as application Data, code and image Data, update Data, the Network Configuration Data, as well as the five Data programming tasks [11]. As shown in the Figure 4.

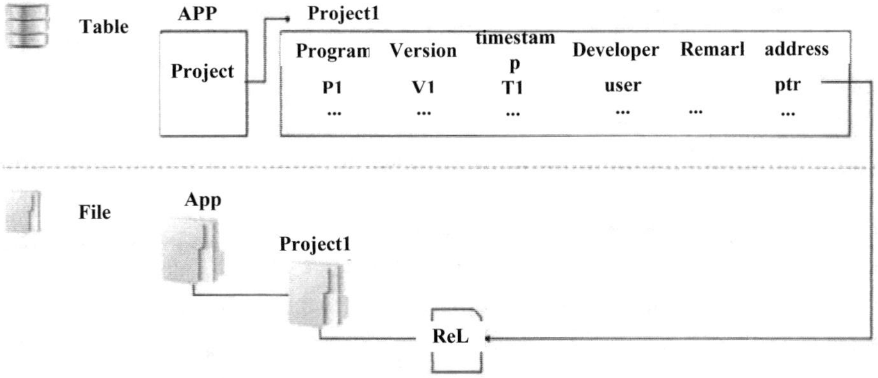

This chapter, based on the efficient reprogramming method proposed RSM system, gives the overall architecture and describes the Web, server and WSN layers in the system architecture. Then, based on the efficient re-programming method proposed in this paper, the overall architecture of RSM system is given, and the Web layer, Server layer and WSN layer in system architecture are described. The key modules of the function and important implementation details. Finally, the application of the infrared sensor in the intelligent laboratory application to reprogram the case, introduced through the registration of wireless sensor networks, add applications, and submit reprogramming tasks using RSM system for reprogramming the relevant details.

4. Discussions

With the development of wireless sensor network applications, enrich the application demand, often due to the application requirements of wireless sensor network application to modify, add, and need to replace the sensor application executable code updates. In the academic world, the application of wireless sensor network will be called wireless sensor network application reprogramming. Due to heavy programming need many sensor collaborations of a large number of wireless data transmission, which can result in sensors consume large amounts of energy and time, the lifetime of wireless sensor network application and availability impact cannot be ignored. There is work to reduce the reprogramming overhead by reducing the size of the updated data and reducing the amount of traffic that is distributed to update the data process. However, the development of wireless sensor network application increased network platform scale and enhancing the heterogeneity, the application requirements are more complex and frequent change, when your own work efficiency is still inadequate in the face of the new situation. This paper revolves around "to provide high efficiency wireless sensor network application programming method" this goal, to support the application of efficient reprogramming technology framework and involved in the framework of a programming language, language processing system, research and exploration of batch data distribution agreement.

In terms of technical framework, this paper presents a framework for the application of high efficiency reprogramming technology based on wireless sensor network syste. This paper based on the analysis of the process for reprogramming heavy programming model, and then combined with the application of wireless sensor network platform and application characteristics include programming language, language processing system is proposed, and the bulk data used to support efficient distribution agreement reprogramming the technical framework of wireless sensor network applications.

5. Conclusion

In programming language, this paper designs a logic programming language ReLog for wireless sensor network. In terms of the language processing system, this paper discusses the ReLog language processing system in reducing the size of the updated data and reducing the execution overhead of the code. The ReLog language processing system includes both the compiler and the virtual machine. In the aspect of bulk data distribution protocol, this paper discusses the specific work of the volume data distribution protocol T2C to reduce the traffic volume of the updated data distribution process. This paper presents the design and prototype implementation of the reprogramming service management (RSM) system. To improve the description of the ReLog language, consider giving the ReLog language a custom function interface. The ReLog language processing system requires the ability to handle user-defined functions accordingly. We need to make a theoretical analysis of the conclusions in the volume analysis data distribution protocol proposed in this paper. we currently need a theoretical analysis of the conclusions in the volume analysis.

Acknowledgement

Project Fund: Collaborative Education by the Ministry of Education. Fund No. 202101196010, Application of Network Crawber in Big Data Teaching.

References

[1] He D, Chen C, Chan S, et al 2013 Security Analysis and Improvement of a Secure and Distributed Reprogramming Protocol for Wireless Sensor Networks *IEEE Transactions on Industrial Electronics* **60**(11) 5348-54

[2] Li J W, Li S N, Zhang Y, et al 2016 An Analytical Model for Coding-Based Reprogramming Protocols in Lossy Wireless Sensor Networks *IEEE Transactions on Computers* **66**(1) 24-37

[3] Dong W, Liu Y, Chen C, et al 2013 R2: Incremental Reprogramming Using Relocatable Code in Networked Embedded Systems *IEEE Transactions on Computers* **62**(9) 1837-49

[4] Qu H, Yi Z, Yang S X 2013 Efficient shortest-path-tree computation in network routing based on pulse-coupled neural networks *IEEE Transactions on Cybernetics* **43**(3) 995-1176

[5] Park T, Kim S Y, Kwon G I 2013 Multi-block FEC for reprogramming wireless sensor networks *Electronics Letters* **49**(14) 905-7

[6] Ahmed A A 2015 A comparative study of QoS performance for location based and corona based real-time routing protocol in mobile wireless sensor networks *Wireless Networks* **21**(3) 1015-31

[7] Pilloni V, Franceschelli M, Atzori L, et al 2016 Deployment of Applications in Wireless Sensor Networks: A Gossip-Based Lifetime Maximization Approach *IEEE Transactions on Control Systems Technology* **24**(5) 1828-36

[8] Hu S, Han J, Wei X, et al 2015 A multi-hop heterogeneous cluster-based optimization algorithm for wireless sensor networks *Wireless Networks* **21**(1) 57-65

[9] Degirmenci G, Kharoufeh J P, Baldwin R O 2013 On the performance evaluation of query-based wireless sensor networks *Performance Evaluation* **70**(2) 124-47

[10] Stolikj M, Meyfroyt T M M, Cuijpers P J L, et al 2015 Improving the Performance of Trickle-Based Data Dissemination in Low-Power Networks. In: Abdelzaher T., Pereira N., Tovar E. (eds) *Lecture Notes in Computer Science* **8965** 186-201

[11] Kumar A K, Sivalingam K M, Kumar A 2013 On reducing delay in mobile data collection based wireless sensor networks *Wireless Networks* **19**(3) 285-99

Improvement Strategy for Voltage Quality of Microgrid Based on Master-Slave Control

Shijun Li[1,*], Jinjie Gu[1] and Biao Li[1]

[1]College of Electrical & Information Engineering, Hunan Institute of Engineering, Xiangtan, Hunan, China

[*]corresponding author's e-mail: lsj181627@qq.com

Abstract. Due to the nonlinear and unbalanced load of micro-grid, the output voltage quality on the main inverter side is easily affected in the micro-grid based on Master-Slave control.So, this paper put up a strategy that compensating the harmonic current and negative-sequence current by subordinate inverter. This strategy makes the current that flow through main inverter only have fundamental positive sequence component, so as to achieve the aim of improving the quality of the output voltage. Finally, the simulation model is built in the PSCAD simulation environment, which shows the effectiveness of the strategy.

1. Introduction

In recent years, as a small power distribution system that can realize self-control and self-management, micro-grid has attracted much attention [1-2]. There are two working modes in Miro-grid: grid-connected operation mode and off-grid operation mode[3].In off-grid operation mode, without the support of large power grid [4], the presence of nonlinear unbalanced mixed load in micro-grid will lead to output voltage distortion and imbalance of inverter in Micro-grid. If the Distributed Generation (DG) in the Micro-grid is not controlled, the voltage imbalance and harmonics will do great harm to the electrical equipment and Micro-grid. Therefore, when the DG in the Micro-grid completes the form conversion of electric energy, it should also ensure that it can still provide high-quality electric energy under the nonlinear unbalanced mixed load[5-6].

This paper firstly analyzes the reasons for the output voltage distortion and imbalance of the main inverter of the Micro-grid, and obtains a quantitative and intuitive conclusion. Secondly, based on the mechanism of output voltage deterioration of the main inverter, this paper put up a strategy that compensating the harmonic current and negative-sequence current by subordinate inverter. This strategy makes the current that flow through main inverter only have fundamental positive sequence component, so as to achieve the aim of improving the quality of the output voltage, and the effectiveness was finally verified through simulation.

2. Micro-grid structure and degradation mechanism analysis of volta-ge quality

The Micro-grid structure with multiple distributed power sources is shown in Fig .1, where K represents the static switch connected to the distribution grid. After being transformed into alternating current by the inverter, each distributed power supply is connected to the alternating current bus and forms a Micro-grid system together with the load. The micro-grid system is connected to the distribution grid through the static switch K, which can realize the switch between the grid-connected operation mode and off-grid operation mode. This paper mainly studies off-grid operation mode.[7]

The voltage and frequency of the Micro-grid based on master-slave inverter control is maintained by one of the DGs, called main inverter, such as the gas turbine in Fig. 1. Because of its output power is adjustable within a certain range, and the output power is stable and easy to control, it is often used as the main inverter, which can maintain the stable operation of the Micro-grid for a long time in off-grid mode. Other DGs, such as pv, fan, etc. would take this voltage and frequency as the reference to inject power into the Micro-grid, so they are known as subordinate inverter. Normally, the function of subordinate inverter is only to inject power into Micro-grid, without improving the voltage of the Micro-grid, so this paper mainly research on how to improve the control strategy of subordinate inverter, so as to improve the power quality of the Micro-grid voltage.[8,9]

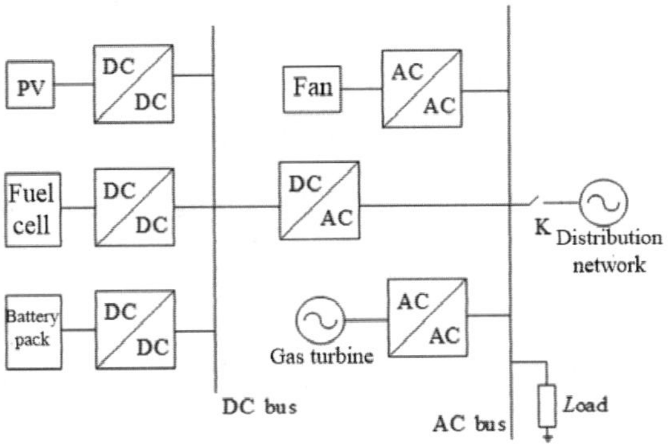

Figure 1. Micro-grid structure with multiple distributed generation

The voltage and frequency of the Micro-grid based on master-slave inverter control is maintained by one of the DGs, called main inverter, such as the gas turbine in Fig. 1. Because of its output power is adjustable within a certain range, and the output power is stable and easy to control, it is often used as the main inverter, which can maintain the stable operation of the Micro-grid for a long time in off-grid mode. Other DGs, such as pv, fan, etc. would take this voltage and frequency as the reference to inject power into the Micro-grid, so they are known as subordinate inverter. Normally, the function of subordinate inverter is only to inject power into Micro-grid, without improving the voltage of the Micro-grid, so this paper mainly research on how to improve the control strategy of subordinate inverter, so as to improve the power quality of the Micro-grid voltage.

3. The compensation strategy with subordinate inverter to improve the voltage quality of Micro-grid

3.1. Mechanism analysis of voltage quality improvement on the output side of the inverter

Master-slave control, that is, a main inverter with larger capacity adopts a control strategy to keep constant voltage and constant frequency so as to maintain the frequency and voltage stability of the Micro-grid, which can be equivalent to the controlled voltage source, while the subordinate inverter uses constant power control mode to inject power into the Micro-grid, which can be equivalent to the controlled current source.[10] Its main circuit is shown in fig 2.

Among them, the i_o is the load current, composed of fundamental wave current i_{of} and harmonious wave current i_{oh}. By controlling the subordinate inverter, the harmonious wave current in the load i_{oh} and the fundamental negative sequence current i_{of}^- can be provided by subordinate inverter, and the main inverter only provide the fundamental positive sequence current i_{of}^+. Zero sequence current i_{of}^0 would be provided by the fourth bridge arm, shared by the main and subordinate inverter. we can know that through controlling the subordinate inverter, the main inverter equivalent output impedance will not contain harmonics, negative sequence and zero sequence voltage drop, making the output voltage of main inverter three-phase symmetrical, and does not contain harmonics.

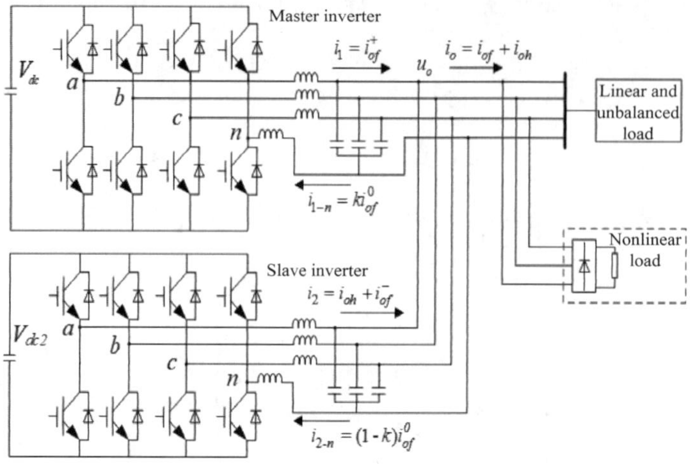

Figure 2. Main circuit topology

3.2. Compensation control strategy by the subordinate inverter

Under the condition of non-linear unbalanced mixed load, if we want the inverter to output the three-phase symmetric pure sine wave without being affected by the load, all the ways we can choose are essentially reducing the harmonic current, negative sequence current and zero sequence current to make its characteristics closer to the ideal power supply.

Therefore, master and subordinate inverters are used to control the output current of the master inverter to maintain the sinusoidal characteristics of the fundamental wave, while the harmonic current in the load and the negative sequence current of the fundamental wave are provided by the subordinate inverter, and the zero sequence current is shared by the fourth bridge arm of the master and subordinate inverter. The overall control block diagram is shown in Fig. 3

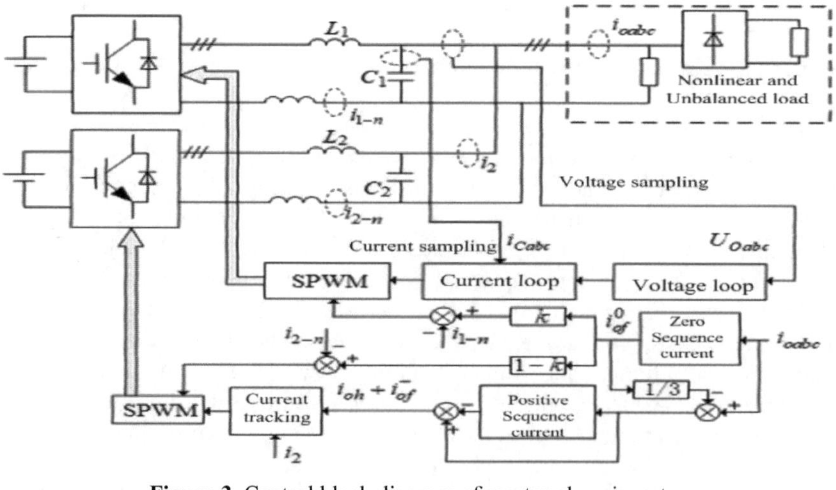

Figure 3. Control block diagram of master-slave inverter

L_1、L_2、C_1 and C_2 are respectively the filtering inductance and capacitance of the main and subordinate inverters; $i_{1\text{-n}}$ and $i_{2\text{-n}}$ are the current flowing into the fourth bridge arm of the master and subordinate inverter respectively. i_{Cabc} is the capacitance current of main inverter; i_2 is the output current from the inverter; i_{oabc} is the load current; U_{Oabc} is the output voltage of main inverter.

When the output side of the inverter is loaded with nonlinear unbalanced mixed load and the nonlinear load is three-phase uncontrollable rectifier, the load current

$$i_{oabc} = i_{of}^{+} + i_{of}^{-} + i_{of}^{0} + \sqrt{2}\sum_{h=6k\pm1}^{n} I_h \sin(hwt+\varphi_h)\,,\ k=1,2,3,\ldots \tag{1}$$

According to the load current characteristics, if we want to ensure that the output current of the main inverter is the positive sequence current of the fundamental wave, the harmonic current and negative sequence current in the load are provided by the subordinate inverter. The sum of the harmonic current and negative sequence current components extracted from the load current is taken as the reference instruction of the slave inverter, and the control block diagram of the slave inverter obtained is shown in Fig 4 .

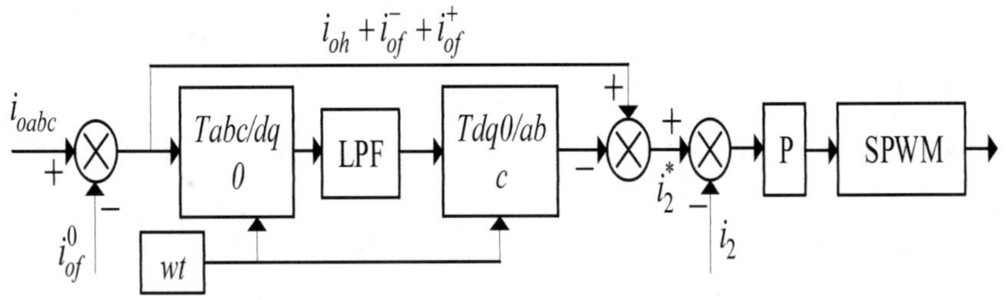

Figure 4. Control block diagram of slave inverter

After removing the zero sequence components, the result of the load current flowing through the coordinate transformation is as follows

$$\begin{aligned}\begin{bmatrix} i_d \\ i_q \\ i_0 \end{bmatrix} &= T(\theta=wt)_{abc/dq0}\begin{bmatrix} i_{oa}-i_{ofa}^{0} \\ i_{ob}-i_{ofb}^{0} \\ i_{ob}-i_{ofb}^{0} \end{bmatrix} = \sqrt{2}I_{of}^{+}\begin{bmatrix} \cos(\theta_1) \\ -\sin(\theta_1) \\ 0 \end{bmatrix} + \sqrt{2}I_{of}^{-}\begin{bmatrix} \cos(2\omega t+\theta_2) \\ -\sin(2\omega t+\theta_2) \\ 0 \end{bmatrix} \\ &+\sqrt{2}\sum_{h=6k+1}^{n} I_h\begin{bmatrix} \cos((h-1)wt+\theta_h) \\ -\sin((h-1)\mathrm{wt}+\theta_h) \\ 0 \end{bmatrix} + \sqrt{2}\sum_{h=6k-1}^{n} I_h\begin{bmatrix} \cos((h+1)wt+\theta_h) \\ -\sin((h+1)\mathrm{wt}+\theta_h) \\ 0 \end{bmatrix}\end{aligned} \tag{2}$$

I_{of}^{+} and I_{of}^{-} are the effective values of positive and negative sequence currents of fundamental waves respectively, I_h is the effective values of harmonic currents corresponding to harmonic sub harmonic, and the expression of $T(\theta)_{abc/dq0}$ is as follows

$$T(\theta)_{abc/dq0} = \frac{2}{3}\begin{bmatrix} \cos\theta & \cos(\theta-120°) & \cos(\theta+120°) \\ \sin\theta & \sin(\theta-120°) & \sin(\theta+120°) \\ 1/\sqrt{2} & 1/\sqrt{2} & 1/\sqrt{2} \end{bmatrix} \tag{3}$$

4. Simulation and analysis

In order to verify the effectiveness of the control strategy, a simulation model of the master and slave inverter with nonlinear unbalanced mixed load was built in the PSCAD environment. The master and subordinate inverter both adopt three -phase four-wire system, the main inverter connected to the load through LC filter, and was put into use from the 1s time, and to observe the output voltage and current waveform of both main and subordinate inverter in the three working conditions: with nonlinear load, the unbalanced load and nonlinear mixed load. [10]

4.1. Mechanism analysis of voltage quality improvement on output side of inverter

Fig 5 shows the load voltage waveform of the master-slave inverter, the output current waveform of the main inverter and the output current waveform of the subordinate inverter when the main-slave inverter runs with nonlinear load. From 0s to 1s, it can be seen that the load voltage and the output current of the master-slave inverter are seriously distorted and the harmonic content is large. At the 1s, harmonic compensation is input from the inverter. Through the effective control of the subordinate inverter, most of the harmonic current in the load is provided by the subordinate inverter. As can be seen from figure 5(d), the voltage distortion rate before compensation is 8.85%, and the voltage distortion rate after compensation is 1.1%. The harmonic in the output current of the load voltage and main inverter is effectively suppressed, and the harmonic current in the load is mainly provided by the subordinate inverter.

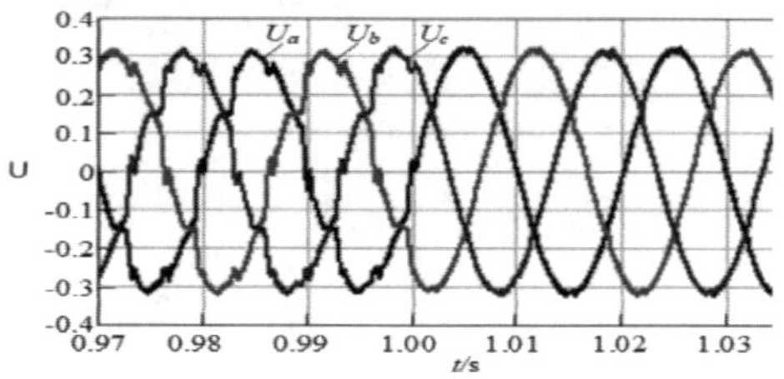

(a) Load voltage waveform

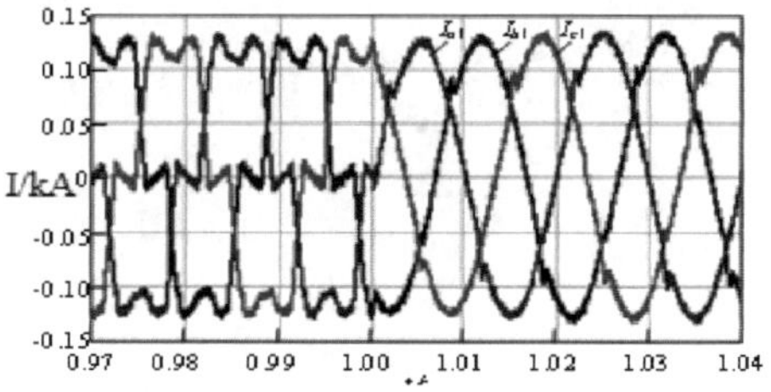

(b) Output current waveform of main inverter

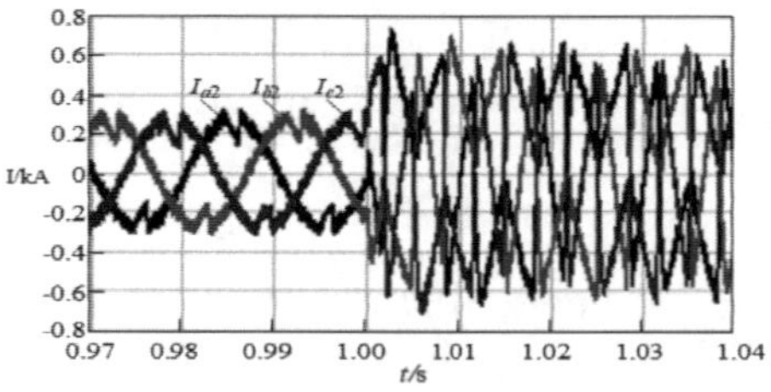

(C) Output current waveform of Slave inverter

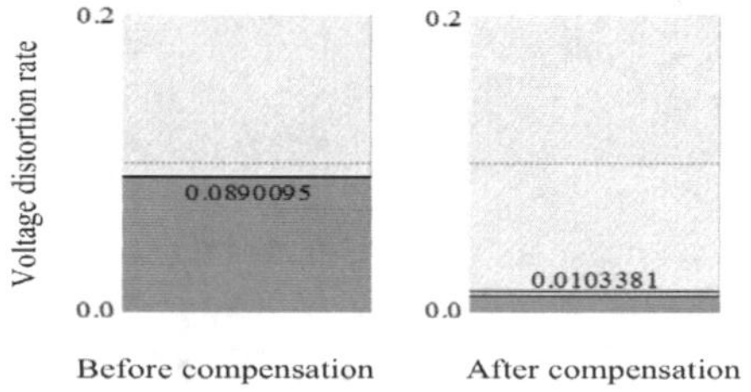

(d) Load voltage distortion rate

Figure 5. The output waveform of the master-slave inverter under nonlinear load

5. Conclusion

On the basis of analyzing the mechanism of voltage distortion on the output side of the Micro-grid main inverter, this paper proposes a Master-slave Inverter Control Strategy for nonlinear unbalanced mixed load through theoretic analysis and derivation, and draws the following conclusions through PSCAD simulation verification:

1) the output voltage distortion of the inverter is caused by the harmonic voltage drop generated by the harmonic current on the output impedance of the inverter. The voltage drop caused by the negative sequence and zero sequence current on the output impedance of the inverter is the cause of the unbalanced output voltage of the inverter.

2) by controlling the subordinate inverter, the negative sequence, zero sequence and harmonic current in the main inverter are reduced, and the voltage distortion and imbalance on the output side of the main inverter are effectively suppressed.

Acknowledgments

This paper is supported Project supported by Scientific Research Fund of Hunan Provincial Education Department (19A108), and provincial natural science foundation of Hunan (2019JJ40059) AND collaborative education project of industry university cooperation of the Ministry of Education

(201902143006) (2019003020) (201902145014), and Hunan teaching reform project(HNJG-2021-1350) ,and Doctoral Talent Research Fund(20RC002), and Xiangtan enterprise science and technology special expert project(CG-TPZJ20211019), and Xiangtan science and technology innovation project((CG-YB20211019).

References

[1] Dipankar De and Venkataramanan. A Proportional+ Multiresonant Controller for Three-Phase Four-Wire High-Frequency Link Inverter.IEEE Transactions on Power Electronics,2010, 25(4):899-906.

[2] Lim K,Choi JJang J,et al.P+ multiple resonant control for output voltage regulation of microgrid with unbalanced and nonlinear loads. 2014.

[3] Karimi H, Mokhtari H.Harmonic and Negative-Sequence Current Control in an Islanded Multi-Bus MV Microgrid. IEEE Transactions on Smart Grid, 2013, PP(1):1-1.

[4] Hassan Moussa,Ahmed Shahin, Jean-Philippe Martin, et al. Harmonic Power Sharing with Voltage Distortion Compensation of Droop Controlled Islanded Microgrids. IEEE Transactions on Smart Grid, 2017, PP (9):1-1.

[5] Q.-C. Zhong, F. Blaabjerg, J. Guerrero, and T. Hornik, "Reduction of voltage harmonics for parallel-operated inverters equipped with a robust droop controller," in Proc. IEEE Energy Convers. Congr. Expo. Phoenix,AZ, 2011, pp. 473–478.

[6] Baghaee H R, Mirsalim M, Gharehpetian G B, et al. Nonlinear Load Sharing and Voltage Compensation of Microgrids based on Harmonic Power-Flow Calculations using Radial Basis Function Neural Networks. 2016.

[7] Yong Li, Zhiwen Zhang, Rehtanz C, et al. Study on steady- and transient-state characteristics of a new HVDC transmission system based on an inductive filtering method. IEEE Transactions on Power Electronics, 2011, 26(7): 1976-1986

[8] Liang C G,Xu J Z,Luo L F,et al.Harmonic Elimination Using Parallel Delta-Connected Filtering Windings for Converter Transformers in HVDC Systems. IEEE Transactions on Power Delivery,2016,10(11)1109-1118

[9] Yong Li，Longfu Luo，Rehtanz.C,et al. Simulation of the electromagnetic response characteristic of an inductively filtered HVDC converter transformer Using field-circuit coupling. IEEE Transactions on Industrial Electronics,2012,59(11):4020- 4031.

[10]Li Biao,JIAN Yong,CAO Xueyuan,etal. Strategy for improving voltage quality of microgrid based on master-slave control.Electric Power Engineering Technology,2020,39(6):117-123

Design and Fabrication of QN9080DK Based Bluetooth 5.0 Fingerprint Lock Controller

Yong Guo[1,*], Cheng Xiong[2], Yu Wang[1], Shikai Wu[1], Le Luo[1] and Qiang Fu[1]

[1]School of Electronic Engineering, Chengdu Technological University, Chengdu, Sichuan, China

[2] School of Optical-Electrical and Computer Engineer, University of Shang Hai for Science and Technology, Shang Hai, China

[*]corresponding author's e-mail: robinfit01@163.com

Abstract. The QN9080DK is a development board from NXP specifically for the wireless access market, which has an integrated QN9080 Bluetooth system-on-a-chip (SOC) with an Arm CortexM4F core and a dedicated fusion sensor coprocessor. In this paper, with the help of MCUXpresso integrated development environment (IDE), SDK, CrotexM4F core, Bluetooth 5.0, serial communication, Arduino software and hardware and so on, a Bluetooth fingerprint lock controller is designed and fabricated. The lock controller has the functions of user Bluetooth password access and unlocking, fingerprint entry, fingerprint unlocking and fingerprint permission deletion, etc. After the joint debugging of system software and hardware, the designed Bluetooth Low Energy (BLE) 5.0 fingerprint smart electronic lock controller meets the expected requirements.

1. Introduction

As the tide of smart home[1-3] continues to surge forward, the Bluetooth electronic lock, as a gateway to smart home, has become more and more abundant in its application scenarios because it can be unlocked using a mobile phone terminal or remotely[4-6]. There have been numerous papers on the design of smart electronic locks, including smart Bluetooth door lock design based on the internet of things (IoT)[7], smart door locks using wireless near field communication (NFC)[8] and WiFi[9] technologies, and smart door lock systems that combine fingerprint, facial recognition, and radio frequency identification (RFID) wireless technologies[10-12]. This design combines the QN9080 SOC with Arduino, supplemented by serial port and Bluetooth communication technology, to realize a new Bluetooth electronic lock controller design. The Bluetooth electronic lock controller has the functions of fingerprint identification, Bluetooth password transmission, voice broadcasting and wireless unlocking, which can meet the needs of unlocking in users' daily life. Bluetooth electronic lock controller not only avoids the inconvenience of carrying physical keys, but also can set the authority and time period of unlocking, which has excellent unlocking experience while enhancing security, all these advantages will make the Bluetooth electronic lock controllers more and more popular among users.

2. Modular design of the lock controller

The Bluetooth electronic lock controller studied in this paper adopts a modular design idea, i.e., the lock controller is split into different functional modules according to the specific functional requirements of the Bluetooth electronic lock controller, and finally the Bluetooth electronic lock controller is implemented with the joint debugging of each module.

2.1. BLE communication module
Since the snake-shaped Bluetooth antenna has been integrated into the QN9080 core board on the QN9080DK development board, in order to implement the BLE communication function of the BLE communication module, it is only necessary to program the QN9080 on MCUXpresso. According to the official SDK, call the corresponding Bluetooth protocol functions to realize the BLE communication function, i.e., the function calls are generally completed at the Bluetooth user layer to complete the BLE communication function.

2.2. Fingerprint unlocking module
The hardware of the fingerprint unlocking module is a combination of AS608 optical fingerprint module and Arduino. This module has the following features: the input time is less than 1 second, the resolution is up to 500 dpi, 300 fingerprints can be input, the communication adopts data packet sending and receiving mode, and supports two access modes of USB and UART at the same time, which are very suitable for door lock fingerprint unlocking scenarios. The Arduino has a perfect open source function library for AS608, which is safe, stable and has low error probability of fingerprint verification, which not only reduces the difficulty of development, but also saves time costs. In the software of this module, the fingerprint unlocking function is completed by modifying and improving some basic functions in the open source function library and combined with the door lock control logic.

2.3. Electric lock driver module
The driver unit of the electric lock generally uses an electromagnet or a direct current (DC) speed reduction motor. If a DC speed reduction motor type electric lock is adopted, a motor driver chip is required to control the forward and reverse rotation of the motor to open or close the lock. The electric lock driver module uses two L9110 H-bridge chips to drive the electric lock. The control ports of the DC motor of the electric lock are connected to the I/O ports of the QN9080DK, and the forward and reverse rotation of the motor are controlled by programming the output high and low voltage levels of the I/O ports, so as to realize the functions of unlocking and locking.

2.4. Voice module
The voice prompt module adopts an external module with STC15W08AS as the core controller. This module is equipped with a TF card for storing sound recordings, and has a 3.5mm headphone jack with a maximum power of 3W. The QN9080 communicates with the external module to complete the function of playing, stopping and selecting the voice output. Due to the high integration of the audio control module, the codes of this part are very concise. The voice module works stably and reliably, and the contents of the voice prompt are: "The door lock has been opened." and "Unlocking failed, please try again."

2.5. Display module
The display module uses a 0.96-inch full color OLED screen, which is used to prompt the user to record the fingerprint, delete the fingerprint, and display the user ID when the user fingerprints are verified. This module adopts Serial Peripheral Interface (SPI) communication protocol, and Arduino writes the contents to be displayed and the corresponding operation instructions to the display module through SPI communication protocol. In order to simplify the source code writing of OLED control and communication, the OLED screen display module control functions in the Arduino function library are used to control the screen display related content, and the main library function are Adafruit_SSD1331.h, Adafruit_GFX.h and SPI.h.

2.6. Serial communication module
During the operation of the fingerprint lock controller, communication is needed between the Arduino and the QN9080 to complete the information interaction. The communication is divided into two modes, one is: Arduino responds to the QN9080 with the result of fingerprint unlocking according to the high and low levels of the specific I/O ports. The QN9080 determines whether the fingerprint is unlocked successfully or not according to the status of the I/O ports, and plays the corresponding voice prompt.

The other one is: during the password inputting or the mode selection, the QN9080 communicates with the Arduino through the serial port to transmit the content that needs to be displayed on the OLED screen, and the QN9080 also controls the Arduino to register and delete the fingerprint. The operation mode of the serial communication is 115200 baud rate, 8 data bits, 1 stop bit, and no parity bit.

3. Main design process and experimental verification

After completing the modular design of the fingerprint lock controller, the above mentioned functional modules need to be jointly debugged to complete the system design of the QN9080DK-based BLE fingerprint lock controller, and then through the joint debugging of hardware and software to complete the experimental verification of the entire fingerprint lock controller, the specific design and experimental process are as follows.

3.1. BLE fingerprint lock controller system composition

The QN9080DK is used as the control core for the system integration of each functional module, and the completed hardware structure is shown in Figure 1, which mainly consists of QN9080DK development board, Arduino control board, fingerprint unlock module, voice module and display module, and the system introduced a DC speed reduction motor to simulate an electronically controlled lock. At the software level, the fingerprint lock controller uses the FreeRTOS embedded real-time operating system, which abstracts each functional module into different tasks, and burns the program into the QN9080 after completing the corresponding code writing and logic design. FreeRTOS real-time operating system can well manage the processes required by Bluetooth control protocols such as master-slave selection, broadcasting, scanning mode selection and user data processing, which makes the original complex and cumbersome software design process more concise and efficient, and also facilitates users to carry out secondary development on this basis. Figure 2 shows the program control scheme of the fingerprint lock controller. Through the program control logic, the basic software design of the Bluetooth fingerprint lock controller is completed.

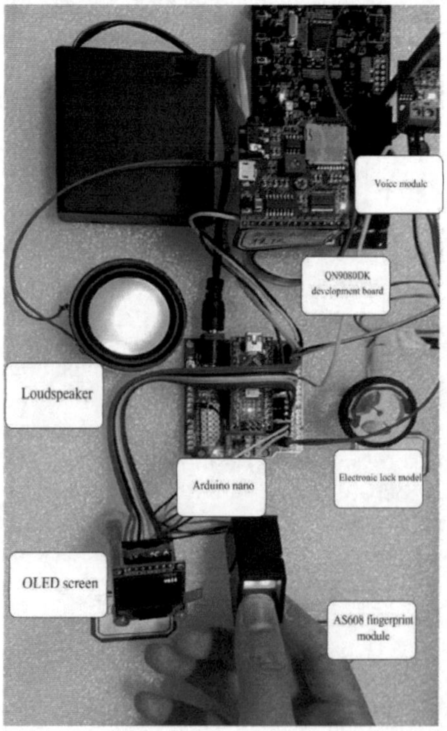

Figure 1. Physical view of the Bluetooth fingerprint lock controller.

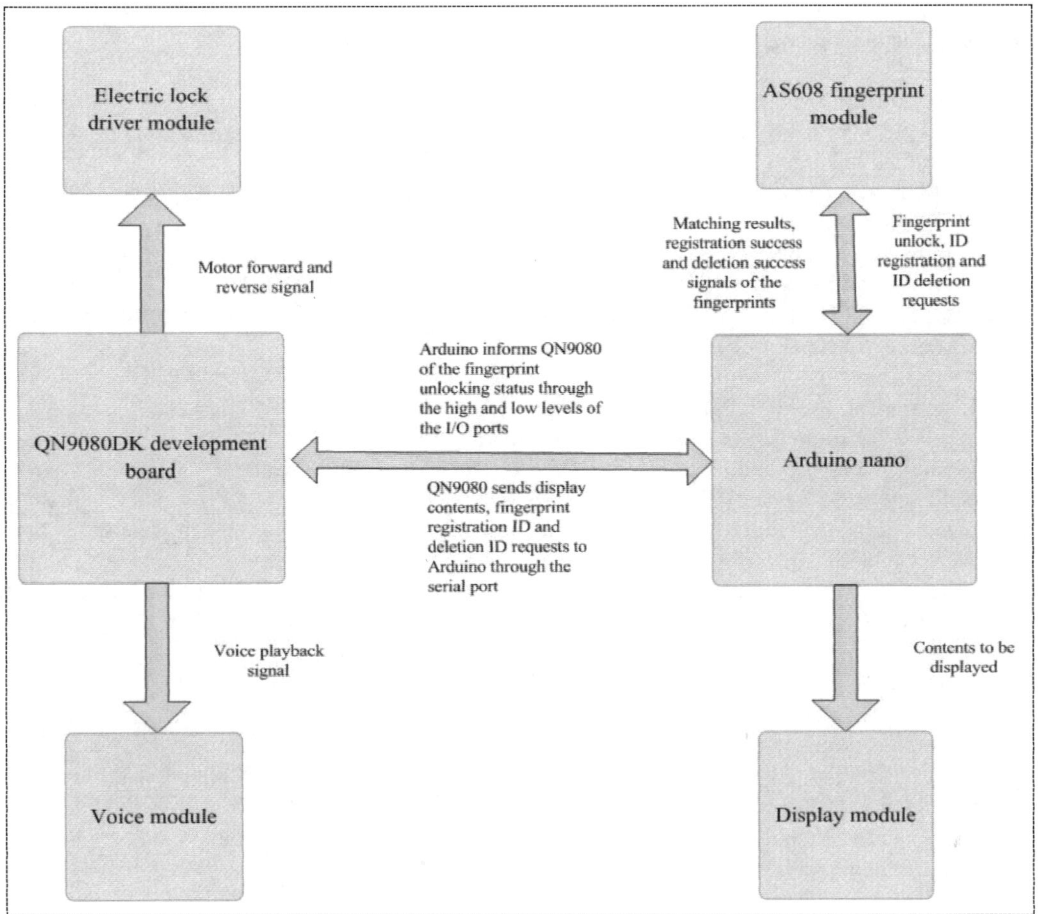

Figure 2. Program control diagram of fingerprint lock controller.

3.2. Joint debugging of the BLE fingerprint lock controller system

During the joint debugging of the fingerprint lock controller system, the fingerprint unlocking, password unlocking, ID registration and ID deletion operations were randomly performed. The OLED display was normal, the voice module broadcast was normal, and the DC speed reduction motor of the simulated electric lock operated correctly. The test results show that the design of the BLE fingerprint lock controller meets the requirements of the design specification.

4. Conclusion

In this paper, a Bluetooth 5.0 fingerprint lock controller is designed and implemented using QN9080DK as the core board and a modular design idea. In the design process, both hardware design and software development are all presented in a modular design. For the functional modules such as fingerprint recognition, voice broadcast, Bluetooth communication, information display, electric lock driver and serial communication, each module is designed, fabricated and debugged individually, and then the modules are combined together for joint debugging of the system to finally achieve the design task requirements. The designed Bluetooth fingerprint lock controller has functions such as wireless password unlocking, fingerprint recognition unlocking, voice broadcast prompt, electric control lock drive and OLED human-computer interaction, which meets the basic functional requirements of smart home access control.

Acknowledgments
This work was supported by the 2021 National Innovation and Entrepreneurship Training Program for College Students (Project No.202111116019), and the Talent Training Quality and Teaching Reform Project of Chengdu Technological University.

References
[1] Agarwal K, Agarwal A and Misra G 2019 Third International Conf on I-SMAC (IoT in Social, Mobile, Analytics and Cloud) (I-SMAC) Palladam India pp 629-633
[2] Khunchai S and Thongchaisuratkrul C 2020 International Conf on Power, Energy and Innovations (ICPEI) Chiangmai Thailand pp 105-108
[3] Fongbedji M, Krami N and Bouya M 2020 IEEE 2nd International Conf on Electronics, Control, Optimization and Computer Science (ICECOCS) Kenitra Morocco pp 1-4
[4] Tran Q. T, Tran D. D, Doan D and Nguyen M S 2020 International Conf on Advanced Computing and Applications (ACOMP) Quy Nhon Vietnam pp 170-174
[5] Swati and Gupta R P 2018 International Conf on Advances in Computing, Communication Control and Networking (ICACCCN) Greater Noida India pp 80-85
[6] Yu Y 2018 IEEE International Conf on Consumer Electronics (ICCE) Las Vegas NV USA pp 1-2
[7] Shanthini M, Vidya G and Arun R 2020 Third International Conf on Smart Systems and Inventive Technology (ICSSIT) Tirunelveli India pp 92-96
[8] Pacheco J and Miranda K 2020 IEEE International IOT, Electronics and Mechatronics Conf (IEMTRONICS) Vancouver BC Canada pp 1-5
[9] Yashwant S K, Krishna P V, Kumar B N V S B, Chandan G and Prasad J V D 2020 2nd International Conf on Innovative Mechanisms for Industry Applications (ICIMIA) Bangalore India pp 718-721
[10]Gudavalli D K P, Monica I S and Vidya Sagar M E C 2019 Global Conf for Advancement in Technology (GCAT) Bangalore India pp 1-6
[11]Masykuroh K, Syifa F T, Setiyanto G R, Ramadhani A D, Kurnianto D and Iryani N 2021 IEEE International Conf on Communication, Networks and Satellite (COMNETSAT) Purwokerto Indonesia pp 342-347
[12]Tiwari S, Thakur S, Shetty D and Pandey A 2018 Second International Conf on Inventive Communication and Computational Technologies (ICICCT) Coimbatore India pp 979-984

Journal of Physics: Conference Series **2173** (2022) 012067 doi:10.1088/1742-6596/2173/1/012067

The Vertex Degree Distances of One Vertex Union of the Cycle and the Star

Haiyan Xie, Nini Xue* and Jingjing Yang

School of Science, Dalian Maritime University, Dalian, Liaoning, China

*corresponding author's email: 1298215387@qq.com

Abstract. The degree distance of a graph is a graph invariant that is more sensitive than the Wiener index. In this paper, we calculate the vertex degree distances of one vertex union of the cycle and the star, and the degree distance of one vertex union of the cycle and the star. These results lay a foundation for further study on the extremal values of the vertex degree distances, and the distribution of the vertices with the extremal values in one vertex union of the cycle and the star.

1. Introduction

Topological indices and graph invariants based on the distances between the vertices of a graph are widely used in theoretical chemistry to establish relations between the structure and the properties of molecules[11,12]. The Wiener index is a well-known topological index which equals the sum of distances between all pairs of vertices of a molecular graph[1,8,10,13]. The degree distance of a graph is a graph invariant that is more sensitive than the Wiener index[2-6]. It is defined in the following way: given a graph G ,the vertex degree distance (VDD for short) of $v \in V(G)$ is defined as $D'(v) = d(v)\sum_{u\in V(G)} d(u,v)$, where *d(v)* is the degree of *v*, *d(u,v)* is the distance between *u* and *v*. The degree distance of *G* is defined as $D'(G) = \sum_{v\in V(G)} D'(v)$.

This paper focuses on a special type of the unicyclic graph, one vertex union of the cycle and the star, vertex distance[7,9].

Denote an m-cycle as $C_m (m \geq 3)$ and an n-star as S_n (n≥3): $V(C_m) = \{u_0, u_1, u_2, \ldots, u_{m-1}\}$, $V(S_n) = \{v_0, v_1, v_2, \ldots, v_{n-1}\}$, where v_0 is the center of S_n . If $|V(C_m) \cap V(S_n)| = 1$, then the union of C_m and S_n is called one vertex union of the cycle and the star. There are two different structure of one vertex union of the cycle and the star: One is a graph on which a point on a circle coincides with the center of the star (denote as $CS_{mn}^{(1)}$, see Figure.1 (a)),called the first kind of the union graph. The other is a graph that a point on the cycle coincides with a leaf of the star (denote as $CS_{mn}^{(2)}$, see Figure.1 (b)), which is called the second kind of the union graph.

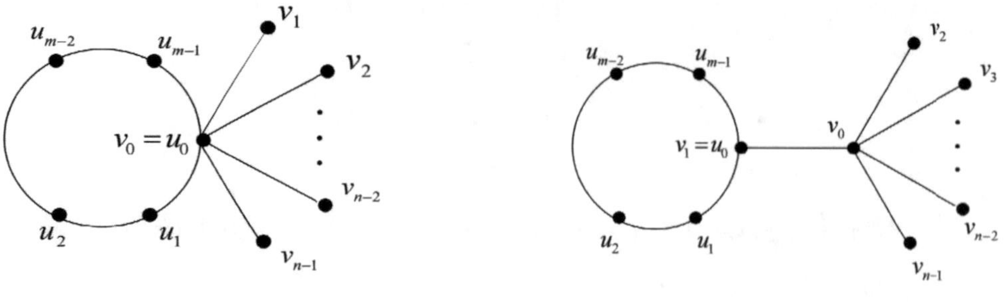

(a)The first kind (b) The second kind

Figure 1. One vertex union of thecycle and thestar.

2. VDD of One vertex union of the cycle and the star of the first kind

Theorem 1.If m is even, then $\forall x \in V(CS_{mn}^{(1)})$,

$$D'(x)=\begin{cases} \frac{1}{4}m^2(n+1)+n^2-1, & x=u_0=v_0; \\ 2(n-1)(i+1)+\frac{1}{2}m^2, & x=u_i(1\le i\le \frac{m}{2}-1); \\ \frac{1}{2}m^2+(m+2)(n-1), & x=u_{\frac{m}{2}}; \\ 2(n-1)(m-i+1)+\frac{1}{2}m^2, & x=u_i(\frac{m}{2}+1\le i\le m-1); \\ \frac{1}{4}m^2+m+2n-4, & x=v_i(1\le i\le n-1). \end{cases} \tag{1}$$

Proof. For even m, $\forall x \in V(CS_{mn}^{(1)})$, we have the following results.

(1) The vertex degree distance of $u_0\ (=v_0)$ is

$$d'(v_0)=d(v_0)\left(\sum_{k=1}^{n-1}d(v_k,v_0)+2\sum_{k=1}^{\frac{m}{2}-1}d(u_k,v_0)+d(v_{\frac{m}{2}},v_0)\right)=(n+1)\left((n-1)+2\sum_{k=1}^{\frac{m}{2}-1}k+\frac{m}{2}\right)=\frac{1}{4}m^2(n+1)+n^2-1 \tag{2}$$

(2) The degree distance of the vertex $u_i(1\le i\le \frac{m}{2}-1)$ on the cycle C_m is

$$d'(u_i)=d(u_i)\left(\sum_{k=0,k\ne i}^{m-1}d(u_k,u_i)+\sum_{k=1}^{n-1}d(v_k,u_i)\right)=2\left(\sum_{k=1}^{m-1}d(u_k,u_0)+(n-1)(i+1)\right)$$

$$=2\left(2\sum_{k=1}^{\frac{m}{2}-1}k+\frac{m}{2}+(n-1)(i+1)\right)=\frac{m^2}{2}+2(n-1)(i+1) \tag{3}$$

(3) According to symmetry, the degree distance of the vertex $u_i(\frac{m}{2}+1\le i\le m-1)$ on the cycle C_m is

$$d'(u_i)=d'(u_{m-i})=\frac{m^2}{2}+2(n-1)(m-i+1) \tag{4}$$

(4) The degree distance of the vertex $u_{\frac{m}{2}}$ on the cycle C_m is

$$d'(u_{\frac{m}{2}})=d(u_{\frac{m}{2}})\left(\sum_{k=0,k\ne i}^{m-1}d(u_k,u_{\frac{m}{2}})+\sum_{k=1}^{n-1}d(v_k,u_{\frac{m}{2}})\right)=2\left(\sum_{k=1}^{m-1}d(u_k,u_0)+(n-1)\left(\frac{m}{2}+1\right)\right)=\frac{m^2}{2}+(m+2)(n-1) \tag{5}$$

(5) The vertex degree distance of the vertex $v_i\ (1\le i\le n-1)$ is

$$d'(v_i)=d(v_i)\left(d(v_0,v_i)+\sum_{k=1,k\neq i}^{n-1}d(v_k,v_i)+2\sum_{k=1}^{\frac{m}{2}-1}d(u_k,v_i)+d(v_{\frac{m}{2}},v_i)\right)=\frac{m^2}{4}+m+2n-4 \tag{6}$$

Thus, the proof is completed.

Theorem 2. If m is odd, then $\forall x\in V(CS_{mn}^{(1)})$,

$$D'(x)=\begin{cases}\frac{1}{4}(m^2+4n-5)(n+1), & x=u_0=v_0;\\ 2(n-1)i+\frac{1}{2}m^2+2n-\frac{5}{2}, & x=u_i(1\le i\le \frac{m-1}{2});\\ 2(n-1)(m-i)+\frac{1}{2}m^2+2n-\frac{5}{2}, & x=u_i(\frac{m+1}{2}\le i\le m-1);\\ \frac{1}{4}m^2+m+2n-\frac{17}{4}, & x=v_i(1\le i\le n-1).\end{cases} \tag{7}$$

Proof. For odd m, $\forall x\in V(CS_{mn}^{(1)})$, we have the following results.

(1) The vertex degree distance of $u_0\ (=v_0)$ is

$$d'(v_0)=d(v_0)\left(\sum_{k=1}^{n-1}d(v_k,v_0)+2\sum_{k=1}^{\frac{m-1}{2}}d(u_k,v_0)\right)=(n+1)\left((n-1)+2\sum_{k=1}^{\frac{m-1}{2}}k\right)=\frac{1}{4}(m^2+4n-5)(n+1) \tag{8}$$

(2) The degree distance of the vertex $u_i(1\le i\le \frac{m-1}{2})$ on the cycle C_m is

$$d'(u_i)=d(u_i)\left(\sum_{k=0,k\neq i}^{m-1}d(u_k,u_i)+\sum_{k=1}^{n-1}d(v_k,u_i)\right)=2(n-1)i+\frac{m^2}{2}+2n-\frac{5}{2} \tag{9}$$

(3) According to symmetry, the degree distance of the vertex $u_i(\frac{m}{2}+1\le i\le m-1)$ on the cycle C_m is

$$d'(u_i)=d'(u_{m-i})=2(n-1)(m-i)+\frac{m^2}{2}+2n-\frac{5}{2} \tag{10}$$

(4) The vertex degree distance of the vertex $v_i(1\le i\le n-1)$ is

$$d'(v_i)=d(v_i)\left(d(v_0,v_i)+\sum_{k=1,k\neq i}^{n-1}d(v_k,v_i)+2\sum_{k=1}^{\frac{m-1}{2}}d(u_k,v_i)\right)=\frac{m^2}{4}+m+2n-\frac{17}{4} \tag{11}$$

Thus, the proof is completed.

According to equation (1) and equation (7), we have the following results.

Corollary 1[5]

$$D'(CS_{mn}^{(1)})=\begin{cases}\frac{1}{2}(m^3+2m^2n-2m^2+6n^2+6mn-6m-16n+10), & \text{if } m \text{ is even};\\ \frac{1}{2}(m^3+2m^2n-2m^2+6n^2+6mn-7m-18n+12), & \text{if } m \text{ is odd}.\end{cases} \tag{12}$$

3. VDD of One vertex union of the cycle and the star of the second kind

Theorem 3.If m is even, then $\forall x\in V(CS_{mn}^{(2)})$,

$$D'(x)=\begin{cases}\frac{3}{4}m^2+6n-9, & x=u_0=v_1;\\ 2(n-1)i+\frac{1}{2}m^2+4n-6, & x=u_i(1\le i\le \frac{m}{2}-1);\\ \frac{1}{2}m^2+m(n-1)+4n-6, & x=u_{\frac{m}{2}};\\ 2(n-1)(m-i)+\frac{1}{2}m^2+4n-6, & x=u_i(\frac{m}{2}+1\le i\le m-1);\\ (\frac{1}{4}m^2+m+n-2)(n-1), & x=v_0;\\ \frac{1}{4}m^2+2(m+n)-5, & x=v_i(2\le i\le n-1).\end{cases} \quad (13)$$

Proof. For even m, $\forall x\in V(CS_{mn}^{(2)})$, we have the following results.

(1) The vertex degree distance of $u_0\ (=v_1)$ is

$$d'(u_0)=d'(v_1)=d(v_1)\left(d(v_1,v_0)+\sum_{k=2}^{n-1}d(v_k,v_1)+2\sum_{k=1}^{\frac{m}{2}-1}d(u_k,v_1)+d(v_{\frac{m}{2}},v_1)\right)=\frac{3m^2}{4}+6n-9 \quad (14)$$

(2) The degree distance of the vertex $u_i(1\le i\le \frac{m}{2}-1)$ on the cycle C_m is

$$d'(u_i)=d(u_i)\left(\sum_{k=0,k\ne i}^{m-1}d(u_k,u_i)+d(u_k,v_0)+\sum_{k=2}^{n-1}d(v_k,u_i)\right)=2(n-1)i+\frac{m^2}{2}+4n-6 \quad (15)$$

(3) According to symmetry, the degree distance of the vertex $u_i(\frac{m}{2}+1\le i\le m-1)$ on the cycle C_m is

$$d'(u_i)=d(u_{m-i})=2(n-1)(m-i)+\frac{m^2}{2}+4n-6 \quad (16)$$

(4) The degree distance of the vertex $u_{\frac{m}{2}}$ of the cycle C_m is

$$d'(u_{\frac{m}{2}})=d(u_{\frac{m}{2}})\left(\sum_{k=0,k\ne i}^{m-1}d(u_k,u_{\frac{m}{2}})+d(v_0,u_{\frac{m}{2}})+\sum_{k=2}^{n-1}d(v_k,u_{\frac{m}{2}})\right)=\frac{m^2}{2}+m(n-1)+4n-6 \quad (17)$$

(5) The degree distance of the vertex v_0 is

$$d'(v_0)=d(v_0)\left(\sum_{k=1}^{n-1}d(v_k,v_0)+2\sum_{k=1}^{\frac{m}{2}-1}d(u_k,v_0)+d(v_{\frac{m}{2}},v_0)\right)=\left(\frac{m^2}{4}+m+n-2\right)(n-1) \quad (18)$$

(6) The degree distance of the vertex $v_i(2\le i\le n-1)$ is

$$d'(v_i)=d(v_i)\left(d(v_0,v_i)+\sum_{k=1,k\ne i}^{n-1}d(v_k,v_i)+2\sum_{k=1}^{\frac{m}{2}-1}d(u_k,v_i)+d(v_{\frac{m}{2}},v_i)\right)=\frac{m^2}{4}+2(m+n)-5 \quad (19)$$

Thus, the proof is completed.

Theorem 4.If m is odd, then $\forall x\in V(CS_{mn}^{(2)})$,

$$D'(x)=\begin{cases}\frac{3}{4}m^2+6n-\frac{39}{4}, & x=u_0=v_1;\\ \frac{1}{2}(m^2+4ni-4i+8n-13), & x=u_i(1\le i\le\frac{m-1}{2});\\ 2(n-1)(m-i)+\frac{1}{2}m^2+4n-\frac{13}{2}, & x=u_i(\frac{m+1}{2}\le i\le m-1);\\ \frac{1}{4}\left(m^2+4(m+n)-9\right)(n-1), & x=v_0;\\ \frac{1}{4}m^2+2(m+n)-\frac{21}{4}, & x=v_i(2\le i\le n-1).\end{cases} \tag{20}$$

Proof. For odd m, $\forall x\in V(CS_{mn}^{(2)})$, we have the following results.

(1) The degree distance of the vertex u_0 $(=v_1)$ is

$$d'(v_1)=d(v_1)\left(d(v_1,v_0)+\sum_{k=2}^{n-1}d(v_k,v_1)+2\sum_{k=1}^{\frac{m-1}{2}}d(u_k,v_1)\right)=3\left(1+2(n-2)+2\sum_{k=1}^{\frac{m-1}{2}}k\right)=\frac{3m^2}{4}+6n-\frac{39}{4} \tag{21}$$

(2) The degree distance of the vertex $u_i(1\le i\le\frac{m-1}{2})$ on cycle C_m is

$$d'(u_i)=d(u_i)\left(\sum_{k=0,k\ne i}^{m-1}d(u_k,u_i)+d(u_k,v_0)+\sum_{k=2}^{n-1}d(v_k,u_i)\right)=2\left(2\sum_{k=1}^{\frac{m-1}{2}}k+(i+1)+(n-2)(i+2)\right)$$

$$=2(n-1)i+\frac{m^2}{2}+4n-\frac{13}{2} \tag{22}$$

(3) According to symmetry, the degree distance of the vertex $u_i(\frac{m+1}{2}\le i\le m-1)$ on the cycle C_m is

$$d'(u_i)=d(u_{m-i})=2(n-1)(m-i)+\frac{m^2}{2}+4n-\frac{13}{2} \tag{23}$$

(4) The degree distance of the vertex v_0 is

$$d'(v_0)=d(v_0)\left(\sum_{k=1}^{n-1}d(v_k,v_0)+2\sum_{k=1}^{\frac{m-1}{2}}d(u_k,v_0)\right)=(n-1)\left((n-1)+2\sum_{k=1}^{\frac{m-1}{2}}(k+1)\right)$$

$$=\frac{1}{4}\left(m^2+4(m+n)-9\right)(n-1) \tag{24}$$

(5) The degree distance of the vertex $v_i(2\le i\le n-1)$ is

$$d'(v_i)=d(v_i)\left(d(v_0,v_i)+\sum_{k=1,k\ne i}^{n-1}d(v_k,v_i)+2\sum_{k=1}^{\frac{m-1}{2}}d(u_k,v_i)\right)=1+2(n-2)+2\sum_{k=1}^{\frac{m-1}{2}}(k+2)$$

$$=\frac{m^2}{4}+2(m+n)-\frac{21}{4} \tag{25}$$

Therefore, the equation(20) is established.
According to equation (13) and equation (20), we have the following results.
Corollary 2

$$D'(CS_{mn}^{(2)}) = \begin{cases} \frac{1}{2}(m^3 + 3m^2n - 3m^2 + 6n^2 - 12mn - 14m + 9), & \text{if } m \text{ is even;} \\ \frac{1}{2}(m^3 + 2m^2n - 2m^2 + 6n^2 + 14mn - 23m - 11n + 20), & \text{if } m \text{ is odd.} \end{cases} \quad (26)$$

4. Conclusion

The one vertex union of the cycle and the star is a typical unicyclic graph. By calculating the vertex degree distances of one vertex union of the cycle and the star, we can further study the distribution of the maximum and maximum vertex degree distances and the extreme graphs of the maximum vertex degree distances of the class of graphs.

Acknowledgments

This research was financially supported by National Natural Science Foundation of China (61803065).

References

[1] Dankelmann P, Gutman I, Mukwembi S and Swart H C 2009 On the degree distance of a graph *Discrete Appl. Math.* **157** 2773–77
[2] Dobrynin A and Kochetova A 1994 Degree distance of a graph: A degree analogue of the Wiener index *J. Chem. Inf. Comput. Sci.* **34** 1082–86
[3] Du Z and Zhou B 2010 Degree distance of unicyclic graphs *Filomat* **24** 95–120
[4] Du Z and Zhou B 2011 Maximum degree distance of graphs with exactly two cycles *Bull. Math. Soc. Sci. Math. Roumanie (N.S.)* **54** 119–136
[5] Feng L, Liu W, Ilic´ A and Yu G 2013 The degree distance of unicyclic graphs with given matching number *Graphs Combin.* **29** 449–462
[6] Gutman I 1994 Selected properties of the Schultz molecular topological index *Journal of Chemical Information and Computer Sciences* **34** 1087–89
[7] Hou Y and Chang A 2006 The unicyclic graphs with maximum degree distance *J. Math. Study* **39** 18–24
[8] Mukwembi S and Munyira S 2013 Degree distance and minimum degree *Bull. Austral. Math. Soc.* **87** 255–271
[9] Tomescu I 1999 Some extremal properties of the degree distance of a graph *Discrete Appl. Math.* **98** 159–163
[10]Tomescu I 2008 Unicyclic and bicyclic graphs having minimum degree distance *Discrete Appl. Math.* **156** 125–130
[11]Wiener H 1947 Structural determination of paraffin boiling points *J. Am. Chem. Soc* **69** 17–20
[12]Xu K, Liu M, Das K C, Gutman I and Furtula B 2014 A survey on graphs extremal with respect to distance–based topological indices *MATCH Commun. Math. Comput. Chem.* **71** 461–508
[13]Dobrynin A and Kochetova A 1994 Degree distance of a graph: A degree analog of the Wiener index *Journal of Chemical Information and Computer Sciences* **34** 1082–86

Radar Cross Section Reduction for the Support System of Electrically Large Objects

Wei Du*, Peng Cui, Qian Zhao and Lewu Deng

AVIC Cheng'Du Aircraft Industrial(Group)Co.Ltd, Chengdu, Sichuan, China

*corresponding author's e-mail: dwbuaa@163.com

Abstract. In the tests of the electromagnetic scattering characteristics of electrically large objects, due to the large object size and the installation height, the physical size of the system is large, and the scattering is strong. In low-frequency bands, cancellation by filtering cannot be achieved, which affects the test accuracy. This study proposes a radar cross section (RCS) reduction design for the target supporter based on a low-scattering structure, and low-scattering models with a flat shape were formed in an array, which reduced the horizontally polarized scattering within ±45° at low frequencies and ensured that the vertical polarization did not deteriorate. Integrated with the corresponding driving equipment, the proposed design can reduce the 360° scattering intensity and improve the cancellation by filtering.

1. Introduction

Currently, the measurement of the radar cross section (RCS) [1] of electrically large objects in China is mainly achieved with the installation method of synchronous motion of support and lifting, and the objects' jacking interface is usually used, which ensures safety in the posture adjustment and lifting. However, due to the oversized target supporter and the coupling problem with the target, too much clutter is introduced in the testing process, and the scattering intensity cannot be completely cancelled in the low-frequency bands.

Currently, a bead-shaped target supporter with horizontal isotropy is commonly used, which ensures that the scattering intensity is constant during 360° rotation. However, in monostatic RCS testing, the circular structure's main lobe direction is forward, which is not consistent with the low forward scattering characteristics of the plane, and the long length of the target supporter limits the scattering characteristics. Therefore, the scattering of the target supporter in low frequencies is strong, and the horizontally polarized scattering is significantly higher than the vertical polarization [6].

In view of the above problem, this study proposes an RCS reducing design for the target supporter based on a low-scattering structure. Low scattering models with a flat shape reduced the horizontally polarized scattering at low frequencies and ensured the vertical polarization performance. Integrated with the corresponding driving equipment, the proposed design can reduce the 360° scattering intensity and improve the cancellation by filtering.

2. Design of the low scattering structure

The scattering response of an object to an incident electromagnetic wave is the reflection and diffraction of the electromagnetic wave when it encounters obstacles in space [5]. Specifically, reflection can be divided into surface reflection and multiple reflections, and diffraction includes edge diffraction, crawling waves, etc.

A low-scattering structure design should be able to suppress various types of scattering. According to the mechanism of electromagnetic wave scattering, the principle of suppression involves absorption and concentration. Absorption refers to the use of wave-absorbing materials to attenuate electromagnetic waves by converting electromagnetic energy into heat. Concentration refers to the reflection of electromagnetic waves into other nonthreatening areas and the diversion of strong scatterings, such as specular reflection and edge diffraction from the focus angle domains through the geometric design.

A structure with a large metal surface area, according to the principle of geometrical optics, reflects an electromagnetic wave in the direction opposite that of the incident ray symmetrical to the surface normal. Under irradiation from electromagnetic waves with vertical polarization relative to the incidence plane, a strong surface wave can be generated on the surface. The surface wave can travel along the metal surface. If there are sudden changes in the shape of the metal surface, the traveling wave can be reflected, contributing to the backward RCS. For a flat shape, when a vertically polarized wave is incident, the surface can generate a strong traveling wave current, and when the current encounters the discontinuity at the rear edge of the carrier, reflection could occur.

On the premise that the surface is smooth with no truncation, certain measures must be taken to suppress the scattering of traveling waves on the surface to further reduce the RCS. Concentration measures, such as geometric modification to make angles of the leading and trailing edges equal, can be used. Through the adjustment of the angle of the trailing edge, the scattering of traveling waves in VV polarization can be effectively reduced so that the scattering of the trailing edge coincides with the leading-edge diffraction peak, thereby deviating the scatterings from the forward focus area and reducing the forward scattering of traveling waves.

3. Array design of the low-scattering structure

3.1. Bead-shaped support structure

In this study, 1:1 element model with a height of 1 m is used in the design of the low-scattering target supporter and the subsequent optimization to characterize the scattering characteristics under different polarizations and to compare the RCS reduction results of various schemes, with the goals of shortening the calculation time and improving the design efficiency.

The RCS of the metal structure and the loading structure of the 1 m high bead structure at low and high frequency bands are calculated as follows:

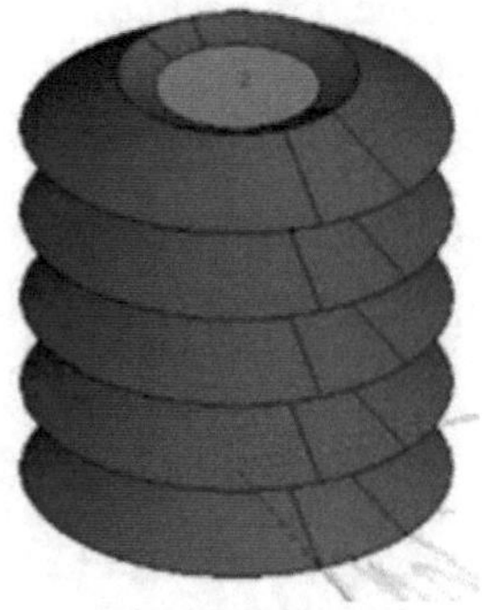

Figure 1. Simulation model of the 1 m bead-shape structure

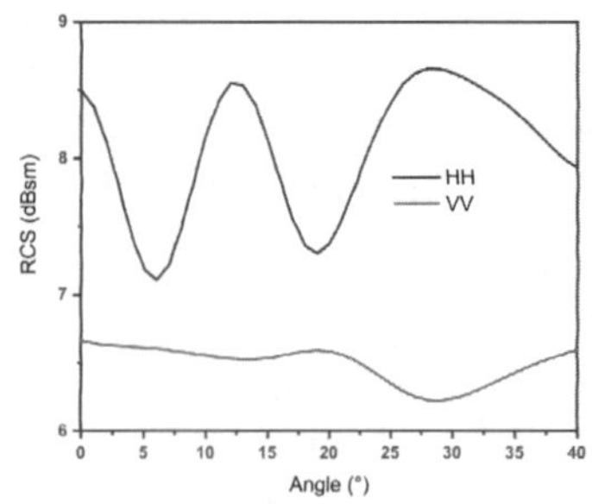

Figure 2. RCS curve of the all-metal bead-shaped structure at 1.7 GHz

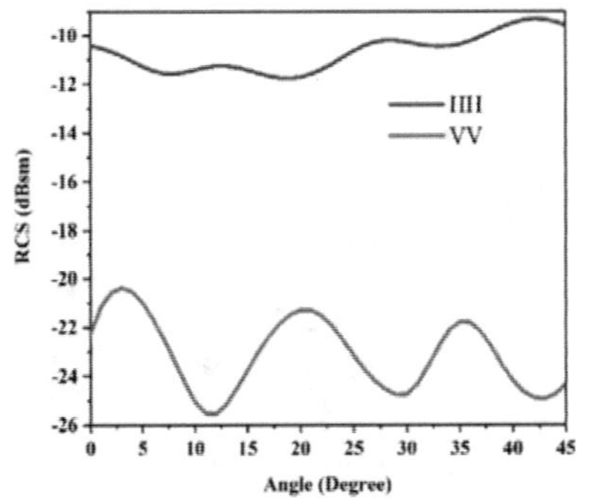

Figure 3. RCS curve of hard foam material at 1.7 GHz

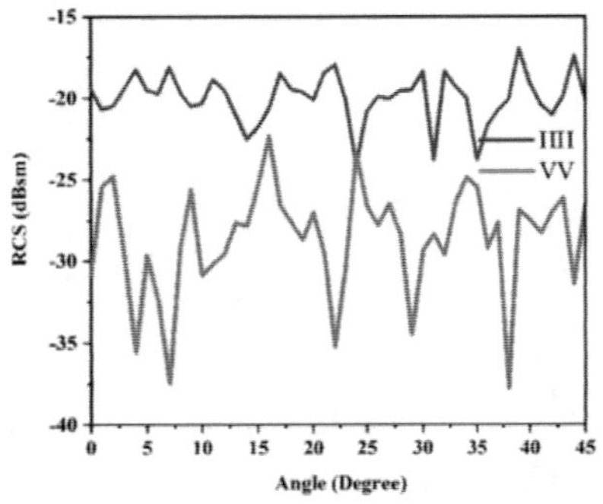

Figure 4. RCS curve of hard foam material at 9.41 GHz

The support system used in the current RCS measurement system has the advantage of horizontal isotropy. The scattering characteristics are basically the same for 360° of rotation, and the structure is simple. However, the scattering intensity in the focus area is relatively high. Especially for low-scattering objects in low frequency, the scattering of the structure is strong.It will lead to errors in the identification of scattering centers in the process of multiple filtering based on scattering centers and leading to a poor cancellation by filtering.

3.2. Structure design of the low-scattering element

There is a clear direction in the geometric design and the final indices of any low-scattering object with a complex structure [4]. Therefore, compared with omnidirectional-scattering reduction, it is more feasible to directionally reduce the RCS of the object support system. Mechanical structures can be used to directionally match the external sleeve of the target supporter with the test object to realize a large difference in scattering between the measured object and the support system in the whole angle domain and to achieve better a cancellation by filtering [7].

Although typical structures and RCS estimation have been extensively studied [2][3], low-scattering characteristics cannot be ensured when the object is large. The model proposed in this study has a flat shape and is aimed at RCS reduction in the forward ±45°. The design of 70° of both leading and trailing edges ensures that the forward reflection peak is near ±55°, and the scattering of the vertical polarization of the downward wave coincides with the edge scattering peak. Under horizontal polarization, traveling wave current is generated by the tail truncation effect and flows along the trailing edge, producing strong traveling wave scattering in the forward ±55°. The simulation results are as follows:

Figure 5. Metal model of the low-scattering structural element.

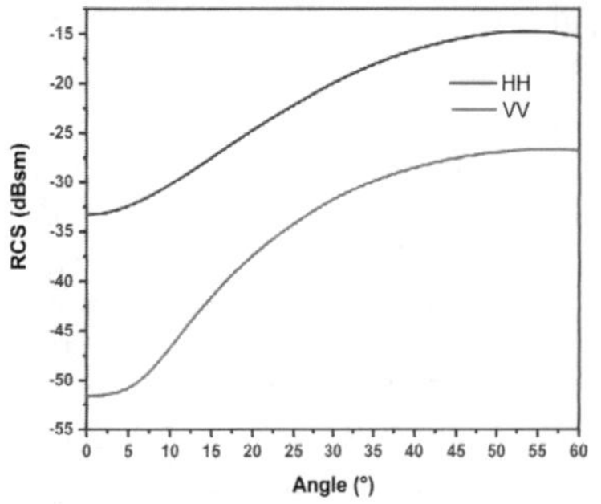

Figure 6. RCS curve of the metal model with low scattering at 1.7 GHz

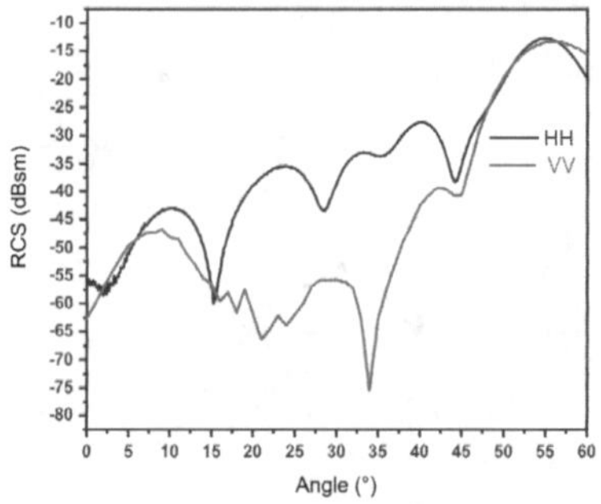

Figure 7. RCS curve of the metal model with low scattering at 9.41 GHz

3.3. Low-scattering array

For directional RCS reduction, the following scaled low-scattering array is designed. Since the RCS value is highly correlated with the angle, it is suggested that real-time angle correction be carried out according to the mechanical structure to ensure minimum scattering in the incident direction.

The model elements are stacked, and the edges and the top and bottom surfaces are smoothed. A center hole is created for the target supporter to pass through. A 1 m model is used in this study for qualitative calculation.

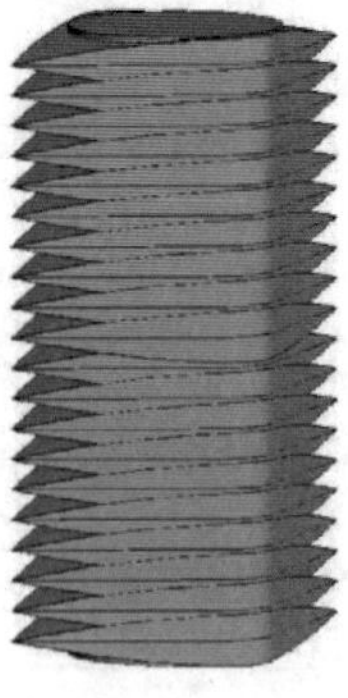

Figure 8. Low-scattering array model

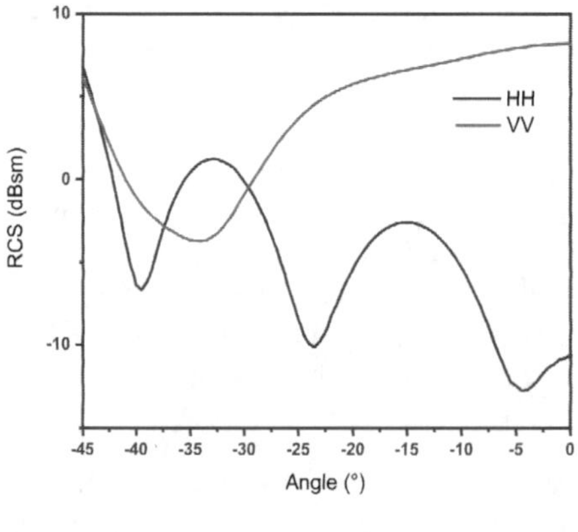

Figure 9. RCS curve of the all-metal low-scattering array model at 1.7 GHz

The above results are compared with those of the bead-shaped structure with two types of materials (i.e., metal and hard foam), as shown in Table 1.

Table 1. Comparison of the optimal RCS of different models.

Model	HH (dBsm)	VV (dBsm)
Low-scattering array PEC_1m	-13 (minimum)	-4 (minimum)
Modified low-scattering array PEC_1m	-6 (minimum)	-17 (minimum)
Modified low-scattering array (hard foam)_1m	-23 (minimum)	-53 (minimum)
Bead-shaped structure PEC_1m	8	6.7
Bead-shaped structure (hard foam)_1m	-11	-24

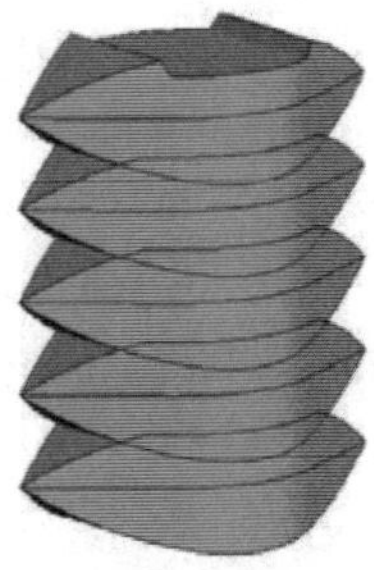

Figure 10. Modified all-metal low-scattering array.

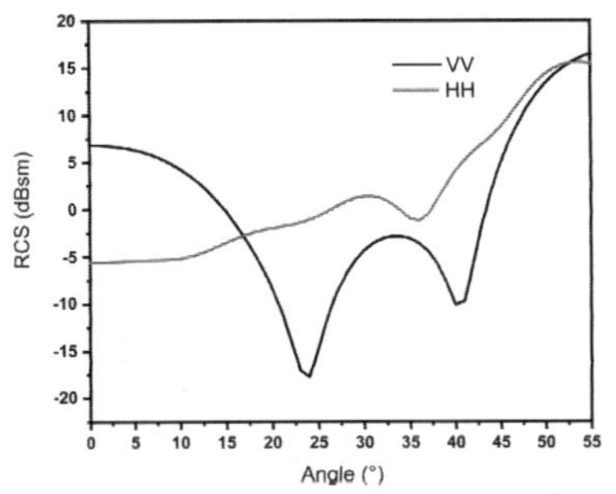

Figure 11. RCS curves of the modified all-metal low-scattering array at 1.7 GHz(all-metal)

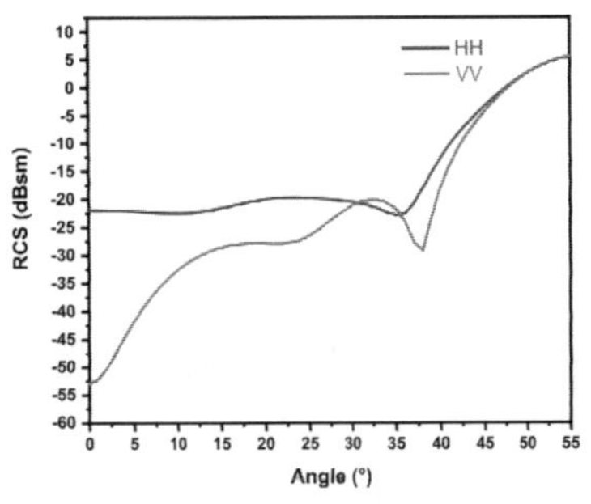

Figure 12. RCS curves of the modified all-metal low-scattering array at 1.7 GHz(hard absorbing foam)

The comparison of the arrays before and after modification shows that the RCS of the array changes with the height of the array element. The HH polarization RCS is positively correlated with the element height, and the VV polarization RCS is negatively correlated with the element height.

For low frequencies, under horizontal polarization, the result of the modified 1 m all-metal low-scattering array is consistent with that of the bead-shaped structure, while under vertical polarization, it is similar to that of the all-metal bead-shaped structure. When the hard absorbing foam material is used, the scattering intensity is significantly lower. The minimum horizontal polarization is -23 dB, and the minimum vertical polarization is -53 dB. Compared with that of the bead-shape structure with the same material, the RCS for the proposed model is greatly reduced.

4. Conclusions

In this study, a new design for RCS reduction in the support system of electrically large objects was proposed for electromagnetic scattering measurement. A typical low-scattering model with a flat shape was used to form an array to realize adjustable RCS reduction in the support system. The results of 1:1 model simulation showed that the structure has low scattering intensity in the forward ±45°, and the RCS under both horizontal and vertical polarizations is better than that of the current bead-shaped structure. In actual application, according to the type of the test object, the element size can be adjusted so that the scattering intensity under different polarizations matches the test requirements. Integrated with the corresponding driving equipment, the optimal scattering angle always faces the reflector in the test process of horizontal and vertical polarizations, thereby reducing the scattering intensity at 360° and improving the cancellation by filtering and the imaging accuracy. Thus, the proposed model has strong adaptability and is of great implementation value.

References

[1] Huang P K, Yin H C, Xu X J 2005 *Radar Target Characteristics(Leida Mubiao Texing)* (Beijing: Publishing House of Electronics Industry) p
[2] Crispin J W, Maffett A L 1965 Radar cross-section estimation for simple shapes *Proceedings of the IEEE* **53**(8) 833 - 48
[3] Cown B J, Ryan C E 1989 Near-field scattering measurements for determining complex target RCS *IEEE Transactions on Antennas & Propagation* **37**(5) 576-85
[4] Li C, Ai J Q, Long W G 2018 Shape Design of Low RCS Supporter for Embedded Cavities *Telecommunication Engineering* **58**(06):105-10
[5] Wang M G 1985 *Geometrical Theory of Diffraction* (Xi'an: XiDian University Press) p
[6] Sang J H.2013 *Low-Observable Technologies of Aircraft* (Beijing: Aviation Industry Press) p
[7] Gao X, Chai J Z, Wang J 2012 Low-profile low-RCS design of pylons for the test of stealth aircraft *Aircraft Engineering* 000(003) 59-62

Measurement and Analysis Method of Light Intensity Distribution for Street Lamps

Wei Du[1], HaoJie Cheng[2,4], Lin Wang[2,4], DiGuan Feng[3] and XiaoKe Wang[1]

[1]Art School (Architecture School), Nantong University, Nantong City, 226007, China
[2]School of Optoelectronic Science and Engineering and Collaborative Innovation Center of Suzhou Nano Science and Technology, Soochow University, Suzhou, 215006 China
[3]Nantong Shanggong Industrial Design Co. Ltd, Nantong City, 226007, China
[4]Key Lab of Advanced Optical Manufacturing Technologies of Jiangsu Province and Key Lab of Modern Optical Technologies of Education Ministry of China, Soochow University, Suzhou, 215006 China

[*]corresponding author's e-mail: weidu@126.com

Abstract. In order to evaluate the actual illumination effect of high-pressure sodium lamp on the road, a high-pressure sodium lamp illuminance test system is developed in this paper. In the testing of sodium lamp illumination, light intensity distribution is the most important evaluation index. It is also one of the most difficult physical parameters to test due to its large amount of data and high accuracy requirements. This thesis investigates a fast and high-precision testing method for light intensity distribution of high-pressure sodium lamps, so as to improve the measurement of sodium lighting. The system performs light intensity distribution testing in the laboratory, studies the test method of light intensity illuminance distribution of high-pressure sodium lamps, uses Matlab to optically model high-pressure sodium lamps, uses various methods to model the data obtained from the experiments, and analyzes the resulting illuminance data, the system is able to quickly test and evaluate the illuminance of high-pressure sodium lamps after installation.

1. Introduction

Urban road lighting is not only directly related to people's travel quality and safety, but also indirectly reflects the overall style and image of the city. In road lighting, if the lighting effect is not good, zebra crossing is easy to occur, which requires that the illuminance formed by lamps on the road surface has good uniformity. High-pressure sodium lamps are widely used in expressways, trunk roads, secondary trunk roads and branch roads, so it is particularly important to test the illuminance of sodium lamps and establish the illuminance light field model. In tunnel lighting, in addition to the uniformity of illumination, the brightness and color temperature of lamps are required to be higher. At present, there are metal halide lamps, incandescent lamps, LED lamps [1], etc., which are commonly used in road lighting. Therefore, sodium lamps are tested and other lamps are evaluated and analyzed based on this. It can not only provide guidance for the secondary optical design of sodium lamp, it can also provide basic reference in optical design for other better light sources to be applied in the field of road lighting.

2. Testing principle and method

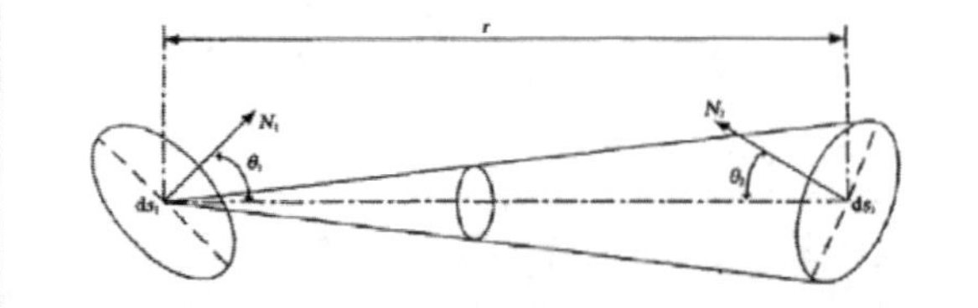

Figure 1. Schematic diagram of principle of surface light source test.

Figure 1 is a schematic diagram of the principle of surface light source testing. The surface s_1 of the light source and the illuminated surface s_2 each have a microelement with an area of ds_1 and ds_2. The included angles between their connecting lines and their normal lines are θ_1 and θ_2, respectively, and the brightness of the surface light source is L. According to the theorem, the luminous flux emitted by ds_1 is

$$d\emptyset_1 = Lds_1dw_1cos\theta_1 \tag{1}$$

In which ds_1 is the area of its microelement, dw_1 is ds_2 is the solid angle at the center of the light-emitting surface source. And because:

$$dw1 = ds_2\frac{cos\theta_2}{r^2} \tag{2}$$

It can be deduced from equations 1 and 2:

$$d\emptyset_1 = Lds_1cos\theta_1\frac{ds_2cos\theta_2}{r^2} \tag{3}$$

Therefore, the illuminance of ds_1 on the detection surface is:

$$dE_2 = \frac{d\emptyset_1}{ds_2} = Lcos\theta_1\frac{ds_1cos\theta_2}{r^2} \tag{4}$$

The above is the illumination formula generated by a small area element at a certain point in space. The illuminance generated by a limited size area light source s_1 can be obtained by integrating the above formula:

$$E_2 = \iint_{S_1} dE_2 = \iint_{S_2} Lds_1cos\theta_1\frac{ds_1cos\theta_2}{r^2} \tag{5}$$

The surface light source is regarded as the light source of cosine radiator, that is, the brightness in all directions is the same.

In the experiment, the illuminance data were collected, and sodium lamps were regarded as cosine radiators scattering uniformly in all directions. The center distribution method [2] was used to collect the illuminance data: in the area of illuminance measurement, the test area was generally divided into rectangular grids, and the grids should be square, and the illuminance was measured at the center of the rectangular grids. This method is suitable for horizontal illumination, Measurement of vertical illuminance or vertical illuminance in camera direction. The average illuminance formula is:

$$E_{av} = \frac{1}{M \times N}\sum E_i \tag{6}$$

The total uniformity can be obtained after the illumination data is obtained in the experiment. Because the higher the average brightness of the light field, the higher the contrast sensitivity of the eyes, and the higher the reliability of the observer's perception of obstacles inside. On the contrary, if the brightness in a certain area of the road is very low, the observer's ability to perceive objects will be greatly affected. In this area, the contrast sensitivity of human eyes decreases, instantaneous maladjustment will occur, so the minimum brightness value of the road surface should be guaranteed, that is, the light emitted by lamps should reach the appropriate total uniformity [3]. Total luminance (or illuminance) uniformity U_0 is defined as the ratio of the minimum luminance (or illuminance) L_{min} to the average luminance (or illuminance) L_{av} of the road surface, that is

$$U_0 = \frac{L_{min}}{L_{av}} \tag{7}$$

U_0 should not be less than 0.4 in order to maintain an acceptable perceptibility. The smaller the maximum and minimum illumination values along the center line of the longitudinal axis, the better the comfort of human eyes. Longitudinal uniformity is defined as the ratio of the minimum luminance (luminance) L_{min} of the road surface to the maximum luminance (luminance) L_{max} of the road surface on the line of the longitudinal axis, that is, $U_L=L_{min}/L_{max}$. Generally, it is recommended that the minimum value of U_L of main roads be about 0.7 to ensure sufficient visual comfort level. For secondary roads, U_L can be reduced to about 0.5 [4].

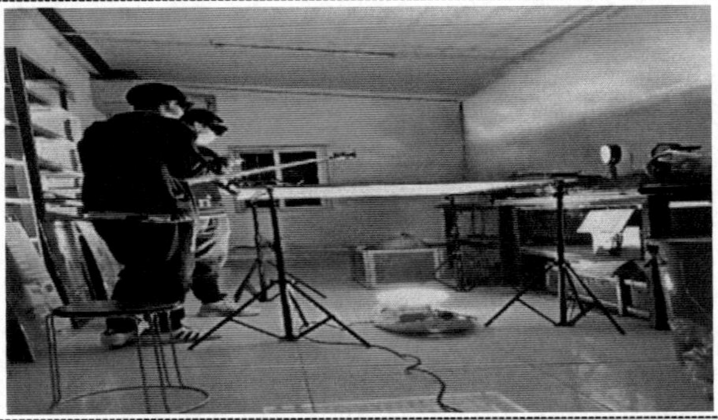

Figure 2. Illumination data acquisition

Experiments were carried out by using the above methods. This test is a high-pressure sodium lamp with rated voltage of 220V, working current of 3A, luminous flux of 26000lm, lamp cap specification of E40 and power of 250W. Figure 2 is the scene diagram of the illumination data collected in the laboratory. In the experiment, 225 sampling points were set up with 5.75cm lateral intervals and 15 intervals. The longitudinal interval is 5.75cm, with 15 intervals. Four groups of data were collected in the experiment, the heights are 0.75m,1m,1.5m and 2m respectively. After lighting the high-pressure sodium lamp for 5 min, keep it constant before measuring; When measuring the illuminance of vertical plane, normal line and working surface, place the light receiver of the light meter above the specified direction, and measure the illuminance at the measuring point of the corresponding target.

3. Light field fitting algorithm

3.1. linear fitting algorithm of 3D surface

The illuminance data of sodium lamp collected in the experiment is three-dimensional data. Three algorithms are used to model the experimental data, and the advantages and disadvantages of the light field models obtained by the three algorithms are compared and analyzed. Firstly, three-dimensional surface linear fitting is used to fit the three-dimensional data based on the least square method and nonlinear regression principle. The least square method (also known as the least square method) is a mathematical optimization technique [5].It seeks the best function matching of data by minimizing the sum of squares of errors. The unknown data can be simply obtained by using the least square method, and the data can be supplemented due to insufficient experimental data, and the sum of squares of errors between these obtained data and actual data can be minimized. The smaller the error, the more accurate the data can be. .Commonly used processing methods include linear iterative method of regression function, piecewise regression method, iterative least square method and so on.

3.2. Local weighted regression algorithm

The general idea of local weighted regression (Lowess) is: take a point x as the center, intercept a piece of data with the length of frac forward and backward, and do a weighted linear regression with the weight function W for this piece of data, remember $(x, \widehat{y})$ is the center value of the regression line, where it is the corresponding value of the fitted curve [6]. For all n data points, n weighted regression lines can be made, and the connection line of the center value of each regression line is the Lowess curve of this data. For the prediction problem, the simplest linear regression in regression is to fit the trend of data with a linear method. But for periodicity, fluctuating data can't be simply fitted in a linear way, otherwise the model will deviate greatly, and Lowess can better deal with this problem. You can fit a line that conforms to the overall trend, and then make a prediction. Lowess can also solve the smoothing problem better. When doing data smoothing, there will be some trend or seasonal data. For such data, we can't use simple mean plus or minus three times the standard deviation to eliminate outliers, and we need to consider the trend and other conditions. With Lowess, a trend line can be fitted, which is taken as the baseline, and the real outlier point [7]is far away from the baseline.

3.3. polynomial surface fitting

Let the fitting polynomial be:

$$y = a_0 + a_1x + a_2x^2 + \ldots + a_kx^k \tag{8}$$

You can get the sum of distances from each point to this curve, that is, the sum of deviation squares is as follows:

$$R^2 = \sum_1^n \left[y_i - \left(a_0 + a_1x + a_2x^2 + \ldots + a_kx^k\right)\right]^2 \tag{9}$$

Polynomial surface fitting is a linear model for nonlinear function of data, which can be operated as efficiently as linear model, and at the same time, the model can be applied to a wider range of data [8].

Three algorithms are used to build light field models for experimental data respectively, and the obtained modeling models are as follows. Figure 3 shows the 3D model obtained by 3D linear fitting. The 3D modeling effect obtained by linear fitting of curved surface is not very good, the obtained 3D model is not smooth enough, and the illumination model does not conform to the light intensity distribution of the light source model. Figure 4 For the three-dimensional model obtained by Lowess algorithm, quadratic programming model (Quadratic) is adopted, and its span value is set at 30%. The light field model obtained by Lowess algorithm [9] is obviously better than that obtained by previous three-dimensional linear fitting, and the illumination model is closer to a light intensity distribution that accords with the light source model. Figure 5 Polynomial surface fitting model. In that polynomial surface fit, after fitting the surface [10] with different orders for many times, it is finally concluded that the coefficient of determination (R-square) obtained when the X and Y variables are all third order is the highest, reaching 97.61%, which also shows that the accuracy of this model is very high and close to the light field model.

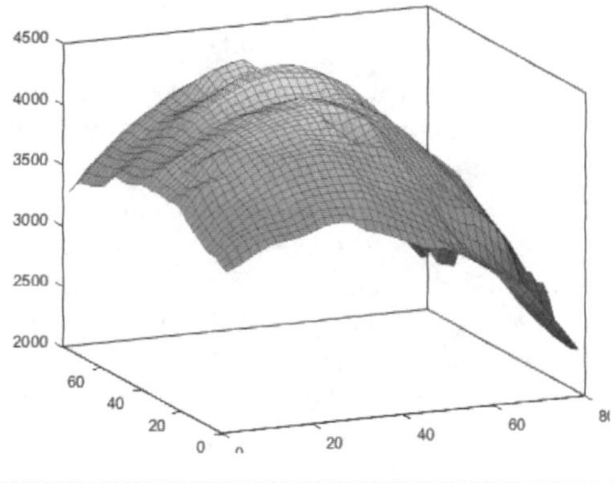

Figure 3. Linear Fitting of Surface to 3D Model.

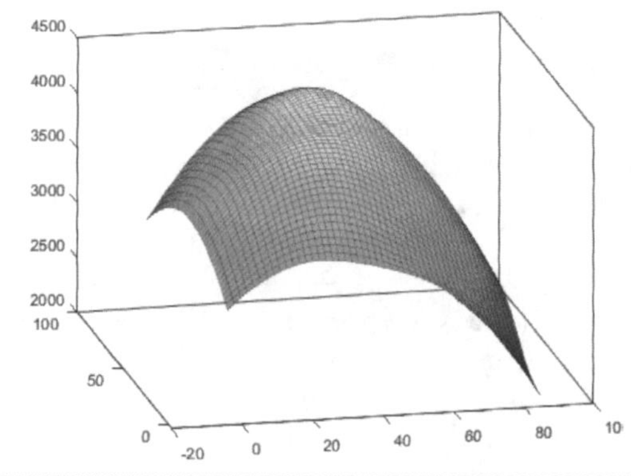

Figure 4. Lowess 3D Model.

Journal of Physics: Conference Series **2173** (2022) 012069 doi:10.1088/1742-6596/2173/1/012069

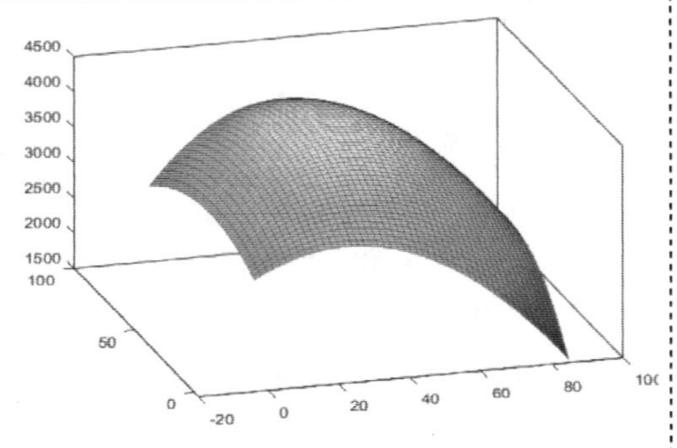

Figure 5. Polynomial Surface Fitting Model.

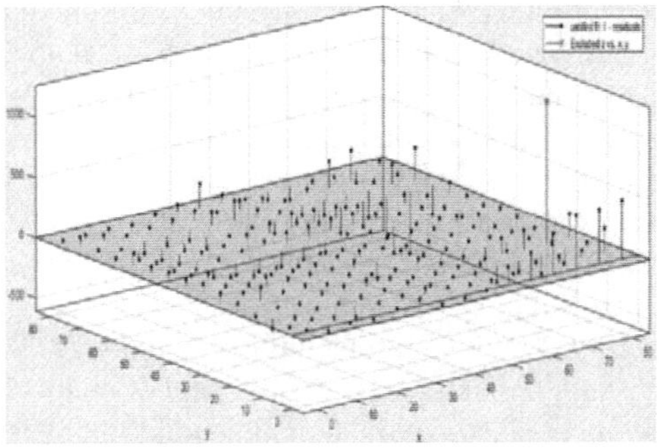

Figure 6. Distribution of Residual Scatters.

According to the light field modeling models obtained by the three algorithms, the modeling results can be analyzed: when the test height H=1.5m, the model obtained by the three-dimensional surface algorithm has a low degree of agreement with the ideal model, the coefficient of determination (R-square) obtained by the Lowess algorithm is 94.01%, the root mean square difference (Rmse) is 120.7, and the coefficient of determination (R-square) obtained by the polynomial fitting algorithm is Figure 6. The mean square error (Rmse) of residual scatter distribution map [11] obtained by polynomial surface fitting is 76.48. Compared with the first two algorithms [12], the mean square error is smaller and accords with normal distribution, so a good modeling model can be obtained. Therefore, in the subsequent data analysis, the polynomial fitting algorithm is used to establish the light field model for the experimental data, and the analysis and evaluation are carried out.

4. Testing and data analysis and evaluation

Road lighting indicators include average brightness (illuminance), uniformity of brightness (illuminance), glare limitation and inducement. In China's standards, it is difficult to fully adopt the brightness standard, so two sets of indicators of brightness and illumination are given [13]. Figure. 7 shows the distribution of plane light intensity data collected in the experiment. According to the 225 sampling data points set in the experiment, the intensity and illumination data of the whole plane are obtained after polynomial surface fitting. From the analysis of the plane light intensity distribution model in Figure 7, it can be concluded that the color in the model gradually deepens from blue to yellow, the color depth also represents the illumination of the light field at the corresponding position [14]; The illumination of the light field along the X and Y axes shows a trend of first increasing and then decreasing, with the center of the high-pressure sodium lamp as the highest illumination field and the lowest illumination on both sides. In the plane model diagram, the overall distribution of light intensity can be seen more intuitively in one plane, it can be seen from the figure that the light intensity distribution is relatively uniform on both sides of the lamp post, with the tail of the lamp post on the left and the front of the lamp post on the right, and the distribution model [15] is obtained.

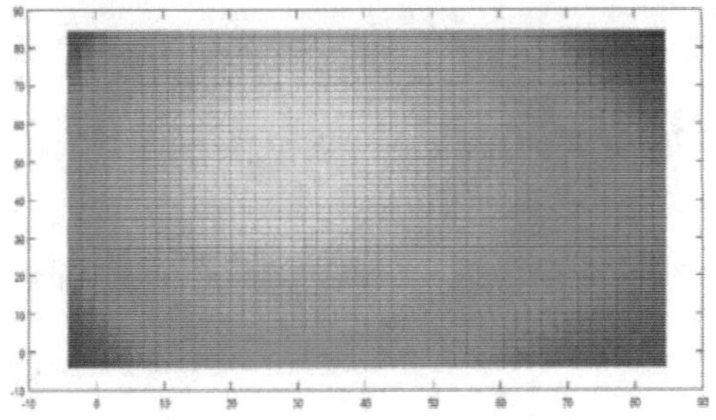

Figure 7. PPlane light intensity distribution model.

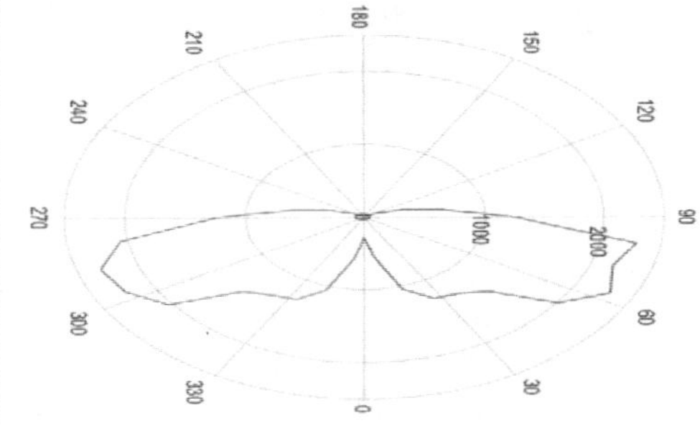

Figure 8. Dlight intensity distribution trend.

Figure 8 shows the distribution trend of plane light intensity, from which we can see the spatial distribution of plane light intensity. From the light intensity distribution trend chart, it can be seen that the light emitted by the sodium lamp light source is roughly symmetrical at both sides of the center [16], the light intensity of the lamp post and the tail part is slightly different, and the overall light is uniform. By calculating the road lighting evaluation parameters of each group of data, the following parameters can be obtained: Field illuminance, average illuminance, lateral luminous angle, longitudinal luminous angle, longitudinal uniformity and total uniformity [17]. According to the measured experimental data, the longitudinal uniformity is about 0.68, the total uniformity is about 0.38, the lateral luminous angle and the longitudinal luminous angle are 62.38 and 65.94 respectively, and the overall uniformity is insufficient, with the total uniformity of 0.44 slightly higher than the evaluation standard of 0.4, which meets the lighting requirements. However, its longitudinal uniformity of 0.68 is slightly lower than the evaluation standard of 0.7, which will affect people's visual comfort.

5. Summary

In this paper, the light intensity distribution of high-pressure sodium lamps is tested by methods. The collected illumination data of high-pressure sodium lamps are modeled by three-dimensional surface linear fitting algorithm, local weighted regression algorithm and polynomial surface fitting algorithm on Matlab software, and the obtained light field model [18] is analyzed and evaluated for its determination coefficient and root mean square difference. The plane light intensity distribution is obtained from the experimental data, and the transverse and longitudinal luminous angles, longitudinal uniformity and total uniformity are calculated by the light field model data, which are analyzed and evaluated according to the road lighting standard.

According to the test and calculation data, the advantages and disadvantages of sodium lamps are compared with other light sources such as LED and metal halide lamps in the market: at present, high-pressure sodium lamps and metal halide lamps are widely used in road lighting. The luminous rays of the high-pressure sodium lamp light source are emitted everywhere, and the brightness is the brightest directly under the lamp, so the utilization ratio of road lighting is not high. In addition, the high-pressure sodium lamp is not designed to distribute light for the road surface, and a large part of light cannot reach the road surface, resulting in the longitudinal uniformity and total uniformity only slightly lower than the road lighting standard; The color rendering index of high-pressure sodium lamp is poor, only 20 to 40, and it feels dim, and the color of objects is difficult to distinguish under the light [19]; Metal halide lamps have the characteristics of high luminous efficiency, long life, good color rendering, compact structure and stable performance, but they contain mercury and are not environmentally friendly. Compared with high-pressure sodium lamps and other light sources, LED light source, as the third generation light source, has the advantages of higher luminous efficiency, longer service life, faster response speed and so on. With the development of technology, these advantages are gradually expanding. Therefore, in the field of road lighting, LED street lamps are an inevitable trend. In view of the existing market of sodium lamp street lamps, how to upgrade the traditional high-pressure sodium lamp street lamps to LED street lamps in a low-cost way is the direction of further research.

References

[1] Xie Di, Du Tong and Wang Zhenlong 2016 Street Illuminance Measurements and Data Processing System *J. Electrical Switch* **54** pp 9-10
[2] Chen Huihai, Wei Yadong, He Peirong, Lin Junli, Wu Guixin,and Lu Qiqing 2014 Design of the Illumination Test System of LED Street Light *J. Building Energy Efficiency* **42** pp 66-68
[3] Cao Fan 2011 *Research on Photometric Measurement Method of LED Street Luminaries (*Dalian Polytechnic University) pp 15-17
[4] Eberhart Martin , Loehle Stefan and Offenhäuser Philipp 2021 3-D visualization of transparent fluid flows from snapshot light field data *J. Experiments in Fluids* **62** pp 20-22
[5] Xuan Li 2012 *Study on Mesopic Vision and Measurements of Road Illuminance under LED Light Sources (*China Jiliang University) pp 5-6
[6] Chen Weihua 2012 *The Design of LED Lamp Light Intensity Detection (*Hangzhou Dianzi University) pp 8-9
[7] Chen Yingying et al 2021 Real-time Pre-rectification of Aberrations for 3D Light-field Display

Based on a Constructed Virtual Lens and Ray Path Tracing *J. Optics Communications pp* 22-25
[8] Yun Zhang 2007 *Research on Illumination Systems used for LED Energy-saving Road Lighting and Illuminance Distribution* (Tianjin University) pp 83-87
[9] Shi Lei, Xie Fengzhen, Zhang Xianwei and Hu Yibo 2020 Research on Fine Modeling of 3D Data Based on Multi-software Processing *J. Scientific and Technological Innovation* **23** pp 95-96
[10]Temmler André et al 2021 Laser Micro Polishing of Tool Steel 1.2379 (AISI D2): Influence of Intensity Distribution, Laser Beam Size, and Fluence on Surface Roughness and Area Rate *J. Metals* **11** pp 33-36
[11]Li Cunlong, Chen Weimin, Lei Xiaohua, Lai Wei, Zeng Haipeng and Wang Fuquan 2010 Optimization of Test Step Angle for the Photometry and Goniophotometry of LED Road Lights Based on Correlation Analysis *J. Optical technology* **36** pp 52-56
[12]Li Hanshan,Yue Shiqiang and Zhang Xiaoqian 2021 Measurement Model and Method of Multiple Projectile Dispersion Position Based on Dual Light Field Intersection Imaging *J. Measurement* **186** pp 24-26
[13]Ge Shaobo et al 2021 Novel Bilayer Micropyramid Structure Photonic Nanojet for Enhancing a Focused Optical Field. *J. Nanomaterials (Basel, Switzerland)* **11** pp 31-33
[14]Wang Minshuai ,Huang Xiaojing ,Cai Xiaomei ,Yang Lan ,Zheng Kai and Wang Xing-yan 2015 Second Optical Design and Optimization for High-power LED Street Lamp *J.Journal of Jimei University (Natural Science Edition)* **20** pp 20-35
[15]Sun Haojie , Li Mengyuan , Zhuang Jie, Li Baicheng, Zhang Dawei and Yang Weiqiao 2014 Research on LED Light Intensity Distribution Test Methods *J. Optical instruments* **36** pp 5-8
[16]Zhao Jing 2014 Design of Test Platform for LED Far Field Intensity Distribution *J. Automation Technology and Application* **33** pp 78-81
[17]You Ji 2018 A Comparative Analysis Between LED and High Pressure Sodium Lamp in Road Lighting Based on DIALuxevo *J. China Lighting Appliance* **8** pp 30-33
[18]González Laprea Jesús et al 2021 Photoelectric Effect Measurements on a Conventional Neon Bulb *J. American Journal of Physics* **89** pp 69-74
[19]Anatolij D. Semenov and Anton V. Volkov and Alena Yu. Shumkina 2020 Neural Network Identification of a Nonlinear Model of a High-pressure Sodium Lamp *J. MATEC Web of Conferences* **329** pp 59-62

Multi-protocol Integration and Intercommunication Technology Based on OPC UA and MQTT

Zhen Wang*, Dantao Han, Yanjie Gong and Yanling Zhao

Instrumentation Technology and Economy Institute (ITEI), Beijing 100055, China

*Corresponding author's e-mail: wangzhen@tc124.com

Abstract. OPC UA and MQTT are widely recognized key technologies for realizing the interconnection and intercommunication of industrial equipment as well as data transmission to the cloud. Based on OPC UA and MQTT, this paper presents a new method to solve the problem of multiple protocols coexistence and information islands in industry. Using the unified name ID to create mappings between OPC UA, MQTT and data acquisition drive, a shared database as data consistency communication medium to support concurrent data reading and writing of OPC UA and MQTT, a data acquisition drive as trigger to realize OPC UA and MQTT data synchronization update. Thus the difficulties of multi-protocol device integration and intercommunication and data uploading to the cloud can be tackled.

1. Introduction

To meet the market needs for diversity, rapidness and miniaturization, intelligent manufacturing theory and technology have been constantly and gradually advanced, and an increasing number of enterprises have realized the importance of integration and intercommunication and data uploading to the cloud, followed by the emergence of a batch of new technologies, such as the IOT, reconfigurable manufacturing system and industry 4.0, etc. Among them, OPC UA(Open Platform Communications Unified Architecture) is one of the most widely used specifications for industrial systems and device communications. It provides a consistent and complete address space and service model, which can solve the previous problem that information of the same system cannot be accessed in the same way. Meanwhile, it provides a fast modeling method, realizes data semantic expression, and reduces the difficulty of data intercommunication and data fusion. However, the traditional C/S communication architecture adopted by OPC UA severely restricts its application in industry, and the problem becomes even more prominent when combined OPC UA with the increasingly widely used cloud technology.

To solve the problem that OPC UA lacks the function of asynchronous data communication and data uploading to the cloud, OPC UA International Standards Organization proposed the PUB/SUB[1] technical standard based on UDP and MQTT[2]. MQTT protocol supports PUB/SUB, which has been widely used in the Internet of Things (IoT) and other fields. The technical idea of combining OPC UA and MQTT to enable OPC UA to support PUB/SUB is a consensus, but implementation methods are diverse. Previous study described the specific method of realizing OPC UA PUB/SUB based on MQTT and its configuration tools [3]. While in another research, the principle of OPC UA PUB/SUB is introduced in detail and the performance of combining OPC UA and MQTT is evaluated and discussed [4]. Furthermore, a study on gateway of OPC UA and MQTT integration is conducted in article [5], introducing a bidirectional mapping technology of OPC UA and MQTT data. Additionally, article [6] and [7] introduced the integration of OPC UA PUB/SUB and TSN and industrial applications of MQTT respectively.Considering the actual needs of industrial application, this paper proposes a multi-protocol integration and intercommunication technical method based on OPC UA and MQTT.

The rest of this paper is organized as follows. Section 1 discusses the OPC UA industrial application background and some existing problems, indicating that OPC UA supports PUB/SUB function is a general trend; Section 2 presents the methods and key technologies of combining OPC UA and MQTT, including the architecture and data mapping mechanism. Section 3 concludes the paper and proposes suggestions for future research.

2. Method

OPC UA and MQTT protocols each have their own communication models. MQTT uses broker agent technology to achieve communication with cloud platforms, which can support one-to-many and many-to-many communications; While OPC UA by using C/S architecture and TCP communication technology can support one-to-one communication only. These features determine that the two application scenarios are quite different, but combining with MQTT can completely enable OPC UA to achieve one-to-many communication.

Data integration and intercommunication is the foundation of industrial digitalization, and it is an import way to deal with problem of multiple protocol coexistence and information islands. This paper proposes an integration and intercommunication technology based on convergence of MQTT and OPC UA. Key technologies and difficulties include the principle of data mapping between OPC UA, MQTT and multi-protocol data acquisition drive module, and the establishment of data sharing mechanism. As mentioned above, OPC UA and MQTT have different communication models, and data acquisition drive must support diverse protocols. In this case, how to establish an appropriate mapping relationship between the above three is the focus of this article. In addition, a data sharing mechanism is established based on database technology to realize integration of OPC UA and MQTT, using data acquisition drive as a trigger to achieve data synchronous update of OPC UA and MQTT. The following elaborates on the research content and key points.

2.1. Architecture

OPC UA protocol implements periodic data collection mainly through monitoring and polling methods, so as to support one-to-one data monitoring. The new method proposed in this article integrating OPC UA and MQTT can enable devices to support OPC UA and PUB/SUB data asynchronous communication as well. The architecture is shown in Figure 1, and the main contents include:

a. Parameter configuration module. By defining a unified name ID, it establishes the mapping relationship among MQTT protocol, OPC UA protocol, data acquisition drive protocol and database storage to form a universal parameter configuration file,

b. Configuration loading module. It receives and parses the configuration file, and sends messages to MQTT loading module, OPC UA loading module, database module, and data acquisition drive module to update parameter information,

c. MQTT loading module. After getting configuration parameter update instruction, it generates a JSON model file to realize MQTT data communication,

d. OPC UA loading module. After getting configuration parameter update instruction, it generates an OPC UA address space model to realize OPC UA semantic communication,

e. Data acquisition drive module. After getting configuration parameter update instruction, it updates the communication protocol parameters of data acquisition drive, and starts to collect external device data and store it in a shared database,

f. Database module. After getting configuration parameter update instruction, it establishes a shared database to receive the actual data collected by data acquisition drive module, giving MQTT and OPC UA modules both read and write access to shared data using the unified name ID.

This method realizes data consistency of OPC UA and MQTT based on unified name ID and shared database. Meanwhile, the shared database is updated in real time through data acquisition drive module, which triggers the real-time data update of MQTT and OPC UA.

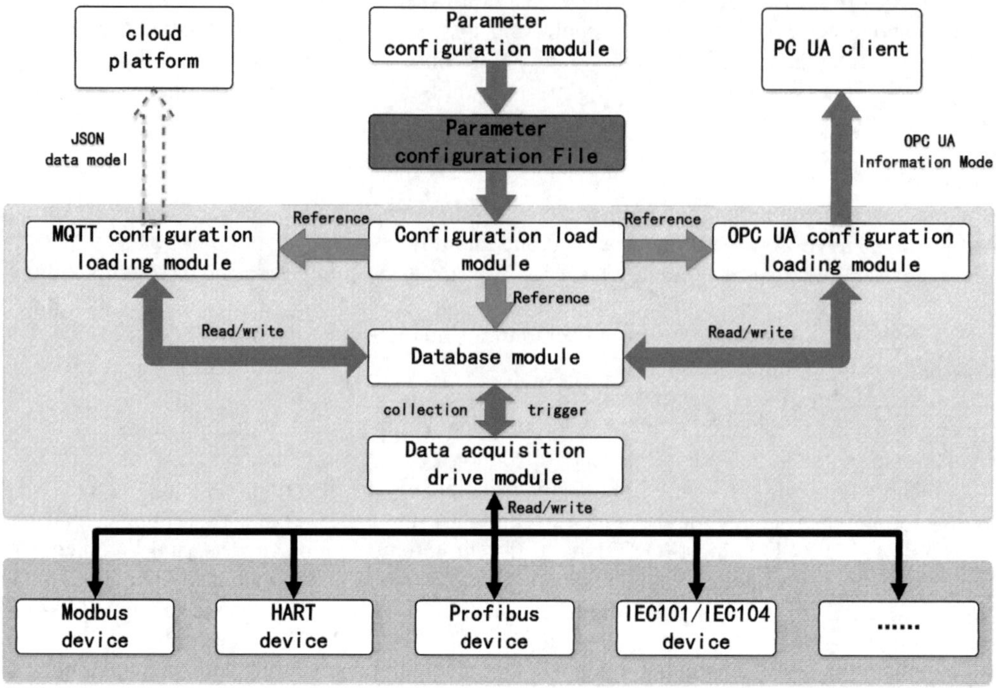

Figure 1. Architecture

2.2. Protocol mapping mechanism

Using the universal format xml file as carrier, the configuration file includes unified name ID, drive parameters and protocol parameters information. Figure 2 shows the configuration file template and parameters mapping architecture, mainly including:

a. Unified name ID. Using name and description information in the configuration file as the unified access identifier to realize data sharing among MQTT, OPC UA, database and acquisition drive,

b. Drive parameters. The drive parameters in the configuration file include communication port type and configuration items (baud rate, parity, data bit, stop bit etc.), data type, data address, access cycle, etc. It realizes initialization and normal operation of data acquisition drive module, and can be flexibly expanded according to different industrial communication protocols,

c. Protocol parameters. The protocol parameters include the necessary parameter information to realize functions of reading and writing data of MQTT and OPC UA. The configuration file is the core foundation in this method. Through configuration loading, MQTT, OPC UA, data acquisition drive and database parameter configuration file are finally formed and passed to each functional modules to complete their initialization and operation. The content of configuration file and mapping process are shown in Figure 2.

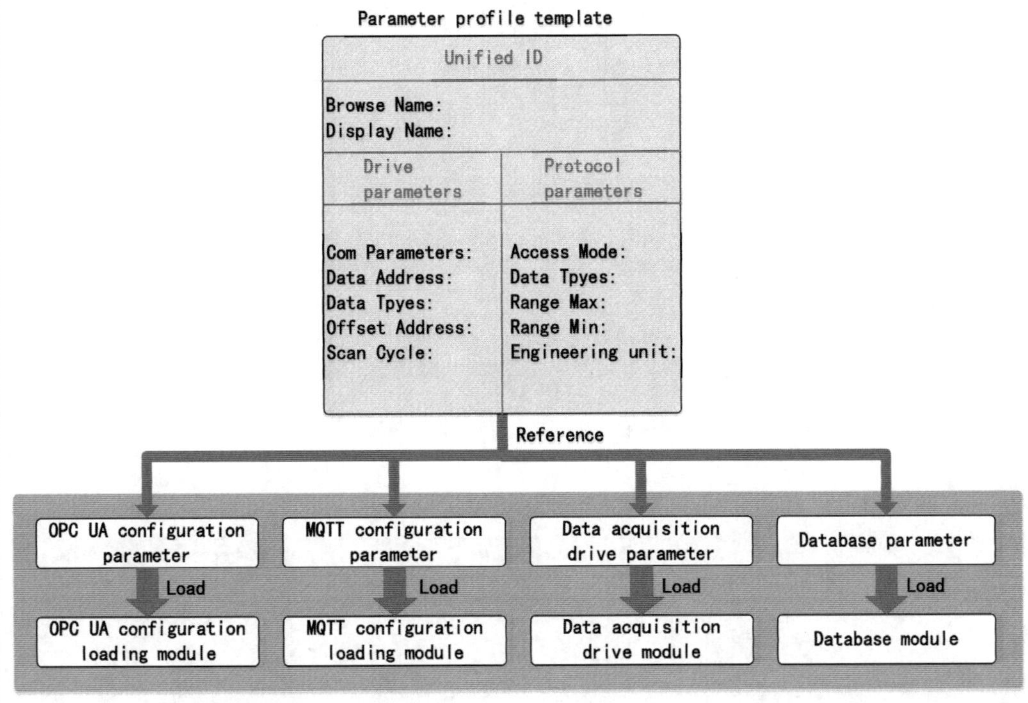

Figure 2. Parameter profile template and mapping architecture

2.3. Configuration parameter mapping example

Taking modbus data collection as an example to verify the technical principle. To realize data collection and storage, OPC UA display and cloud synchronization response, the mapping process should be configured as Figure 3 shows, in which four sub-configuration files are generated through the configuration loading module, and then respectively passed to OPC UA, MQTT, data acquisition drive and shared database, and a synchronization mechanism is established. OPC UA and MQTT synchronization data update is triggered by real-time update of data acquisition drive.

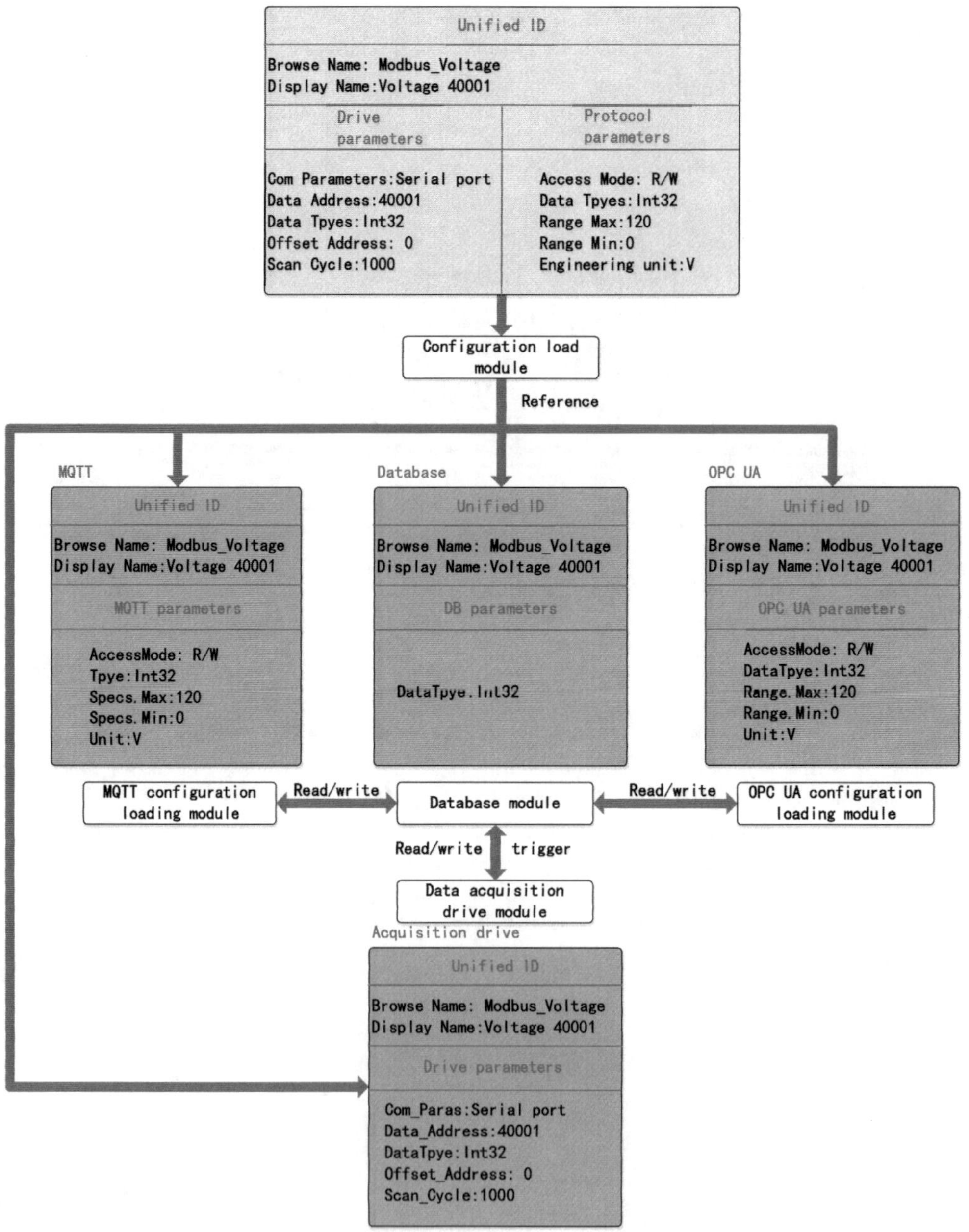

Figure 3. Configuration parameter mapping example

3. Summary

This paper presents a method to integrate OPC UA and MQTT protocols. The configuration file is formed based on the xml model, and the mapping relationship among OPC UA protocol, MQTT protocol and data acquisition drive is established through a unified name ID. In the operation process, a shared database is the foundation for data collection drive to store data and OPC UA and MQTT to access data using the unified name ID, hence data sharing is realized. Furthermore, the real-time data collected by data acquisition drive is used as a trigger to achieve data synchronization between OPC UA and MQTT. This method has been tested on Alibaba Cloud platform and realized data reading and writing as well as trigger of events. Later, OPC UA and MQTT convergence will be further optimized to improve data access efficiency, and the concept of OPC UA information model will be integrated into MQTT data transmission.

Acknowledgment

This research was fully supported by a grant from the National Key Research and Development Program "New system architecture and key techniques of instrument based on semantic interaction integration" (Grant No. 2018YFB2003200).

References

[1] OPC Foundation, "OPC Unified Architecture Specification, Part14:PUB/SUB," 2018.
[2] MQTT https://mqtt.org/.
[3] Liu, Z. , and P. Bellot . "A configuration tool for MQTT based OPC UA PUB/SUB." 2020 RIVF International Conference on Computing and Communication Technologies (RIVF) IEEE, 2020.
[4] Raddatz, H. , et al. "Evaluation and Extension of OPC UA Publish/Subscribe MQTT Binding." 2020 IEEE Conference on Industrial Cyberphysical Systems (ICPS) IEEE, 2020.
[5] Peniak, P. , E. Bubenikova , and J. Spalek . "Model of Integration Gateway for Communication of OPC/MQTT Devices." 2020 Cybernetics & Informatics (K&I) 2020.
[6] J. Pfrommer, A. Ebner, S. Ravikumar, and B. Karunakaran, "Open source opc ua PUB/SUB over tsn for realtime industrial communication," in 2018 IEEE 23rd International Conference on Emerging Technologies and Factory Automation (ETFA), vol. 1. IEEE, 2018, pp. 1087–1090.
[7] P. Nenninger, M. Gierl, and R. Kriesten, "A contribution to publish-subscribe based communication in industrial applications," in 2019 IEEE 10th Annual Information Technology, Electronics and Mobile Communication Conference (IEMCON). IEEE, 2019, pp. 0424–0429.

Research on University Computer Teaching Based on Data Mining Learning Model

Guibing Xu

Nanchang institute of technology, Nanchang, Jiangxi, China

e-mail: xuguibing188@163.com

Abstract. Under the background of the development of big data, the research of learning analysis technology has shifted from theoretical fields to practical application fields such as teaching big data analysis and learning behavior prediction. However, it is worth noting that the wide application of big data technology has led to a large increase in text data, and how to quickly obtain accurate information has become a decisive factor in improving students' learning ability. Therefore, this article combines the current status of computer teaching in colleges and universities to construct a data mining learning model to evaluate the effects of current college computer teaching methods. On this basis, it helps computer teachers to choose the most reasonable teaching style and teaching methods, and guide students to better master computer skills in order to achieve the best computer teaching effect.
Keywords. Data mining learning model; Colleges and universities; Computer teaching

1. Introduction

Data mining is an effective means of knowledge mining in the context of big data, which can select potentially undiscoverable data from a massive information database. It can be simply understood as finding valuable information that can be reused on the basis of existing data and information, and analyzing the specific relationship between these information[1]. For computer teaching in colleges and universities, the information collected through big data and other means is often high-dimensional[2]. If effective measures are not taken for processing, the information will gradually accumulate or even exceed the load. Once information errors occur, it will easily lead to computer teaching reforms. The direction has deviated. If the data mining learning model is introduced for information processing, it will be able to find some hidden value information in a more targeted manner, so as to make a correct assessment of the current computer teaching work, find the main factors affecting the employment of computer majors, and serve as a reform of computer teaching. Work provides scientific decision-making.

2. The construction of a data mining learning model for computer teaching in colleges and universities

The basic flow of the data mining algorithm, as shown in Figure 1, mainly includes problem definition, data collection, data preprocessing, data mining, interpretation results, evaluation results, etc.

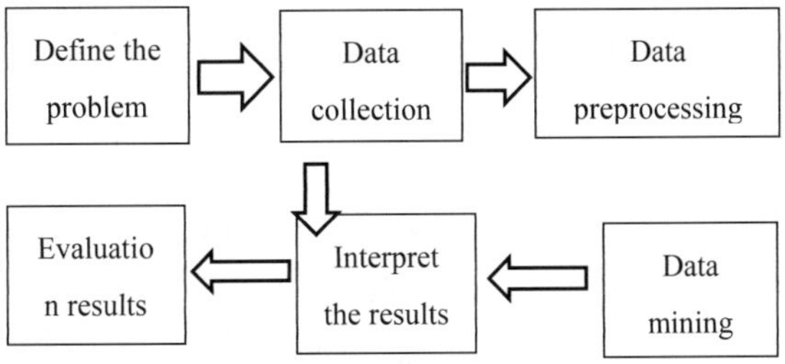

Figure 1. Basic flow of data mining algorithm

Among them, the definition problem refers to the identification of what kind of information data mining needs to process in the evaluation of computer teaching in colleges and universities, and the selection of data mining algorithms in this research[3].

Data collection and preprocessing are mainly to capture the information needed for research, and carry out targeted preprocessing, including screening, removal, conversion, etc., so that the calculation of the algorithm will not be affected by irregular data[4].

Data mining is based on the previous link, analyzing the characteristics of the data obtained, and then selecting the appropriate algorithm according to the data, so as to evaluate the computer teaching data of colleges and universities, forming a data decision tree, and realizing data classification to complete data mining[5].

The interpretation and evaluation of the results are after the completion of the data mining work, but because the results of the data mining are not required for research, it is necessary to interpret the evaluation result information obtained so as to convert it into information that can be easily understood by everyone[6].

3. Selection of data mining learning analysis methods and algorithm

3.1. ID3 algorithm

The ID3 algorithm is a kind of decision tree. The principle of its algorithm operation is to get the best conclusion with the least amount of calculation. ID3 algorithm is based on information theory, and uses information entropy and information gain as measurement standards to realize the inductive classification of data[7]. Among them, information entropy mainly represents the probability of discrete random events. If a system is more orderly, its corresponding information entropy will be lower, otherwise the information entropy will be higher. Therefore, in the operation of ID3 algorithm, the order status of system data is mainly measured by information entropy.

Assuming that there is a random variable X with a range of values, the probability of each type is $\{x_1, x_2, \cdots, x_n\}$expressed as$\{p_1, p_2, \cdots, p_n\}$, then the entropy definition of X is expressed as:

$$H(X) = -\sum_{i-1}^{n} p_i \log_2 p_i \quad (1)$$

Like the above formula, if a variable changes more, it means that the amount of information it carries is also larger. As far as the classification system is concerned, it is expressed by the C variable, the value range is $\{C_1, C_2, \cdots, C_n\}$, and the probability of each category is expressed as $\{P(C_1), P(C_2), \cdots, P(C_n)\}$, then the entropy definition of the classification system is expressed as:

$$H(C) = -\sum_{i=1}^{n} P(C_i) \log_2 p(C_i) \quad (2)$$

The information gain is for each feature. Looking at a feature t, what is the amount of information when the system has it and without it, and the difference between the two is represented by the information gain. The calculation formula is:

$$IG(S|T) = Entropy(s) - \sum_{value(T)} \frac{|S_v|}{S} Entropy(S_v) \quad (3)$$

In the formula, S represents the set of samples, and represents the set of all values of T, v represents an attribute value in the set, and represents the set of V in S, and represents the number of samples in S.

The operation of the ID3 algorithm is mainly to calculate the information gain when each non-leaf node of the decision tree is classified. If the value of the information gain is larger, the sample is more representative.

3.2. BP neural network model

BP neural network refers to adding several layers of neurons between the input layer and the output layer[8]. If the input layer is x_i, then the output is expressed as:$O_i^{(1)} = x(i), i = 1,2,\cdots, n$.

Assuming that the weight coefficient of the hidden layer in the BP neural network is, then the input and output formulas are respectively expressed as:

$$net_i^{(2)}(k) = \sum_{j=1}^{m} w_{ij}^{(2)} O_j^{(1)}(k) \tag{4}$$

$$O_i^{(2)}(k) = f\left[net_i^{(2)}(k)\right] \tag{5}$$

Assuming that the actual output and network output of the p-th sample are expressed as $O_p(k+1)$ and $O'_p(k+1)$ respectively, then the error calculation formula is expressed as:

$$E_p = \frac{1}{2}\left[O_p(k+1) - O'_p(k+1)\right]^2 \tag{6}$$

On this basis, adjust $w_{ij}^{(2)}(k)$ and $w_{li}^{(3)}(k)$. η is the learning rate and α is the momentum factor. The results are as follows:

$$w_{ij}^{(2)}(k) = \alpha\, w_{ij}^{(2)}(k-1) + \eta \frac{\partial E}{\partial w_{ij}^{(2)}(k)} \tag{7}$$

$$w_{li}^{(3)}(k) = \alpha\, w_{li}^{(3)}(k-1) + \eta \frac{\partial E}{\partial w_{li}^{(3)}(k)} \tag{8}$$

4. Empirical analysis of university computer teaching based on data mining learning model

For the evaluation of computer teaching in colleges and universities, this article selects the online learning course of "Computer Application Fundamentals" of XX College for analysis and prediction. First, the Apriori algorithm is used to perform information pruning processing according to the impact factor and support. On this basis, the ID3 algorithm is used to measure the information entropy to form a data decision tree to achieve the goal of data mining. details as follows:

4.1. Data acquisition and processing

To capture the information data of the "Computer Application Fundamentals" course of XX College, it is necessary to establish a corresponding database and enter the data. Specifically, eight databases were obtained, including professional teacher information table, teacher education qualification table, teacher title table, student information table, curriculum table, teacher evaluation table, teacher class table, and student employment table. On this basis, the data is integrated, entity identification, redundancy processing and data conflict handling are carried out.

4.2. Evaluation index system of computer teaching in colleges and universities

Regarding the composition of evaluation indicators for computer teaching in colleges and universities, it is aimed at the needs of teaching and comprehensively considers a series of related indicators that affect the teaching effect. Therefore, the teaching evaluation indicators are divided into teacher quality indicators and teacher quality indicators. Among them, the quality of teachers is mainly the evaluation of teachers' teaching ability, involving hard indicators such as academic qualifications, professional titles, and teaching achievements. The quality of teachers involves teachers' teaching skills, including indicators such as classroom participation, course difficulty, teaching interaction, and comprehensive evaluation of teaching. In the evaluation index system of computer teaching in colleges and universities, irrelevant items such as gender and age equivalence are eliminated, and the data information of teacher quality index and the corresponding attributes of the evaluation of teacher quality are obtained. Preprocess the collected data, perform coding conversion, and form corresponding values. The specific information is shown in Table 1.

Table 1. Computer teaching evaluation information

Name	Attribute value	Integerization
Education	Specialist	1
	Undergraduate	2
	Postgraduate	3
	PhD	4
Job title	Assistant	1
	Intermediate	2
	Deputy Senior	3
	Positive	4
Whether there is interaction	Yes	1
	No	2
Degree of difficulty of the course	Easy	1
	Acceptable	2
	Harder	3
	Difficulty	4
Interactive teaching	Excellent	1
	Good	2
	Middle	3
	Difference	4
Classroom participation	Positive	1
	Negative	2
Course assessment method	Closed book	1
	Unbook	2
	Practice	3
Comprehensive evaluation of teaching	Excellent	1
	Good	2
	Middle	3
	Difference	4

On this basis, the data is mined and processed to analyze the relationship between various data, to make a scientific and reasonable evaluation of computer teaching in colleges and universities, and to provide data decision support for teaching reform.

4.3 Evaluation results of computer teaching in colleges and universities

4.3.1 Data mining results With the continuous expansion of colleges and universities, the quality of students is uneven. In order to make better use of data mining technology to improve students' learning ability, to guide students to acquire professional computer knowledge and skills, so as to better achieve employment. Therefore, it is necessary to find out the association of each data in these information sets. Based on the above analysis, the Apriori algorithm is used to generate association rules. There are 45 rules in total, some of which are shown in Table 2.

Table 2. Teaching Association Rules

serial number	Association rules	Support	Confidence
001	002->0x101	26.5%	81.2%
201	201->0x801	17.8%	81.2%
301	301->0x501	18.1%	70.9%
401	401->0x801	14.1%	72.8%
601	601->0x201	40.8%	59.0%
701	701->0x401	27.9%	71.1%
801	001->0x801	14.4%	71.1%

As shown in the above table, 002->0x101 means that the degree is a bachelor’ s degree, and the comprehensive evaluation is excellent. The support is 26.5%, which means that teachers who are evaluated as excellent account for 26.5% of the total teachers, and the confidence level is 81.2%, which means that the comprehensive evaluation is excellent. As a percentage of the total number of teachers.

4.3.2. BP neural network model evaluation results Use the BP neural network model to evaluate the teaching quality of colleges and universities, set the target training error value to 0.0001, and then select the last 8 sets of data for testing. The results are shown in Table 3. It can be seen that the error is small, and the teaching quality constructed by the BP neural network can be seen The evaluation model has good fault tolerance and generalization ability, and basically achieved the expected effect.

Table 3. BP neural network model evaluation results

Number	Expected value	Output value	Error
001	0.7878	0.7784	0.31%
201	0.7899	0.7823	0.23%
301	0.8122	0.8080	0.53%
401	0.7799	0.7832	0.87%
501	0.8034	0.8087	0.68%
601	0.7901	0.7882	0.65%
701	0.8033	0.8101	0.58%
701	0.7913	0.7889	0.76%

In summary, compared with the BP neural network model evaluation method, the interference of human factors on the results is reduced to a certain extent, and the error of the results obtained is smaller. However, compared with data mining methods, if the sample is enlarged, the reliability of the results will be greatly reduced. Data mining methods can perform typical analysis on detailed samples, and the results have a higher reference for work improvement.

5. Conclusions and recommendations

5.1. Conclusion

Integrating all the data mining results, the following conclusions are obtained:

First, in the computer evaluation of colleges and universities, the interactivity of teaching has a vital influence on the improvement of the effect of computer courses in colleges and universities, and it is one of the key influencing factors of the comprehensive evaluation of computer teaching. Combining the learning characteristics of some freshmen, it is found that their willingness to learn independently is relatively weak, and they lack the ability to actively explore and learn. By integrating teaching interaction into the teaching of computer courses in colleges and universities, actively guiding students to interact with each other and inspiring students to participate in computer learning Positivity can also just guide students to understand professional knowledge and skills, inspire students' initiative and creativity in learning, and can also improve students' problem-solving ability.

Second, classroom participation can also have a greater impact on computer teaching in colleges and universities, and will directly affect the comprehensive evaluation results of computer teaching. Because some freshmen have just entered the campus and are separated from the heavy high school learning tasks, they are easily attracted by the novel campus life and other entertainment, and they are not active in computer teaching. By setting up diversified computer teaching content through classroom participation, more students are attracted to actively participate in the course interaction, so that teachers' teaching is no longer one-way, but two-way learning interactive communication, and strengthen the interactive effect of computer teaching.

Third, under normal circumstances, professional teachers with a high degree of education in computer teaching in colleges and universities are often more popular with students. Based on the above data mining results, it is found that teachers with the same teaching experience and highly educated teachers are often able to explain some difficult-to-understand problems more profoundly and easily to understand, simplifying multi-level complex problems, so that students are more willing to take electives Courses for highly educated teachers. In addition, teachers with high academic qualifications often have higher learning ability and are willing to learn information-based teaching methods, creating a more relaxed and lively learning range, so as to lead more students to actively participate in the teaching practice of the course, so as to achieve the improvement of teaching effect.

In summary, through data mining on the evaluation of computer teaching in colleges and universities, it is found that the current single teaching method can no longer improve the actual teaching effect. It is necessary to improve the computer courses in colleges and universities in terms of teaching interaction, course participation, teacher ability and so on. Enriching the content of computer classrooms makes it easier for students to accept and understand knowledge.

5.2. Suggestions

Based on this, in view of the current shortcomings of computer teaching in colleges and universities, we should combine the above-mentioned data mining results and propose targeted improvements.

First, change the concept of computer teaching. In computer teaching in colleges and universities, teachers should combine the goals of national modern curriculum construction, change teaching concepts, teach students in accordance with their aptitude, and stimulate students' enthusiasm for computer learning[9-10]. Computer teachers should pay attention to cultivating students' independent learning and practical ability, strengthen teaching exchanges and interaction, guide students to actively explore new knowledge and skills, meet social employment needs, and cultivate high-quality professional computer talents.

Second, enrich the content of computer teaching and mobilize students' enthusiasm for work as much as possible[11]. Teachers of computer teaching reform in colleges and universities should improve their lesson preparation ability and course information processing ability, enrich the knowledge in books, guide students to clearly grasp the importance of computer skills for future study and work, and stimulate students' desire to explore new knowledge , So as to actively participate in autonomous learning and strengthen the quality and effect of computer teaching[12].

Third, introduce diversified computer teaching methods. In the new era, computer teaching in colleges and universities should pay attention to cultivating students' comprehensive ability, not only to strengthen the cultivation of students' innovative ability, but also to cultivate students' innovative thinking[13]. On this basis, computer teaching in colleges and universities can introduce open teaching methods, fully integrate multi-channel computer learning capabilities, break the limitations of traditional computer classrooms, and improve students' computer skills.

Fourth, create a good interactive atmosphere for teaching. In order to further strengthen the effect of computer teaching, computer teaching in colleges and universities should pay attention to creating a good teaching interaction atmosphere[14]. Teachers should fully respect the dominant position of students and guide students to actively participate in interactive communication. At the same time, teachers can use group discussions and other forms to guide students to actively participate in learning interaction, exchanges and cooperation, etc., to improve students' initiative in learning, and stimulate students' learning innovation and creativity[15].

References

[1] Shi Z 2017 Research on the Application of Computer Data Mining Technology in University Scientific Research Management System *Science and Technology Information* **15**(17) 195-6
[2] Yuan S, Ye C 2018 Research on the Curriculum Construction of Big Data in Computer Science and Technology in Applied Undergraduate Universities *Science and Technology Vision* **13** 198-9
[3] Sui Yongbo, Cao Xu 2018 A preliminary study on the application of data mining-based learning analysis in teaching——Taking computer courses in colleges and universities as an example *Journal of Changchun University of Traditional Chinese Medicine* **34**(05) 988-91
[4] Pu Y 2018 Analyze the optimization path of computer basic courses in colleges and universities based on computational thinking *Journal of Jiangxi Vocational and Technical College of Electric Power* **31**(12) 87-8
[5] Ren P, Ruan C 2019 Research on the Path of Data Mining in the Precision Employment of College Graduates——Taking Computer Major as an Example *Journal of Shanxi Datong University (Natural Science Edition)* **35**(02) 26- 9
[6] Sun B 2019 Empirical research model analysis of university computer teaching based on data mining learning analysis *Modern Electronic Technology* **42**(19) 127-31
[7] Hu R, Qian L 2019 Research on the Teaching Mode of "Micro-course" of Computer Culture Foundation in Colleges and Universities *Software Guide* **18**(11) 203-5
[8] Wang Z, Zhang T, Duan L, Zhu H 2020 Research on Innovative Teaching Strategies of Computer Majors in Colleges and Universities under the Background of Big Data *Education Modernization* **7**(19) 88-89+100
[9] Guo J, Bai J, Shi L 2020 Analysis and discussion of basic computer scores in medical universities based on association rules: Taking Baotou Medical College of Inner Mongolia University of Science and Technology as an example *Curriculum Education Research* **16** 242
[10]Wang H 2021 Research on Big Data Technology in Computer Information Teaching in Colleges and Universities *Network Security Technology and Application* **02** 83-5
[11]Shi W 2021 The reform of the computer professional curriculum system in colleges and universities under the guidance of professional competence training *Journal of Hubei Open Vocational College* **34**(04) 132-3
[12]Lu X 2021 Analysis of the Application of Computer Data Mining Technology in University Archives Information Management System *Computer Knowledge and Technology* **17**(15) 41-2+47
[13]Lu H 2021 Research on the Evaluation of University Computer Laboratory Management Level Based on Data Mining *Modern Electronic Technology* **44**(13) 115-9
[14]Shang Y 2021 Analysis of university computer data processing technology based on big data *Electronic Testing* **17** 132-3+99
[15]Yang X, Wu S, Xia H, Zhou J, Yu X, Fan Y 2021 Research on the teaching reform of data mining and analysis courses in the era of big data *Computer Age* **09** 107-11

The Design of Computer Network Virus Defense System Based on Data Mining Technology

Hui Xu*

School of Computer and Information Engineering,Nanchang Institute of Technology, Nanchang, Jiangxi 330044, China

*corresponding author's e-mail: 408566853@qq.com

Abstract. With the development of modern society in China, the application of computer technology is becoming more and more extensive. Today, with the gradual popularization of network technology, related computer network viruses are also constantly appearing. These viruses often pass through the loopholes of the network system and now the computer system, and damage the data and malicious tampering in the system.In order to better protect against the harm of computer network viruses, the related data mining technology has emerged.This paper discusses the type of computer network virus and explores the computer network virus defense system based on data mining technology.

Keywords. Data mining technology, Network virus defense systems

1. Introduction

In the era of Internet information, the network security use of computer users has become the focus of social development.[1] The generation and spread of network viruses will seriously threaten the network security of computer users, lead to the paralysis of user computer systems, and cause certain economic losses of users.[2] In view of this, it is necessary to scientifically and effectively use data mining technology to improve the defense level of computer network for users, and avoid various threats and damage to users' computer network, so as to promote the stable and sustainable development of the whole computer network industry, and protect the personal interests of users[3].

2. Data mining

Data mining refers to the process of algorithmic searching for information hidden in it from a large number of data.Data mining is often related to computer science, and achieves these goals through statistics, online analysis processing, intelligence retrieval, machine learning, expert systems (relying on past rules of thumb), and pattern recognition.In recent years, data mining has attracted great attention in the information industry, mainly due to the existence of a large amount of data that can be widely used, and the urgent need to convert such data into useful information and knowledge.The information and knowledge obtained can be used in a wide variety of applications, including business management, production control, market analysis, engineering design and scientific exploration, etc.Data mining is a technology to find its rules from a large number of data by analyzing each data [4] .

There are mainly three steps: data preparation, rule search and law representation.Data preparation is to select the required data from the relevant data sources and integrate it into the dataset for data mining; the rule search is to find out the rules contained in the data set; the law representation is as understandable as possible (such as visualization).[5] The tasks of data mining are association analysis, cluster analysis, classification analysis, anomaly analysis, specific group analysis, and evolution analysis[6].

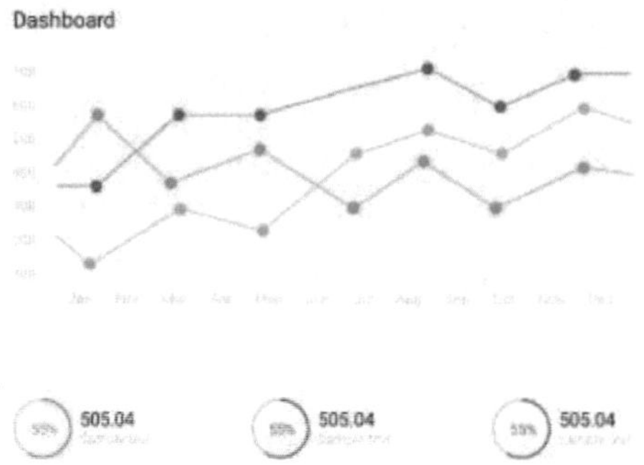

Figure 1 Schematic diagram of the development of data mining technology

In recent years, data mining has attracted great attention in the information industry, mainly due to the existence of a large amount of data that can be widely used, and the urgent need to convert such data into useful information and knowledge[7].The information and knowledge obtained can be used in a wide range of applications, including business management, production control, market analysis, engineering design, and scientific exploration.Data mining utilizes ideas from some fields: ① sampling, estimation, and hypothesis testing from statistics; search algorithms, modeling techniques, and learning theories for ② artificial intelligence, pattern recognition, and machine learning.[8] Data mining has also rapidly embraced ideas from other areas, including optimization, evolutionary computing, information theory, signal processing, visualization, and information retrieval[9].Some other areas also play an important supporting role.In particular, database systems require efficient storage, indexing, and query processing support.Techniques derived from high-performance (parallel) computing are often important in processing massive datasets.Distributed technology can also help process massive amounts of data and is critical when data cannot be processed together.

3. Computer virus type

Computer virus refers to a set of computer instructions that are compiled or inserted in a computer program to destroy computer functions or destroy data, affect the normal use of the computer, and are capable of self-copying.

Computer viruses are transmissible, covert, infectious, latent, excitable, expressive, or destructive.Life cycle of computer virus: development period of infection incubation period detection period of digestive period extinction period.

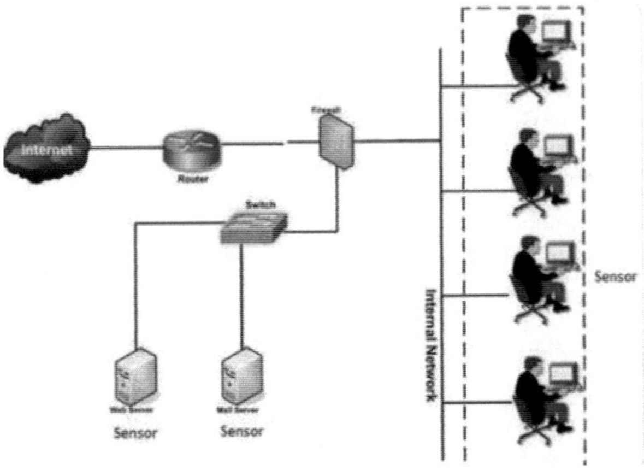

Figure 2. Computer virus invading a computer

Any virus that invades the system will have a different impact on the system and the application.Light people will reduce computer work efficiency, occupy system resources, heavy people can lead to data loss, system collapse. The programmability of a computer virus, representing it as with any other legal programs, is an executable program, but it is not a complete program, but a program parasitic on other executables that destroys only when other programs run.Once the virus is executed after entering the computer, it will search for other qualified environments, determine the target, and then copy itself, so as to reach the purpose of self-reproduction. Therefore, infectivity is an important condition to judge the computer virus[10].

Computer viruses are man-made, destructive, infectious, and latent, and programs designed to disrupt computer information or systems.It does not exist independently, but is hidden in other executable programs.After the virus in the computer, the light speed of the machine, the heavy dead system damage; therefore, the virus brings great losses to the user, usually, we call this destructive program a computer virus[11].

Computer viruses can be classified into guide viruses, file viruses and hybrid viruses according to existing media, three source viruses, embedded viruses and operating system viruses, DOS system virus, Windows system virus and UNIX system[12].Today's computer viruses are constantly innovating, including some unique new viruses that cannot be classified by regular types, such as Internet viruses (spread through the Internet, some viruses carrying more and more data), email viruses, etc.

Computer viruses are recognized as the number one enemy of data security. Since 1987, the computer virus received widespread attention worldwide, and China also first discovered a computer virus in 1989.At present, the new virus is developing towards being more destructive, more secretive, with a higher infection rate and a faster transmission speed.Therefore, we must continue to study the basic common sense of computer viruses and strengthen the prevention of computer viruses.

4. Preliminary design of the system design1. Vision of the system design

4.1. The vision of system design

In the increasingly rapid development of computer networks today, in order to better resist the intrusion of computer viruses, data mining technology has been gradually applied in the defense system of computer network viruses.To strengthen the data mining technology and improve the virus defense index, it is necessary to strengthen the analysis of the speed of the transmission of network viruses.Generally speaking, viruses transmitted through the network often need to scan the data information of the computer system when invading and destroying the computer system, while the data generated in the scanning operation is collected by the data mining technology, and the application of no data mining technology provides more technical support.Data mining steps are complicated, and multi-function modules can be divided according to their practical characteristics.Data mining system is mainly composed of data source module, data preprocessing module, rule library, data mining module, etc. Data mining system generated under these modules can defend and analyze viruses.

4.2. System design module

Data source module.The data source module contains a packet grab, which refers to the packet sent to the host system on the network.The original network package in the data source module contains some data structure of specific data, which will be handed over to the system preprocessing module for analysis after the packet grab program has intercepted it.[13]

Preoutsourcing module, the data package in the preoutsourcing module usually needs to be classified and then transferred to the data that can be recognized by the system, such as the data port information, IP address, source file IP address and other information should be sorted and classified in the processing module.Pre-processed data can reduce the late data analysis and processing time, and further improve the data identification.

The rule library module.The rule library module can identify, analyze, and mine the data after the network virus appears in the computer system.The rule set acquired in data mining has been adjusted to the network viruses, the record can be used in data mining work, the record can help the system in potential virus analysis, and play a certain defensive role.As the most important part of the data mining technology, the rule library module can also improve the data mining information database and provide virus analysis data for the system.

Data mining module.The Data mining module is the most critical part of the whole data mining technology.[14] The module mainly includes the event library and the data mining algorithm. The event library is formed by the data mining method to collect the relevant data. In the event library, we can sort out and classify the data, so as to get a clearer data results.

5. Current status of computer virus defense system in China

Due to the computer popularization time in China is relatively short, the relevant technical measures are not perfect and mature enough, so there are many problems, mainly manifested in two aspects: First, because the network is virtual, so it is difficult to control in life.Whether at home or abroad, once the virus appears, it may lead to a large area of computer paralysis, and even cause serious property losses.Second, there are some hackers who are familiar with the network and will attack the network for a variety of purposes.

6. Conclusion

In the current cloud computing environment, it has become more and more urgent to strengthen the protection of computer network security vulnerabilities, and to explore the more effective safeguard measures can protect the integrity of the information security of computer network users.Today's computer network has been increasingly more and more commonly applied, the challenge in security protection is more severe, must effectively deploy internal network, internal threats in computer network virus increasingly rampant today, in order to eliminate and avoid these viruses to computer system, we must constantly develop related data mining technology to analyze and kill the virus, in order to reduce the possibility of virus now computer system, bring security for computer system, and at the same time to create a pure network working environment for people.

Acknowledgement
Project Fund: Collaborative Education of the Ministry of Education, Fund No. 202002158020, Research on python Mixed Innovative Teaching for Big Data.

References
[1] Yaoyao X 2021 The Design of the Computer Network Virus Defense System Based on the Data Mining Technology *Network Security Technology and Applications* **11** 45-6
[2] Haifeng F 2020 The Application of Data Mining Technology in the Development of Computer Network Virus Defense System *Electronic Technology and Software Engineering* **17** 236-7
[3] Yanshi J 2020 The Application of Data Mining Technology in Computer Network Virus Defense China High-tech **17** 55-6
[4] Weihuang B 2020 The Design of the Computer Network Virus Defense System Based on the Data Mining Technology *Electromechanical information* **23** 140-1
[5] Yibai G, Yuche H 2020 The Application of Data Mining in Computer Network Virus Defense *Electronic Technology and Software Engineering* **04** 250-1
[6] Yuliu, Ke Z 2020 The Application of Data Mining Technology in the Development of Computer Network Virus Defense System *Journal of Beijing Printing Institute* **28**(01) 133-5
[7] Xiaolei G 2019 The Development and Analysis of the Computer Network Virus Defense System Based on Data Mining *Network Security Technology and Applications* **09** 50-2
[8] Yongjun T 2019 The Application of Data Mining in Virus Defense of Computer Network Scientific and technological innovation and productivity **07** 61-3
[9] Linasun Y 2019 Application of Data Mining Technology in the Development of Computer Network Virus Defense System *Software Guide* **18**(08) 166-9 + 173
[10] Zhangjun H 2018 *A Network Virus Attack and Defense Strategy Based on Incomplete Information Game Theory* (Xiamen University)
[11] Jingzidu H, Liu Xuan G 2018 The Design and Implementation of the Network Virus Defense System Based on Data Mining Technology *Information and Computer (Theory Edition)* **06** 55-7
[12] Li Y 2016 Exploration of Computer Network Virus Defense Technology Based on Data Mining Technology *Modern Electronics Technology* **39**(21) 120-2 + 126
[13] Yueyao T 2013 Strengthen the computer defense system to resist the network virus attack, the combination of—— tradition and innovation integration *Information System Engineering* **09** 82
[14] Pengwei L, Wenying G 2012 Study of network virus invasion defense system technology *Coal Technology* **31**(09) 219-21

Database Programming Technology Based on Computer Software Engineering

Weiwei Gong*

Jiangxi College of Applied Science and Technology, Nanchang, Jiangxi 330044, China

*Corresponding author's e-mail: qgjxzbt@163.com

Abstract. Based on practical application, this paper further discusses the database programming technology in computer software engineering. Computer technology has developed to a certain extent and is still active in various fields. However, because the demand for computers in various industries and the requirements for technical performance are different, software engineers are required to develop software systems suitable for enterprises according to their own production characteristics. Because the efficiency and quality of computer software can not reach at present, the programming technology level of database may have a certain impact on the software system.In order for database programming technology to play a full role in various fields, it is necessary to increase investment in database programming technology.This paper analyzes the database technology of computer software engineering in detail. In the process of establishing the actual database programming system, we make full use of the file creation and file access of the database to improve the database programming technology in the current computer software engineering, and then improve the stability of computer software. This paper analyzes computer software engineering, summarizes database programming program, fully realizes the application value of program technology in actual production, and combines database programming technology with the design of computer software engineering project, so as to promote the continuous innovation and development of computer software technology in China.
Keywords. Computer software; Engineering; Database; Programming

1. Introduction

To establish a complete and comprehensive database system, we first need to have good operation stability in the network environment, so that we can update and improve our own system software on the basis of computer network.It is based on the computer network operating.[1] environment that we can fully realize the systematic management and control of the network environment.[2] Before designing the framework of database programming technology of computer software engineering, designers need to comprehensively and systematically analyze and extract the influencing factors existing in the computer network environment, and extract and convert the value of valuable data information.[3] Secondly, it can also optimize the design of software functions, further reduce the proportion of potential security risks in the network environment, and make system structure adjustment and system optimization schemes in time.[4] Through the reasonable design of each system, the function of the software is realized by using the synergy between the systems.[5] After comprehensively analyzing the problems existing in the network operation environment and putting forward the corresponding solutions, it can also improve the information security in the database and effectively improve the reliability of software operation.[6] As shown in the Figure 1.

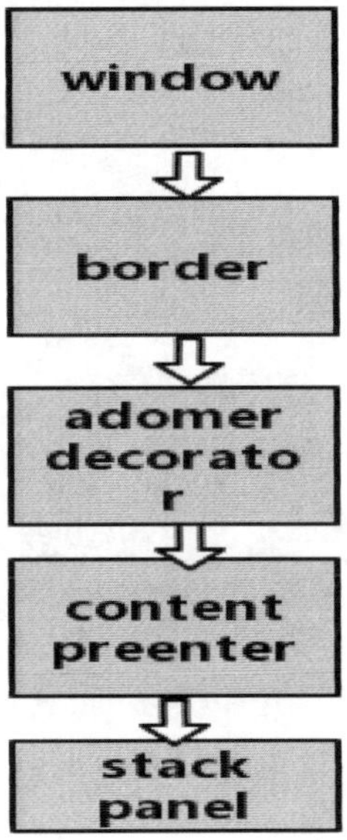

Figure 1. Database system model overview

2. Computer software database programming technology

Computer software database programming technology covers all kinds of information technology and media based on programming technology, which ensures the normal operation of information processing system and the processing of different tasks.[7] Database programming technology promotes the creation and development of computer database and plays a vital role in its basic characteristics.[8] In order to ensure that the effectiveness and performance of software engineering design meet the requirements of relevant design, it is necessary to determine its use direction and performance requirements before programming the database.[9] This requires software engineers to communicate with users and carry out actual software programming after fully understanding users' needs. [10] Firstly, after the needs of users, it is necessary to establish a basic structural framework, which can meet the needs of users, supplement the corresponding content in real time, and increase the functional content of measurement and drawing software, so as to ensure that the functional software of measurement and drawing can resist the performance of common threats in the network environment.[11] In principle, the security of data information software must be guaranteed.At the same time, you must ensure that the information in the database is kept up to date.[12]

In order to ensure the scientificity of database establishment, this can be realized through the generation of each partition, that is, different computer software plug-ins will create a relatively independent database, and the synergy between functional plug-ins is fully considered.[13] In the process of establishing the database, the utilization and role of these functions will be considered, so that when the software is used, their overall performance can be valued.In this way, the establishment of partitions can better organize the preparation of the database and effectively avoid the internal conflict in the software database.For external installation software, the generation of database is relatively easy, and can be selectively installed according to the needs of users.It also supports user-defined installations to effectively avoid potential database security problems.[14] This external installation software also has strong response ability and relative plasticity to the network environment.As shown in the Figure 2.

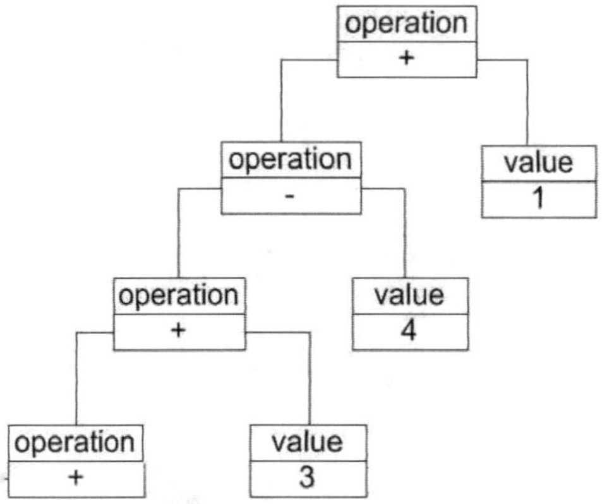

Figure 2. Scope of application of computer software engineering

In the functional analysis of customer specifications, it is necessary to clarify the purpose of software design and optimize the mode of computer system engineering.The second is the standardization of software development technology.We must focus on the software development and input of computer system design, ensure the utilization value of computer data, and improve the goal of technical application of software engineering.As long as the software technology is analyzed, we must focus on the application of computer technology and the management of computer hardware.Through the analysis of software system specifications and equipment, the development of software system is the core goal of design optimization.

3. Database programming system mode

Computer software database programming technology through the research on file encryption protection technology, it can be concluded that in the process of related technology operation.Using file encryption technology can not only protect data information, but also protect users' personal privacy information. However, once transmission leakage occurs, the security performance of the whole database protection will be reduced.Therefore, the security performance of data file encryption should be systematically analyzed and studied, so as to further strengthen the use frequency of computer software engineering.The general database file encryption technology has the following two principles: first, through the continuous encryption guarantee for user needs, the continuous password replacement and verification code input, and the file encryption processing, which can further improve the overall operation speed of database files and optimize the database solution,Fully realize the transformation from database file analysis to login password and verification code.In the process of file encryption, we should choose file encryption between different systems in combination with the selection of data information.In case of functional conflict, I can optimize the system in combination with the system framework, and optimize the data use system by pointing out the innovation of the system design system.

In the design of storage functions, specific and in-depth analysis should be carried out in combination with the use performance of data.Generally speaking, when using the software, users should first observe whether the running state of the computer software is normal, and then design and formulate the prevention scheme according to the software technology.Show the final value of database storage model design, especially in the process of software system design, the key should be to select the software function system, combined with the engineering project design, carefully design the scheme optimization, increase the value of data system design, and provide reference for project design and innovative work task form.Combined with the selection of programming technology, in the process of software engineering stability analysis, it is necessary to optimize the system operation state and resource optimization, timely analyze the restrictive problems existing in the system operation, and show the value of the system operation.In the analysis of system function problems, combined with the restrictive problems existing in the programming stage, observe the stable operation of the system, combined with the data transmission problem, use the data transmission information to optimize the data control process, improve the software function value and ensure the effectiveness of data processing.As shown in the Figure 3.

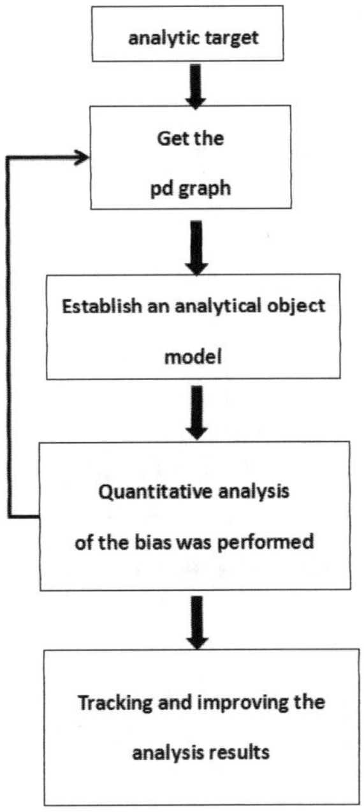

Figure 3. Schematic diagram of the computer software engineering

At the same time, in the transformation of system structure, the system resources are combined with the stability of system operation to realize the development of computer software working form, optimize the use function of software system in network environment, avoid harm to network environment and ensure the final use value of data resource database.[15] Secondly, during the operation of the system, by constructing the data storage mode, we can quickly screen the information data, optimize the data processing process and show the effectiveness of the software system combined with the information use and classification methods.Third, after the software is put into use, it should optimize the new data resources in combination with the operation situation, observe whether there are problems in the operation of the database, and observe whether there are unstable factors in the system in combination with the differential functions of various modes, so as to better optimize and improve the system project.[16]

4. Conclusion

With the continuous development and improvement of database programming technology, people have widely recognized the defects existing in computer software information, and use computer database programming technology to integrate and optimize computer software.Greatly improve the diversification and scientization of computer data resources.The research on database programming technology of computer software engineering can help the computer database technology to better transform the model, and make the database programming technology promote the continuous development and progress of China's information industry.Database programming technology directly affects the efficiency and quality of computer software, and its stability in the network environment will also have different effects.This paper mainly makes a systematic analysis and Research on the core technology of database programming technology, which can effectively avoid the data maximization of computer software technology and engineering database programming in database programming.The reliable operation of computer software in the network environment further improves the application of computer in different majors.

References

[1] Guojun Y 2020 Database programming technology based on computer software engineering *Digital World* **12** 53-4

[2] Yihui C 2020 Analysis of database programming technology based on computer software engineering *Computer Programming Skills and Maintenance* **11** 88-9 + 97

[3] Zhen L 2020 Discussion on database programming technology based on computer software engineering *Information and Communication* **09** 95-7

[4] Jianfeng Z 2020 Analysis of database programming technology strategy based on computer software engineering *Computer Products and Circulation* **04** 21 + 34

[5] Hongyan D 2020 Database programming technology based on computer software engineering *Science and Technology Innovation* **06** 90-1

[6] Siqi S 2020 Database programming technology of computer software engineering *Electronic Technology and Software Engineering* **01** 135-6

[7] Xiaoling H, Lilu F, Xiaohong Q 2020 Important role and principles of computer software database design *Electronic Technology and Software Engineering* **02** 186-7

[8] Mingwu F 2019 Analysis of database programming technology based on computer software engineering in the new era *Hubei Agricultural Mechanization* **21** 157-8
[9] Qingping Z 2019 Analysis of database programming technology based on computer software engineering *Science and Technology Wind* **23** 110
[10] Yusuo L 2019 Programming technology of computer software engineering based on database *Electronic Technology and Software Engineering* **16** 172-3
[11] Xiaoxin W 2019 Database programming technology based on computer software engineering *Electronic Testing* **01** 86-7 + 98
[12] Chengyu Z 2018 Database programming technology based on computer software engineering *Automation and Instrumentation* **06** 186-8 + 191
[13] Yunxiang S 2018 Discussion on database programming technology strategy based on computer software engineering *Computer Knowledge and Technology* **14**(16) 16-7
[14] Liping F 2017 Database programming technology based on computer software engineering *Electronic Technology and Software Engineering* **01** 164-5
[15] Xiaoxue J 2012 *Website construction of basic courses of computer software technology* (Dalian Maritime University)
[16] Aibing D 2012 *Research on the application of computer software interface technology in ship comprehensive design software* (Shanghai Jiaotong University)

Journal of Physics: Conference Series **2173** (2022) 012074 doi:10.1088/1742-6596/2173/1/012074

Analysis of Group Delay and Phase Delay Algorithms for Short Baseline Interferometry

Dong Li, Yiwen Jiao*, Feilong Mao, Zefu Gao, Pengcheng Ge and Kuanfei Sun

School of Electronic and Optical Engineering, Space Engineering University, Beijing, 101400, China

*Corresponding author's email: jiaoyiwen1985@163.com

Abstract. As a fast direction finding technology, interferometry has the advantages of simple algorithm and high sensitivity, so it is widely used in various direction finding systems. Short baseline interferometer Angle measurement is the basic principle of phase Angle measuring method, it is using the delay difference between different receiving station in the system to obtain the target azimuth information, has the advantage of high accuracy of Angle measuring, but interferometer is inevitable during the high frequency signal direction finding Angle measuring fuzzy problems exist, in order to solve this problem in general with phase delay method improve the interferometer Angle measurement accuracy, The group delay method is used to deblur. In this paper, group delay and phase delay algorithms are simulated and analyzed, and the simulation results prove the effectiveness of the proposed method.

1. Introduction

Radio direction finding methods can be divided into ultra-long wave, long wave, medium wave, short wave, ultra-short wave and microwave direction finding methods according to different frequency bands. According to the type of work: fixed direction finding, mobile lateral direction, or fixed point direction finding, vehicle-mounted, shipborne, airborne direction finding; It can also be divided into phase interferometer direction finding, specific amplitude direction finding, time difference method direction finding, spatial spectrum estimation direction finding, Doppler direction finding and other methods according to the different principle of direction finding. In the direction finding system of phase interferometer, the time delay difference between the echo signals received by different antennas is used to measure the Angle[1-4]. The accuracy of the time delay difference is directly related to the accuracy of the Angle measurement, and it has a vital influence on the accuracy of the whole direction finding system. Therefore, to ensure the accuracy and accuracy of the measurement of the delay difference is the key to the direction finding system of the phase interferometer. In addition, the ambiguity of the signal phase difference also has a key influence on the accuracy of the Angle measurement. When the RF signal is a single point frequency signal, the phase-delay method is adopted, and the system measurement accuracy is high but the measurement value is easy to be blurred. When the RF signal is a broadband signal, the group time-delay method is adopted, and the system measurement accuracy is low but the fuzzy effect is good. Based on this, the multipoint frequency signal is used as the RF signal, and the ambiguity of phase delay is extended by group time[5-7].

2. Basic working principle of short baseline interferometer

Radio waves propagate uniformly along a straight line in space and incident into the antenna array from different directions. The phase of the wireless signals received by each array element is different. By comparing the phase difference between two arrays, the azimuth information of the signals can be obtained and the direction of the incoming wave can be determined[8].

That is to say, the phase interferometer direction finding determines the target's incident direction by comparing the phase difference of the same signal received by two antennas with a certain distance. We call line arrays baseline, the longer the baseline phase interferometer direction finding precision is higher, but the longer the baseline is prone to phase ambiguity, especially when the radiation signal belongs to the high frequency signal, due to the incoming signal frequency, incidence Angle and antenna spacing and the effect of the antenna are more likely to be greater than the actual phase 2 PI and phase discriminator is 2 PI for cycle, The phase difference of the signal passing through the phase detector is constantly folded within the range of $[-\pi,\pi]$[9-10]. At this point, the phase ambiguity problem is generated, so that the Angle of the incident signal cannot be correctly measured. Aiming at the problem of phase ambiguity in phase interferometer direction finding, according to different application conditions, there are many methods to solve the phase ambiguity, such as long and short baseline method, virtual baseline method, three-dimensional baseline method, etc., each method has its applicable conditions[11].

The short-baseline interferometer consists of two channels, as shown in Figure 1. Here, we assume that the Angle measuring model of the system is a far-field model, that the front wave of the electromagnetic wave radiated by the radiation source reaching the interferometer is a plane wave, and that the RF signals received by the two antennas of the interferometer are parallel signals. As can be seen from the figure, the geometric delay of electromagnetic waves radiated by the radiation source reaching the two antennas is:

$$\tau = \frac{l\cos\theta}{c} \tag{1}$$

l is the baseline length, namely the distance between two antennas, and c is the speed of light, then the direction Angle of incoming wave is:

$$\cos\theta = \frac{c\tau}{l} \tag{2}$$

The derivative of Equation (2) is obtained as follows:

$$\delta\theta = -\frac{c\delta\tau}{l\sin\theta} \tag{3}$$

According to Equation (3), the Angle measurement accuracy of the interferometer is related to the measurement accuracy of the time delay difference, the size of the incident Angle and the length of the baseline. The higher the measurement accuracy of the time delay difference, the larger the incident Angle and the longer the baseline, the higher the Angle measurement accuracy.

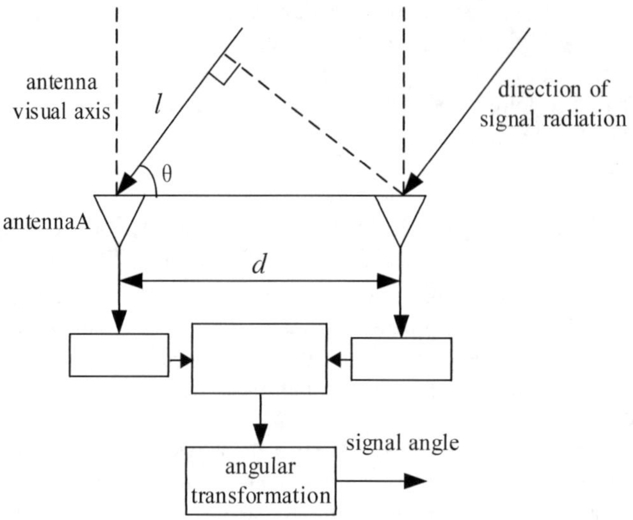

Figure 1. Schematic diagram of short baseline interferometer

3. Mathematical model of direction finding signal

Assuming that the RF signals received by the two antennas of the interferometer are multi-point frequency signals, the mathematical model of the RF signals received by the two antennas A and B is as follows:

Antenna A: $s(t) = e^{j2\pi f_c t}$

Antenna B: $s(t-\tau) = e^{j2\pi f_c(t-\tau)}$

$s(t)$ is the radio frequency signal, f_c is the carrier frequency, and τ is the delay of signal arriving at antenna B. The radio frequency signal received by the two antennas is down-converted. The local oscillator signal of down-converted is: $f(t) = e^{-j2\pi f_0 t}$, and f_0 is the frequency value of the down-converted frequency converter. After down-converted, the signals of the two antennas are expressed as:

Antenna A: $s_0(t) = s(t)\cdot f(t) = e^{j2\pi(f_c - f_0)t}$

Antenna B: $s_1(t-\tau) = s(t-\tau)\cdot f(t) = e^{j2\pi(f_c-f_0)t - j2\pi f_c\tau}$

The signal received by antenna B is compensated with time delay, and the mathematical expression of the compensated signal is shown as follows, where τ_0 is the compensation delay, and the magnitude of τ_0 is estimated from the group delay result:

$$s_1(t-\tau+\tau_0) = e^{j2\pi(f_c-f_0)(t+\tau_0)-j2\pi f_c\tau} \tag{4}$$

Since the principle of direction finding by phase interferometer is to determine the incident direction of the target by comparing the phase difference of the same signal received by two antennas with A certain distance, the signal received by antenna A and the signal received by antenna B after time delay compensation are conjugate multiplied:

$$\begin{aligned} s_0(t)\cdot s_1^*(t-\tau+\tau_0) &= j2\pi[(f_c-f_0)t-(f_c-f_0)(t-\tau_0)+f_c\tau] \\ &= j2\pi[(f_c-f_0)\tau_0+f_c\tau] \end{aligned} \tag{5}$$

The phase difference of the signal after conjugate multiplication is extracted, and the following is obtained:

$$\varphi = 2\pi(f_c - f_0)\tau_0 + 2\pi f_c \tau = 2\pi f_c(\tau - \tau_0) - 2\pi f_0 \tau_0 \tag{6}$$

From the above equation, it can be seen that $2\pi f_0 \tau_0$ is the phase difference caused by the down conversion. Therefore, this phase difference should be removed before the calculation of the time delay difference. After removing the phase difference caused by the local vibration frequency during the down conversion, we can get:

$$\varphi = 2\pi f_c(\tau - \tau_0) \tag{7}$$

Thus, the result of residual delay calculated through phase delay is as follows:

$$\tau - \tau_0 = \frac{\varphi}{2\pi f_c} \tag{8}$$

The final result of the delay difference is equal to the compensation delay plus the residual delay value:

$$\tau = \tau_0 + \frac{\varphi}{2\pi f_c} \tag{9}$$

The above equation is the time delay difference between the received signals of two array elements in direction finding of short baseline interferometer.

4. Simulation analysis and verification

Parameter setting of simulation experiment is shown in Table 1.

Table 1. Parameter setting of simulation experiment

Sampling rate	SNR	The length of the baseline	Bandwidth	Local oscillator signal frequency	Minimum frequency of RF signal	Maximum frequency of RF signal
5.08GHz	30dB	2m	24M	2GHz	2.004GHz	2.028GHz

The radio frequency signal is divided into 241 parts according to the frequency interval of 100kHz, and the signal is divided into several segments before the extraction of phase difference, and the signal points are divided into several segments, and then the Fourier transform is performed piecewise.

In the direction finding system of short-baseline interferometer, both phase delay algorithm and group delay algorithm are used to calculate the two-channel signal delay. The group delay is used to calculate a signal delay value. The phase difference of each point frequency of multi-point frequency signal is extracted in the frequency domain, and the Fourier transform of $s_0(t)$ and $s_1(t-\tau)$ can be obtained:

$$\begin{aligned} &s_0(t) \to S(\omega) \\ &s_1(t-\tau) \to S(\omega)\cdot e^{-j\omega\tau} \end{aligned} \tag{10}$$

Conjugate and multiply $S(\omega)$ and $S(\omega)\cdot e^{-j\omega\tau}$ to obtain:

$$Y(\omega) = |S(\omega)|^2 \cdot e^{j\omega\tau} \tag{11}$$

From the above equation, it can be deduced that the phase difference corresponding to each point frequency of the multi-point frequency signal is:

$$\varphi = \omega_0 \tau = angle(Y(\omega_0)) \tag{12}$$

Using phase interferometer direction finding process, the signal phase difference between two antenna receives is 2π cycle, because the phase discriminator is 2π model to measure the phase difference between the two signals, therefore measure the phase difference in the range of $[-\pi, \pi]$, if the actual phase difference is beyond this range, namely when the baseline length is greater than the wavelength, Multi-valued fuzzy appears in the phase calculation of direction - finding system.

In order to eliminate the phase ambiguity, a phase interferometer is used to solve the ambiguity. The traditional interferometer direction-finding system usually uses a combination of long and short baselines to solve the phase ambiguity. However, this algorithm requires that the length of the "short baseline" is less than half of the signal wavelength. When the signal frequency band is very high and the wavelength is very short, it is undoubtedly difficult to adopt the long baseline method in physical implementation. The virtual baseline method solves the problem of the long and short baseline method by establishing virtual array elements, but it has some limitations on the antenna array, so it cannot be applied to the circular array. Because $\omega\tau \in [-\pi, \pi]$ and therefore more frequency signal phase difference is easy to change in place, in order to solve this problem, need to get the phase difference of winding, draw the solution before and after the winding phase frequency curve (interference fringe), finally, the solution after winding interference fringes using the least squares fitting, fitting the linear intercept is more frequency in the early phase of the signal, From $\varphi = \omega\tau = 2\pi f \tau$, it can be seen that the slope of the line is $2\pi\tau$. Divide the slope by 2π to get the delay difference of the fitted multi-point frequency signal. Figure 2 shows the phase frequency curve of the signal before unwinding, Figure 3 shows the phase frequency curve of the signal after unwinding, Figure 4 shows the least square fitting of the phase difference of the multipoint frequency signal.

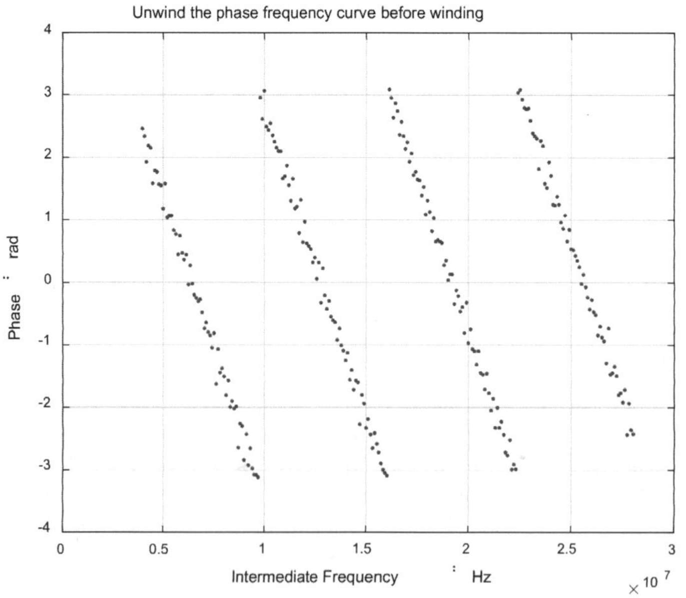

Figure 2. The phase frequency curve of the signal before unwinding

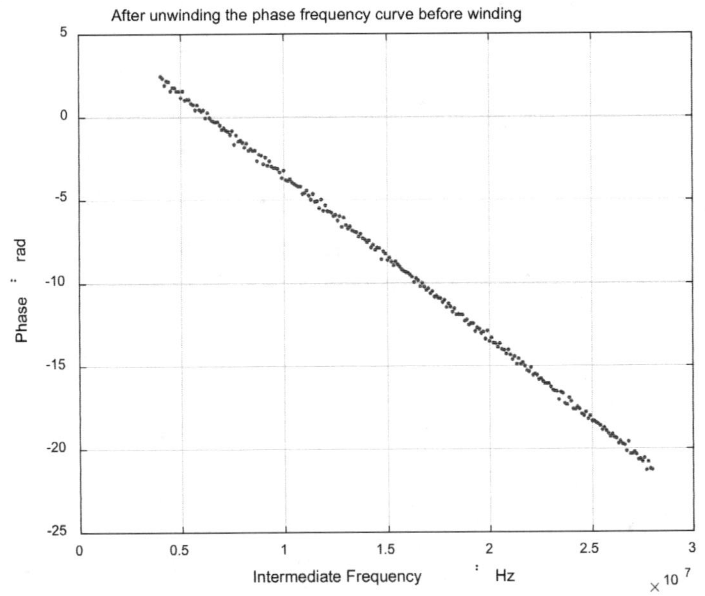

Figure 3. The phase frequency curve of the signal after unwinding

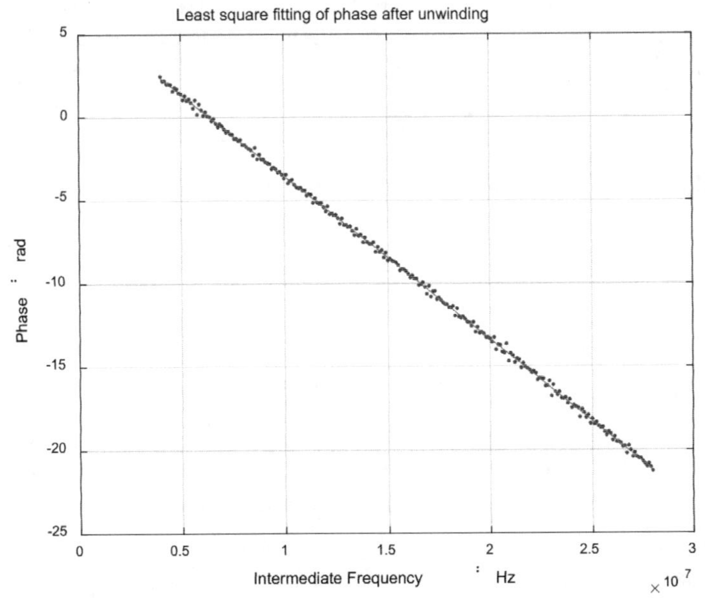

Figure 4. The least square fitting of the phase difference of the multipoint frequency signal

After the delay value is calculated by the above group delay method, the group delay result is approximately equivalent to several sampling points, and then the delay compensation is carried out for the signal with delay, that is, the sampling points calculated by the group delay are inserted or deleted. After the signal is compensated, the residual delay of the signal is calculated by the phase delay algorithm, and the final delay of the signal is equal to the compensation delay plus the residual delay. In the phase delay algorithm, after the signal is compensated for the delay, Fourier transform is performed on the signal again, and then the conjugate is multiplied to obtain the phase difference of the frequency at a certain point. From the phase difference, the time delay is obtained, which is the residual time delay. After obtaining the signal delay, it is substituted into Equation (2) to obtain the azimuth Angle of the incoming wave of the emitter when the baseline length is known.

Table 2. Time delay measurement results

Group delay	Residual time delay	Time delay	The Angle
-1.9692*10^-9	-1.0275*10^-14	-1.9685*10^-9	107.1742

In order to verify that the phase delay algorithm has higher Angle measurement accuracy than the group delay algorithm, the group delay and phase delay were simulated for 2000 times respectively, and 2000 phase results and delay results were obtained respectively. Then the variance of the phase and delay was calculated respectively, and the variance results were shown in the table below.

Table 3. The results of variance

	Phase variance	Time delay variance
Group delay	6.6076*10^-7	1.439*10^-24
The phase delay	6.6076*10^-7	4.1676*10^-27

As can be seen from the above table, if the IF signal is evaluated by the two algorithms, the delay variance of the phase delay is smaller than that of the group delay and the accuracy is higher.

5. Conclusion

In the short-baseline interferometry system, multipoint frequency signal is usually selected as the RF signal. The maximum span of the signal frequency, i.e. the bandwidth, determines the measurement accuracy, while the minimum interval of the point frequency ensures the effect of understanding the ambiguity. Therefore, the group delay algorithm is used to solve the ambiguity of phase delay when measuring the Angle. The Angle measurement accuracy of the system is not only related to the signal frequency but also to the length of the baseline. The longer the baseline, the higher the accuracy.

Acknowledgments

This work was supported by the Space Engineering University and the School of Electronic and Optical Engineering.

References

[1] Si Weijian, Cheng Wei. Research and Implementation of Rotating Interferometer Defuzying Method [J]. Journal of Projections, Missiles and Guidance,2010.

[2] He Ming, Li Yunhao, Tang Bin. A Novel Direction Finding Fuzzy Algorithm for Radar Signal Rotating Interferometer [J]. Telecommunication Technology,2013.

[3] Si Weijian, Chu Ping, Sun Shenghe. Research on Defuzzification Technology Based on Semicircle Array [J]. Systems Engineering and Electronics,2008.

[4] Yang Jing, Luo Bing, Xia Shaojun. Research on Fuzzy Algorithm for Dissolving Irregular Baseline of Phase Interferometer [J]. Aerospace Electronic Warfare,2012.

[5] Jiang Xuejin, Gao Xia, Shen Yang. A multi-baseline phase interference Instrument Design Method [J]. Electronic Information Countermeasure Technology, 2008 (04) : 39-45.

[6] He Qing, Liu Yuan 'an, Li Shulan, Yu Cuiping. Improvement of dry Analysis of Involved Antenna Array and Its Direction Finding Error [J]. Beijing Journal of University of Posts and Telecommunications,2011,34(06):121-124.

[7] Guo Binxing, Zhang Yongjie. The invention relates to a rotating phase interferometer measurement Fuzzy Algorithm of Angle Solution [J]. Projectiles, Rockets and Guidance Newspaper, 2015, 35 (02) : 185-189.

[8] Zhao Chunlei, Wang Jian. Phase stem based on virtual baseline Optimization Design of Intervention Instrument Array [J]. Ship Electronic Countermeasure, 2008. 2019 and (6) : 38-41.
[9] Zhang Jinxiu, Tao Haihong, Wang Yuan. One is based on a double basis Improved Interferometer Positioning Algorithm Based on Rotation [J]. Beijing Theory Journal of Engineering University of China,2018,38(03):320-324.
[10] Xiao Xiangbin, Du Ye, Ren Feng. Ternary array phase interferometer Fuzzy Baseline Design [J]. Information Recording Material Material, 2019, 20 (01) : 88-89.
[11] LIANG Shuanggang. Research on Positioning Technology Based on Phase Interferometer Direction Finding Algorithm [D]. Northwestern Polytechnical University,2006.

A Hybrid Optimization Method for Manufacturing Cell Scheduling with Random Interruptions Based on Improved Wolf Pack Algorithm and Simulation

Zian Zhao and Hong Zhou*

School of Economics and Management, Beihang University, Beijing 100191, China

*corresponding author's e-mail: h_zhou@buaa.edu.cn

Abstract. Cell scheduling is an important issue in the cell manufacturing system. Considering the problems of machine interruption and energy consumption, this paper establishes a stochastic optimization model to minimize the makespan as well as the cost of energy consumption during machine idling and the interruption cost. A hybrid algorithm, based on the wolf pack algorithm (WPA) and simulation, is proposed to solve the problem. In order to deal with the problems of slow convergence speed and easily falling into a local optimum which often happens for the standard WPA, the iterative local search strategy, Cauchy mutation strategy, and Gaussian disturbance are introduced into the search process of WPA to improve its performance. The solution with stochastic parameters is evaluated via simulation. Numerical experiments demonstrate that the proposed hybrid algorithm shows a good performance in both solution quality and convergence speed, and a satisfactory solution to the problem can be reached within a reasonable number of iterations.

1. Introduction

The cellular manufacturing system combines the advantages of just-in-time manufacturing, lean manufacturing, and group technology, which has gradually becomes an important organization form in modern manufacturing systems[1].

Cell scheduling is one of the most important management problems in the cellular manufacturing system, and has received extensive attentions and investigations. For example, Feng et al.[2] establish a mixed integer programming model for cell scheduling and use a genetic algorithm to solve it, while Sinaki et al.[3] solve the established weighted multi-objective cell scheduling model with Lingo. For a long time, research on cell manufacturing systems has usually focused on the performance indicators related to efficiency and output. But recently, energy related issues gradually come into the spotlight. According to estimates by the International Energy Agency (IEA)[4], the industrial sector will account for the largest share of energy consumption (nearly 50%) by 2040, and energy consumption in manufacturing has become an increasingly important issue. Therefore, it is necessary to fully consider the issue of energy consumption to achieve sustainable production. Iqbal et al.[5] establish an energy-saving model that minimizes the idle time of machines, and solve it through a two-stage method to obtain the best scheduling arrangement. Liu et al.[6] find that turning off and reopening the machine will consume more energy, and sustainable development can be achieved by reasonably turning off the machine with a long idle time. It is not unusual in the manufacturing that processing is interrupted and delayed due to machine failures. The handling of machine failures is a difficult problem in manufacturing process. Li et al.[7] develop a scheduling optimization method that simultaneously considers the arrival of dynamic tasks and machine failures, including machine idling or standby. Sarker et al.[8] propose a hybrid evolutionary algorithm to solve the job scheduling problem considering machine fault maintenance. Based on the above considerations, a cell scheduling optimization problem is studied in this paper, which minimizes the cost of energy consumption during machine idleness, maintenance cost, , as well as the makespan.

The rest of this paper is organized as follows. First, section 2 provides a stochastic optimization model for the problem under study. Then in section 3, a hybrid optimization method is proposed that integrates an improved wolf pack algorithm with a discrete event simulation procedure. Numerical experiments are conducted in section 4 to demonstrate the efficacy of the model and the algorithm. At the end of the paper, the main conclusions are summarized in section 5.

2. Problem formulation

2.1. Problem description

The problem studied in this paper focuses on the job scheduling in a cellular manufacturing system (CMS), i.e. to determine the order of the jobs being processed on the machines in different cells. Multiple objectives are considered. As a common performance index for cell scheduling problems, the makespan is first chosen to be minimized. Secondly, the problem assumes that the operations on machines are mainly machining, thus the tools are often worn out and need to be replaced during the machining process, inevitably resulting in the interruptions of the machining process. Hence the cost of machine interruptions is taken into account. In addition, the cost for energy consumption during machine idling is also considered in this paper. Thus the total cost is chosen as another objective to be minimized.

Assuming that the cell configuration is determined in advance, the intra-cell movement time is ignored; each machine can process at most one operation at a time, each operation can be processed by at most one machine, and the machine setup times are included in the processing times; the time interval of machine interruption and the time of failure follow a negative exponential distribution; the machine will consume energy during idle time, and the interruption process will incur a cost.

2.2. Model

The notations used in this article are summarized as follows.

Parameters:

n : Number of jobs;

m : Number of machines;

c : Number of cells;

j : Index of jobs $j \in J = \{1,2,...,n\}$;

k : Index of machines $k \in M = \{1,2,...,m\}$;

u : Index of cells $u \in U = \{1,2,...,c\}$;

n_j : Number of operations for job j ;

O_{ij} : The i-th operation of job j $i \in I_j = \{1,2,...,n_j\}$;

p_{ijk} : Processing time of operation O_{ij} on machine k ;

d_j : Due time of job j ;

$t_{uu'}$: Inter-cell movement time for jobs moving from cell u to u';

C_1 : Energy consumption cost per unit time when the machine is idle;

BR : Repairing cost per unit time of machine breakdown;

$$y_{ku} = \begin{cases} 1, \text{if machine } k \text{ belongs to cell } u \\ 0, \qquad \text{otherwise} \end{cases} ;$$

s_{ij} : Starting time of operation O_{ij} ;

C_{ij} : Completion time of operation O_{ij} ;

$$F_{ijk} = \begin{cases} 1, \text{If operation } O_{ij} \text{ is interrupted on machine } k, \sum_{j\in J}\sum_{i\in n_j} P_{ijk} \times x_{ijk} \geq \theta_k \\ 0, \quad \text{otherwise} \end{cases};$$

θ_k : The time between failures of machine k, which follows the negative exponential distribution;

ω_k : The failure time of machine k, which follows the negative exponential distribution ;

λ_k : Expected time for machine k failure;

Variables:

$$x_{ijk} = \begin{cases} 1, \text{if operation } O_{ij} \text{ is assigned to machine } k \\ 0, \quad \text{otherwise} \end{cases};$$

$$z_{jj'k} = \begin{cases} 1, \text{if job } j' \text{is processed immediatly after job } j \text{ on machine } k \\ 0, \quad \text{otherwise} \end{cases};$$

The two objectives of the model, noted as f_1 and f_2, are to minimize the makespan as well as the total cost including the cost of interruptions due to machine failure and the cost of energy consumption during machine idling.

$$f_1 = \max_j C_{n_j j} \tag{1}$$

$$f_2 = \sum_{k\in M}\left[\sum_{\substack{j,j\in J \\ j=j'}} \left(\sum_{i\in I_j} s_{ij'} \times x_{ij'k} - \sum_{i=I_j} C_{ij} \times x_{ijk}\right) \times z_{jj'k}\right] \times C_1 + BR \times \sum_{k\in M}\sum_{j\in J}\sum_{i\in I_j} F_{ijk} \times \lambda_k \tag{2}$$

The first part of the objective function (2) represents the cost of energy consumption, while the second part represents the cost of machine interruption.

To simplify the calculation, this paper adopts the form of weighted sum to handle the multiple objectives. Considering the uncertain failure time and random interruptions, the expected value of the objective function will be optimized. Thus the model can be represented as follows.

$$\min \quad E(f) = w_1 \times E(f_1) + w_2 \times E(f_2) \tag{3}$$

subject to

$$\sum_{k\in M} x_{ijk} = 1, \forall i \in I_j, j \in J \tag{4}$$

$$s_{i'j'} \geq (s_{ij} + p_{ijk}) \times z_{jj'k} \times x_{ijk} \times x_{i'j'k}, \forall k \in M, \{i,i'\} \in I_j, \{j,j'\} \in J \tag{5}$$

$$C_{ij} = s_{ij} + \sum_{k\in M} p_{ijk} \times x_{ijk} + \sum_{k\in M} F_{ijk} \times \omega_k, \forall i \in I_j, j \in J \tag{6}$$

$$s_{(i+1)j} - C_{ij} \geq x_{(i+1)jk} x_{ijk'} y_{ku} y_{k'u'} t_{uu'}, \forall i \in I_j, j \in J, \{k,k'\} \in M, k \neq k', \{u,u'\} \in U \tag{7}$$

$$C_{n_j j} \leq d_j, \forall j \in J \tag{8}$$

$$\sum_{i\in I_j} x_{ijk} + \sum_{i\in I_j} x_{ij'k} \geq z_{jj'k} \times 2, \forall k \in M, \{j,j'\} \in J \tag{9}$$

$$x_{ijk}, z_{jj'k} \in \{0,1\}, \forall i \in I_j, \{j, j'\} \in J, k \in M \tag{10}$$

Formula (4) means that each operation can only be assigned to one machine. Formula (5) means that on any machine, an operation can be started only after the previous operation on this machine has completed. Formula (6) indicates that the total time of a job includes process time, inter-cell movement time, and machine repair time. Equation (7) implies that the inter-cell movement time is considered when adjacent operations are performed on the machines belonging to different cells. Formula (8) ensures that the job are completed within the due date. Constraint (9) implies that $z_{jj'k}$ is 1 only when job j and job j' are processed on the same machine k.

3. Hybrid optimization algorithm

A hybrid optimization algorithm, noted as HWPA-Sim, is proposed in this section, which is based on the integration of an improved wolf pack algorithm and simulation. Wolf Pack Algorithm (WPA) solves the problem by simulating the process of prey hunting by wolves in the nature [9]. The wolf hunt for prey together through three intelligent behaviours: scouting behaviour, summoning behaviour, and besieging behaviour. After each generation of besieging behaviour, the wolves will be updated through the survival of the fittest mechanism.

The standard wolf pack algorithm needs a large number of iterations in search for a satisfactory solution to some problems, which results in a low efficiency. To overcome the difficulty, some efforts have been made recently, such as adding improved search strategies and leader strategies [10-12]. Similar to most meta-heuristic algorithms, WPA also faces the problem of being prone to falling into a local optimum. In response to these problems, this paper proposes an improving strategy for WPA, which introduces the iterative local search, Cauchy mutation strategy, and Gaussian disturbance into the wolf pack algorithm. In addition, the logarithmic movement factors are employed to replace random factors, and the adaptive step length strategy, as proposed by Guo [13], is adopted. Due to the uncertainty in the interruption and failure time, the objective function is evaluated through a discrete event simulation process.

3.1. Encoding

A three-layer encoding scheme, as proposed by Feng et al. [2], is adopted in this paper, which can effectively express the relationship between cell, machines, and job sequences.. The value in the first layer represents the job index j corresponding to the operation O_{ij}, and the second layer represents the machine performing the operation O_{ij}, and the third layer specifies the cell where operation O_{ij} is processed.

Operation O_{ij}	5	2	5	6	3	4
Machine k	4	5	3	1	4	4
Cell u	2	1	1	1	2	2

Figure 1. The three-layer individual coding.

3.2. Initialization

The algorithm is initialized by generating a set of individuals. The first layer of the individual is obtained by randomly generating the job sequence. Then the corresponding available machine is found for each operation to obtain the second layer of the individual. Finally, find out the index of the cell where the machine is located to fill in the third layer.

3.3. Improving strategies for WPA

3.3.1. Mobility factor. In the standard wolf pack algorithm, the uncertain change of the random factor (rand) in each intelligent behaviour will increase the uncertainty of the algorithm's mid-to-late evolution, thus the speed of convergence and the accuracy of the final solution are very unstable. Based on the experimental results of Duan [14], this paper chooses the logarithmic mobility factor, defined as formula (11), to replace the random factor.

$$\sigma = \log_{\text{genmax}} \text{gen} \qquad gen = 1, 2, \ldots, gen\max \tag{11}$$

In the formula, $genmax$ represents the largest iterations, and gen represents the current iterations.

3.3.2. Adaptive step length. In order to speed up the convergence speed of the algorithm, this paper adopts the adaptive step length proposed by Guo [13], so that each wolf's advancement length can be dynamically adjusted according to the current position of the lead wolf X_{lead} and its own position x_i. The step length is defined as

$$step_{\bullet} = \log_{gen\max} gen \times \left\| x_i - X_{lead} \right\|_2 \tag{12}$$

By adopting the adaptive step length, the three intelligent behaviour movements of scouting behaviour, summoning behaviour, and besieging behaviour can be defined as

$$x_i^p = x_i + \sin(2 \times \pi \times \frac{p}{h}) \times step_a \times \log_{gen\max} gen \times \left\| x_i - X_{lead} \right\|_2 \tag{13}$$

$$x_i(k+1) = x_i(k) + \frac{(X_{lead} - x_i(k))}{\left| X_{lead} - x_i(k) \right|} \times step_b \times \log_{gen\max} gen \times \left\| x_i - X_{lead} \right\|_2 \tag{14}$$

$$x_i(k+1) = x_i(k) + (2 \times \lambda - 1) \times step_c \times \left| G_k - x_i(k) \right| \times \log_{gen\max} gen \times \left\| x_i - X_{lead} \right\|_2 \tag{15}$$

3.3.3. Iterative local search. One of the shortcomings of WPA is that it is easy to fall into the local optimum. To avoid or mitigate this problem, an iterative local search procedure is added in the standard WPA after each wolf scouting behaviour. With this procedure, an intermediate state is obtained by perturbing the local optimum, and a new optimum can be reached through the search from the intermediate state. Thus the algorithm can quickly jump out of the local optimum. The steps of the procedure are as follows.

Step 1. The best solution x_i^{p*} is perturbed at the end of each wolf scouting behaviour, and the new state x_i^{p**} is obtained by the update $x_i^{p**} = x_i^{p*} \times rand$.

Step 2. if $f(x_i^{p^{**}}) < f(x_i^{p^{*}})$, then $x_i^{p^{*}} = x_i^{p^{**}}$, $f(x_i^{p^{*}}) = f(x_i^{p^{**}})$.

Step 3. Repeat the above steps until the boundary condition is reached; otherwise, go to Step 4.

Step 4. Output the current optimal solution and end the iterative local search procedure.

3.3.4. Cauchy mutation strategy. In an effort to avoid the scouting behaviour falling into the local optimum, the Cauchy mutation strategy is introduced [15], i.e. the individuals without improvement after the current scouting is given the Cauchy disturbance. Specifically, if no improvement is found in scouting when the wolf i moves along h directions, it will be mutated with a certain probability P_{ed}. In this paper, the individual is updated according to Eq. (16) after Cauchy mutation.

$$x_{next}^*(t) = x_i^*(t) \times (1 + cauchy(0,1)) \tag{16}$$

where $x_{next}^*(t)$ is the new individual obtained after Cauchy perturbation, and Cauc$hy(0,1)$ is the Cauchy operator.

3.3.5. Gaussian disturbance. The traditional wolf pack algorithm uses the position of the lead wolf as the prey position during the besieging behaviour, which results in a deviation and affects the algorithm's ability to explore the optimal solution. This paper employs the idea of the whale pack algorithm [16] to disturb the lead wolf with a certain probability P_{ga} in the besieging behaviour. The position of the prey is obtained with a small disturbance of lead wolf, which improves the algorithm's ability to explore the optimum. Thus the position of the prey is determined by formulas (17) (18).

$$NG_{best}^{t} = G^{t} \times (1 + Gaussian(\mu, \sigma)) \tag{17}$$

$$G^{t+1} = \begin{cases} NG_{best}^{t}, f(NG_{best}^{t}) < f(G^{t}) \\ G^{t}, \quad otherwise \end{cases} \tag{18}$$

where G^{t} and NG_{best}^{t} are the position of the lead wolf and the prey obtained after Gaussian perturbation in the current iteration, respectively.

3.4. HWPA-SIM algorithm structure

In summary, the structure of the hybrid algorithm HWPA-SIM proposed in this paper is shown in Figure 2.

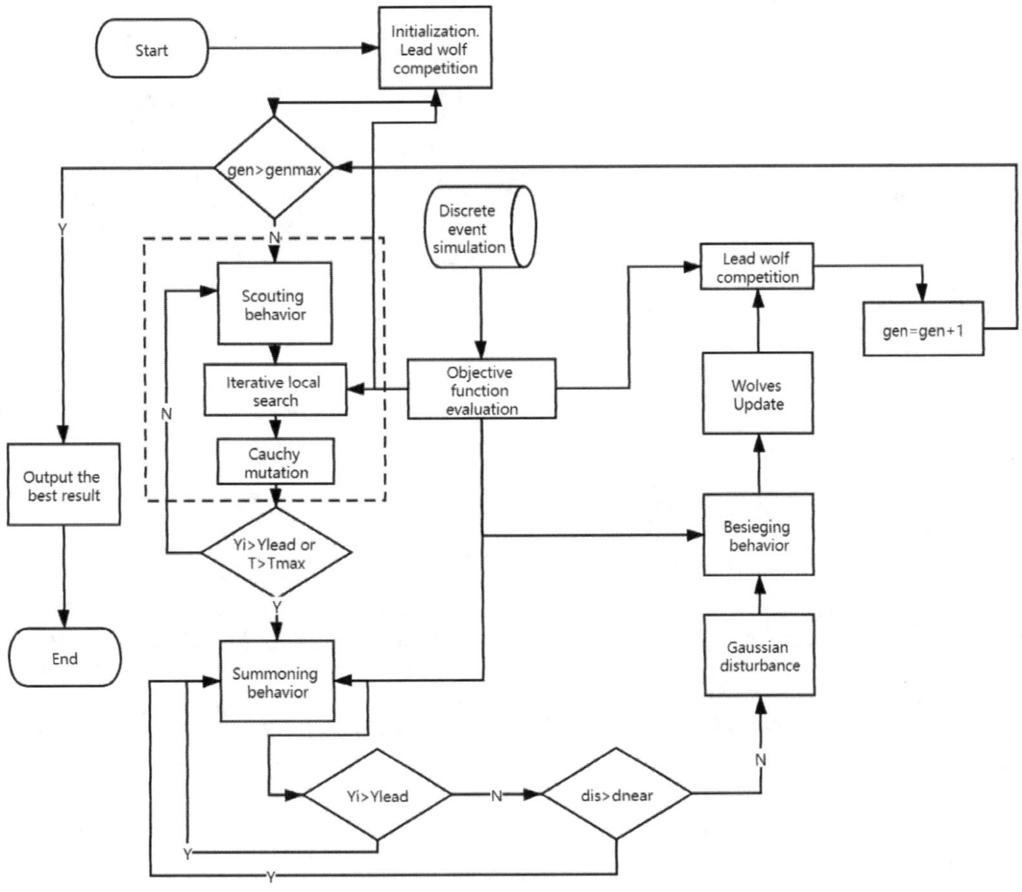

Figure 2. Flow chart of HWPA-SIM.

4. Simulation experiment and results

The simulation experiments are carried out with MATLAB on the computer with 2 GHz Inter® Core™ i7 CPU and 8 GB RAM. The instance includes 6 jobs, 10 machines, and 2 cells. The machine sets of the two cells are $U_1=\{M_1,M_3,M_5,M_7,M_9\}$ and $U_2=\{M_2,M_4,M_6,M_8,M_{10}\}$, respectively. Each job requires 6 operations, and some operations can be completed on different machines in the cell. The process plans for the jobs are shown in Table 1, including the alternative machines and the corresponding processing times for each operation. The process time is randomly generated from the interval [1,10]. The time between failures of machine follows the negative exponential distribution with the parameter $\lambda=1/5$; the failure time of machine follows the negative exponential distribution with parameter $\mu=1/3$; the weights w_1 and w_2 are set 0.5. The unit time cost is 1 for energy consumption, and also 1 for machine failure. The parameter settings of the improved wolf pack algorithm are shown in Table 2.

Table 1. Process plan of jobs.

Job	Operations on alternative machines [machine1(process time1)/machine2(process time2)/...]					
	1	2	3	4	5	6
1	[8(9)]	[4(3)]	[1(6)]	[10(4)/6(4)]	[10(5)/7(4)]	[6(8)]
2	[3(4)]	[7(5)/3(6)]	[7(5)]	[8(8)/6(6)]	[6(3)]	[1(3)/2(5)]
3	[9(8)]	[8(7)/7(8)]	[3(3)]	[9(3)/8(4)]	[7(7)/6(6)]	[4(4)]
4	[3(5)]	[7(4)]	[8(4)/2(6)]	[6(5)]	[9(1)/8(5)]	[4(9)/6(8)]
5	[5(7)/7(5)]	[6(3)]	[7(7)/9(5)]	[7(6)]	[6(3)]	[3(5)/5(6)]
6	[6(8)/1(6)]	[3(3)]	[4(5)/1(9)]	[8(3)]	[9(1)]	[8(4)/10(6)]

Table 2. Parameter values of HWPA-SIM.

HWPA-SIM Parameters	Value
Number of wolves	100
The maximum number of iterations	80
Maximum number of scouts	30
Scout wolves scale factor	4
Distance determination factor	800
Step factor	1000
Elimination factor	10
Search method upper bound	15
Search method lower bound	8
Cauchy mutation rate P_{ed}	0.5
Gaussian disturbance rate P_{ga}	0.5

Figure 3 shows the legend of the mean value and the best value of the objective function for 10 runs of the HWPA-SIM. It is easy to see that the best value of the objective function drops from the initial 129.5990 to the minimum value of 92.4979 from the 12th generation, which shows a satisfied convergence performance. More specifically, the energy consumption is reduced by 34.06% and the makespan is reduced by 15.51%, showing that this model can greatly reduce both the energy consumption during machine idle time and the make span. Hence the hybrid algorithm performs well in both calculation efficiency and effectiveness.

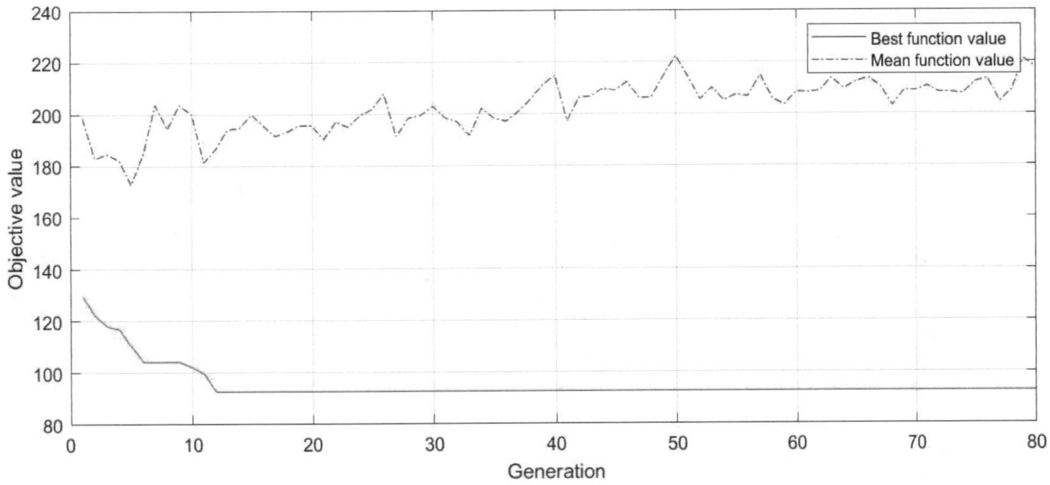

Figure 3. The mean value and optimal value of the objective function solved by HWPA-SIM.

Figure 4 shows the gantt chart of the best scheduling plan with the makespan of 51.0073, where each bar represents the operation of job *j* on machine *i*.

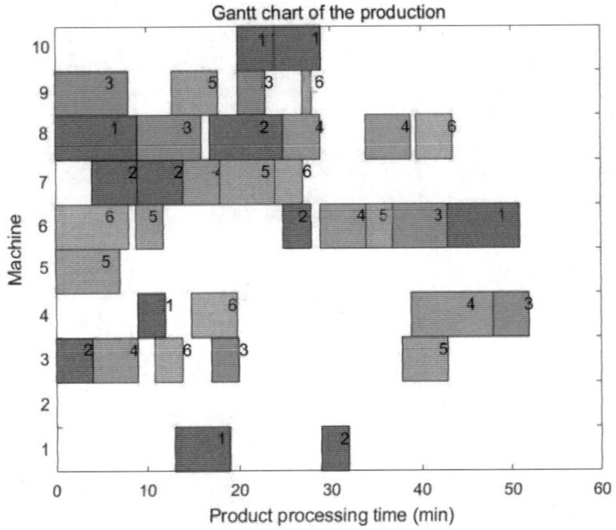

Figure 4. Scheduling plan of the problem using HWPA-SIM.

5. Conclusions

In this paper, a stochastic optimization model is established for the cell scheduling problem whose objective is minimizing the makespan as well as the total cost of machine interruption and energy consumption during machine idle time. Furthermore, a hybrid algorithm is proposed, which takes the improved WPA as the main optimization engine. Iterative local search, Cauchy mutation strategy, and Gaussian disturbance are introduced into the standard WPA to enhance its optimization ability, while the logarithmic movement factor and adaptive step length are adopted to speed up the convergence speed. Furthermore, the discrete event simulation procedure is combined with the optimization process to evaluate the objective function. Simulation experiments show that the proposed method can obtain a satisfactory scheduling which achieves the effective trade-off among the makespan, the cost of energy consumption, and the cost of machine interruption.

Acknowledgments
This work is supported by the National Key R&D Program of China (Grant No. 2018YFB1701600).

References
[1] Wu X, Chu C H, Wang Y, et al. 2007. A genetic algorithm for cellular manufacturing design and layout, *Eur. J. Oper. Res.*, **181**, 156-167.
[2] Feng Y, Li G, Sethi S P S 2018. A three-layer chromosome genetic algorithm for multi-cell scheduling with flexible routes and machine sharing, *Int. J. Prod. Econ.*, **196**, 269-283.
[3] Sinaki R Y, Sadeghi A, Suer G, et al. 2019. A Weighted Multi-Objective Mathematical Model for Cell Scheduling and Environmentally Sustainable Supply Chain Network, *Procedia Manuf.*, **39**, 1559-66.
[4] International Energy Agency (IEA), 2017. International Energy Outlook 2017. Retrieved from. https://www.eia.gov/pressroom/presentations/mead_91417. pdf.
[5] Iqbal N, Aziz M H, Jahanzaib M, et al. 2016. Integration of cell formation and job sequencing to minimize energy consumption with minimum make DATA SET, *Proc. Inst. Mech. Eng., Part B*, **231**(14), 2636-51.
[6] Liu Y, Dong H, Lohse N, et al. 2016. A multi-objective genetic algorithm for optimization of energy consumption and shop floor production performance, *Int. J. Prod. Econ.*, 179, 259-272.
[7] Li Y, He Y, Wang Y, et al. 2020. An optimization method for energy-conscious production in flexible machining job shops with dynamic job arrivals and machine breakdowns, *J. Cleaner Prod.*, **254**, 120009.
[8] Sarker R, Omar M, Hasan K S M, et al. 2013. Hybrid Evolutionary Algorithm for job scheduling under machine maintenance, *Appl. Soft Comput.*, 1**3**(3), 1440-47.
[9] Wu H S, Zhang F M, Wu L S 2013. New swarm intelligence algorithm—wolf pack algorithm, *Syst. Eng. Electron.*, 1**3**(3), 1440-47.
[10]Li G L, Wei Z H, Xu L 2015. Wolf pack algorithm based on modified search strategy, *J. Comput. Appl.*, **35**(6), 1633-36, 87.
[11]Zhou Q, Zhou Y Q 2013. Wolf colony search algorithm based on leader strategy, *Appl. Res. Comput.*, **30**(9), 2629-32.
[12]Chen X Y, Tang C J, Wang J, et al. 2018. Improved Wolf Pack Algorithm Based on Differential Evolution Elite Set, *IEICE Trans. Inf. Syst.*, **E101D** (7), 946-49.
[13]Guo L T 2018. Improved wolf pack algorithm based on adaptive step length and adjustable scouting direction, *J. Zhejiang University (Science Edition)*, **45**(3), 284-293.
[14]Duan Y M 2014. Research on bee colony algorithm, *XiDian University*.
[15]Sapre S, Mini S 2019. Opposition-based moth flame optimization with Cauchy mutation and evolutionary boundary constraint handling for global optimization, *Soft Comput.*, **23**, 6023-41.
[16]Mirjalili S, Lewis A 2016. The Whale Optimization Algorithm, *Adv. Eng. Softw.*, **95**(5), 51-67.

Optimization of the Production Preparation Process by Simulation Modeling Methods

G A Pollack, A G Palei* and O S Buslaeva

School of economyics and management ,South Ural State University (National Resarch University), 80 Lenin avenue, Chelyabinsk 454080, Russia

*corresponding author's e-mail: paleiag@susu.ru

Abstract. The article describes a stochastic multiphase multichannel queuing system with an unlimited input capacity for the coordination of commercial business deals in an industrial enterprise. To build the model, it is proposed a conceptual map was used. The model allows experiments in the virtual world to optimize the parameters of a real process. As a result of experiments on the model, "bottlenecks" were identified in the process of negotiating deals. Recommendations are proposed to reduce the coordination time, which allows increasing the number of agreed business deals by 3 times.

1. Introduction

The technological stage coordination of the business deals at an industrial enterprise consists of two steps: obtaining primary information from the customer and approval of individual items and the order as a whole by the services of the enterprise. In the approval process, information on the possibility and timing of delivery, technical requirements, product range, number of products supplied, terms of delivery and payment, as well as price is specified.

For Chelyabinsk Pipe-Rolling Plant JSC known as ChelPipe, the actual approval time is on average 10-15 days, which is 2 times longer than the deadlines set by the regulations. This time is 35% of the order passage time. Such delays have a significant impact on the total time of the order execution, which confirms the relevance of the problem of studying and optimizing the ordering process for different values of incoming parameters.

The problem of optimizing the order approval process is relevant since, at this stage, there is a close interaction with a client and clarification of all conditions. The speed of receiving all necessary information by the client depends on the quality of the plant services.

It is necessary to establish the reasons that have a significant impact on the time of approval of the business deals, and to find the optimal solution to eliminate them.

The process is described by a complex, weakly structured system.

In such systems, when making management decisions, situations arise when the reasons and possible ways to correct an undesirable situation are unclear, and the consequences of an ineffective management decision are difficult to assess. One of the ways to study processes in poorly structured systems is to use simulation models.

From the point of view of computer implementation, simulation is a comprehensive method for the study of complex systems, which includes the construction of conceptual, mathematical and software models. In other words, this is the construction of a digital analogue of a real process, with the help of which targeted virtual simulation experiments of various natures can be repeatedly carried out in virtual space.

The paper proposes the use of cognitive maps to build a simulation model of a complex weakly structured system.

The cognitive map allows you to identify the main elements of the system, their characteristics and the interaction between the elements [4,5].

As a result of the analysis the business deals approval process a cognitive model was built. It was revealed that 6 plant services are involved in the process of coordinating the business deals: managers, production loading department (PLD), technical department (TD), production planning department (PPD), price department (PD), the management of the purchase of metal (MPM).

The cognitive maps in the BPMN (Business Process Model and Notation) are shown in figure 1. The cognitive model is a subjective model of a situation (process), reflecting the expert's knowledge about the laws of its development. The process of coordinating the position of each individual business deals is dynamic and not predetermined, therefore it is proposed to use a diagram in which the relationships between the objects are probabilistic in nature.

The probability of a business deals position returning to another department are calculated based on real data and expert estimates.

The connections between the process objects (figure 1) show subjective assessments of the probability of transferring a business deals position between departments and/or the customer.

The values of the probabilities were either obtained by expert means or calculated based on a photograph of the working day of managers involved in the execution of the business deals.

2. Implementation of the model

In the context of computer implementation, simulation modeling is an integrated technique for studying complex systems, which includes the construction of conceptual, mathematical and software models. In other words, this is the construction of a digital analog of a real technological process, with the help of which it is possible to carry out targeted simulation experiments of a different nature in a virtual space.

As a result of the analysis of the real business deals process, a conceptual model was built at the enterprise (Figure 1). The model made it possible to identify the main elements of the simulated system, their characteristics and the interaction between the elements.

The process consists of 14 tasks and includes the work of seven blocks. Upon receipt of an order from the client, each business deals item is agreed upon to determine the possibility of production. Moreover, each position can be repeatedly sent between departments for adjustment and re-coordination.

The process consists of 14 tasks and includes the work of seven blocks. When the order is received from the client, the manager draws up the business deals and sends it for approval to the Production Loading Department (PLD), which checks the correctness/sufficiency of the data. In case of any deficiencies, the order item is corrected. If there are no shortcomings, the position is approved and the PLD sends the order to the Technical Department (TD) to determine the possibility of its production.

If the enterprise has the opportunity to manufacture the product and the price is determined, the order is returned to the PLD again to coordinate the position (business deals to the "Approved" status).

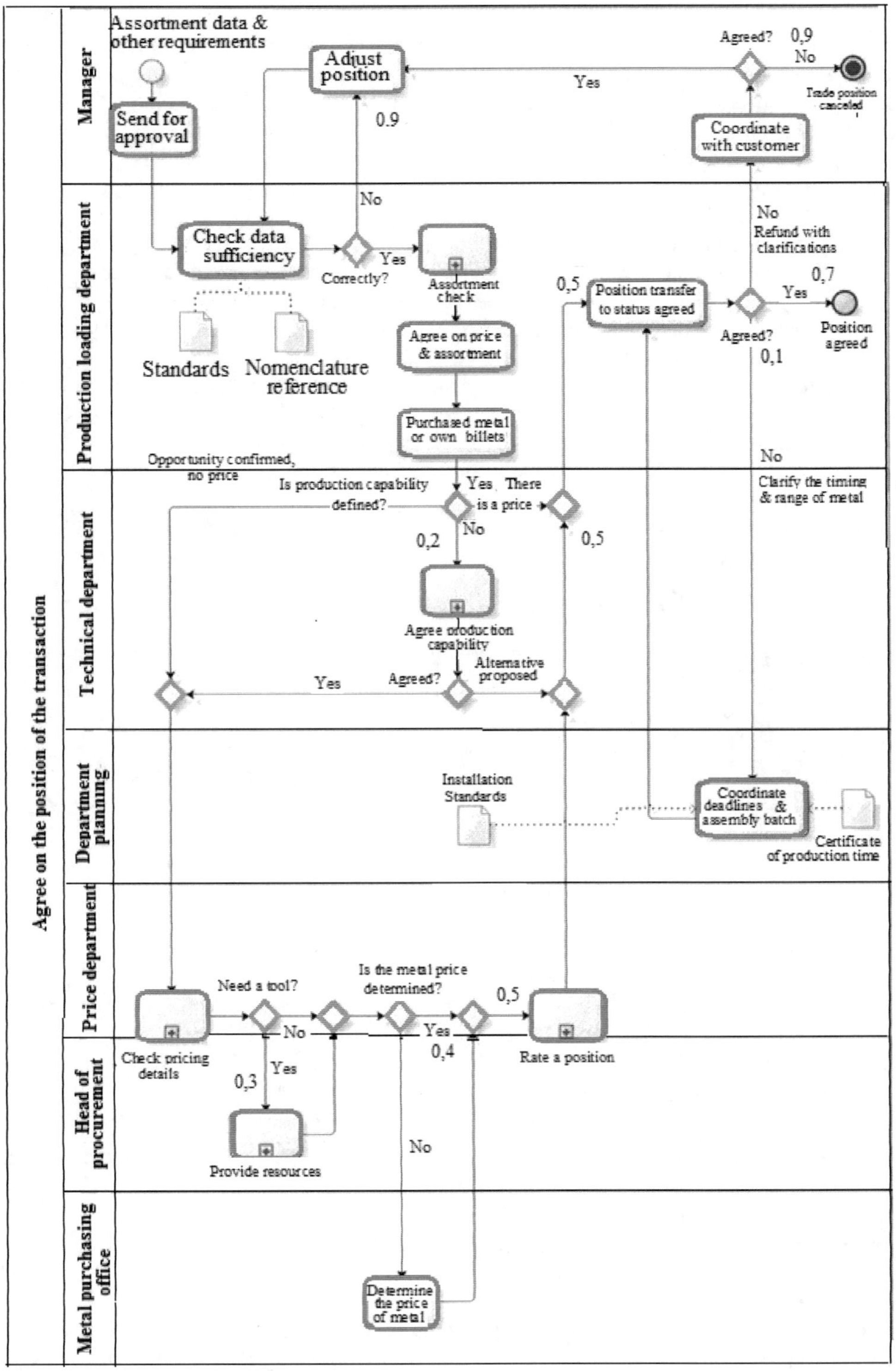

Figure 1. The cognitive map in the BPMN

If the price is not available, the order is sent to the Price Department (PD) to check the availability of all parameters necessary for quotation.If an additional tool is needed for production, the PD sends an order to the head of the Management of Purchasing raw materials to provide the necessary resources.If it is necessary to determine the metal price, the OC sends the order to the Metal Purchasing Department (MPD). When all necessary data are received, the PD evaluates the position and sends it to the PLD node to agree on the position.

If the position is agreed and has a reference price (RP), the approval process is completed. If it is necessary to clarify the metal for the order and its timing, it is sent to the Production Planning Department (PPD) to agree on the timing, installation, metal, and production batch. If it is necessary to coordinate some specifications with the customer, the order is returned to the Manager node.

In case of a negative response, the order item is canceled. Otherwise, it is sent for correction and re-approval circle. Thus, the order can be returned to the same node several times.

Figure 1 shows that the order approval process is a multiphase multichannel queuing system of closed type with unlimited input capacity [1,2,3].

The simulation model (Figure 2) is built using the AnyLogic 8.3 modeling tool. When building the model, the discrete-event library was used, which assumed the representation of the simulated system as a process, that is, the sequence of operations performed with the agents.

The model in AnyLogic is set graphically as a sequence of blocks, each of which is a separate operation.

Figure 1 shows that the order approval process is a multiphase multichannel queuing system of closed type with unlimited input capacity.

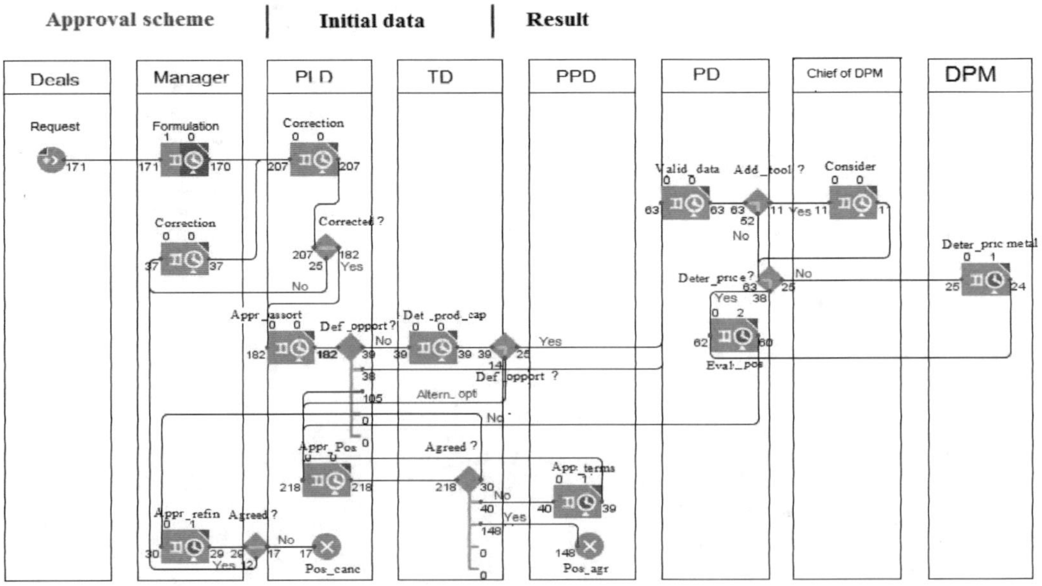

Figure 2. The simulation model

The adequacy of the model was tested on a simulation experiment with data determined by the regulations. During the experiment, the model was trained by selecting the values of the probabilities of transferring transaction positions between departments to obtain an average approval time close to the corresponding regulatory parameters.

During the simulation experiment (figure 3) with the actual data, it was determined that the operation "Check the correctness of filling / sufficiency of data" is performed repeatedly and requires the greatest amount of time to complete, which violates the time schedule.

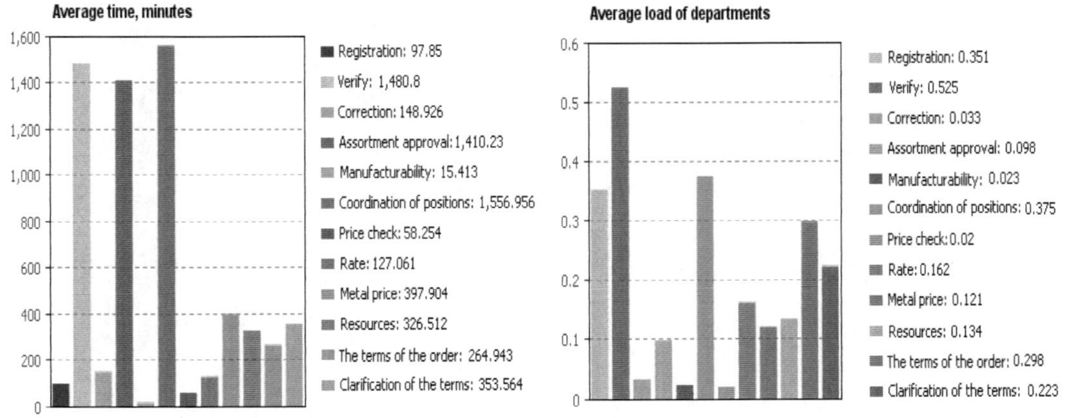

Figure 3. The results of the real data experiment

To conduct an experiment on real data, work time study was carried out in each department. It was revealed that the processing time for order items is close to the exponential distribution law.

This operation is performed by the production loading department (PLD), which has maximum workload and a large number of unprocessed, expired transactions. We can conclude that this unit is the "bottleneck" of the whole technological process of application execution. The measures on automation of these operations were formulated and proposed.

The obtained optimal estimates were used to conduct an experiment to maximize the number of approved requests.

A comparison of the results of the experiments is shown in Table 1.

Table 1. A comparison of the results

Options	Time-limit	Timing	Optimiza-tion
Average approval time, (days)	5	10	6
Number of Agreed Business Deals	148	32	84

3. Conclusion

Thus, the implementation of the optimal variant of the simulated process organization allowed

- 1. The proposed approach to modeling and analysis of a poorly structured system allows us to develop its formal simulation model, which can be used to optimize the organization of the process under study.
- 2. The implementation of the optimal variant of the process under study (coordination of the business deals at an industrial enterprise) will increase the number of agreed transactions by 3 times and reduce the duration of approval of transactions compared to the actual one by 40%.

The modeling, carried out in this way as an integral part of the digital double of the whole process, makes it possible to analyze the current state of the product, planning activities for the organization of product sales, which allows real-time control of all factors affecting the cost and quality of the product before its production.

References

[1] F. Guala, "Models, Simulations, and Experiments",Model-Based Reasoning: Science, Technology, Values, L. Magnani and N. Nersessian (eds.), New York: Kluwer, 2002, pp. 59–74,.

[2] A.M. Low, M.G. McComas, "How to build valid and credible simulation models," Proceedings of the 2001 Winter Simulations Conference, 2001.

[3] L. Jay; B. Behrad and Kao, Hung-An "A Cyber-Physical Systems architecture for Industry 4.0-based manufacturing systems". Manufacturing Letters.3:

18–23.doi:10.1016/j.mfglet.2014.12.001 [January 2015].
[4] A.G Palei,. G.A. Pollak, N.V. Kalashnikova and E.A. Konova "Building Models of Economic Systems Using Cognitive Methods", Innovation management and education excellence through vision 2020 : Proceedings of the 31st International Business Information Management Association Conference (IBIMA), Milan:2018, pp. 4062- 4070.
[5] Z.K. Avdeeva,. "Kognitivnoe modelirovanie clozhnykh sistem" [Cognitive modeling for solving problems of managing poorly structured systems], Large systems management, 2007, pp. 26-39.

Hybrid MMC Redundancy Configuration Calculation Method Based on Heuristic Algorithm

Xiaojian Lu[1,a], Zi'an Zeng[2,b], Xichao Lv[1,c], Xusong Luo[1], Gongliao Zhang[1] and Youping Fan[2]

[1]Liuzhou Bureau, UHV Transmission Company, Liuzhou, Guangxi, China
[2]School of Electrical Engineering and Automation, Wuhan University, Wuhan, Hubei, China

[*]E-mails: [a]409436177@qq.com, [b]2020202070059@whu.edu.cn, [c]lvxichao1man@163.com

Abstract. Most of the existing redundant configuration methods of modular level converters (MMC) only consider the reliability of the MMC, and rarely consider the utilization of insulated gate bipolar transistors (IGBT). A heuristic algorithm-based calculation method for hybrid MMC redundancy configuration is proposed in order to better meet the engineering practicality. Firstly, the topology of the hybrid MMC is introduced, and the correlation between submodules is introduced into the calculation of its reliability. Secondly, the index of IGBT utilization is proposed to judge the economic advantage. The effects of correlation and service life on the reliability of the hybrid MMC and the IGBT utilization are analyzed. Then, heuristic factors are constructed with reliability and IGBT utilization, and the optimal redundancy configuration scheme of hybrid MMC is obtained by heuristic algorithm. Finally, based on the parameters of Kunliulong multi-terminal hybrid DC transmission project, the calculation is verified in Matlab and compared with the results of the difference method to illustrate the feasibility and superiority of the proposed redundancy configuration calculation method.

1. Quotes

In the context of global energy interconnection, China is vigorously developing clean energy and renewable energy, optimizing the energy structure, and achieving clean energy development by transforming the development, production, allocation and utilization of energy; secondly, power electronics are fully applied to the flexible DC transmission process to maximize the power system's ability to transmit electricity.

In recent years, Modular Multilevel Converter (MMC) has been widely used in flexible DC transmission. Compared with the traditional two-level converter, the MMC output has higher waveform quality, better fault handling capability, and better converter expansion capability [1-4].

In order to ensure the MMC can operate safely and stably, the reliability of the MMC needs to be evaluated and the redundant configuration of the MMC needs to be designed. A number of studies on MMC reliability analysis have been done in the literature. The literature [5] used the k/n(G) mathematical model to analyze the reliability of MMC with different redundancy numbers. The literature [6] analyzed the effect of different redundancy configurations on the reliability of MMC based on the literature [5], considering the effect of control and protection system and valve cooling system on the reliability of MMC. In the literature [7-8], the first-order difference of the reliability function for hybrid MMC is given for the hybrid MMC redundancy configuration scheme. Literature [9] proposed that MMCs are not completely independent from each submodule due to the influence of loop flow when they are operating, and the correlation between MMCs needs to be considered to calculate their reliability. The literature [10-11] proposed a method to calculate MMC reliability considering correlation based on Copula function.

However, most of the literature mentioned above, analyzes and models the reliability of half-bridge MMCs. There are few analyses for hybrid MMC reliability and redundancy configurations. Moreover, the factor of economy is not considered when designing the redundancy configuration, which does not meet the practical requirements of engineering.

In order to solve the above deficiencies, this paper firstly establishes a sub-module reliability calculation model from the hybrid MMC topology. Considering the correlation between submodules, the Gumbel-Copula function is introduced to obtain the hybrid MMC reliability calculation model after coordinate transformation. Next, IGBT utilization is proposed to measure the economy of MMC design, and the effects of correlation and service life on the reliability of hybrid MMC and IGBT utilization are analyzed. Finally, based on the heuristic algorithm, a redundant configuration method for hybrid MMC is proposed. And the simulation verification in Matlab, the calculation results are reasonable, can be more suitable for the actual calculation of the optimal redundancy configuration number of MMC in the project, and better save the cost of the project.

2. Hybrid MMC sub-module topology

The hybrid MMC is composed of six bridge arms. Each bridge arm contains N_h half-bridge sub-modules (HBSM) and N_f full-bridge sub-modules (FBSM). In the hybrid topology, for the half-bridge power module, only zero or positive voltage can be output, and the current flowing through the power module needs to be bi-directional to keep the module voltage stable. For full-bridge power modules, zero, positive, and negative capacitive voltages can be output, and module voltage stability can be achieved even if the current is in a single direction.

Each submodule consists of an IGBT module, capacitors, voltage equalization resistors, and a submodule control system. The detailed structure is shown in Figure 1.

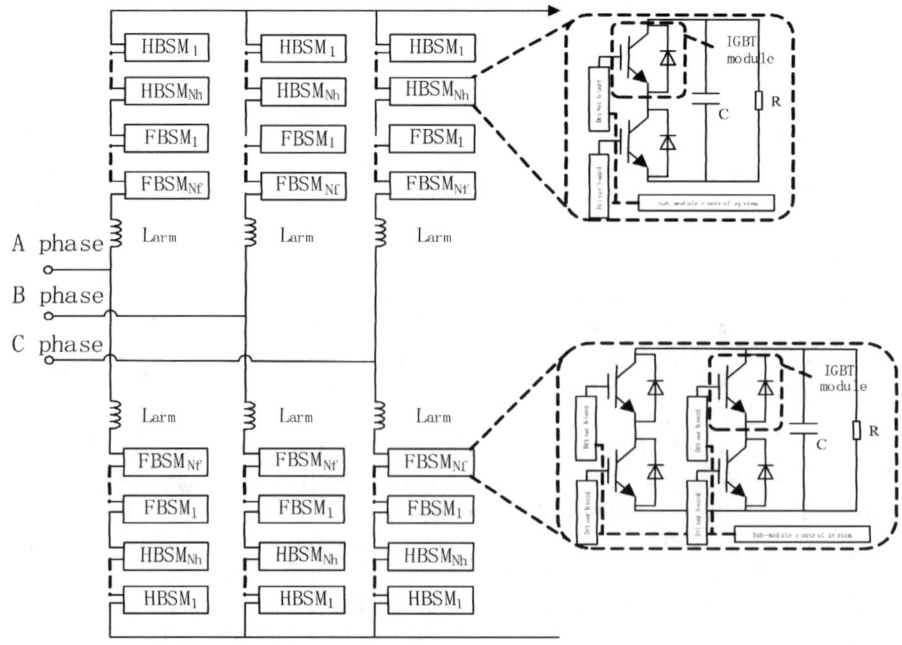

Figure 1. Schematic diagram of hybrid MMC structure

3. Hybrid MMC reliability calculation method

3.1. Sub-module reliability calculation model

For most power electronics, the exponential distribution is generally used statistically to describe its lifetime because of the memoryless nature of the exponential distribution. For non-repairable components, its failure rates all satisfy the bath curve [12].

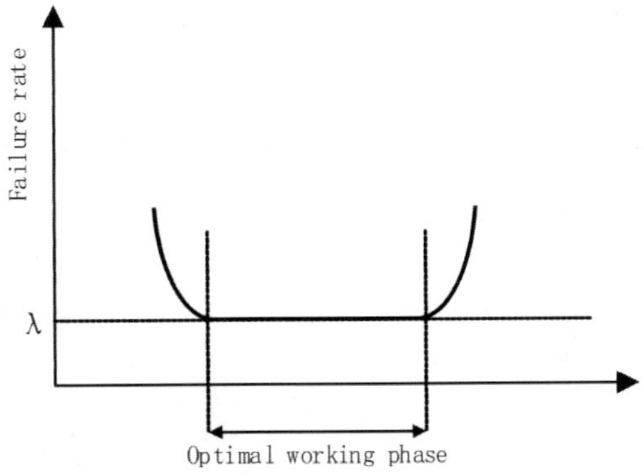

Figure 2. Bath Curve

It is assumed that when the component is in the optimal operating phase, its failure rate remains a constant, which is used to denote λ. At this point the reliability of the component R(t) can be expressed as

$$R(t)=e^{-\lambda t} \tag{1}$$

In this paper, assuming that all sub-modules satisfy the above conditions, the reliability functions of HBSM and FBSM can be described by the following equations according to the reliability calculation method of series systems.

$$\begin{aligned} R_h &= R_{igbt}^2 R_c R_r R_{cl} = e^{-\left(2\lambda_{igbt}+\lambda_c+\lambda_r+\lambda_{cl}\right)t} = e^{-\lambda_h t} \\ R_f &= R_{igbt}^4 R_c R_r R_{cl} = e^{-\left(4\lambda_{igbt}+\lambda_c+\lambda_r+\lambda_{cl}\right)t} = e^{-\lambda_f t} \end{aligned} \tag{2}$$

Where λ_{igbt}, λ_c, λ_r, and λ_{cl} denote the failure rates of IGBT module, capacitor, equalization resistor, and control system, respectively. λ_h and λ_f are the failure rates of HBSM and FBSM, respectively.

3.2. Copula Theory

In practical engineering, each submodule is not really completely independent, so the correlation between submodules must be taken into account when calculating the reliability of MMC bridge arms. However, it is very difficult to solve a multidimensional non-normal joint density function, and the Copula function provides a way to find the multidimensional non-normal joint density function by edge distribution.

The Copula function is essentially only a linking function that combines the edge density functions of each subvariable, and the Copula function [13-14] describes the correlation between these subvariables.

Suppose that $(T_1,T_2...T_N)$ is an N-dimensional random vector and $F_1(T_1)$, $F_2(T_2)$ ··· $F_N(T_N)$, is its edge distribution function. According to the theorem of Sklar [15-16], the N-dimensional Gumbel-Copula function [17] is expressed as follows.

$$\begin{aligned} & C^{N}_{1/(1-\theta)}\left(F_1(T_1)_1, F_2(T_2)\cdots F_N(T_N)\right) \\ & = P\left(T_1 \le t, T_2 \le t \cdots T_N \le t\right) \\ & = \exp\left\{-\left[\left(-\ln F_1(T_1)\right)^{\frac{1}{1-\theta}} + \left(-\ln F_2(T_2)\right)^{\frac{1}{1-\theta}} + \cdots\left(-\ln F_N(T_N)\right)^{\frac{1}{1-\theta}}\right]^{1-\theta}\right\} \end{aligned} \tag{3}$$

Where $\theta \in (0,1)$ is used to indicate the correlation between submodules, and the value of θ is proportional to the correlation of submodules.

3.3. Reliability analysis of hybrid MMC

Because of the highly symmetrical structure of the hybrid MMC, the reliability of the entire MMC can be expressed in terms of the reliability of the individual bridge arms.

Assume that a single bridge arm consists of N_h half-bridge submodules, N_{h0} redundant half-bridge submodules, N_f full-bridge submodules, and N_{f0} redundant full-bridge submodules. the number of half-bridge submodule failures at time T is ih and the number of full-bridge submodule failures is if.

From the literature [18], the reliability of the submodules after combining the N-dimensional Gumbel-Copula function and performing the coordinate transformation is expressed by the following equation, when the submodules are in a mutually independent state.

$$\begin{aligned} R_H &= \sqrt[N_h+N_{h0}+N_f+N_{f0}]{\frac{\exp\left\{-\left[(N_h+N_{h0})(\lambda_h t)^{1/(1-\theta)} + (N_f+N_{f0})(\lambda_f t)^{1/(1-\theta)}\right]^{1-\theta}\right\}}{\exp\left[-(\lambda_f-\lambda_h)t(N_f+N_{f0})\right]}} \\ R_F &= \sqrt[N_h+N_{h0}+N_f+N_{f0}]{\frac{\exp\left\{-\left[(N_h+N_{h0})(\lambda_h t)^{1/(1-\theta)} + (N_f+N_{f0})(\lambda_f t)^{1/(1-\theta)}\right]^{1-\theta}\right\}}{\exp\left[-(\lambda_h-\lambda_f)t(N_h+N_{h0})\right]}} \end{aligned} \tag{4}$$

Considering the specificity of the hybrid MMC structure, the redundancy replacement method during its normal operation can be divided into two cases, as shown in Figure 3.

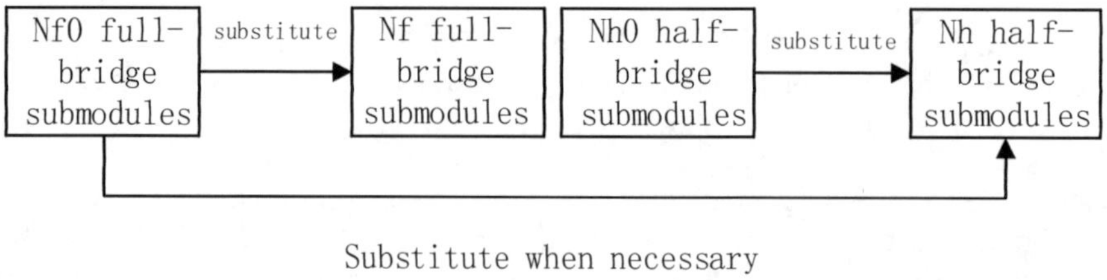

Figure 3 Schematic diagram of the hybrid MMC redundancy replacement method

- (1) When $i_h < N_{h0}, i_f < N_{f0}$ faulty half-bridge/full-bridge submodules are all replaced by redundant half-bridge/full-bridge submodules, at this time the bridge arm reliability can be expressed as

$$R_1 = \sum_{i_h=0}^{N_{h0}}\left[\sum_{i_f=0}^{N_{f0}} C^{i_h}_{N_h+N_{h0}} R_H^{N_h+N_{h0}-i_h}\left(1-R_H\right)^{i_h} C^{i_f}_{N_f+N_{f0}} R_F^{N_f+N_{f0}-i_f}\left(1-R_F\right)^{i_f}\right] \tag{5}$$

- (2) When $i_h > N_{h0}, i_f < N_{f0} + N_{h0} - i_h$, the number of redundant half-bridge sub-modules is not enough to meet the demand of fault replacement, and the full-bridge sub-modules will temporarily change their working mode to act as half-bridge sub-modules, at which time the bridge arm reliability can be expressed as

$$R_2 = \sum_{i_h=N_{h0}+1}^{N_{h0}+N_{f0}} \left[\sum_{i_f=0}^{N_{h0}+N_{f0}-i_h} C_{N_h+N_{h0}}^{i_h} R_H^{N_h+N_{h0}-i_h} \left(1-R_H\right)^{i_h} C_{N_f+N_{f0}}^{i_f} R_F^{N_f+N_{f0}-i_f} \left(1-R_F\right)^{i_f} \right] \tag{6}$$

Combining the equations, we can obtain the reliability R equation for the hybrid MMC bridge arm.

$$R\left(N_{h0},N_{f0}\right) = R_1\left(N_{h0},N_{f0}\right) + R_2\left(N_{h0},N_{f0}\right) \tag{7}$$

It is easy to see the relationship between the reliability of the hybrid MMC bridge arm and the number of redundancies of the submodules from the equation. However, too many redundant configurations do not mean that the reliability of the hybrid MMC can be improved without limit, but will increase the economic investment.

Therefore, considering the optimal redundancy configuration for hybrid MMC requires a combination of both reliability and economics.

4. Hybrid MMC redundancy configuration method

4.1. IGBT utilization of hybrid MMC

The most expensive and core device in the hybrid MMC sub-module is the IGBT, and this paper uses the utilization rate of the IGBT in the MMC module to evaluate the economics of the MMC.

The ratio of the number of IGBTs effectively utilized by the redundant modules of the hybrid MMC configuration to the total number of IGBT redundant modules is defined here as the IGBT utilization rate.

The effective utilization number Neff of IGBT can be defined by the following equation.

$$N_{eff} = \left(2i_h + 4i_f\right) \times R\left(N_{h0},N_{f0}\right) \tag{8}$$

where the number of failed half-bridge submodule failures is ih, the number of full-bridge submodule failures is if, and R is the hybrid MMC bridge arm reliability.

According to the definition above, the expression of the IGBT utilization function can be obtained as follows

$$e\left(N_{h0},N_{f0}\right) = N_{eff} / \left(2N_{h0} + 4N_{f0}\right) \tag{9}$$

where N_{h0} denotes the number of redundant half-bridge submodules and N_{f0} denotes the number of redundant full-bridge submodules.

Based on this, this paper uses a heuristic algorithm to calculate the full-bridge and half-bridge submodule redundancy configurations. A brief description of the heuristic algorithm is given in the following section.

4.2. Heuristic-based algorithm for redundant configuration calculation

In real engineering, most linear programming problems have the quality of nonlinear constraints. The heuristic algorithm[19] can solve this problem perfectly. It is done by comparing the heuristic factors of each first-order neighborhood by iteration after iteration, so that the result gradually approximates the optimal solution. The difference in the selection of heuristic factors is the different heuristic algorithms.

The specific steps of the heuristic algorithm are as follows.

Step 1: First set up an initial feasible solution $\boldsymbol{x}=(x_1, x_2,\ldots, x_n)$ for the model; and construct a set of first-order neighborhoods $\boldsymbol{y}_j=(y_1, y_2,\ldots, y_n)$, j=1,2, ...,n of x

$$\begin{cases} y_j = x_j + 1 \\ y_k = x_k, k = 1,2,\ldots,n, k \neq j \end{cases} \tag{10}$$

Step 2: Determine whether the solution $\boldsymbol{x}$ satisfies the constraints of the system at this point. If not satisfied, calculate the corresponding heuristic factor[20-22] G_j for all first-order neighborhoods $\boldsymbol{y}_j$ of $\boldsymbol{x}$ and find the neighborhood $\boldsymbol{y}_{\max}$ corresponding to the largest heuristic factor. If satisfied, skip to Step 4.

Step 3: Update the feasible solution so that $\boldsymbol{x} = \boldsymbol{y}$max. skip to Step 2 to continue the calculation.

Step 4: At this point, $\boldsymbol{x}$ is the optimal solution of the model.

For the hybrid MMC redundancy module configuration problem, first establish its redundancy model.

$$\begin{cases} \max e\left(N_{h0}, N_{f0}\right) \\ \text{s.t. } R\left(N_{h0}, N_{f0}\right) \geq R_0 \\ N_{h0_0} = 1, N_{f0_0} = 1 \end{cases} \tag{11}$$

The ratio of the increment of IGBT utilization to the increment of MMC system reliability is chosen as the heuristic factor and calculated as follows:

$$G_j = \frac{e(\mathbf{y}_j) - e\left(N_{h0}, N_{f0}\right)}{R(\mathbf{y}_j) - R\left(N_{h0}, N_{f0}\right)} \tag{12}$$

Where R_0 is the reliability threshold of the system, N_{h0}, N_{f0} are the number of redundant modules for half bridge, full bridge respectively. N_{h00}, N_{f00} are the initial values of the set number of redundancy N_{h0}, N_{f0}, $e(N_{h0}, N_{f0})$ is the IGBT utilization function, $R(N_{h0}, N_{f0})$ is the reliability function of the hybrid MMC.

Then follow the steps of the heuristic algorithm and iterate step by step until finding the optimal redundancy configuration scheme for hybrid MMC that satisfies the model requirements. The specific flow chart is shown in Figure 4.

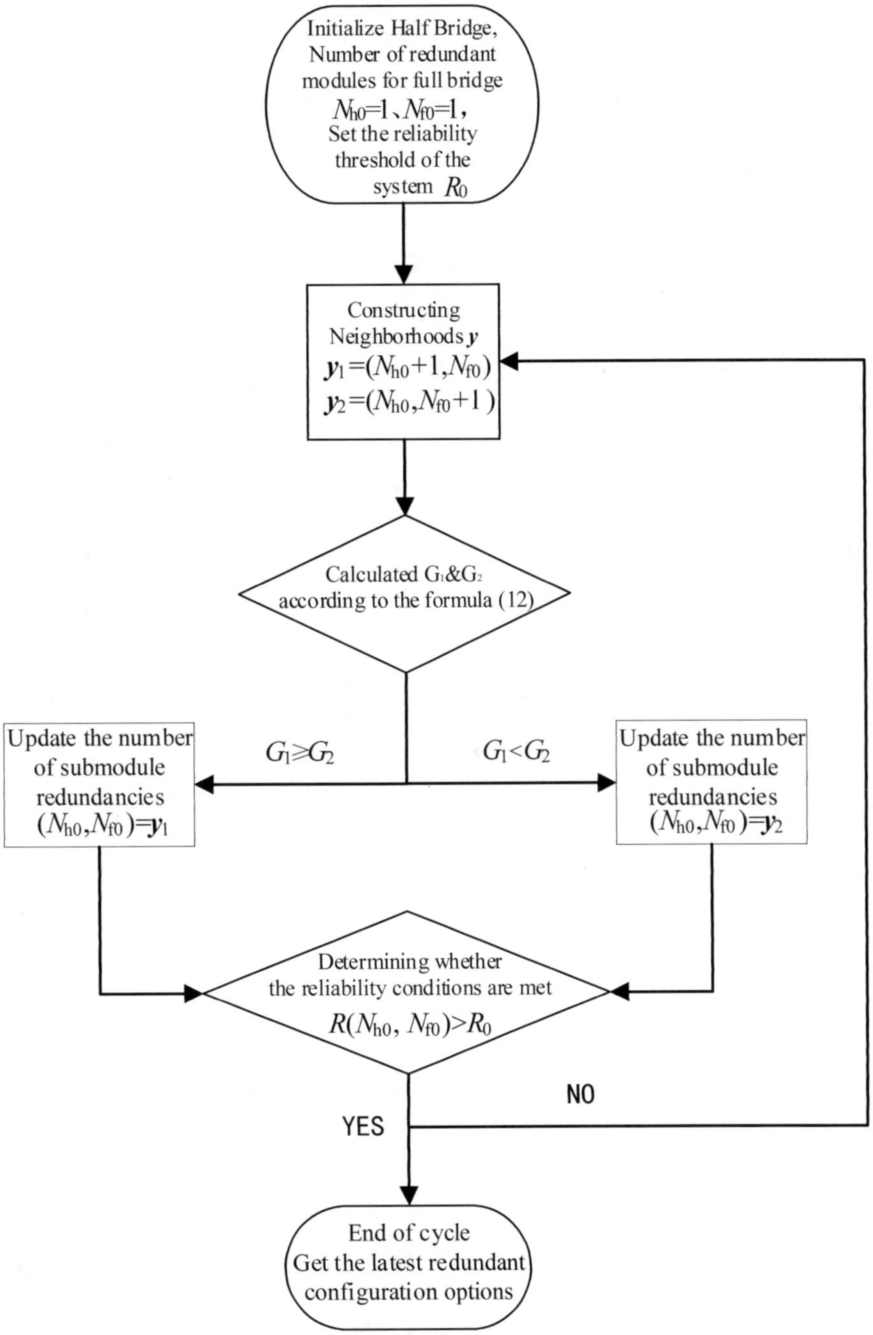

Figure 4. Flow chart of redundant configuration calculation based on heuristic algorithm

5. Analysis of calculation cases

In order to verify the feasibility of the redundancy configuration method and reliability calculation method described in this paper, this section relies on a multi-terminal hybrid DC transmission project for analysis.

The project uses a hybrid MMC converter with a DC voltage of 400 kV, regardless of redundancy, and the bridge arm is configured with 160 full-bridge submodules and 40 half-bridge submodules. The failure rates of each part of the submodules are shown in Table 1.

Table 1. MMC component failure rate

Part Name	Number of full bridge sub-modules included	Number of half-bridge sub-modules included	Failure rate/(times/ year)
IGBT Modules	4	2	0.0003504
Capacitance	1	1	0.0000876
Uniform voltage resistor	1	1	0.0023214
Communication Module	4	2	0.000876
Integrated Circuit Modules	1	1	0.00142788

According to Equation (2), the failure rates $\lambda_h = 0.0063$ and $\lambda_f = 0.0087$ can be obtained for HBSM and FBSM, respectively.

Let the number of redundancies of HBSM and FBSM be 15 and 5, respectively. By substituting the failure rate into Eqs. (4)-(9), the curves of hybrid MMC reliability and utilization with correlation θ and service life t are obtained as shown in Figs. 4 and 5.

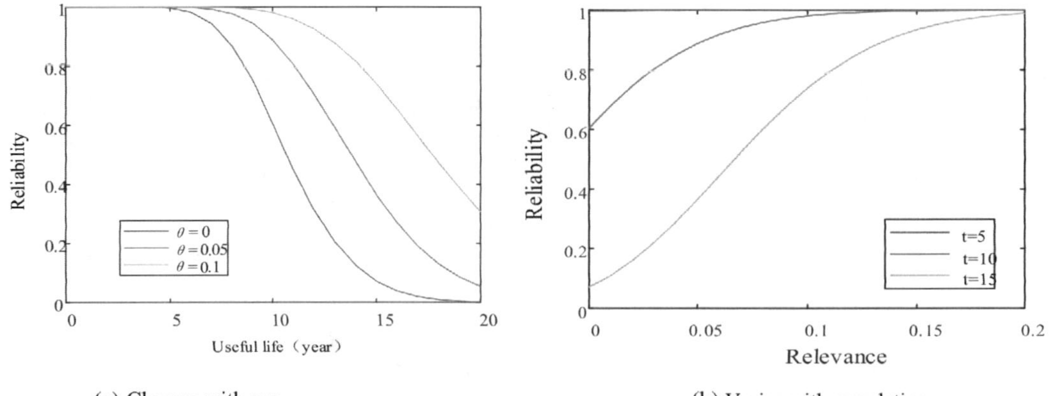

(a) Change with age (b) Varies with correlation

Figure 5 Reliability curve with correlation and service life

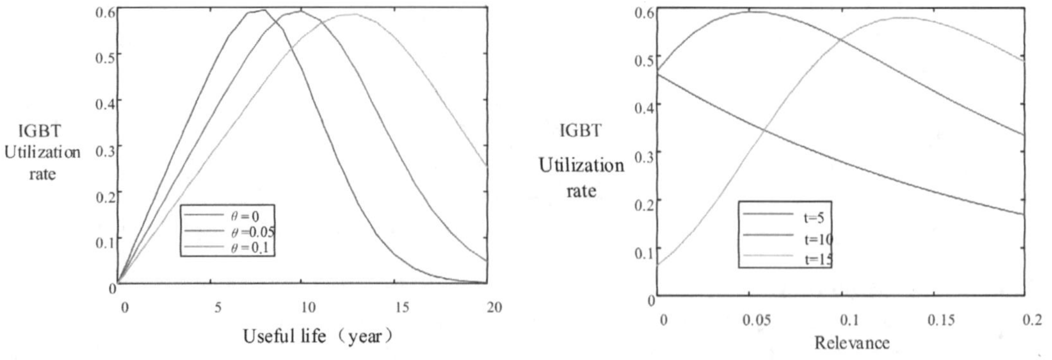

(a) Change with age (b) Varies with correlation

Figure 6. IGBT utilization curve with correlation and service life

The graph shows that the reliability decreases with increasing service life t and increases with increasing correlation θ. The IGBT utilization tends to increase and then decrease with increasing service life t and correlation θ. As the correlation/life increases, the peak IGBT utilization is delayed.

Next, the optimal redundancy configuration of the hybrid MMC is analyzed based on the above data.

Relying on computer simulations to randomly generate hybrid MMC lifetime data satisfying E(λ_h= 0.0063)、E(λ_f= 0.0087), and exponential distribution, the mean value of correlation θ was obtained by the great likelihood estimation method, $\theta = 0.06$, and set the service life $t = 10$ years and the reliability threshold $R_0 = 92\%$.

In order to be able to filter out the optimal redundancy configuration scheme in this example, the maximum number of redundancies for HBSM and FBSM are set to 25 and 20, respectively. The distribution of hybrid MMC reliability and IGBT utilization under different redundancy configurations are plotted in MATLAB according to Eqs. (4)-(9), as shown in Figs. 6 and 7.

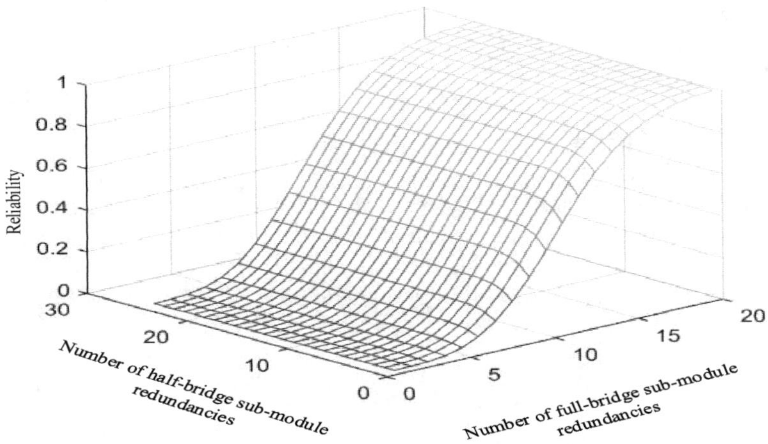

Figure. 7 Reliability with the number of submodule redundancy surface

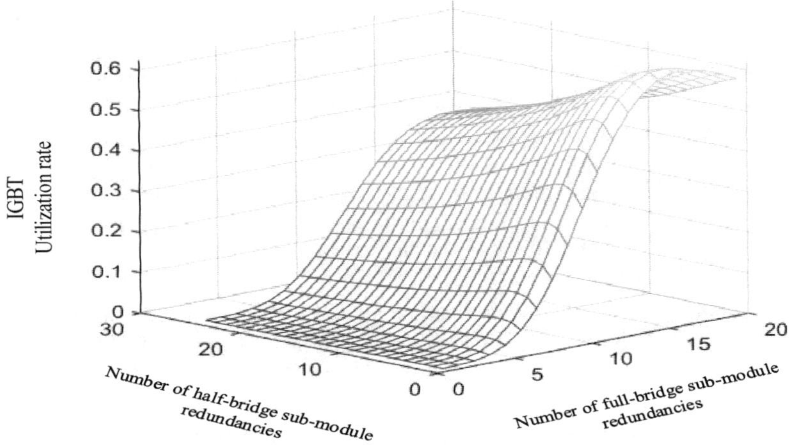

Figure. 8 Reliability surface with the number of submodule redundancies

The trend of IGBT utilization and hybrid MMC reliability with the number of redundancies is different as seen in the figure. Reliability tends to increase with the number of sub-module redundancy.

the degree of reliability improvement is higher for FBSM than HBSM, and up to a certain point, the reliability improvement is limited. As the number of sub-module redundancy increases, IGBT utilization takes the lead in increasing and then decreasing. Similar to reliability, FBSM has a higher impact on IGBT utilization than HBSM.

The heuristic algorithm proposed in this paper and the difference method in the literature [7] are used to analyze the redundant configuration of the hybrid MMC in the arithmetic example, respectively. The obtained results are shown in Table 2

As can be seen from Table 2, the results obtained by the difference method and the results of the algorithm in this paper meet the requirements in terms of reliability, and the difference is not large. In the aspect of IGBT utilization, there is a certain difference between the results of the two methods, and the heuristic algorithm of this paper is obviously better than the difference method. Moreover, the computation time of the heuristic algorithm is also less than that of the difference method.

Table 2. Comparison of the results of the difference method and heuristic algorithm

	Nh0	Nf0	R	e	Calculation time
Differential method	8	16	95.11%	54.91%	1.893s
Heuristic algorithms	2	16	93.65%	62.40%	1.262s

When the redundancy number of FBSM is 16 and the redundancy number of HBSM is 2, the whole hybrid MMC system can meet the reliability requirements and the IGBT utilization can be taken to the maximum value to satisfy the system design requirements. Therefore, the hybrid MMC redundancy configuration results obtained by the heuristic algorithm proposed in this paper are feasible and faster to compute.

6. Conclusion

In this paper, the IGBT utilization rate of hybrid MMC is used as an index to measure the economy in order to consider both the reliability and economy factors of hybrid MMC. In the process of calculating the reliability and IGBT utilization of hybrid MMC, the correlation between submodules is also considered to make the calculation results more consistent with the engineering reality. From the results, it can be seen that correlation and service life have significant effects on the reliability and IGBT utilization of hybrid MMC.

In this paper, a heuristic algorithm-based redundancy configuration method for hybrid MMCs is proposed. Starting from the redundancy configuration problem of hybrid MMC, the heuristic factor is constructed with the ratio of reliability and IGBT utilization. The results show that this proposed algorithm can better solve the hybrid MMC redundancy configuration method with shorter computation time compared to the difference method.

The hybrid MMC redundancy configuration method proposed in this paper is applicable to any hybrid MMC. only the corresponding sub-module failure rate needs to be adjusted. The results of this paper also provide a more reliable reference for MMC redundancy design.

References

[1] Knaak H J. Modular multilevel converters and HVDC/FACTS: A success story[C]// Proceedings of the 14th IEEE European Conference on Power Electronics and Applications. Birmingham, England: IEEE, 2011.

[2] Rohner S, Bernet S, Hiller M, et al. Modulation, losses, and semiconductor requirements of modular multilevel converters [J]. IEEE Transactions on Industrial Electronics, 2010, 57(8):2633-2642.

[3] Saeedifard M, Iravani R. Dynamic performance of a modular multilevel back-to-back HVDC system [J].IEEE Transactions on Power Delivery, 2010, 25(4) : 2903-2912.

[4] Harnefors L, Antonopoulos A, Norrga S, et al. Dynamic analysis of modular multilevel converters [J]. IEEE Transactions on Industrial Electronics, 2013, 60(7): 2526-253

[5] Ding M, Wang JJ, Song Q. Reliability modeling and redundancy analysis of converter valves in

flexible DC transmission system based on k/n(G) model[J]. Power Grid Technology, 2008, 32(21): 33-36.
[6] Wang X.L., Guo J.L., Pang F., et al. Structural reliability analysis of modular multilevel converters[J]. Chinese Journal of Electrical Engineering, 2016,36(7): 1908-1914.
[7] Xu JC, Zhao PH, Jiang W, et al. Reliability analysis and redundancy configuration method of hybrid MMC with DC fault ride-through capability[J]. Chinese Journal of Electrical Engineering, 2016, 36(4): 953-960.
[8] Xu Jianzhong, Zhao Penghao, Zhao Chengyong. reliability analysis and redundancy configuration of MMC with hybrid sub-module topologies [J]. IEEE Transactions on Power Electronics, 2016, 31(4):2720-2729.
[9] Wang Biyang, Wang Xifan, Bie Zhaohong, et al. Reliability model of MMC considering periodic preventive maintenance [J]. IEEE Transactions on Power Delivery, 2017, 32(3): 1535-1544.
[10]Hao Jing, Jianzhong Xu, Ying Xu, et al. Reliability analysis method of MMC considering submodule correlation[J]. Chinese Journal of Electrical Engineering,2017, 37(13): 3835-3842.
[11]Zheng W D, Xu Q D, Shao Z G, et al. Reliability modeling and analysis of MMC system based on k/n(G) model[J]. Journal of Hunan University (Natural Science Edition),2020,47(12):108-116.
[12]NARESKY J. Numerical approach to electronic reliability [J]. Proceedings of the IRE, 1959, 47(5): 946-956.
[13]He, Cheng-Ming, Wu, Wei, Meng, Qing-Jun. Copula-based reliability model of mechanical system and its application[J]. Journal of Military Engineering,2012,33(3): 379-384.
[14]Tang JY, Zhao YX, He P, et al. A new Copula theory for the calculation of correlation reliability of mechanical systems[J]. Mechanical Science and Technology, 2009,28(4):532-535.
[15]Zhou JY, Xie RY, Han WQ, et al. Copula reliability model for symmetric structural systems[J. Journal of Military Engineering,2011, 32(2): 204-209.
[16]Han WQ, Zhou JY, Zhu FX. Failure probability calculation based on Copula failure-related structural system[J]. Mechanical Design and Manufacture, 2011(2): 183-185.Han Wenqin, Zhou Jinyu, Zhu Fuxian. Failure probability calculation in structural systerms with correlated failure modes based on Copula[J]. Machinery Design & Manufacture,2011(2): 183-185.
[17]Nelsen R B. An introduction to Copulas [M]. 2nd ed. New York: Springer, 2006.
[18]Xu JC, Wang L, Wu DY, et al. Reliability modeling of hybrid modular multilevel converters using submodule correlation coordinate transformation decoupling[J]. Journal of Electrotechnology,2019,34(18):3821-3830.
[19]Zhao Zhicao, Song Baowei, Zhao Xiaozhe, et al. Optimization of system reliability redundancy based on Copula function[J]. Systems Engineering Theory and Practice,2014,34(8):2149-2153.
[20]Guo Bian, V. Rajendra Prasad, Frank A. Tillman, et al. Optimal Reliability Design:Fundamentals and Applications [M]. Guo Jinli, Ge Chunning, Translation. Beijing:Science Press, 2011.
[21]Sharma U, Venkateswaran K V. A direct method for maximizing the system reliability [J]. IEEE Transactions on Reliability, 1971,20(20):256-259.
[22]Lai F, Zhu C, Wang W. Cost modeling and optimization of redundant systems [J] Systems Engineering and Electronics Technology, 2000,22(9): 84-89.

Optimization Algorithm Design of Laser Marking Contour Extraction and Graphics Hatching Based on Image Processing Technology

Haoyang Luo[1,a], Jun Luo[2,b,*], Ruimin Li[3,c] and Meiqi Yu[1,d]

[1]Jiangxi Normal University, Nanchang, Jiangxi, China
[2]School of Engineering, Guangzhou College of Technology and Business, Guangzhou, Guangdong, China
[3]University of Electronic Science and Technology of China, Chengdu, Sichuan, China

E-mails: [a]luohaoyang@jxnu.edu.cn, [b]luojuncn@qq.com, [c]2019050306021@std.uestc.edu.cn, [d]201826001001@jxnu.edu.cn

Abstract. On the premise of solving the problem in a specific situation, this paper aims at the laser marking machine for the design and optimization of the outline extraction of the two-dimensional composite curve figure and the figure hatching scheme. The scanning line algorithm based on computer graphics and the polygon hatching are based on the algorithm theory. Based on the foundation, an improved polygon scan line algorithm model is established, and the contour extraction and graphic shadows of two-dimensional polygons are realized by MATLAB programming. The simulation results show that our algorithm model can meet our requirements well, and at the same time has a certain improvement effect. At the same time, compared with traditional two-dimensional polygon scanning and hatching algorithms, our algorithm model has strong analysis and processing capabilities and strong applicability. Compared with traditional two-dimensional polygon scanning and shading algorithms, model analysis can still run stably in a high refresh rate environment. We hope to further optimize the algorithm in the later stage, so that it can be better used in related engineering fields.
Keywords. Laser marking; Polygon hatching scanline algorithm; Shortest path planning

1. Introduction

1.1. Background of the question
Laser marking has become a widely used processing method because of its high efficiency, non-contact operation, no consumables and slight influence on products surface. The two main hatching methods of laser marking are direction parallel hatch and contour parallel hatch. Direction parallel, known as "zigzag hatch", is based on line segments moving parallel to the initial selected reference direction and connecting it to alternate from left to right (or right to left) to obtain a coherent path. And contour parallel hatch is based on the equidistant line segments and generates the equidistant curve in a helical way.

The hatching entity must be closed curvilinear polygons, and multiple mutually-nested objects may be filled inward or outward with combining profile lines by direction parallel or contour parallel hatch. As shown in figure 1. The hatch curves should be no missing or duplicate areas for the hatching process. To meet the requirements of efficient laser marking, the hatching curves should be parallel to contour lines, evenly distributed and automatically generated quickly. As shown in figure 2. Efficiency is an important index to generate hatched graph.

Figure 1. Graphical representation of zigzag parallel and contour parallel hatch

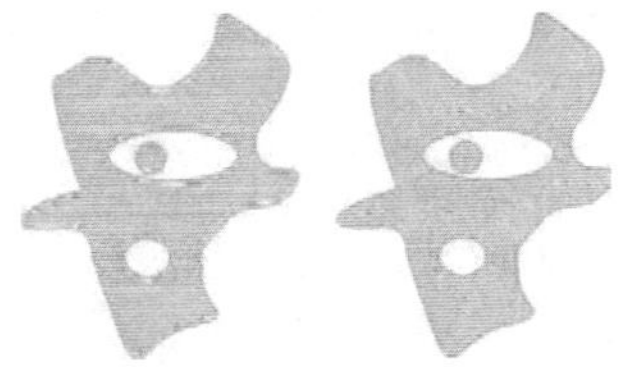

Figure 2. zigzag parallel and contour parallel hatch of multi-layer curves (From outside to inside of nested areas, only odd layers are hatched.)

1.2. Restatement of the question

The problems studied in this paper are as follows:

Please research the characteristics of zigzag parallel and contour parallel hatch, establish mathematical model of hatching, design the algorithm and discuss its operating efficiency, and finally answer the following questions:

1.Considering direction parallel hatch horizontally, realize the zigzag parallel and contour parallel hatch of the single-layer contour pattern under two sets of input parameters according to the coordinate data of the figure in Attachment 1:

(1) Internal contraction of boundary distance $1mm$, hatch line spacing $1mm$;

(2) Internal contraction of boundary distance $0.1mm$, hatch line spacing $0.1mm$.

Calculate the total length of hatching and count the number of horizontal lines of zigzag parallel hatch and the number of circles of contour parallel hatch. And count the average elapsed time of the program and calculate the time-consuming ratio for the program running under conditions of parameter groups (2) and (1).

2.Considering direction parallel hatch horizontally, realize the zigzag parallel and contour parallel hatch of the multi-layer contour pattern under two sets of input parameters according to the coordinate data of the figure in Attachment 2:

(1) Internal contraction of boundary distance $1mm$, hatch line spacing $1mm$;

(2) Internal contraction of boundary distance $0.1mm$, hatch line spacing $0.1mm$.

Calculate the total length of hatching and count the number of horizontal lines of zigzag parallel hatch and the number of circles of contour parallel hatch. And count the average elapsed time of the program and calculate the time-consuming ratio for the program running under conditions of parameter groups (2) and (1).

2. Analysis of the questions

For problem 1 and 2, according to the coordinates given in Appendix 1 and 2, we can draw two two-dimensional planar polygons using MATLAB. According to the requirements of the title, we need to fill the two graphics with line cutting and circular cutting. In order to solve the problem, we have made some necessary mathematical assumptions, which will be given in Chapter 3 of the article. Based on the assumption that we set up, whether the graph is nested or not has no influence on the program design. Therefore, we generalize the solutions of question 1 and question 2 into the realization of two hatching algorithms: line tangent and ring tangent[1].

(1) Contour parallel hatch: According to the known conditions, use polyshape function to draw the required hatching figure in MATLAB. To realize circumcision, that is to complete the hatching of isometric curve in the graph, that is to use the polybuffer function to indent and fill the isometric curve according to the step size (1 / 0.1)[2]. Finally, we can get the hatching figure we need and complete the ring cutting step.

(2) Direction parallel hatch: According to the known conditions, draw the required hatching figure with polyshape in MATLAB. To realize circumscribe, we first need to get the inflection point of the parallel line. The inflection point can be regarded as the intersection point of the parallel line and the

indented graph[3]. Therefore, we use polybuffer to indent the original figure and get the intersection point, and then judge the division of the parallel line according to the number of intersection points of a parallel line. If there are two intersections (such as four), the parallel lines will be broken, and two parallel lines will be formed according to the intersection points. At the same time, the parallel lines will be divided according to the step size (1 / 0.1). Then, the intersection points will be sorted, and the intersection points will be connected in order to form a cutting parallel line. Then we will get our final hatching result.

3. Establishment and solution of the question

3.1. Assumptions

In order to facilitate the solution of the model and ensure the strong generalization of the model, we set up the following assumptions[4]:

1. Ignore the moving time during the period of segment by segment rendering.
2. Ignore the effects of mark materials and processing technology on incision types.
3. The hatching curve is always parallel to the figures' boundary lines and uniformly distributed within the graph for the generation of hatched figures.
4. No omissions or repeated hatching areas during hatching.
5. The closet points of connection exist in the data set when achieving hatching with area segmentation algorithm.

3.2. Algorithm flow chart

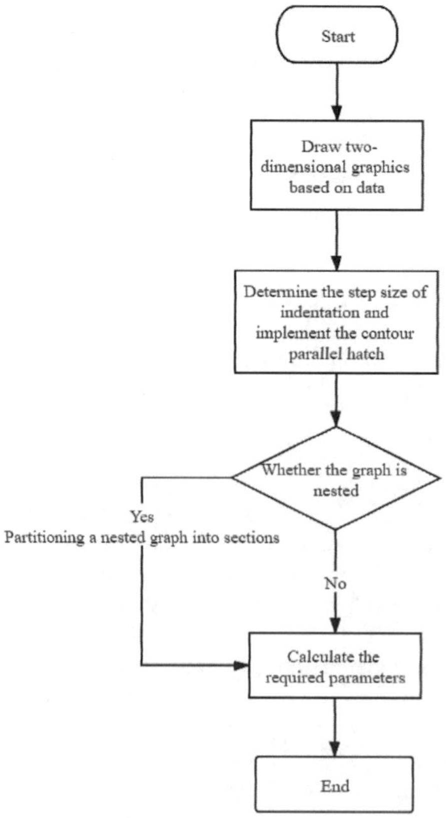

Figure 3. Flow chart of zigzag parallel

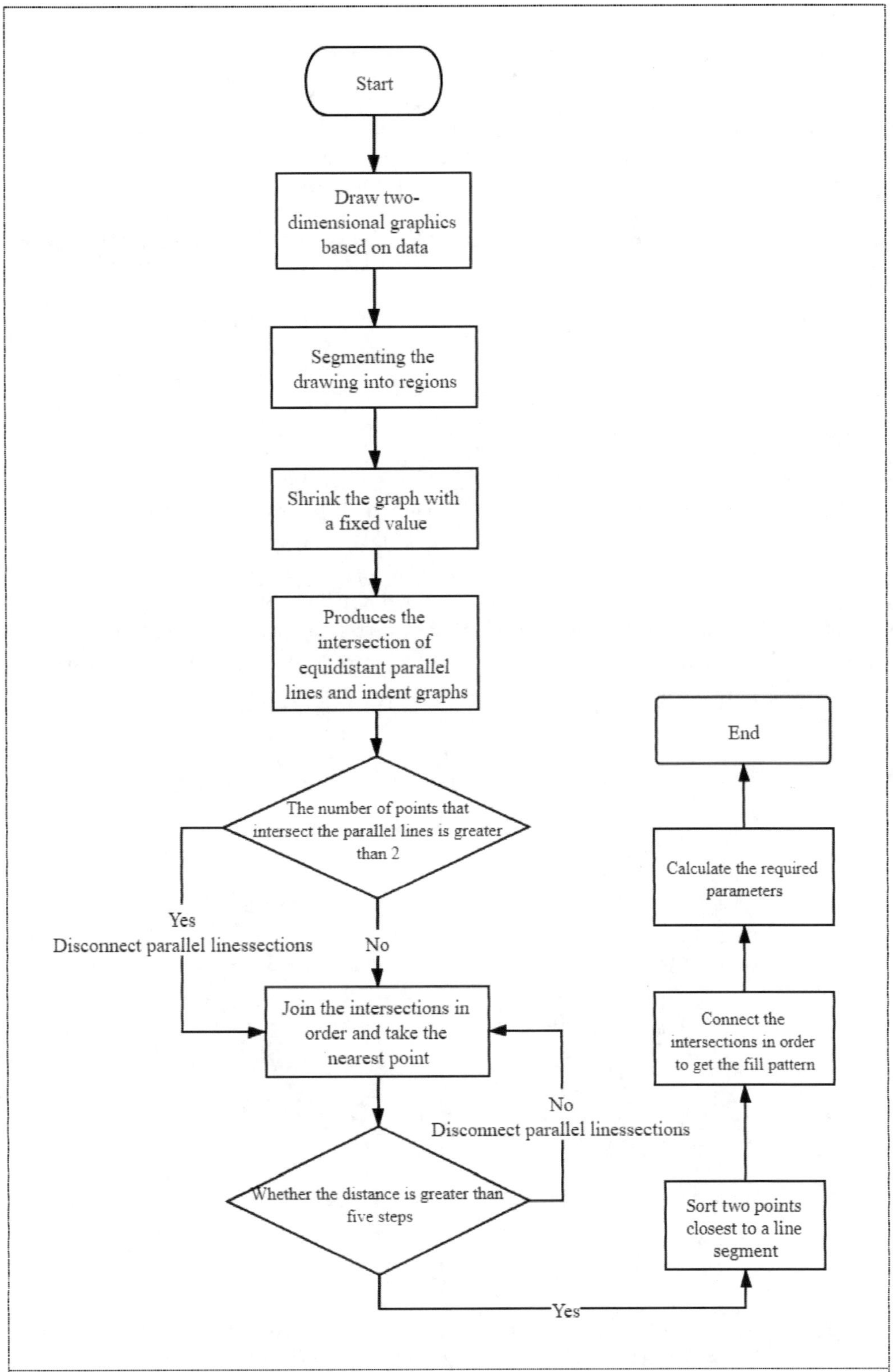

Figure 4. Flowchart of the contour

3.3. The result of problem 1

According to figure 3, programming with MATLAB, we get the following results:

(1) Under the condition of 1 *mm* internal contraction boundary distance and 1 *mm* hatching line spacing, the total length of line cutting hatching curve is 196.4336 *mm*. The total length of ring cutting hatching curve is 896.8205 *mm*, while the number of horizontal lines is 50, the number of turns is 9. And the average running time is 654 *ms* and 1293 *ms* respectively.

(2) Under the condition of 0.1 *mm* internal contraction boundary distance and 0.1 *mm* hatching line spacing, the total length of row cutting hatching curve is 989.5264*mm*. The total length of ring cutting hatching curve is 10017.094*mm*, while the number of horizontal lines is 525, the number of turns is 86. And the average running time is 2600 *ms* and 7857 *ms* respectively. The specific results are shown in figure 5.

Under the two parameters, the running time ratio (direction parallel: contour parallel) is 0.5058 and 0.3309, respectively.

3.4. Results of Problem 2

According to figure 4, programming with MATLAB, we get the following results:

(1) Under the condition of 1 *mm* internal contraction boundary distance and 1 *mm* hatching line spacing, the total length of zigzag parallel hatching curve is 189.0599 *mm*. The total length of contour parallel hatching curve is 773.0452 *mm*, while the number of horizontal lines is 50, the number of turns is 6. And the average running time is 1304 *ms* and 1266*ms* respectively.

(2) Under the condition of 0.1 *mm* internal contraction boundary distance and 0.1 *mm* hatching line spacing, the total length of direction parallel hatching curve is 1306.2959 *mm*, the total length of ring cutting hatching curve is 9060.1234 *mm*. The number of horizontal lines is 525, the number of turns is 60. And the average running time is 3045*ms* and 6559*ms* respectively. As shown in figure 6.

The running time ratio (zigzag parallel: contour parallel) is 1.0300 and 0.4642, respectively.

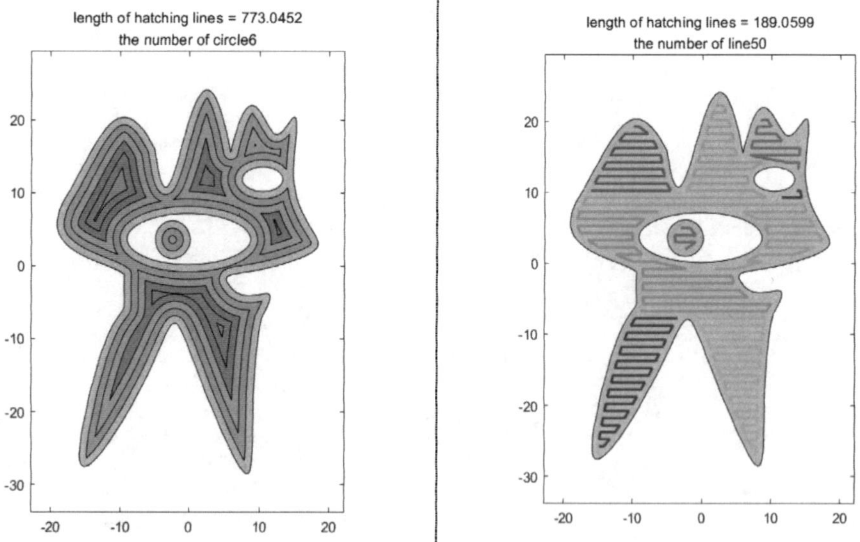

(a)	(b)
Figure 5 (a) parameter group (1) multi-layer direction parallel hatching	Figure 5 (b) parameter group (1) multi-layer contour parallel hatching

Figure 5. parameter group (1) multi-layer direction parallel and contour parallel hatching

Journal of Physics: Conference Series **2173** (2022) 012078 doi:10.1088/1742-6596/2173/1/012078

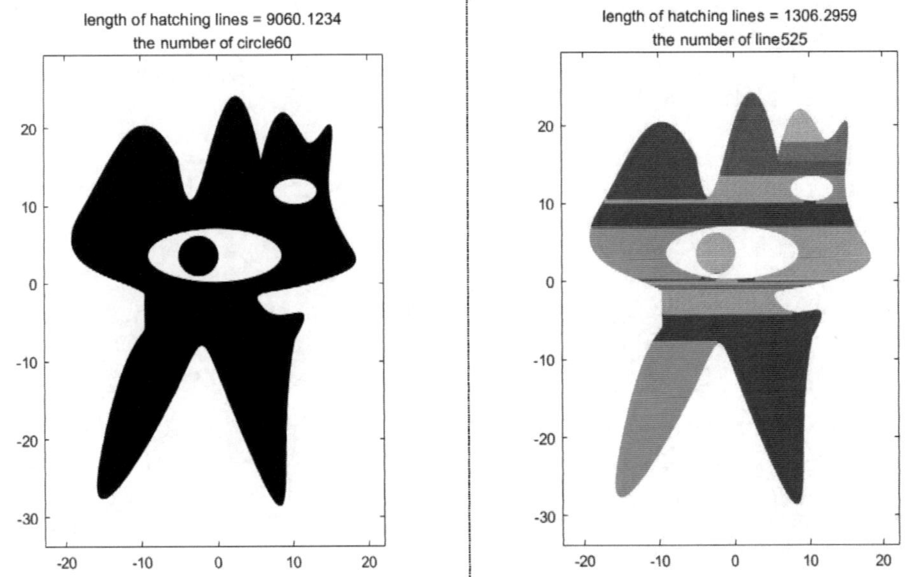

(a) (b)

Figure 6 (a) parameter group (2) multi-layer contour parallel hatching

Figure 6 (b) parameter group (2) multi-layer direction parallel hatching

Figure 6. parameter group (2) multi-layer direction parallel and contour parallel hatching

4. Evaluation and extension of the model

In this chapter, we will evaluate and popularize the model, hoping that our model can help professionals in related fields to solve problems.

4.1. Advantages of the Model

(1) The improved algorithm model of polygon hatching scanline greatly simplifies the calculation of intersection in scanning and segmentation, and solves the problem of intersection point better.The model is efficient and easy to program. Only a small amount of comparison judgment and simple addition and multiplication are needed.

(2) Our solution to the same problem model is not limited to a single algorithm, but to extract and combine the advantages of a variety of practical algorithms to process data to obtain the highest accuracy.

(3) The algorithm model we built is not limited to the intuitive data image representation. In solving the model, we correctly use some algorithms with abstract thinking. The purpose is to output the most effective data with the simplest ideas.

(4) Compared with the traditional scanning and hatching algorithm for two-dimensional polygon, the model analysis has a strong processing ability and high applicability. Compared with the traditional scanning and hatching algorithm for two-dimensional polygon, the model analysis can still run stably in a high refresh rate environment[5].

4.2. Generalization of the Model

(1) The algorithm model established by us has strong universality, easy programming, fast operation speed and high efficiency. It can be used as a reference for universities to carry out laser marking,

laser engraving, 3D printing and other teaching or extracurricular academic projects of students.

(2) Our model can solve the problem of finding the intersection point between the line cutting scanning line and the boundary of the cavity contour quickly, accurately and stably when generating the center path of the line cutting tool in the cavity machining of mechanical parts, which can bring some inspiration to the design optimization in related engineering fields[6].

Acknowledgements

Fund Project: Construction of electronic information experimental teaching demonstration center project of Guangdong undergraduate university teaching quality and teaching reform project in 2020

References

[1] Wang W, Wu X 2001 Image processing technology in laser marking *Electromechanical engineering technology* **33**(8) 30-1

[2] Ren A 2005 *Computer Graphics* (Beijing: Beijing University of Aeronautics and Astronautics Press) p 66-70

[3] Zhang C M, Yang X, Li X 2005 Computer Graphics (Beijing: Science Press) p 46-51

[4] Xie M, He X 2008 Application of polygon hatching scanning algorithm in line cutting tool path generation *Journal of Jiamusi University* (Natural Science Edition) **26**(2) 162-5

[5] Liu X 2008 Review of rapid prototyping technology development *Agricultural Equipment and Vehicle Engineering* **2** 10-3

[6] Chen X, Mcmains S 2005 Proc. of the ASME 2005 Int. Design Engineering Technical Conferences and Computers and Information *Engineering Conference. Volume 2: 31st Design Automation Conference, Parts A and B* (Long Beach, California: ASME) pp 565-75

Structural Optimization Design Based on Improved Ant Colony Algorithm

Jian Jun Deng*, **Lin Zhang and Da Wu**

The Air Defense and Anti-missile Institute, Air Force Engineering University, Xian, Shaanxi, China

*corresponding author's e-mail: jianjundeng1@163.com

Abstract. Ant colony algorithm is a kind of bionic intelligent algorithm, which is suitable for combinatorial optimization. The principle and calculation flow of ant colony algorithm is discussed in this paper, how to use improved ant colony algorithm to solve the optimization problem of truss structure is analyzed, the optimization problem of 13-bar truss structure is calculated. The results show that ant colony algorithm has certain advantages in solving the structural optimization problem with discrete design variables.

1. Introduction

The variables in structural system design are usually discrete variables. Traditional methods mostly use mathematical programming methods, these methods have great limitations when applied to engineering design problems, and tend to converge to a local optimal solution. The initial value have a great impact on the result of the solution, it is difficult to get the global optimal solution. Ant colony algorithm is a kind of random search algorithm, like other simulated evolutionary algorithms, it seeks the optimal solution through a group of candidate solutions in the evolution process. Numerous studies have proved that the ant colony algorithm has excellent ability to find better solutions. The algorithm uses the principle of positive feedback, which can speed up the evolution process to a certain extent. It is also a parallel algorithm, which is carried out continuously information exchange and transfer between different individuals, therefore it is being better to find optimal solutions. Ant colony algorithm has advantages in solving complex optimization problems and it is a promising method. Using this method to solve TSP problems, allocation problems and so on, good experimental results has been obtained. In this paper, an improved ant colony algorithm is used in the structural optimization design to overcome the defects caused by the mathematical programming method used in the structural optimization design[1].

2. Basic ant colony system model

The behavior of a single ant is extremely simple, but the ant colony group composed of such a single simple individual shows extremely complex behavior. Ant individuals transmit information through a substance called pheromome. Ants can leave pheromones along their moving routes, and there will be more ants where the pheromone concentration is high. In equal time, the pheromone concentration is higher in the shorter path, so the number of ants choosing the shorter path will also increase. If there are more ants walking along a path, the following ants will be more likely to choose this path. This leads to an increase in the number of ants taking the short route while those taking the other route slowly disappear. It is through this pheromone communication that ant colonies finally choose the optimal path to search for food, which is the biological principle of ant colony algorithm[2]. The algorithm principle is showed by following flowchart figure.

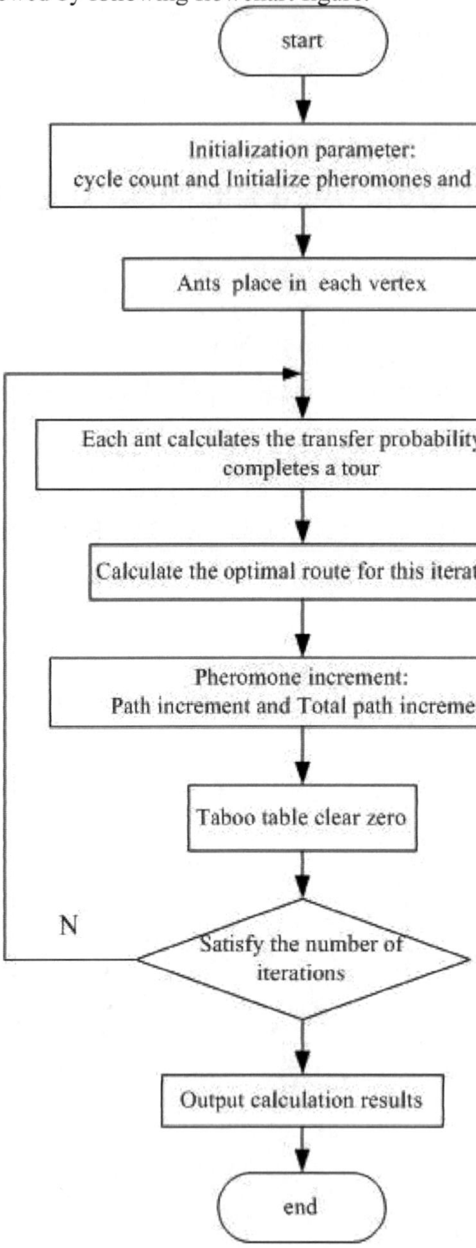

Figure 1. Calculation principle of ant colony algorithm

3. Improved ant colony algorithm

3.1. The main differences between ant colony system and improved ant colony algorithm[3]

3.1.1. Using the better solution. When choosing the next city, the improved ant colony algorithm makes more use of the current better solution.

3.1.2. Selective addition of pheromone. Add pheromone on the edge to which the global optimal solution belongs.

3.1.3 .Reduce pheromone appropriately. Every time the ant moves from city *i* to city *j*, the pheromone on edge *ij* will be reduced appropriately.

3.2. Application of improved ant colony algorithm to solve the optimization problem of truss structure
Regarding the nodes of the truss structure as the city in the TSP problem, and the optimal cross-sectional area of the truss as the distance between cities, the quality problem of the truss structure can be transformed into a path optimization problem. The calculation of the truss structure in this paper adopts the improved ant colony algorithm.

4. The thirteen-bar truss structure optimization design based on the improved ant colony algorithm

The above-mentioned improved ant colony algorithm is used in the optimization design of the thirteen-bar truss structure system as shown in Figure 2.

The truss structure is the problem of plane truss, there are a total of 20 unknown quantities, it is solved by used matrix displacement method, each stem cell matrix is listed, overall stiffness matrix is integrated, the load array and the displacement array are listed, overall stiffness equation is formed, the displacement boundary conditions is introduced, the boundary conditions is that the horizontal and vertical displacement of the upper and the lower fixed support are both 0.

Its basic parameters are: elastic modulus E=70560Mpa,ensity$\rho_1 = 2.810^3$kg/m^3, tensile allowable stress: σ_{max}=421MPa and the maximum allowable displacement μ_{max}= 10mm, the length of line is 2m, the included acute angle between the rods is 45 degrees. Find the optimal cross-sectional area of thirteen-bar truss to minimize the total mass of the truss system. Penalty function method was used to deal with constrained condition.

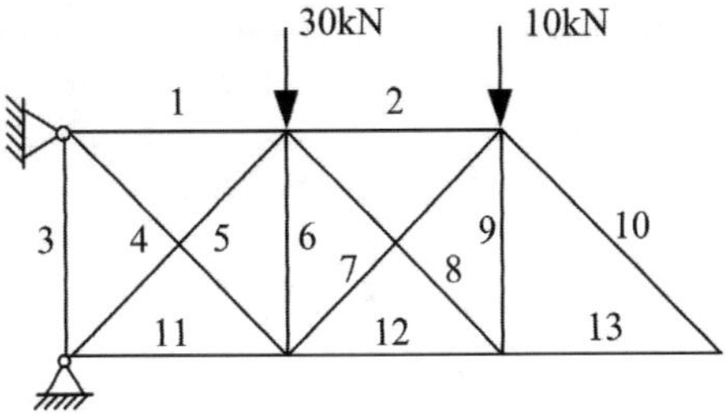

Figure 2. Results of structure calculation

Suppose the domain of design variables (optimal cross-sectional area) S=(9.21, 10.75, 13.32, 15.46, 17.52, 19.68, 21.83, 24.59, 28.26, 32.18, 38.55, 41.77, 49.24, 52.99, 63.65, 70.37, 75.43, 80.67, 84.49, 88.70, 97.95, 104.63, 109.36, 117.26, 122.48, 134.96, 173.57, 188.50, 203.85) (unit cm^2).

The mathematical expression of the problem is:

Objective function: $\min f(x) = \sum_{j=1}^{n} \rho_1 \cdot x_j \cdot l_j$;

In the formula,
x_j: the sectional area of the j-th support frame, j=1,2,...n;
l_j: the length of the j-th support frame, j=1,2,...n;
ρ_1: material density;
According to the constraints of structural design,

$\sigma_j \leq [\sigma]$ j=1,2,...n;

$\sigma_{cj} \geq [\sigma_c]$ j=1,2,...n;

$\mu_j \leq \mu_{\max}$ j=1,2,...n_0;

σ_j : Stress of the j-th support frame;

$[\sigma]$: Maximum allowable stress;

μ_j : Displacement of the *j*-th node, j=1,2,...n_0, n_0 is the number of nodes;

$\mu_{\max}$: Maximum allowable displacement.

Since ant colony algorithm itself can only deal with unconstrained problems, constraints need to be converted into unconstrained problems through certain rule. In this paper, the penalty method is used to deal with the constraint conditions. Constructing penalty function as follows:

$$\phi(\boldsymbol{x}, \gamma^{(k)}) = f(\boldsymbol{x}) + \gamma^{(k)}\left(\sum_{i=1}^{n} (\max(0, \sigma_i(\boldsymbol{x})))^2 + \sum_{i=1}^{n} (\max(0, \sigma_{ci}(\boldsymbol{x})))^2 + \sum_{j=1}^{n_0} (\max(0, \mu_j(\boldsymbol{x})))^2\right) \quad (\boldsymbol{k} = 0,1,2\cdots) \tag{1}$$

$g_i(x)$ is the constraint condition of stress, $\mu_j(x)$ is the constraint condition of displacement, $\gamma^{(k)}$ is the penalty factor, $\gamma^{(0)}=10^{-10}$.using variable scale method(DFP)in unconstrained optimization method to solve penalty function.

The improved ant colony algorithm is used for simulation. There is no theoretical basis for the selection of parameters in the algorithm[4]. If the value of α is too large, the ant colony search will fall into the local optimum too early. The higher the value of β is, the more likely the ant will choose the shortest path at a local point. If it is too large, the random performance and global search ability of the algorithm will be affected; if it is too small, the random performance and global search ability of the algorithm will be improved, but the convergence speed of the algorithm will be reduced. If $1-\rho$ is too large, the random performance and global search ability of the algorithm will be affected; if it is too small, the random performance and global search ability of the algorithm will be improved, but the convergence speed of the algorithm will be reduced. According to experience and through several simulation running tests, various simulation operating parameters are given as Table 1. the number of ants in this example is 20. The calculation results of the optimization solver are shown in Table 2, the calculation process is shown in Figure3.

Table 1. Calculation parameter setting

n	α	β	ρ	Q	q_0	τ_0	n_c
20	0.5	2	0.9	5000	0.95	300	200

Table 2. Results of calculate

Design variables(cm2)	Parameters before optimization	Parameters after optimization
X1	203.85	203.85
X2	104.63	109.36
X3	173.57	122.48
X4	41.77	49.24
X5	84.49	75.43
X6	63.65	49.24
X7	70.37	52.99
X8	28.26	38.55
X9	188.50	173.57
X10	134.96	117.26
X11	188.50	188.50
X12	97.95	84.49
X13	24.59	17.52
Total mass of structure (kg)	1629.73	1576.28

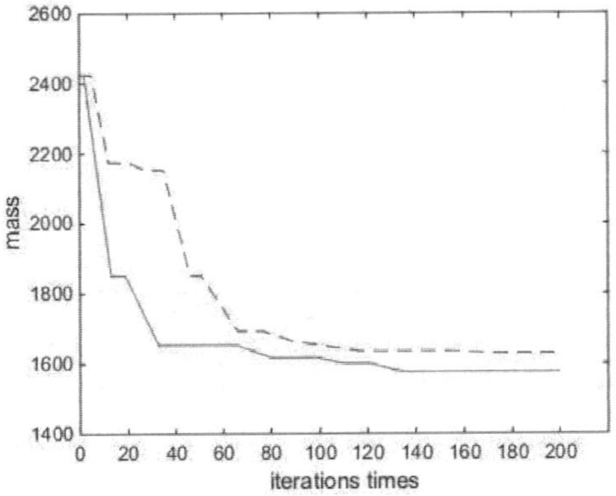

Figure 3. Calculation process

5. Conclusion

In this paper, an improved ant colony algorithm is used to optimize the thirteen-bar truss structure. From the simulation results, it is in line with the actual situation and the feasibility of the method is verified. The optimized design can improve the stress distribution, optimize the structure layout, reduce its own weight, and save material consumption, which is of great significance for reducing the manufacturing cost of the mechanical structure. Ant colony algorithm has something in common with genetic algorithm and simulated annealing algorithm[5], that is, it does not have very high requirements on the mathematical model of the optimization object, then optimization solution can be carried out, it is a good algorithm with potential.

References

[1] Zhang Z Q, Ren Z D, Li H N, Li N and Wang W 2016J. Journal of Architecture and Civil Engineering **33** 76
[2] Jiang G C, Xiao Y J and Xu X F 2020J. Shanghai Energy Conservation **12** 1464
[3] Wang L and Shi X 2019J. Journal of Nanjing University of Science and Technology **43** 700
[4] Li G, Li H ,Zhang S Y, Luo Q H and Yang H F 2018 J. China Sciencepaper **13** 1909
[5] Gong Y B 2014J. Mechanical Science and Technology for Aerospace Engineering **33** 807

Research on Programming Model and Compilation Optimization Technology of Multi-Core GPU

Liyan Chen

School of Computer and Information Engineering, Nanchang Institute of Technology, Nanchang, Jiangxi 330044, China

*corresponding author's e-mail: 76025835@qq.com

Abstract. GPGPU (General Purpose Computing on Graphics Processing Units) has been widely applied to high performance computing. However, GPU architecture and programming model are different from that of traditional CPU. Accordingly, it is rather challenging to develop efficient GPU applications. This paper focuses on the key techniques of programming model and compiler optimization for many-core GPU, and addresses a number of key theoretical and technical issues. This paper proposes a many-threaded programming model ab-Stream, which would transparentize architecture differences and provide an easy to parallel, easy to program, easy to extend and easy to tune programming model. In addition, this paper proposes memory optimization and data transfer transformation according to data classification. Firstly, this paper proposes data layout pruning based on classification memory, and then proposes Ta T (Transfer after Transformed) for transferring Strided data between CPU and GPU. Experimental results demonstrate that proposed techniques would significantly improve performance for GPGPU applications.
Keywords. Many-core GPU; Programming model; Compiler optimization

1. Introduction

With the increasingly refined scientific and technological research, scientific computing to large-scale parallel computing put forward higher demand. GPU (Graphic Processing Unit) is the processor for throughput design and on-chip integration of many parallel computing components. In recent years, the use of unified architecture of the GPU and the use of high-level language programming development platform, making the GPU common computing domain GPGPU (General Purpose computing on Graphics Processing Units) into the rapid development period [1]. However, GPU architecture and programming model are different from the traditional CPU architecture and programming model, the development of efficient GPU applications is still very challenging. Therefore, this paper is aimed at the unified architecture of GPU architecture. It deeply studies the programming model and compile and optimization key technologies for public core GPU. It mainly includes: public thread programming model for many-core GPU, GPGPU application mapping to calculate granularity parallel method, GPU storage structure optimization technology, CPU+GPU heterogeneous system load balancing strategy, the many- core GPU for the public thread programming model framework prototype system design and implementation.

At present, the many-core processor has been widely used in the computer from the giant machine to the various computing areas; computing technology has been fully into the nuclear era. Intel demonstrated its 80-core processor prototype in the fall of 2006, marking the industry's move toward multi-core to multi-core development [2]. IBM also introduced the 80-core processor Cyclops-64 to support the construction of thousands of megabytes of giant machines. In 2007, NVIDIA introduced its new GPU architecture CUDA (Compute Unified Device Architecture), the latest support CUDA architecture GPU contains 1024 computing cores, which has been put into practical use, but how efficient use of hundreds of computing core is still facing challenge [3]. Therefore, the programming model and compilation optimization technology of the public core architecture will be an important development direction for future high performance computing technology.

2. Methodology

This paper uses the literature analysis method to define the related concepts involved in this study, and clarifies the meaning of these concepts in this study. It analyzes the basic theory used in this study and its practical guiding significance to this course. And using empirical research to analyze key techniques of programming model and compiler optimization for many-core GPU, and addresses a number of key theoretical and technical issues [4]. As a result of the unified architecture GPU powerful computing power and the use of high-level language programming a variety of development platform, unified architecture GPU has been widely used in high-performance computing and scientific computing, including molecular dynamics, signal processing, graphic transformation, data discovery, medical diagnosis, biomedicine, video conversion, financial analysis, astrophysics, genetic structure prediction, weather forecasting, oil exploration, earthquake prediction and many other fields. In other words, the use of GPU parallel computing power has become an important way to solve intensive computing science and engineering problems.

3. Results and discussion

3.1. Ab-stream public thread programming model

Corresponding to the many-core processor architecture, this paper proposes the following principles that the public thread programming model framework ab-Stream must meet: Easy-to-parallel, ab-Stream programming model must be able to fully exploit the advantages of many-core computing resources in the many-core processor; Easy programming, because the C language programming model has been accepted by the industry and widely used, so the best ab-Stream programming model is a kind of C language programming mode; Easy to expand, the ab-Stream programming model must be able to provide a well-expandable interface to the emerging many-core architectures [5]. In view of the above criteria, mapping rules of the public thread programming model ab-Stream is shown in Figure 1, at the same time, the overall framework design of the ab-Stream is shown in Figure 2.

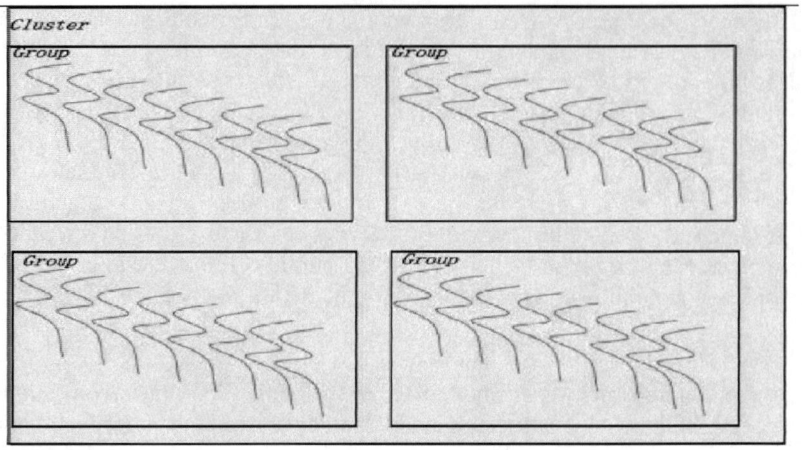

Figure 1. Mapping rules of ab-Stream

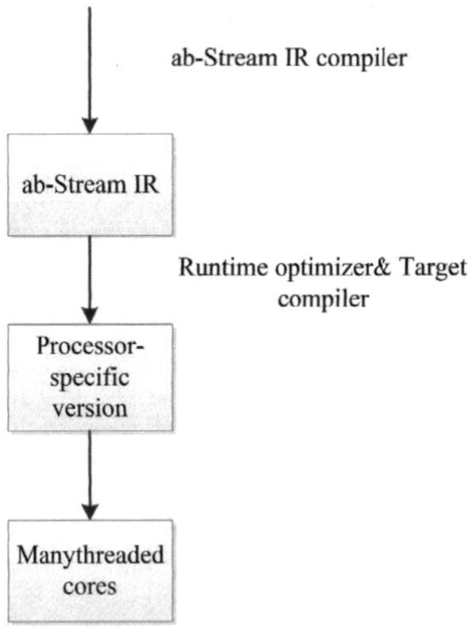

Figure 2. Overall frame of ab-Stream

It can be seen from the above that the ab-Stream programming model can well shield the kernel architecture differences and provide programmers with a programming model framework that is easy to parallel, easy to program, easy to extend and easy to tune.

3.2. Public thread architecture storage optimization technology

The many-core architecture is typically a storage-constrained, high-performance computing architecture, so the storage architecture optimization technology of the architecture is critical to the performance of the program [6]. In order to broaden the application of nuclear processors and ease the storage wall, the many-core processor architecture is a multi-level storage structure. The typical storage hierarchy is shown in Figure 3.

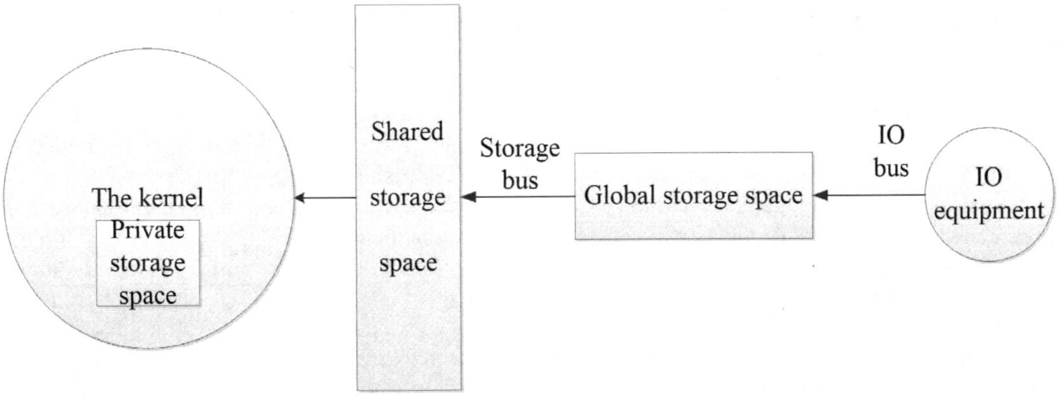

Figure 3. Storage hierarchy

According to the multi-level storage structure of the design rules, off-chip large-capacity global storage space usually access latency is large, the on-chip access delay is small and the private storage space capacity is very limited. Mainstream core processors have adopted a multi-level storage tier design strategy [7]. The typical many-core processor storage level and its related attributes are shown in Table1.

Table 1. The typical many-core processor storage level and its related attributes

Memory	Typical capacity	Position	Implementation technology	Delay
Global storage space	>1G	Outside	CMOS DRAM	>300Cycles
Shared storage space	<16M	Inside	CMOS SRAM	<100Cycles
Private storage space	<16K	Inside	Multi-port conventional memory CMOS	<50Cycles

The support of multi-level storage hierarchy and peer-to-peer mode diversity enables the developer to efficiently exploit the time and spatial locality of the application. Through the effective use of on-chip memory space, it can alleviate the global memory pressure and hide the global memory access delay. And each type of storage space has its own unique advantage and special purpose [8]. As a result, storage optimization techniques for the many-core architecture are critical to the overall performance of the application based on procedural features.

3.3. Storage technology optimization for CUDA

The CUDA architecture contains a variety of storage resources and storage levels to support a wide range of GPGPU application mappings, and different memory spaces have different attributes to meet the needs of different types of applications. Although the existing CUDA storage optimization has formed some general optimization criteria, the choice of diversified CUDA storage space in different types of CUDA application mapping is still relatively random and does not form a more systematic and comprehensive experience and guidelines of data structure [9]. Therefore, based on the practical application, this paper summarizes a set of effective data access optimization methods through a large number of engineering practices: data access optimization technology based on data classification [10].

Data classification storage is a very important task of memory, but also a very tricky and heavy task. According to the data type, we can make data analysis and sorting classification, and according to the appropriate criteria for matching [11]. When the data classification is completed, the next will be more arduous work, that is, to find the most suitable for storing each type of data storage space, and then the data movement. Therefore, the completion of this simple and routine work requires the accumulation of a lot of engineering experience and costly human resources costs

Data classification storage is important because it is the best to meet user needs, but also makes storage costs are minimized. The advantages of data classification and storage are as follows: Reduce overall storage costs. According to the data localized storage data can be integrated to play a multi-level storage architecture performance advantages and cost advantage. Performance optimization: Classification of storage can make different characteristics and use of storage space to maximize the overall benefits. Improve data availability: The classification store can store data according to the data locality, which can reduce the repeated storage and access of data, reduce the storage access time, and improve the reusability of the data, and maximize the bandwidth utilization of the storage system [12]. The basic process of data sorting is shown in Figure 4.

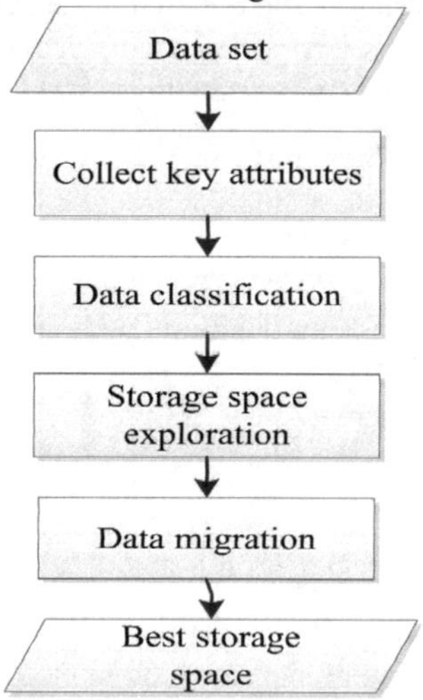

Figure 4. The basic process of data sorting

The data classification method proposed in this paper is based on the CUDA storage hierarchy. According to the CUDA application, the key attributes of the data structure includes the access rights, the owner, the life cycle, the size and other key data attributes to sort the data, and then explicitly migrate each type of data to a memory space that maximizes the storage bandwidth to optimize the storage of data [13]. The specific operation of the process is shown in Figure 5.

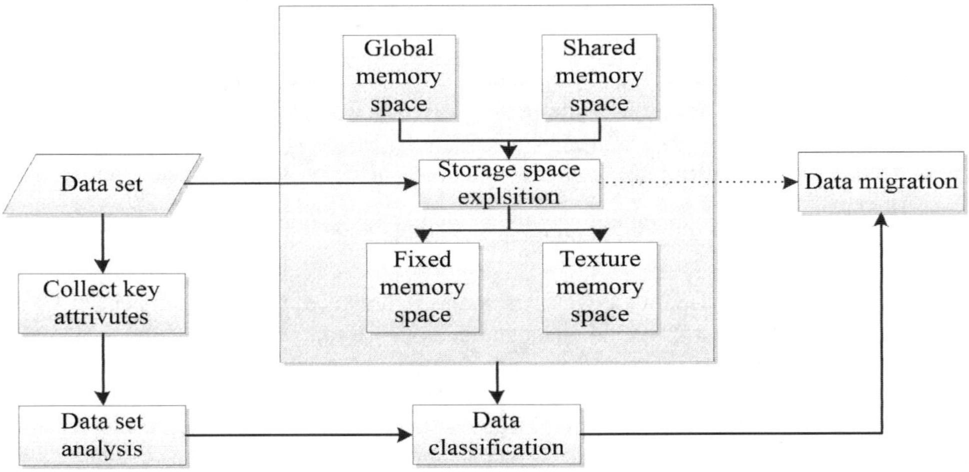

Figure 5. The specific operation of the process

3.4. Transmission optimization technology for Strided data

In order to formalize the transmission method of Strided data structure, defining the array A [D1] [D2] ... [Dn], Strided data vector data Strided A, which consists of the internal vector of array A [GB1: GE1], [GB2: GE2], ..., [GBn: GEn], is transferred from CPU memory to Strided data vector in GPU memory space structural variables Strided M [*] [HB1: HE1] [HB2: HE2] ... [HBn: HEn] storage space [14].The process is shown in Figure 6.

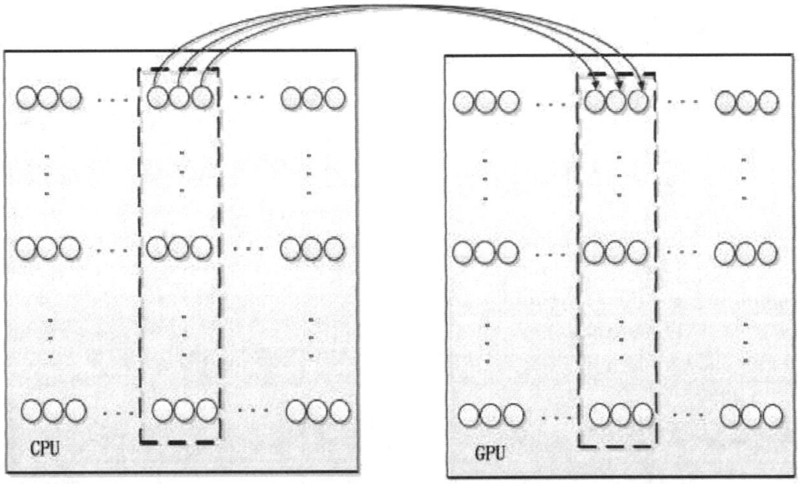

Figure 6. Strided data vector data is transferred from the CPU to the GPU

According to the CUDA data transmission API rules, one of the most simple Strided data vector data structure transmission methods that selective copy (Selective Copy, SC) is starting up cuda Memcpy for many times and constantly recalculates. And calculates the next piece of data stored in the next section. The method needs to constantly calculate the data source address offset scoffed and the target address offset stiffest to locate dst and src in order to facilitate the accurate transmission of data during transmission.

For the SC mode in the transmission Strided data vector data structure not only calculates the positioning complex, and needs additional startup overhead, there is a good method: a time-effective Strided data vector data structure transfer method that is redundant copy. This method will be a non-continuous Strided data vector data space first copying to the target storage space, which will be concerned about the Strided data vector data and do not care about other additional data as a continuous storage space, which copies to the target storage space. Then the selection of the Strided data vector data of interest is transferred to the next calculation process, but the extra data of non-attention needs to occupy a certain amount of storage space [15]. The RC mode is simple to operate and does not require multiple start of the cuda Memcpy, which reduces the overhead of additional startup.

4. Conclusion

In recent years, GPGPU has been widely used in high-performance computing. However, GPGPU is storage-constrained high-performance architecture, and GPU architecture and programming model is different from the traditional CPU architecture and programming mode, the development of efficient GPU applications is very challenging. Based on the challenges and constraints of GPGPU in the field of general computing, this paper has made the stage research results, and provides an effective solution for GPGPU to further expand the application space and promote the development of high performance computing field. This paper focuses on the GPU-oriented public thread programming model and its key support technology: GPGPU application mapping, based on CUDA storage level optimization technology, as well as the prototype system design and implementation of some phased results, for further research related to the good foundation. Combined with the basis of this paper and the future direction of development of the GPU, there are the following issues worthy of further study: Research to optimize the system for the operation of the many-core GPU; Research on compiling optimization technology for many-core.

Acknowledgement

Project fund: General Projects of Science and Technology Projects of Jiangxi Provincial Department of Education, Project No. GJJ212114, Optimization of resource scheduling algorithm based on responsive web front end.

References

[1] Kan G, Liang K, Li J, Ding L, He X, Hu Y, Mark A-B, Adel H 2016 Accelerating the SCE-UA Global Optimization Method Based on Multi-Core CPU and Many-Core GPU *Advances in Meteorology* **3**(4) 56-9

[2] Simon M S, James P, Richard B S, Amaurys A I 2015 High performance in silico virtual drug screening on many-core processors *International Journal of High Performance Computing Applications* **29**(2) 22-9

[3] Dorgham O M, Laycock S D, Fisher M H 2012 GPU accelerated generation of digitally reconstructed radiographs for 2-D/3-D image registration *IEEE Transactions on Biomedical Engineering* **59**(9) 32-4

[4] Smelyanskiy M, Holmes D, Chhugani J, Larson A, Carmean D M, Hanson D, Dubey P, Augustine K, Kim D, Kyker A, Lee V W, Nguyen A D, Seiler L, Robb R 2009 Mapping high-fidelity volume rendering for medical imaging to CPU, GPU and many-core architectures *IEEE Transactions on Visualization and Computer Graphics* **15**(6) 22-8

[5] Changkyu K, Jatin C, Nadathur S, Eric S, Anthony D N, Tim K, Victor W L, Scott A B, Pradeep D 2011 Designing fast architecture-sensitive tree search on modern multicore/many-core processors *ACM Transactions on Database Systems (TODS)* **36**(4) 54-5

[6] Giles M B, Mudalige G R, Sharif Z, Markall G, Kelly P H J 2012 Performance Analysis and Optimization of the OP2 Framework on Many-Core Architectures *The Computer Journal* **55**(2) 77-9

[7] Pennycook S J, Hammond S D, Mudalige G R, Wright S A, Jarvis S A 2012 On the Acceleration of Wavefront Applications using Distributed Many-Core Architectures *The Computer Journal* **55**(2) 23-9

[8] Gábor J T, György C 2014 Initial condition for efficient mapping of level set algorithms on many-core architectures *EURASIP Journal on Advances in Signal Processing* **1** 32-7

[9] Gan X B, Shen L, Tan Q Y, Liu C, Wang Z Y 2011 Performance Evaluation and Optimization on GPU *Advanced Materials Research* **1231**(219) 44-7
[10] Ikebata A, Xiao F 2016 GPU-accelerated large-scale simulations of interfacial multiphase fluids for real-case applications *Computers and Fluids* **3**(5) 34-9
[11] Wang B, Qi Z, Ma R, Yao J, Guan H 2016 A user mode CPU–GPU scheduling framework for hybrid workloads *Future Generation Computer Systems* **2**(3) 34-9
[12] Gizem E, Bülent Ç 2017 Accelerating local search algorithms for the travelling salesman problem through the effective use of GPU *Transportation Research Procedia* **22**(2) 44-7
[13] Hwu W W, Chang P P 1989 Achieving high instruction cache performance with an optimizing compiler *ACM SIGARCH Computer Architecture News* **17**(3) 47-53
[14] Sheng L, Dai Z B 2017 Research on the Maritime Communication Cryptographic Chip's Compiler Optimization *Polish Maritime Research* **24**(2) 22-8
[15] Li L, Meng Q Y 2014 Application of Computer Technology in the Design of the Third Party E-Commerce Platform *Applied Mechanics and Materials* **3277**(608) 33-8

Solving Smallest Enclosing Circle Problem by Wading across Stream Algorithm

Shang Gao[1,*] and Hao Pan[2]

[1]School of Computer, Jiangsu University of Science and Technology, Zhenjiang 212003, Jiangsu, China
[2]Suzhou Institute Of Construction and Communications, Suzhou 215104, Jiangsu, China

[*]corresponding author's e-mail: gao_shang@just.edu.cn

Abstract. The feature of the smallest enclosing circle problem is analysed. According to idea of wading across the stream by feeling the way, a kind of fast efficient wading across stream optimization algorithm is put forward. For solving smallest enclosing circle problem, the improved wading across stream algorithm shrink the search space gradually. It is proved that wading across stream algorithm is a simple and effective algorithm.

1. Introduction

The Smallest Enclosing Circle Problem is, simply stated, the problem of finding the smallest circle that completely contains a set of points. The minimal enclosing circle is useful when people want to plan the location of a shared facility. In 1991, Welzl [1] puts forward the new method based on some early research methods. Welzl used randomized incremental algorithm for smallest enclosing ellipse problem, and this algorithm is extended to the smallest enclosing ball problem and the smallest enclosing ellipsoid problem. Mark deBerg [2] improved the method was called the randomized incremental algorithm. In 2005, Frank Nielsen [3] proposed a deterministic algorithm based on dual decision-making, which was called dual decision algorithm. In 2000, Wang Wei [4] et al proposed a fast algorithm for discrete point set minimum bounding circle, known as the most distant point first approximation algorithm. Li Hongjun [5] analysed and compared various algorithms. Wang Hongfa [6] Found Minimum Contain Circle of Plane Finite Points By Genetic Algorithm. In this paper, a wading across stream algorithm is proposed to solve this problem.

2. The smallest enclosing circle problem

The smallest-circle problem or minimum covering circle problem is a mathematical problem of computing the smallest circle that contains all of a given set of points in the Euclidean plane.

If n is the number of points. The intuitive approach is that all the circles generated by three points and by two points are enumerated, then find the smallest one. The number of circles is $C_n^2 + C_n^3$, so the complexity of this algorithm is $O(n^3)$.

Most of the geometric approaches for the problem look for points that lie on the boundary of the minimum circle . The smallest enclosing circle has the following properties:

Properties I: Given N points in the plane, the smallest enclosing circle is unique.

If there are two the smallest enclosing circle C_1 and C_2, which are the same radius and enclose all points as shown in Figure 1(a). Then all points are in the intersection part of two circle as shown in

Figure 1. Obviously, you can find the smaller circle C , which encloses intersection part as shown in Figure 1(b).

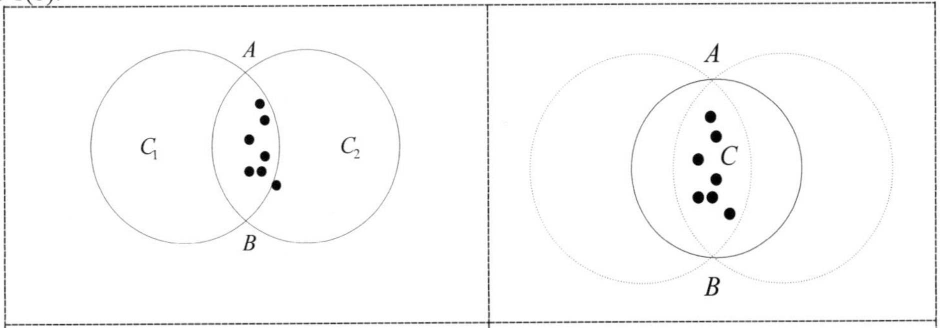

(a) two smallest enclosing circles C_1 and C_2 (b) a smaller enclosing circle C

Figure 1. .The two smallest enclosing circle contained all points

Properties II: The minimum covering circle of a set S can be determined by at most three points in S which lie on the boundary of the circle. If it is determined by only two points, then the line segment joining those two points must be a diameter of the minimum circle. If it is determined by three points, then the triangle consisting of those three points is not obtuse.

The centre of smallest-circle is (x, y), the mathematical model is as follows

$$\min \max_{1\le i\le n} \sqrt{(x-x_i)^2+(y-y_i)^2} \tag{1}$$

Or

$$\begin{aligned} &\min v \\ &s.t. \quad \sqrt{(x-x_i)^2+(y-y_i)^2} \le r \quad (i=1,2,\cdots,n) \end{aligned} \tag{2}$$

3. The idea of wading across stream algorithm

The idea of fast random optimization algorithm is from the idea of wading across the stream by feeling the way. The fast random optimization algorithm, which is called the wading across stream algorithm [7][8][9] , acts a solution as a start point, then searches several random solutions near the start point, and find the best solution of these solutions. This best solution is taken as the next start point, and then searches several random solutions near this start point, and so on. The iterative process is shown in Figure 2. The optimization problem contains continuous optimization problem and the discrete optimization problems. we use wading across stream algorithm to solve these two kinds problems separately as follow.

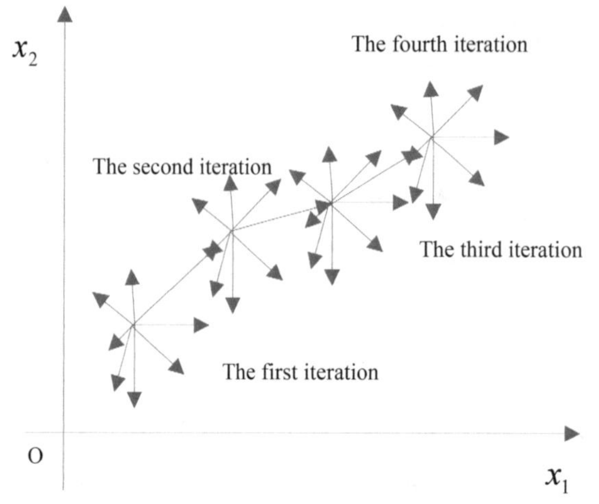

Figure2. Iterative process of wading across the stream algorithm

The wading across stream algorithm (WSA) and simulated annealing algorithm (SAA) are somewhat similar, but the effect of WSA are better than SAA, and the complexity of WSA is simpler than SAA. The simulated annealing algorithm is from the one solution to search one solution, but wading across stream algorithm is from the best solution of several solutions to search one solution.

4. Solve smallest enclosing circle problem by wading across stream algorithm

4.1. Wading across stream algorithm

The framework of solving smallest enclosing circle problem by wading across stream algorithm (WSA) is as follows:

Step 1. Set the parameters of algorithm: the number of solution m, neighborhood radius L, and the maximum number of iterations $n_{\max}$.

Step 2. Calculate the mean value of n points $(x_i, y_i), i = 1,2,\cdots,n$ as (x_0, y_0), $f^* = \max_{1\le i\le n}\sqrt{(x_0 - x_i)^2 + (y_0 - y_i)^2}$. Set $k = 0$.

Step 3. Generate m neighborhood solutions (x_{01}, y_{01}), (x_{02}, y_{02}), …, (x_{0m}, y_{om}) in the neighborhood of (x_0, y_0) with radius L. Set $x_{0i} = x_0 + L \cdot r_{xi}$, $y_{0i} = y_0 + L \cdot r_{yi}$, where r_{xi}, r_{yi} is the uniform distribution random number in the interval [-1, 1].

Step 4. Calculate the objective function $f'_1, f'_2, \cdots, f'_m$ of m neighborhood solutions. Find the best solution (x_0^*, y_0^*) and its objective function f^*. Set $(x_0, y_0)=(x_0^*, y_0^*)$. Save the best solution and set $k = k + 1$.

Step 5. If the given stopping condition (up to the required number of iterations nmax) is not met, save the best solution and go to step 3.

The time complexity of the algorithm is estimated as follows: Because it take more time to calculate the value of the objective function, the time complexity is about $O(m \cdot n_{\max})$.

4.2. The improved wading across the stream algorithm

With the number of iterations increases, more and more solutions close to the optimal solution. If the neighbourhood radius remains unchanged, the convergence speed will become slow. The improved method reduce the neighbourhood radius gradually when the number of iterations increases. The simple method is as the following $L_k = \alpha L_{k-1}$, where α is the contraction coefficient.

In general, α is about 0.99. The smaller α is , the faster search speed is. The improved wading across stream algorithm (IWSA) is as follows:

Step 1. Set the parameters of algorithm: the number of solution m, neighbourhood radius L, and the maximum number of iterations $n_{\max}$, contraction coefficient α.

Step 2. Calculate the mean value of n points $(x_i, y_i), i = 1,2,\cdots,n$ as (x_0, y_0), $f^* = \max_{1\le i\le n} \sqrt{(x_0 - x_i)^2 + (y_0 - y_i)^2}$. Set $k = 0$.

Step 3. Generate m neighbourhood solutions (x_{01}, y_{01}), (x_{02}, y_{02}), ..., (x_{0m}, y_{om}) in the neighbourhood of (x_0, y_0) with radius L. Set $x_{0i} = x_0 + L\cdot r_{xi}$, $y_{0i} = y_0 + L\cdot r_{yi}$, where r_{xi}, r_{yi} is the uniform distribution random number in the interval [-1, 1].

Step 4. Calculate the objective function $f'_1, f'_2, \cdots, f'_m$ of m neighbourhood solutions. Find the best solution (x_0^*, y_0^*) and its objective function f^*. Set $(x_0, y_0) = (x_0^*, y_0^*)$. Save the best solution and set $k = k+1$ and $L_k = \alpha L_{k-1}$.

Step 5. If the given stopping condition (up to the required number of iterations nmax) is not met, save the best solution and go to step 3.

5. Example analysis

5.1. Comparison of two algorithms

Example 1: The coordinates of 12 points are as shown in table 1（[6]）. The radius is 14.7085, and the center coordinates is (18.9786, 16.2653). the iterative process are shown in figure 2

Table 1. coordinates of 12 points

point	x	y
1	15.85	6.05
2	29.82	22.11
3	4.84	20.32
4	25.47	29.46
5	33.65	17.31
6	7.30	10.43
7	28.15	6.67
8	20.85	11.47
9	22.83	2.07
10	26.28	23.15
11	14.39	30.24
12	30.24	25.23

The parameters of WSA and IWSA are set as follows: $m = 20$, $n_{\max} = 100$, $L = L_0 = 1$, $\alpha = 0.9$. Their iterative process is shown in Figure 3.

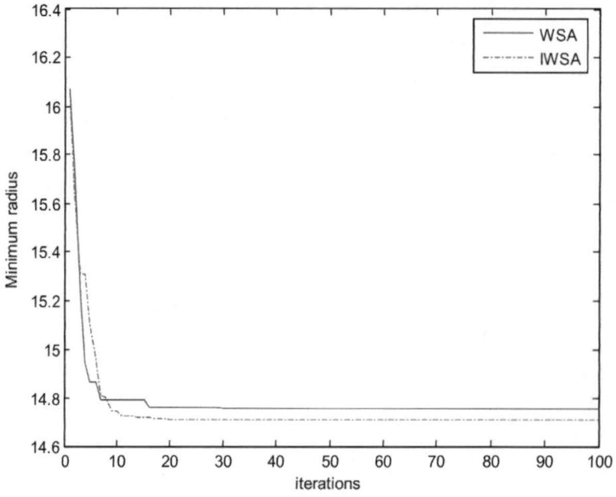

Figure 3. Iiterative process of two algorithms

From Figure3, the convergence speed of the improved algorithm is faster than that of the original basic algorithm, and the effect is better.

5.2. Parameter selection test

The shrinkage coefficient has an impact on the algorithm, where the shrinkage coefficient is 1, 0.8, 0.6 and 0.4. The contraction coefficient is 1, which is the original basic stone crossing algorithm. The iterative process is shown in Figure 4. It can be seen from Figure 4 that it is better to take moderate shrinkage coefficient. If the contraction coefficient is too small, the iteration area is relatively small and may fall into local extreme value. As shown in Figure 4, the effect of $\alpha = 0.4$ is not good. Figure 5 shows the iterative process when the contraction coefficient is taken as 1, 0.9, 0.8, 0.7 and 0.6. It can be seen from Figure 5 that there is little difference in the effect when the contraction coefficient is between 0.6 and 0.9.

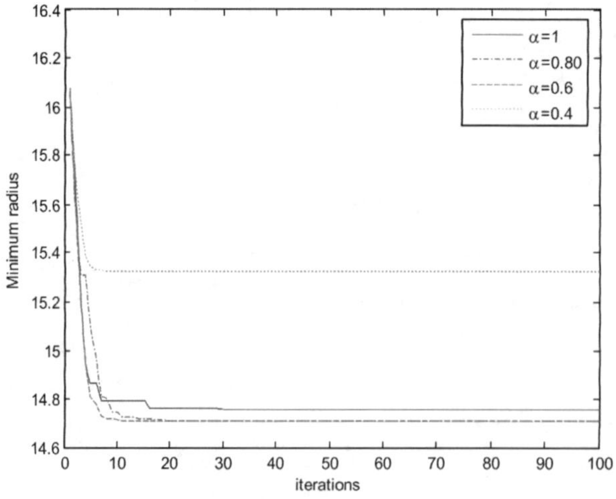

Figure 4. Comparison of different shrinkage coefficients

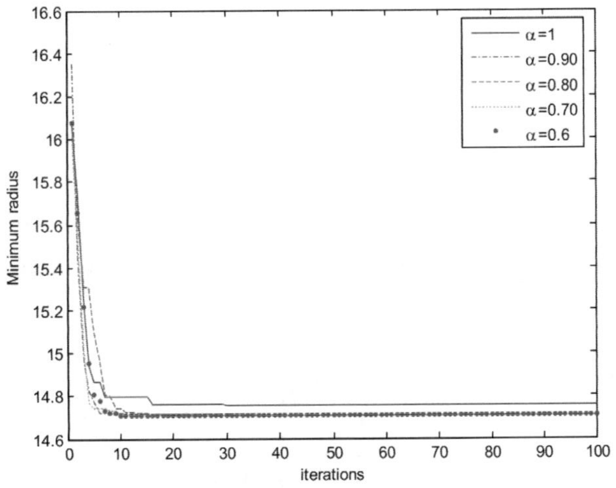

Figure 5. iterative process of different contraction coefficients

6. Conclusion

The smallest enclosing circle problem is an old problem. With the rapid development of computer technology, the smallest enclosing circle problem is applied more and more widely, such as design, application in industrial robot, collision detection, machine learning, computer graphics and other fields. Research results are more and more. This paper presents IWSA to solve of the smallest enclosing circle algorithm, and give its proof. The algorithm is simple and effective.

Acknowledgments

This work was supported by the National Natural Science Foundation of China under Grant and the National Natural Science Foundation of China under Grant 62176107.

References

[1] W. Emo, "Smallest enclosing disks (balls and ellipsoids)", in Maurer H. (Ed.), New Results and New Trends in Computer Science, Lecture Notes in Computer Science, Heidelberg: Springer-Verlag, vol.555, 1991, pp.359-370.
[2] Mark de Berg, Marc van Kreveld, Mark Overmars, et al, Computational Geometry: Algorithms and Applications. Springer-Verlag, 3rd rev. ed. 2008.
[3] N Frank, N Richard, "A fast deterministic smallest enclosing disk approximation algorithm", Information Processing Letters, 2005, vol.93, no.6, pp.263-268.
[4] W. Wang, W.P Wang, J. Y. Wang, An Algorithm for Finding the Smallest Circle Containing all Points in a Given Point Set", Journal of Software, 2000, vol.11, no.9, pp.1237-1240.(inChinese)
[5] H. J. Li, X. P. Zhang, "Analysis and improvement of smallest enclosing disk algorithm on discrete set of points", JOURNAL OF GRAPHICS, 2012,vol.3, no.2, pp.34-38. (inChinese)
[6] H. F. Wang, J. Liu, "Finding Minimum Contain Circle of Plane Finite Points with Genetic Algorithm", Journal of Nanchang University(Engineering & Technology), 2007, vol.29, no.4,pp.384-386. (inChinese)
[7] Shang Gao, Hualong Yu, Ling Qiu, Cungen Cao. The wading across stream algorithm. international Journal of Computers and Applications, 2014,36(4):127-132.
[8] Shang Gao, Cungen Cao, Solving Continuous Optimization Problem by Hybrid Wading across Stream Algorithm- Estimation Distribution Algorithm . Computer Science,2016,43(1):81-84. (inChinese)
[9] Shang Gao, Ling Qiu,Cungen Cao, Solving Continuous Optimization Problem by Wading Across Stream Algorithm. Journal of Nanjing Normal University (Natural Science Edition), 2015,38(1):108-112 (inChinese)

An Improved Particle Swarm Algorithm with the Elite Retain Strategy for Solving Flexible Jobshop Scheduling Problem

Mingliang Wu[1,2,*], Dongsheng Yang[1,2] and Tianyi Liu[1,2]

[1]Intelligent Electrical Science and Technology Research Institute, Northeastern University, Shenyang, Liaoning, China
[2]Key Laboratory of Integrated Energy Optimization and Secure Operation of Liaoning Province, Northeastern University, Shenyang, Liaoning, China

[*]corresponding author's e-mail: wuml_neu@163.com

Abstract. As a combination optimization problem that an enterprise often encounters in the process of jobshop production management, the flexible job shop scheduling problem has become one of the hot issues that scholars at home and abroad have continuously studied and explored in recent years. With the development of computer technology and the rise of interdisciplinary, intelligent optimization algorithms have gradually become the primary method to solve such problems. Particle swarm algorithm is an evolutionary search calculation method that simulates birds gathering flight, constantly changing their position and speed during movement, and finally reaching the optimal state. However, the application of this algorithm in the field of solving jobshop scheduling is not particularly mature, and there are some shortcomings in its application. Aiming at the disadvantages of particle swarm algorithm that it is easy to fall into the optimal local solution and the slow convergence speed of the algorithm in the later stage, this paper proposes an elite retention strategy and integrates it into the PSO to enhance its later solution ability.

1. Introduction

With the rapid development of modern science and technology and the continuous growth of the global economy, the market competition environment for enterprises has become more and more fierce. If enterprises want to develop better in the fierce competition environment, they need corporate leaders to stand [1-3]. From a higher perspective, make a practical and feasible long-term development plan for the enterprise. In the face of product diversification, small batches, and more targeted customer needs in the market, improving enterprises' production capacity and efficiency is an important direction [4].

The rational allocation of production resources is a significant challenge that modern enterprises need to face. Therefore, whether it is in theoretical scientific research or the actual production process of an enterprise, the flexible jobshop scheduling problem (FJSP) is a scheduling optimization problem that an enterprise needs to solve frequently [5]. This scheduling problem does not only appear in the production process of the enterprise jobshop, but also frequently appears in the fields of aerospace science and engineering, transportation, network communication, Etc [6]. Therefore, studying how to solve the FJSP has become a current hot issue, attracting more and more domestic and foreign experts and scholars. As the most typical shop scheduling problem, the FJSP has the characteristics of more complex modeling, more incredible calculation difficulty, and more constraints. It is challenging to solve such problems using traditional accurate calculation methods, and the time cost will be huge [7]. Therefore, scholars at home and abroad have shifted the focus of research on solving such problems to intelligent optimization algorithms. For example, intelligent optimization algorithms such as genetic algorithm, artificial immune algorithm, ant colony algorithm, dragonfly algorithm, and simulated annealing algorithm are applied to solve jobshop scheduling. In the problem, a new idea is provided to solve the problem of jobshop scheduling [8-12]. Although this type of algorithm cannot guarantee to find the optimal solution to the scheduling problem, compared with the precise calculation method, it will be found that this type of algorithm runs faster and has higher efficiency, and has a more critical role in the actual production scheduling of the enterprise [13,14]. However, the traditional swarm intelligence optimization algorithm has limitations in processing FJSP and is prone to the premature phenomenon [15]. In order to avoid this difficulty, this paper proposes an elite retention strategy (ERS) and integrates it into PSO to enhance its optimization ability.

The rest of this article is organized as follows. First, introduce FJSP in Section 2. Then the basic process of PSO and the details of ERS are introduced in Section 3. Then, perform experiments and conduct experimental analysis in Section 4. Finally, the full text is summarized in Section 5.

2. The description of FJSP

The FJSP is an essential part of the production process of an enterprise. Having an efficient, scientific, and reasonable jobshop scheduling plan can improve the efficiency of jobshop production activities and ensure the effective operation of business activities [16]. The FJSP can be described as follows: a set of several jobs to be processed in the jobshop, and these jobs to be processed are reasonably distributed on each processing machine according to the pre-defined processing sequence. Processing, the process of finally meeting the required scheduling goals. Without loss of generality, the constraints of FJSP can be described as:

- (1) There can be multiple machines for each process, but only one machine can be selected for processing.
- (2) Once each process starts processing, it cannot be interrupted during processing.
- (3) A machine can only have one job processed on it at the same time.
- (4) Operations of one same job exist sequence.
- (5) Only one machine can be selected for each operation.
- (6) All jobs can be processed, and all machines are optional when the task starts.
- (7) The processing progress is non-preemptive, and any preemptions or cancellations are not permitted.

The scheduling objectives of the FJSP are usually selected to maximize the benefits of the enterprise, but there may be many scheduling objectives for an enterprise. This is because each department of the enterprise will set different needs according to the functions and interests of their departments—scheduling goals to maximize the benefits of the department of the enterprise. For example, to shorten the time and increase the output, the production department will set the minimum and maximum completion time as the scheduling target. The sales department will be very concerned about whether the delivery can be delivered to the customer on time and take the delivery date as the scheduling target. Therefore, the development of an enterprise needs to select an appropriate scheduling target based on the actual situation of the enterprise. Classically, this paper utilizes the maximum completion time to test the performance of the PSO, which can be given as:

$$f_1 = min(Makespan) = min(CT_k) \tag{1}$$

where *Makespan* represents the maximum completion time of all machine, it is the main indicator to measure the completion time of all jobs. CT_k is expressed as:

$$CT_k = \sum_{i=1}^{n} \sum_{j=1}^{K_i} (S_{ijk} R_{ijk} + T_{ijk} R_{ijk}) \tag{2}$$

where S_{ijk}, T_{ijk} implies the starting time, and the processing time of the *jth* operation of the *ith* job, respectively. R_{ijk} denotes whether the machine *k* processes operation O_{ij}, 1 means processed, 0 means unprocessed. K_i represents the number of operations of job J_i. *n* represents the number of job.

3. The improved particle swarm optimization

3.1. The origin of particle swarm optimization

The particle swarm optimization (PSO) algorithm is a global random search algorithm based on swarm intelligence proposed by Kennedy and Eberhart, inspired by the results of artificial life research, by simulating the migration and swarming behavior of birds in the foraging process. All kinds of organisms in nature have certain group behaviors. One of the leading research fields of artificial life is exploring the group behaviors of natural organisms to construct its group model on the computer. The group behavior of birds and fish in nature has always been the research interest of scientists. The biologist Craig Reynolds proposed a very influential bird grouping model in 1987. In his simulation, each individual follows:

- 1) Avoid collisions with individuals in the neighborhood;
- 2) Match the speed of individuals in the neighborhood;
- 3) Fly to the center of the flock, and the entire flock flies to the target.

3.2. Algorithm principle

PSO draws inspiration from this model and is used to solve optimization problems. In PSO, the potential solution of each optimization problem is a bird in the search space, called a particle. All particles have a fitness value determined by the optimized function, and each particle also has a speed that determines the direction and distance of their flight. Then the particles follow the current optimal particle and search in the solution space.

PSO is initialized as a group of random particles (random solution), and then the optimal solution is found through iteration. In each iteration, the particle updates itself by tracking two extreme values; the first is the optimal solution found by the particle itself, and this solution is called the individual extreme; the other extreme is the optimal solution currently found by the entire population Solution, this extremum is the global extremum. In addition, instead of the entire population, only a part of it can be used as the neighbors of the particle. Then the extremum in all the neighbors is the local extremum.

Suppose that in a D-dimensional search space, there is a tribe composed of N particles, where the ith particle represents a D-dimensional vector.

$$X_i = (x_{i1}, x_{i2}, \dots, x_{iD}), i = 1,2, \dots, N \tag{3}$$

The flight speed of the ith particle is also a D-dimensional vector.

$$V_i = (v_{i1}, v_{i2}, \dots, v_{iD}), i = 1,2, \dots, N \tag{4}$$

The optimal position of the ith particle searched so far is called the individual extremum, denoted as:

$$Pbest_i = (p_{i1}, p_{i2}, \dots, p_{iD}), i = 1,2, \dots, N \tag{5}$$

The optimal position searched by the entire particle swarm so far is the global extremum, denoted as:

$$Gbest = (p_{g1}, p_{g2}, \dots, p_{gD}) \tag{6}$$

In each iteration, the particles update their speed and position according to the following Equation.

$$v_i = v_i + c_1 * rand * (Pbest_i - x_i) + c_2 * rand * (Gbest - x_i) \tag{7}$$

$$x_i = x_i + v_i \tag{8}$$

where rand is the randoom number in [0,1], c_1 and c_2 are the learning factor.

3.3. The elite retain mechnism

In order to enhance the power of the PSO, the elite retain mechanism (ERM) is proposed and combined with PSO. The number of elite solutions retained can be mathematically calculate as:

$$N_p = \frac{\sum_{i=1}^{N} f(X_i)/N - f(X_{best})}{\sum_{i=1}^{N} f(X_i)/N} \times N \tag{9}$$

where N represents the population size, X_{best} denotes the individual with the smallest makespan, N_p implies the number of elite solutions retained. In each iteration, we use the first N_p elites of the current generation to randomly replace N_p individuals in the next generation to achieve the purpose of retaining excellent genes. The flowchart of the improved PSO is placed in Fig. 1

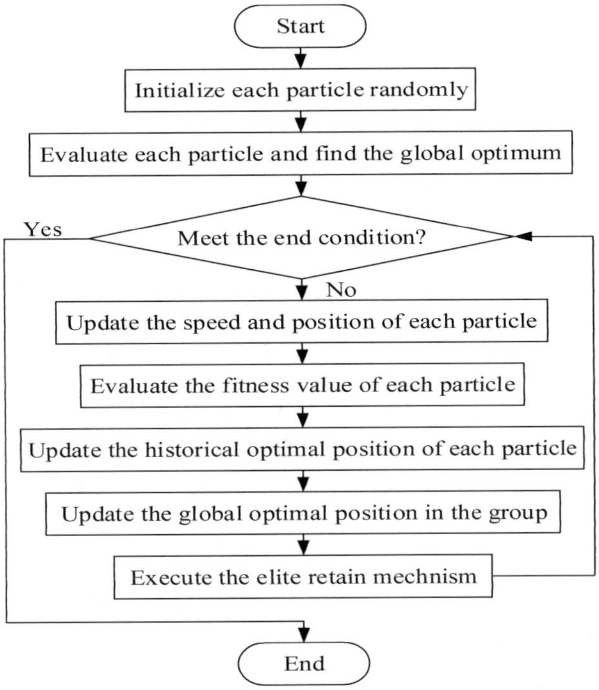

Figure 1. The flowchart of our PSO

4. Results and discussion

In this paper, the makespan is employed to test the advantage of our PSO. Meanwhile, 6 classic meta-heuristic algorithms that included dragonfly algorithm (DA), salp swarm algorithm (SSA), sine cosine algorithm (SCA), whale optimization algorithm (WOA), grasshopper optimisation algorithm (GOA), multi-verse optimizer (MVO) are introduced as comparison algorithm [2,17-21]. 10 typical FJSP instances that come from Fattahi are utilized as the benchmark in testing. We set the maximum number of iterations as 100, population size as 100, *c1* and *c2* of of PSO as 2. The parameters of other algorithms follow the recommended values of the original literature. The algorithm terminates when the number of iteration reaches the maximum generation count. Perform a total of ten experiments and record the average result in TABLE I (the results with * are the best solution results). We can see that our PSO can always obtain the shortest completion time. Moreover, the gap between other algorithms and our algorithm is very obvious. It clearly reminds us of the power and advantages of improved PSO.

The convergence curves of FJSP1-10 are placed in Fig. 2-4. In FJSP 2, 3, 6, 7, 9, 10, the lower black convergence curve representing improved PSO shows that improved PSO is in the leading position almost all the time. For FJSP 4, 8, the improved PSO can turn the situation around in the middle of the iteration. For FJSP 1, 5, the improved PSO can finally gain the advantage in the later iteration. All in all, the improved PSO has good global search ability and local search ability that can handle the above all FJSP instances.

The Gantt chart of FJSP1 and FJSP10 by our PSO are put in Fig. 5 (The horizontal axis represents the completion time, and the vertical axis represents the machine number). For example, in part(a) of

Fig. 5, at zero time, machine one starts the first process the first operation of the third job, machine two starts the first process the first operation of the fourth job, and machine three starts the first operation of the first job. While waiting for the third job's first operation to be processed, machine four starts to process the second operation of the third job. Waiting for the first operation of the fourth job to be processed, machine five starts to process the second operation of the fourth job, and so on. The Gantt chart in Fig. 5 has very few gaps, which proves that the scheduling scheme obtained by our PSO is efficient and the best choice with minimum completion time.

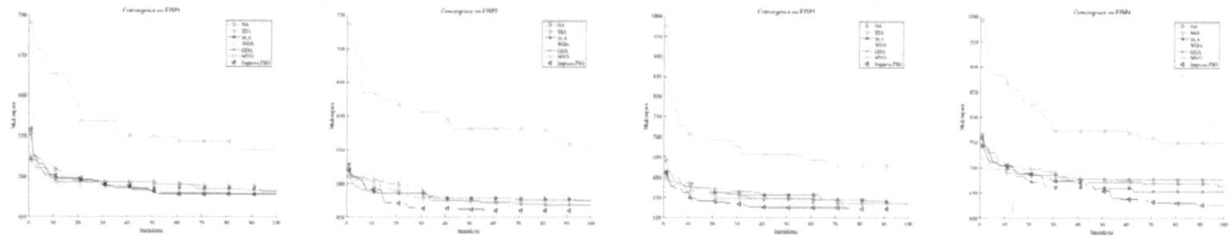

(a) Instance.1 (b) Instance.2 (c) Instance.3 (d) Instance.4

Figure 2. The convergence curve of *Instance. 1-4*

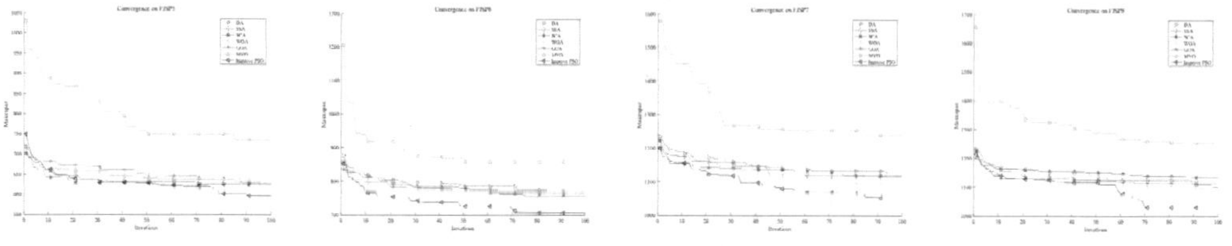

(a) Instance.5 (b) Instance.6 (c) Instance.7 (d) Instance.8

Figure 3. The convergence curve of *Instance. 5-8*

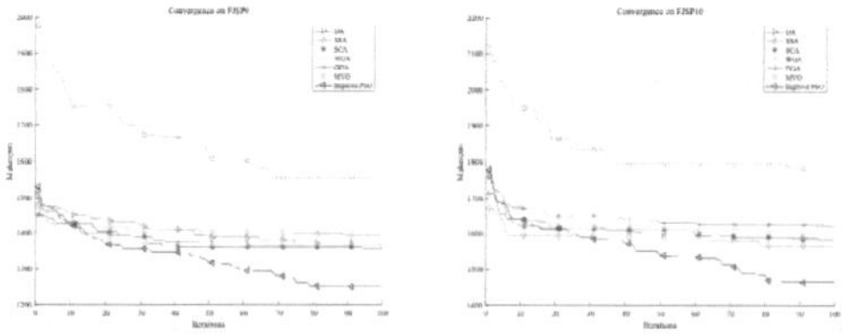

(b) Instance.9 (c) Instance.10

Figure 4. The convergence curve of Instance. 9-10

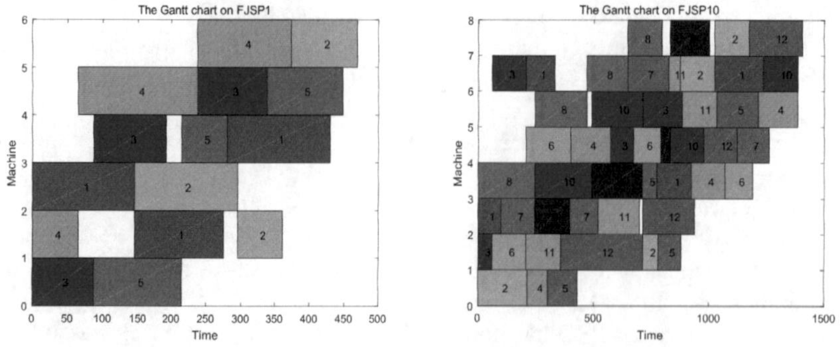

(a) The Gantt chart of FJSP1 by our PSO (b) The Gantt chart of FJSP10 by our PSO

Figure 5. The Gantt chart of FJSP1 and FJSP10 by our PSO

Table 1. The obtained average results

Algorithm	Problem									
	FJSP 1	F JSP2	FJSP 3	F JSP4	FJSP 5	F JSP6	FJSP 7	F JSP8	FJSP 9	F JSP1 0
DA	488	475.4	540.2	664.4	627	771.6	1125	1100.8	1361.4	1580.8
SSA	483	468.8	536.8	661	627.4	769.8	1115.4	1122	1395.6	1567.8
SCA	481	475	539.8	653.4	623.6	756.4	1119.2	1135.6	1356.2	1583.4
WOA	478.6	468.2	531.8	677.2	624.6	764.6	1117.4	1114	1364.4	1621.6
GO A	533.6	547.2	627.4	749	734.4	859	1239	1252.6	1555.8	1752.2
MV O	481.8	468	533.6	651.4	637.6	762.8	1093.8	1098.4	1373	1603.8
PSO	477.8*	460*	522.2*	626.8*	596.6*	705.8*	1053.2*	1028.8*	1251.6*	1464.8*

5. Conclusion

FJSP is a challenging and important NP-hard problem in the manufacturing field. Obtaining an efficient scheduling plan is very important for enterprises. All aspects of life are inseparable from the value created by FJSP research for people. In order to achieve a better solution to FJSP and or to obtain a scheduling plan with a smaller makespan, this paper proposes an elite retention strategy (ERS) and integrates it into the PSO. Ten FJSP instances are used as benchmarks to test the performance of our proposed algorithm. All experimental results show that our algorithm presents a significant advantage in comparison with the other six classic meta-inspired algorithms. Both convergence curve analysis and Gantt chart analysis can reflect the power and potential of the improved PSO. On this basis, further development and improvement of PSO will continue in the future.

Acknowledgement

This research work is supported by the National Key Research and Development Project under Grant 2018YFB1700500.

Reference

[1] X. Li and L. Gao, "An effective hybrid genetic algorithm and tabu search for flexible job shop scheduling problem," International Journal of Production Economics, vol. 174, pp. 93-110, 2016.

[2] S. Saremi, S. Mirjalili, and A. Lewis, "Grasshopper optimisation algorithm: theory and application," Advances in Engineering Software, vol. 105, pp. 30-47, 2017.

[3] J. Tian-Hua, "Flexible job shop scheduling problem with hybrid grey wolf optimization algorithm," Control and Decision, vol. 33, no. 3, pp. 503-508, 2018.

[4] J. Vallikavungal Devassia, M. Angelica Salazar-Aguilar, and V. Boyer, "Flexible job-shop scheduling problem with resource recovery constraints," International Journal of Production Research, vol. 56, no. 9-10, pp. 3326-3343, 2018.

[5] H. Wang, Z. Jiang, Y. Wang, H. Zhang, and Y. Wang, "A two-stage optimization method for energy-saving flexible job-shop scheduling based on energy dynamic characterization," Journal of Cleaner Production, vol. 188, no. JUL.1, pp. 575-588, 2018.

[6] Y. Demir and S. K. isleyen, "Evaluation of mathematical models for flexible job-shop scheduling problems," Applied Mathematical Modelling, vol. 37, no. 3, pp. 977-988, 2013.

[7] J. Li, X. Yuan, L. Bo, and Z. Zeng, "A pseudo-dynamic search ant colony optimization algorithm with improved negative feedback mechanism to solve tsp," 2018.

[8] F. Pezzella, G. Morganti, and G. Ciaschetti, "A genetic algorithm for the flexible job-shop scheduling problem," Computers & operations research, vol. 35, no. 10, pp. 3202-3212, 2008.

[9] A. Bagheri, M. Zandieh, I. Mahdavi, and M. Yazdani, "An artificial immune algorithm for the flexible job-shop scheduling problem," Future Generation Computer Systems, vol. 26, no. 4, pp. 533-541, 2010.

[10]L. Wang, G. Zhou, Y. Xu, S. Wang, and M. Liu, "An effective artificial bee colony algorithm for the exible job-shop scheduling problem," The International Journal of Advanced Manufacturing Technology, vol. 60, no. 1-4, pp. 303-315, 2012.

[11]D. Yang, M. Wu, Z. Yang, Y. Guo, and W. Feng, "Dragony algorithm for solving flexible jobshop

scheduling problem," in 2020 35th Youth Academic Annual Conference of Chinese Association of Automation (YAC), pp. 867-872, IEEE, 2020.

[12]P. J. Van Laarhoven, E. H. Aarts, and J. K. Lenstra, "Job shop scheduling by simulated annealing," Operations research, vol. 40, no. 1, pp. 113-125, 1992.

[13]M. Nouiri, A. Bekrar, A. Jemai, D. Trentesaux, A. C. Ammari, and S. Niar, "Two stage particle swarm optimization to solve the flexible job shop predictive scheduling problem considering possible machine breakdowns," Computers & Industrial Engineering, vol. 112, no. oct., pp. 595-606, 2017.

[14]L. Sun, L. Lin, M. Gen, and H. Li, "A hybrid cooperative coevolution algorithm for fuzzy flexible job shop scheduling," IEEE Transactions on Fuzzy Systems, vol. 27, no. 5, pp. 1008-1022, 2019.

[15]G. Zhang, X. Shao, P. Li, and L. Gao, "An effective hybrid particle swarm optimization algorithm for multi-objective exible job-shop scheduling problem," Computers & Industrial Engineering, vol. 56, no. 4, pp. 1309-1318, 2009.

[16]M.Nouiri, A. Bekrar,A. Jemai, S. Niar, and A.C.Ammari,"An effective and distributed particle swarm optimization algorithm for flexible job-shop scheduling problem," Journal of Intelligent Manufacturing, vol. 29, no. 3, pp. 603-615, 2018.

[17]S. Mirjalili, "Dragony algorithm: a new meta-heuristic optimization technique for solving single-objective, discrete, and multi-objective problems," Neural Computing & Applications, vol. 27, no. 4, pp. 1053-1073, 2016.

[18]S. Mirjalili, A. H. Gandomi, S. Z. Mirjalili, S. Saremi, H. Faris, and S. M. Mirjalili, "Salp swarm algorithm: A bio-inspired optimizer for engineering design problems," Advances in Engineering Software, vol. 114, pp. 163-191, 2017.

[19]S. Mirjalili, "Sca: a sine cosine algorithm for solving optimization problems," Knowledge-based systems, vol. 96, pp. 120-133, 2016.

[20]S. Mirjalili and A. Lewis, "The whale optimization algorithm," Advances in engineering software, vol. 95, pp. 51-67, 2016.

[21]S. Mirjalili, S. M. Mirjalili, and A. Hatamlou, "Multi-verse optimizer: a nature-inspired algorithm for global optimization," Neural Computing and Applications, vol. 27, no. 2, pp. 495-513, 2016.

Research and Implementation of an OpenMV-Based Target Edge Detection and Tracking System

Yong Guo[1,*], Jun Yin[2], Yu Wang[1], Shikai Wu[1], Le Luo[1] and Qiang Fu[1]

[1]School of Electronic Engineering, Chengdu Technological University, Chengdu, Sichuan, China
[2]School of Mechanical and Electrical Engineering, Chengdu University of Technology, Chengdu, Sichuan, China

[*]corresponding author's e-mail: robinfit01@163.com

Abstract. The improvement of ARM microcontroller unit (MCU) processing capability has greatly broadened the application fields of machine vision, and the huge benefits brought by machine vision can be seen in scenarios such as medicare and health, industrial production, military and national defense. Edge detection is a common processing method in image processing, which can provide basic guarantee for subsequent target tracking and other controls. In this paper, OpenMV is used as a hardware platform to extract the edges of a target in combination with specific algorithms, and then the common PID algorithm in the field of automatic control is written in the OpenMV integrated development environment (IDE) using Python language to achieve low-latency tracking of dynamic target objects. In addition, this thesis uses the color threshold tool in the IDE to capture the color threshold of defective product as an eigenvalue, and then calls the Canny operator in the function library and functions such as kernel filtering to process the image. Through actual testing and comparison, it can be found that the image edge is retained more completely after the kernel filtering process. After the general area of the defective products is calibrated on the object to be detected, the defective products can be screened.

1. Introduction

With the rapid development of digital image processing technology, the subject of machine vision is widely used in scenarios such as industrial production [1], smart cities [2], and modern agriculture [3], especially in the detection of product defects and the monitoring and tracking of moving objects. Specifically, machine vision can not only detect product defects in fruit appearance [4], flat steel surface [5], PCB defects [6], text printing quality [7], optical component surface defects [8], and particleboard surface defects [9], etc., but also monitor and track moving objects [10-11], and even combine machine learning techniques for intrusion detection [12-13]. Edges are where the pixel values of an image vary most significantly, and are also one of the distinguishing features of an image. Unprocessed digital images often contain rich information and can be roughly divided into two major parts: contours (high frequency) and details (low frequency). However, when embedded MCUs implement various controls by recognizing image information, too much detail will not only occupy the limited resources of the MCU, but also the image-independent information will cause various interferences.

In this thesis, a typical edge detection and tracking control system is designed using OpenMV as the hardware platform and relying on its IDE. Edge detection is often used as the initial processing step of machine vision. After edge detection, key information of image can be extracted and other color details in low frequency parts can be masked. After the preliminary edge extraction, it can not only remove the noise interference of the image itself, but also reduce the resource consumption of the MCU, which provides a good technical basis for the subsequent operations of defective detection and automatic control. The object tracking in this paper is realized by reading the color threshold and under the control of PID algorithm. This algorithm can effectively overcome the problems of inertia and lack of precision in the position-based control algorithm. Usually, the three criteria for measuring the pros and cons of image edge detection are: high signal-to-noise ratio, high positioning accuracy, and unique response to a single boundary. Therefore, this thesis will compare kernel filtering and Canny operator for processing the same images, will find the difference between the two image processing, and select a more appropriate scheme of the defective products calibration application. This design overcomes the drawbacks of traditional screening, and has broad application prospects in the field of industrial production and testing.

2. Fundamental principles and methods

The OpenMV module adopts a firmware library development method, and many frequently used functions can be directly called in conjunction with the supporting documentation. It is worth noting that since the IDE is coded in Python, familiarity with the differences between the Python language and other programming languages is required.

2.1. STM32H765VI main controller chip

The core of the OpenMV module is the STM32H765VI high-performance ARM processor produced by STMicroelectronics, which is a high-performance Cortex-M7 MCU with DSP and DP-FPU (Double Precision-Floating Point Unit). Due to this MCU has faster digital signal processing capability, higher operating frequency, more peripheral modules and lower power consumption, these provide a reliable guarantee that the OpenMV module can implement face detection, color recognition, object tracking and other functions.

2.2. OV7725 photosensitive element

Photoelectric sensors are a class of devices which can convert optical signals into electrical signals based on the photoelectric effect. The camera of the OpenMV hardware platform uses an OV7725 CMOS sensor, which is controlled by the Serial Camera Control Bus (SCCB) interface, and can provide full-frame, sub-sampled or windowed 8-bit/10-bit images in a variety of formats. The device is capable of running at 60 frames per second (60FPS) in VGA format with resolutions up to 300K pixels, and the OpenMV module runs at 30FPS operating rate in most cases. We found that the images captured by the OV7725 had distortion effects in our practical tests.

2.3. Canny algorithm

Canny operator, a classical algorithm, is used to realize edge detection, which is divided into the following four steps:

Gaussian blur (filtering): Gaussian filtering can effectively remove the noise in the image and smooth the image.

Gradient calculation: the classical edge detection operators include Roberts, Prewitt, Sobel, etc. The digital image can obtain the first order derivatives (rate of change) in the horizontal and vertical directions after processed by these operators, and then the gradient value G and the direction angle Θ of one pixel can be obtained after calculating the x-direction gradient and y-direction gradient according to the following equation (1).

$$G=\sqrt{G_x^2+G_y^2}\ ,\quad \theta=\arctan(Gy/Gx) \tag{1}$$

Non-maximum signal suppression (NMS): After obtaining the gradient value and direction angle of the image, the edge information will be misjudged if the edge is determined based on this criterion only, and it is difficult to meet the key technical requirement of "uniqueness of the response to a single boundary" mentioned in the previous section, thus the use of NMS can make the image edge extraction more accurate.

Dual-threshold detection: Dual-threshold refers to the maximum and minimum values, and the remaining pixels after non-maximum suppression processing can often accurately represent the important information of image edges. When processing some images whose edge features are not particularly obvious, because these edge features are close to the random noise signals, it will lead to the misjudgment of the signal. The solution to this problem is to use a weak edge pixel and its 8 neighboring pixels to connect to the edge, as long as one of pixels is a strong edge pixel, then the weak edge point can be recognized as a real edge and thus retained.

2.4. Incremental PID control algorithm

This scheme uses an incremental PID algorithm to achieve dynamic tracking of the objects. The difference between this algorithm and the positional PID is that it outputs an increment of the control quantity, which is only related to the past two states T(n-1) and T(n-2), and has a much smaller impact on the system. When the camera captures an object in the viewing area that is similar to the preset color threshold, the STM32H765VI MCU will output PWM waves with different duty cycles according to the program design, and control the steering engine on the PTZ to rotate at different speeds to make the target object in the center of the image as much as possible.

3. Main design and experimental procedure

The design process of this subject consists of two key control processes as shown in Figure 1 and Figure 2 below:

3.1. Edge detection implementation procedure

The flow chart of edge detection is shown in Figure 1, the core algorithms used in this process have been introduced in Section 2 fundamental principles and methods and will not be repeated here.

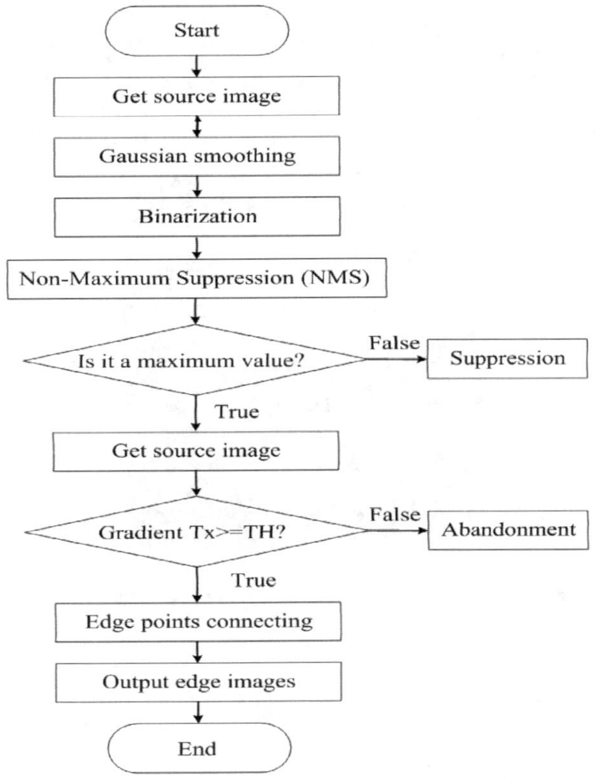

Figure 1. Implementation flow chart of edge detection.

3.2. Dynamic tracking realization process

Firstly, we need to use the color threshold tool in the IDE to determine the color threshold, and then assign the value to a specific variable in the program. Figure 2 shows the flow control of dynamic tracking, in which the control of steering engine rotation is completed on the PTZ.

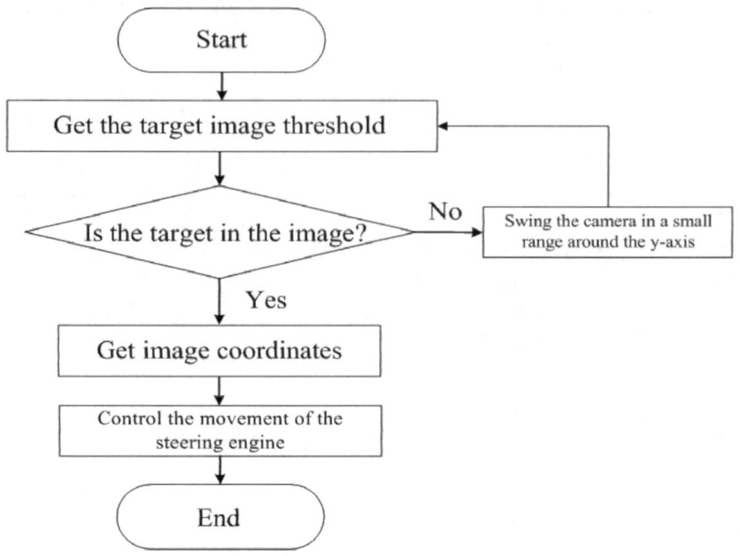

Figure 2. Dynamic tracking flow chart.

The physical diagram of the object tracking PTZ designed in this research is shown in figure 3. The system is composed of a 5V to 3.3V DC-DC voltage conversion module, an OpenMV module, and a 2D PTZ. Next, the specific implementation process of the edge detection experiment, the defective product calibration experiment and the target dynamic tracking experiment will be described respectively.

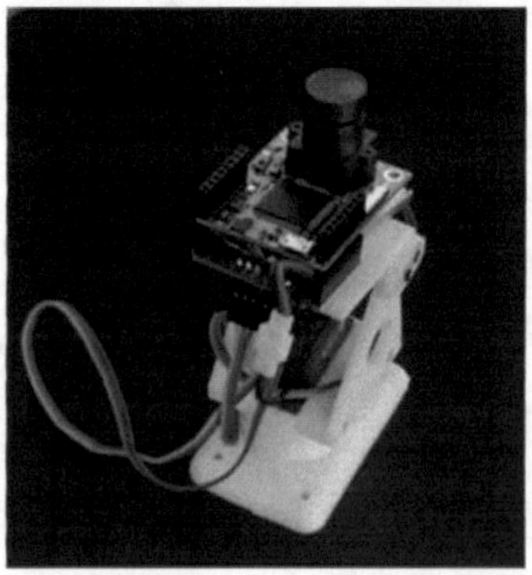

Figure 3. Physical view of dynamic object tracking PTZ.

3.2.1. Edge detection experiment

Figure 4 (a) shows the original keyboard image, and it can be seen from the image that when a camera captures such comparatively large images, the images near the edge of the camera objective are prone to distortion. According to the optical principle, the various pixel points on the image have different distances to the lens, which can produce a fisheye effect (barrel aberration). This kind of problem can be corrected by using an orthoscopic camera lens or the function image.lens_corr（1.8）. Figure 4 (b) shows a image processed by Canny operator, the processed image format has been changed to a grayscale image, although from the output image we can found that the useless information has been effectively suppressed, after actual testing, it is found that the information of letters on the keyboard is difficult to obtain, but the image effect of kernel filtering is better.

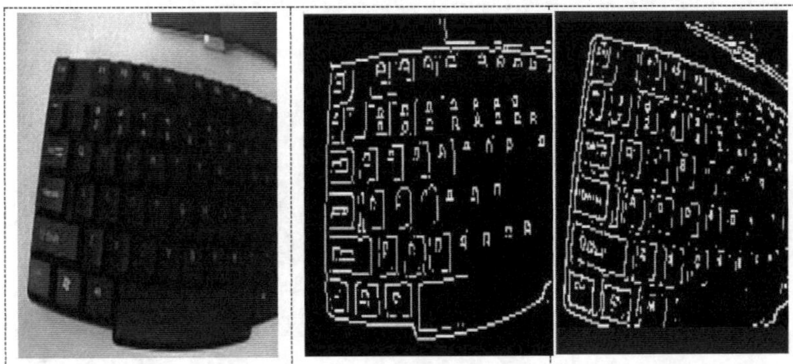

(a) (b) (c)

Figure 4. Edge detection experiment of keyboard graph. (a) Original keyboard image. (b) Canny operator processed image. (c) Kernel filtering processed image.

The original image in the figure 5 (a) is much more informative, and the image after edge detection has been changed from the original RGB three-primary colors to a binary image, and redundant details are effectively filtered out. This set of experimental results also shows that the effect of image edge detail preservation processed by kernel filtering is better than that of Canny operator processing.

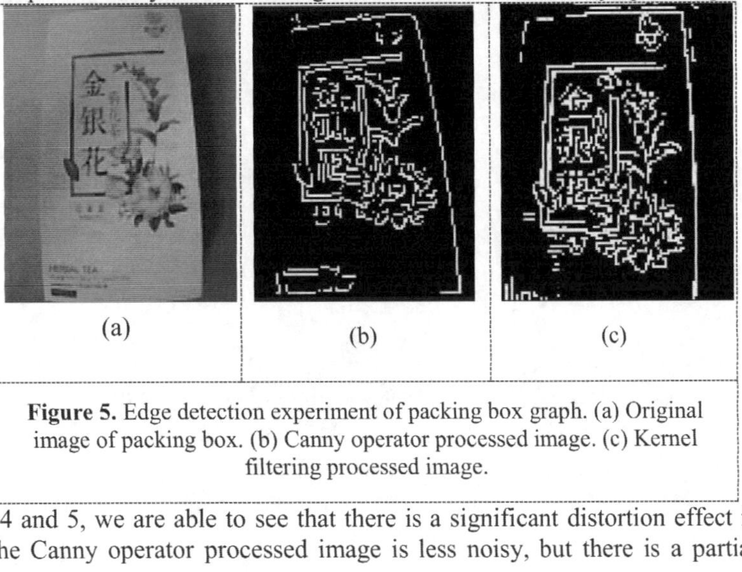

(a) (b) (c)

Figure 5. Edge detection experiment of packing box graph. (a) Original image of packing box. (b) Canny operator processed image. (c) Kernel filtering processed image.

From Figure 4 and 5, we are able to see that there is a significant distortion effect in the OV7725 image sensor, the Canny operator processed image is less noisy, but there is a partial loss of edge information.

3.2.2. Defective product calibration test experiment

In this experiment, one side of the Rubik's cube is selected as the product testing object, and there are 9 small squares in that side of the cube. The small square in the center of the image is severely worn and the surrounding 8 normal squares are significantly different (in color and area size), and the Rubik's cube is determined as a defective product by the testing algorithm. Figure 6 shows the processing process of the defective product screening:

Figure 6 (a) shows the original image, and the color image contains orange, red, blue, white, and yellow squares, but the orange square in the center of the image is badly worn and clearly distinguishes itself from the other normal squares in terms of image shape and area.

Figure 6 (b) shows the original image processed by Sobel operator, and the color threshold of defective products is preset before screening. When the system detects an image close to the threshold, it will calibrate the contour of the defective product.

Figure 6 (c) shows that the defective orange square is detected based on image (b) and the position of the defective square is marked.

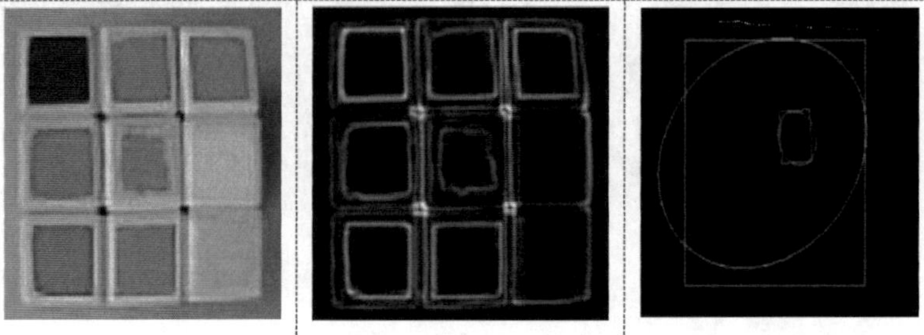

(a) (b) (c)

Figure 6. Experimental diagram of defective products detection. (a) Original image. (b) Image processed by Sobel operator. (c) Defect marking image.

3.2.3. Target dynamic tracking experiment

(a) (b) (c)

Figure 7. Experimental diagram of dynamic object tracking. (a) The target object is directly in front of the camera. (b) The target object moves to the lower right area. (c) The object moves directly above the camera.

In Figure 7 (a), the target object is located directly in front of the camera and there is no obvious jitter occurs during this period, indicating that the PID parameter settings are reasonable. When the target object in Figure 7 (b) moves to the lower right area, the steering engine controlling the X axis swings to make the camera move to the right, and the steering engine of the Y axis moves downward to make the camera move downward. The object in Figure 7 (c) moves directly above the camera, at this moment, only the steering engine controlling the Y axis moves to make the target appear in the image area of the camera as much as possible.

4. Conclusion

In summary, it can be seen that both Canny algorithm and kernel filtering algorithm can be used to extract the edge contour of the target objects, the image processed by Canny operator is less noisy but there is a partial loss of edge information, while the effect of image edge detail retention by kernel filtering is better than the result of Canny processing. The steering engine of this system can move in one dimension or double steering engine can move in two dimensions at the same time, and the camera stops swinging after capturing the image, which shows that this system effectively utilizes the PID control algorithm to achieve high accuracy and low latency tracking of moving objects in the two dimensional plane.

Acknowledgments
This work was supported by the 2021 National Innovation and Entrepreneurship Training Program for College Students (Project No.202111116019), and the Talent Training Quality and Teaching Reform Project of Chengdu Technological University.

References
[1] Ge Q, Fang M and Xu J 2018 IEEE International Conf. on Mechatronics and Automation (ICMA) Changchun China pp 832-836
[2] WANG X, CHEN F and LIANG X 2020 Chinese Automation Congress (CAC) Shanghai China pp 5209-5213
[3] Pei Y, Ye J, Jiang Y, Lian M, Han X and Yu G 2019 IEEE International Conf. on Signal, Information and Data Processing (ICSIDP) Chongqing China pp 1-5
[4] Prabhu A, Sangeetha K V, Likhitha S and Shree Lakshmi S 2021 International Conf. on Intelligent Technologies (CONIT) Hubli India pp 1-10
[5] Luo Q, Fang X, Liu L, Yang C and Sun Y 2020 IEEE Transactions on Instrumentation and Measurement **69** pp 626-644
[6] Khalilian S, Hallaj Y, Balouchestani A, Karshenas H and Mohammadi A 2020 International Conf. on Machine Vision and Image Processing (MVIP) Iran pp 1-5
[7] Liu X, Xia M, Xia N, Tang S, Yao A and Li W 2021 4th International Conf. on Advanced Electronic Materials, Computers and Software Engineering (AEMCSE) Changsha China pp 955-959
[8] Lin Y, Xiang Y, Lin Y and Yu J 2019 2nd International Conf. on Information Systems and Computer Aided Education (ICISCAE) Dalian China pp 415-418
[9] Prasitmeeboon P and Yau H 2019 5th International Conf. on Engineering, Applied Sciences and Technology (ICEAST) Luang Prabang Laos pp 1-4
[10]Wangsheng Z and Xiuqin P 2021 IEEE International Conf. on Power, Intelligent Computing and Systems (ICPICS) Shenyang China pp 731-738
[11]Wang S, Liu Y, Wang Y and Li L 2021 IEEE International Conf. on Artificial Intelligence and Computer Applications (ICAICA) Dalian China pp 1143-1147
[12]Nayak D U, Mohite A P, Nair P P and Bide P J 2021 Asian Conf. on Innovation in Technology (ASIANCON) PUNE India pp 1-6
[13]Guan L, Li X, Yang H and Jia L 2020 International Conf. on Sensing, Diagnostics, Prognostics, and Control (SDPC) Beijing China pp 291-295

Application of Improved Ant Colony Algorithm in Aerobat Structure Optimization

Jian Jun Deng*, **ChengTao Zhang and Ning Wang**

The Air Defense and Anti-missile Institute, Air Force Engineering University, Xian, Shaanxi, China

*corresponding author's e-mail: jianjundeng1@163.com

Abstract. This paper analysis the characteristics of two kinds of ant colony algorithm, some kind of aircraft structure dynamic optimization design model is established, based on the analysis of the model, two kinds of ant colony algorithm are combined with optimization calculation method respectively, the two kinds of ant colony algorithm is studied in the aircraft structure optimization design, the calculation results indicate that improved ant colony algorithm is scientific and feasible, it can effectively solve the optimization problems of aerobat and other structures.

1. Introduction

Simulating the inherent laws of natural phenomena in nature is one of the main ideas for constructing optimization algorithms, such as particle swarm optimization algorithm, genetic algorithm, these algorithms provide a general framework for solving complex system optimization problems. The algorithm does not depend on the strict mathematical relationship, also does not depend on problems of specific areas, which has been widely applied in the research of a variety of disciplines, a train of thought is provided to solve the problem of complex system optimization. Ant colony algorithm is similar to it, inspired by the research results of real ant colony behavior, Dorigo•M first proposed this kind of simulated evolutionary algorithm based swarm [1]. It is a random search algorithm, which makes full use of the similarity between the process of ant colony searching for food and the famous traveling salesman problem. This algorithm is called "human worker ant colony algorithm", which simulates the process of ants searching for food to solve the traveling salesman problem. A series of good experimental results are obtained by using this method to solve the traveling salesman problem, assignment and scheduling problem. The advantages of ant colony algorithm in solving complex optimization problems have proved that it is a promising method.

Ant algorithm like other intelligent bionic algorithm, it find the optimal solution through a set of candidate solutions in the process of evolution, does not require any prior knowledge. The mechanism of understand solution space include the memory of the ants, communicate with each other and cluster activities, it is a kind of parallel algorithms in essence, information exchanges and transfers are carried out continuously between each other, which is conducive to finding better solutions[2].

In this paper, ant colony algorithm is introduced into structural optimization design in order to provide a new idea for structural optimization in theory and practical application. The results show that the proposed method can effectively solve the optimization problem of aerobat structure cabin, and has strong practicability and universality.

2. Discussion of characteristics about basic ant colony system model and improved ant colony algorithm

2.1. Advantages of basic ant colony system

Advantages of basic ant colony system are robustness , distributed computing and easy to combine with other methods

2.2.Shortcomings of basic ant colony system

Shortcomings of basic ant colony system are long time to search, easy to stagnate and it is difficult to deal with continuous space optimization problems.

2.3.The main differences between basic ant colony system and improved ant colony algorithm

2.3.1. Different state transition rules. In the improved ant colony algorithm, ants use pseudo-random ratio selection rules to select the next city[3].

2.3.2. Local pheromone updating strategies are different. The function of local pheromone update is to make the selected path have less influence on the subsequent ants, so that the ants have a stronger ability to explore the unselected path. In the improved ant colony algorithm, pheromone addition is selective. When an ant moves from one node to another node, part of pheromone on the path is eliminated accordingly. thus, a negative feedback mechanism is introduced through local adjustment of pheromone; the probability of the selected path being selected again is reduced.

2.3.3.Different global pheromone update strategies. In the improved ant colony algorithm, only the pheromone on the shortest path currently searched is updated[4].

Basic ant colony system model and improved ant colony algorithm are difficult to deal with continuous space optimization problems. Because the choices made by each ant at each stage are always limited, it requires discrete solution space, so it is suitable for discrete optimization problems such as combinatorial optimization, but cannot be directly applied to continuous space optimization problems.

3. The aerobat structure optimization design based on the improved ant colony algorithm

A simplified model of an aerobat structure is showed in Figure 1， When the aerobat is flying, if the frequency of external exciting force is the same as the natural frequency of the structure, resonance will occur, which can result in unacceptable deformation or even damage of the structure. Therefore, in the aerobat structure design, the natural vibration frequency should be limited according to the external frequency, such as the requirement of natural vibration frequency ω_I not less than a certain allowable frequency $\underline{\omega}$, ω_I is the first natural frequency.

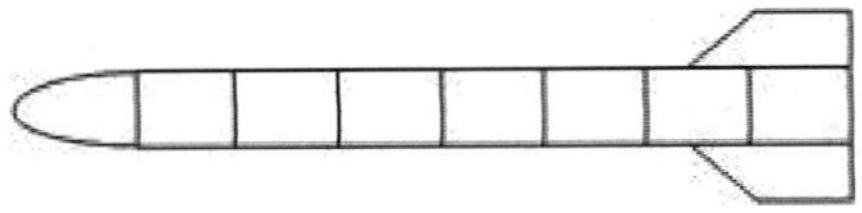

Figure 1. Aerobat structure model

Since the natural frequency of the structure is related to the stiffness of the structure, the stiffness of the structure should be adjusted during the design to meet the above condition. At the same time, the weight of the structure is guaranteed to be the lightest. Assuming that the aerobat is divided into r sections, and the structural mass and stiffness of each section are evenly distributed, that is, the cross-sectional area of each section is constant, and then the structural weight of the aerobat is:

$$W = g\sum_{i=1}^{r} \rho_i l_i A_i + g\sum_{i=1}^{2} \rho_i b_i t_i \lambda_i \tag{1}$$

Then the mathematical model of the problem is:

$$\text{search } A=[A_1, A_2, \cdots A_r]^T \text{ and } v=[b, t, \lambda]^T$$

where A_i is the cross-sectional area of the i-th section of the cabin; b is average chord of the wing , t is average thickness of the wing, λ is spread of the wing.

To minimize $W = g\sum_{i=1}^{r} \rho_i l_i A_i + g\sum_{i=1}^{2} \rho_i b_i t_i \lambda_i$ and satisfy

$$\omega_1^2 \geq \underline{\omega}^2 \quad (2)$$

$$\underline{A}_i \leq A_i \leq \overline{A}_i \quad (i=1,2,\cdots,r) \quad (3)$$

$$\underline{\lambda} \leq \lambda \leq \overline{\lambda} \quad \underline{t} \leq t \leq \overline{t} \quad \underline{b} \leq b \leq \overline{b} \quad (4)$$

The formula(2)-(4)are the upper and lower limits of the design variables, this problem is a r dimensional nonlinear programming problem with $2r+7$ inequality constraints.

The idea of the algorithm: because ant colony algorithm is difficult to deal with continuous variable optimization problems, according to the value range of variables, each design variable is separated and dispersed by 0.5mm from the lower limit to the upper limit, and the discrete state of each variable forms a spatial grid. The direction of ant movement is determined according to the pheromone quantity left by ants on the grid nodes. After repeated cycles, the solution ends when the grid spacing is less than the given value, the geometry constraint was solved. Penalty function method was used to transform constrained nonlinear programming into unconstrained nonlinear programming, penalty function is constructed as follows, which solves the inequality constraint problem of natural frequency.

$$\phi(\boldsymbol{x}, \gamma^{(k)}) = \boldsymbol{f}(\boldsymbol{x}) + \gamma^{(k)}[\max(0, \omega_1(\boldsymbol{x})]^2 \quad \boldsymbol{k} = (0,1,2\cdots) \quad (5)$$

$\omega_1(x)$ is the constraint condition of natural frequency, $\gamma^{(k)}$ is the penalty factor.

The transfer matrix method is an effective method to solve the natural frequency of the chain structure, the research object of this paper is the chain structure with two branches, so it is suitable to use the transfer matrix method to calculate the natural frequency of the model.

In this paper, the component separation surface is taken as the node to establish the centralized mass model. There are 9 nodes in the model, and two wings are located away from the axis of the aerobat. As the branch system of the centralized mass model, the connection stiffness between the nodes must be consistent with the stiffness and distribution of the whole aerobat.

Natural frequency solution steps is as follows:

a. Establish the model of each component and determine the state vectors at both ends of the component.

b. Deduce the lattice between each node, deduce the field matrix between state vectors at both ends of the elastic element, and find the transfer matrix of the typical segment.

c. Establish the total transfer matrix to represent the relationship between state vectors at both ends of the model.

d. Boundary conditions are introduced, both ends of the object studied in this paper are free ends, and the boundary conditions are that both of the bending moment and stress of the free end are 0.

e. Establish frequency equation to solve natural frequency of the model.

In order to prevent the algorithm from falling into the local optimum, the pheromone update strategy is adjusted in the algorithm. When all the ants have completed a trip, randomly select the length of the loop path L_k of one ant, compare the length of the loop path $L_i(i \neq k)$ of other ants with the path, and update the pheromone of the L_k path and the pheromone on the edge shorter than the path, so that the algorithm can effectively overcome the search stagnation and avoid the algorithm falling into local optimization quickly, it is helpful to shorten the search time and improve the ability to search the optimal solution[5].

4. Simulation analysis

In the calculation, the head is equivalent to cabin, r=8, l=0.5m, inner diameter of cabin d=265mm, outer diameter D(310mm,300mm,304mm), b(210mm,200mm,205mm), t(22mm,15mm,18mm), λ (280mm,260 mm,265mm), the numbers in brackets are the upper and lower limits of parameters and the initial values respectively. The outer diameter D^2, the wingsurface segment λ, the relative thickness of the airfoil t and the average aerodynamic chord b were taken as the design variables. $\omega = 1200 rad/s$, elastic modulus E=40180MPa, ρ_i=1.78× 10^3kg/m^3. The basic ant colony algorithm and the improved ant colony algorithm are used for simulation respectively. The simulation running parameters are given as follows: by number of design variables n=11,on the basis of experience ,α=2,β=3, α is the relative importance of the trajectory,β is the relative importance of visibility, evaporation coefficient of the pheromone ρ=0.5,q=0.8, Q=200, the initial pheromoneτ_0=500, and the number of ants in this example is 30.The parameter value and minimum mass of each section before and after optimization are shown in the following table.

From the calculation results, the aerobat structure optimized by the improved ant colony algorithm is 1.31kg less than the basic ant colony algorithm, which shows that the improved ant colony algorithm has a better optimization effect than the basic ant colony algorithm.at same time, It can be seen from Figure 2 that the improved ant colony algorithm is not only more optimized than the basic ant colony algorithm, but also has better convergence speed than the basic ant colony algorithm.

Table 1. Optimization calculation results of aerobat structure

Initial value of design variables(mm)	Parameters before optimization(mm)	Parameters after optimization(mm)
304	300	300
304	303.5	302.5
304	304	303.5
304	306.5	306
304	305	305
304	303.5	303.5
304	300.5	300.5
304	300	300
205	208.5	207.5
18	21	20.5
265	276.5	274
Total mass of structure (kg)	234.13	232.82

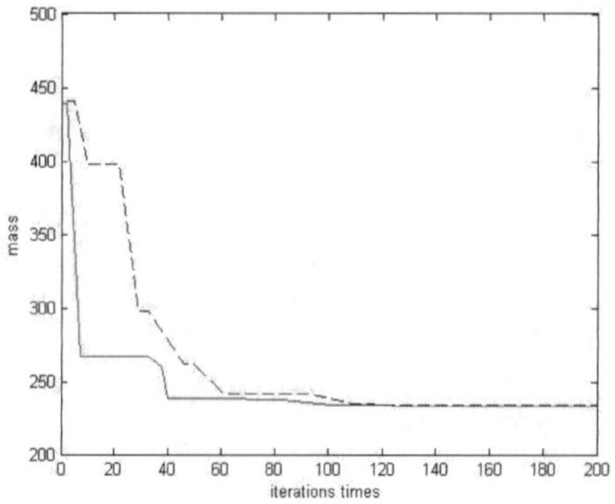

Figure 2. Optimization design history

5. Conclusion

The improved ant colony algorithm is used to optimize the aircraft structure in this paper, the simulation results show that the method is feasibility, it has certain application value to the design of aerobat structure. For the optimization problem of structural parameters, as long as the optimization objective function can be expressed explicitly, the optimization problem can be solved quickly[6]. it's not like genetic algorithm and simulated annealing algorithm based on analysis method and the solid mathematical foundation, the algorithm convergence problems remain to be further research, the results is not stable, restricted by the number of ants, easily plunged into local optimal solution, through proper improvement and mixed with other algorithms, ant colony algorithm is a very promising algorithm.

References

[1] Zhang J B, Zuo T, Hu X Y and Wu Y W 2021*J. Chinese High Technology Letters* **31** 288
[2] Tao Y M 2020*J. Software Guide* **19** 276
[3] He Y H 2021*J. Computer and Digital Engineering* **49** 920
[4] Ma R G, Cui H, Xue S J, Guo L and Yuan Chao 2016*J. Journal of Xidian University(Natural Science)* **43** 185
[5] Jiang L,Wang J, Liang C Y and Zhao S P 2019 *J. Computer Engineering and Applications* **55** 244
[6] Zhao Y Q,Yan X Ge Z H and Li H Q 2020*J. Journal of Chongqing Institute of Technology* **34** 1

Journal of Physics: Conference Series **2173** (2022) 012085 doi:10.1088/1742-6596/2173/1/012085

An Inverse Solution Algorithm for Industrial Robot

Peng Wang[1,4], Yiren Zhou[2,*] and Nan Yan[3]

[1]School of Mechanical Engineering, Tianjin University, Tianjin, China
University in Zhejiang, Shaoxing, Zhejiang, China
[2]Nanchang University of Technology, Nanchang, Jiangxi, China
[3]Mechanical & Electrical Engineering College, Ningbo Polytechnic, Ningbo, Zhejiang, China
[4]International Institute for Innovative Design and Intelligent Manufacturing of Tianjin

[*]corresponding author's e-mail: renyiz@126.com

Abstract. The inverse solution process of industrial robot is complex. Because the traditional matrix algorithm involves a lot of matrix operation, its process is complex. Numerical algorithms usually do not yield all the solutions, generally, the selection of initial value and the search algorithm have great influence on the convergence and accuracy. The geometric algorithms is simple and intuitionistic with simple expression and small computation. In this paper, a hybrid algorithm combining the geometric method and the inverse transformation matrix method is proposed for puma 560 industrial robots. The geometric algorithms is used to solve the first three joints and the matrix algorithm is used to obtain the final three joint angles. This paper introduces the inverse solution algorithm for this kind of robot in detail, and verifies the effectiveness of the proposed algorithm through matlab simulation.

1. Introduction

Kinematics of industrial robots, as an important research field of robotics, has a profound impact on robot control, robot dynamics and trajectory planning, and is directly related to robot control, trajectory planning and off-line programming accuracy, which is the basis of research on robot dynamics, trajectory planning and motion simulation[1]. It describes and studies the motion characteristics of the robot from the perspective of geometry or mechanism [2], without considering the effect of the force or torque that causes these motions, which can be divided into forward kinematics and inverse kinematics. given the geometric parameters of the connecting rod and joint variables, forward kinematics is to find the position and orientation of the end-effector of the robot relative to the reference coordinate system for a given robot, Inverse kinematics is to solve the joint variables of the robot that can achieve the desired pose by knowing the geometric parameters of the connecting rod and the pose of the end-effector of the robot relative to the reference coordinate system.

The commonly used robot kinematics methods [3] are mainly analytical and numerical methods, including vector algebra, geometry, matrix. For example, the matrix method [3-5] requires the establishment of the corresponding coordinate system on each joint, and the forward kinematics model is established by means of coordinate transformation. In the inverse kinematics, the inverse transformation algorithm is usually adopted, and the solution process is complex. Numerical method [6-8] directly solves the constraint equations, and can obtain the real solutions of any mechanism by iterative operation, but usually cannot get all the solutions. Generally speaking, the selection of initial value and search algorithm have a great influence on the convergence and accuracy.

The geometric algorithm [9-10] of robot inverse solution has been studied by scholars. This paper presents a geometric method and inverse transformation method. The puma560 robot was simulated by matlab to verify the effectiveness of the algorithm.

2. Hybrid inverse solution algorithm

2.1. Mathematical model of robot kinematics

The research object of this paper is puma560 industrial robot, whose structural characteristic is that the last three joint axes are intersect at one point, and it is spherical wrist. The characteristics of this mechanism meet the PIEPER principle, and the robot coordinate system is established according to the robot coordinate system establishment rules proposed by Craig J.J, As shown in Figure 1, 6R robot DH parameters shown in Table 1 .

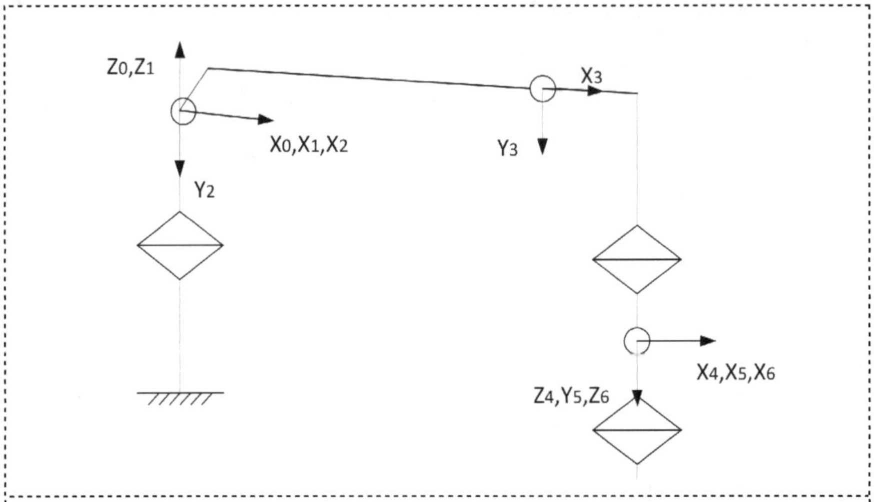

Figure 1. structure sketch of robot configuration

Table 1. 6R robot DH parameters

Joints	the angle of connecting rod (a_{i-1}/rad)	the length of connecting Rod(α_{i-1}/rad)	the offset of connecting Rod (d_i/mm)	joint Angle (θ_i/mm)
1	0	0	0	θ_1
2	-90	0	0	θ_2
3	0	a_2	d_3	θ_3
4	-90	a_3	d_4	θ_4
5	90	0	0	θ_5
6	-90	0	0	θ_6

Based on the geometric model of the robot, the inverse solution of the robot first obtains the homogeneous transformation matrix of the end link with respect to the coordinate system of the base link. the matrix can be expressed as follows:

$$ {}^0_6T = \begin{bmatrix} r_{11} & r_{12} & r_{13} & p_x \\ r_{21} & r_{22} & r_{23} & p_y \\ r_{31} & r_{32} & r_{33} & p_z \\ 0 & 0 & 0 & 1 \end{bmatrix} = {}^0_1T(q_1)\,{}^1_2T(q_2)\,{}^2_3T(q_3)\,{}^3_4T(q_4)\,{}^4_5T(q_5)\,{}^5_6T(q_6) \quad (1) $$

In equation 1, p_x, p_y and p_z are the coordinates of the origin of the attached coordinate system of the robot's end link. r_{ij} is the element of rotation matrix, i, j=1, 2, 3; ${}^{0}_{I}T(q_I)$ is the matrix of the coordinate system of the i-1 connecting rod relative to the coordinate system of the i connecting rod, and is a function of the angle of the joint θ_i.

2.2. Solution of angle and joint

First of all, the angle of joint θ_1 is easy to be solved.

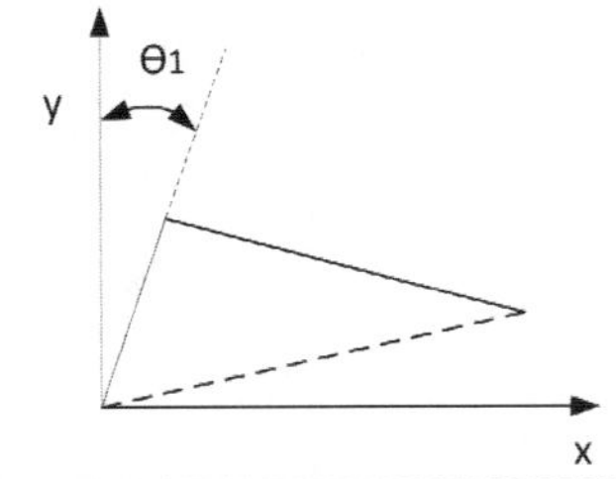

Figure 2. Top view simplification

In Figure 2, the Angle of joint θ_1 needs to be combined with the joint Angle at the previous moment to select the optimal Angle, which can be deduced as follows:

$$q_{pre} = \text{atan}\,2\left(p_y, p_x\right) \tag{2}$$

$$\theta_{\text{var}} = \text{acos}\left(\frac{d_3}{\sqrt{p_y^2 + p_x^2}}\right) \tag{3}$$

In the formula, θ_{pre} and θ_{var} is the middle value of angle of joint θ_1. Joint Angle θ_1 has two possible solutions as follows:

$$\theta_1 = \theta_{\text{pre}} + \theta_{\text{var}} \tag{4}$$

$$\theta_1 = \theta_{\text{pre}} - \theta_{\text{var}} - \text{pi}/2 \tag{5}$$

The big arms and small arms of the three robot can be converted into the form of a triangle, as shown in Figure 3:

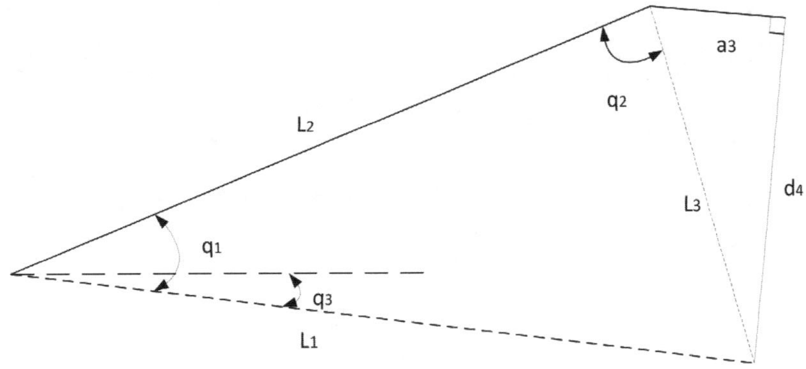

Figure 3. Simplified geometric structure of robot

The parameter L_2 in the Figure 3 is the length of the robot arm, and the numerical value is equal to parameter a_2 of the connecting rod. The parameter L_3 is obtained through the geometric relationship between the connecting rod parameters a_3 and d_4. The geometric meaning of L_1 is the straight-line distance between the origin of the end coordinate system and the origin of the coordinate system attached to connecting rod 2. L_1 and L_2 and L_3 form a triangle, where the Angle between L_2 and L_1 is q_1, the Angle between L_2 and L_3 is q_2, and the Angle between L_1 and the horizontal line is q_3.

For robot configuration 1, the angles q_1 and q_2 are

$$q_1 = \text{acos}\left(\frac{L_2^2 + L_1^2 - L_3^2}{2L_1^2L_2^2}\right) \tag{6}$$

$$q_2 = \text{acos}\left(\frac{L_2^2 + L_3^2 - L_1^2}{2L_2^2L_3^2}\right) \tag{7}$$

the Angle q_3 of robot is:

$$q_3 = \text{atan}\ 2\left(p_z, \sqrt{p_x^2 + p_y^2 + p_z^2}\right) \tag{8}$$

There are two ways of looking at the Angle of joint 2 as follows.

If $\theta_2 = -(q_1 + q_3)$, then:

$$\theta_3 = \pi - q_2 - \text{atan}\ (d_4, a_3) \tag{9}$$

If $\theta_2 = q_1 - q_3$, then

$$\theta_3 = q_2 + \text{atan2}\,(d_4,\ a_3) - \pi \tag{10}$$

Rewrite Eq. 1 as follows.

$$[{}^0_3T(q_2)]^{-1}\,{}^0_6T = {}^3_4T(q_4)\,{}^4_5T(\theta_5)\,{}^5_6T(\theta_6) \tag{11}$$

$$\begin{bmatrix} c_1c_{23} & s_1c_{23} & -s_{23} & -a_2c_3 \\ -c_1s_{23} & -s_1s_{23} & -c_{23} & a_2s_3 \\ -s_1 & c_1 & 0 & -d_3 \\ 0 & 0 & 0 & 1 \end{bmatrix} \begin{bmatrix} r_{11} & r_{12} & r_{13} & p_x \\ r_{21} & r_{22} & r_{23} & p_y \\ r_{31} & r_{32} & r_{33} & p_z \\ 0 & 0 & 0 & 1 \end{bmatrix} = {}^3_6T \tag{12}$$

${}^3_6T$ given by the following formula.

$${}^3_6T = {}^3_4T\,{}^4_6T = \begin{bmatrix} c_4c_5c_6 - s_4s_6 & -c_4c_5s_6 - s_4c_6 & -c_4s_5 & a_3 \\ s_5c_6 & -s_5s_6 & c_5 & d_4 \\ -s_4c_5c_6 - c_4s_6 & s_4c_5s_6 - c_4c_6 & s_4s_5 & 0 \\ 0 & 0 & 0 & 1 \end{bmatrix} \tag{13}$$

The elements (1, 3) and elements (3, 3) on both sides of the Eq. 12 are equal, respectively.

$$r_{13}c_1c_{23} + r_{23}s_1c_{23} - r_{33}s_{23} = -c_4s_5 \tag{14}$$

$$-r_{13}s_1 + r_{23}c_1 = s_4s_5 \tag{15}$$

From Eq. 14 and Eq. 15, As long as $\sin\theta_5 \neq 0$, θ_4 can be obtained:

$$\theta_4 = \text{Atan }2\left(-r_{13}s_1 + r_{23}c_1, -r_{13}c_1c_{23} - r_{23}s_1c_{23} + r_{33}s_{23}\right) \tag{16}$$

When θ_5 =0, the manipulator is in a singular configuration

Rewrite Eq. 1 as follows:

$$_4^0T(\theta_4)^{-110}{}_6T = {}_5^4T(\theta_5){}_6^5T(\theta_6) \tag{17}$$

$_4^0\text{T}(\theta_4)$ is given by the following Eq. 6:

$$\begin{bmatrix} c_1c_{23}c_4+s_1s_4 & s_1c_{23}c_4-c_1s_4 & -s_{23}c_4 & -a_2c_3c_4+d_3s_4-a_3c_4 \\ -c_1s_{23}s_4+s_1c_4 & -s_1c_{23}s_4-c_1c_4 & -s_{23}s_4 & a_2c_3s_4+d_3c_4+a_3s_4 \\ -c_1s_{23} & -s_1s_{23} & -c_{23} & a_2s_3-d_4 \\ 0 & 0 & 0 & 1 \end{bmatrix} \tag{18}$$

In Eq. 13, c_i represents for $\cos\theta_i$, c_{ij} represents for $\cos(\theta_i+\theta_j)$, In the same way , s_j represents for $\sin(\theta_j)$, s_{ij} represents for $\sin(\theta_i+\theta_j)$, i, j=1,2,3.4,5,6.

$_6^4T$given above as follows:

$$_6^4T = {}_5^4T{}_6^5T = \begin{bmatrix} c_5c_6 & -c_5s_6 & -s_5 & 0 \\ s_6 & c_6 & 0 & 0 \\ s_5c_6 & -s_5s_6 & c_5 & 0 \\ 0 & 0 & 0 & 1 \end{bmatrix} \tag{19}$$

If the elements (1,3) and (3,3) on both sides of Eq. 17 are equal, then:

$$r_{13}\left(c_1c_{23}c_4+s_1s_4\right)+r_{23}\left(s_1c_{23}c_4-c_1s_4\right)-r_{33}\left(s_{23}c_4\right)=-s_5 \tag{20}$$

$$r_{13}\left(-c_1s_{23}\right)+r_{23}\left(-s_1s_{23}\right)+r_{33}\left(-c_{23}\right)=c_5 \tag{21}$$

From Eq. 20 and Eq. 21, we can get the result of θ_4 is

$$\theta_5 = \text{atan }2\left(s_5, c_5\right) \tag{22}$$

Applying the above method again, it can be calculated that Eq. 1 can be written as follows.

$$\left({}_5^0\text{T}\right)^{-1}{}_6^0T = {}_6^5T(\theta_6) \tag{23}$$

As mentioned above, if the elements (1, 1) and (3, 1) on both sides of (19) are equal, then:

$$s_6 = -r_{11}\left(c_1c_{23}s_4-s_1c_4\right)-r_{21}\left(s_1c_{23}s_4+c_1c_4\right)+r_{31}\left(s_{23}s_4\right) \tag{24}$$

$$c_6 = r_{11}\left[\left(c_1c_{23}c_4+s_1s_4\right)c_5-c_1s_{23}s_5\right]+r_{21}\left[\left(s_1c_{23}c_4-c_1s_4\right)c_5-s_1s_{23}s_5\right]-r_{31}\left(s_{23}c_4c_5+c_{23}s_5\right) \tag{25}$$

From Eq. 24 and Eq. 25, we can get the result of θ_6 is:

$$\theta_6 = \text{Atan }2\left(s_6, c_6\right) \tag{26}$$

Since signs appear in Eq. 11 and Eq. 15, there are four possible solutions to these equations. In addition, four other solutions can be obtained by operating the "rotation" of the brachial joint. For the four solutions calculated above, the "flip" of the wrist joint can be obtained:

$$\theta_4' = \theta_4 + 180 \tag{27}$$

$$\theta_5' = -\theta_5 \tag{28}$$

$$\theta_6' = \theta_6 + 180 \tag{29}$$

When all eight solutions are calculated, some (or even all) of them are eliminated due to the limitation of joint motion range. Of the remaining valid solutions, the one closest to the current operator is usually chosen.

3. Algorithm simulation and analysis

Because the collected signal is non-linearity and non-constant Q, it needs domain knowledge and expert experience to extract features of the signal. CEEMDAN not only can adaptively decompose vibration signals into different frequency bands, but also solve the phenomenon of mode mixing by adding specific white noise. LSTM not only can extract the inherent features of the rolling bearing vibration signal adaptively, but also make use of the memory feature of the network, which can combine the running state of the front and back time to make the classification more accurate. Therefore, CEEMDAN-LSTM model by connecting CEEMDAN and LSTM is proposed to realize fault diagnosis. The overall structure of the model is shown in Figure 3.

In order to verify the effectiveness of the robot's hybrid inverse solution based on geometric method and the inverse transformation matrix algorithm, a spiral was first planned in cartesian space, then the joint angle was obtained by inverse solution and then the positive solution was returned to cartesian space for verification. The simulation experiment is carried out under the platform of matlab 2017a, and the algorithm has been realized through matlab programming. The test environment is as follows: the operating system is Windows 7; Memory: 8GB; CPU is Inter Core i7-4790 3.60 Ghz.

The spiral path is:

$$\begin{cases} x = 500\cos(\pi t) \\ y = 500\sin(\pi t) \\ z = 0.1t - 10 \end{cases}$$

Where t is the parameter, and t=0: 0.02: 4

End link orientation of robot is α=0.01: 0.01: 2.01, β=0.005: 0.005: 1.005, γ= 0.015: 0.015: 3.015 which corresponding to the position form the combination of 201 pose.

For the puma560 connecting rod parameters: a_2 = 670 mm, a_3 = 50 mm, d_3=30 mm, d_4 = 450 mm

Firstly, for the 201 points, the inverse geometric solution algorithm was used to calculate the value of the joint angles, and then the actual pose of the end link was obtained by the positive kinematics solution of the robot.

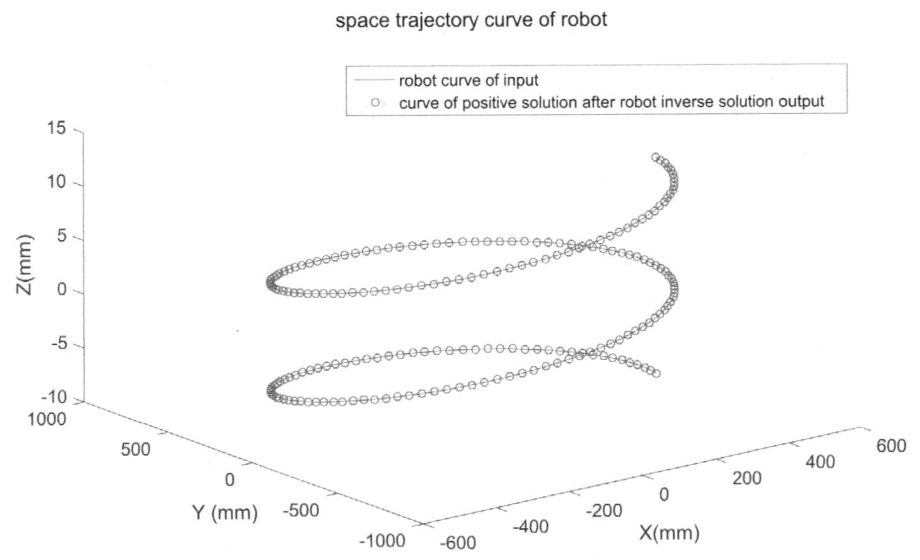

Figure 4. Positive curves and actual curves of the proposed robot algorithm

In Figure 4, the inverse solution proposed in this paper also shows that the difference of the trajectory is very small, almost the same.

From Figure 5 and Figure 7, the trajectory of the simulation is highly coincident with the desired trajectory, the position accuracy of the inverse solution proposed in this paper reaches level 10^{-13}.However, the position accuracy of the inverse transformation matrix method is 10^{-5}. From Figure 6 and Figure 8, the orientation error of the inverse solution proposed in this paper is also minimal, and reaches level 10^{-15}, However, the orientation accuracy of the inverse transformation matrix method is 10^{-7}. the inverse solution proposed in this paper better meets the requirements of use. The hybrid algorithm has certain advantages and practicability.

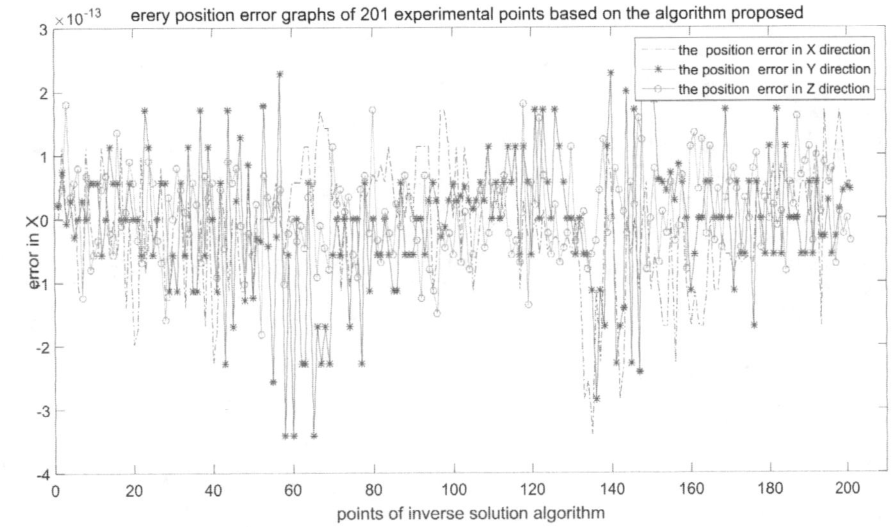

Figure 5. every position error graph of 201 experimental points based on the algorithm proposed

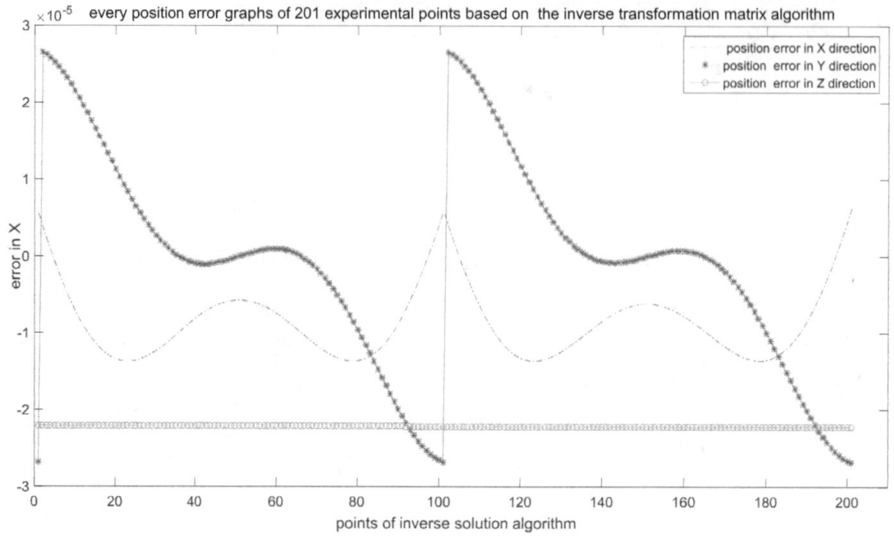

Figure 6. every position error graph of 201 experimental points based on the inverse transformation matrix algorithm

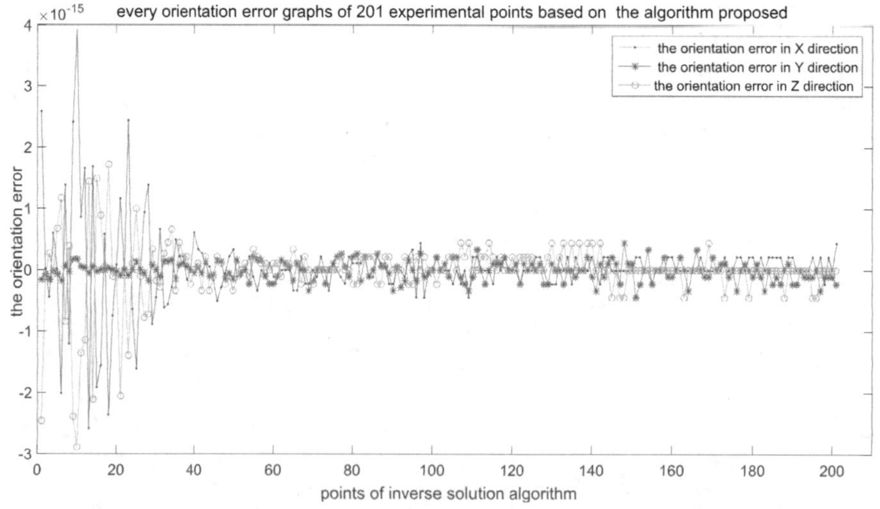

Figure 7. every orientation error graph of 201 experimental points based on the algorithm proposed

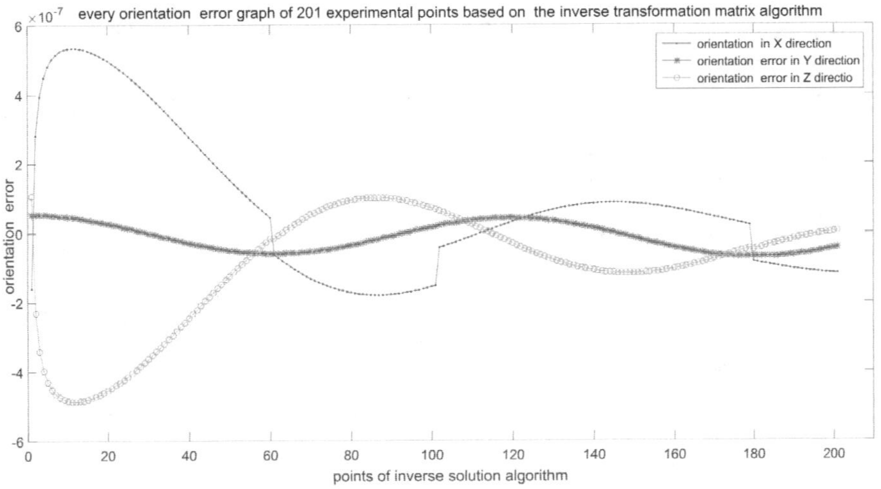

Figure 8. every orientation error graph of 201 experimental points based on the inverse transformation matrix algorithm

4. Conclusion

As conclusion, the mathematical model of 6R joint robot was established. A hybrid inverse solution algorithm based on geometric method and the inverse transformation matrix method was proposed. The traditional matrix method to solve the inverse problem of industrial robots have complex expressions. As it presented in this paper, the geometric method is used to solve the first three joints and the last three joints of the mechanical arm, and the matrix method is used to obtain the final inverse solution. The hybrid algorithm has certain advantages and practicability。

Acknowledgments

This work is supported by The Innovation Fund of Industry, education and research of the Science and Technology Development Center of the Ministry of Education(2018C01068), Science and Technology Research Fund of Jiangxi Provincial Department of Education(GJJ191006),2020 Nanchang Engineering Research Center and Key Laboratory Foundation(2020-NCZDSY-005), Ningbo Science and Technology Innovation 2025 Major Special Project,(No. 2019B10075), Ningbo Science and Technology Innovation 2025 Major Special Project (No. 2018B10010), 2021 Nanchang Key Laboratory of Intelligent Manufacturing Electronic Assembly Innovation and Development (2021-NCZDSY-018)

References

[1] Ye S, Liu D 2014 Robot Inverse Kinematics Algorithm Including Multi-Solutions and Singularity Analysis *Machine tool & hydraulics* **3** 27-9

[2] Cai Z 2015 *Robotics (Second Edition) (Ji Qi Ren: Di Er Ban)* (Beijing: Tsinghua University Publishing House)

[3] Li X, Guo Y, et al 2013 Inverse Kinematics Solution and Verification of Modular 6-DOF Manipulator *Transactions of the Chinese Society for Agricultural Machinery* **4** 246-51

[4] Dong M, Zhou Y, Zhang M 2000 A New Solution for Inverse Kinematics Problems of PUMA Robot *Modular Machine Tool & Automatic Manufacturing Technique* **10** 19-21

[5] Zhao K, LI Y, Li C 2002 A New Solution For Inverse Kinematics Of Standford Robot Based On There Intersect Joints *Journal of Taiyuan University of Technology* **6** 637-9

[6] Liu D, Wang T 2007 Algorithm Combined Analytical and Numerical Solutions for Robot Inverse Kinematics *Journal of Beijing University of Aeronautics and Astronautics* **6** 727-30

[7] Goldenberg A A, Benhabib B, Fenton R G 1985 A Complete Generalized Solution to the Inverse

Kinematics of Robots *IEEE Journal on Robotics & Automation* **1** 14-20
[8] Wang X, Yang G, et al 2010 A Inverse Kinematics Algorithm of the General 6 DOF Robot Based on Newton-Raphson Iteration *Transducer and Microsystem Technologies* **10** 116-8
[9] Wang W, Yu H, Zhou W 2017 Geometric Inverse Algorithm for 6R Robot *Science Technology and Engineering* **6** 83-7
[10] Lu Z, Zheng S 2017 Inverse Kinematics Algorithm of 6-DOF Robots based on Geometric Method and Screw Theory *Journal of Mechanical Transmission* **6** 111-4

SNR Classification Based Multi-Estimator IRM Speech Enhancement Algorithm

Xinqiang Li, Xingmian Wang, Yanan Qin, and Jing Li*

College of Electronic and Information Engineering, Shandong University of Science and Technology, Qingdao, China

*Corresponding author's e-mail: peilianglj@163.com

Abstract. Deep neural network(DNN)-based ideal ratio mask(IRM) estimation methods are often adopted in speech enhancement tasks. In the previous work, IRM estimation was usually realized by a single DNN-based IRM estimator without considering the SNR levels, which had a limited performance in real applications. Therefore, a two stage speech enhancement method is proposed in this paper. Firstly, a DNN-based SNR classifier is employed to classify the speech frames into three classes according to different SNR thresholds. Secondly, three corresponding DNN based IRM estimators related to the three SNR classes are trained respectively, from which the amplitude spectrum is corrected. Finally, speech enhancement is realized by doing IDFT to the corrected speech spectrum combined with the phase information of noisy speech. Experiment results show that the algorithm proposed in this paper has better performances in the evaluation of short time objective intelligibility(STOI), perceptual evaluation of speech quality(PESQ) and segmental signal-to-noise ratio improvement(SSNRI) scores.

1. Introduction

Speech enhancement is an important issue in speech signal processing, with the purpose of improving speech intelligibility and also reducing the signal distortion. Traditional unsupervised speech enhancement algorithms, such as Wiener filtering[1], spectral-subtractive algorithm[2] and statistical-model-based methods[3], were usually based on stationary noise assumptions, and could not achieve good performance in non-stationary noise conditions. Non-negative matrix factorization (NMF) is another speech enhancement method[4], which is based on matrix decomposition and has a good performance when it comes to structured interferences, e.g. music or a competing speaker, but the high computational cost limits its application in real-time speech enhancement tasks.

For supervised speech enhancement methods, such as algorithms using DNN as the learning machine, have shown much potential in single-channel speech enhancement tasks, especially under non-stationary noise environments.

DNN-Based IRM algorithms[5-8] is also supervised speech enhancement algorithms, which have been proved as an effective solution on speech enhancement. For DNN-Based IRM algorithms, a DNN is usually used as the IRM estimator to predict the value of IRMs, from which amplitude spectrums of the noisy speeches will be corrected, and then, speech enhancement can be realized by obtaining the IDFT of the corrected amplitude spectrum combined with the phase information, we call these algorithms as single-estimator IRM algorithms. But for the single-estimator IRM algorithms, it's hard for the specific DNN to fits in all the speeches because the SNR levels may change abruptly. To solve the problem, the local SNR(the average SNR of a speech frame,usually around 20ms) classification based multi-estimator IRM algorithm is proposed in this paper.

2. Speech enhancement based on local SNR classification

The idea of the IRM based speech enhancement is to use the estimated T-F masks to correct the amplitude spectrums of the noisy speech frames, as shown in equation (1), and then do IDFT to the corrected spectrums combined with the phase information to obtain the enhanced speech.

$$\hat{y}_{t,f} = \hat{irm}_{t,f} \cdot y_{t,f} \tag{1}$$

Where $y_{t,f}$ is the spectrum of the original noisy speech, $\hat{y}_{t,f}$ is the corresponding corrected spectrum, t is the frame index, f is the discrete frequency index and $\hat{irm}_{t,f}$ is the mask predicted by the IRM estimator. The T-F mask denoted by equation (2) is the training target according to literature [5].

$$irm_{t,f} = \left(\frac{s_{t,f}^{\ 2}}{s_{t,f}^{\ 2} + n_{t,f}^{\ 2}} \right)^{\beta} \tag{2}$$

Where $s_{t,f}$ and $n_{t,f}$ are the spectrums of clean speech and noise respectively, $irm_{t,f}$ is the ideal mask using as the training target and the value of β is usually set to 0.5, so equation (2) fits in the optimal estimation of the power spectrum.

To observe the performance of single IRM estimator algorithm described in literature [5] in different SNR conditions, speech enhancement tests were carried out for speech samples with the average SNR level of -5dB, so the enhanced speech frames can be obtained.

Then we classified all the noisy speech frames into two sets with the SNR level 0dB as the threshold as the following:

$$\begin{cases} \text{SET1: frames with SNR} \leq \text{0db} \\ \text{SET2: frames with SNR} > \text{0db} \end{cases}$$

and we analyzed the mask distortion of the enhanced frames corresponding to the two set respectively by equation (3).

$$\delta_t = \frac{1}{N} \sum_{f=0}^{N-1} \frac{\left| irm_{t,f} - \hat{irm}_{t,f} \right|}{irm_{t,f}} \tag{3}$$

Where δ_t is the mask distortion, N is the length of mask vector $IRM_t = (irm_{t,0}, irm_{t,1}, \ldots, irm_{t,N-1})$. The distortion of the test samples is shown in figure 1.

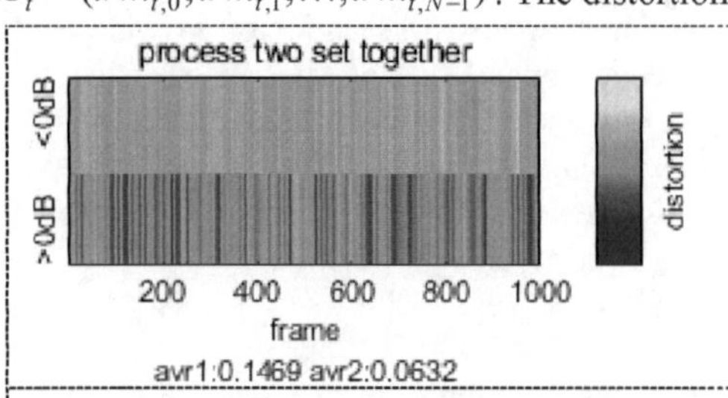

Figure 1. IRM distortion of the single-estimator IRM algorithm.

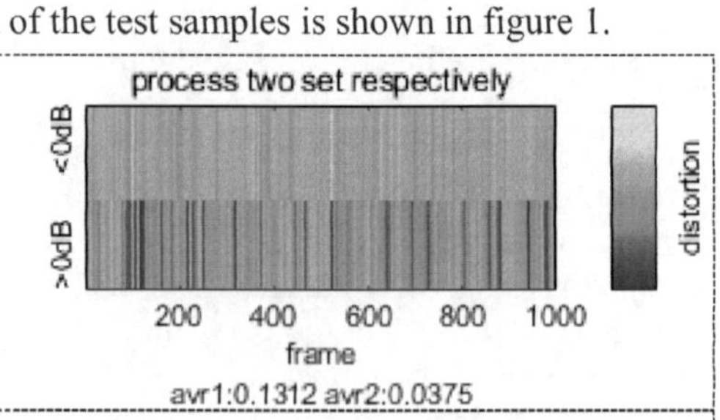

Figure 2. IRM distortion by employing two IRM estimators.

In figure 1, the top half give the mask distortions of enhanced speech frames corresponding to SET1, avr1 is the mean distortion of them, and the bottom half gives the mask distortions of the enhanced frames corresponding to SET2, avr2 is the mean distortion of them. It can be seen that the mask distortion is worse in the speech segments with lower SNR levels.

As a comparison, we trained two DNN IRM estimators with the same two sample sets: SET1 and SET2 respectively, those were, one for SET1 and another one for SET2. The distortion denoted in equation (3) was evaluated and given by figure 2.

Comparing figure 1 with figure 2, it can be seen that mask estimation distortion can be greatly reduced by classifying speech frames into different sets according to the SNR levels and then training different IRM estimators respectively, and this is the idea of local SNR classification based multi-estimator IRM speech enhancement algorithm.

3. Method

3.1. SNR classification

Considering the complexity of the algorithm, the noisy speech frames are divided into three categories according to local SNR levels by two thresholds, namely low SNR category(SNR1), medium SNR category(SNR2) and high SNR category(SNR3).

The number of frames and the classification accuracy of SNR classifiers are considered when we decide the thresholds. A higher accuracy sometimes means uneven frame quantities of each SNR category. We have tried different thresholds, for example, we tried the thresholds of 0dB and 5dB and achieved the highest classification accuracy of 95.4%, but the frame quantity of SNR3 were too less to train the corresponding IRM estimator.

Finally we select the thresholds for each category as given in table 1.

Table 1. The thresholds of SNR categories.

SNR category	SNR levels(dB)
SNR1	<-8
SNR2	>=-8 and <2
SNR3	>=2

Table 2. The coding of the SNR categories.

SNR category	code
SNR1	(1,0,0)
SNR2	(0,1,0)
SNR3	(0,0,1)

The SNR classifier consists of a neural network with softmax layer, as shown in figure 3.

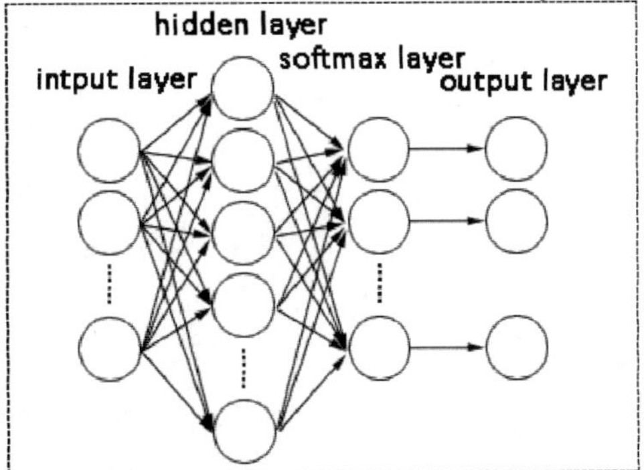

Figure 3. Structure of SNR classifier.

The network includes an input layer, a hidden layer, a softmax layer and an output layer. The input layer contains 100 neurons, consisting of Mel frequency cepstrum coefficient(MFCC), relative spectral transformed perceptual linear prediction cepstral coefficient(RASTA-PLP) and their first and second order difference. The hidden layer contains 500 neurons, the activation function is tan-sigmoid function.

The output result is a vector of 3 elements, and here, hot coding is used to represent the SNR categories of the speech frames, and the corresponding hot coding table is shown in table 2.

Softmax layer maps the input of each neuron to (0,1), and the final output of the network is a vector like (a, b, c), where a, b, and c are all decimal values from 0 to 1, and the largest element of the three elements is set to 1, and the other two elements are set to 0.

In this paper, 200 utterances in TIMIT data base and another 50 utterances are mixed with white noise at -5dB and used as the training speech and the test speech respectively to test the SNR classification method. The experimental results show that the classification accuracy is about 86% with the threshold values of - 8dB and 2dB.

3.2. multi-estimator IRM Algorithm

The local SNR classification based multi-estimator IRM speech enhancement algorithm is proposed in this section. The block diagram of the proposed method is given by figure 4.

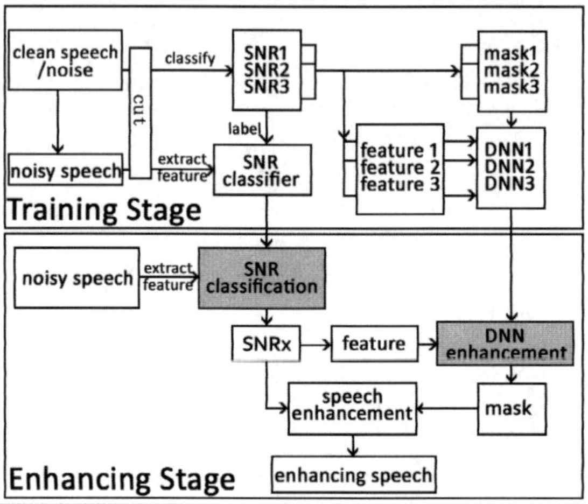

Figure 4. Block diagram of the proposed method.

In the training stage, firstly we need to train the SNR classifier. So the testing frames are classified into three SNR categories according to the thresholds in table 1, from which we get the labels of all the test frames. After that, the SNR classifier is trained with the robust parameters introduced in section 3.1 as the input, and the training label is the hot codes based on SNR categories as shown in table 2.

Secondly, we need to train the three IRM estimators, each of them is trained respectively according to the three categories mentioned above. The input data is the robust parameters extracted from the noisy speech frames, and the training target is the original masks of the noisy speech according to equation (2). Finally we get three trained IRM estimators denoted as DNN1, DNN2, DNN3 corresponding to the three SNR categories separately.

Here, the robust feature parameters of the three IRM estimators include MFCC, RASTA-PLP, amplitude modulation spectrum(AMS), the first and second order difference data of MFCC and the first order difference data of RASTA-PLP.

4. Results and Discussions

4.1. Experimental Settings

In this paper, we select 2144 utterances(about 1 hour) in TIMIT database as the clean speech, and the training set is obtained by mix the clean speech with white noise, pink noise and babble noise selected from noisex-92 noise databases, each with SNR levels - 5dB, 0dB and 5dB separately.

The test set is obtained by mixing 330 clean utterances in TIMIT database with the same noises adopted in the training set respectively, that is, white noise, pink noise and babble noise and each with SNR levels - 5dB, 0dB and 5dB.

The structure of the three DNN-based IRM estimators is illustrated in figure 5. The three hidden layers use the log-sigmoid function as the activation function, and each hidden layer contains 400 neurons.

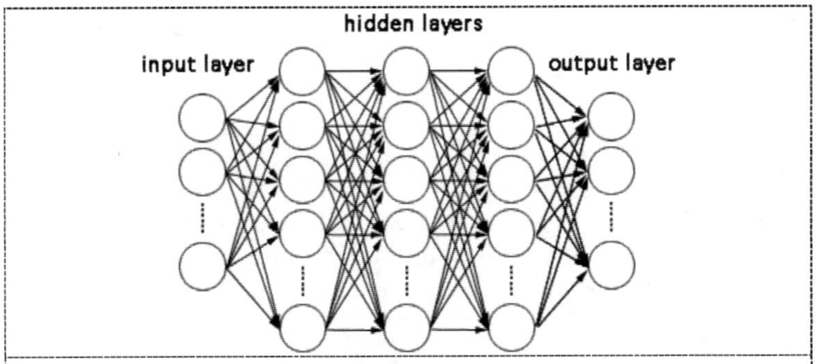

Figure 5. The structure of the IRM estimators.

The output data is an mask vector with the dimension of 160, which corresponds to the frequency range 0 ~ 4kHz. In this paper, the training epoch is set to 50, the mean square error function is used as the cost function, and the training method is the gradient descent method with adaptive learning rate.

As a comparison, the single-estimator IRM algorithm is used as the baseline algorithm, and the speech enhancement performance is evaluated by the measurements of short time objective intelligibility(STOI), perceptual evaluation of speech quality(PESQ) and segmental signal-to-noise ratio improvement(SSNRI).

4.2. *STOI Scores*

In this section, the STOI scores of our algorithm are evaluated under the SNR levels of -5dB, 0dB and 5dB. Comparision of the scores between the single-estimator IRM algorithm(named as SE-IRM) and the proposed multi-estimator IRM algorithm(named as ME-IRM) are carried out, and the results are shown in table 3. It can be seen that under the conditions of - 5dB, 0dB and 5dB, the method proposed in this paper achieved better performance than the single-estimator IRM algorithm. It is worth noting that the babble noise group get the lowest STOI scores at all the SNR conditions when employing SE-IRM algorithm but have highest improvements when employing ME-IRM algorithm with the STOI values of 0.0866, 0.0622 and 0.0435. And the similar STOI scores between different noise conditons in each SNR level means better adaptability of the proposed method than SE-IRM method.

Table 3. The STOI scores.

	pink			white			babble		
	noisy	SE-IRM	ME-IRM	noisy	SE-IRM	ME-IRM	noisy	SE-IRM	ME-IRM
-5dB	0.5653	0.6835	0.7370	0.5422	0.6875	0.7364	0.5557	0.6631	0.7497
0dB	0.6824	0.7800	0.8206	0.6574	0.7741	0.8187	0.6676	0.7577	0.8199
5dB	0.7959	0.8543	0.8820	0.7764	0.8497	0.8747	0.7722	0.8338	0.8773

4.3. *PESQ Evaluation*

In this section, the PESQ scores of our algorithm and the SE-IRM method are calculated as illustrated in table 4.

Table 4. The PESQ scores.

	pink			white			babble		
	noisy	SE-IRM	ME-IRM	noisy	SE-IRM	ME-IRM	noisy	SE-IRM	ME-IRM
-5dB	1.4465	1.8204	2.0939	1.2827	1.7155	2.0350	1.5743	1.8372	2.1263
0dB	1.7891	2.2144	2.4485	1.5553	2.0532	2.3685	1.9040	2.1822	2.3898
5dB	2.1489	2.5554	2.7488	1.8746	2.3656	2.5798	2.2499	2.5044	2.7011

From table 4, we can see that the SE-IRM method can improve the speech quality of noisy speeches, but the ME-IRM proposed in this paper has better performance compared to the SE-IRM method. Notice that the single-estimator IRM algorithm and the proposed method both have lower PESQ scores for white noises, but still the ME-IRM method has the highest improvement compared to the SE-IRM method.

4.4. SSNR Improvement

The segmental signal-to-noise ratio(SSNR) is also often adopted to evaluate the performance of speech enhancement algorithms. The SSNR improvement(SSNRI) scores of the proposed method and the baseline method are calculated as table 5 illustrates. We can see that the proposed ME-IRM method has better performance than the SE-IRM method in all conditions.

Table 5. SSNR improvement related to noisy speech.

	pink			white			babble		
	noisy	SE-IRM	ME-IRM	noisy	SE-IRM	ME-IRM	noisy	SE-IRM	ME-IRM
-5dB	-	11.0638	11.9448	-	8.5279	9.0975	-	9.4012	10.7923
0dB	-	8.8292	9.8105	-	6.9399	7.6776	-	7.2807	8.6360
5dB	-	6.5591	7.5469	-	5.1890	6.0822	-	5.0424	6.7164

5. Conclusions

A local SNR classification based multi-estimator IRM speech enhancement algorithm is proposed in this paper. In this algorithm, a speech SNR classifier is developed and by which speech frames are classified into three sets, and three corresponding IRM estimators are trained to estimate the IRM values, then the spectrums are corrected by employing the IRMs, and finally speech enhancement is realized by doing IDFT to the corrected spectrums. The experiment results show that the algorithm we proposed has better performance than the single-estimator IRM algorithm. However, our algorithm may introduce the problem of data discontinuity in the reconstruction stage in a certain extent, especially when the local SNR classification accuracy is not high enough. So the further research is needed on this issue.

References

[1] Yang, Yan , and C. Bao . "RS-CAE based AR-Wiener Filtering and Harmonic Recovery for Speech Enhancement." IEEE/ACM Transactions on Audio, Speech, and Language Processing PP.99(2019):1-1.
[2] Siam, A. I. , et al. "A Novel Speech Enhancement Method Using Fourier Series Decomposition and Spectral Subtraction for Robust Speaker Identification." Wireless Personal Communications 108.2(2019).
[3] Faraji, N. , and A. Kohansal . "MMSE and MAP estimators for speech enhancement in additive noise assuming a t-location-scale clean speech prior." IET Signal Processing 12.4(2018):532-543.
[4] Suh, et al. "Localized user-driven topic discovery via boosted ensemble of nonnegative matrix factorization." Knowledge & Information Systems (2018).
[5] Wang, Y. , A. Narayanan , and D. L. Wang . "On Training Targets for Supervised Speech Separation." IEEE/ACM Transactions on Audio, Speech, and Language Processing 22.12(2014):1849-1858.
[6] Zhao, Y. , Z. Q. Wang , and D. L. Wang . "Two-Stage Deep Learning for Noisy-Reverberant Speech Enhancement." IEEE/ACM Transactions on Audio, Speech, and Language Processing 27(2018):53-62.
[7] Wang, X. , and C. Bao . "Mask estimation incorporating phase-sensitive information for speech enhancement." Applied Acoustics 156(2019):101-112.
[8] Li, R. , et al. "A multi-objective learning speech enhancement algorithm based on IRM post-processing with joint estimation of SCNN and TCNN." Digital Signal Processing 101.

A Trust Index-Based Adaptive Kriging Method for the Cooling Fan Optimization

Li Lu[1,*], Yizhong Wu[1], Zhehao Xia[1] and Qi Zhang[1]

[1]National CAD Supported Software Engineering Centre, School of Mechanical Science and Engineering, Huazhong University of Science and Technology, 1037 Luoyu Road, 430074 Wuhan, People's Republic of China

[*]Corresponding author's e-mail: luli1992@hust.edu.cn

Abstract. In the cooling fan optimization, there are many local minima near the optima, which improves the accuracy requirement of the Kriging model. Due to unexpected prediction errors caused by some deceptive samples, the model exploration capability of the traditional method is not enough. To overcome this problem, an adaptive Kriging method based on the trust index is proposed in this paper. By considering the sample distribution and region nonlinearity, the trust index is used to evaluate the reliability of the samples, which can enhance the sampling strategy for new candidates. Several classic test functions with many local minima are employed to verify the effectiveness of the proposed method. Further, the method is used to optimize the cooling fan, and the simulation result shows that the performance of the optimization objective is significantly increased.
Keywords. Adaptive Kriging, Trust index, Global Exploration, Cooling Fan

1. Introduction

In engineering optimization, most problems are black-box, and the simulation cost is usually expensive. Therefore, researchers propose the optimization method base on the surrogate models to overcome the difficulty[1; 2]. By using the surrogate model technique such as Kriging[3], radial basis function[4], and artificial neural network[5], the optimization cost of the black box problem is effectively reduced. The key to building an accurate surrogate model is the additional sample searching strategy called 'filling criteria'[6]. In studies of filling criteria, efficient global optimization (EGO) is one of the most widely used algorithms. The EGO believes that the prediction value and error of the Kriging model obey the normal distribution. The filling criteria of EGO, named Expected Improvement function(EI), is designed based on this viewpoint. However, the prediction ability of the Kriging model is affected by some deceptive samples, resulting in unexpected prediction errors when solving complex problems[7]. Therefore, many studies are carried out to improve EI by balance global and local search capabilities[8-10]. Meanwhile, since the intelligent algorithm has good exploration ability, they are combined with the surrogate model-based method to form the classic optimization framework[11-13]. By constructing fidelity surrogate models with the appropriate intelligent algorithm, the design cycle of the product is dramatically shortened.

In this paper, the trust index is defined and used to rebalance the global and local search capability. Further, particle swarm optimization is employed to search the additional sample with the trust index-based filling criteria. There are two aspects considered by the trust index, sample distribution and region nonlinearity. For the first aspect, the trust index represents the reliability of the samples. Controlling by the adaptive threshold, the domain is partitioned into reliable and unreliable regions. The algorithm will prioritize searching for the additional sample in the unreliable region. As for the second aspect, based on the trained Kriging model, the trust index represents the region nonlinearity. During the additional sample searching, the region with higher nonlinearity will be concerned. Overall, by introducing the index, the exploration ability of the adaptive method is improved.

The other parts of this paper are arranged as follows. In the first section, the trust index is defined, and the quantification method of the trust index is described. In section two, combining with the algorithm flow, the usage of the trust index in refining the global characteristics and searching the global optima are elaborated. In the third section, the effectiveness of the proposed method is verified by classical test functions. In section four, the proposed method is used to optimize the shape of the cooling fan. The optimization and origin design scheme are compared by computational fluid dynamics(CFD) simulation.

2. The trust index-based evaluation method for training sample

In this section, the trust index is divided into two parts for describing. In the first part, the definition of the trust index is about the sample itself. In the second part, the definition is about the relationship between the search particle and its nearest sample. Herein, the particle is the definition in the particle swarm optimization, which is used to qualify the second factor of the trust index. The samples are evaluated by normalizing and weighting these two factors. Therefore, the distrust index is also considered and explored while search the optimal solution.

The trained Kriging model can predict the response value and relative error of unknown points. However, due to the differences in nonlinearity regions and samples distribution, there are many discrepancies in the prediction effect. To better express the views of this paper, the following test example, named the Griewank function, is used to explain the definition and quantification process of the trust index. The samples are obtained by the Latin hypercube sampling.

Evaluate those random samples by expensive evaluation and construct the surrogate model. Through the prediction function of the Kriging method, the prediction error of the surrogate model in the feasible domain is displayed, as shown in Figure 1. It can be illustrated that the prediction error is directly proportional to the distance around the sample, and the error rapidly increases when far away from the samples. Therefore, add samples from sparse areas with more significant errors has become the basic idea of most research.

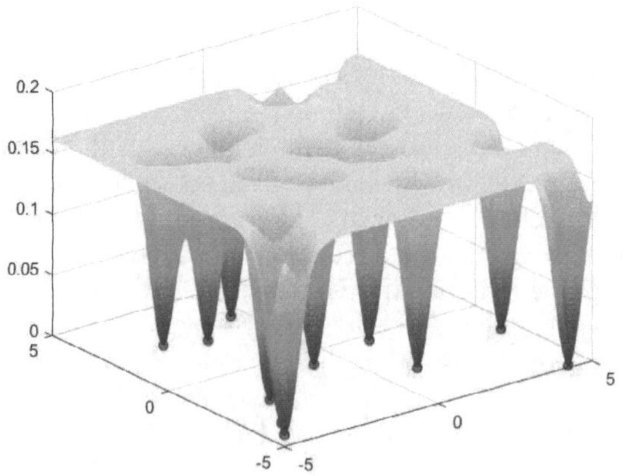

Figure 1. Mesh plot of the MSE of trained Kriging model

In each iteration, analyze the predictor error near the samples to generate the trust space, as shown in Figure 2. The initial radius $\alpha(x_i)$ of the trust space is set to 0.01 of the minimum parameter interval length, and the space size is further adjusted by evaluating the reliability of prediction points. When the space reliability is less than the threshold, the radius is determined. The radius, which represents the size of trust space, is the first aspect of the trust index.

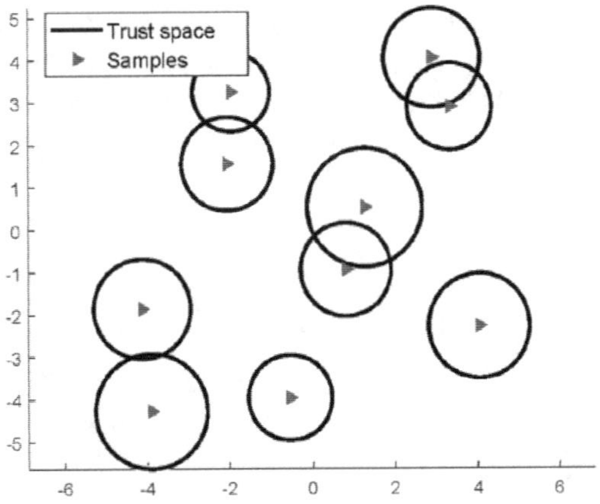

Figure 2. The trust space around the samples

The trust index generation steps for each sample are described as follows.

1) According to the radius of the trust space, evaluation points are generated on the circle boundary. The constructed Kriging model is adopted for the evaluation, so the expensive evaluation problem does not need to be considered. As the prediction result of points can be treated as the normal distribution $Y(x) \quad N\left(\hat{Y}(x), \hat{s}^2(x)\right)$ [14], the reliability of evaluation points can be calculated with the formula(1)

$$R(x) = \frac{k\ \hat{s}(x)}{\left|\hat{Y}(x)\right|} \leq P \tag{1}$$

where k is the value of the inverse function of cumulative probability function of set probability, and the value is set as 2. P represents the variation deviation threshold of trust space, and the value is 0.2.

Through the adjusted parameters k and P , the trust space can be maximized as possible. For example, according to the two parameters set above, the probability relative error of about 84%($\Phi(2)$) points are not larger than 0.2. In this paper, the global characteristics of the objective function are only roughly constructed, so the threshold condition is not be set strictly.

2) If all boundary points in the current space meet the condition(1), expand the current space radius α with $\alpha_{itr} = m\alpha_{itr-1}$ where m is the expansion factor. After completing the space expansion, return to step 1 and recalculate the reliability of the circle boundary. If the result still meets the condition(1), expand the space size again. If the condition does not be satisfied, stop the expansion of the current space. When the size of all spaces is finalized, go to step 3.

3) When all the size of spaces are determined, merged the overlapping spaces. Weighting according to the size of the original space to obtain the equivalent center. In this paper, since the size of trust spaces is the weight in the optimization process, the equivalent center is calculated as the centroid

$$x_{eq} = \frac{\sum_{i=1}^{n} x_i\ \alpha_i}{\sum_{i=1}^{n} \alpha_i} \tag{2}$$

At this point, the trust index of the sample itself is constructed. During optimization, the size of the trust space is reconstructed according to the Kriging model in every iteration. Adding samples in the relatively least trusting space helps reduce the global prediction error.

As mentioned earlier, another factor in the trust index definition is the region nonlinearity between search particle and their nearest sample. Another simple example is used to illustrate the idea of the area nonlinearity qualification. The zero-section of the levy function is intercepted and shown in Figure 3. The figure shows that the original function has strong nonlinearity at both ends of the feasible domain while weak in the middle. Meanwhile, samples are carried out by employing uniform sampling. These samples are used to train the Kriging model, and the prediction error curve is plotted.

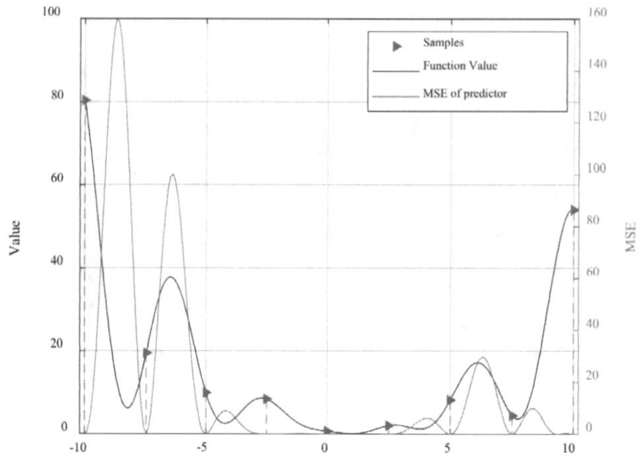

Figure 3. Prediction error of Kriging model of levy function with uniform sampling

In the middle region with weak nonlinearity, the prediction error of the Kriging model tends to be close to 0, and the prediction of the model is accurate. However, at both ends region, the situation is different. Firstly, the original function between the three sample points on the left is highly nonlinear, while the prediction error of the Kriging model fluctuates wildly among the sample points, indicating that the current samples in this area are not enough to build an accurate surrogate model. Meanwhile, the two samples on the far right in the figure are also in the nonlinear region, but the objective function change between the two samples tends to be linear. As shown in the figure, the prediction error of the Kriging model is smaller than that of the left area. Therefore, this paper relies on the built Kriging model to evaluate the degree of nonlinearity between the search particle and sample. For the area with lower reliability, the weight is increased when searching the additional samples. Through the gradient information of the Kriging model, the space between the search particle and sample is equally divided, and the total value of the gradient change is obtained by stepwise accumulation. The calculation method of the trust index is as follows:

$$R_{\vec{x}_{obj}-\vec{x}} = \frac{\sum_{i=1}^{q}\left(\left\|\nabla f(x_i)\right\| - \left\|\nabla f(x_{i-1})\right\|\right)}{\left\|\vec{x}_{obj} - \vec{x}\right\|} \tag{3}$$

where q is the number of equally divided spaces, $\vec{x}_{obj}$ is the vector of the target sample. x_i is the i-th division joint on the vector between the search particle and the target point.

So far, the qualification method of two factors of the trust index is explained. In the next section, the algorithm flow of the trust index-based Kriging method is elaborated.

3. Algorithm flow of trust index-based adaptive Kriging method

This paper aims to maintain a rough grasp of the global characteristics when searching for local optima. Therefore, the process of global exploration is added in parallel. At the end of each iteration, the samples are filtered to avoid crowded sampling. The fitness function of global exploration is constructed as follows, and the $\vec{x}_{near}$ represents the nearest sample of the particle.

$$min\,mize\ -\left(\left(\left\|\vec{x}_{near}-\vec{x}\right\|-\alpha_{near}\right)\cdot R_{\vec{x}_{near}-\vec{x}}\right). \tag{4}$$

In each iteration, the search for the optima and the global exploration point is performed in parallel, and the sample set is gradually updated until the optimal result meets the convergence condition. The complete algorithm flow is described as follows, and the flow chart is shown in Figure 4.

1) Using the Latin hypercube sampling method to generate 5d initial samples according to the dimension d of the problem.

2) Construct the Kriging model of the optimization objective basing on the initial samples.

3) According to the content of section 2, the trust index of samples is evaluated during the global exploration process.

4) Activate the particle swarm optimization algorithm to optimize two computational goals in parallel. One of the goals is finding out the optima of the optimization problem, and the other is sampling in the lowest trust index region. As mentioned in step 3, the trust index between the particle and the sample is evaluated. Further, It combined with the size of trust spaces to evaluate the fitness function(4) of each particle.

5) Check whether the iteration reaches the set maximum number of iteration steps, or the optimal value converges in the last ten iterations. If the stop condition is satisfied, output the optimal result. Otherwise, put the two samples found in step 4 into the sample set and go to step 6.

6) Evaluate the congestion degree of the samples. When the spacing between the samples is smaller than the threshold of congestion degree, which is set as the minimum index radius α, take the better performance one to update the sample set, return to step 2.

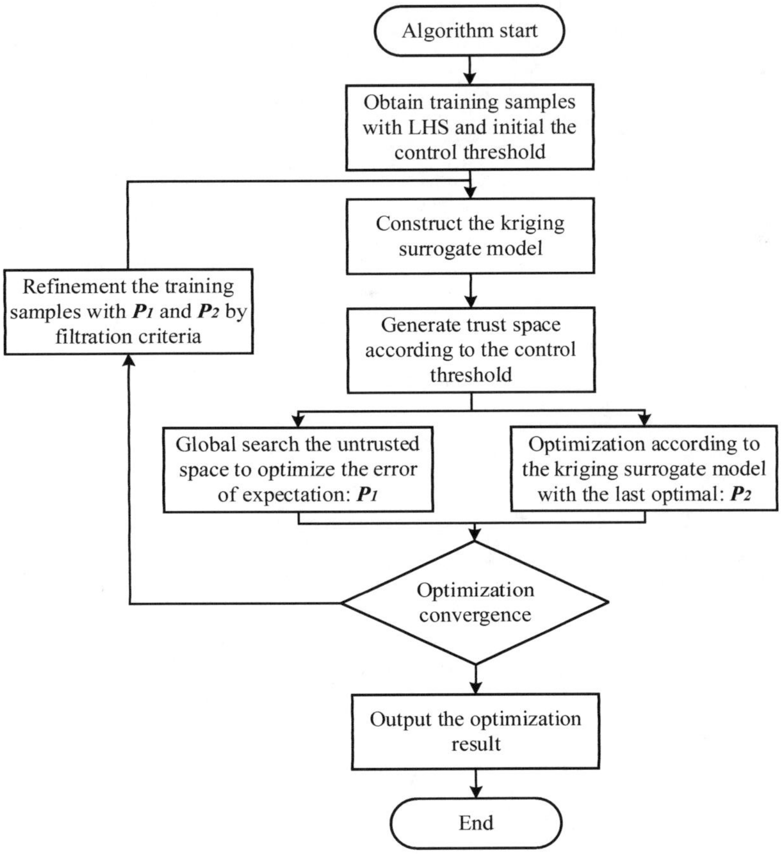

Figure 4. The algorithm flow chart of the trust index-based Kriging method

4. Numerical examples

In order to verify the feasibility of the method proposed in this article, several numerical test cases were selected[7; 9; 15-18]. These commonly used test functions contain multiple local minima, high nonlinearity, and other characteristics. The test results are shown in Table 1. In order to prove that the proposed method improves optimization efficiency, the article focuses on the following points. All examples are run 30 times to prove the stability of the proposed method. The mean, variance, and worst solution of the results are displayed to prove that the results of proposed method can converge to the same result. At the same time, the results are compared with the classic algorithm with the EI function to compare the accuracy of the solution.

As listed in Table 1, the proposed method can obtain accurate optimization results for all the test examples. Since this paper uses the heuristic algorithm to optimize the black-box problem, the actual value cannot be obtained. Although some deviations in the optimization results, the deviation is slight and within the allowable error range. Meanwhile, the adaptive Kriging method, which uses the EI function as filling criteria, is employed for comparison. As listed in the table, the optimization results obtained by the proposed method are more accurate in most cases. Further, in the test of Shubert and Cross In Tray function, the classic method fails to obtain the optimal solution.

Table 1. The test result of numerical examples (Num=30)

Function	Calibration Optima	EI function	Optimization Result		
			Mean	Variance	Worst
Branin	0.3979	0.3978	0.3977	3.3121e-6	0.3915
Six-Hump Camel	-1.0316	-1.0317	-1.0317	1.7245e-6	-1.0303
Griewank	0	0.0001	0.0012	1.1359e-6	0.0017
Bohachevsky	0	0.0209	0.0010	2.7831e-6	0.0043
Levy	0	0.0152	0.0020	5.0631e-6	0.0058
Goldstein Price	3	3.0034	3.0012	7.2343e-6	3.0087
Shubert	-186.7309	-164.8672	-186.7414	6.2978e-5	-186.7719
Cross In Tray	-2.0626	-2.0257	-2.0626	1.3778e-10	2.06263

The relative error of the 30 times optimization is calculated and shown as a box diagram in Figure 5. Since the optima of some test functions are 0, the relative error cannot be calculated. Therefore, the relative error calculation in these test cases is based on 1e-5, set for visualization artificially. These three cases are displayed on the left, and the others are displayed on the right.

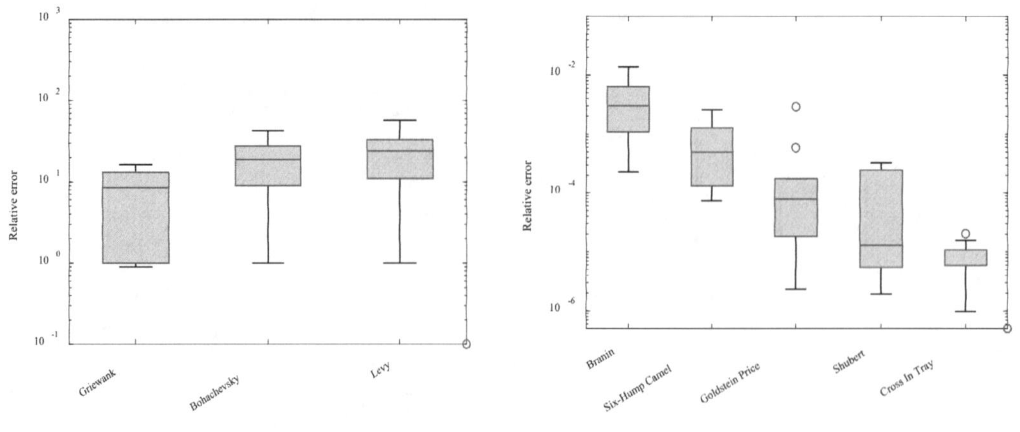

(a) (b)

Figure 5. The relative error of test functions

Journal of Physics: Conference Series **2173** (2022) 012087 doi:10.1088/1742-6596/2173/1/012087

5. Cooling fan optimization

The object of optimization is the fan of the cooling system in the automobile. In engineering design, the design of the cooling fan usually adopts the method of practical invention. Due to computational fluid dynamics(CFD) popularity, the optimization method has gradually been converted to a combination of alternative models and simulation. The design variables are shown in Figure 6; five variables are controlled to get the optimal bending shape of the fan blade.

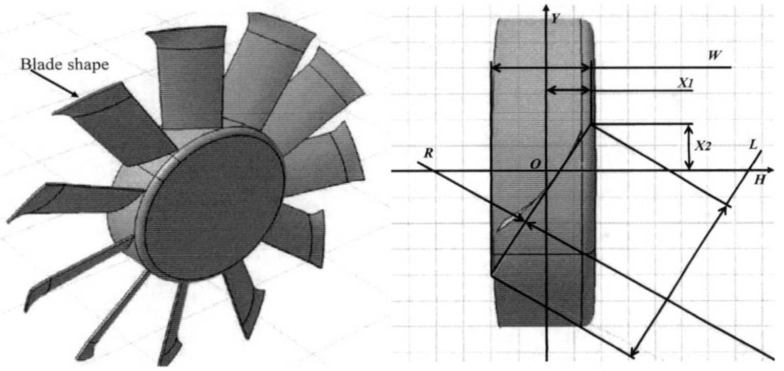

(a) (b)

Figure 6. The design variables of the cooling fan

where x_1 and x_2 are the abscissa and ordinate of the endpoint of the control curve, respectively. w represents the width of the fan blade. R and L are the radius and chord length of the arc where the two endpoints of the control curve are located.

Three computational domains are constructed: the inlet, outlet, and rotation to efficiently simulate the cooling fan working condition. The gas flow in the rotating domain is more complicated than the others, while the flow in the inlet and outlet domains is relatively simple. Therefore, to accelerate the CFD iteration process, the mesh size of the rotation domain is set to 5mm, while the mesh size of the inlet and outlet domains is relatively sparse, with a length of 15mm. According to the actual situation, the initial calculation domain inlet is set as the pressure stagnation inlet with a value of -60 Pa. the calculation domain outlet is set as the atmospheric pressure outlet with a value of 0 Pa. And, the rotation speed of the rotating calculation domain is set as 1200rpm.

According to the method process proposed in this article, the initial sample uses LHS to obtain 5x5=25 samples. CFD simulation is performed according to the set simulation parameters, and the mass flow rate is read while simulation convergence. When the algorithm calculates the next point, The simulation is performed according to the same process. Finally, the algorithm converged to obtain the optimal result. The data comparison between the original design and the optimization result is shown in Table 2. In addition, the value in brackets is the prediction result of the Kriging model. It can be seen that the relative error from the actual result is 0.48%, indicating that the surrogate model construction is accurate near the optima.

Table 2. Comparison between the original design and optimization

Type	Design parameters (x1,x2,w,L,R)	The mass flow rate of the inlet
Original design	21.36, 18.86, 46.58, 78.78, 335.19	0.588
optimization	26.55, 21.74, 46.64, 65.39, 406.01	0.618(0.621)

In order to compare the simulation results more intuitively, we display the cross-section of the velocity vector field of the simulation domain in Figure 7. The graph shows that the optimized design also has a uniform distribution of flow field and a higher mass flow rate in the fan domain.

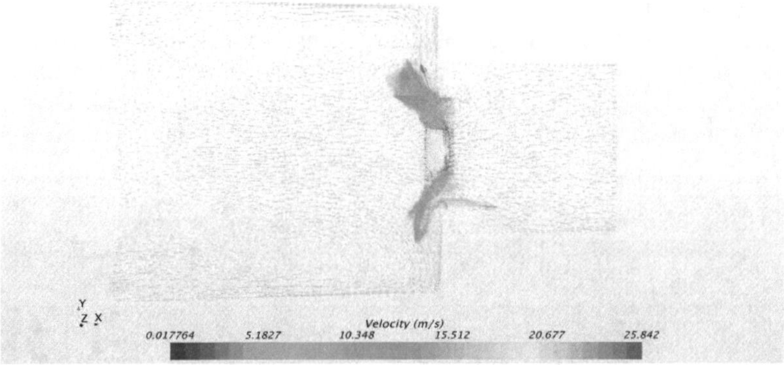

(a)

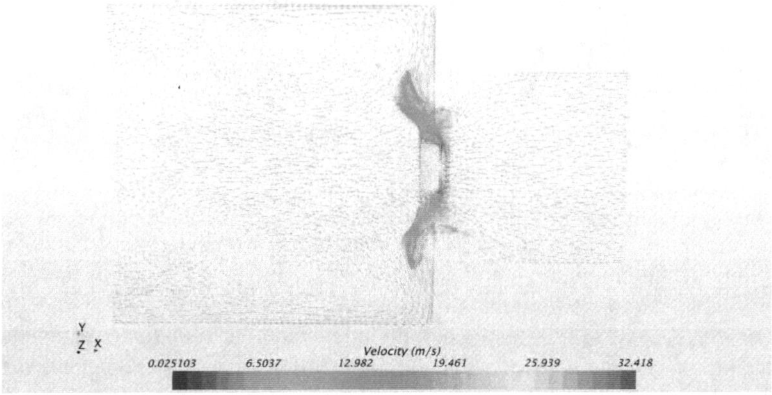

(b)

Figure 7. The final vector field of the original design and optimization

6. Conclusion

In this paper, the trust index-based adaptive Kriging method, which concentrates on the distribution and the nonlinearity of samples, is proposed. The evaluation method of the trust index is formed with the prediction function of the Kriging model. During the optimization process, the sample distribution and region nonlinearity are considered to improve the filling criteria. Several classical test functions and an engineering example, which contain several local minima, are employed and verified the effectiveness of the proposed method. The test results show that the method proposed in this paper can solve complex optimization problems. In the cooling fan optimization, the obtained scheme dramatically increases the mass flow rate of the cooling fan.

Acknowledgments

This paper is supported by the National Natural Science Foundation of China(Grant NO.51575205) and the National Key Research and Development Project(Grant No. 2019YFB1706501). These supports are gratefully acknowledged.

Reference

[1] Gu JC, Zhang H, Zhong XG (2020). Hybrid meta-model-based global optimum pursuing method for expensive problems. Struct Multidiscip O 61:543-554 doi:https://doi.org/10.1007/s00158-019-02373-w
[2] Cai XW, Gao L, Li XY (2020). Efficient Generalized Surrogate-Assisted Evolutionary Algorithm for High-Dimensional Expensive Problems. Ieee T Evolut Comput 24:365-379 doi:https://doi.org/10.1109/TEVC.2019.2919762.
[3] Chen LM, Qiu HB, Gao L, Jiang C, Yang Z (2020). Optimization of expensive black-box problems

via Gradient-enhanced Kriging. Comput Method Appl M 362 doi:https://doi.org/10.1016/j.cma.2020.112861

[4] Yi J, Shen YC, Shoemaker CA (2020). A multi-fidelity RBF surrogate-based optimization framework for computationally expensive multi-modal problems with application to capacity planning of manufacturing systems. Struct Multidiscip O 62:1787-1807

[5] Nobahari A, Vahedi A, Takorabet N (2020). ANN-based optimization approach devoted to the sizing of arbitrary rotor pole geometries of permanent magnet motors for electric motorcycle. Int J Appl Electrom 64:S155-S166 doi:https://doi.org/10.3233/JAE-209512

[6] Jones DR, Schonlau M, Welch WJ (1998). Efficient global optimization of expensive black-box functions. J Global Optim 13:455-492 doi:https://doi.org/10.1023/A:1008306431147

[7] Li Y, Shi J, Cen H, Shen J, Chao Y (2021). A Kriging-based adaptive global optimization method with generalized expected improvement and its application in numerical simulation and crop evapotranspiration. Agricultural Water Management 245:106623 doi:https://doi.org/10.1016/j.agwat.2020.106623

[8] Liu H, Ong Y-S, Cai J (2018). A survey of adaptive sampling for global metamodeling in support of simulation-based complex engineering design. Struct Multidiscip O 57:393-416 doi:https://doi.org/10.1007/s00158-017-1739-8

[9] Xing J, Luo Y, Gao Z (2020). A global optimization strategy based on the Kriging surrogate model and parallel computing. Struct Multidiscip O 62:405-417 doi:https://doi.org/10.1007/s00158-020-02495-6

[10] Zhou Y, Haftka RT, Cheng G (2016). Balancing diversity and performance in global optimization. Struct Multidiscip O 54:1093-1105 doi:https://doi.org/10.1007/s00158-016-1434-1

[11] Fan CD, Hou B, Zheng JH, Xiao LY, Yi LZ (2020). A surrogate-assisted particle swarm optimization using ensemble learning for expensive problems with small sample datasets. Applied Soft Computing 91 doi:https://doi.org/10.1016/j.asoc.2020.106242

[12] Dong HC, Dong ZM (2020). Surrogate-assisted grey wolf optimization for high-dimensional, computationally expensive black-box problems. Swarm Evol Comput 57 doi:https://doi.org/10.1016/j.swevo.2020.100713

[13] Briffoteaux G, Gobert M, Ragonnet R, Gmys J, Mezmaz M, Melab N, Tuyttens D (2020). Parallel surrogate-assisted optimization: Batched Bayesian Neural Network-assisted GA versus q-EGO. Swarm Evol Comput 57 doi:https://doi.org/10.1016/j.swevo.2020.100717

[14] Echard B, Gayton N, Lemaire M (2011). AK-MCS: An active learning reliability method combining Kriging and Monte Carlo Simulation. Structural Safety 33:145-154 doi:https://doi.org/10.1016/j.strusafe.2011.01.002

[15] Dong HC, Song BW, Dong ZM, Wang P (2016). Multi-start Space Reduction (MSSR) surrogate-based global optimization method. Struct Multidiscip O 54:907-926 doi:https://doi.org/10.1007/s00158-016-1450-1

[16] Saad A, Dong Z, Buckham B, Crawford C, Younis A, Karimi M (2019). A new Kriging–Bat Algorithm for solving computationally expensive black-box global optimization problems. Eng Optimiz 51:265-285 doi:https://doi.org/10.1080/0305215X.2018.1461853

[17] Wang LQ, Shan SQ, Wang GG (2004). Mode-pursuing sampling method for global optimization on expensive black-box functions. Eng Optimiz 36:419-438 doi:https://doi.org/10.1080/03052150410001686486

[18] Zhou G, Duan LB, Zhao WZ, Wang CY, Ma ZD, Gu JC (2016). An enhanced hybrid and adaptive meta-model based global optimization algorithm for engineering optimization problems. Sci China Technol Sc 59:1147-1155 doi:https://doi.org/10.1007/s11431-016-6068-4

Research on Bearing Fault Diagnosis Method under Variable Operating Conditions Based on MWDCNN

Lifan Kong[1], Taiyong Wang[1], Peng Wang[1,3,*] and Yiren Zhou[2]

[1]School of Mechanical Engineering, Tianjin University, Tianjin, China
[2]Nanchang University of Technology, Nanchang, Jiangxi, China
[3]International Institute for Innovative Design and Intelligent Manufacturing of Tianjin University in Zhejiang, Shaoxing, Zhejiang, China

[*]corresponding author's e-mail: pengwang@tju.edu.cn

Abstract. Aiming at the problem that the current convolutional neural network model cannot adaptively learn more effective features, which leads to the low accuracy of bearing fault diagnosis under unknown operating conditions, a multi-scale weight distribution convolutional neural network model is proposed. The model is composed of three parts: feature extraction layer, multi-scale feature connection layer and classification layer. The weight distribution mechanism is introduced in the feature extraction layer. In order to verify the feasibility of the model, experiments were carried out using the Case Western Reserve University (CWRU) data set, which proved that the model can effectively extract features under variable operating conditions and has good fault recognition capabilities.

1. Introduction

As an important part of the rotating machinery system, the rolling bearing plays a role in carrying and transmitting loads. Its operating state determines the reliability of the rotating machinery. Therefore, it is of great significance to study the fault diagnosis methods of rolling bearings [1]. The current intelligent fault diagnosis methods are mainly based on traditional fault diagnosis and deep learning fault diagnosis. Traditional fault diagnosis is to manually extract the main features of the original signal from the time domain, frequency domain, and time-frequency domain, and then input it into the existing model. Learn, and finally use the model for fault diagnosis and identification. This method requires expert experience and advanced signal processing methods, such as time-domain statistical characteristics[2], time-frequency spectrum binarization[3], variational modal decomposition[4], wavelet transform[5], etc., with certain Limitations, thus limiting the further development of traditional fault identification methods. The deep learning fault recognition uses the deep network structure to directly extract the features of the input samples, get rid of the dependence on manual feature extraction and expert knowledge, and realize the end-to-end fault diagnosis. Jing et al.[6] directly used the frequency data of the vibration signal as the input of the CNN, and compared it with the three types of data from the original data, the frequency spectrum and the combined time-frequency data. The results showed that CNN can adaptively learn the features from the frequency data. Zong et al.[7] used sample segmentation, sample expansion, and scaled conjugate gradient (SCG) algorithm to extract the features of the original vibration signal for fault diagnosis. Compared with the existing models, it has better feature learning and classification performance. Feng et al.[8] improved the convolutional neural network with regularized weights and weighted loss functions to identify rolling bearing faults under unbalanced samples. Jin Yufeng et al.[9] proposed a rolling bearing fault diagnosis model based on the combination of residual network and attention mechanism to solve the problems of complex feature extraction and difficulty in classifier training in the process of bearing fault diagnosis.

Although the above methods have good fault diagnosis capabilities, there are still some different problems. 1) Not able to adapt to the changing working environment; 2) Ignoring the problem of different weights of valid features and invalid features in the process of feature extraction; 3) Insufficient use of shallow features and deep features. In response to the above problems, this paper proposes a multi-scale weight distribution convolutional neural network model (MWDCNN): use the convolutional neural network to extract features from the original signal, assign weights to the features extracted in different layers, and then combine the shallow layers It is fused with the features after the weight distribution in the deep layer, and finally the multi-classifier is used for pattern recognition to realize the bearing fault diagnosis.

2. Multi-scale weight distribution convolutional neural network

2.1. Principle of weight distribution

According to the difference between the effective and invalid feature weights in the feature extraction process, redistributing the weights of the features can extract features more effectively. This paper divides the weight distribution into two parts: channel-based weight distribution and space-based weight distribution, as shown in Figure 1.

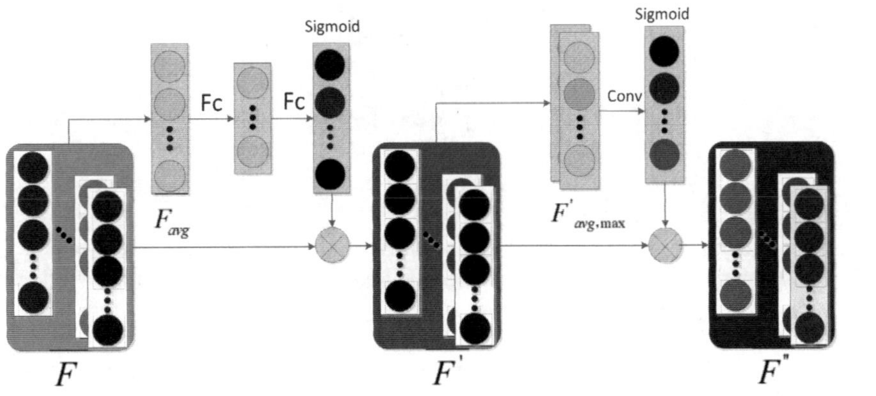

Figure 1. Schematic diagram of weight distribution

The formula for channel-based weight distribution is as follows:

$$F' = \sigma(W_1(W_0(F_{avg})) \otimes F \tag{1}$$

Where: σ is the sigmoid function; W0 and W1 are the weights of the two fully connected layers; $F_{avg} \in R^{C\times1\times1}$; F、$F' \in R^{C\times H\times W}$; F is the input feature map; F' is the channel-based weighted feature map; F_{avg} is the feature based on the spatial global average pooling Figure; *R* represents the dimension of the feature map; *H* is the height, *W* is the width, and *C* is the number of channels.
The formula for space-based weight distribution is as follows:

$$F'' = \sigma(f^{3\times3}F'_{avg,\max}) \otimes F' \tag{2}$$

Where: σ is sigmoid function; $F'_{avg,\ max}$ is a feature map based on spatial tie pooling and maximum pooling combined; $F'' \in R^{C\times H\times W}$; $f^{3\times3}$ is a 3×3 convolutional layer; $F_{avg} \in R^{C\times1\times1}$; F' is a channel-based weighted feature map; $F'_{avg,\max} \in R^{2\times H\times W}$; F^{*} is a space-based weighted feature map.

2.2. Multi-scale weight distribution convolutional neural network model

The model proposed in this paper consists of three parts: a feature extraction layer, a multi-scale feature connection layer, and a classification layer, as shown in Figure 2.

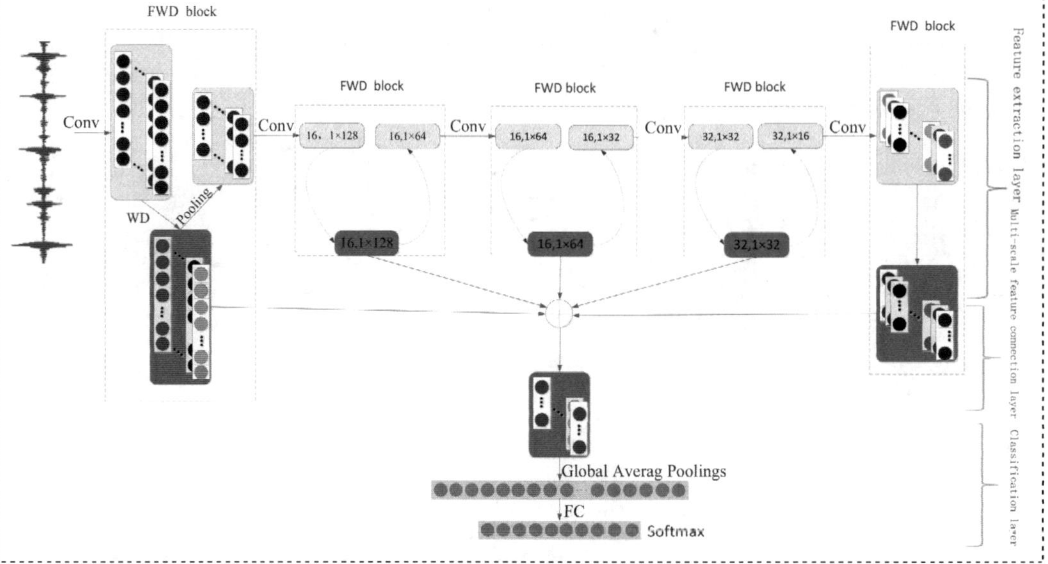

Figure 2. Multi-scale weight distribution convolutional neural networks model

The feature extraction layer is composed of 5 feature weight distribution modules (FMD block). The module uses the weight distribution principle to adaptively increase the weight of effective features according to the importance of the feature, and suppress the less useful features in the task. , And then improve the model's ability to extract features. The feature weight distribution module is composed of three parts: convolutional layer, weight distribution, and pooling layer. First, the input signal is feature extracted through the convolutional layer, and then the weight is assigned to the extracted features, the dimension of the output feature remains unchanged, and finally after the average pooling process, the output is used as the input of the next feature weight assignment module. This can extract high-weight features more effectively.

The multi-scale feature connection layer merges the shallow features with the deep features, and uses the number of channels to merge the features output by the feature weight distribution module to realize the reuse of effective features, and the features can complement each other to make the model Has better generalization performance.

The classification layer performs global average pooling on the output features of the multi-scale feature connection layer, and then inputs it to the fully connected layer, and finally uses the softmax function for fault diagnosis.

3. Experimental verification

3.1. Data sources

This article uses Case Western Reserve University (CWRU) rolling bearing data for experiments and analysis , The test object of this experiment is the drive end bearing, and the bearing model is deep groove ball bearing SKF6205.

A total of 4 data sets have been established in the experiment, as shown in Table 1. In the experiment, 2048 data points are used for fault diagnosis each time. Under variable conditions, the training set uses three types of loads, and the test set uses a single load type. 1728 samples are taken for each bearing state in the training set, and 192 samples are taken in the test set. Among them, the training set samples are all overlapped, and there is no overlap between the test set samples, in order to fully reflect the performance of the diagnostic method.

Table1. Description of experimental data set

Working conditions	data set	Training set load/HP	Test set load/HP
Variable conditions	A	0/1/2	3
	B	0/1/3	2
	C	0/2/3	1
	D	1/2/3	0

3.2. Parameter settings

The parameters of the MWDCNN model are shown in Table 2. In the training process, the number of samples processed in each batch is 16, the optimizer is Adam, the learning rate is 0.0005, and the number of iteration steps is 20.

Table 2. Parameter settings of MWDCNN

Internet	Nuclear size/step size	Number of cores	Output	Zero padding
convolution 1	112×1/16×1	16	128×16	yes
Weight distribution 1	-	-	128×16	-
Pooling 1	2×1/2×1	16	64×16	no
convolution 2	3×1/1×1	16	64×16	yes
Weight distribution 2	-	-	64×16	-
Pooling 2	2×1/2×1	16	32×16	no
convolution 3	3×1/1×1	32	32×32	yes
Weight distribution 3	-	-	32×32	-
Pooling 3	2×1/2×1	32	16×32	no
convolution 4	3×1/1×1	32	16×32	yes
Weight distribution 4	-	-	16×32	-
Pooling 4	2×1/2×1	32	8×32	no
convolution 5	3×1/1×1	32	8×32	no
Weight distribution 5	-	-	8×32	-
Pooling 5	2×1/1×1	32	4×32	no
Multi-scale feature connection layer	-	-	4×128	-
Global average pooling	-	-	128	-
Fully connected	-	-	128	-
Softmax	-	-	10	-

3.3. Experimental results and analysis under variable conditions

First, use the data sets A, B, C, and D under variable load conditions for training and testing to verify the fault recognition accuracy of the MWDCNN model under variable load conditions, and generate a confusion matrix to represent the experimental results , As shown in Figure 3. Although the diagnosed fault sizes are different, the faults are all located as outer ring faults, indicating that the MWDCNN model has learned the nonlinear characteristics of these three types of faults (outer ring faults, inner ring faults, and rolling element faults). Load adaptation capability.

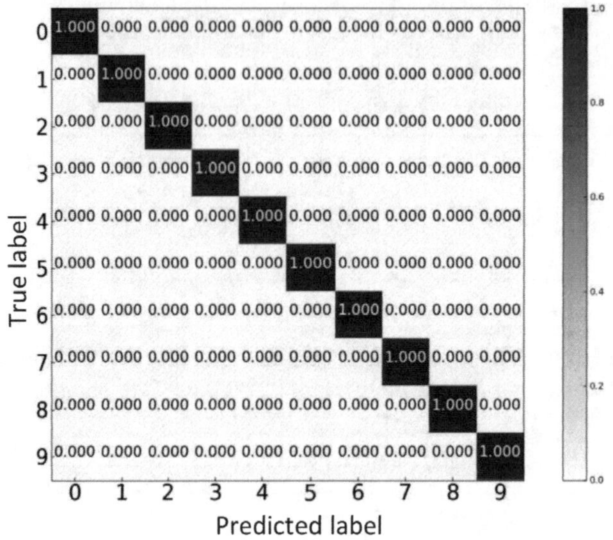

(a)Data set B experiment results

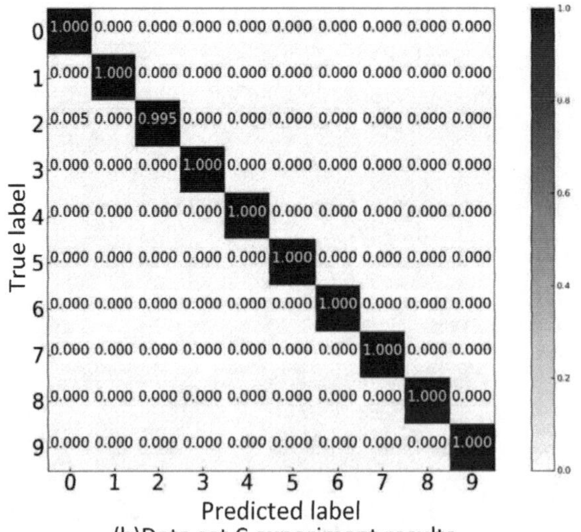

(b)Data set C experiment results

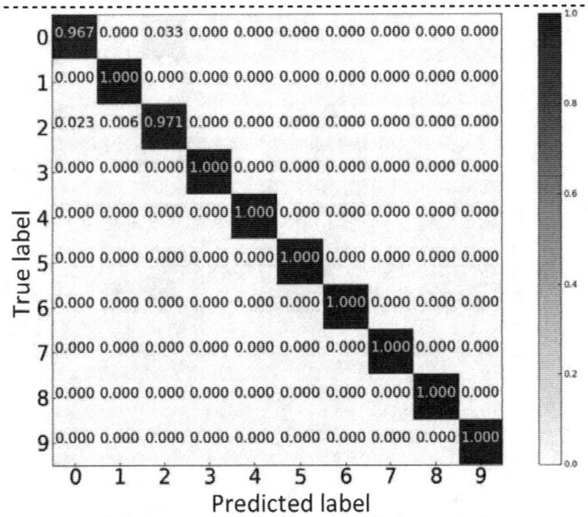

(c) Experimental results of data set D

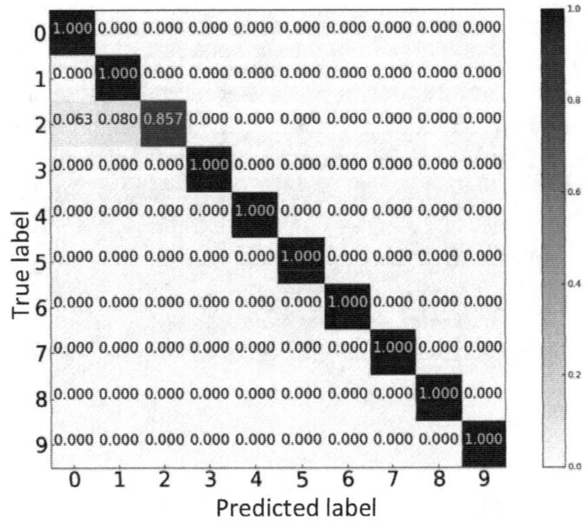

(d)Experimental results of data set E

Figure 3. Experimental results of testing datasets

In order to verify that the application of the feature weight distribution module and multi-scale feature connection in the convolutional neural network can improve the fault recognition ability and domain adaptation ability of the model under load changes, verify the superiority of the MWDCNN model, and other related models comparing. The experimental results are shown in Figure 4. It can be seen from the figure that the accuracy of the model proposed in this paper can reach 98.64% or more in each data set test, which is the model with the best test effect, which further verifies that the model still has good fault recognition under variable load conditions. Capability and domain adaptability. At the same time, it can be seen that CNN has the worst performance in each experiment, which shows that in the case of variable load, the multi-scale feature connection and feature weight distribution module can improve the ability of the model to extract features.

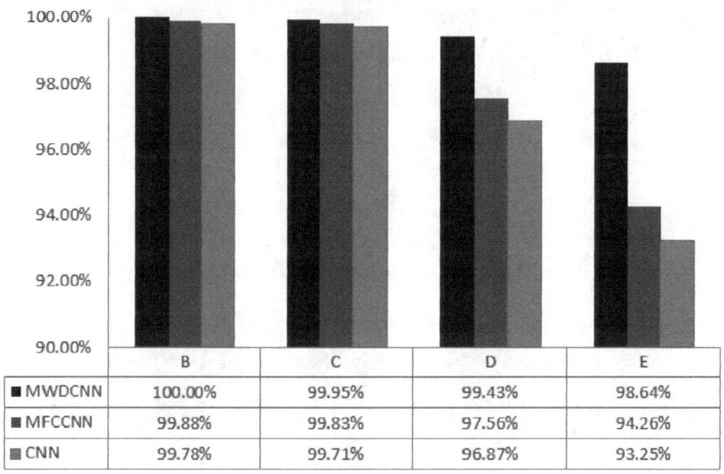

Figure 4. Comparison of model accuracy under variable load

Use data set A1 and add different degrees of Gaussian white noise to the test samples to perform fault diagnosis under variable noise conditions. At the same time, compare with other related models to verify the accuracy of fault recognition of this model under variable noise conditions. Experimental results As shown in Figure 5. Obviously, the MWDCNN model proposed in this paper is better than other models. When the signal-to-noise ratio is greater than 0dB, the accuracy of the model can reach more than 95.53%. This shows that the model can effectively extract features, has strong anti-noise performance, and can adapt to changing working environments.

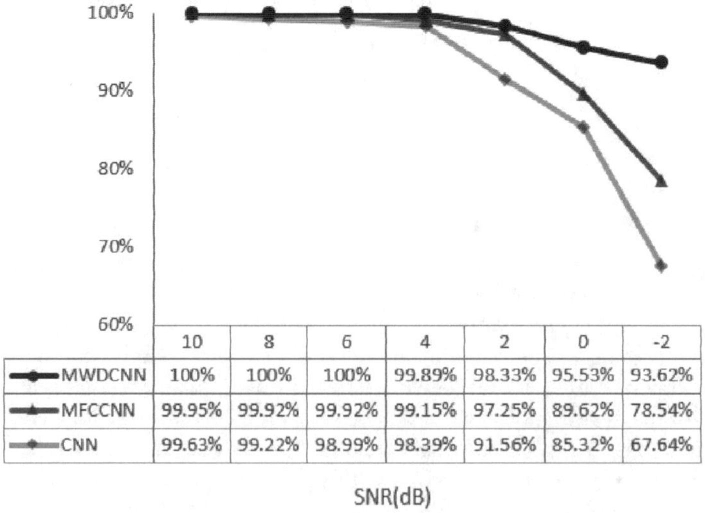

Figure 5. Comparison of model accuracy under variable noise

4. Conclusion

Compared with other methods, the multi-scale weight distribution convolutional neural network model proposed in this paper has the following three advantages: 1) The weight distribution method enables the model to better extract effective feature information and reduces the weight of invalid features. , Improve the accuracy and stability of fault recognition; 2) Multi-scale connection of the weighted shallow features and deep features to achieve the complementary advantages between effective features, improve the model's ability to extract features and the model's generality Chemical performance; 3) The model has anti-noise performance and high variable load adaptability.

Acknowledgments

This work is supported by The Innovation Fund of Industry, education and research of the Science and Technology Development Center of the Ministry of Education(2018 C 01068), Science and Technology Research Fund of Jiangxi Provincial Department of Education(GJJ191006),2020 Nanchang Engineering Research Center and Key Laboratory Foundation(2020-NCZDSY-005).

References

[1] Tian Y L, Liu X Y 2019 A deep adaptive learning method for rolling bearing fault diagnosis using immunity *Tsinghua Science and Technology* **24**(6) 750-62
[2] Wang D, Tsui K L 2017 Statistical Modeling of Bearing Degradation Signals *IEEE Transactions on Reliability* **66**(4) 1331-44
[3] Wang P, Wang T, Dong J 2016 Weak Feature Extraction of Vibration Signal Based on Binaryzation of EEMD Time-Frequency Map *Journal of Tianjin University* (Science and Technology) **49**(07) 667-73(in Chinese)
[4] Jiang X, Wang J, Shi J, et al 2019 A coarse-to-fine decomposing strategy of VMD for extraction of weak repetitive transients in fault diagnosis of rotating machines *Mechanical Systems and Signal Processing* **116** 668-92
[5] Chen B J, Shen B M, Chen F F, et al 2019 Fault diagnosis method based on integration of RSSD and wavelet transform to rolling bearing *Measurement* **131**(1) 400-11
[6] Jing L, Zhao M, Li P, et al 2017 A convolutional neural network based feature learning and fault diagnosis method for the condition monitoring of gearbox *Measurement* **111** 1-10
[7] Meng Z, Guo X, Pan Z, et al 2019 Data segmentation and augmentation methods based on raw data using deep neural networks approach for rotating machinery fault diagnosi *IEEE Access* **7** 79510-22
[8] Feng J, Yaguo L, Na L, et al 2018 Deep normalized convolutional neural network for imbalanced fault classifycation of machinery and its understanding via visuali-zation *Mechanical Systems and Signal Processing* **110** 349-67
[9] Jin Y, Yao M, Liu X, et al 2020 Rolling Bearing Fault Diagnosis Model Combining with Residual Network and Attention Mechanism *Mechanical Science and Technology for Aerospace Engineering* **39**(06) 919-25(in Chinese)

Balanced Optimization Design of Aluminum Alloy Impeller Rapid Casting Production Line

Wenqing Zheng*, Siqi Zhang, Yang Fan, Aonan Song, Bowen Li and Yanqiu Lu

School of Transportation Engineering, Shenyang Jianzhu University, Shenyang , Liaoning, China

*Corresponding author email: 24255822@qq.com

Abstract. The improvement and design of the production line is the key to improve the overall efficiency and reduce the total cost of each enterprise. The main purpose of the current rapid casting aluminum alloy impeller is through the determination of reasonable distribution center, production facilities, and the location of the relationship between the warehouse, so as to improve the overall efficiency of the logistics industry, promote interest rate. Reasonable aluminum alloy impeller rapid casting scheme, guiding the future transportation power, logistics innovation is an important factor. The manufacturing industry will continue to advance, and the optimization of production lines for more advanced innovations will be a critical research goal in the future.

1. The status quo in China and abroad

In the 1950s, Japanese engineers introduced the concept of logistics engineering, and promoted the development of logistics engineering, and put forward the famous concept of lean production and continuous improvement.

Garrido J M. (2009) [1] published The book Introduction to Flexsim, which introduced a lot of basic knowledge and usage of Flexsim. Bart Beemsterboer, Martin Land, Ruud Teunter (2017) [2] Used markov decision process model to prove that the optimal strategy of this kind of hybrid system changes with the change of batch size. Richard Adamson, Martin Hobbs (2017) [3] Discussed the development and application of integrated real-time production scheduling and control strategy for multiple cryogenic air separation units (ASU) and compressor units. Eghtesadifard M, Khalifeh M, Khorram M (2020) [4] discussed that assembly line balancing (ALB) refers to how tasks are allocated on an assembly line in order to achieve the desired productivity and optimization goals.

Xiang Zhengjie (2018) [5] took the optimization of storage and transportation field of X coal mine as the research object, and introduced ECRS method and process program analysis method in detail. Zhou Kangqu et al. (2019) [6] took the optimization of improving the production capacity of a micro-tiller in a company as the research object, and analyzed the operation content and process flow of each station. Huang Kai, Yu Jianbo (2019) [7] Took the lean improvement research on the machining line of air conditioning compressor as an example, and obtained lower cost and working time by analyzing the process flow of the production line and improving unnecessary costs in the process. Gu Jia, Wu Cong, Xiong Genliang, Yang Shaolin (2016) [8] Briefly introduced the knowledge related to system simulation and Flexsim software, and modeled the actual assembly line based on Flexsim simulation software and analyzed its existing bottleneck.

2. Research content and methods

2.1. Current Status of Rapid Casting Production Line for Aluminum Alloy Impeller of R Company
This design takes the rapid casting production line of aluminum alloy impeller of R company as the research object, and the production line of this company is a single-product production line. R company is a professional enterprise that manufactures parts with rapid casting technology. The parts produced are made of a variety of materials, mainly aluminum alloy, titanium alloy, stainless steel and non-metallic parts.

This design selects R company's aluminum alloy rapid casting workshop production line for balance optimization study. In the process of field observation of the rapid casting production line of aluminum alloy impeller, it is found that some stations have the phenomenon of operators idle and waiting, while some stations have been operating under high load. Therefore, it can be preliminarily judged that the rapid casting line has the problem of unbalanced production line.

2.2. 0-1 integer programming model of production line balance
The line balance problem is divided into two main categories, as shown in Table 1,

Table 1. Classification of production line balance

regimentation	The connotation of the balance problem
I The production cycle is known, find the minimum total number of stations required.	In the early design and configuration stage of the production line, the management can predict consumer demand for products based on the results of market research, so as to determine the beat of the production line. By reorganizing and distributing all operating elements, each station can be operated within the time range required by the production cycle, so as to achieve the optimization goal of the least number of stations. That is, when the production efficiency is constant, the production cost of the production line is reduced.
II The total number of stations is known, find the minimum production cycle required.	After the production line has been built, the equipment and operator situation is basically fixed, at which point the production line needs to be adjusted to minimize the production beat. That is, to ensure that production costs remain the same to improve the efficiency of the production line.

Because the number of seats of the company is small in this design, the close relationship between the working elements is tight, the time consumption of the job elements is quite different, it is difficult to make large changes to the number of seats on the original basis. Therefore, the design of this design is based on the Class II prerequisites for the production line optimization.

Before establishing a mathematical model of a line balancing problem, make assumptions about the model in the following areas, including:

- (1) The types of products produced on each production line are known and only one;
- (2) The working time of each operation in the production line is fixed;
- (3) The number of workers on each station is known and the same;
- (4) the process beat on each production line is less than or equal to the production beat;
- (5) Each job element is the smallest individual and cannot be further decomposed;
- (6) The allocation of operations in the workstation must meet the constraints of the priority relationship.

Preparation for modeling:

(1) Model symbol description

CT (Cyctime): production beats for fast casting lines;

M: Number of seats;

m: The serial number of the station, k=1,2,3,... K;

k: The number of job elements;

i,j: The serial number of the job element;
t_i: Standard working time required to complete the i job element: i-1, 2,3,... m;
Tasks={1,2,3,... m) represents a collection of job elements that are already on the production line;
Pred: A collection that represents a tight front-to-back relationship between the various job elements;
Pred,j, job element i must be arranged before job element j;
(2) Description of the decision variable
X_{ik}:X_{ik} is a 0-1 variable, where:

$$X_{ik} = \begin{cases} 0 \text{ means that job element i is not assigned to station k} \\ 1 \text{ means that the job element i is assigned to the station k} \end{cases}$$

$A_k$is the indicator variable of station k, where:

$$A_k = \begin{cases} 0 \text{ means that the kth station is not occupied} \\ 1 \text{ means that the kth station is occupied} \end{cases}$$

(3) The 0-1 integer planning model description of the line balance optimization design problem
a. The establishment of a 0-1 integer planning model for the balanced optimization design of Class I production lines

Each job must be assigned to a workstation that meets the constraints:

$$\sum_{K=1}^{K} X_{ik} = 1, i = 1,2,3, \dots m \tag{1}$$

For each station, the work time must not be greater than the production beat of the quick casting line, i.e. to meet the constraints:

$$\sum_{i=1}^{m} X_{ik} t_i \leq CT, k = 1,2,3, \dots k \tag{2}$$

For each job element pair that has a priority relationship, the constraints of the actual priority relationship must be met, i.e. the job element j is scheduled with a station number larger than the station number that the job element i is scheduled for. That is, to satisfy the constraints:

$$\sum_{k=1}^{K} K(K_{jk} - K_{ik}) \geq 0, (i,j) \in Pred \tag{3}$$

Give a workstation a 0-1 variable, which is 1 when a job element is scheduled in a workstation in a quick casting line, and 0 to 0. That is, to satisfy the constraints:

$$\sum_{K=1}^{K} X_{ik} \leq mA_k, \quad k = 1,2,3, \dots k \tag{4}$$

2.2.1. The first kind of mathematical model Through model operation, the output result of 0-1 integer programming model of the balanced optimal design problem of class I production line is no different from that of the original workstation arrangement, which does not play an optimization role. Therefore, class I model is not used in this project. Therefore, the modeling process is not described.

2.2.2. The second kind of mathematical model The mathematical model of the 0-1 integer planning model for the class II line balance optimization design problem is:

Target function:

$$Minz = CT \tag{5}$$

Constraints:

$$\begin{cases} \sum_{k=1}^{k} X_{ik} = 1; (i = 1,2,3,\dots m) \\ \sum_{i=1}^{m} X_{ik} t_i \leq CT\,; (k = 1,2,3,\dots k) \\ \sum_{k=1}^{K} K(K_{jk} - K_{ik}) \geq 0\,, ((i,j) \in Pred) \end{cases} \tag{6}$$

Lingo, an acronym for Linear Interactive and General Optimizer, is an interactive linear and universal optimization solver, launched by Lingo Systems Inc., and can be used to solve nonlinear planning, as well as for solving some linear and nonlinear equations, and is a powerful and optimal choice for solving optimized models.

2.2.3. Lingo programming and solving On the basis of the 0-1 integer planning model of the balance problem of class I rapid casting production line, the constraints and target functions of the mathematical model are entered in Lingo software, and the results of Lingo software are as follows Figure 1:

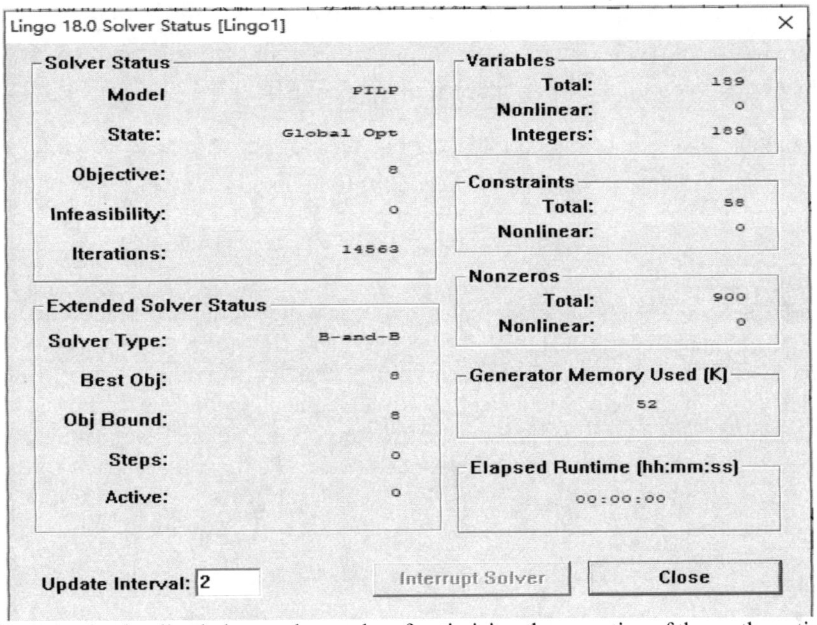

Figure 1. Class I production line balances the results of optimizing the operation of the mathematical model

As can be seen from the results of the output operation, the results obtained do not change from the original workstation arrangement. That is, eight seats for R company's rapid casting production line is already the best solution, no longer need to build on this design.

On the basis of the 0-1 integer planning model of the balance problem of the aluminum alloy impeller assembly line ii, the target function and constraints of the mathematical model are written to lingo software correspondingly and run.

2.2.4. The optimization results The results are listed in Table 2 below, which is divided by the optimized line station division that can be obtained by running the Lingo software:

Table 2. Category II Balance Optimization Design Problem Solved After Line Station Division

The station serial number	The name of the station	The operation serial number	The content of the job	Seat time (second)
1	Additive manufacturing	1、2、3、4	Mold to eliminate internal stress, tree formation, configuration of paint, add defoaming agent stirring	232.91
2	Cleaning	5、6	Clean the prototype and hang it in a stand	184.07
3	Coating	7、8、9	Coated shell, dewaxed, roasted	241.23
4	Foundry	10、11	Pour and clear the shell	231.44
5	Machining	12、13、14	Cut the gate, polish the gate, blast the shot	219.38
6	First inspection	15	Inspection 1	223.11
7	Appearance treatment	16、17、18	Crude car, fine car, T6 heat treatment	247.91
8	Retest the incorporated storage	19、20	Inspection 2, storage	214.00

Results can be obtained from the results of the operation of the software and from Table 2, when the specified number of hours is 8, the production line can be optimized to a smaller production beat of 247.91s.

According to the line balance rate formula, the line balance rate is divided under the new production line station$\frac{1794.07}{247.91\times 8}\times 100\% = 90.4\%$, which belongs to the production line with a higher equilibrium rate.

3. Conclusion

This design takes R company's aluminum alloy impeller rapid casting production line as the research object. First of all, on the theoretical basis of logistics engineering, the relevant data of the existing rapid casting production line in the literature are analyzed. Secondly, using the 0-1 integer planning theory, which is widely used in the field of logistics, a 0-1 integer planning model is established for the balance of production lines under two different constraints. And on the basis of the mathematical model, the Lingo software is used to run it to find the optimal solution. The running results of Lingo show that the 0-1 integer programming model of class II based on the balance problem of rapid casting production line has better optimization effect.

Acknowledgments

I would like to thank my thesis advisor: Mrs. Zhang Siqi from Shenyang Jianzhu University school of Traffic Engineering. Mrs. Zhang Siqi made guiding opinions and recommendations on the research direction of my paper, and gave directions to the difficulties and doubts I encountered in time during the process of writing the paper. This project "Balance Optimization Design of Aluminum Alloy Impeller Rapid Casting Production line" is studied on the basis of her project 2021CSLKT3-250 "Balance Optimization Design of Aluminum alloy production line based on Lean Production". At the same time, I would like to thank my friends and classmates for their great support and help in the preparation of the paper, which brought me great inspiration.

4. References

[1] Garrido J M. 2009. Introduction to Flexsim[M].Object Oriented Simulation. Springer US: 31-42.
[2] Bart Beemsterboer,Martin Land, Ruud Teunter. Flexible Lot Sizing in Hybrid Make-to-order//Make-to-stock Production Planning [J].European Journal of Operational Research, Volume 260, Issue 3, 2017: 1014–1023.
[3] Richard Adamson,Martin Hobbs, Andy Silcock, Mark Willisa. Integrated Real-time Production Scheduling of A Multiple Cryogenic Air Separation Unit and Compressor Plant [J]. Computers & Chemical Engineering, 2017,9: 25–37.
[4] Eghtesadifard M,Khalifeh M,Khorram M. A systematic review of research themes and hot topics in assembly line balancing through the web of science within 1990–2017[J]. Computers & Industrial Engineering, 2020, 139: 106-182.
[5] Mr. Xiang. Research on the optimization of process of X coal storage and transportation field based on ECRS method. Modern trade and industry; 2018,(3):28-33.
[6] Zhou Kangdao; Yang Kun; Think about it. HTBS micro-till packaging line simulation and optimization; Industrial engineering; 2019,(22)1:77-85.
[7] Gu Jia, Wu Cong, Xiong Genliang, Yang Shaoliang. Simulation and optimization of the assembly line of the air conditioner in-house machine. . . mechanical design and manufacturing2015, 7:20-30.
[8] Huang Kai, Yu Jianbo, Lean Improvement Research of Air Conditioning Compressor Processing Line , Process and Manufacturing 2019, 8 (1): 56-59.
[9] Zhang Guohui, Zhang Lingjie. Flexsim-based mixed-flow line simulation and optimization. Logistics Technology 2017, 34 (23) :83.86.
[10]Li Jingfeng, Dong Shaohua. The study of mixed genetic algorithm on the balance of mixed-flow assembly line. Journal of Beijing University of Science and Technology, 2016, 34 (8): 952.958.

AUTHOR INDEX